Unified SuperStandard Theories for Quaternion
Universes & The Octonion Megaverse

Stephen Blaha Ph. D.
Blaha Research

Pingree-Hill Publishing

MMXX

Rev. 00/00/01   March 25, 2020

To My Children & Grandchildren

How happy is he born and taught,
That serveth not another's will;
Whose armor is his honest thought,
And simple truth his utmost skill!

The Character of a Happy Life,
Sir Henry Wotton (1568-1639)

Some Other Books by Stephen Blaha

All the Megaverse! Starships Exploring the Endless Universes of the Cosmos using the Baryonic Force (Blaha Research, Auburn, NH, 2014)

SuperCivilizations: Civilizations as Superorganisms (McMann-Fisher Publishing, Auburn, NH, 2010)

All the Universe! Faster Than Light Tachyon Quark Starships & Particle Accelerators with the LHC as a Prototype Starship Drive Scientific Edition (Pingree-Hill Publishing, Auburn, NH, 2011).

Unification of God Theory and Unified SuperStandard Model THIRD EDITION (Pingree Hill Publishing, Auburn, NH, 2018).

The Exact QED Calculation of the Fine Structure Constant Implies ALL 4D Universes have the Same Physics/Life Prospects (Pingree Hill Publishing, Auburn, NH, 2019).

Unified SuperStandard Theory and the SuperUniverse Model: The Foundation of Science (Pingree Hill Publishing, Auburn, NH, 2018).

Quaternion Unified SuperStandard Theory (The QUeST) and Megaverse Octonion SuperStandard Theory (MOST) (Pingree Hill Publishing, Auburn, NH, 2020).

United Universes Quaternion Universe - Octonion Megaverse (Pingree Hill Publishing, Auburn, NH, 2020).

Available on Amazon.com, bn.com Amazon.co.uk and other international web sites as well as at better bookstores (through Ingram Distributors).

CONTENTS

FIGURES and TABLES

INTRODUCTION

In previous books this author has derived the Unified SuperStandard Theory in our 3 + 1 dimension space-time from Complex General Relativity and Quantum Field Theory (suitably extended.) This book creates a deeper basis[1] for our theory in Complex Quaternion space for our universe, and beyond that in Complex Octonion space for the Megaverse of universes. The general conclusions of our development are:

Our universe is a 32 Complex Quaternion Dimensions Space. It is equivalent to QUeST, which has a 3 + 1 Complex Quaternion Dimensions Space-Time and Internal Symmetries.

Upon restriction of QUeST to real-valued coordinates, QUeST Becomes the Unified SuperStandard Theory.

The Megaverse is a 32 Complex Octonion Dimensions Space. It is equivalent to MOST, which has a 7 + 1 Complex Quaternion Dimensions Space-Time and Internal Symmetries.

The 7 + 1 Complex Quaternion Dimensions Space-Time of the MOST Megaverse contains 3 + 1 Complex Quaternion Dimensions QUeST Universes' Space-Times.

QUeST → real-valued coordinates Unified SuperStandard Theory

---

[1] QUeST and MOST are defined in Blaha (2020a) and (2020b).

We show that the Unified SuperStandard Theory is an exact consequence of Quaternion and Octonion space.*The generality of these results, and the author's previous derivation of the exact value of the Fine Structure Constant α, and approximate derivations of the Weak and Strong coupling constants (shown later), shows that the Internal Symmetries of the Standard Model, and our Unified SuperStandard Theory, are now understandable.*

Thus particle physics can be explained by Quaternion and Octonion space characteristics. We show this in some detail in the following chapters.

1. Unified SuperStandard Theory in 3 + 1 Real Dimensions

This chapter presents a summary of the main consequences of our Unified SuperStandard Theory in 3 + 1 real dimensions. We will see that it is the limit of the Quaternion Unified SuperStandard Theory (QUeST) in 32 Complex Quaternion Dimensions when it is restricted to 3 + 1 real-valued space-time coordinates. Chapters 26 – 58 present the details of the Unified SuperStandard Theory in 3 + 1 real dimensions.

The internal symmetries (interactions) of the theory, the vector boson spectrum, and the fermion spectrum are presented. The reader will see later that the QUeST predicts all these results in complex 32 dimension quaternion space. ***Thus QUeST is the ground of the Unified SuperStandard Theory in 3 + 1 real dimensions that we derived previously based on Complex General Relativity and Quantum Field Theory.***

1.1 Axioms and Derivation of the Unified SuperStandard Theory in 3 + 1 real dimensions

The axioms and derivation of the Unified SuperStandard Theory is presented in Appendix 1-A together with a historical perspective on the derivation of theories of physical reality.

An important aspect of derivations for all physical theories is the dimension of space-time. Appendix 1-B shows that the dimension must be greater than or equal to four for physical processes to execute in parallel.

1.2 Unified SuperStandard Theory Interactions

The vector boson symmetries[2] in the Unified SuperStandard Theory were found to be

$$[U(1)\otimes SU(2)\otimes SU(3)\otimes U(1)\otimes SU(2)\otimes SU(3)\otimes U(4)^4]^4\otimes U(4) \qquad (1.1)$$

The symmetries are of similar form in each of the four layers. In addition the symmetries of the "normal" sector and the Dark sector are duplicates.

QUeST has the same internal symmetries and adds an additional U(2) symmetry for each of the four layers. Thus $U(2)^4$. This symmetry maps each normal fermion to its corresponding Dark equivalent, and vice versa, layer by layer. We discuss this symmetry in more detail later.

The symmetries of the normal sector are

$$[U(1)\otimes SU(2)\otimes SU(3)\otimes U(4)^2]^4\otimes U(4) \qquad (1.2)$$

where the lone U(4) is that of the Species group described later.

The symmetries of the Dark sector are similar (but different0

$$[U(1)\otimes SU(2)\otimes SU(3)\otimes U(4)^2]^4\otimes U(4) \qquad (1.3)$$

where the lone U(4) is again that of the Species group.

Each of the four layers of fermions experiences it own set of interactions. They must be different since inter-layer interactions are not seen. Similarly the normal and Dark sectors of each layer must also have different sets of interactions. The U(2) groups added by QUeST must also yield ultra-weak interactions since interactions between normal and Dark matter are not seen.

The Layer groups act to interconnect the four layers for the normal and Dark layers separately. The U(2) groups serve to connect the normal and Dark sectors (in principle). Thus the symmetries result in an interconnected spectrum of fermions.

---

[2] Most of these symmetries are broken symmetries.

1.3 Unified SuperStandard Theory Vector Bosons

Fig. 1.1 displays the Unified SuperStandard vector bosons except for the Species group which all particles experience, and except for the U(2) groups added by QUeST.

The "Normal" Vector Boson "Periodic" Table

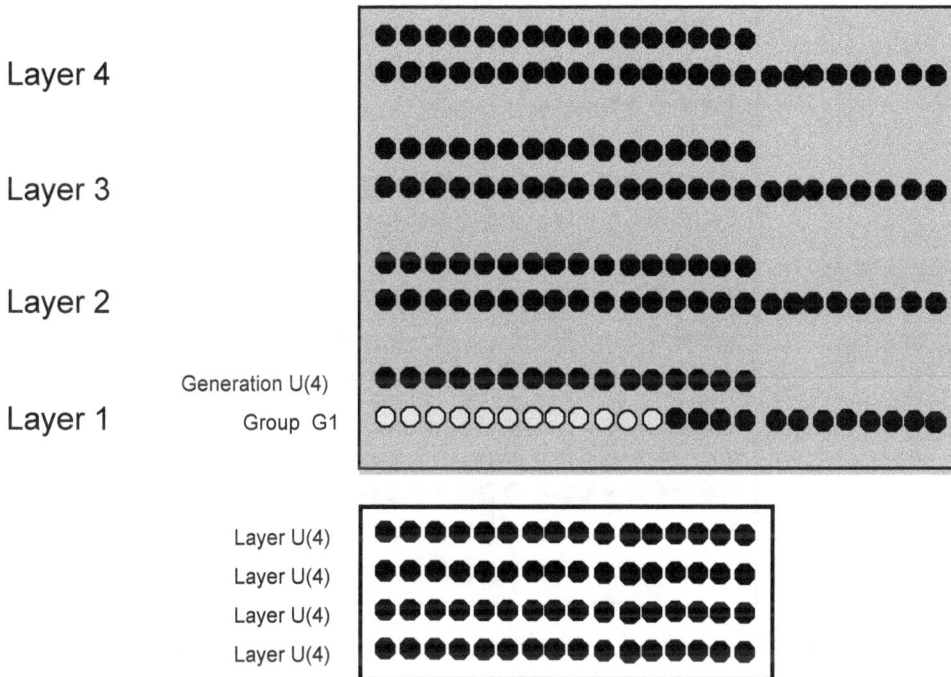

Figure 1.1. The vector bosons. Each circle represents a group generator. The known vector bosons are in the lowest row with a white interior. Yet to be found vector bosons are solid black. The Layer groups straddle all four layers. G1 is SU(3)⊗SU(2)⊗U(1). The list of groups for the higher three levels is the same as those of the first layer. There are 224 normal vector bosons not counting Species and U(2) groups.

The "Dark" Vector Boson "Periodic" Table

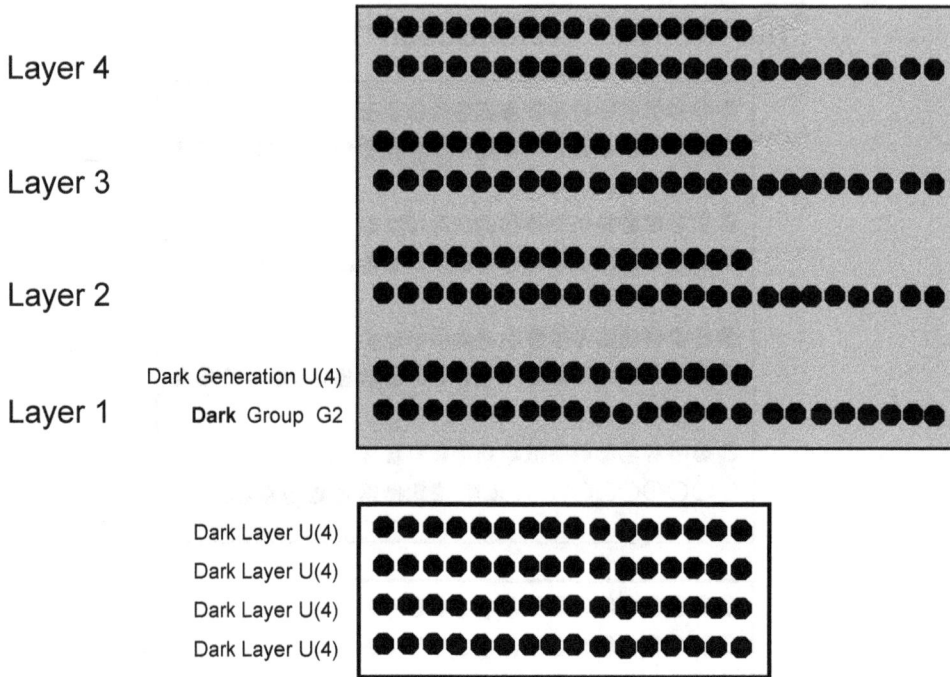

Layer 4

Layer 3

Layer 2

Dark Generation U(4)

Layer 1   **Dark** Group G2

Dark Layer U(4)
Dark Layer U(4)
Dark Layer U(4)
Dark Layer U(4)

Figure 1.2. The vector bosons. Each circle represents a group generator. The known vector bosons are in the lowest row with a white interior. Yet to be found vector bosons are solid black. The Layer groups are distributed by layer symbolically although they each straddle all four layers. G2 is $SU_D(2) \otimes U_D(1) \otimes SU_D(3)$. The list of groups for the higher three levels is the same as those of the first layer. There are 224 Dark vector bosons not counting Species and U(2) groups.

1.4 Fermion Mass Spectrum

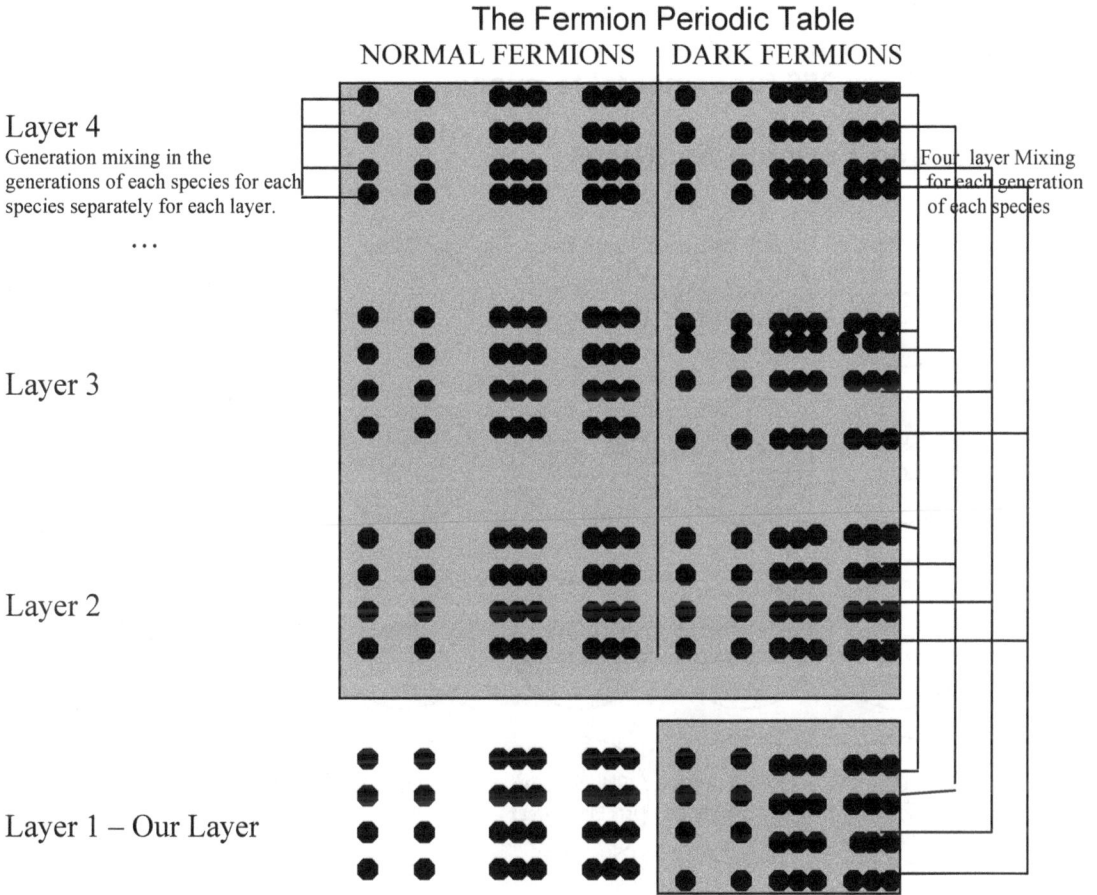

The Fermion Periodic Table

NORMAL FERMIONS | DARK FERMIONS

Layer 4
Generation mixing in the
generations of each species for each
species separately for each layer.

...

Four layer Mixing
for each generation
of each species

Layer 3

Layer 2

Layer 1 – Our Layer

Figure 1.3. Fermion particle spectrum and partial example of pattern of mass mixing of the Generation group and of the Layer group. Dark sector parts of the periodic table are shaded gray. Unshaded parts are the known fermions with an additional, as yet not found, 4th generation shown. The lines on the left side (only shown for one layer) display the Generation mixing within each layer's species. The Generation mixing applies within each layer using a separate Generation group for each layer. The lines on the right side show Layer group mixing with the mixing amongst all four layers for each of the four generations individually. There are four Layer groups. There are 256 fundamental fermions.

1.5 Groups and Fermion Splittings

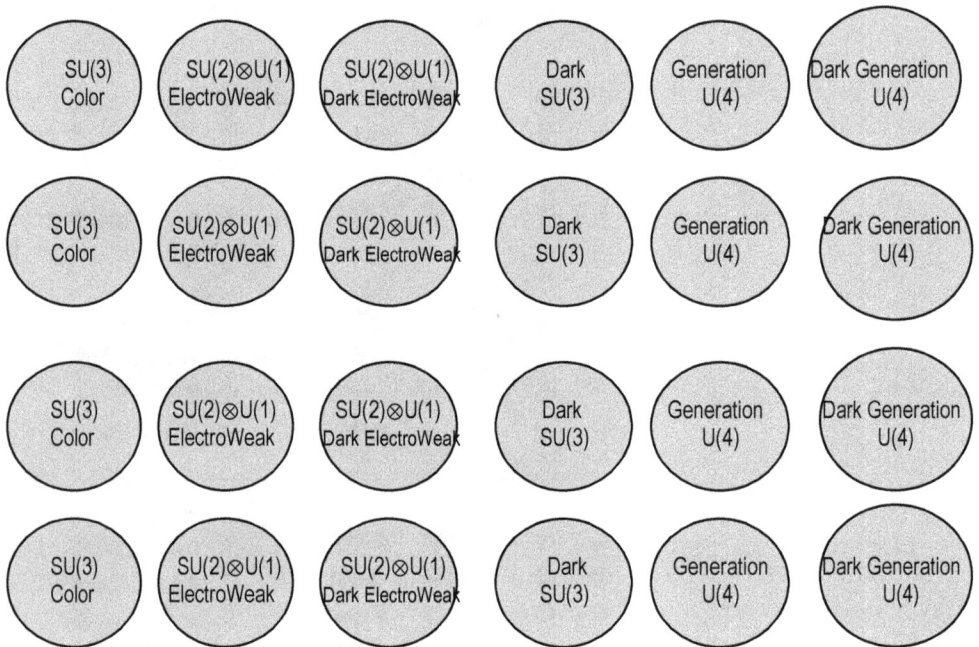

Figure 1.4. The set of four layers of internal symmetry groups corresponding to four generations in four layers of spin ½ fermions and the four layers of vector bosons. In addition there are the Normal and Dark Layer groups, and the Species group, which are *not* displayed.

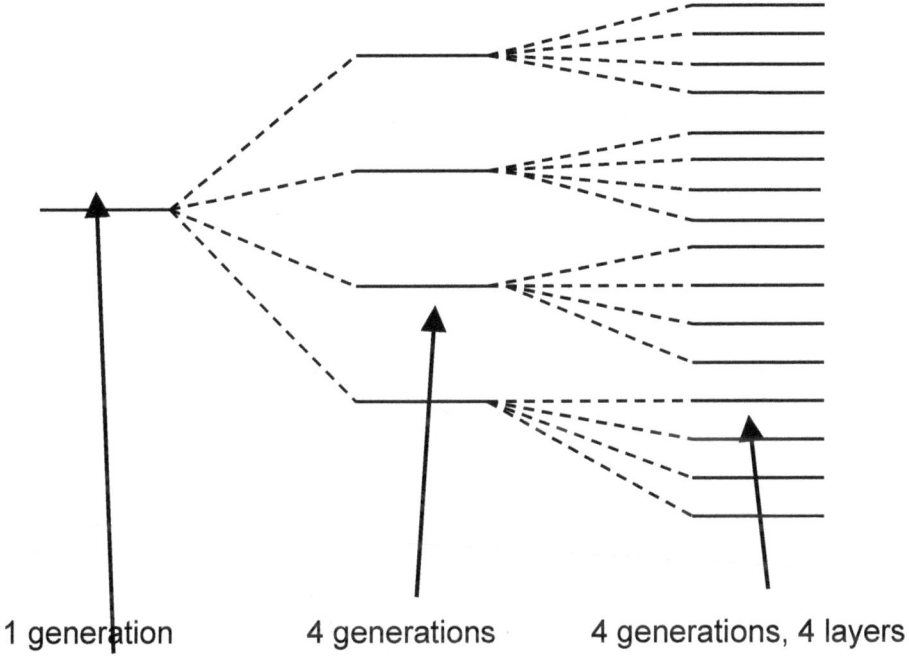

1 generation   4 generations   4 generations, 4 layers

Figure 1.5 The "splitting" of a single generation fermion into four generations and then into four layers.

Appendix 1-A. Unified SuperStandard Theory 3 + 1 Derivation

In this appendix we will outline the process deriving our Unified SuperStandard Theory from fundamental axioms.[3] Our theory has the somewhat unique feature of being derivable from first principles. Most other fundamental theories are constructed in an ad hoc fashion, often based on symmetry considerations, with features being added rather than derived.

The goal of Unified SuperStandard Theory was to derive it in the manner of Euclid with a clear connection between the steps of the derivation just as Euclid developed geometry from a progression of theorems.

We present a deeper foundation for the derivation of the Unified SuperStandard Theory that does not conflict with the derivation presented in the first edition (Blaha (2017f)). It embodies a 'simpler' set of primitive terms and axioms, and 'explains' parts of the derivation in a cleaner way as well as providing *a new basis* for particle phenomena such as 'spooky' Quantum Entanglement.

We begin this appendix with a brief historical view of the progress of fundamental physics from the cosmogonies of the Pre-Socratic Philosophers to modern views of Physics. Remarkably the conceptual thought processes of the Pre-Socratics is surprisingly similar to those of contemporary physicsts: similar questions, similar proposals and similar qualitative views of reality – all based on limited empirical knowledge, limited mathematics, and limited means of proving assertions.

The great strength of the Pre-Socratics was perspicacity, a grasp of Logic and the ability to see beneath appearances.

---

[3] Appendix C outlines the general process of deriving a unified theory of everything from a set of axioms. It is described in detail in Blaha (2017c) and (2017d) as well as in versions in earlier books.

1-A.1 The Historical Path from Pre-Socratic Cosmogony to Current Attempts at a "Theory of Everything"

Pre-Socratic efforts to understand physical reality and the Cosmos could be viewed as occurring in two phases: an initial phase where all happened as the result of the direct intervention of the Gods, and a second phase where physical materials and processes 'caused' natural phenomena and the invocation of a deity (deities) was not required.

1-A.1.1 Gods-Based Cosmogony

Prior to the 6th Century BC, Man's understanding of the Cosmos and physical phenomena were based on theological concepts – all events were based on the history and actions of the gods. In Mediterranean civilizations cosmology was well described by Hesiod's *Theogony,* and on Orphic and Zoroastrian cosmology that partly originated in Egypt, the Middle East and Persia. The origin of the universe was a history of the gods and their actions beginning from a primeval Chaos. Physical phenomena were the results of the actions of the gods.

1-A.1.2 Philosophic-Materialistic Cosmogonies

In the 6th Century BC a number of philosophers – notably Pherecydes – developed a materialistic view of the creation and Physics of the universe. Pherecydes postulated a primitive kronos that became the god Chronos – Time – from whom emanated a 'cosnic egg' and thence the gods Night and Eros. These gods created the constituents of the universe: earth, sky-air, and water. A reading of the historical fragments[4] describing this cosmology shows that much of the logic of their description of physical history and phenomena consisted of replacing theological terms with physical materialistic terms.[5] Thus there was a 'smooth' transition from theological descriptions to materialistic descriptions.

---

[4] See Kirk (1962) for a detailed study of Pre-Socratic Philosophy.

[5] This author pointed out this similarity of description in a paper (unpublished) in 1964. The similarity is reminiscent of the architectural similarity of wooden Greek temples which used pins to hold beams together and later stone Greek temples such as those of the Parthenon which placed decorative marble 'pegs' in the temple in imitation of the earlier wooden temples' pegs.

Various philosophers developed 'Theories of Everything' from this point on. The best known theory is the four element theory: fire, wind, water and earth of the 5th Century BC.

Miletus was a hotbed of Philosophic thought with Thales (and Homer) in the 6th Century BC holding that water was the sole element from which all elements were generated. He was followed by Anaximander who posited a sophisticated concept – 'the Indefinite Apeiron' – a single primary element from which the four elements were generated as well as intermediate forms of these elements. Anaximander also proposed 'plural worlds of infinite extent' – a concept familiar in various 'modern' forms since the 16th Century including our Megaverse proposal.

Anaximenes of Miletus suggested that the apeiron was air. The philosopher Diogenes also postulated that all was ultimately composed of air.

A succession of philosophers made noteworthy suggestions:

2. Theophrastus proposed that all matter is in eternal motion – dynamics.
3. Heraclitus proposed the unity of all things with fire as the primary element.
4. Democritus and Leucippus proposed matter is made of atoms of an infinite variety of shapes (since there was no reason why they should all have the same shape).
5. Xanophanes suggested that all things were made of Earth and water.
6. Pythagoras suggested all was based on Number as an extension of apeiron to unity with all other numeric quantities (multiplicity) generated from it.

Thus the concepts of materialistic Physics have a clear beginning in the Pre-Socratics. Notably, philosophers were quick to seize on the notion of a primary form of matter from which the varieties seen in Nature were derived. The Indefinite Apeiron of Anaximander is an especially noteworthy conceptual precursor to our current theories of elementary particles.

A reading of the fragments of the Pre-Socratic philosophers' writings raises a number of significant questions that do not appear to have been seriously considered:

1. Why and how did the transition from a theocratic description of the universe to a materialistic description take place? It is thought that Pherecydes was the first (or the leading person) in proposing this change of *zeitgeist*. But the source of the concept is elusive since the observation of Nature was rudimentary and the leap of thought – contrary as it was to prevailing religious thinking – is substantial – inspired!

2. After the transition to a materialistic/scientific thinking pattern, how did the observed multifold aspects of Nature become viewed as based on appearances of a few substances (fire, air, water, and earth or one or more of these substances)? We can see that transformations between substances such as water upon heating becoming 'air' but to have a 'Theory of Everything' based on a few, or one, fundamental substance is a leap of thought reminiscent of the 20th century attempts at a Theory of Everything such as String theory.

No ready answer to these questions can be found – mostly due to the loss of books when the Great Library of Alexandria was burned. The burning of the Library and the destruction of the first group of 'scientists' engaged in the study of natural phenomena[6] (together with the murder of Archimedes in Syracuse) shows that war can have a major retrograde effect on civilization.[7]

If one must have an answer one can only suppose that these advances – and they were major conceptual advances – were the result of 'inspired' creative thought. A possible view of this form of creativity is perhaps embodied in the story of Newton and a falling apple. The natural view is to see the fall of a physical apple. However, a mind can reconstruct this view to see an instance of a force of nature – gravity. What one sees, can be interpreted differently in the 'mind's eye.' And lead to a new, different perspective.

---

[6] Such as a prototype steam engine.
[7] Of course today's world is different with major science and engineering efforts for war. Today we face the issues of the obliteration of civilizations.

1-A.1.3 Atomistic Theory

From the time of the Pre-Socratics to the mid-19th Century the nature of matter was a subject of dispute with some arguing matter was continuous and others arguing that matter was ultimately composed of atoms. The dispute was poignantly brought to a head by the success of the atomistic theory of Boltzmann and others. Their success was controversial and vigorously condemned by opponents for decades.

By 1900 the atomic theory of matter was largely accepted and the door was opened to the successes of quantum theory.

1-A.1.4 Twentieth Century Physics

With the acceptance of the atomistic theory of matter a series of developments principally related to atomic spectra and black body radiation led to the development of quantum mechanics in the 1920's, and quantum field theory in the 1930's and 1940's. The great experiments from the 1930's through the 1970's led to the Standard Model of elementary particles and their interactions.

Starting from the nineteen-teens a series of efforts were made to create a unified theory of everything with noteworthy initial efforts by Weyl, Einstein and Eddington. The efforts took on a new life with the development of unified quantum field theories and of string theories in the 1970's and afterwards.

In an effort to avoid the *ad hoc* nature of these theories which assumed group symmetries and/or strings, this author developed a theory derived from fundamental principles in a manner reminiscent of Euclidean geometry that culminates in this book since it describes the known properties of matter, proposes new possible features, and opens the door to extensions to new phenomena should they be discovered in future experiments.

This appendix now begins the derivation by specifying the requirements of the theory, defining primitive terms and then stating the theory's basic axioms à la Euclid. There are two advantages of a derivation from sound axioms: 1) the resulting derivation shows what must be included in the complete theory; and 2) the derivation automatically excludes possibilities that a creative theorist might think to include in the theory.

1-A.2 Underlying Physical Principles of Creation

1-A.2.1 Creation by a Cosmic Pseudo-Euclid Entity

The derivation of the Unified SuperStandard Theory seems to be best presented within the framework of a putative being who may or may not be a 'real' entity. Of this 'being', which might have been called the Cosmic Pseudo-Euclid Entity, but we will call the Entity, we can say: 1) It must be 'outside of' (independent of) time, 2) It must be 'immaterial' (not composed of anything), 3) It must be 'unchanging' in itself and 4) It originates the derived theory presented here.

In earlier editions we have called this Entity the 'Unmoved Mover' that implements the theory as Reality and causes all events to happen in a manner consistent with the Unified SuperStandard Theory.[8]

In this section we start with a discussion of the basic prerequisites of Creation (a general theory of everything) and then proceed to specify the axioms of creation that yield the derivation.

1-A.2.2 Creation Dynamics Rationale

The derivation of the theory for our universe (and for other possible universes[9] within a Megaverse) has certain fundamental prerequisites if one wishes to have dynamical physical processes, as we know them, to occur within interesting universes. We take these prerequisites for granted normally and proceed to concoct theories of physics as if they may be pulled out of a hat like the proverbial rabbit. However any attempt to create a universal physical theory must meet these prerequisites to build a theory from fundamental primitive terms and axioms in a manner similar to Euclid's construction of geometry.

---

[8] The role and features of the Entity would suggest that it is God to many and is discussed earlier in the God Theory section. However, since we address only Physics issues, we shall leave its nature an open issue in the Unified SuperStandard Theory derivation.

[9] We might call other universes *exoverses*.

1-A.2.3 Fundamental Prerequisites for a Fundamental Theory of Physics

We can list fundamental prerequisites based on a general knowledge of the necessary nature of a fundamental theory of Physics. This approach presumes a general knowledge of the theory that we wish to construct illustrating the maxim, "Our ends determine our beginnings."

A. A time variable must exist that may have various forms.

B. We wish to have a dynamical fundamental theory that evolves in time. Thus there must be a mechanism(s) that allow dynamical processes to exist that may, or may not, run in parallel.

C. Multiple parallel processes can execute.

D. There must be a space with a coordinate system(s), and distance measure, within which processes can execute.

E. There must be particles upon which dynamical processes execute.

F. There must be a space of functionals that support the creation of particle states and help determine their properties. The particle functional space frees particles from a complete dependence on coordinate space.

G. There must be a space of 'waves' of free field fourier expansions for all the fundamental particles absent interactions..

H. There must be an order in the 'created' dynamical theory that is embodied in a form of a computational language[10] with a Chomsky-like *Grammar* using a

---

[10] The possibility that the universe is one enormous Word was explored in *Cosmos and Consciousness* (Blaha (2003)) in physical, philosophical, and religious contexts. A few years ago around 2012 the author found a book with a similar title by R. M. Bucke published in 1901 entitled *Cosmic Consciousness* on the evolution of Man to a new level

finite set of terminal and non-terminal *symbols* that constitutes an alphabet (vocabulary).[11] The ordering in the form of a language with grammar *Production Rules* ensures the consistency of the generated theory.

I. Creation should opt for Vitamorphic[12] universes that support life in some form. Recent studies have shown that evolution favors the development of increasingly intelligent life. Thus the ultimate appearance of intelligent life at places within universes appears to be natural—making the *Anthropic Principle* an evolutionary consequence[13] of the *Vitamorphic Principle*.

These prerequisites would seem to be necessary and sufficient for the specification of primitive terms and axioms for a fundamental theory.

1-A.3 Basis of SuperStandard Axioms

In the following sections we present a revised set of 'primitive' terms and axioms for our theory. A comparison of this new set of axioms with those provided in earlier editions would show that they are equivalent except for a few new axioms. They are also more simply stated, have fewer overlaps between axioms, and cleanly lead to our theory of elementary particles.

The goal of this edition is to derive the Unified SuperStandard Theory in the manner of Euclid with a clear connection between the steps of the derivation just as Euclid developed geometry from a progression of theorems.

---

of consciousness. The content of this book is inrelated to Blaha (1998) – first edition - and (2003) as well as Blaha's other books.

[11] Particle Computer Languages are described in Blaha (2005b) and (2005c) as well later in this volume and in other books by the author.

[12] The *Vitamorphic Principle* states that universes should support some form of life realizing that there are many varieties of life and borderline forms of life. A 'tight' definition of life has not been satisfactorily constructed. There are many borderline entities that may or may not be called life. We take 'Vitamorphic' to mean 'life enabling' in English. Vitamorphism is not a concept without meaning—a universe (Megaverse) consisting of only inert matter without energy present would be non-Vitamorphic. The Anthropic Principle, briefly put, states that intelligent human-like life should exist.

[13] One can well wonder whether the emergence and dominance of Mankind has eliminated the possibility of the emergence of other intelligent species on earth from the many semi-intelligent species that exist now and in the past.

1-A.4 Primitive Terms and Axioms

Primitive terms can be as simple as those of Euclid or they can be more complex. The level of simplicity depends on the nature of the theory and the Physical Laws that emerge from it. In the case at hand, a fundamental unified theory, the constructs that emerge in the construction of the theory are mathematically complex. Consequently, the choice of primitive terms and axioms may be expected to be mathematically complex as well, unless one wishes to expand the primitive terms into a more detailed, term by term description in simpler, more basic primitives. We will not pursue that alternative here since the terms that we use are 'self-explanatory' to the Elementary Particle Physics theorist knowledgeable about quantum field theory and particle symmetries.

1-A.5 Mathematics and Conceptual Prerequisites

Due to the complexity of the Theory we have chosen to specify mathematics prerequisites and use them in the derivation rather than devoting parts of the derivation to mathematical preliminaries. Therefore we use complex variable theory, Riemannian coordinates, group theory, classical and Quantum Logic, functionals, Chomsky-like computational languages, and so on without bringing in unnecessary supporting details from them.

We also assume certain physical concepts such as distance, quantum features, second quantization, covariance under a group transformation, and spatial curvature.

The axioms use some of these prerequisite concepts treating them as primitive terms for the derivation.

1-A.6 Primitive Terms for the Unified SuperStandard Theory

The set of primitive terms of the theory are:

Qubits
Qubes
Qubas
Core
Grammar

Terminal and Nonterminal Symbols
Production Rules
Speed of Light
Spatial Dimensions
Space and Time Coordinates
Covariance under group transformations
Asynchronous processes
Parallel Processes
Reference Frame
Complex Lorentz Group
General Coordinate Transformations
Gravity
Universe
Particle Masses
Fermions
Bosons
Particle States
Particle Rest State
Particle Momenta
Spin
Canonical Quantization
Quantum Process
Quantum Entanglement
Second Quantization
Quantum Field Theory
Quantum States
Asymptotic Particle States
Internal Symmetries
Coupling Constants
Discrete Symmetries
Yang-Mills Local Gauge Theory
Functionals
Functional space

In choosing these primitives, we understand that they generally embody a significant theoretic description or body of knowledge. We do not include names used in the mapping to reality (such as quark) in the list of primitives since the mapping to reality is a separate issue in our view.

1-A.7 Axioms for the Unified SuperStandard Theory

The set of axioms that we list below is supplemented by the Decision Axioms of Appendix C.1.3. The 'new' physical axioms are

PARTICLE AXIOMS

1. All matter and energy is composed of particles.
2. Each fundamental particle has a physico-logic structure within it that we designate its core.
3. Particles form an alphabet with a finite number of characters and combine in ways specified by the quantum probabilistic production rules of a quantum computational grammar.[14]
4. A core is a particle functional that combines with a free field fourier coordinate expansion in an inner product to produce a free second quantized particle field.
5. There is a 4-dimensional space of particle functionals, called *particle functional space*, with the distance measure, eq. 1-A.1 below, specifying the transformation group of particle functionals.
6. Particle functional space consists of a single point.
7. The core of a fermion functional is called a *qube*. Fundamental bosons have a core consisting of a boson functional called a *quba*.
8. Qubes have a bare mass. Qubas have zero mass.

SPACE AXIOMS

9. The dimensions of a coordinate space-time are determined by the number of fundamental[15] interactions, and the requirement that all parallel processes, with parts perhaps separated by distances, can occur synchronously.
10. Spatial coordinates are inherently complex-valued.
11. Space has one complex-valued component that plays the role of time. Physical phenomena dynamically evolve based on the time variable.
12. The infinitesimal distance ds between two space-time points is given by

---

[14] See Blaha (2005b).
[15] Interactions that would exist in the absence of fermion particles.

$$ds^2 = dt^2 - d\mathbf{x}^2 \qquad\qquad (1\text{-}A.1)$$

where $d\mathbf{x}$ is a vector of the spatial coordinates. Transformations between coordinate systems preserve the value of ds and define a transformation group. (The Complex Lorentz Group)

13. Physically acceptable reference frames have real-valued coordinates. These coordinates can be obtained by group transformations from complex-valued coordinate systems. Physical space-time measurements are made in a real-valued coordinate system.

14. The speed of light is the same in all reference frames.

15. Free fundamental leptons must have a real-valued energy.

16. Gravity may cause space-time to be curved. (Complex General Coordinate transformations[16])

DYNAMICS AXIOMS

17. The complete theory has a lagrangian formulation. If the lagrangian is truncated to quadratic form (with interactions set to zero) then symmetries appear that are the source of particle symmetry groups that persist with broken symmetry after interactions are reintroduced. The lagrangian specifies a set of production rules of a type 0 Chomsky language generalized to include production rules for the generation of all strings of symbols (particles) from any strings of symbols (including the *head symbol*.)[17]

18. The lagrangian of the theory must be invariant under coordinate system transformations.

---

[16] If the metric tensor of space-time is analogous to one of the metric tensors of the superfluid phases of ^{3}He, then space-time might have several metric tensors in 'various regions.' If the space-time metric tensor is analogous to the ^{3}He-B superfluid phase metric tensor, which has an effective gravity with a complex metric tensor, the space-time metric tensor would be the familiar one of General Relativity. However if the space-time metric tensor is analogous to the metric tensor of superfluid ^{3}He-A, which exists at higher pressure and temperature, then the space-time metric tensor might be similar to the Penrose twistor theory metric tensor. In this case the corresponding General Relativity may have a twistor-like metric tensor: perhaps in the early universe, and/or inside black holes, and/or in small universes with higher pressure and temperature than our universe. We will assume the conventional metric for Complex Special and General Relativity.

[17] Chapter 8 of Blaha (2018b) discusses computational languages for particles in detail.

19. Dynamical particle equations must be covariant under group transformations.
20. All interactions have a local Yang-Mills gauge theory formulation.
21. The vector bosons, and the interactions among them, are determined by terms in complete lagrangian, some of whose parts are obtained from the Riemann-Christoffel Curvature Tensor.

QUANTIZATION AXIOMS

22. All fields must be canonically quantized.
23. Fermion and Boson vacua can be defined that are valid in all coordinate systems.
24. The number of particles in an asymptotic state of any given type is invariant in all reference frames.
25. Quantum processes starting in an initial quantum state, with parts separated by a distance after a time, can have the parts synchronously change each other instantaneously. (Quantum Entanglement)

1-A.8 The Derivation of the Unified SuperStandard Theory

The derivation of the Unified SuperStandard Theory has been a multi-year process undertaken by the author. Much of the derivation appears in Blaha (2015a), (2016f), (2017b), (2017c), (2017d), (2018a), (2018b), and (2018c). Earlier work, upon which these books are based, is referenced in these books and listed in the References in this book.

In the following we describe the derivations presented in detail in earlier books with some changes. The goal is to show a clear logical development of the Unified SuperStandard Theory from first principles in a manner reminiscent of the derivation of Euclidean geometry. This derivation will be seen to be primarily based on a 'simple' concept—the space-time in our universe. The derivation explains the construction process of the physical theory. The manner of the derivation embodies the mapping of the theory to physical reality. The question of the 'Unmoved Mover' (Appendix C) is necessarily beyond the scope of Physics. (See the God Theory section of this book.)

1-A.9 The 'Time' Order of the Derivation

An examination of the derivation presented here will show that the derivation has an implicit ordering which is analogous to the ordering of theorems in Euclidean Geometry. It begins with the most 'primitive' aspects of the theory and progressively develops and adds features.

Thus it is not time-ordered like the dynamical progression of the universe from the Big Bang, or like the dynamical evolution of physical processes in time. To some extent the derivation consists of a derivation of the basic nature of the fundamental fermion and boson spectrum, and the Standard Model interactions, from a Complex Lorentz group analogy; a derivation of the number of fermion generations, and the Generation group interaction, from the form of the free lagrangian fermion terms and their associated conservation laws; a derivation of the fermion layers, and the Layer groups interactions, from conservation laws implicit in the free lagrangian plus ElectroWeak terms; and a derivation of boson layers from Complex Lorentz group considerations.

1-A.10 Emergence vs. Derivation

Emergence is an interesting approach that is being studied in Physics and Biology. The use of emergent concepts to 'derive' the features of complex phenomena from simpler constructs is conceptually different from a derivation along Euclidean lines. However the procedure is quite similar. The most significant difference is that emergence suggests that complex phenomena somehow mask an underlying simpler *dynamical* structure at their base. The dynamics of the basic structure then generate complex phenomena. The author believes the emergent approach introduces a deeper layer of Physics for which no experimental evidence exists.

An examination of the derivation of the Unified SuperStandard Theory presented here shows that the primitives of the derivation – terms and axioms – may be sufficiently 'masked' so as to justify describing our derived model as quasi-emergent, since it is based primarily on the Complex Lorentz group and Complex General Relativity.

Appendix 1-B. Constraint on Space-Time Dimensions

Generally the dimension of space-time is taken as given and inquiries are not made as to the origin and purpose of the number of dimensions. In this chapter we show a fundamental justification exists for the number of space-time dimensions. The justification is simply the physical requirement that (asynchronous) physical processes can execute in parallel.

1-B.1 Determination of the Number of Space-Time Dimensions

The *a priori* determination of the dimensions of a space-time is guesswork in the absence of a fundamental principle(s).

This chapter derives the dimensions of our 4-dimensional space-time using one principle that yields the same space-time dimension of four. The principle (axiom) specifies the requirement that space-time must allow physical processes to run in parallel following the principles of Asynchronous Logic.[18] This requirement is physically necessary. Spatially separated quantum entangled processes must have parts that proceed in parallel at a distance. This principle is also required in order to have spatially separated classical processes to proceed in parallel.

1-B.2 Synchronization of Non-Local Physical Processes

In earlier books we discussed the central role of Logic and the need for synchronization of non-local physical processes. The need for synchronized non-local physical processes requires the introduction of a new principle: the Principle of Asynchronicity.[19] When processes take place in parallel whether it is Quantum

---

[18] This approach is developed in (2012a) and (2015a).
[19] Much of this chapter is covered in Blaha (2011c) and so is printed in smaller type. Some might argue that it should be called the principle of synchronicity since the goal is synchronization of the parts of an evolving process. We

Mechanical entangled processes at small/large distances, or in high order Feynman diagrams (or their old fashioned time ordered perturbation theory predecessor) the synchronicity of a process is a physical requirement. It is implicitly resolved by physical laws that prevent asynchronicities (situations when parallel processes get "out of sync" resulting in the failure of an entire physical process to complete properly.) The Principle of Asynchronicity is described in the following pages. Asynchronicity can be briefly described as:

> In computation asynchronicity issues can arise. For example parallel computations or computer processes on a chip or set of chips have to be carefully managed for a parallel computer process to complete properly. In the case of computer chip design (VLSI chips and so on) techniques have been developed for the design of chips based on multi-valued logic. One conceptual approach uses 4-valued logic to define clock-less computer logic circuits. The 4-valued logic developed by Fant (2005) has the four logic values TRUE, FALSE, NULL, and INTERMEDIATE. It is an extension of Boolean Logic that can accommodate time asynchronicities in asynchronous computer circuits. It enables circuits to avoid the use of system clocks to implement synchronization.[20] Thus the synchronization is explicit in 4-valued logic and non-logical constructs are not needed.[21] Concurrent transitions are coordinated solely by logical relationships with no need for any time constraints or relationships.

Realizing that the Standard Model, and physical theories that are ultimately derived from it, such as Quantum Mechanics, potentially contain asynchronicities, we suggest that a Principle of Asynchronicity is necessarily embodied in the fundamental theory of Physics.

---

chose to follow the terminology in the field of Asynchronous Logic as exemplified by Fant (2005) – a classic in that field.

[20] Remarkably Bjorken (1965) pp. 220-226 presents an analogy between Feynman diagrams and electrical circuits where momenta map to currents, coordinates to voltages, Feynman parameters to resistance, and free particle equations of motion to Ohm's Law plus the equivalent of Kirchhoff's Laws. Thus Feynman diagrams and computer circuits are analogous.

[21] A two-valued asynchronous logic is also possible – just as the Dirac equation can be expressed as two 2-dimensional equations. See Fant (2005) and Bjorken (1965).

1-B.3 Four-Valued Asynchronous Logic

The basic defining features of asynchronous circuits and Asynchronous Logic are:

1. An *asynchronous circuit* is a circuit in which the component parts are autonomous and can act in parallel at various rates of time evolution. They are not controlled by a clock mechanism but proceed, or wait for signals indicating that they can proceed.

2. *Asynchronous logic* is the logic used in the design of asynchronous circuits. The logic embodies the asynchronicity, and so the circuits built using the logic do not use a clock to control the execution speed of the various parts of an asynchronous circuit. Consequently logic elements do not necessarily have a distinct true or false state at any given point in time. 2-valued Boolean Logic is not sufficient and so asynchronous logic is multi-valued. The logic contains states that allow for "stop and go" states within an executing asynchronous circuit.

In Fant's asynchronous 4-valued logic the four possible truth values of a state are:

True – status is true and all data is current
False – status is false and all data is current
Intermediate – status is indefinite with some data current
NULL – status is indefinite with no data present – results in a suspension of processing of the circuit part in a NULL state until current data becomes present

"Data" is the information flowing through all or part of a circuit. Using these truth-values the evolution in time of the parts of an asynchronous circuit are effectively synchronized by the logic without the use of a clock mechanism. (A clock mechanism effectively is a subsidiary time constraint or set of time constraints.) See Fant (2005) for further details.

An implicit aspect of asynchronous logic is the coordination of spatially separated parts of a circuit. Since spatial separations in a circuit can be mapped to time delays using the speed of data propagation between parts, spatial asynchronicities are subsumed under time asynchronicities. This is particularly true for computer chips that are kept small to minimize delays.

1-B.4 Principle of Asynchronicity

An obvious feature of elementary particle phenomena is the coordination of the parts of a physical process in time and space. Complex Feynman diagrams embody the coordination of the parts of interacting particles. Quantum entanglement phenomena embody the coordination of the parts of a physical phenomenon separated by large distances and perhaps times. Examples of these types, which could be multiplied indefinitely, lead to a Principle of Asynchronicity.

Principle (Axiom): Nature requires an asynchronicity principle. Aasynchronicity is coordinated by 4-valued physico-logical structures for matter.

Elaboration: Elementary particle physical phenomena must support extended coordinated physical phenomena in space and time. The fundamental laws of particle physics must be such as to permit coordinated physical phenomena with coordination between the parts of a physical phenomenon at small/large distances and small/large time intervals. The coordination must be embodied within physical laws. We make it an axiom of our theory.

Asynchronicities are common in the many sub-circuits of a computer chip. Asynchronicities are also common in the many interaction sub-regions of a set of interacting particles. Page 7 of Fant (2005) has a diagram of a circuit with a set of sub-circuits with five time slices of the interacting sub-circuits showing five states of the "'data' wave front" at five points in time. This diagram is similar to the time-sliced diagram of an interacting system of particles in "old fashioned" time-ordered perturbation theory. Page 29 of Blaha (2005b) displays a similar diagram (Fig. 5.1.4) in a description of a Standard Model Quantum Language Grammar – a language representation of particle physics. Blaha's diagram[22] is remarkably similar to Fant's diagram in overall features as one might expect since both address time asynchronicity.

The asynchronicity that appears in perturbation theory diagrams is intimately related to the appearance of antiparticles in diagrams. As noted earlier antiparticles are interpretable as negative energy particles traveling backwards in time. The time orderings, which are implicit in the Feynman diagram approach and explicit in old fashioned perturbation theory, evidence time asynchronicity and the effects of the dynamics. They coordinate asynchronicities such that correct results follow from perturbative calculations.

1-B.5 Dimension of the Megaverse

Having found the dimension of our universe is 4, we now turn to the dimension of the enclosing Megaverse. The dimension must be greater than four. But its value cannot be specified by a fundamental principle except perhaps minimality. Later we will see the choice of a dimension of eight yields a Megaverse set of internal symmetries that are a simple superset of the dimensions of our 4-dimensional QUeST. It does add additional Dark sectors in a way that is compatible with the Unified SuperStandard Theory.

Other choices such as a 6-dimensional alternative are either more complex or have too many internal symmetries and fundamental particles.

On this basis we choose an 8-dimensional Megaverse. It gives a Megaverse that can have an infinite number of universes and yet is not "overly" wasteful of particles and interactions.

---

[22] Created without knowledge of Fant's work.

2. New Deeper Axioms for Unified SuperStandard Theories in Quaternion and Octonion Space-times

Previous derivations of the Unified SuperStandard Theory were based on Complex General Relativity and Quantum Field Theory. This book presents a deeper derivation that unites space-time coordinates and internal symmetry coordinates. Our motivation is the close analogy between Lorentz subgroups and Standard model groups suggesting a possible common origin, which **we found in a close examination of the symmetries of the Unified SuperStandard Theory in 3+ 1 real-valued coordinates.** We seek a corresponding structure in the form of a larger space-time. The basis for our approach is the trend of types of coordinates:

Real → complex → quaternion → Complex quaternion → octonion → Complex octonion

The Standard Model is based on real-valued coordinates. The Unified SuperStandard Theory, and its precursors by the author, was based on complex coordinates (specifically Complex Lorentz group transformations). Recently the author has developed biquaternion (complex quaternion), and bioctonion (complex octonion), SuperStandard Models in Blaha (2020a) and Blaha (2020b) with features that *directly lead to the Unified SuperStandard Theory upon restriction to real-valued coordinates.* This book expands on these books.

We create a Unified SuperStandard Theory in 32 complex quaternion coordinates (called QUeST) for our universe, and a Unified SuperStandard Theory in 64 complex octonion coordinates for the Megaverse of universes (called MOST). These theories were outlined in Blaha (2020a) and (2020b).

These new theories mesh well with the universe and Megaverse, which we described previously. Upon restriction to 3 + 1 real-valued coordinates QUeST becomes the Unified SuperStandard Theory of our universe described in earlier books

and later in this book. Upon restriction to 7 + 1 real-valued coordinates MOST becomes the Unified SuperStandard Theory of the Megaverse described in earlier books and later in this book

2.1 Axioms for Unified SuperStandard Theories

AXIOMS

1. Biquaternion space is the basic space-time of our universe. Bioctonion space is the basic space-time of the Megaverse. These spaces factorize into a coordinate space-time and an internal symmetry space-time.

2. Physical processes can execute in parallel.

3. Matter and energy are particulate.

4. Space--times are locally Lorentzian.

5. All calculations are finite.

6. Particle theory can be defined in any curved space-time.

7. Each particle has a wave function determined by a functional inner product defining the particle state. The functionals form a set without a distance measure.

2.2 General Implications of the Axioms

In this section we describe some of the implications of each of the axioms.

1. Biquaternion space is the basic space-time of our universe. Bioctonion space is the basic space-time of the Megaverse. These spaces factorize into a coordinate space-time and an internal symmetry space.

The factorization into a space-time and an internal symmetry space must be a form of spontaneous symmetry breaking of yet unknown origin. It appears to be related to a breakdown of the vacuum.

2. Physical processes can execute in parallel.

Physical processes are known to be able to execute in parallel at any distance of separation. As Fant has shown parallel execution requires a minimal number of dimensions: 4. Consequently the dimension of space-time must be 4 or greater. The biquaternion space-time of QUeST is 4-dimensional allowing parallel process execution.

The bioctonion space-time of MOST is 8-dimensional and also allows parallel process execution. The choice of eight dimensions is natural since it allows 4-dimensional universes within it. It also has a form that allows a clean formulation. Lastly, as will be seen later, it conforms to the pattern of interplay between Lorentz symmetry and internal symmetry found in the Unified SuperStandard Theory. See Appendix 1-B for an extended discussion of parallelism and dimension.

This axiom leads to a view of the origin of the dimensions.

3. Matter and energy are particulate.

The most direct method of specifying a theory of matter and energy is through the Use of Quantum Field Theory. Thus Quantum Field Theory is implied.

4. Complex Space-times are locally Lorentzian.

A locally complex Lorentzian space-time leads to Complex General Relativity. In flat space-time Complex General Relativity becomes Complex Lorentz group. (In point of fact the Complex Poincaré group follows.

5. All calculations are finite.

Given the need for Quantum Field Theory it becomes necessary to find a formulation that yields finite values for calculations in perturbation theory. The only approach that eliminates high energy divergences, and yet preserves the results found in perturbation theory calculations that agree with (primarily QED) experiments, is Two-Tier Quantum Field Theory. This is discussed in detail in earlier books starting in 2002. Thus only our Two-Tier formalism satisfies this axiom.

6. Particle theory can be defined in any curved space-time.

In the 1970s we developed a formalism that allows the definition of particle states in any space-time in such a way that its physical content is preserved when transformed to any coordinate system.[23] This PseudoQuantum Quantum Field Theory satisfies this axiom.

7. Each particle has a wave function determined by a functional inner product defining the particle state. The functionals form a set without a distance measure.

This axiom is satisfied by our formulation of quantum functionals in Blaha (2019f) and earlier books. Our formulation eliminates the superficial violation of the Theory of Relativity by "spooky" quantum entangled processes with parts separated by a physically "large" distance.

The seven axioms imply the Unified SuperStandard Theories and its deeper biquaternion and bioctonion hypercomplex formulations.

---

[23] S. Blaha, Il Nuovo Cimento **49A**, 35 (1979).

3. Quaternion Unified SuperStandard Theory (QUeST) Formulation Compared to Unified SuperStandard Theory

This chapter[24] lists some of the highlights of the Unified SuperStandard Theory (presented in Blaha (2018e) and (2019g)) and briefly identifies the differences from our new 32 complex quaternion dimensional Quaternion Unified SuperStandard Theory (QUeST) formulation presented in this book and in Blaha (2020a) and (2020b).

QUeST has the known features of the Standard Model and of the Unified SuperStandard Theory. QUeST adds some additional, beneficial, features to the Unified SuperStandard Theory

3.1 COMPARISON

1. The number of spatial dimensions was determined to be the number of generators in the primary set of interactions of the space. In the case of an *empty* universe the primary set of interactions is the U(2) qubit transformations group. The number of U(2) generators is four and thus the dimension of space is 4 complex dimensions. Also and more importantly, considerations of Asynchronous Logic, and the requirement that physical processes must be able to proceed in parallel, require the number of spatial dimensions to be four. The book justifies four complex space-time dimensions with a Lorentz metric yielding Complex Lorentz group symmetry.

Change: The dimensionality is set by the quaternion foundation assumptions..

2. Boosts of the Complex Lorentz group transform a Dirac-like equation with a Landauer mass into four different forms (called species). Each form maps to a type of

---

[24] This chapter is extracted from Blaha (2020) for the reader's convenience.

fermion: neutral leptons (neutrinos), charged leptons, up-type quarks, and down-type quarks. Neutral leptons and down-type quarks are tachyons. Some evidence exists for tachyonic neutrinos. Complex Lorentz boosts lead to the Complex Lorentz group factorization: $SU(2) \otimes U(1) \otimes SU(3) \otimes SU(2) \otimes U(1)$. We map $SU(2) \otimes U(1) \otimes SU(3)$ to fermion particle functional space to obtain the internal symmetry group for ElectroWeak and Strong Interactions: $SU(2) \otimes U(1) \otimes SU(3)$. The remaining factors $SU(2) \otimes U(1)$ we map to the internal symmetry group for Dark Matter, which we take to be the Dark ElectroWeak Interaction (unconnected to normal matter interactions).

> Change: We find an additional SU(3) Dark Strong Interaction group giving an $SU(2) \otimes U(1) \otimes SU(3) \otimes SU(2) \otimes U(1) \otimes SU(3)$ symmetry.

3. Parity Violation as seen in the Weak Interactions follows directly from the forms of the four types of fermions predicted by the Complex Lorentz Group.

> Change: None

4. The existence of four conserved (and partially conserved) quantum numbers such as baryon number and lepton number indicates that there is a U(4) group whose **4** representation causes each species to have four generations—three of the generations are known. We suggest that a fourth generation of much higher mass fermions exist.

> Change: Normal matter has a U(4) Generation group and Dark matter has a separate U(4) Dark Generation group.

5. In each generation there are four partially conserved quantum numbers. Thus we find that there is another U(4) group (called a Layer group) for each generation yielding the combined Layer groups $[U(4)]^4$. The **4** representation of each U(4) results in a fermion spectrum of four layers of four generations or 192 fermions in all. We see only one layer at present. The additional three layers of fermions remain to be found at much higher masses. The symmetry group of the Unified SuperStandard Theory is

$$[SU(2)\otimes U(1)\otimes SU(3)\otimes SU(2)\otimes U(1)\otimes U(4)\otimes U(4)]^4\otimes U(4)$$

where the last factor is for the broken Species group, which follows from Complex General Relativity.

Change: Changed symmetry. Additional Dark Strong SU(3) group and separate Generation and Layer groups for the Dark sector to avoid generating normal particle and Dark particle interactions. The result:

$$[U(1)\otimes SU(2)\otimes SU(3)\otimes U(1)\otimes SU(2)\otimes SU(3)\otimes U(4)^4]^4\otimes U(4)$$

6. Assuming all particles are massless at the Big Bang, and all particle types have an equal proportion of the total mass-energy then, we find that the 192 fermions and 192 vector bosons yield a Dark Matter percentage of 83.33% (experimentally the estimates are 84.5% and 81.5%). The proportion of Dark Mass-Energy is found to be 91% of the universe's mass-energy. Experimentally the proportion has been estimated to be 95%. These results agree well with experiment. See chapter 14 of Blaha (2019g) and (2018e) for details.

Change: Now 224 Dark fermions and 224 "normal" fermions (of which 24 are known) totaling 448 fermions.

7. The instantaneous quantum effects between space-like separated parts of a quantum state ('spookiness') are taken to be a feature of fundamental importance. The only sensible way to implement this feature in quantum theory is to assume that the wave function of every particle is the inner product of a particle functional and a wave (Fourier) coordinate expansion. Particle functionals exist in a space with no distance measure. The space of coordinate expansions also has no distance measure. Other functionals in a state (and their implicit coordinate Fourier expansions) change *instantaneously* when one of the functionals comprising a state changes since coordinate space distance is irrelevant.

Change: None

8.   Fermion particle functionals are called *Qubes*. They exist 'within' every fermion. They have a mass that we take to be the Landauer mass—the minimal energy of a qubit. Boson particle functionals are called *Qubas*. They are assumed to be massless in the absence of all interactions to preserve free vector boson and spin 2 boson gauge symmetry. Free Higgs particles are assumed to be massless for consistency.

Change: None

9. To have a completely finite theory with no infinities (including no fermion triangle infinities) we introduced Two-Tier Coordinates that replaced normal point like coordinates with a type of 'fuzzy' coordinates.

$$X^\mu = x^\mu + i Y^\mu(x)/M_c^2.$$

Change: None

10. Since the Unified SuperStandard Theory lagrangian would require higher order derivatives to account for quark confinement (linear potential terms) and for MoND-like deviations from conventional gravity, and since such terms would be outside a canonical lagrangian formulation, we introduced two fields for each particle (fermions and bosons) in a formulation we call PseudoQuantum Theory. PseudoQuantum theory enables a canonical lagrangian formulation. It has other advantages such as a clean separation of vacuum expectation values from quantum fields for Higgs particles. It also supports second quantization in arbitrary coordinate systems while maintaining the same particle interpretation of states in all coordinate systems.

Change: None

11. The book also describes Higgs symmetry breaking and the use of the Faddeev-Popov Mechanism in detail for the theory.

Change: None

12. Since a Complex Special Relativity requires a Complex General Relativity we considered Complex General Relativity and showed that it could be 'factored' into General Relativity and a new U(4) group that we called the Species group. Since Complex General Relativity must support interactions with all types of matter we specified a Species group interaction with all matter. Further, we assumed that the Species vector bosons acquired masses through the Higgs Mechanism. The Higgs Mechanism caused Species group contributions to each fermion mass. Such a mass term would require each fermion particle mass to be both inertial *and* gravitational *solving the mystery of the equality of inertial and gravitational mass.*

Change: None

13. We showed that the implicitly higher derivative Riemann-Christoffel curvature tensor for all interactions leads to new interactions beyond The Standard Model. In addition to yielding quark confinement and MoND-like modifications of gravity, it may help understand the missing nucleon spin issue, discrepancies in proton radius measurements, vector meson dominance (VDM), and so on.

Change: Addition of Dark SU(3) terms and Dark Generation and Layer groups to the Riemann-Christoffel tensor for each layer.

14. We defined an Interaction Rotations group that caused rotations among all the vector boson interactions of The Unified SuperStandard Theory. We found that rotations that respected Superselection rules such as the Charge Superselection rule could have physical significance. One example is ElectroWeak Theory, which is an application of Interaction Rotation transformations.

Change: The Interaction Rotations group is now not compelling since it is not in QUeST.

15. Since the number of fundamental fermions (192) and fundamental vector interaction bosons (192) is equal we considered Supersymmetric features of the Unified SuperStandard Theory.

Change: The number of fermions and vector bosons is changed. (Item 6.) SuperSymmetry not indicated.

16. The discovery of two new particles that do not appear to be within the framework of The Standard Model, as it is currently known, raises the possibility that they may be within the expanded fermion spectrum in The Unified SuperStandard Theory. Towards that end we present a *preliminary* assignment of the locations of the new fermions within the spectrum of the Unified SuperStandard Theory.

Change: The additional Dark Strong SU(3) group implies Dark quarks are triplets.

17. We showed that the coupling constants of the Standard Model including the Fine Structure Constant of QED are determined by eigenvalue functions. As a result they differ and in a way that appears contrary to the hypotheses of Grand Unified Theories (GUTs) even taking account of running coupling constant considerations.

Change: None

Appendix 3-A. Gauge Groups Based on Particle Numbers

In this Appendix[25] we show the origin of the Generation and Layer groups in particle number operators since they are not well known. Chapters 43 – 45 describe the in more detail. Particle interactions followed directly in the Unified SuperStandard Theory by analogy with Complex General Relativity subgroups yielding

$$SU(2) \otimes U(1) \otimes SU(3) \otimes SU(2) \otimes U(1) \otimes SU(3) \qquad (3\text{-}A.1)$$

where the latter three factors are the Dark interactions.

They have a SU(10) covering group that contains this direct product of groups. The groups in eq. 3-A.1 are particle interaction groups in the Unified SuperStandard Theory.

Unlike other attempts to develop a formulation of the Standard Model (or generalizations) the Unified SuperStandard Theory was originally directly based on a theory foundation consisting of Complex General Relativity and Quantum Field Theory. Later we will show a deeper basis in Quaternions and Octonions.

To those who might prefer to base a theory on real General Relativity we note that proofs in Quantum Field Theory *require* the Complex Lorentz Group.[26] Thus the Complex Lorentz group is unavoidable for a properly (and rigorously) formulated Quantum Field Theory. Since the formulation of the Complex Lorentz Group in flat space-time can only be as the limit of Complex General Relativity, the choice of a foundation of Complex General Relativity is required.

---

[25] This appendix is an extract from Blaha (2020 for the readers convenience.
[26] Streater (2000).

Since particles are countable, and thus have discrete particle numbers, Quantum Field Theory brings particle numbers, and particle number laws such as particle conservation laws, into consideration.

Blaha (2019e) and earlier books showed that Complex Lorentz boosts generate four types of fermion particles that we call *particle species*. We map these four species to charged leptons (such as electrons), neutral leptons (such as neutrinos), up-type quarks (such as the u quark), and down-type quarks (such as the d quark).

3-A.1 Basis of the Generation Group

We define two particle number operators for normal up-quark particles and down-quark particles, B_{uq} and B_{dq}. Similarly we define two particle number operators for normal species "e" (electron) particles and species "v" particles, B_e and B_v. Similarly we define Dark matter equivalents:[27] B_{De}, B_{Dv}, B_{Duq}, and B_{Ddq}.

In the absence of interactions these fermion particle number operators are conserved. Each set are "diagonal" operators within a U(4) group. Thus we have a normal U(4) Generation Group and a Dark U(4) Generation group.

On this basis we find there are four generations of each species in the normal and in the Dark matter sectors. One generation of normal fermions with large masses has not as yet been found.

The gauge vector bosons of the Generation Group also have large masses. If the conservation of the fermion particle numbers is broken then we view it as a consequence of Generation Group symmetry breaking.

3-A.2 Basis of the Layer Group

The set of particle number operators can be further refined if we take account of the fourfold fermion generations. To further refine the set of particle number operators we temporarily neglect all interactions that would violate conservation laws for the set.

---

[27] By analogy, we assume that there are four species of Dark matter: charged Dark leptons, neutral Dark leptons, Dark up-type quarks, and Dark down-type quarks. Thus we are led to the Dark particle numbers: Dark Baryon Numbers, and Dark Lepton Numbers shown above.

We therefore subdivide the above particle number set into four particle numbers per generation. For the ith generation we define

L_{ie} – The "e" species particle number for the ith generation
L_{iv} – The v species particle number for the ith generation
L_{iuq} – The up-quark species particle number for the ith generation
L_{idq} – The down-quark species particle number for the ith generation

L_{iDe} – The Dark "e" species particle number for the ith generation
L_{iDv} – The Dark v species particle number for the ith generation
L_{iDuq} – The Dark up-quark species particle number for the ith generation
L_{iDdq} – Dark down-quark species particle number for the ith generation

for each generation i = 1, 2, 3, 4. Individual fermions have positive L_{ia} = +1 values and anti-fermions have negative L_{ia} = –1 values for species a = 1, 2, 3, 4 (with the three color subspecies of quarks treated as part of one species.)

At this point we have four particle number operators for each generation. We define a group framework for each set of particle numbers. The simplest way is to assume that each generation consists of four layers with the particles in each generation in a U(4) fundamental representation.[28] Then each generation has a U(4) Layer group with the generation's four number operators (above) as its diagonal operators. We call this group the Layer Group of the ith generation L_{ia}. With four generations we obtain four U(4) Layer groups for normal matter. In addition there are four U(4) Dark Layer groups. See Fig. 2.4.

The consequence of this expansion of particle numbers and groups is that the set of fermions increases fourfold. We now have four layers, with each having four generations, Experimentally, we know of three generations of fermions—the lowest generations of the lowest level. The remaining generation and three levels of fermions are of much higher mass and yet to be found.

---

[28] See Fig. 2.3 for a depiction of the "splitting" of fermions: first into generations, then into layers.

See Blaha (2019g) and (2018e) for a detailed discussion of the Layer Groups. We note in passing that the symmetries of these number operators are badly broken. Yet the underlying group structure remains.

Quaternion/Octonion Derivation of Universe and Megaverse Internal Symmetries and Particle Spectrums

4. Quaternion Universe–Octonion Megaverse Scenario

4.1 Hypercomplex Number Based Higher Dimensions

The Unified SuperStandard Theory was based on complex-valued coordinates. This choice enabled us to understand the reason behind the four types of fermions found in nature: neutral fermions, charged fermions, up-type quarks, and down-type quarks. A study of Complex Lorentz group subgroups showed that they were similar to the factors of the Standard Model symmetry group. This similarity motivated this author to consider a space with dimensions that corresponded to space-time and to internal symmetry dimensions.

In choosing a higher dimension space for a larger theory of elementary particles the use of coordinate systems based on hypercomplex number systems seemed reasonable. The hope was that just as complex coordinates led to a deeper understanding of the internal symmetries of the Standard Model, the use of hypercomplex coordinates might lead to a further understanding of the origin of both space-time and internal symmetries.

The pattern of rising hypercomplexity is:

Real → Complex → Quaternion → Biquaternion → Octonion → Bioctonion

The Unified SuperStandard Theory took particle theory from real-valued coordinates to complex-valued coordinates. Complex quaternion (biquaternion) and complex octonion (bioctonion) extensions took us to QUeST and MOST. They used larger spaces to unite space-time symmetry and internal symmetry.

Since the requirement of parallel physical processes made the minimal space-time dimension 4 and since the Megaverse must include universes as subspaces, we were led to a 4-dimensional complex quaternion formulation for our universe and an 8-

dimensional complex octonion formulation for the Megaverse. They were considered in Blaha (2020a) and (2020b). Generalizations to higher dimensional quaternion and octonion spaces were considered to accommodate the four layers suggested by the Unified SuperStandard Theory.

In this book we consider these higher dimensional theories and find that the complex quaternion higher dimensional theory QUeST) leads directly to the 3 + 1 dimensional Unified SuperStandard Theory upon restriction of the theory to real-valued coordinates. Similarly the Megaverse MOST leads to a reasonable Megaverse theory upon restriction to a generalization of the Unified SuperStandard Theory in 7 + 1 real space-time dimensions.

Fig. 4.1 below summarizes QUeST and MOST dimensions and particles.

Space	Complex Quaternion[29]	Complex Octonion[30]
Number of Quaternions[31] per Space-Time Coordinate	2	4
Number of Quaternion/Octonion Space-Time Dimensions per layer	7 + 1	7 + 1
Total Number of Dimensions (real) per layer	64	128
Internal Symmetry Dimensions (real) per layer	52	104
Space-Time Dimensions (real) per layer	8	16
Number of Fermions Per Layer[32]	64	128
Total Number of Fermions	256	512

Table 4.1. Comparison of QUeST Universe and MOST Megaverse Features.

---

[29] From chapter 6.

[30] From chapter 10.

[31] The 3 quaternion case is not considered here. It seems unsatisfactory because it does not lead to the desired internal symmetries or fermion spectrum. Cases where the number of quaternions exceed 4 appear to have much too many internal symmetries and particle spectra.

[32] The fermion number counts quark triplets: three types for each quark species.

4.2 Glossary of Table 4.1 Entries

Number of Quaternions – The dimension of a coordinate measured in multiples of four..
:

Number of Space-Time Dimensions – The number of time and spatial quaternion or octonion coordinates per layer.

Total Number of Dimensions – The total number of coordinates counting all time and spatial coordinates within quaternions (octonions) per layer.

Internal Symmetry Dimensions – The number of dimensions devoted to being internal symmetry representation coordinates per layer.

Space-time Dimensions – The number of coordinates that become space-time coordinates per layer.

Number of Fermions per Layer – The number of fundamental fermions counting individual quarks in quark triplets per layer.

Number of Fermions – The sum over layers of the fundamental fermions counting individual quarks in quark triplets.

Quaternion Universes and 32 Dimension Complex Quaternion Unified SuperStandard Theory

The chapters in this section describe the form of complex quaternion universes. For good reason it appears that all universes have this form; the same internal symmetries, the same fundamental fermion spectrum, the same vector boson spectrum and interactions, and the same coupling constants. Thus all universes may be similar. And the form of the Megaverse may be determined.

5. Quaternion Unified SuperStandard Theory (QUeST)

Quaternions and octonions are hypercomplex numbers with special properties that make them similar to complex numbers.[33] Quaternions and octonions are both normed division algebras over the reals (*hypercomplex* number systems) with salutary properties for quantitative studies in quantum field theory and perturbation theory. Some of their new features are listed on the cover page.

Quaternions have significant properties that distinguish them:

1 .They are associative.

2. They are one of the two finite dimensional division rings having the real numbers as a proper subring. (The other is octonions—considered in chapter 8.)

3. They are non-commutative. (This is not a roadblock for quantum field theory which is also non-commutative in general.)

These features support the development of physics theories.[34]

5.1 Some Basic Quaternion Features

A quaternion is a 4-tuple of real numbers. A complex quaternion is a 4-tuple of complex numbers:

---

[33] Much of this chapter appears in Blaha (2020a) and (2020b).

[34] There is an extensive literature on quaternions starting with the original work of Hamilton. Some recent, relevant papers are: S. L. Adler, "Generalized Quantum Dynamics", IASSNS –HEP-93/32 (1993); S. De Leo, arXiv:hep-th/9506179 (1995); Rolf Dahm, arXiv:hep-th/9601207 (1996); S. De Leo, arXiv:hep-th/9508011 (1995); S. L. Adler, arXiv.hep-th/9607008 (1996) and references therein.

$$x = a + bi + jc + kd \ = a + \mathbf{v} \tag{5.1}$$

where a, b, c, d are real or complex numbers, and $\mathbf{v}$ is a 3-vector. The symbols i, j, and k are fundamental quaternion units. A quaternion norm is defined by

$$\|\mathbf{x}\| = \text{sqrt}(aa^* + bb^* + cc^* + dd^*) \tag{5.2}$$

and the norm of $\mathbf{v}$ is

$$\|\mathbf{v}\| = \ \text{sqrt}(bb^* + cc^* + dd^*) \tag{5.3}$$

An important identity is

$$e^x = e^a (\cos (\|\mathbf{v}\|) + \mathbf{v}/\|\mathbf{v}\| \sin(\|\mathbf{v}\|))s \tag{5.4}$$

.It is used to define boosts in quaternion space.

5.2 Motivation and Procedure

Our goal is to create a larger dimension space within which we can derive our space-time and the Unified Superstandard Theory in such a way as to understand the similarity of Lorentz subgroups and Standard Model internal symmetry groups. The development of a deeper form of the Unified SuperStandard Theory will lead to refinements in the theory.

There are two possible procedures to follow in developing the deeper basis:

1. One can develop the Quantum Mechanics and Quantum Field Theory in a quaternion space and then extract the dynamics, fermion spectrum, gauge fields, and so on of our familiar space-time.

2. One can define a quaternion space, and then using its coordinates, directly extract the space-time, internal symmetries, fermion spectrum, gauge field spectrum and dynamics. Quaternion algebra may not be used. A quaternion then specifies a 4-vector of coordinates.

We have chosen the latter approach as it will directly lead to the Unified SuperStandard Theory in 3 + 1 real-valued dimensions..

In developing the deeper space, upon which we will build, we will take guidance from the derivation of the Unified SuperStandard Theory. That theory assumes a complex 4-dimensional space-time upon which Complex General Relativity is constructed. It then proceeds to complex flat space-time and Complex Relativity. Next it restricts the complex coordinates of the theory to real-valued coordinates (excepting quarks and color gluons that remain complex as described later).

We follow this procedure emulating the approach of the 4-dimensional Unified SuperStandard Theory.

5.3 Definition of Quaternion Space – Complex Quaternion (Biquaternion) Space

Following the stated procedure we define a complex or biquaternion space[35] with one "time" biquaternion and seven "spatial" b iquaternions believing the 7+1 space-time of our experience is a reflection of this deeper level.

Time Biquaternion
$$t = (a + bi + jc + kd) + I(a' + b'i' + j'c' + k'd')$$

Spatial Biquaternions
$$x = (a_x + b_x i + jc_x + kd_x) + I(a'_x + b_x'i' + j'c_x' + k'd_x')$$
$$y = (a_y + b_y i + jc_y + kd_y) + I(a'_y + b_y'i' + j'c_y' + k'd_y')$$
$$z = (a_z + b_z i + jc_z + kd_z) + I(a'_z + b_z'i' + j'c_z' + k'd_z')$$
$$x1 = (a_{x1} + b_{x1}i + jc_{x1} + kd_{x1}) + I(a'_{x1} + b_{x1}'i' + j'c_{x1}' + k'd_{x1}')$$
$$y1 = (a_{y1} + b_{y1}i + jc_{y1} + kd_{y1}) + I(a'_{y1} + b_{y1}'i' + j'c_{y1}' + k'd_{y1}')$$
$$z1 = (a_{z1} + b_{z1}i + jc_{z1} + kd_{z1}) + I(a'_{z1} + b_{z1}'i' + j'c_{z1}' + k'd_{z1}')$$
$$w1 = (a_{w1} + b_{w1}i + jc_{w1} + kd_{w1}) + I(a'_{w1} + b_{w1}'i' + j'c_{w1}' + k'd_{w1}')$$

where I is an additional "imaginary" number with $I^2 = -1$ that explicitly *makes biquaternions (complex quaternions) from real quaternions. We will not use the algebra of quaternions but simply treat the quaternion coordinates as coordinates in a space.*

---

[35] We will use biquaternion to mean complex quaternions from this point on.

Fig 5.1 symbolically depicts the space with a black circle for each real-valued coordinate. We initially treat the real-valued and imaginary parts of a complex quaternion as a set of all real-valued coordinates.

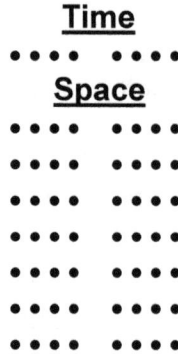

<u>Time</u>
```
• • • •   • • • •
```
<u>Space</u>
```
• • • •   • • • •
• • • •   • • • •
• • • •   • • • •
• • • •   • • • •
• • • •   • • • •
• • • •   • • • •
• • • •   • • • •
```

Figure 5.1. Eight-Dimensional biquaternion space with coordinates represented by • 's. There are 64 real-valued coordinates.

This biquaternion space has 64 real dimensions (32 complex dimensions.).

5.3.1 Dimension Functionals

The dimension array in Fig. 5.1 can be represented in terms of *Dimension Functionals*. We can define a dimension row functional D_i that takes a column dimension as its argument and generate the dimension array $[d_{ij}]$ of Fig. 5.1 with

$$d_{ij} = D_i(d_j)$$

where $i = 1, 2, ..., 8$ and $j = 1, 2, ---, 8$. Dimension functionals can also be used in the case of octonion coordinates in chapter 9.

This formalism illustrates the possibility of dimensions playing the role of functionals. We will discuss functionals in more detail in chapter 7.

5.4 Biquaternion Lorentz Group

Our definition of time and space biquaternion coordinates purposefully resembles those of our real space-time. One might ask why should there be a Lorentz-like group for biquaternion space. The only apparent reason is the need for a special speed c in our space-time that enables one to boost from a rest frame of a mass m particle with energy m to a moving frame of greater energy.[36] Without c the group of the above coordinates would presumably be the biquaternion U(8) group. In this group, transformations preserve the norm of a state so that a "boost-like" transformation does not exist.

The biquaternion Lorentz transformations do have a unique speed c (the speed of light) and specify a unique rest frame for any particle—both sublight particles and tachyon particles.[37] . Thus we select the Biquaternion Lorentz group SU(7,1) for biquaternion space.

Flat space biquaternion Special Relativity generalizes directly to a biquaternion General Relativity which may be constructed directly (mindful of quaternion non-commutativity).

The flat space biquaternion Lorentz group transformations have constant biquaternion matrix elements that are analogous to those of the Lorenz group.. (See Appendix 12-A for Lorentz group boosts.)

Motions in complex quaternion space are of four types: motions where the norm of the velocity is real-valued and less than c, motions where the norm of the velocity is real-valued and greater than c, motions where the norm of the velocity is complex-valued and whose absolute value is less than c, and motions where the norm of the velocity is complex-valued and whose absolute value is greater than c,

Boosts generating these cases lead to the separation of fermions into four species as we show in Appendix 12-A for the Unified SuperStandard Theory and its complex quaternion basis QUeST (as well the complex octonion Megaverse MOST.)

---

[36] A similar consideration applies to tachyons.
[37] Tachyons can be transformed by a Complex Lorentz transformation to and from a rest frame. See Blaha (2018f) and Appendix 12-A below.

5.5 Extracting the Symmetries and Particle Spectra

As stated earlier in section 5,2 we will directly describe the symmetry structure implied by the form of the biquaternion coordinate system while mindful of sections 5.3 and 5.4.

The Unified SuperStandard Theory developed the group structure from which the particle species were derived from a subset of Lorentz boost transformations. It used boosts with complex exponentiation similar to quaternion exponentiation in eq. 5.4. Complex boosts mapped a system at rest to a system in motion with a real energy and complex 3-momenta in general. Biquaternion boosts play a similar role.

The 4-dimensional subspace for the Unified SuperStandard Theory complex coordinate boosts' was

<u>**Time**</u>

•

<u>**Space**</u>

• •

• •

• •

Figure 5.2. Four-Dimensional subspace for Unified SuperStandard derivation of particle spectra with coordinates represented by • 's. Note the time coordinate wasreal-vlaued.

Following the same line of reasoning we now specify a biquaternion subspace analogously restricted to that of Fig. 5.2 to define the relevant set of coordinates for determining particle symmetries and spectra. Note time is a "real" quaternion in this subspace. The spatial coordinates consist of complex quaternions.

Time
• • • •
Space
• • • •   • • • •
• • • •   • • • •
• • • •   • • • •
• • • •   • • • •
• • • •   • • • •
• • • •   • • • •
• • • •   • • • •

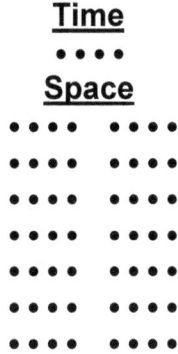

Figure 5.3. 8-dimensional biquaternion subspace for symmetries and particle spectra.

We expect its 30 complex coordinates will split into a 4-dimensional complex coordinate space[38] which will support Complex Lorentz transformations, and the remaining 26 complex coordinates will represent internal symmetry space. Internal symmetry space has an initial SU(26) group before breakdown.

The mechanism for this symmetry breakdown may involve vacuum energy effects in the biquaternion universe.

The remaining four time dimensions (coordinates) shown in Fig. 5.1 will be shown to correspond to a U(2) group that maps between normal fermions and Dark fermions on a one-to-one basis. Experiment indicates this group's interactions must be extraordinarily weak.

In the next chapter we analyze the symmetries of the 26 complex dimensional subspace.

---

[38] The 7 + 1 biquaternion space yields a 3 + 1-dimensional complex coordinate space which then becomes the 3 + 1 dimensional real-valued space of our experience.

6. Internal Symmetries of the Quaternion Unified SuperStandard Theory (QUeST)

The coordinate space picture of QUeST described in chapter 5 enables us to simply find the internal symmetries and particle spectra of QUeST. They will turn out to be those of the Unified SuperStandard Theory.

The QUeST coordinate subspace for the determination of internal symmetries is depicted in Fig. 6.1..It is based on the discussion of Fig. 5.3. The complex space-time 4-vector is separated from the internal symmetry coordinates.

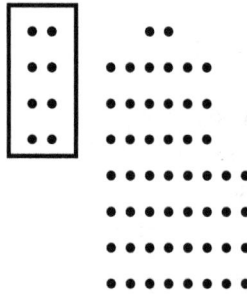

Figure 6.1. Internal Symmetry determination subspace. Eight real-valued space-time coordinates (giving a complex 4-vector) are separated from the 60 real-valued coordinates for internal symmetries.

The internal symmetry coordinates number 52 real coordinates or 26 complex coordinates. These coordinates serve as coordinates[39] of the fundamental

---

[39] Simple counting of fundamental representation dimensions shows this to be true:: $2 + 3 + 2 + 3 = 10$ respectively. The set of 10 complex coordinates support transformations to a factorized block-diagonal form. The 10 complex coordinates can be transformed into fundamental representations of the above factor groups.

representations of each of the factors of *by the definitions of the sets of internal symmetry coordinates as fundamental representations of the subgroups.*

$$SU(2) \otimes U(1) \otimes SU(3) \otimes SU(2) \otimes U(1) \otimes SU(3) \otimes U(4)^4 \qquad (6.1)$$

6.2 One Layer QUeST

The 26 complex coordinates form an SU(26) internal symmetry space. This space undergoes a breakdown. The resulting factorized form of the internal symmetries *of one layer*[40] is given in eq. 6.1.

The 7 + 1 dimension QUeST leads to a one layer form of the Unified SuperStandard Theory. In this theory there is one Layer group with a singlet representation. When we generalze 7 + 1 QUeST to 32 complex quaternion dimensions we will have a QUeST with four layers corresponding to the Unified SuperStandard Theory in 3 + 1 dimensions.

The one layer QUeST is represented by Fig. 6.2. The seemingly duplicate factors in eq. 6.1

$$SU(2) \otimes U(1) \otimes SU(3) \ \otimes SU(2) \otimes U(1) \otimes SU(3) \qquad (6.2)$$

are for the "normal" and the Dark sectors as shown in Fig. 6.2. Fig. 6.2 reflects a transformation of the dimensions (coordinates) of Fig. 6.1 into representations of internal symmetry groups.

6.2.1 The U(2) Rotation Group Between Normal and Dark Sectors

The omitted four real-valued dimensions (coordinates) in Fig. 6.1 (compared to the complete set of dimensions in Fig. 5.1) play the role of a U(2) group. It appears that the only role this group could play is to transform between the Normal and Dark sectors. It transforms each normal fermion to its Dark counterpart and *vice versa.*

Experiment suggests this group would then be necessarily broken with large mass vector bosons and with very ultra weak interactions.

---

[40] Strictly speaking this is a simplification since the Layer group mixes all layers.

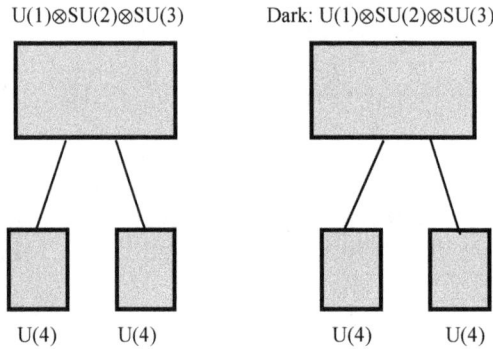

Figure 6.2. Schematic of the internal symmetry groups of eq. 6.1. The two large blocks are each 5 complex coordinate representations of SU(2)⊗U(1)⊗SU(3).

6.2.2 Generation and Layer Groups

The lower U(4) groups in Fig. 6.2 are the Generation and Layer number groups. One pair of each number group is for each of the two U(1)⊗SU(2)⊗SU(3) factors above. Layer and Generation groups are part of the Unified SuperStandard Theory. See chapters 44 and 45 for detailed discussion as well as Appendix 3-A.

6.3 The Breakdown of the Symmetry of the 20 Real Coordinate Group SU(10)

The 20 real coordinate subspace of Fig. 6.1 contains a representation of

$$SU(2) \otimes U(1) \otimes SU(3) \otimes SU(2) \otimes U(1) \otimes SU(3) \qquad (6.3)$$

which clearly could be transformed to a block diagonal in the group factors.

However it is instructive to consider the generation of the component factors using a Lorentz boost-like framework. Separating the internal symmetry part of Fig. 6.1 into six sets of Lorentz-like coordinates with a real energy and complex momenta we obtain Fig. 6.2.

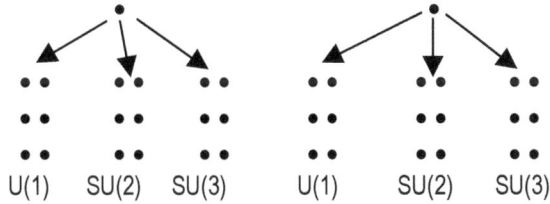

Figure 6.3. Each of the two real-valued time coordinate is linked to three complex spatial coordinates. Lorentz transformations applied to each 4-vector so constructed yields a factor of eq. 6.3. The factors so generated commute to give eq. 6.3 with commuting subgroup factors listed in the figure.

The use of coordinates with real-valued energy and complex-valued spatial momentum is shown in Appendix 12-A to lead to the four fermion species using Lorentz boosts. Here we see a similar effect in the generation of particle interactions.

6.4 Map to Unified SuperStandard Theory Groups

The map of the four complex coordinates to Complex space-time is direct. Complex Lorentz group transformations relate coordinate systems.

The complete set of internal symmetries for one layer is given by eq. 6.3 and the U(2) rotation group described above in section 6.2.1. The map follows from the observations:

1. The factors of eq. 6.3 can be separated into separate factors for the normal and Dark sectors.

2. Since the set of normal and Dark fermions are split into four species we can unambiguously associate the factors of eq. 6.3 with its factors.

3. The SU(3) factors have $\underline{3}$ representations which we can associate with the up-quark and down-quark, normal and Dark species. Thus the SU(3) subgroups are normal and Dark Strong interaction subgroups.

4. The SU(2)⊗U(1) factors map to ElectroWeak interactions for the normal and Dark sectors.

5. The four U(4) factors map to the Generation and Layer number groups for the normal and for the Dark sectors.

Thus we have a map to the interactions of the *one layer* Unified SuperStandard Theory. Although we consider only one layer the considerations apply to all four layers.

6.5 Four Layer QUeST Exactly

The QUeST figures above do not display the Internal Symmetry, Generation and Layer groups of all four layers of the Unified SuperStandard Theory. The groups of the layers are not the same. Each layer has its own set of groups;

$$SU(2)⊗U(1)⊗SU(3)⊗SU(2)⊗U(1)⊗SU(3)⊗U(4)^4 \qquad \text{See Fig. 6.2}$$

and

$$U(2) \qquad \text{Section 6.2.1}$$

The overall internal symmetry is the internal symmetry group of the Unified SuperStandard Theory augmented by $U(2)^4$ giving

$$[SU(2)⊗U(1)⊗SU(3)⊗SU(2)⊗U(1)⊗SU(3)⊗U(4)^6]^4 \qquad (6.2)$$

The one layer theory is described by $7 + 1$ dimension complex quaternion QUeST. The four layer theory is described by a 32 dimension complex quaternion QUeST. Thus it consists of four "copies" of the coordinates:

Figure 6.4. 32 dimension QUeST schematic.

which yield four duplicates of the internal symmetry schematic in Fig. 6.2 displayed in Fig. 6.5,

and four U(2) rotation groups – one for each layer,

and a complex Quaternion space-time consisting of 3 + 1 complex-valued quaternion coordinates, which map to 3 + 1 complex coordinates, and then to the 3 + 1 real-valued coordinates of the Unified SuperStandard Theory.

The sum total of real-valued dimensions is 256, as is the sum of the dimensions of the above parts which are constructed from these dimensions.

Four layer QUeST *is needed* to obtain the 3 + 1 complex quaternion formulation of the Unified SuperStandard Theory which we call QUeST.

4 layer QUeST → 4 layer real-valued coordinates Unified SuperStandard Theory

U(1)⊗SU(2)⊗SU(3)   Dark: U(1)⊗SU(2)⊗SU(3)

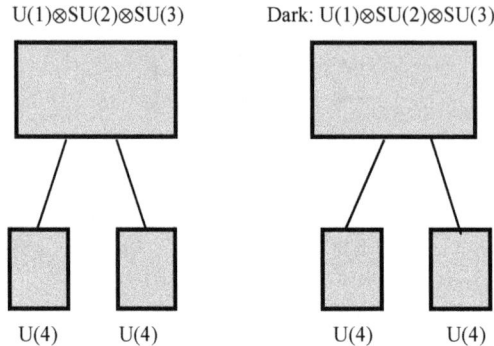

U(4)   U(4)   U(4)   U(4)

Figure 6.5. One of the four schematics of the four layer QUeST.

6.6 The QUeST Universe

The QUeST universe has a 3 + 1 complex quaternion space-time of 32 dimensions counting the dimensions within each of the four quaternion coordinates. The universe has a normal part and a Dark part. The internal symmetries include those of the Unified SuperStandard Theory and four U(2) groups that support transitions between normal and Dark matter and fields. There is one U(2) group for each layer. The U(2) interactions are presumably ultra-weak since Dark matter has not been detected experimentally.

The fermion and vector boson spectrums that emerge are those of the Unified SuperStandard Theory. They are displayed in Figs. 1.1 – 1.4. There are 64 fundamental fermions in each layer yielding 256 fermions in four layers.

The Unified SuperStandard Theory in 3 + 1 real-valued dimensions emerges when QUeST is restricted to real coordinates.

Our universe is a 32 Complex Quaternion Dimensions Space. It is equivalent to QUeST, which has a 3 + 1 Complex Quaternion Dimensions Space-Time and Internal Symmetries.

Upon restriction of QUeST to real-valued coordinates, QUeST Becomes the Unified SuperStandard Theory.

QUeST → real-valued coordinates Unified SuperStandard Theory

7. Particle-Dimension Duality

An examination of the total number of dimensions and the number of fundamental fermions (per layer) as we saw in Table 4.1 shows that these quantities are equal for the QUeST case (64) and the MOST case (128) seen later. The equality of these parameters raises the possibility that the number of internal symmetry dimensions equals the number of fundamental fermions in some deep sense. This possibility is enhanced because the number of particles depends on the singles nature of leptons and the triplet nature of quarks.

When one considers the measurement of distance in a space one sees the measuring process requires the use of particles. Without particles one cannot distinguish space-time points from each other. And the distance between space-time points is determined either directly or indirectly by bosons such as light transmitted (and possibly reflected) between fermion "clumps.'

Without particles distance, space-time is meaningless. Without space-time particles will "clump' and dynamics is impossible.

Thus a relation between particles and dimensions is required for Physics.

7.1 Particles Mapped to Quantum Functionals

A particle has a quantum field. We have shown in earlier books that quantum fields may be factored into an inner product of a quantum functional and a wave in coordinate space-time.[41] Every particle has a set of internal symmetry and space-time quantum numbers. We can map the set of fundamental fermions to a set of quantum functionals. These functionals have transformation properties under internal symmetries. They are thus representations of the internal symmetry groups.

---

[41] See Blaha (2019e) and (2020) as well as earlier books by the author.

There is no distance in the set of functionals. As we showed in Blaha (2020) and earlier books factoring quantum fields into inner products of functionals and waves eliminates the instantaneity problem of Quantum Entanglement.

7.2 Coordinates and Functionals

The lack of distance in the set of functionals parallels the lack of distance between coordinates. For example there is no distance between the "x" coordinate and the "y" coordinate in a coordinate system. Functionals have features in common with coordinates.

Quantum functionals can also have commutation relations similar to coordinates: Defining a quantum functional conjugate momentum:

$$\pi_k = d/df_k \tag{7.1}$$

For quantum functional f_k we obtain the commutation relation

$$[f_k, \pi_j] = i\delta_{kj} \tag{7.2}$$

where k and j represent internal symmetry indices. Eq. 7.2 mirrors the form of quantum mechanical commutation relations adding to the similarity between coordinates (dimensions) and quantum functionals.

Biquaternion and bioctonion coordinates can be mapped to quantum functionals since they have the same role, and number, as the coordinates of internal symmetries. We set

$$x_m = R_{mn}f_n \tag{7.3}$$

where R is a transformation.

7.3 Particle-Dimension Duality

Thus in QUeST and MOST we find a particle-dimension (coordinate) duality.[42] The point of view offered by this duality suggests that fermion particles are in some sense "interchangeable" with internal symmetry coordinates—with particle functionals as an intermediary.

7.4 Particle – Functional – Dimension Triality

The above discussion shows a close analogy – a *triality* – between fermion particles, dimensions (coordinates) and quantum functionals.

---

[42] Although the discussion is for one layer of particles it can be generalized to four layers of fermions with the introduction of additional sets of coordinates.

8. Functionals Implement Quantum Entanglement

In 1935 Einstein, Podolsky, and Rosen[43] (EPR) proposed a definition of Physical Reality and then proceeded to consider a quantum state composed of two separated quantum systems. In this chapter we will propose an extension of their definition of Physical Reality, and then show, by counterexample, that the clarification is needed.

Then we will consider their separated quantum systems example within the framework of our quantum functional formalism. We will show our formalism eliminates instantaneous quantum action-at-a-distance. (No spookiness.)

8.1 Definition of Physical Reality

EPR suggested the definition of Physical Reality:

EPR Criterion for Physical Reality

If, without in any way disturbing a system, we can predict with certainty (i.e., with probability equal to unity) the value of a physical quantity, then there exists an element of physical reality corresponding to this physical quantity.

EPR suggested there might be other ways of defining physical reality in a manner consistent with Physical Theory. It appears that a more comprehensive definition of physical reality is required to justify a correspondence with Physical Theory.

The criterion for a correspondence between physical reality and a physical theory must include the requirement that a physical theory's complete set of observables must correspond to physical reality. Predicting with certainty the value of one physical

---

[43] Einstein A, Podolsky B, and Rosen N, "Can Quantum-Mechanical Description of Physical Reality Be Considered Complete?", Phys. Rev. **47**, 777 (1935).

quantity leaves open the question of other possible physical quantities. In the case of conventional simple quantum theory there are two observables Q and P with commutation relation [Q, P] = i. Within this framework the EPR definition appears acceptable. However one should not hang the definition of physical reality on one possible quantum mechanics formalism. Physical reality must be independent of all valid formulations. A simple counter example shows that the EPR definition is not sufficient in the case of an alternate quantum theory.

Counter Example

Consider the formalism of Blaha (2016f) which was used to create a larger quantum theory that encompassed both quantum physics and classical physics as extremes. The theory supports both classical and quantum physics and an intermediate range between them. In this theory (called PseudoQuantum Mechanics) the range from quantum to classical physics is specified by an angle θ.

The theory had two commuting variables x_1 and p_1, which we augment with two new variables x_2 and p_2, defined by

$$x_i = (m\omega/\hbar)^{-\frac{1}{2}} Q_i \tag{8.1}$$
$$p_i = (m\omega\hbar)^{\frac{1}{2}} P_i$$

for i, j = 1, 2 where

$$P_2 = -i \, d/dQ_1 \tag{8.2}$$
$$Q_2 = i \, d/dP_1$$

with the commutation relations:

$$[Q_i, P_j] = i(1 - \delta_{ij}) \tag{8.3}$$

for i, j = 1, 2.

Next we define raising and lowering operators

$$a_i = 2^{-\frac{1}{2}}(Q_i + iP_i) \tag{8.4}$$
$$a_i^\dagger = 2^{-\frac{1}{2}}(Q_i - iP_i)$$
$$Q_i = (a_i + a_i^\dagger)/\sqrt{2}$$
$$P_i = (a_i - a_i^\dagger)/(\sqrt{2}i)$$

with

$$[a_i, a_j^\dagger] = (1 - \delta_{ij}) \qquad (8.5)$$
$$[a_i, a_j] = 0$$
$$[a_i^\dagger, a_j^\dagger] = 0$$

for i, j = 1, 2.

We now define an alternate set of raising and lowering operators that will use an angle θ to provide a continuous transition from classical to quantum (and vice versa)

$$b_1 = Q_1\cos\theta + iP_2\sin\theta \qquad (8.6)$$
$$b_2 = -Q_2\sin\theta + iP_1\cos\theta$$

$$b_1^\dagger = Q_1\cos\theta - iP_2\sin\theta \qquad (8.7)$$
$$b_2^\dagger = -Q_2\sin\theta - iP_1\cos\theta$$

Their commutation relations are

$$[b_1, b_1^\dagger] = \sin(2\theta) \qquad (8.8a)$$
$$[b_2, b_2^\dagger] = -\sin(2\theta)$$
$$[b_1, b_2^\dagger] = [b_2, b_1^\dagger] = 0$$
$$[b_1, b_2] = [b_1^\dagger, b_2^\dagger] = 0$$

The PseudoQuantum Hamiltonian is

$$\hat{H} = p_1 p_2/m + m\omega^2 x_1 x_2 \qquad (8.8b)$$
$$= \tfrac{1}{2}\omega(\{a_1, a_2^\dagger\} + \{a_2, a_1^\dagger\})$$
$$= \omega(P_1 P_2 + Q_1 Q_2)$$

Note: all physical quantities commute for $\theta = 0$ by eq. 8.8a. But they are quantum for $\theta = \pi/4$. Thus this theory encompasses both classical and quantum theory. Blaha (2016f) develops a harmonic oscillator example (and other examples) showing both its classical and quantum limits as well as intermediate oscillator states.

The example illustrates the point that predicting the value of one physical quantity with certainty does not make it physically real. All physical quantities of a system must also have commutation relations in agreement with Physical Reality. The above counterexample illustrates this point by defining a sharp physical quantity with variable commutation relations. The choice of the parameter θ leads to a quantum theory or a classical theory or an intermediate theory.

A more satisfactory criterion for Physical Reality is:

<u>**New Criterion for Physical Reality**</u>
If, without in any way disturbing a system, we can predict with certainty (i.e., with probability equal to unity) the value of a physical quantity and determine the commutation relations of all observables in the Physical Theory, then there exists an element of physical reality corresponding to this physical quantity and to the enveloping Physical Theory.

The commutation relations of the physical quantity are critical to the reality question. The question of physical reality is dependent on the complete formulation. Thus the question of physical reality depends on more than the value of each physical quantity. The reality question requires a total specification of commutation relations and eigenvalues.

8.2 Quantum Entanglement and Action-at-a-Distance

EPR considered the quantum entanglement of two systems and showed that instantaneous action-at-a-distance (spookiness) resulted. In this section we will show that our quantum functional formalism, which generalizes quantum theory, eliminates the problem of instantaneous action-at-a-distance.[44] We will show the solution provided by quantum functionals using the same example as EPR.

The key feature of quantum functionals is their ubiquitous presence at every point of space-time. In a multi-system quantum state all functionals are directly, instantaneously linked no matter what the separation of the constituent systems. When a reduction of the state of one system occurs due to a measurement, all other systems are instantly updated since the space-time separation of the individual systems is not

---

[44] This section presents the solution for quantum spookiness that we proposed in Blaha (2019g) and (2018e).

relevant. The linkage of all quantum functionals is relevant. A reduction of one system immediately impacts the other related systems.

8.3 The EPR Two System State Example

EPR considered a state consisting of two systems that might become separated spatially. We can represent the state as

$$\Psi = \Sigma_n \psi_{1n}(x_1)\psi_{2n}(x_2) \tag{8.9}$$

We can represent a measurement (reduction of state) with a projection Π_{1a} of system "1" to a state ψ_{1a} with

$$\psi_{1a} = \delta_{ab} \, \Pi_a \, \psi_{1b} \tag{8.10}$$

Then

$$\Psi_{projected} = \Pi_{1a} \, \Sigma_n \psi_{1n}\psi_{2n} \ = \psi_{1a}(x_1)\psi_{2a}(x_2) \tag{8.11}$$

The effect of the measurement of system "1" is *instantaneous* of system "2" because the quantum functionals f_{1n} and f_{2n}, and the projections Π_{1n} and Π_{2n} of both systems are not separated by distance. By eq. 7.3

$$\psi_{1n}(x) = f_{1xn}(\Pi_{1xn}\Phi \ ) \ = (f_{1xn}, \, \Pi_{1xn}\Phi) \tag{8.12}$$

$$\psi_{2n}(y) = f_{2ny}(\Pi_{2yn}\Phi \ ) \ = (f_{2yn} \, , \, \Pi_{2yn}\Phi) \tag{8.13}$$

with $x = x_1$ and $y = x_2$. *The quantum functional and the projection select the wave and its coordinate parameterization. The coordinates in the wave are merely place holders.*

Therefore the relative distance between the coordinates x_1 and x_2 is not relevant for the change of state of system "2". The quantum functionals and projections give the instantaneity of the change in ψ_{2a} upon the measurement of system "1".

The EPR Spookiness is resolved by quantum functionals. There is no conflict with the Theory of Special Relativity.

Octonion Megaverse

The chapters in this section describe the form of the complex octonion Megaverse that contains universes. It appears that the Megaverse has a superset of universe features: internal symmetries, fundamental fermion spectrum, vector boson spectrum and interactions, and coupling constants.

9. Bioctonion Megaverse

In chapters 5 and 6 we developed a biquaternion theory called QUeST that could be used to derive the group structure, and the fundamental fermion and vector boson spectrums. It led to the Unified SuperStandard Theory and accounted for the close similarity between the internal symmetries of the Standard Model sector and the subgroups of the Lorentz group. They both exhibit U(1), SU(2) and SU(3) symmetries.

In this chapter we define a bioctonion space and use it to define a Megaverse fundamental basis of a somewhat more general Unified SuperStandard Theory, called MOST. MOST develops a more robust set of internal symmetries and fundamental particles. It creates a new view of Dark matter that appears to help explicate the lack of interactions between normal matter and Dark matter.

In chapters 5 and 6 we found the Unified SuperStandard Theory, which was based on the Complex Lorentz group, could be based on a biquaternion space for our universe.

If we assume the existence of a Megaverse[45] containing our universe, and other universes, then we can define a *Megaverse Octonion SuperStandard Theory* (MOST) that becomes a more general basis of the Unified SuperStandard Theory.

Remarkably MOST, when "restricted" to our universe, yields QUeST.

9.1 Octonion Features

Octonions have significant properties that enable them to be used in a quantum field theory development:

1. An octonion is an 8-tuple of real numbers. A complex octonion is an 8-tuple of complex numbers.

---

[45] The Megaverse is described in some detail in Blaha (2017c), (2017f), and (2018e) together with evidence for its existence. See chapter qqqq for possible experimental evidence.

2. They are nonassociative.
3. They are one of the two finite dimensional division rings having the real numbers as a proper subring. (The other is quaternions—considered in chapter 5.)
4. They are non-commutative. (This is not a roadblock for quantum field theory which is also non-commutative in general.)

These features support the development of physics theories.

We can represent a bioctonion b as

$$b = b_{real} + Ib_{imaginary}$$

where b_{real} and $b_{imaginary}$ are real-valued octonions.

9.2 Motivation and Procedure

Our goal again is to create a larger dimension space within which a universe based on QUeST can exist, and where we can ultimately derive our space-time and the Unified SuperStandard Theory. Again we use the similarity of Lorentz subgroups and Standard Model internal symmetry groups in our development.

Again there are two possible procedures to follow in developing the deeper basis:

1. One can develop the Quantum Mechanics, Quantum Field Theory, ... in an octonion space and then extract the dynamics, fermion spectrum, gauge fields, and so on of our familiar space-time.

2. One can define an octonion space and then use its coordinates to directly extract the space-time, internal symmetries, fermion spectrum, gauge field spectrum and dynamics.

We have chosen the latter approach as it will more directly lead to a Unified SuperStandard Theory for the Megaverse.

In developing the deeper space, upon which we build, we will take guidance from the derivation of the Unified SuperStandard Theory. This theory assumes a complex 4-dimensional space-time upon which Complex General Relativity is constructed. It then proceeds to complex flat space-time and Complex Relativity.

After defining features of Complex Lorentz transformations the Unified SuperStandard Theory used Lorentz boosts to derive the Dirac forms of the four fermion species. The boosts were required to boost a fermion from a rest state to a state with a real-valued energy, and real or complex-valued 3-momenta. Thus a real time – complex-valued spatial part is required for the proper definition of species.

The Unified SuperStandard Theory then showed Lorenz subgroups mapped to Standard Model internal symmetry subgroups.

9.3 Definition of Bioctonion Space

Following the above stated procedure we define a bioctonion[46] (complex octonion) space with *one* "time" biquaternion and *seven* "spatial" bioctonions as a generalization of the 3 + 1 space-time of our experience. The choice of 8 bioctonion dimensions seemed natural but was not required by a principle. It does lead to a larger set of internal symmetries with *one layer* QUeST symmetries as a subset.[47] We will use the symbol • to represent each of the bioctonion space coordinates in Fig. 9.1.

We have chosen a complex 8-dimensional bioctonion space-time as the one MOST layer Megaverse space-time. There are 128 real coordinates in the bioctonion Megaverse space from which complex 8-dimensional space-time (4-dimensional complex space-time) is extracted. It embeds our universe's 4-dimensional complex space-time as a subspace-time.

Fig 9.1 symbolically depicts the space with a "dot" • for each real-valued coordinate. *Again we treat the bioctonion space as a higher dimensional space and do not use details of octonion algebra in our development.*

---

[46] We use bioctonion synonymously with complex octonion in this and subsequent chapters.
[47] We start the complex octonion discussion by developing a one layer MOST. Then we develop a four layer MOST which contains four layer QUeST-based universes.

Time

• • • • • • • •   • • • • • • • •

7-Space

• • • • • • • •   • • • • • • • •
• • • • • • • •   • • • • • • • •
• • • • • • • •   • • • • • • • •
• • • • • • • •   • • • • • • • •
• • • • • • • •   • • • • • • • •
• • • • • • • •   • • • • • • • •
• • • • • • • •   • • • • • • • •

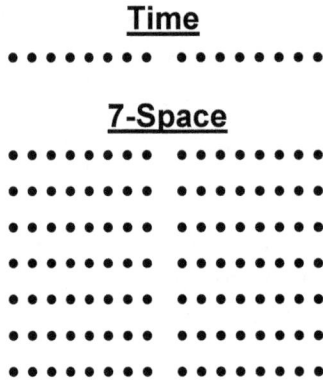

Figure 9.1. Eight-Dimensional (7 + 1) bioctonion space with coordinates represented by • 's.

This bioctonion space has 128 real dimensions (64 complex dimensions.).

9.4 Bioctonion Lorentz Group

Our definition of time and space bioctonion coordinates purposefully resembles those of our real space-time. One might ask why there should be a Lorentz-like group for bioctonion space. up.

The only apparent reason is the need for a special speed c that, in our space-time, its existence enables one to boost from a rest frame of a mass m particle with energy m to a moving frame of greater energy.[48] Without c the group of the above coordinates would presumably be the bioctonion U(8) group. U(8) transformations preserve the norm of a state so that a "boost-like" transformation does not exist.

The bioctonion Lorentz transformations do have a unique speed c (the speed of light) and specify a unique rest frame for any particle—both sublight particles and tachyon particles.[49] . Thus we select the bioctonion Lorentz group for bioctonion space.

---

[48] A similar consideration applies to tachyons.
[49] Tachyons can be transformed by a Complex Lorentz transformation to and from a rest frame. See Blaha (2018f) and Appendix 12-A below.

Flat space bioctonion Special Relativity generalizes to a bioctonion General Relativity which may be constructed directly (mindful of octonion non-commutativity).

Flat space bioctonion Lorentz group transformations have constant bioctonion matrix elements that are analogous to those of the Lorenz group.

9.5 Extracting the Symmetries and Particle Spectra

As stated earlier in section 9.2 we will directly describe the symmetry structure implied by the form of the bioctonion coordinate system while mindful of sections 9.3 and 9.4.

The Unified SuperStandard Theory developed the group structure, from which the particle species were derived, from a subset of Lorentz boost transformations. Complex boosts mapped a system at rest to a system in motion with a real energy and complex 3-momenta in general. Bioctonion boosts play a similar role.

The 4-dimensional representation of the Unified SuperStandard Theory complex coordinates subset is given in Fig. 9.2.

Time
•
Space
• •
• •
• •

Figure 9.2. Four-Dimensional space for Unified SuperStandard derivation of particle spectra with coordinates represented by • 's.

Following the same line of thought, which is described in more detail earlier, we now specify a bioctonion subspace restricted to that of Fig. 9.3 to define the relevant set of coordinates for determining particle symmetries and spectra.

Time
• • • • • • • •

7-Space

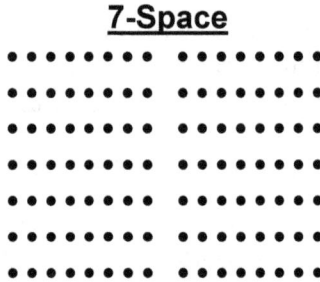

Figure 9.3. 8 complex dimension bioctonion subspace for symmetries and particle spectra.

The subspace's 60 complex coordinates will split into an 8-dimensional complex coordinate space which will support 8-dimensional Complex Lorentz transformations, and a 52 complex coordinates internal symmetry space. The neglected eight dimensions support a U(4) rotations group transforming between normal and Dark sectors. (There are four sectors in MOST necessitating the U(4) rotations that transform among them.)

The mechanism for this symmetry breakdown may be due to vacuum energy effects in the bioctonion Megaverse.

In the next chapter we analyze the symmetries of the 52 complex dimensional subspace. We find it contains the Standard Model symmetries and the Generation and Layer number symmetries. The number symmetries are inherently part of the set of internal symmetries. We will also see later that the Dark fermion sectors of the theory may have spinors that occupy different parts of the overall 16 component spinors of complex 8-dimensional space-time if the U(4) group does not transform between sectors. In that case it would only have a singlet representation.

10. Symmetries of the Megaverse Octonion SuperStandard Theory (MOST)

The coordinate space picture of the Megaverse described in chapter 8 enables us to simply find the internal symmetries and particle spectra of MOST. They will turn out to be a superset of those of the Unified SuperStandard Theory.

The MOST Megaverse subspace for the determination of the Internal Symmetries is depicted in Fig. 10.1. It is based on the discussion of Fig. 9.3. The space-time complex 8-vector is separated from the internal symmetry coordinates.

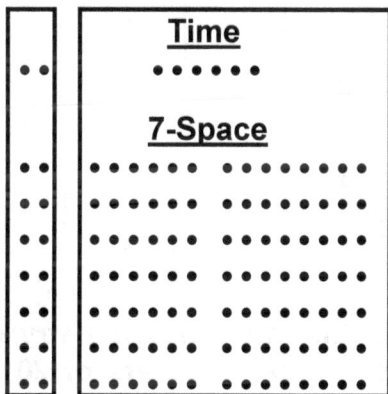

Figure 10.1. Internal Symmetry determination subspace. It is Based on Fig. 9.3; The 16 real space-time coordinates are separated from 104 real coordinates for internal symmetries.

The internal symmetry coordinates above number 104 real coordinates or 52 complex coordinates. These coordinates serve as the coordinates of the fundamental representations of each of the factors of

$$[SU(2){\otimes}U(1){\otimes}SU(3){\otimes}SU(2){\otimes}U(1){\otimes}SU(3)]^2{\otimes}U(4)^8 \qquad (10.1)$$

The factorized internal symmetry emerges from another breakdown(s) which corresponds to the subgroup structure of the Lorentz group. Eq. 10.1 evidently follows from the structure of the bioctonion Lorentz transformations.

The U(4) Generation and Layer groups are represented in Fig. 10.1. We depict the pattern of symmetry implied by Fig. 10.1 and eq. 10.1 in Fig. 10.2 and 10.3.

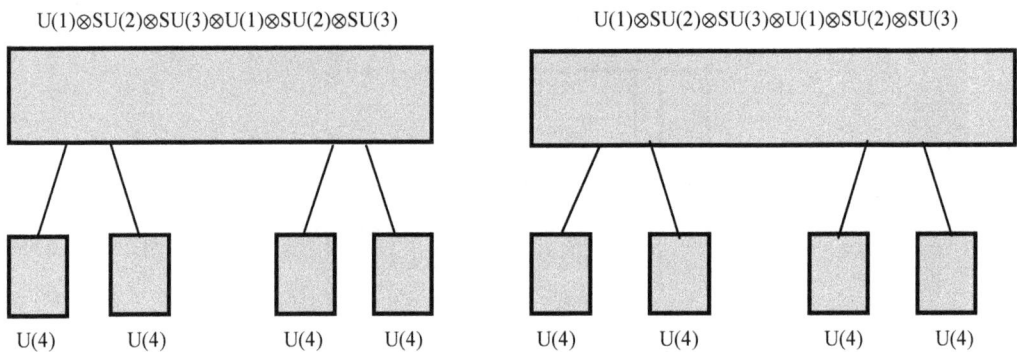

Figure 10.2. Schematic of the internal symmetry groups' coordinates of Fig. 10.1. The two "large" blocks are sets of 20 real-valued coordinates furnishing representations of the listed groups. The lower U(4) groups are the Generation and Layer number groups. The total number of real-valued coordinates is 104.

Each U(1)⊗SU(2)⊗SU(3)⊗U(1)⊗SU(2)⊗SU(3) block in Fig. 10.2 has a 10 complex coordinates (20 real-valued coordinates) representation. The blocks are subdivided in Fig. 10.3 into sets of 10 real-valued coordinates supporting representations of U(1)⊗SU(2)⊗SU(3). There are three Dark blocks. The first block contains the representations of the known parts of the Standard Model. The internal symmetry groups of each part are listed in Fig. 10.4.

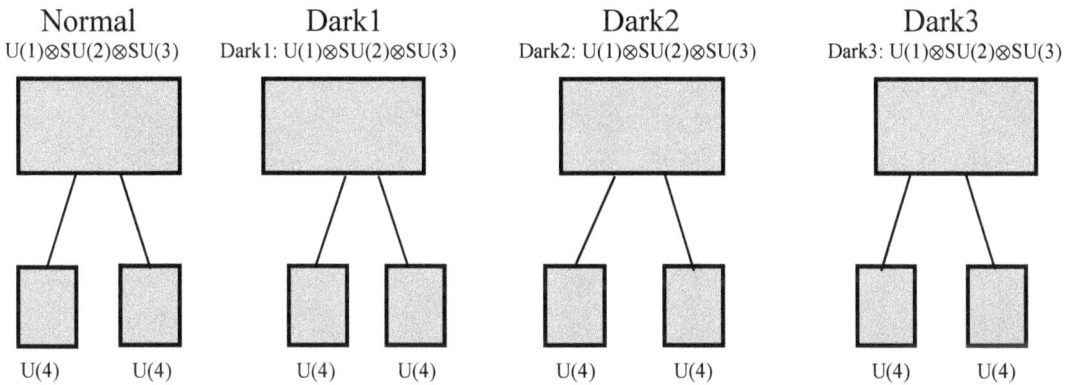

Figure 10.3. Schematic of the internal symmetry groups of eq. 10.1. The lower U(4) groups are the Generation and Layer number groups. One pair of each number group is for each of the four U(1)⊗SU(2)⊗SU(3) factors above. The result is the total Internal Symmetry group of the enlarged Unified SuperStandard Theory.

The factorization of each of the four blocks is accomplished by following the procedure given in section 6.4 for each block.

10.1 One Layer MOST

The above sections specify the *one layer* MOST. The symmetries of the three other layers are the same but their groups are individual to each layer. The groups of each layer are each flagged with a different layer index.

The overall one layer MOST internal symmetry is specified by Fig. 10.3 plus a U(4) group that rotates among the four normal and Dark parts shown in Fig. 10.3.

$$[SU(2) \otimes U(1) \otimes SU(3)]^4 \otimes U(4)^8 \tag{10.2}$$

There is also a space-time sector with 8 (= 7 + 1) complex coordinates.

Comparing Figs.6.3 and 10.3 we see that a one layer MOST has a "Normal" part and one Dark part while the MOST Megaverse adds two more Dark parts.

Thus the internal symmetry groups of one Layer MOST are:

<u>"Normal" Gauge Groups</u>
SU(3)⊗SU(2)⊗U(1)
Generation Group U(4)
Layer Group U(4)

<u>Dark1 Gauge Groups</u>
SU(3)⊗SU(2)⊗U(1)
Generation Group U(4)
Layer Group U(4)

<u>Dark2 Gauge Groups</u>
SU(3)⊗SU(2)⊗U(1)
Generation Group U(4)
Layer Group U(4)

<u>Dark3 Gauge Groups</u>
SU(3)⊗SU(2)⊗U(1)
Generation Group U(4)
Layer Group U(4)

PLUS

A U(4) group that rotates among the four normal and Dark sectors

Figure 10.4. One layer MOST vector bosons list from eq. 10.2. The four layer MOST quadruples the above list: with one distinct set for each layer. In one layer the total number of vector bosons of the above list is 192 for one layer. Thus four layers yield a total count of 768 vector bosons in MOST (not counting the Species group). We require each level has a separate U(4) rotation group.

10.2 Four Layer MOST

The four layer MOST is described by a 32 dimension complex octonion QUeST. Thus it consists of four "copies" of the coordinates:

Figure 10.4. The 32 dimension MOST schematic. Four layer MOST has 512 real-valued dimensions.

which yield four duplicates of the internal symmetry schematic in Fig. 10.3,

and four U(4) rotation groups – one for each layer,

and a 8 complex quaternion space-time consisting of 7 + 1 complex-valued quaternion coordinates.

The sum total of real-valued dimensions is 512 as is the sum of the dimensions of the above parts constructed from the dimensions.

In short we obtain the four layer MOST which may contain QUeST universes.

10.3 Fermion and Gauge Vector Boson Spectrums

The fermion and vector boson spectrums that emerge in MOST are those of an enlarged QUeST and Unified SuperStandard Theory. They are displayed below. MOST has an additional two Dark sectors beyond QUeST and the Unified SuperStandard Theory.

Vector Bosons

From Fig. 10.4 we find MOST has 192 vector bosons in one layer. Thus four layer MOST has a total count of 768 MOST vector bosons. There are two additional Dark vector boson sectors beyond QUeST and the Unified SuperStandard Theory.

Fermions

There are 512 fundamental fermions in MOST, which includes two additional Dark fermion sectors. Fig. 10.5 shows the MOST fermion spectrum.

```
        1      2      3      4
      ....   ....   ....   ....
      ....   ....   ....   ....
      ....   ....   ....   ....
      ....   ....   ....   ....

      ....   ....   ....   ....
      ....   ....   ....   ....
      ....   ....   ....   ....
      ....   ....   ....   ....

      ....   ....   ....   ....
      ....   ....   ....   ....
      ....   ....   ....   ....
      ....   ....   ....   ....

      ....   ....   ....   ....
      oooo   ....   ....   ....
      oooo   ....   ....   ....
      oooo   ....   ....   ....
```

Figure 10.5. Schematic spectrum of the fermions of MOST. Each fermion is represented by a •. Quark triplets are represented by a single •. Four sets of four species in four generations which are in turn in 4 layers. Open symbols ∘ represent known fermions. There are 512 fundamental fermions taking account of quark triplets.

10.6 The MOST Megaverse

Four layer QUeST universes can exist in a four layer MOST Megaaverse.

The MOST Megaverse has a 7 + 1 complex quaternion space-time of 64 dimensions counting the dimensions within each of the four quaternion coordinates. The universe has a normal part and a Dark part. The internal symmetries include those of the Unified SuperStandard Theory plus two additional Dark sectors and four U(4) groups that support transitions between normal and Dark matter and fields. There is one U(4) group for each layer. The U(4) interactions are presumably ultra-weak.

The fermion and vector boson spectrums that emerge are those of listed in section 10.5.

The Megaverse is a 32 Complex Octonion Dimensions Space. It is equivalent to MOST, which has a 7 + 1 Complex Quaternion Dimensions Space-Time and Internal Symmetries.

The 7 + 1 Complex Quaternion Dimensions Space-Time of the MOST Megaverse contains 3 + 1 Complex Quaternion Dimensions QUeST Universes' Space-Times.

11. Darkness Superselection or SuperWeak Interaction: Why Normal and Dark Particles Don't Interact

Despite numerous searches no interaction between normal and Dark matter has been found except for gravitation. Consequently a sizeable number of physicists has concluded that Dark matter does not exist.

Our study of MOST raises contrary possibilities:.

1. A superselection rule for a conserved Darkness number that precludes all interactions between Dark and normal matter except gravitation. We suggest a quantum number, *Darkness* denoted D, exists, similar to charge, with integer values: 1 for normal matter, and 2, 3, and 4 for the Dark matter sectors of MOST. When telescoped downward to QUeST the Darkness quantum number D is 1 for normal matter and 2 for Dark matter. The columns in Fig. 11.1 are labeled with the MOST Darkness quantum number. Interaction vector bosons also have the Darkness quantum number as shown in Fig. 11.2.

2. A superweak interaction between normal and Dark matter: a U(2) interaction in QUeST; and a U(4) interaction in MOST that makes normal ←→ Dark transitions undetectable.

11.1 Superselection Rule: Darkness Restricted Interactions

A superselection rule for D implies it is a conserved quantum number with no interactions between sectors with different values of D. Thus fermion – vector boson interactions have the form

$$\overline{\psi}_k A_k \psi_k \tag{11.1}$$

for k = Darkness number.

No vector interactions exist between normal and Dark matter. D conservation implies the lack of matter – Dark matter interactions. Gravitation being universal does exist between all forms of matter. The gravitational spinor connection interaction can be defined to exclude interactions between D sectors.

The picture implied by D having a superselection rule appears to agree with current experiment.

Darkness D:

	1	2	3	4
	••••	••••	••••	••••
	••••	••••	••••	••••
	••••	••••	••••	••••
	••••	••••	••••	••••
	••••	••••	••••	••••
	••••	••••	••••	••••
	••••	••••	••••	••••
	••••	••••	••••	••••
	••••	••••	••••	••••
	••••	••••	••••	••••
	••••	••••	••••	••••
	••••	••••	••••	••••
	••••	••••	••••	••••
	○○○○	••••	••••	••••
	○○○○	••••	••••	••••
	○○○○	••••	••••	••••

Figure 11.1. Schematic of the spectrum of fermion layers of MOST labeled with Darkness obtained from Fig. 10.5. The empty circles are known fermions. Quark triplets are represented by a ○ or a •. Darkness 1 includes the known fermions of the Standard Model. Darkness 1 and 2 are the fermions of the Unified SuperStandard Theory and QUeST in our universe. Darkness 1 through 4 appear in MOST in the Megaverse.

11.2 Other Superselection Rules Based Quantum Numbers

Each Darkness sector has its own charge and other internal symmetry quantum numbers. The Darkness superselection rule implies superselection rules for:

Charge
Other Internal Symmetry Quantum Numbers
    SU(3) Numbers
    $SU(2) \otimes U(1)$ Numbers
    Generation Numbers
    Layer Numbers
    Spin
    Total Angular Momentum

The spin superselection rule follows from conservation of angular momentum separately in each sector. See section 11.3.

The restriction of the above quantities to different Darkness sectors where they have their own specific conservation laws removes the special significance of Charge and other quantum numbers. They are not of universal significance. They are of sector-restricted significance.

The Darkness superselection rule, which precludes inter-sector interactions, implies the inter-penetrability of matte and energy. The absence of interactions enables matter and energy of all sectors to coincide at the same point; point by point, throughout the universe and Megaverse.

11.3 MOST Darkness Structured Spinors

The case of the particle spin superselection rule is of importance in MOST and QUeST. First we note that total angular momentum must be separately conserved in each sector.

Secondly, we note the 7 + 1 (eight-dimensional) Megaverse space-time has 16 component Dirac fermion spinors. Given the split of the fermion spectrum into four sectors (Fig. 11.1) it is reasonable to decompose 16-spinors into four 4-dimensional Dirac spinors using spin projection operators Π_k.

$$\Pi_1 = (I, 0, 0, 0) \tag{11.2}$$
$$\Pi_2 = (0, I, 0, 0)$$
$$\Pi_3 = (0, 0, I, 0)$$
$$\Pi_4 = (0, 0, 0, I)$$

with fermion wave functions for each Darkness sector

$$\psi_i(x) = \Pi_i \psi(x) \tag{11.3}$$

using 4×4 matrices composed of 0's and the identity matrix I.

If we define all vector interactions to have a spin of the type of its Darkness sector then normal entities only interact with normal entities, and Dark entities of each sector only interact with other Dark entities in its sector. Total angular momentum is then conserved by interactions. (eq. 11.1)

Thus the absence of interactions between normal and Dark matter is enhanced.

Darkness D:	1	2	3	4
	Gauge Groups	Gauge Groups	Gauge Groups	Gauge Groups
	Gauge Groups	Gauge Groups	Gauge Groups	Gauge Groups
	Gauge Groups	Gauge Groups	Gauge Groups	Gauge Groups
	Gauge Groups	Gauge Groups	Gauge Groups	Gauge Groups
Generic Field:	A_1	A_2	A_3	A_4

Figure 11.2. The list of MOST vector bosons based on Fig. 11.4. There are four layers shown vertically. D = 1 includes the known vector bosons of the Standard Model. D = 1 and 2 are the vector bosons of the Unified SuperStandard Theory and QUeST in our universe. D = 1 through 4 appear

in MOST in the Megaverse. The generic vector boson field for each value of D appears below the list.

11.4 QUeST 8-Spinors

In QUeST there is a norm matter sector with $D = 1$ and a Dark matter sector with $D = 2$. In chapters 5 and 6 we did not consider QUeST spin. We now suggest that spin "filters down" from MOST to QUeST. We choose to have 8-spinors with the upper four spinor components for normal $D = 1$ matter and the lower four components for Dark $D = 2$ matter.

Then using the above discussion restricted to $D = 1$ and $D = 2$ we obtain the scenario of section 11.1 and the Darkness superselection rule, and superselection rules for the quantum numbers in section 11.2, with the $D = 1$ and $D = 2$ Darkness sectors. The QUeST vector interactions have the form

$$\Sigma_k \; \overline{\psi}_k A_k \psi_k \tag{11.4}$$

for normal ($k = 1$) and Dark ($k = 2$) interactions.

Total angular momentum is separately conserved in each Darkness sector. Normal matter and Dark matter do not interact.

This discussion can be directly used to generalize the Unified SuperStandard theory.

11.5 Superweak Interactions Between Normal and Dark Matter

Both QUeST and MOST each have four rotation groups between normal and Dark matter sectors: four U(2) groups for QUeST and four U(4) groups for MOST.

They imply that the Superweak interactions are the most likely possibility.

12. Megaverse-Universe Connection

MOST describes the unified elementary particle theory of the Megaverse. Universes within the Megaverse, which are subspaces of the Megaverse, are described by QUeST. Table 4.1 compared MOST and QUeST features.

QUeST is the 4-dimensional subset of MOST. As chapters 6 and 9 show QUeST has a subset of MOST fermions, vector bosons and interactions. MOST has four sectors of differing Darkness. The MOST 16-spinor, which is composed of four 4-spinors becomes an 8-spinor composed of two 4-spinors. The below figure depicts the relation between MOST and QUeST. (QUeST and the Unified SuperStandard Theory have the same features. They are both enhanced with some new features (shown in the dotted box) emanating from MOST as shown below.) *assuming Darkness has a Superselection rule.*

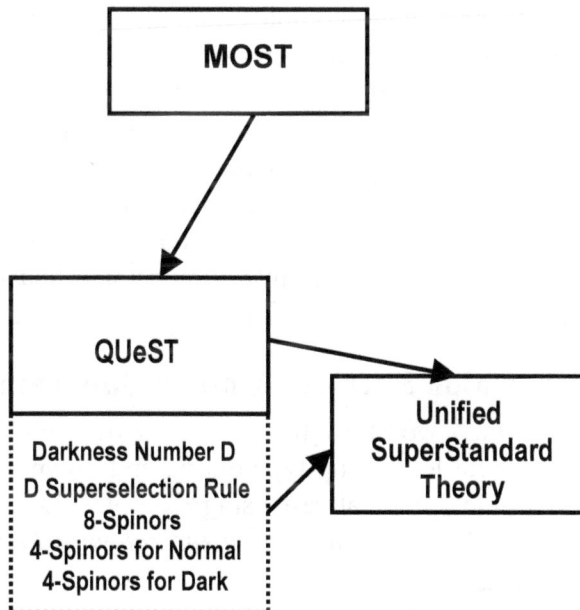

Appendix 12-A. Fermion Species of QUeST, MOST, and the Unified SuperStandard Theory

In Blaha (2018e) (and earlier books) we showed that the overall form of the fundamental fermion spectrum consisted of four species that were determined by the restriction of Complex Lorentz boosts to boosts from a particle's rest frame to a coordinate system with a real-valued energy and time. There are four types of these boosts: two yield a normal particle and two yield a tachyon. Each of the two boosts of each type boost to a coordinate system with real-valued spatial coordinates and to a coordinate system with complex-valued spatial coordinates.

We find that this general form of the four boosts, and fermion species, occurs in the Unified SuperStandard Theory (four complex dimensions), four complex quaternion dimension QUeST, and eight complex octonion dimension MOST. This appendix describes the boosts and species in each theory.

In complex quaternion (and octonion) space we must restrict the "time" quaternion (octonion) to be "real-valued" for the purpose of determining the fermion and vector boson spectrums. In this appendix we summarize results of earlier books.

We find that each of the three theories has a "charged" lepton species, a neutral lepton species, an up-type quark species, and a down-type quark species. The octonion quark species, because they are in an 8-dimensional space, have a distinctive higher dimensional feature with a six-dimensional "internal" momentum.

12-A.1 Four Fermion Species – Unified SuperStandard Theory

In this section[50] we show the free form of each of the four fermion species (charged lepton, neutral lepton, up-type quark, and down-type quark) can be generated from a spin ½ particle spinor at rest using Complex Lorentz group boosts. *A crucial part of this study is the requirement that the relevant Lorentz boosts transform from*

---

[50] See Blaha (2018e) and earlier books by the author for more complete details.

fermion rest frames to "moving" frames with a real-valued energy and a real-valued time. The four species fermion spectrum form is a direct consequence as this section shows.

12-A.1.1 Matrix Representation of Complex Lorentz Group L_C Boosts

We begin with Complex Lorentz Group (L_C) boosts because they will be crucial in the determination of the equations of motion of various types of spin ½ particles. Defining

$$\omega \mathbf{w} = \mathbf{u_r}\omega_r + i\mathbf{u_i}\omega_i \tag{12-A.1}$$

we see an L_C boost can be expressed in the form

$$\Lambda_C(\mathbf{v_c}) = \exp[i\omega\hat{\mathbf{w}}\cdot\mathbf{K}] \tag{12-A.2}$$

where

$$\omega = (\omega_r^2 - \omega_i^2 + 2i\omega_r\omega_i \,\hat{\mathbf{u}}_r\cdot\hat{\mathbf{u}}_i)^{\frac{1}{2}} \tag{12-A.3}$$

$$\hat{\mathbf{w}} = (\omega_r\hat{\mathbf{u}}_r + i\omega_i\hat{\mathbf{u}}_i)/\omega \tag{12-A.4}$$

Since $\hat{\mathbf{u}}_r\cdot\hat{\mathbf{u}}_r = 1 = \hat{\mathbf{u}}_i\cdot\hat{\mathbf{u}}_i$

$$\hat{\mathbf{w}}\cdot\hat{\mathbf{w}} = 1 \tag{12-A.5}$$

The (generally) complex relative velocity is

$$\mathbf{v_c} = \hat{\mathbf{w}} \tanh(\omega) \tag{12-A.6}$$

We now analytically continue to complex ω and complex unit vectors $\hat{\mathbf{w}}$. The resulting complex generalization will be the matrix form of proper L_C boosts:

$$\Lambda_C(\mathbf{v_c}) = \exp[i\omega\hat{\mathbf{w}}\cdot\mathbf{K}] \equiv \Lambda_C(\omega, \hat{\mathbf{w}})$$

$$= \begin{bmatrix} \cosh(\omega) & -\sinh(\omega)\hat{w}_x & -\sinh(\omega)\hat{w}_y & -\sinh(\omega)\hat{w}_z \\ -\sinh(\omega)\hat{w}_x & 1 + (\cosh(\omega) - 1)\hat{w}_x{}^2 & (\cosh(\omega) - 1)\hat{w}_x\hat{w}_y & (\cosh(\omega) - 1)\hat{w}_x\hat{w}_z \\ -\sinh(\omega)\hat{w}_y & (\cosh(\omega) - 1)\hat{w}_x\hat{w}_y & 1 + (\cosh(\omega) - 1)\hat{w}_y{}^2 & (\cosh(\omega) - 1)\hat{w}_y\hat{w} \\ -\sinh(\omega)\hat{w}_z & (\cosh(\omega) - 1)\hat{w}_x\hat{w}_z & (\cosh(\omega) - 1)\hat{w}_y\hat{w}_z & 1 + (\cosh(\omega) - 1)\hat{w}_z{}^2 \end{bmatrix}$$

$$(12\text{-A}.7)$$

Since analytic continuations are unique, the above form for $\Lambda_C(\mathbf{v_c})$ is well-defined and unique. It spans the complete set of proper L_C boosts.

12-A.1.2 Left-handed Part of LC

If we let

$$\hat{\mathbf{u}}_i = \hat{\mathbf{u}}_r \equiv \hat{\mathbf{u}} \qquad\qquad (12\text{-A}.8)$$

so that the vector $\hat{\mathbf{u}}_i$ is parallel to $\hat{\mathbf{u}}_r$, and

$$\omega_i = \pi/2 \qquad\qquad (12\text{-A}.9)$$

then $\Lambda_C(\mathbf{v_c})$ becomes a Left-handed L_C boost:

$$\Lambda_C(\mathbf{v_c}) = \Lambda_L(\omega_r, \mathbf{u}) \qquad\qquad (12\text{-A}.10)$$

12-A.1.3 Right-handed part of L_C

If we let

$$\hat{\mathbf{u}}_i = -\hat{\mathbf{u}}_r \equiv -\hat{\mathbf{u}} \qquad\qquad (12\text{-A}.11)$$

so that the vector $\hat{\mathbf{u}}_i$ is anti-parallel to $\hat{\mathbf{u}}_r$, and

$$\omega_i = -\pi/2 \qquad\qquad (12\text{-A}.12)$$

then $\Lambda_C(\mathbf{v_c})$ becomes a Right-handed L_C boost:

$$\Lambda_C(\mathbf{v_c}) = \Lambda_R(\omega_r, \mathbf{u}) \qquad (12\text{-A}.13)$$

as described in Blaha (2007b).

12-A.1.4 Free Spin ½ Particles – Leptons & Quarks

In this section we begin by developing dynamical equations for spin ½ particles based on the L_C parts. These spin ½ particles are conventional Dirac particles (Majorana particles are also allowed but not discussed), spin ½ tachyons, and "color" versions of both types totaling four species. We will identify leptons and quarks with these fields.

12-A.1.4.1 First Step - Deriving the Conventional Dirac Equation

In this section we will review a method of obtaining the equation of motion of a particle using a free Dirac equation that is obtained by a Lorentz boost of a spinor wave function of a particle at rest.

In the case of a Lorentz transformation the 4×4 matrix form of a Lorentz transformation of Dirac matrices is

$$S^{-1}(\Lambda(v))\gamma^\nu S(\Lambda(v)) = \Lambda^\nu_{\ \mu}(v)\gamma^\mu \qquad (12\text{-A}.14)$$

where $S(\Lambda(v))$ is

$$S(\Lambda(v)) = \exp(-i\omega\sigma_{0i}v_i/(2|\mathbf{v}|)) = \exp(-\omega\gamma^0\boldsymbol{\gamma}\cdot\mathbf{v}/(2|\mathbf{v}|))$$
$$= \cosh(\omega/2)I + \sinh(\omega/2)\gamma^0\boldsymbol{\gamma}\cdot\mathbf{p}/|\mathbf{p}| \qquad (12\text{-A}.15)$$

with $\omega = \operatorname{arctanh}(|\mathbf{v}|)$, $\cosh(\omega/2) = [(E+m)/(2m)]^{\frac{1}{2}}$ and $\sinh(\omega/2) = |\mathbf{p}|[2m(E+m)]^{-\frac{1}{2}}$. Also

$$S^{-1}(\Lambda(v)) = \gamma^0 S^\dagger(\Lambda(v))\gamma^0 = \exp(\omega\gamma^0\boldsymbol{\gamma}\cdot\mathbf{v}/(2|\mathbf{v}|))$$
$$= \cosh(\omega/2)I - \sinh(\omega/2)\gamma^0\boldsymbol{\gamma}\cdot\mathbf{p}/|\mathbf{p}| \qquad (12\text{-A}.16)$$

We begin by defining a generic positive energy plane wave solution of the Dirac equation for a normal fermion particle at rest with rest mass m, *which we take to be the qube bare mass in the absence of interactions*, as

$$\psi(x) = e^{-imt}w(0) \tag{12-A.17}$$

with $w(0)$ a four component logic spinor column vector. *For a free particle at rest, the rest energy $m = m_0$, the qube mass.* The wave function satisfies the momentum space Dirac equation for a fermion at rest:

$$(m\gamma^0 - m)e^{-imt}w(0) = 0 \tag{12-A.18}$$

Subsequently we will use a similar procedure to construct the free tachyonic Dirac equation.

If we now apply $S(\Lambda(v))$ we find

$$0 = S(\Lambda(v))(m\gamma^0 - m)e^{-imt}w(0) = [mS(\Lambda(v))\gamma^0 S^{-1}(\Lambda(v)) - m]S(\Lambda(v))w(0)$$

A straightforward evaluation shows

$$mS(\Lambda(v))\gamma^0 S^{-1}(\Lambda(v)) = g_{\mu\nu}p^\mu\gamma^\nu = \displaystyle{\not}p \tag{12-A.19}$$

where $p^0 = (p^2 + m^2)^{\frac{1}{2}}$, $\mathbf{p} = \gamma m\mathbf{v}$, and $p = |\mathbf{p}|$. In addition

$$S(\Lambda(v))w(0) = w(p) \tag{12-A.20}$$

is a positive energy Dirac spinor. Therefore the Dirac equation for a fermion in motion in momentum space has the form:

$$(\displaystyle{\not}p - m)e^{-ip\cdot x}w(p) = 0 \tag{12-A.21}$$

where the exponential factor, mt, is also boosted to $p \cdot x$. Eq. 12-A.21 implies the well-known free, coordinate space Dirac equation:

$$(i\gamma^\mu \partial/\partial x^\mu - m)\psi(x) = 0 \tag{12-A.22}$$

12-A.1.4.2 Derivation of the Tachyon Dirac Equation

The Left-handed boost has the form:

$$\Lambda_L(\omega, \mathbf{u}) = \Lambda(\omega + i\pi/2, \mathbf{u}) = \exp[i\omega_L \hat{\mathbf{u}} \cdot \mathbf{K}] \qquad (12\text{-A}.23)$$

where $\omega_L = \omega + i\pi/2$ and

$$\cosh(\omega_L) = i \sinh(\omega) = -\gamma = i\gamma_s \qquad (12\text{-A}.24)$$
$$\sinh(\omega_L) = i \cosh(\omega) = -\beta\gamma = i\beta\gamma_s$$

with, $\beta = v > 1$, $\gamma_s = (\beta^2 - 1)^{-\frac{1}{2}}$, and $\boldsymbol{\omega \geq 0}$. Thus

$$\sinh(\omega) = \gamma_s \qquad (12\text{-A}.25)$$
$$\cosh(\omega) = \beta\gamma_s$$

The corresponding spinor transformation is:

$$S_L(\Lambda_L(\omega, \mathbf{u})) = \exp(-i\omega_L\sigma_{0i}v_i/(2|\mathbf{v}|)) = \exp(-\omega_L\gamma^0\boldsymbol{\gamma}\cdot\mathbf{v}/(2|\mathbf{v}|))$$
$$= \cosh(\omega_L/2)I + \sinh(\omega_L/2)\gamma^0\boldsymbol{\gamma}\cdot\mathbf{p}/|\mathbf{p}| \qquad (12\text{-A}.26)$$

The inverse transformation is

$$S_L^{-1}(\Lambda_L(\omega, \mathbf{u})) = \gamma^2\gamma^0 K^{-1}S_L^{\dagger}K\gamma^0\gamma^2 = \gamma^2\gamma^0 S_L^{T}\gamma^0\gamma^2 = \exp(\omega_L\gamma^0\boldsymbol{\gamma}\cdot\mathbf{v}/(2|\mathbf{v}|))$$
$$= \cosh(\omega_L/2)I - \sinh(\omega_L/2)\gamma^0\boldsymbol{\gamma}\cdot\mathbf{p}/|\mathbf{p}| \qquad (12\text{-A}.27)$$

where the superscript T denotes the transpose and K is the complex conjugation operator (that also appears in the time-reversal operator). Note that S_L is not unitary just as the equivalent spinor Lorentz transformation $S(\Lambda(v))$ is not unitary.

We can now apply a left-handed superluminal transformation to the generic positive energy plane wave solution of the Dirac equation for a particle of mass m at rest. The result is

$$0 = S_L(\Lambda_L(\omega, \mathbf{u}))(m\gamma^0 - m)e^{-imt}w(0)$$
$$= [mS_L\gamma^0 S_L^{-1} - m]e^{-imt}S_L w(0)$$

where $S_L = S_L(\Lambda_L(\omega, \mathbf{u}))$. After some algebra

$$mS_L\gamma^0S_L^{-1} = m[\cosh(\omega_L)\gamma^0 - \sinh(\omega_L)\gamma\cdot\mathbf{p}/|\mathbf{p}|]$$
$$= i\gamma^0E - i\gamma\cdot\mathbf{p} = i\not p \qquad (12\text{-A}.28)$$

using the tachyon energy and momentum expressions

$$\mathbf{p} = m\mathbf{v}\gamma_s \qquad\qquad E = m\gamma_s \qquad (12\text{-A}.29)$$

Also

$$S_Lw(0) = w_T(p) \qquad (12\text{-A}.30)$$

is a tachyon spinor.

The momentum space tachyonic Dirac equation is

$$(i\not p - m)e^{ip\cdot x}w_T(p) = 0 \qquad (12\text{-A}.31)$$

where $p{\cdot}x = Et - \mathbf{p}{\cdot}\mathbf{x}$ after performing a corresponding left-handed superluminal coordinate transformation in the exponential factor. Thus a positive energy wave is transformed into a negative energy wave by the superluminal transformation.

If we apply $i\not p$ we find the tachyon mass condition is satisfied

$$-E^2 + \mathbf{p}^2 = m^2 \qquad (12\text{-A}.32)$$

Transforming back to coordinate space we obtain the *tachyon Dirac equation*:

$$(\gamma^\mu\partial/\partial x^\mu - m)\psi_T(x) = 0 \qquad (12\text{-A}.33)$$

The "missing" factor of i in the first term of eq. 12-A.33 requires the lagrangian to be different from the conventional Dirac lagrangian in order for the lagrangian to be real. The simplest, physically acceptable, free spin ½ tachyon lagrangian density is:

$$\mathscr{L}_T = \psi_T^S(\gamma^\mu\partial/\partial x^\mu - m)\psi_T(x) \qquad (12\text{-A}.34)$$

where

$$\psi_T^S = \psi_T^\dagger i\gamma^0\gamma^5 \qquad (12\text{-A}.35)$$

The corresponding action is

$$I = \int d^4 x \mathcal{L}_T \tag{12-A.36}$$

12-A.1.4.3 Complex Space, and 3-Momentum & Real-Valued Energy Fermions (Quarks)

Spinor boost transformations were used in previous sections to develop the dynamical equations for Dirac fields and tachyon fields. In this section we will use L_C spinor boosts to generate additional fermion field dynamical equations.

The form of the L_C spinor boost transformation corresponding to the coordinate transformation is:

$$S_C(\omega, \mathbf{v_c}) = \exp(-i\omega\sigma_{0k}\hat{w}_k/2) = \exp(-\omega\gamma^0\boldsymbol{\gamma}\cdot\hat{\mathbf{w}}/2)$$
$$= \cosh(\omega/2)I + \sinh(\omega/2)\gamma^0\boldsymbol{\gamma}\cdot\hat{\mathbf{w}} \tag{12-A.37}$$

The inverse transformation is

$$S_C^{-1}(\omega, \mathbf{v_c}) = \gamma^2\gamma^0 K^{-1} S_C^\dagger K \gamma^0\gamma^2 = \gamma^2\gamma^0 S_C^{T}\gamma^0\gamma^2 = \exp(\omega\gamma^0\boldsymbol{\gamma}\cdot\hat{\mathbf{w}}/2)$$
$$= \cosh(\omega/2)I - \sinh(\omega/2)\gamma^0\boldsymbol{\gamma}\cdot\hat{\mathbf{w}} \tag{12-A.38}$$

where the superscript T denotes the transpose and K is the complex conjugation operator (that also appears in the time-reversal operator). Note that S_C is not unitary just as in previous cases considered in this appendix.

We now redo the development of spin ½ dynamical equations of motion of earlier sections for this more general case of complex ω and $\hat{\mathbf{w}}$. Again we apply a boost to a Dirac equation for a positive energy plane wave particle of mass m at rest:

$$0 = S_C(\omega, \mathbf{v_c}))(m\gamma^0 - m)e^{-imt}w(0)$$
$$= [mS_C\gamma^0 S_C^{-1} - m]e^{-imt}S_C w(0) \tag{12-A.39}$$

where $S_C = S_C(\omega, \hat{\mathbf{w}})$. After some algebra

$$mS_C\gamma^0 S_C^{-1} = m[\cosh(\omega)\gamma^0 - \sinh(\omega)\boldsymbol{\gamma}\cdot\hat{\mathbf{w}}] \tag{12-A.40}$$

12-A.1.4.3.1 CASE 1: PARALLEL REAL AND IMAGINARY RELATIVE VECTORS

If the real and imaginary relative vectors parts of $\hat{\mathbf{w}}$, namely $\hat{\mathbf{u}}_r$ and $\hat{\mathbf{u}}_i$, are parallel, then $\hat{\mathbf{u}}_r \cdot \hat{\mathbf{u}}_i = 1$ and

$$\omega = \omega_r + i\omega_i \qquad (12\text{-}A.41)$$

Eq. 12-A.40 can be re-expressed as

$$mS_C\gamma^0 S_C^{-1} = m[\cosh(\omega_r)\cos(\omega_i) + i\sinh(\omega_r)\sin(\omega_i)]\gamma^0 - m[\sinh(\omega_r)\cos(\omega_i) + \\ + i\cosh(\omega_r)\sin(\omega_i)]\gamma\cdot\hat{\mathbf{u}}_r \qquad (12\text{-}A.42)$$

or equivalently

$$mS_C\gamma^0 S_C^{-1} = \cos(\omega_i)\gamma\cdot p_r + i\sin(\omega_i)\gamma\cdot p_i \qquad (12\text{-}A.43)$$

where

$$p_r^0 = m\cosh(\omega_r) \qquad\qquad p_i^0 = m\sinh(\omega_r) \quad (12\text{-}A.44)$$

and

$$\mathbf{p_r} = m\hat{\mathbf{u}}_r\sinh(\omega_r) \qquad\qquad \mathbf{p_i} = m\hat{\mathbf{u}}_r\cosh(\omega_r) \quad (12\text{-}A.45)$$

If $\omega_i = 0$, then we recover the momentum space Dirac equation. If $\omega_i = \pi/2$, then we obtain the left-handed momentum space tachyon equation. Since the range of ω_i is [0, ∞> (due to the cut along the real ω-plane axis) eq. 12-A.43 corresponds to the results of the Left-Handed Lorentz boost part discussed earlier.

12-A.1.4.3.2 CASE 2: ANTI-PARALLEL REAL AND IMAGINARY RELATIVE VECTORS

If the real and imaginary relative vectors parts of $\hat{\mathbf{w}}$, $\hat{\mathbf{u}}_r$ and $\hat{\mathbf{u}}_i$, are anti-parallel $\hat{\mathbf{u}}_r = -\hat{\mathbf{u}}_i$, then $\hat{\mathbf{u}}_r\cdot\hat{\mathbf{u}}_i = -1$ and

$$\omega = \omega_r - i\omega_i \qquad (12\text{-}A.46)$$

We can then express eq. 12-A.40 as

$$mS_C\gamma^0 S_C^{-1} = m[\cosh(\omega_r)\cos(\omega_i) - i\sinh(\omega_r)\sin(\omega_i)]\gamma^0 - m[\sinh(\omega_r)\cos(\omega_i) - \\ - i\cosh(\omega_r)\sin(\omega_i)]\gamma\cdot\hat{\mathbf{u}}_r \qquad (12\text{-}A.47)$$

or

$$mS_C\gamma^0 S_C^{-1} = \cos(\omega_i)\gamma\cdot p_r - i\sin(\omega_i)\gamma\cdot p_i \qquad (12\text{-}A.48)$$

where

$$p_r^0 = m\cosh(\omega_r) \qquad\qquad p_i^0 = m\sinh(\omega_r) \qquad (12\text{-}A.49)$$

and

$$\mathbf{p_r} = m\hat{\mathbf{u}}_r \sinh(\omega_r) \qquad\qquad \mathbf{p_i} = m\hat{\mathbf{u}}_r \cosh(\omega_r) \qquad (12\text{-}A.50)$$

If $\omega_i = 0$, then we again recover the momentum space Dirac equation, If $\omega_i = \pi/2$, then we obtain the right-handed momentum space tachyon equation. (The range of ω_i is again $[0, \infty>$.)

Note: Since the matrix elements in the boost depend on $\gamma = (1 - \beta^2)^{-\frac{1}{2}}$ with a singularities at $\beta = \pm 1$, which in turn corresponds to $\omega = \pm\infty$, there is a branch cut along the ω axis in the complex ω-plane. Therefore we point out again the product of three Left-handed transformations is not equivalent to a Right-handed transformation.

12-A.1.4.3.3 CASE 3: COMPLEXONS: A NEW TYPE OF PARTICLE WITH PERPENDICULAR REAL AND IMAGINARY 3-MOMENTA

If the real and imaginary relative vectors parts of $\hat{\mathbf{w}}$, namely $\hat{\mathbf{u}}_r$ and $\hat{\mathbf{u}}_i$, are perpendicular, $\hat{\mathbf{u}}_r \cdot \hat{\mathbf{u}}_i = 0$, then

$$\omega = (\omega_r^2 - \omega_i^2)^{\frac{1}{2}} \qquad (12\text{-}A.51)$$

Thus ω is either pure real ($\omega_r \geq \omega_i$) or pure imaginary ($\omega_r < \omega_i$).

The momentum space equation generated by the corresponding L_C spinor boost is

$$\{m \cosh(\omega)\gamma^0 - m \sinh(\omega)\boldsymbol{\gamma}\cdot(\omega_r\hat{\mathbf{u}}_r + i\omega_i\hat{\mathbf{u}}_i)/\omega - m\}e^{-ip\cdot x}w_c(p) = 0 \qquad (12\text{-}A.52)$$

Defining the momentum 4-vector

$$p = (p^0, \mathbf{p}) \qquad (12\text{-}A.53)$$

where

$$p^0 = m \cosh(\omega) \qquad\qquad \mathbf{p} = \mathbf{p_r} + i\mathbf{p_i} \qquad (12\text{-}A.54)$$
$$\mathbf{p_r} = m\omega_r\hat{\mathbf{u}}_r \sinh(\omega)/\omega \quad \mathbf{p_i} = m\omega_i\hat{\mathbf{u}}_i \sinh(\omega)/\omega \quad (12\text{-}A.55)$$

and

$$\mathbf{p_r}\cdot\mathbf{p_i} = 0 \qquad (12\text{-}A.56)$$

then we obtain a positive energy Dirac-like equation with complex 3-momentum

$$[p \cdot \gamma - m] e^{-ip \cdot x} w_c(p) = 0$$

or, explicitly,

$$[p^0 \gamma^0 - (\mathbf{p_r} + i\mathbf{p_i}) \cdot \gamma - m] e^{-ip \cdot x} w_c(p) = 0$$

(12-A.57)

with a complex 3-momentum $\mathbf{p}$ and the 4-momentum mass shell condition:

$$p^2 = p^{0\,2} - \mathbf{p_r} \cdot \mathbf{p_r} + \mathbf{p_i} \cdot \mathbf{p_i} = m^2$$

(12-A.58)

Note

$$|\mathbf{v}| = |\mathbf{p}|/p^0 = [(\mathbf{p_r} + i\mathbf{p_i}) \cdot (\mathbf{p_r} + i\mathbf{p_i})]^{\frac{1}{2}}/p^0 = \tanh(\omega)$$

(12-A.59)

and thus the Lorentz factor

$$\gamma = \cosh(\omega)$$

(12-A.60)

Eq. 12-A.57 is the momentum space equivalent of the wave equation

$$[i\gamma^0 \partial/\partial t + i\gamma \cdot (\nabla_r + i\nabla_i) - m]\psi_C(t, \mathbf{x_r}, \mathbf{x_i}) = 0$$

(12-A.61)

where

$$x_c = (t, \mathbf{x_r} - i\mathbf{x_i})$$

(12-A.62)

and where the grad operators ∇_r and ∇_i are with respect to $\mathbf{x_r}$ and $\mathbf{x_i}$ respectively. Since $\hat{\mathbf{u}}_r \cdot \hat{\mathbf{u}}_i = 0$, we see that there is a subsidiary condition on the wave function

$$\nabla_r \cdot \nabla_i\, \psi_C(t, \mathbf{x_r}, \mathbf{x_i}) = 0$$

(12-A.63)

We will call the particles satisfying eqs. 12-A.62 and 12-A.63 *complexons*. In addition we have the anti-commutation relation

$$\{\gamma \cdot \mathbf{p_r}, \gamma \cdot \mathbf{p_i}\} = 0$$

(12-A.64)

which in turn implies

$$\gamma \cdot \nabla_r \gamma \cdot \nabla_i \psi_C(t, \mathbf{x_r}, \mathbf{x_i}) = \gamma \cdot \nabla_i \gamma \cdot \nabla_r \psi_C(t, \mathbf{x_r}, \mathbf{x_i}) = 0 \qquad (12\text{-A.65})$$

Since $\mathbf{p_r} = \mathbf{p_i} = 0$ in the particle rest frame prior to the complex group boost, the boosted particle spin 4-vector s^μ satisfies

$$s^\mu p_r{}_\mu = s^\mu p_i{}_\mu = 0 \qquad (12\text{-A.66})$$

Note that s^μ is itself complex[51] and, if the spin points in the z-direction prior to the complex boost, then the boosted s^μ has the form

$$s^\mu = (-\sinh(\omega)\hat{w}_z, (0,0,1) + (\cosh(\omega) - 1)\hat{w}_z \hat{\mathbf{w}}) \qquad (12\text{-A.67})$$

with $\hat{\mathbf{w}}$ defined earlier: $\hat{\mathbf{w}} = (\omega_r \hat{\mathbf{u}}_r + i\omega_i \hat{\mathbf{u}}_i)/\omega = \mathbf{p}/(m\sinh(\omega))$.

The momentum 4-vector is defined by

$$p = (p^0, \mathbf{p}) \qquad (12\text{-A.68})$$

where

$$p^0 = m \sinh(\omega) \qquad\qquad \mathbf{p} = \mathbf{p_r} + i\mathbf{p_i} \qquad (12\text{-A.69})$$

with

$$\mathbf{p_r} = m\omega_r \hat{\mathbf{u}}_r \cosh(\omega)/\omega \qquad \mathbf{p_i} = m\omega_i \hat{\mathbf{u}}_i \cosh(\omega)/\omega \qquad (12\text{-A.70})$$

and

$$\mathbf{p_r} \cdot \mathbf{p_i} = 0 \qquad (12\text{-A.71})$$

then we obtain the complexon tachyon equation

$$[ip\cdot\gamma - m]e^{+ip\cdot x}w_{cL}(p) = 0 \qquad (12\text{-A.72})$$

with a complex 3-momentum $\mathbf{p}$ and the tachyon 4-momentum mass shell condition:[52]

---

[51] This feature of partons, which is not present in ordinary Dirac particles, might be the source of the discrepancies between theory and experiment in deep inelastic parton spin physics which is based on conventional real parton spins.

$$p^2 = p^{0\,2} - \mathbf{p_r}^2 + \mathbf{p_i}^2 = -m^2 \qquad (12\text{-}A.73)$$

Eq. 12-A.72 is the momentum space equivalent of the wave equation

$$[\gamma^0 \partial/\partial t + \gamma \cdot (\nabla_r + i\nabla_i) - m]\psi_{CL}(t, \mathbf{x_r}, \mathbf{x_i}) = 0 \qquad (12\text{-}A.74)$$

or

$$[\gamma \cdot \nabla - m]\psi_{CL}(t, \mathbf{x_r}, \mathbf{x_i}) = 0 \qquad (12\text{-}A.75)$$

with the subsidiary condition on the wave function

$$\nabla_r \cdot \nabla_i \, \psi_{CL}(t, \mathbf{x_r}, \mathbf{x_i}) = 0 \qquad (12\text{-}A.76)$$

also holds. We note that eq. 12-A.74 is covariant under the real Lorentz group.

12-A.1.5 Up-type Color Quarks

Up-type quarks are assumed[53] to be fermions with complex 3-momenta - complexons, and an internal color SU(3) symmetry, that satisfy $p^2 = m^2$. Their field equation with a color SU(3) index, denoted a, inserted is

$$[i\gamma^0 \partial/\partial t + i\gamma \cdot (\nabla_r + i\nabla_i) - m]\psi_C{}^a(t, \mathbf{x_r}, \mathbf{x_i}) = 0 \qquad (12\text{-}A.77)$$

with the subsidiary condition

$$\nabla_r \cdot \nabla_i \, \psi_C{}^a(t, \mathbf{x_r}, \mathbf{x_i}) = 0 \qquad (12\text{-}A.78)$$

The free field solution is:

---

[52] Note that the presence of the $\mathbf{p_i}^2$ term does not change the tachyon requirement that $\mathbf{p_r}^2 \geq m^2$ as seen in the previous cases.

[53] The complexon theory that we develop and use for quark dynamics in the Standard Model is not required. Our SuperStandard Model could use Dirac fermion dynamics for the up-type quarks and tachyon dynamics for down-type quarks. Then the (broken) Left-handed complex Lorentz boosts would have the basic space-time group rather than L_C. We choose to use complexon dynamics for quarks because they have an internal SU(3)-like structure suggestive of color SU(3). More importantly, their spin dynamics is different and thus may resolve the differences between theory and experiment for the deep inelastic parton spin-dependent structure functions.

$$\psi_C^{\ a}(x) = \sum_{\pm s} \int d^3 p_r d^3 p_i \, N_C(p) \delta(\mathbf{p_r} \cdot \mathbf{p_i}/m^2) [b_C(p,a,s) u_C^{\ a}(p,\, s) e^{-i(p \cdot x + p^* \cdot x^*)/2} +$$
$$+ d_C^{\ \dagger}(p,a,s) v_C^{\ a}(p,\, s) e^{+i(p \cdot x + p^* \cdot x^*)/2}] \quad (12\text{-A.79})$$

The free Feynman propagator arranged into the form of a spectral integral is

$$iS_C^{\ ab}(x,y) = -\delta^{ab} \int dM \, (i\gamma^0 \partial/\partial x^0 - i(\nabla_r - i\nabla_i) \cdot \gamma + m) \delta'(\nabla_r \cdot \nabla_i/m^2) J(\mathbf{x_i} - \mathbf{y_i}, M^2) \triangle_F(x - y, M)$$
$$(12\text{-A.80})$$

where

$$\triangle_F(x - y, M) = (2\pi)^{-4} \int d^4 p_r \, \exp[-ip^0(x^0 - y^0) + i\mathbf{p_r} \cdot (\mathbf{x_r} - \mathbf{y_r})]/(p_r^2 - M^2 + i\varepsilon)$$
$$(12\text{-A.81})$$

and

$$J(\mathbf{x_i}, M^2) = (2\pi)^{-3} \int d^3 p_i \, \delta(M^2 + \mathbf{p_i}^2 - m^2) \, \exp[-i\mathbf{p_i} \cdot (\mathbf{x_i} - \mathbf{y_i})] \quad (12\text{-A.82})$$
$$= (2\pi)^{-2} |\mathbf{x_i} - \mathbf{y_i}|^{-1} \theta(m^2 - M^2) \sin((m^2 - M^2)^{1/2} |\mathbf{x_i} - \mathbf{y_i}|)$$

12-A.1.6 Down-type Color Quarks

Tachyonic complexons with complex 3-momenta, and an internal global SU(3) symmetry, that have mass shell condition $p^2 = -m^{12\text{-A.}}$ Their field equation with a color SU(3) index, denoted a, inserted is

$$[\gamma^0 \partial/\partial t + \gamma \cdot (\nabla_r + i\nabla_i) - m]\psi_{CL}^{\ a}(t, \mathbf{x_r}, \mathbf{x_i}) = 0 \quad (12\text{-A.83})$$

with the subsidiary condition on the wave function

$$\nabla_r \cdot \nabla_i \, \psi_{CL}^{\ a}(t, \mathbf{x_r}, \mathbf{x_i}) = 0 \quad (12\text{-A.84})$$

Its free field left-handed solution is:

$$\psi_{CLL}^{\ +a}(\mathbf{x_r}, \mathbf{x_i}) = \sum_{\pm s} \int d^2 p_r dp^+ d^3 p_i \, N_{CLL}^{\ +}(p) \theta(p^+) \delta((p_i^3(p^+ - p^-)/\sqrt{2} + \mathbf{p_{r\perp}} \cdot \mathbf{p_{i\perp}})/m^2) \cdot \quad (12\text{-A.85})$$
$$\cdot [b_{CLL}^{\ +}(p,a,s) u_{CLL}^{\ a}(p,a,s) e^{-i(p \cdot x + p^* \cdot x^*)/2} + d_{CLL}^{\ ++}(p,a,s) v_{CLL}^{\ +a}(p,a,s) e^{+i(p \cdot x + p^* \cdot x^*)/2}]$$

and its right-handed solution is

$$\psi_{CLR}^{+a}(x_r, x_i) = \sum_{\pm s} \int d^2p_r dp^+ d^3p_i \, N_{CLR}^{+}(p)\theta(p^+)\delta((p_i^3(p^+-p^-)/\sqrt{2} + \mathbf{p}_{r\perp}\cdot\mathbf{p}_{i\perp})/m^2)\cdot$$

$$\cdot[b_{CLR}^{+}(p,a,s)u_{CLR}^{+a}(p,a,s)e^{-i(p\cdot x + p^*\cdot x^*)/2} + d_{CLR}^{++}(p,a,s)v_{CLR}^{+a}(p,a,s)e^{+i(p\cdot x + p^*\cdot x^*)/2}]$$

$$(12\text{-}A.86)$$

The free left-handed Feynman propagator arranged into the form of a spectral integral is

$$iS^+_{CLLF}{}^{ab}(x,y) = -\delta^{ab}\int dM \, C^-R^+(\gamma^0\partial/\partial x^0 + (\nabla_r - i\nabla_i)\cdot\gamma - m)R^-C^+ \delta'(\nabla_r\cdot\nabla_i/m^2)\cdot$$
$$\cdot J_2(\mathbf{x}_i - \mathbf{y}_i, M^2)\triangle_{FT}(x - y, M) \qquad (12\text{-}A.87)$$

with ∇_r and ∇_i derivatives with respect to $\mathbf{x}_r$ and $\mathbf{x}_i$ and where

$$\triangle_{FT}(x - y, M) = (2\pi)^{-4}\int d^4p_r \exp[-ip^0(x^0 - y^0) + i\mathbf{p}_r\cdot(\mathbf{x}_r - \mathbf{y}_r)]/(p_r^2 + M^2 + i\varepsilon)(12\text{-}A.88)$$

and

$$J_2(\mathbf{x}_i, M^2) = (2\pi)^{-3}\int d^3p_i \delta(M^2 - \mathbf{p}_i^2 - m^2) \exp[-i\mathbf{p}_i\cdot(\mathbf{x}_i - \mathbf{y}_i)] \quad (12\text{-}A.89)$$

$$= (2\pi)^{-2}|\mathbf{x}_i - \mathbf{y}_i|^{-1}\theta(M^2 - m^2)\sin((M^2 - m^2)^{1/2}|\mathbf{x}_i - \mathbf{y}_i|)$$

12-A.1.7 Summary: 4 Species of Particles: Leptons and Quarks

12-A.1.7.1 Charged lepton fermions
　　The conventional Dirac equation and solutions.

12-A.1.7.2 Neutral leptons - Neutrinos
　　Simple tachyons with real energy and 3-momentum. Their free field equation is:

$$(\gamma^{\mu}\partial/\partial x^{\mu} - m)\psi_{T}(x) = 0 \qquad (12\text{-}A.90)$$

and their left-handed $\psi_{TL}{}^{+}$ Feynman propagator is:

$$iS^{+}{}_{TLF}(x, y) = \tfrac{1}{2}C^{-}R^{+}\gamma^{0}\!\int d^{4}p(2\pi)^{-4}\, p^{+}e^{-ip\cdot(x-y)}/(p^{2} + m^{2} + i\epsilon) \qquad (12\text{-}A.91)$$

Similarly the light-front Feynman propagator for the right-handed $\psi_{TR}{}^{+}$ tachyon field is

$$iS^{+}{}_{TRF}(x,y) = -\tfrac{1}{2}C^{+}R^{+}\gamma^{0}\!\int d^{4}p(2\pi)^{-4}\, p^{+}e^{-ip\cdot(x-y)}/(p^{2} + m^{2} + i\epsilon) \qquad (12\text{-}A.92)$$

12-A.1.7.3 Up-type Color Quarks

Up-type quarks are assumed[54] to be fermions with complex 3-momenta - complexons, and an internal color SU(3) symmetry, that satisfy $p^{2} = m^{2}$. Their field equation with a color SU(3) index, denoted a, inserted is

$$[i\gamma^{0}\partial/\partial t + i\gamma\cdot(\nabla_{r} + i\nabla_{i}) - m]\psi_{C}{}^{a}(t, x_{r}, x_{i}) = 0 \qquad (12\text{-}A.93)$$

with the subsidiary condition

$$\nabla_{r}\cdot\nabla_{i}\,\psi_{C}{}^{a}(t, x_{r}, x_{i}) = 0 \qquad (12\text{-}A.94)$$

12-A.1.7.4 Down-type Color Quarks

Tachyonic complexons with complex 3-momenta, and an internal global SU(3) symmetry, that have mass shell condition $p^{2} = -m^{2}$. Their field equation with a color SU(3) index, denoted a, inserted is

---

[54] The complexon theory that we develop and use for quark dynamics in the Standard Model is <u>not</u> required. Our SuperStandard Model could use Dirac fermion dynamics for the up-type quarks and tachyon dynamics for down-type quarks. Then the (broken) Left-handed complex Lorentz boosts would have the basic space-time group rather than L_{C}. We choose to use complexon dynamics for quarks because they have an internal SU(3)-like structure suggestive of color SU(3). More importantly, their spin dynamics is different and thus may resolve the differences between theory and experiment for the deep inelastic parton spin-dependent structure functions.

$$[\gamma^0\partial/\partial t + \gamma\cdot(\nabla_r + i\nabla_i) - m]\psi_{CL}{}^a(t, \mathbf{x_r}, \mathbf{x_i}) = 0 \qquad (12\text{-A}.95)$$

with the subsidiary condition on the wave function

$$\nabla_r\cdot\nabla_i \; \psi_{CL}{}^a(t, \mathbf{x_r}, \mathbf{x_i}) = 0 \qquad (12\text{-A}.96)$$

12-A.2 Four QUeST Fermion Species

QUeST is defined in a 8-dimensional complex quaternion space. We have extracted four complex-valued (eight real-valued) coordinates to form a space-time. These coordinates support a Complex Lorentz group just as the United SuperStandard theory. Therefore the considerations of section 12-A.1 above apply without change.

There are four species of fundamental fermions: "charged" lepton species, neutral lepton species, up-type quark species, and down-type quark species. The nature of each species is the same as in the Unified SuperStandard Theory.

The key relation in complex quaternion space is

$$e^x = e^a \; (\cos{(\|\mathbf{v}\|)} + \mathbf{v}/\|\mathbf{v}\| \; \sin(\|\mathbf{v}\|))s \qquad (5.4)$$

in analogy with the similar complex-valued identity used in section 12-A.1.

12-A.3 MOST Fermion Species

The fermion species in MOST number four despite the complex eight-dimensional nature of the space-time extracted from complex octonion space. The eight complex space-time coordinates support a 7+1 Complex Lorentz group.

There are four types of boosts in 7+1 space-time that boost a particle rest state to a state of motion with a real energy and a real-valued or complex-valued momentum (spatial coordinates):

1. A boost from rest to a frame with real-valued energy and momentum with $p^{02} - \mathbf{p}^2 > 0$. A "normal charged lepton-like" fermion.
2. A boost from rest to a frame with real-valued energy and momentum with $p^{02} - \mathbf{p}^2 < 0$. A "tachyonic neutral lepton-like" fermion.

3. A boost from rest to a frame with real-valued energy and a complex-valued spatial momentum with $p^{02} - \mathbf{p}^2 > 0$. An "up-type quark-like" fermion.
4. A boost from rest to a frame with real-valued energy and momentum with $p^{02} - \mathbf{p}^2 < 0$. A "tachyonic down-type quark-like" fermion.

where p^0 is the energy and $\mathbf{p}$ is a spatial 7-vector. The particle states generated by 1 and 2 are "conventional" 8-dimensional analogues of the 4-dimensional case.

The quark-like cases have an 8-dimensional aspect that prompts us to call them *octoquarks*.in 4-dimensional space-time. In 4-space-time a quark momentum is set by the term

$$\omega w = \mathbf{u_r}\omega_r + i\mathbf{u_i}\omega_i \qquad (12\text{-A.1})$$

as shown in section 12-A.1. In 8-dimensional space-time the 8-momentum, *hyper-momentum*,[55] is set by

$$\omega w = \mathbf{u_r}\omega_r + i\mathbf{u_i}\omega_i + j\mathbf{u_3}\omega_i + k\mathbf{u_4}\omega_4 + q\mathbf{u_5}\omega_5 + r\mathbf{u_6}\omega_6 \qquad (12\text{-A.97})$$

where the $\mathbf{u_i}$ are 8-vectors satisfying

$$\mathbf{u_i}\cdot\mathbf{u_j} = \delta_{ij} \qquad (12\text{-A.98})$$

and i, j, k, q, and r are fundamental octonion units.

Note that a fermion spin s satisfies the 8-vector inner product

$$s\cdot p = 0 \qquad (12\text{-A.99})$$

or

$$\mathbf{s}\cdot\mathbf{p} = 0 \qquad (12\text{-A.100})$$

in a manner similar to 4-space-time.

Thus the 7 7-vectors in 8-space-time form an orthonormal set and define quarks with 7-momentum and 7-spin. We call them *octoquarks*.

---

[55] As opposed to complex momentum in 4-space-time.

Normal and tachyonic particles are distinguished in complex 8-dimensional space by the sign of $p^{02} - \mathbf{p}^2$ in a manner similar to 4-space-time. MOST fermions occur in four species just like QUeST fermions and Unified SuperStandard Theory fermions.

There are four species of fundamental fermions in 4-space-time and 8-space-time: "charged" lepton species, neutral lepton species, up-type octoquark species, and down-type octoquark species. The nature of each species (normal or tachyon, and real-valued or complex spatial momentum) is the same as in the Unified SuperStandard Theory.

12-A.4 Subgroups of the Complex Lorentz Group

In Blaha (2018e) and earlier books the author found that the restriction to real-valued time in the set of transformations of the Complex Lorentz group led to the presence of the same set of subgroups: U(1), SU(2), and SU(3) as well as other subgroups: U(1), SU(2), and SU(3). We saw the analogous internal symmetry groups existed in the extended Standard Model.

Thus it appears that one can extract the internal group structure of the complex quaternion and the complex octonion spaces by only considering the set of dimensions limited to the real part of the complex time quaternion and the real part of the complex time octonion respectively.

Coupling Constants

We show that α, the fine structure constant of Quantum Electrodynamics, is exactly determined by the electrodynamic vacuum polarization. Further we show that the Weak and Strong coupling constants are also determined by Weak and Strong vacuum polarization to good accuracy. *Thus all coupling constants appear to be determined by vacuum polarization. Consequently these coupling constants appear to be the same in all universes.*

Together with the previously determined particle spectrums and interactions it follows that *all universes have the same internal symmetry structure. The question of particle masses and symmetry breaking awaits further analysis..*

13. Missing Features of the Unified SuperStandard Theory

The Unified SuperStandard Theory's *internal symmetry* form was derived from a set of axioms. It did not determine the coupling constants or the particle masses of the theory. In this book, and earlier books, we calculated the coupling constants of the Electromagnetic, Weak and Strong interactions of the first layer of fermions. The other fermion layers appear to have similar coupling constants for their internal symmetry group equivalents since all coupling constants appear to be determined by vacuum polarization as we show later.

13.1 Determination of Coupling Constants

The renormalized coupling constants of The Unified SuperStandard Theory can be determined experimentally. However their theoretical calculation was uncertain. This gap in our understanding suggests that there is a major aspect of fundamental physics that is not understood. The fact that we can determine the form – but not the values of coupling constants – so directly from basic principles suggests that a new basic principle(s) is needed to complete The Unified SuperStandard Theory. A similar comment applies to fermion and boson masses – both our mechanism,[56] and the vanilla Higgs Mechanism, arbitrarily fixes particle masses.

The *most* meaningful attempt to determine a coupling constant in a non-trivial 4-dimensional quantum field theory was that of Johnson, Baker and Willey[57] in a 4-dimensional model – massless Quantum Electrodynamics. They developed the theory to the point where if one function, that they called the eigenvalue function, had a zero at

---

[56] Blaha (2015c).
[57] M. Baker and K. Johnson, Phys. Rev. **D8**, 1110 (1973) and references therein. See Appendix C.

the value of the fine structure constant[58] α = 1/137.035999139 (31) at Q^2 = 0. Then the theory would have no infinities.[59]

This author then developed an approximate solution for the eigenvalue function in perhaps the most comprehensive 4-dimensional quantum field theory calculation to all orders in α. The approximate calculation agreed with known exact results for the eigenvalue function to 6th order in e. In 1974 this author[60] did not find an eigenvalue function zero at the known value of the renormalized α for reasons that will be explained later. The remedy for this deficiency, which determined α exactly, will also be described later.

The Johnson, Baker, Willey model QED eigenvalue condition illustrates one possible approach to determining the coupling constants of QED and possibly the other coupling constants of The Unified SuperStandard Theory. It appeared possible that eigenvalue conditions might fix other coupling constant values. In this book we will suggest that the non-aabelian group interactions of the Unified SuperStandard Theory may possess eigenvalues conditions similar to that of massless QED. We will then quite accurately calculate approximate values of other coupling constants.

What other approaches are possible? There is an anthropomorphic approach which posits the necessity of certain ranges of some coupling constants for human life, and life in general, to exist. We are not comfortable with this approach since it seems to "beg the question." The input is equivalent to the output mitigating its character as fundamental.

One could also study the set of coupling constants in a 10-dimensional (or other dimensional) space looking for the set of coupling constant values.

These alternate possibilities are not viable at present. So we will proceed with the eigenvalue function approach, fixing the apparent failure in Massless QED, and developing eigenvalue functions for the coupling constants of other interactions.

---

[58] C. Patrignani et al, (Particle Data Group), Chin. Phys. C**40** 100001 (2016).

[59] An alternate summation in perturbation theory gives the zero at the bare coupling constant α_0. Both alternatives will be discussed later where the renormalized coupling constant will be shown to be the correct zero.

[60] Equation 1 in our paper S. Blaha, Phys. Rev. **D9**, 2246 (1974).

13.2 Origin of Particle Masses

Particle masses were fixed by either the original Higgs Mechanism or by our new mechanism that was based on an extension of Quantum Field Theory to include classical fields, which contain vacuum expectation values, which cropped up in the original Higgs Mechanism and were handled "by hand." (See Blaha (2015c).)

The origin of the constants appearing in either approach to particle masses is unknown at present. They are *ad hoc* parameters inserted by hand. This book will not investigate their origin.

13.3 Principle of Unfolding Depth

There are many situations in Physics where an initial "discovery" leads to deeper and deeper levels of understanding. The clearest example of this process of unfolding depth is the sequence of discoveries of the nature of matter:

1. Four basic elements: earth, air, fire, water
2. The chemical elements of the eighteenth and nineteenth centuries
3. The discovery that the elements are made of atoms by Boltzmann and others in the late nineteenth century
4. The discovery that atoms consist of electrons circling a nucleus in the early twentieth century
5. The discovery that a nucleus is composed of protons and neutrons
6. The discovery that neutrons and protons are composed of quarks

This example of unfolding depth can be multiplied many times in Physics and Chemistry. The case most relevant to the current discussion is that of Electromagnetism where we see:

1. Classical Electromagnetism
2. Renormalizable Quantum Electrodynamics
3. Incorporation of Quantum Electrodynamics within the Weinberg-Salam Model with a complex renormalization program (due to t'Hooft and others).

4. Possible incorporation of the Weinberg-Salam Model within a unified theory such as our Unified SuperStandard Theory with the elimination of renormalization infinities using Two-Tier coordinates.

The above Electromagnetism sequence has the property that the theory in the earlier stages is still correct within its range of validity. So we can still perform computations in Quantum Electrodynamics using Pauli-Villars regularization (stage 2 above) and the Gell-Mann-Low QED renormalization studies are still valid.

Having seen how Physics theories unfold to greater depth in the above examples (which could have been readily expanded to a much larger set of examples), we now propose a simple principle of Unfolding Depth.

Principle: Theories of Physical Phenomena tend to unfold to greater depth with the passage of time in such a way that earlier stages of the unfolding process still retain their validity to a great degree.

14. Vacuum Polarization and Coupling Constants

Recent studies of vacuum polarization have led the author to consider its profound implications for our universe and other universes in the Megaverse.

The vacuum polarization mechanism is based on eigenvalue functions that have a previously unanticipated universality. Eigenvalue functions determine the QED fine structure constant, the coupling constants of the Weak SU(2) and Strong SU(3) gauge theories to good approximation.

The result is that our universe is very similar to other universes with the same Physics and Chemistry, and thus the same prospects for life.

We will also see that a form of vacuum polarization suggests 4D universes are like particles.

15. Universal Coupling Constant Eigenvalue Condition

In a series of remarkable papers Johnson, Baker and Willey[61] developed a finite theory of massless QED (called JBW) without divergences if a certain function $F_1(\alpha)$ of the fine structure constant α called the eigenvalue function were zero. (A zero would imply Z_3, the divergent vacuum polarization constant of the electron, was zero.)

Adler[62] refined the discussion by pointing out that a zero of the eigenvalue function would be an essential singularity with:

$$F_1(\alpha) = 0 \qquad\qquad (15.1)$$
$$d^n F_1(\alpha)/d\alpha^n = 0$$

The calculation of the eigenvalue function was reduced by JBW to the sum of all single loop vacuum polarization diagrams of the general form of Fig. 15.1.

In 1973 the author[63] calculated $F_1(\alpha)$ approximately to all orders in α. A search for an essential singularity proved fruitless. Recently the author noticed that the vacuum polarization of the electron is manifest in experiment with the effective value of α increasing at higher energies. Thus Z_3 is not zero and has a divergent piece.

On this basis the author proposed, in a series of books in 2019, that the *appropriate* eigenvalue condition was

$$F_2(\alpha) = 0$$

where

---

[61] Summarized in some detail in K. Johnson and M. Baker, Phys. Rev. **D8**, 1110 (1973). Also in Blaha (2019b) and (2019c).
[62] S. Adler, Phys. Rev. **D5**, 3021 (1972).
[63] S. Blaha, Phys. Rev. **D9**, 2246 (1974).

$$F_2(\alpha) = F_1(\alpha) - [2/3 + \alpha/(2\pi) - (1/4)[\alpha/(2\pi)]^2] \qquad (15.2)$$

The additional terms are those appearing in the exact low order calculation of $F_1(x)$:

$$F_{1\ low\ order}(\alpha) = 2/3 + \alpha/(2\pi) - (1/4)[\alpha/(2\pi)]^2 \qquad (15.3)$$

In terms of F_2 the renormalization constant Z_3 is

$$Z_3 = 1 + F_1(\alpha)\ln(p/\Lambda) = 1 + F_2(\alpha) + \text{divergent terms} = 1 + \text{divergent terms} \qquad (15.4)$$

The original goal of the JBW Model was to solve massless QED in a manner that made all renormalization constants either 1 or at least finite.

We modified this goal. We shall see that we can obtain a physically better eigenvalue function F_2 that has a zero at the known fine structure constant α. Until now we have not specified the value α that appears in the preceding equations. We now define α as a partially renormalized quantity that is related to the bare fine structure constant α_0 by

$$\alpha = \alpha_0[2/3 + \alpha_0/(2\pi) - (1/4)[\alpha_0/(2\pi)]^2] \qquad (15.5)$$

We will show that the evaluation of the F_2 eigenvalue function gives the known approximate[64] physical value[65] of the fine structure constant:

$$\alpha = 0.007297352\ 5\ \ 693\ (11) \qquad (15.6)$$

The renormalized expressions appearing below are not fully finite. However the intermediate renormalized finite α is physically sensible—more so than the completely finite renormalization constants goal of the JBW Model.

The bare charge constant α_0 is known to approach ∞ at very short distances. The simplest examples of this phenomenon are the physical Coulomb scattering amplitudes

---

[64] The constant α is an irrational number.
[65] 2018 CODATA: P. J. Mohr *et al* CODATA group (2019)

and the first order change in hydrogen-like atomic energy levels.[66] Thus our modified JBW Model with a partial renormalization conforms to physical reality:

$$Z_3 = 1 + \{\alpha F_2(\alpha) + \alpha[2/3 + \alpha/(2\pi) - (1/4)[\alpha/(2\pi)]^2]\}\ln(p/\Lambda)$$
$$= 1 + \alpha\{2/3 + \alpha/(2\pi) - (1/4)[\alpha/(2\pi)]^2\}\ln(p/\Lambda) \qquad (15.7)$$

at α = the physical fine structure constant where $F_2(\alpha) = 0$.

Our approximate 1973 solution, which summed one loop pieces of the vacuum polarization yielded the algebraic equations, is:[67]

$$A_1 = (g + 1)(1 - 2g^2)/[(g + 2)(g - 1)] \qquad (15.8)$$

$$A_2 = [8g^2(2g + 1) - (2g^3 + 2g^2 + g - 2)(g^2 + 2g + 2)]/[2(g^2 - 1)(g^2 - 4)]$$

$$A_3 = -2(1 + 3g + 6g^2 + 2g^3)/[g(g + 1)]$$

$$A_4 = -(g + 2)(1 + 5g + 6g^2 + 2g^3)/[g(g^2 - 1)] - 1/(g + 1)$$

$$\psi = [gA_3 - (4 + 2g)A_1]/[(4 + 2g)A_2 - g A_4]$$

$$(\alpha/2\pi) = [gA_4 - (4 + 2g)A_2]/(A_4A_1 - A_2A_3)$$

$$F_1(g) = (2/3)(1 - 3g^2/2 - g^3) - (\alpha/4\pi)[(2 + 4g + 4g^2)(g - 2) + \alpha\psi g^3]/[(g^2 - 1)(g - 2) + \alpha(2 + 4g + 4g^2)(g - 2) + \alpha\psi g^3]$$

expressed as a function[68] of g (the power of the divergent factor p/Λ) with ψ specifying the gauge, and with the renormalization definitions

---

[66] See E. A. Ueling, Phys. Rev. **48**, 55 (1935) and R. Serber, Phys. Rev. **48**, 49 (1935).

[67] Blaha *op. cit.*

[68] The solution for the eigenvalue function is clearly best expressed in terms of the g factor in the exponents of the divergent renormalization factors. We use $F_1(g)$ and $F_1(\alpha(g))$ interchangeably.

$$\Gamma_\mu(p) = f(\gamma_\mu + 2g\gamma \cdot pp_\mu/p^2)(p/\Lambda)^{2g} \tag{15.9}$$

$$S_F = [f\gamma \cdot p(p/\Lambda)^{2g}]^{-1} \tag{15.10}$$

$$\Gamma_{\mu\alpha}(p) = (f_3/p^2)(\gamma \cdot p\gamma_\mu\gamma_\alpha - \gamma_\alpha\gamma_\mu\gamma \cdot p)(p/\Lambda)^{2g} \tag{15.11}$$

and

$$F_1 = (2/3)(1 - 3g^2/2 - g^3) - f_3/f \tag{15.12}$$

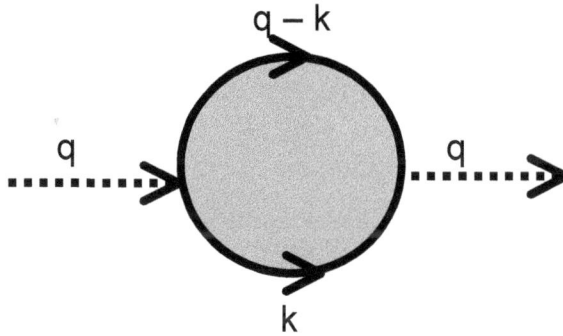

Figure 15.1 One loop vacuum polarization Feynman diagrams with internal free photon propagators.

We thus have an expression for the eigenvalue function F_2 within the framework of massless QED. We shall see that the eigenvalue function generalizes to all Standard Model gauge interactions. We shall also see that it applies to the expansion and contractions of universes upon the introduction of a Dark Energy gauge interaction for universes.

In Blaha (2019b) we generalized eqs. 15.8 to a "universal" eigenvalue function F_2 to include the Weak interaction and the Strong interaction coupling constants by inserting an interaction specific factor in the α_G equation:

$$(\alpha_G/2\pi) = c_G^{-1}[gA_4 - (4 + 2g)A_2]/(A_4A_1 - A_2A_3) \tag{15.13}$$

For a non-abelian group we set

$$c_G^{-1} = [(11/3)C_{ad} - 2C_f/3]/(16\pi)^3 \tag{15.14}$$

where C_{ad} is the dimension of the fundamental representation of the group and C_f is the number of fermions (fermion flavor) of the interaction.

We also extend it to the universe Dark Energy interaction in Blaha (2019c).

In chapter 18 we will see that we obtain the exact QED Fine Structure Constant to the known 13 place accuracy with $F_2 \cong 0$. Thus our "approximate" eigenvalue condition $F_2(\alpha) = 0$ appears to be remarkably accurate. It generalizes directly in chapters 3 and 4 to the Standard Model interactions and the universe scale factor.

Blaha (2019b) describes the universal eigenvalue function F_2 in detail including numerous plots. It also proposes a generalization of F_2 to an "exact" tangent form in chapter 23.

16. Massless Quantum Electrodynamics

Quantum Electrodynamics (QED) has been presented in depth in many papers and textbooks. In its pristine form it contains a massive electron interacting with a massless photon. Due to these interactions the electron mass and the electric charge are renormalized in perturbation theory. The renormalizations lead to infinities that have been of some concern in the early days of QED. Starting with Pauli-Villars regularization, techniques have been developed to control these infinities so the theory makes only finite, measurable experimental predictions. The theoretical predictions of QED have achieved a remarkable degree of consistency with experiment to make QED the most accurate of all theories. Successes include the hydrogen atom, particle magnetic moments, the Lamb shift, and short distance Coulomb scattering – all calculated in QED with up to ten digit accuracy!

16.1 The Importance of the Value of α

Perhaps the most important question in QED is the origin of the value of the QED coupling constant, the Fine Structure Constant denoted α, with the measured value[69] of α = 1/137.035999139 (31) at $Q^2 = 0$. *Many features of Nature and Life have been shown to depend significantly on the value of α.* The form of the universe is largely determined by the value of α and gravity. Chemistry depends to great detail on the value of α. Life is based on Chemistry, as noted by Paracelsus in the 16th Century.

The prodigious importance of the value of α has led to many attempts to determine (or calculate) its origin: attempts ranging from the subatomic to the cosmological. All these efforts have been unsuccessful. Perhaps one of the better known efforts was that of W. Heisenberg who determined α to have a value of about 3. Other efforts, including a past effort of this author, have also been unsuccessful.

---

[69] C. Patrignani et al, (Particle Data Group), Chin. Phys. C**40** 100001 (2016).

In this book we accurately determine the approximate value of α. We shall show that the value of α is determined by an eigenvalue function whose zero removes massless QED divergences. This zero occurs at the measured value of α.

16.2 The QED Divergences Requiring Renormalization

QED has divergences when calculated in perturbation theory. The divergences can be isolated into four divergent quantities:

Z_1 - vertex renormalization factor
$Z_2 = Z_1$ - self-energy renormalization factor
Z_3 - vacuum polarization renormalization factor
δm - self-mass renormalization

The renormalization constants appear in the expressions:

$$S_F(p) = Z_2 S'_F(p)$$
$$D_F(q)_{\mu\nu} = Z_3 D'_F(q)_{\mu\nu}$$
$$V_\mu(q', q) = Z_1^{-1} V'_\mu(q', q)$$

where $S'_F(p)$, $D'_F(q)_{\mu\nu}$, and $V'_\mu(q', q)$ are the divergence-free QED physical propagators and vertex.

In addition the renormalized charge e satisfies

$$e_0 = Z_1 e/(Z_2 Z_3^{1/2}) = Z_3^{-1/2} e \qquad (16.1)$$

where e_0 is the bare electric charge and e is the renormalized charge. The self-mass δm satisfies

$$m = m_0 + \delta m$$

where m_0 is the bare electron mass and m is the physical electron mass.

We now leave the conventional discussion of QED since it is discussed in detail in many textbooks and papers.

16.3 Johnson-Baker-Willey (JBW) Massless QED Model

The JBW model was an attempt to eliminate the divergences of QED at very high energies where the electron mass may be neglected. It was extended to allow for non-zero electron mass. The model is summarized in Phys. Rev. **D8**, 1110 (1973). This paper and references therein present a detailed derivation of the JBW model. The interested reader is referred to the JBW papers, which will only be utilized here for the discussion of the JBW eigenvalue function and its role in eliminating QED divergences as well as in calculating the fine structure constant usually denoted α.

In addition to the JBW series of papers Adler[70] made a significant advance in the understanding of the JBW model and massless QED by showing that the JBW eigenvalue function, if it has a zero at the measured value of α or at α_0 - the bare value of α, will be an essential singularity.

Subsequently this author[71] found an approximation that enabled the JBW eigenvalue function to be calculated to all orders in α. The approximate eigenvalue function did not have an essential singularity although it had poles and branches. It also did not have a zero at α near the experimentally found value of 1/137.035999139 (31) at $Q^2 = 0$. These apparent failures, and the unification of QED and the Weak interactions in the Weinberg-Salam Model, eliminated the interest in the massless QED at that time.

In this book we revive the eigenvalue condition, in a modified form, for divergence-free QED and show that it has a zero near the known value of α. We explain the absence of the essential singularity as due to our approximation. However the approximation does suggest the possibility of an essential singularity in the exact solution. On this basis we show important possible corollary effects of the eigenvalue function zero. And we then extend our discussion to the Weak and Strong interactions where we determine that eigenvalue functions exist that pin down the values of their coupling constants to values near the currently known values.

Thus we have a viable method of determining, at least approximately, the known coupling constants of the Unified SuperStandard Theory.

---

[70] S. Adler, Phys. Rev. **D5**, 3021 (1972).
[71] S. Blaha, Phys. Rev. **D9**, 2246 (1974). Reprinted in Appendix 25-A for the reader's convenience.

17. QED Eigenvalue Function

17.1 Origin of the Eigenvalue Function

The massless QED eigenvalue function of the JBW model was found in a series of papers and summarized in some detail by the paper of K. Johnson and M. Baker,[72] Phys. Rev. **D8**, 1110 (1973). In this section we briefly outline the steps leading to the JBW eigenvalue function based on Johnson and Baker:

1. The electron self-energy and the photon vacuum polarization are calculated using the free electron and photon propagators from the Feynman diagrams of Figs. 2 and 3.

2. The apparent quadratic divergence appearing in the diagrams of Fig. 3 is reduced to a logarithmic divergence in the vacuum polarization.

3. The logarithmic divergence in the vacuum polarization is further reduced to a single power of the logarithm of the ultraviolet cutoff denoted Λ.

4. The coefficient of the divergent logarithmic term is denoted

$$(x/2\pi)F(x)$$

   where x is the bare fine structure constant α_0.

5. If

$$F(x_0) = 0$$

---

[72] The author is pleased to acknowledge discussions with M. Baker and K. Johnson in 1973 that introduced him to their model.

for some value x_0 then a consistent divergence-free (finite) solution of massless QED is found.

6. The F(x) function is reduced to the eigenvalue function $F_1(x)$ which is the sum of logarithmic divergences of all single closed electron loop diagrams. If the eigenvalue function $F_1(x)$ has a zero at x_0 then $F(x_0) = 0$. Consequently the eigenvalue condition becomes

$$F_1(x_0) = 0 \qquad\qquad (17.1)$$

7. Adler[73] made the important observation that a zero in eq. 17.1 would necessarily be an essential singularity:

$$d^n F_1(x)/dx^n|_{x=x_0} = 0 \qquad\qquad \text{for all } n > 0 \quad (17.2)$$

Depending on the summation in perturbation theory of the relevant vacuum polarization diagrams might may occur at the bare coupling constant α_0 or the physical coupling constant $\alpha = 1/137$ ...

8. This author calculated an approximation to F_1 which did not explicitly display an essential singularity and did not have a zero at the physical fine structure constant α.

17.2 Blaha's Approximate Calculation of the Eigenvalue Function

In 1974 this author[74] formulated an approximation to the equations of massless QED and solved them for the vacuum polarization, electron self-energy and the vertex renormalization. The approximation is described in detail in the author's Phys. Rev. D paper in Appendix 25-A.

---

[73] Adler *op. cit.*
[74] Blaha *op. cit.*

The approximate solution for $F_1(x)$ had the encouraging feature that it reproduced the known[75] low order exact calculations of $F_1(x)$:

$$F_{1\text{ low order}}(x) = 2/3 + x/(2\pi) - (1/4)[x/(2\pi)]^2 \qquad (17.3)$$

Our approximate solution, which summed pieces of the vacuum polarization given by the diagrams of Figs. 2 and 3 in Appendix A, yields the algebraic equations:[76]

$$A_1 = (g + 1)(1 - 2g^2)/[(g + 2)(g - 1)] \qquad (17.4)$$

$$A_2 = [8g^2(2g + 1) - (2g^3 + 2g^2 + g - 2)(g^2 + 2g + 2)]/[2(g^2 - 1)(g^2 - 4)]$$

$$A_3 = -2(1 + 3g + 6g^2 + 2g^3)/[g(g + 1)]$$

$$A_4 = -(g + 2)(1 + 5g + 6g^2 + 2g^3)/[g(g^2 - 1)] - 1/(g + 1)$$

$$\psi = [gA_3 - (4 + 2g)A_1]/[(4 + 2g)A_2 - g\,A_4]$$

$$(\alpha/2\pi) = [gA_4 - (4 + 2g)A_2]/(A_4A_1 - A_2A_3)$$

$$F_1(g) = (2/3)(1 - 3g^2/2 - g^3) - (\alpha/4\pi)[(2 + 4g + 4g^2)(g - 2) + \alpha\psi g^3]/[(g^2 - 1)(g - 2) + \\ + \alpha(2 + 4g + 4g^2)(g - 2) + \alpha\psi g^3]$$

as a function[77] of g with ψ specifying the gauge, and with the definitions

$$\Gamma_\mu(p) = f(\gamma_\mu + 2g\gamma \cdot pp_\mu/p^2)(p/\Lambda)^{2g} \qquad (17.5)$$
$$S_F = [f\gamma \cdot p(p/\Lambda)^{2g}]^{-1} \qquad (17.6)$$
$$\Gamma_{\mu\alpha}(p) = (f_3/p^2)(\gamma \cdot p\gamma_\mu\gamma_\alpha - \gamma_\alpha\gamma_\mu\gamma \cdot p)(p/\Lambda)^{2g} \qquad (17.7)$$

---

[75] J. Rosner, Phys. Rev. Lett. **17**, 1190 (1966).
[76] Blaha *op. cit.* The solution for the eigenvalue function is clearly best expressed in terms of the g factor in the exponents of the divergent renormalization factors.
[77] We use $F_1(g)$ and $F_1(\alpha(g))$ interchangeably.

and

$$F_1 = (2/3)(1 - 3g^2/2 - g^3) - f_3/f \qquad (17.8)$$

in the notation of Appendix A. Eqs. 17.4 and 17.5 manifestly cannot lead to a form of F_1 with an essential singularity due to their algebraic form.

The plot of F_1 below did not show a zero of F_1 at the physical fine structure constant. Thus the hopes raised by the JBW model seemed dashed—at least in our approximate solution *then*. Later in this book we revive the hope of a satisfactory eigenvalue function with an eigenvalue at the physical value of the fine structure constant α.

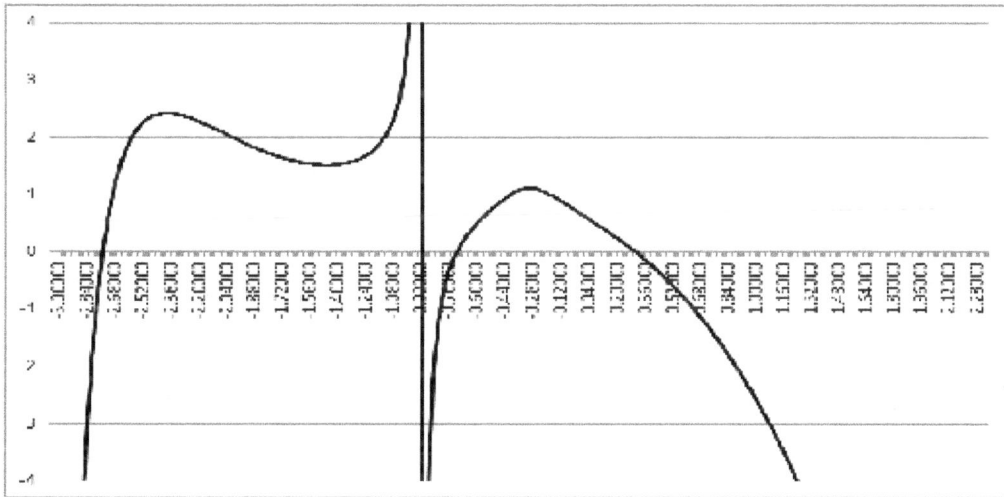

Figure 17.1. A plot of the approximate eigenvalue function $F_1(g)$ (vertical axis) *as a function of g*. Note none of its zeroes correspond to the known physical value of α. (Confirmed by next figure.) It does not have an essential singularity.

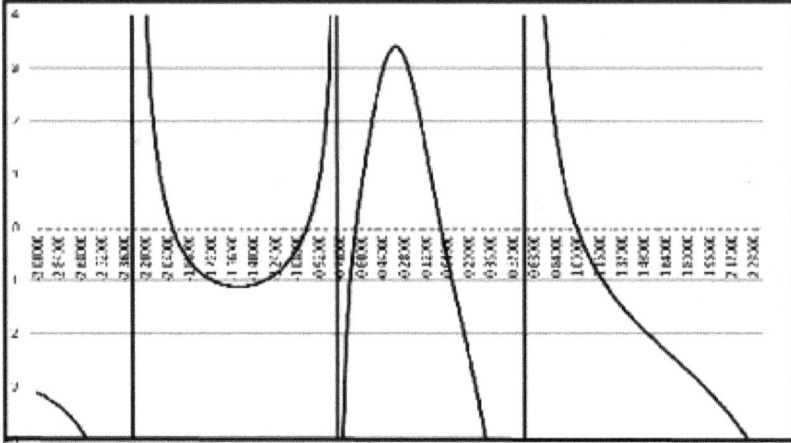

Figure 17.2. A plot of α(g) (vertical axis) *as a function of g* **for the approximate F₁ eigenvalue function. Note that it does include the physical value of α in its range of values but not as a zero of F₁.**

17.3 Some Features Signifying an Essential Singularity of the Eigenvalue Function

A number of important features follow primarily from the existence of an essential singularity in the JBW eigenvalue function:

1. The essential singularity may occur at the value of the bare coupling constant or at the renormalized (physical) coupling constant. This disparity is due to the appearance of differing results depending on the order of summation of perturbation theory for the vacuum polarization.

2. Due to the essential singularity, the large n coefficients ($n \geq k$) of an expansion in α of the eigenvalue function

$$F_1(\alpha) = \sum_{n=k}^{\infty} c_n \alpha^n \qquad (17.9)$$

for some value k may reveal features of F_1 near the point of the essential singularity.

17.3.1 Additional Features of the Eigenvalue Function

The JBW massless QED model has one species of charged fermion. There are more charged fermion species. However their effect is simply to create a multiple of the eigenvalue function:

k Charged Fermion Species → Eigenvalue function = $kF_1(\alpha)$

where k is the number of species. Thus the eigenvalue condition is independent of number of fermion species.

The Strong interactions and other non-abelian interactions are also not necessarily relevant for the calculation of the QED F_1 and its eigenvalue since these interactions are asymptotically free and may be neglected at very high energies.

17.3.2 Recognizing Evidence In an Approximation to an Essential Singularity

In calculating a function approximately, that is known to have an essential singularity, there are several possible types of evidence of the singularity:

1. The singularity will manifest itself directly as an infinite order zero or as in infinite essential singularity value at a point.

2. Since the singularity may reveal itself only approximately, the forms of a revealing approximation may be a flat region of a function (a constant value in a region.) Or it may be an "infinite-valued" point: a pole or a singularity with a branch point.

A simple example illustrating this second possibility is the function:

$$\exp(-1/x) = 1 - 1/x + \dots \tag{17.10}$$

If the approximation finds only the first term of the expansion, then, since constants are trivially infinite order zeroes, a constant region in the approximation might signal an essential singularity. If the approximation has something like the first two terms in the eq. 17.10 expansion, then an infinity in the approximation could signal the presence of the essential singularity.

We shall use this working approach to detecting an essential singularity in the eigenvalue functions we study below.

17.4 New Revised Eigenvalue Function Approximation

As pointed out in our 1974 Phys. Rev. D paper The eigenvalue function F_1 does not have a zero at the known value of the Fine Structure Constant $\alpha = 1/137.035999139$ (31) at $Q^2 = 0$. It is not even close to the value of α. As a result in 1974 we abandoned the effort thinking the approximate calculation was either insufficient to capture the essential singularity or that α was possibly determined by some other, perhaps cosmological, consideration.

We now reconsider our approximation and show that item 2 of section 17.3 (eq. 17.9) leads to a value of α near to the measured value. We define

$$F_2(\alpha) = F_1(\alpha) - [2/3 + \alpha/(2\pi) - (1/4)[\alpha/(2\pi)]^2] \qquad (17.11)$$

where we subtract the known low order terms of F_1 (eq. 17.3) in accordance with eq. 17.9.[78] $F_2(\alpha)$ will be seen below to have a neighborhood where $F_2(\alpha) \approx 0$ with a set of values for α including an approximate value for the known fine structure constant. Since our $F_1(\alpha)$ (and consequently $F_2(\alpha)$) is an approximate solution of the single electron loop QED equations, it is reasonable to expect $F_2(\alpha)$ is not identically zero at the eigenvalue point. However, a possible indication of an essential singularity is a *constant region* of $F_2(\alpha)$, as noted in section 17.3.2 using the example eq. 17.10. *We find such a region exists of constant $F_2(\alpha)$. That region has a value at its "midpoint" closely approximating the known value of α.*

In terms of F_2 the renormalization constant Z_3 is

---

[78] We note that an essential singularity in a function persists if any finite number of terms is subtracted.

$$Z_3 = 1 + F_1(\alpha)\ln(p/\Lambda) = 1 + F_2(\alpha) + \text{divergent terms} = 1 + \text{divergent terms} \quad (17.12)$$

17.5 Revised JBW Model Goal

The original goal of the JBW Model was to solve massless QED in a manner that made all renormalization constants either 1 or at least finite. We shall see that we can obtain an eigenvalue function F_2 that has a zero at the known fine structure constant that we denote α. Until now we have not specified the value α that appears in the preceding equations. We now define α as a partially renormalized quantity that is related to the bare fine structure constant α_0 by

$$\alpha = \alpha_0[2/3 + \alpha_0/(2\pi) - (1/4)[\alpha_0/(2\pi)]^2] \quad (17.13)$$

and specify all appearances of α in eqs. 17.4-17.11 as the α in eq. 17.13 which we will show leads to the known physical value of the fine structure constant: $\alpha = 0.0072973525693$.

Thus the renormalized expressions appearing in eqs. 17.5 - 17.7 are not fully finite. However the intermediate renormalized finite α is physically sensible—more so than the completely finite renormalization constants goal of the JBW Model. The bare charge constant α_0 does approach ∞ at very short distances. The simplest examples of this phenomenon are the physical Coulomb scattering amplitudes and the first order change in hydrogen-like atomic energy levels.[79] Thus our modified JBW Model with a partial renormalization (eq. 17.12) conforms to reality.

$$Z_3 = 1 + \{\alpha F_2(\alpha) + \alpha[2/3 + \alpha/(2\pi) - (1/4)[\alpha/(2\pi)]^2]\}\ln(p/\Lambda) \quad (17.14)$$
$$= 1 + \alpha\{2/3 + \alpha/(2\pi) - (1/4)[\alpha/(2\pi)]^2\}\ln(p/\Lambda)$$

at α = the physical fine structure constant where $F_2(\alpha) = 0$.

---

[79] [79] See E. A. Ueling, Phys. Rev. **48**, 55 (1935) and R. Serber, Phys. Rev. **48**, 49 (1935).

17.5.1 Physical Implications of Intermediate Renormalization

There is no problem with using F_2 as the eigenvalue function since it can be made the factor relating an infinite bare charge to the physical charge (eq. 4.1). (It is an imperfection viewed from the goal of the original JBW model for a divergence-free QED.) However we believe our modified model is reality.

The below plots show the overall form of $F_2(\alpha)$ plotted vs. g and the region where the fine structure value appears. Under these circumstances it appears that the Modified JBW model of QED is correct and determines the fine structure constant if it could be completely solved. *As a result other proposed determinations for the value of α appear to be ruled out: including cosmological and other physical approaches.* Thus the fundamental nature of the universe, and of Life, is fixed by QED.

17.5.2 Intermediate Renormalization of Weak and Strong Coupling Constants

As we noted above in our discussion of the first order change in hydrogen-like atomic energy levels, the bare electron charge e_0 is changed by $\sqrt{Z_3}$ vacuum polarization. The proton bare charge, also e_0, is similarly changed by $\sqrt{Z_3}$ so the charge factor in the level shift is the renormalized charge $e_0^2 Z_3$. Since protons are composed of three quarks, their charges must also be changed by a factor of $\sqrt{Z_3}$. Thus they have infinite bare electric charges.

18. QED α Eigenvalue Calculation

We have examined the values of the quantities in eq. 15.8 looking for an essential singularity (eq. 15.2) or its approximation. Fig. 18.1 below plots $F_2(\alpha)$ as a function of g. It displays a "flat region." While essential singularities usually are thought to imply a transcendental function such as $\exp(1/\alpha)$, a constant function with value zero fulfills the essential singularity conditions in eq. 15.1. Therefore we take the "flat region" to indicate an essential singularity.

Fig. 18.2 shows a "close up" of the flat region[80] where F_2 is approximately zero. Upon close numeric analysis we find the results in Tables 18.1 and 18.2.

g =	-0.0005805369 0000	-0.0005805369 1948	-0.0005805369 5000
α =	0.007297352	*0.0072973525693*	0.007297353
$F_2 \times 10^{10} =$	3.26316 06817671	3.26316 025452474	3.26316 134861337

Table 18.1. Values of g, α and $F_2(\alpha) \times 10^{10}$. F_2 is very close to zero for the displayed range of values and throughout the flat region. F_2 has a local minimum at precisely the known value of α = 0.0072973525693 (11).

g =	-0.00058053700	-0.00058053705	-0.00058053710
α =	0.007297354	0.007297354	0.007297355
$F_2 \times 10^{10} =$	3.26316 299072544	3.26316 29663526	3.26316 408259273

Table 18.2. Other neighboring values of g, α and $F_2(\alpha) \times 10^{10}$ in the flat region *away* from g = 0 (where our approximate F_2 is exactly zero.) F_2 is

---

[80] These figures appeared in Blaha (2019a) and (2019b).

very close to zero for the displayed range of values and throughout the flat region.

Thus we have a very good approximation $F_2(\alpha) \cong 0$ at the experimentally known value that is exact to 13 places with a minimum in $F_2(\alpha)$ as anticipated.

F_2 is nearly zero, as are its derivatives, at the physical Fine Structure Constant. It closely approximates a trivial essential singularity of constant value zero in a neighborhood of the singularity.

Note $F_2(\alpha = 0) = 0$ as well. This zero can be viewed as a type of singularity. If QED could transition from positive α to negative α then it would lead to a catastrophe since like charges would then attract.[81,82]

It is extremely important to note the calculation is strictly QED. Thus α is space and time independent, and not Anthropic.

---

[81] Freeman Dyson has speculated on this possibility.

[82] $F_2(\alpha)$ may have more than one zero. One of the zeroes is at the value of the Fine Structure Constant as we show.

Flat Region Endpoints

$g = -0.00058053691948$, $\alpha_{\text{calculated}}(g) = 0.0072973525693$

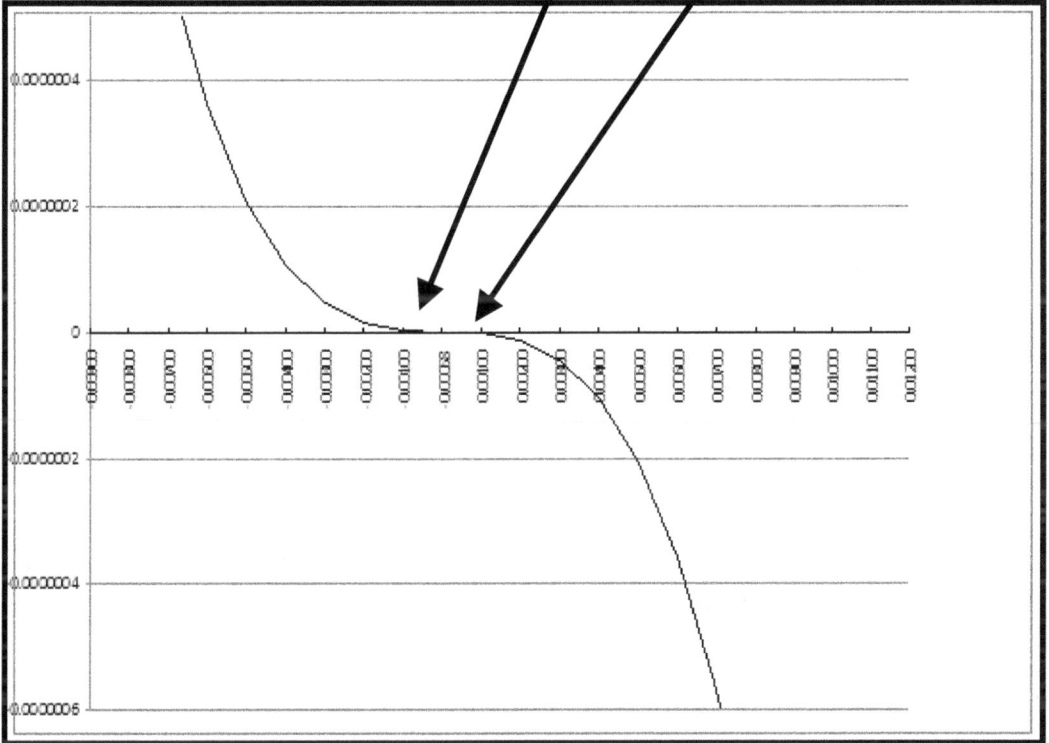

Figure 18.1. Close up plot of our eigenvalue function $F_2(g)$ (vertical axis) vs. g.

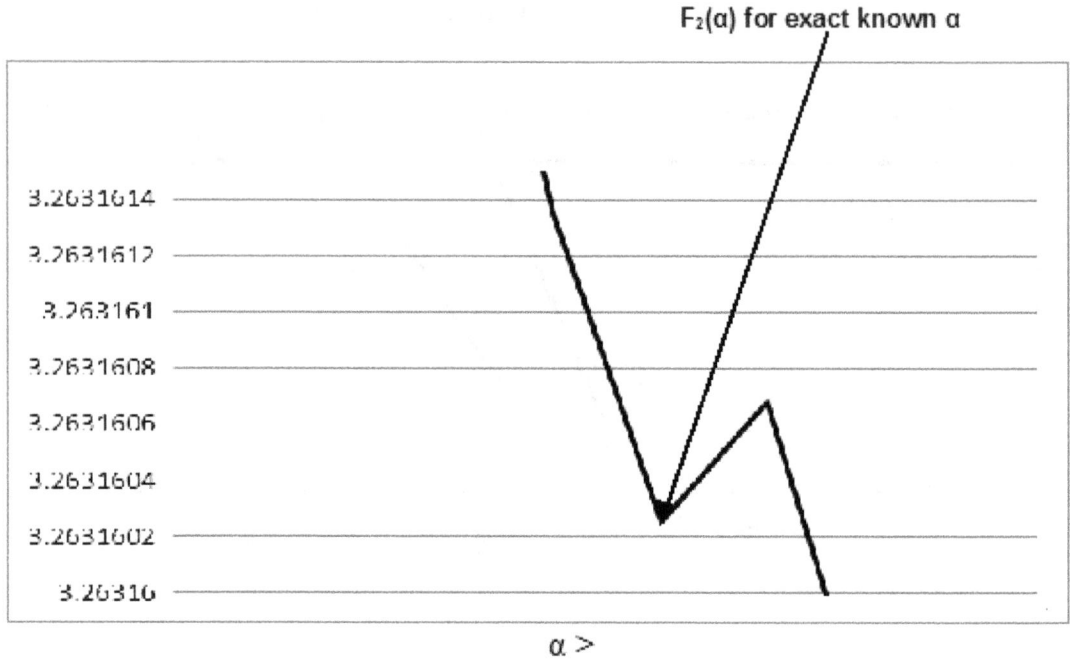

Figure 18.2. Detailed closeup plot of $F_2(\alpha) \times 10^{10}$ data in Tables 18.1 and 18.2. The local minimum of $F_2(\alpha) \times 10^{10}$ at g = -0.00058053691948 corresponds to the exact known value of α = 0.0072973525693.

18.1 Approximate Formula for α in Terms of π and e

Our success in exactly determining α is very encouraging and suggests our approximate $F_2(\alpha)$ calculation in QED may have captured the essence of the full eigenvalue function. However it is of interest to consider α from a different perspective—approximate formulas based on fundamental mathematical constant such as π and e. Speculations of simple formulas for α have been considered for over 100 years.

We shall develop two expressions for α based on

$$e = 2.718281828$$
$$\pi = 3.141592654$$

A remarkably simple expression, yet quite accurate, is

$$\alpha = 1/(16\pi e) \tag{18.1}$$
$$= 0.007318729$$

It is accurate to three places.

A better formula, to which the above is an approximation, is

$$\alpha = 1/[16\pi e(1 + \alpha/e)] \tag{18.2}$$

Solving for α we obtain

$$\alpha = (-e + (e^2 + 1/4\pi)^{1/2})/2 \tag{18.3}$$
$$= 0.007299129$$

giving five place accuracy compared to the known value:

$$\alpha = 0.0072973525693 \ (11)$$

19. Standard Model Coupling Constant Eigenvalues

The remarkable calculation of the QED Fine Structure Constant as evidenced by Table 18.1 leads us to consider the determination of the coupling constants of the other Standard Model interactions; Weak SU(2) and Strong SU(3).

In Blaha (2019b) we generalized the F_2 function (and related functions in eq. 15.8) to the cases of the Weak interaction and the Strong interaction coupling constants by inserting a group theoretic factor in the equation for α_G only:

$$(\alpha_G/2\pi) = c_G^{-1}[gA_4 - (4 + 2g)A_2]/(A_4A_1 - A_2A_3) \qquad (15.13)$$

$$c_G^{-1} = [(11/3)C_{ad} - 2C_f/3]/(16\pi)^3 \qquad (15.14)$$

where C_{ad} is the dimension of the fundamental representation of the group and C_f is the number of fermions (fermion flavor) of the interaction.[83]

The known vector interactions and coupling constants of the Standard Theory are:[84]

- The Strong interaction coupling constant[85] $e_S = 1.22$
- The Weak SU(2) coupling constant $e_W = 0.619$
- The Electromagnetic U(1) coupling constant $e_{QED} = 0.303$

We found good approximations to the SU(2) and SU(3) coupling constants in Blaha (2019b).

---

[83] See Blaha (2019b) for the SU(2) and SU(3) values of c_G and the plots of all Standard Model eigenvalue functions.
[84] All coupling constant values are based on data extracted from C. Patrignani *et al* (Particle Data Group), Chinese Physics **C40**, 100001 (2014).
[85] Based on the running coupling constant value $\alpha_s(M_Z^2) = 0.1193 \pm 0.0016$.

The gauge interaction coupling constants, which we denoted with the label G with $e_G = (4\pi\alpha_G)^{1/2}$, for QED, Weak SU(2) and Strong SU(3) have a remarkable regularity—they double from interaction to interaction:[86]
The deeper significance of this regularity is not known.

Group	Known Coupling Constant e_G	Known $e_G^2/(4\pi)$	Calculated $\alpha_G = e_G^2/(4\pi)$	Calculated[87] Exponent g_G
QED, U(1)	~0.303	$\alpha = 0.0072973525693$	$\alpha = 0.0072973525693$	-0.00058053691948
SU(2)	0.619	0.0305	0.0425	0.54
SU(3)	1.22	0.118	0.086	0.5605

The relative closeness of the calculated values of "fine structure constants" to the experimentally known values is very encouraging—particularly in the case of the Electromagnetic fine structure constant α. It puts to rest other possible explanations for its value.

Our QED calculation of α has no free (adjustable) parameters unlike other attempts in the past. It also is totally based on Quantum Field Theory. The calculation of the non-abelian coupling constants also has no free (adjustable) parameters.

Thus the coupling constant eigenfunctions depend only on inherent perturbation theory based on dynamics. Coupling constant values cannot be "tweaked" to their known values by adjusting input parameters.

The ability of our 1973 calculation of the JBW eigenvalue function together with the new insights into understanding of the precise method to obtain its "fine structure constant" eigenvalues is also encouraging. It opens the possibility that the Standard Model has within itself the mechanism for determining the constants appearing within it. It raises the hope that a similar self-determination mechanism may

---

[86] Chapter 16 shows the universe scale factor g is ½ of the QED Fine Structure g deepening the mystery.
[87] They appear in eqs. 5.5 – 5.8 in Blaha (2019b). See Blaha (2019b) for more details.

also exist within the theory to determine the masses appearing in the Higgs particles sector of the theory.

19.1 Standard Model in All Universes

Since we will see later that 4D universe(s) have the same evolutionary pattern of expansion and contraction, and since the Standard Model[88] coupling constants (which govern all macroscopic and chemical interactions) are determined internally within quantum field theory we can assert that the Physics, Chemistry, and Biology of all 4D universes are the same.

---

[88] One could suggest that the Standard Models of other universes are different. However the simplicity of the group structure would argue otherwise—as would consideration of the case of colliding universes.

20. Non-Abelian Interactions "Fine Structure Constants"

The Unified SuperStandard Theory has non-abelian interactions that go beyond QED by having group transformations that necessitate the introduction of cubic and quartic terms in the lagrangian of the theory. The calculation of non-abelian coupling constants have until now been limited to low order perturbative running coupling constant calculations.[89]

It would have been better to have calculations of coupling constants to all orders—even if only approximately—as we successfully saw in massless QED earlier. In this chapter and the following chapter we will attempt to calculate coupling constants in a manner analogous to that of earlier chapters for QED. Our attempt will be at very high energies where we can assume that all particle masses are negligible. We will make a number of assumptions which are not unreasonable:

1. We will assume that infrared divergences are not relevant.

2. We will assume that cubic and quartic Yang-Mills couplings can be neglected reducing Yang-Mills fields to multiplets of QED-like fields.

3. We will assume that each non-abelian interaction can be considered independent of all other interactions.

4. We will assume that each interacting field of a group is subject to the same vacuum polarization renormalization. This renormalization is assumed to be similar to that of JBW vacuum polarization.

---

[89] See, for example, H. Georgi, H. R. Quinn, S. Weinberg, Phys. Rev. Lett. **33**, 451 (1974).

5. Thus we assume that an eigenvalue function exists for each non-abelian interaction that results from a single logarithmic "divergence" in the vacuum polarization. This divergence stems from one fermion loop vacuum polarization diagrams summed over all fermion species of the fundamental representation of its group.

The justification for this series of assumptions is the similarity of the contributions of the vacuum polarization diagrams if cubic and quartic non-abelian interactions are neglected. The calculations of each Feynman diagram have the same characteristics of massless QED momentum space integrals up to an overall interaction constant factor in each order of perturbation theory.

Based on these assumptions we proceed to calculate non-abelian eigenvalue functions for SU(2), SU(3), and SU(4) in the following chapter.

We do make one further assumption we assume that the coupling constant for each interaction has the factor

$$c_G^{-1} = [(11/3)C_{ad} - 2C_f/3]/(16\pi)^3 \tag{20.1}$$

to all orders where C_{ad} is the dimension of the fundamental representation of the non-abelian group and C_f is the number of fermions (fermion flavor) of the interaction.

The eigenvalue function F_1 for an interaction with coupling constant α can be expressed as a power series in α:

$$F_1(\alpha_G) = \sum_n a_n(\alpha_g)^n = \sum_n a_n(c_G\alpha)^n \tag{20.2}$$

under the assumption that the interaction group constant is approximately c_G to all orders in n where the QED eigenvalue function has the form

$$F_1(\alpha) = \sum_n a_n\alpha^n \tag{20.3}$$

Since non-abelian interactions are known to be asymptotically free—with coupling constants becoming finite at ultra-short distances we will not need to do intermediate renormalization such as we did for electric charge. Thus we will use F_1 as the eigenfunction.

Having established the framework for our approximate calculations of the eigenvalue functions and their eigenvalues we will proceed to calculate them in the next chapter for SU(2), SU(3) and SU(4).

Note that the procedure for obtaining the massless QED eigenvalue function and the fine structure constant will exactly duplicate our calculation in Appendix 25-A.

Chapter 23 provides a simple fit to the F_2 and α functions using tangents.

21. Non-Abelian "Fine Structure Constant" Eigenvalue Conditions

21.1 Coupling Constants for Non-Abelian Interactions

The vector interactions and coupling constants of the Unified SuperStandard Theory are:[90]

- The strong interaction coupling constant[91] $g_S = 1.22$

- The Weak SU(2) coupling constant $g_W = 0.619$

- The Electromagnetic U(1) coupling constant $e = g_E = 0.303$

- The Dark Weak SU(2) coupling constant $g_{DW} = ?$

- The Dark Electromagnetic U(1) coupling constant $g_{DE} = ?$

- The "Normal" U(4) Generation[92] group coupling constant $g_G = ?$

- The "Normal" U(4) Layer group coupling constant $g_L = ?$

- The U(4) Species group coupling constant $g_{Sp} = ?$

---

[90] All coupling constant values are based on data extracted from C. Patrignani *et al* (Particle Data Group), Chinese Physics **C40**, 100001 (2014).

[91] Based on the running coupling constant value $\alpha_s(M_Z^2) = 0.1193 \pm 0.0016$.

[92] We take "Normal" group to refer to the group associated with the known fermions.

In this chapter we will calculate (approximately) g_S, g_W, e, and the "fine structure constant" for SU(4) g_4, which would seem to apply to the U(4) groups listed above, subject to the approximations listed in chapter 20.

We will use the below to calculate "fine structure constant"'s α_G

$$A_1 = (g + 1)(1 - 2g^2)/[(g + 2)(g - 1)] \tag{21.1}$$

$$A_2 = [8g^2(2g + 1) - (2g^3 + 2g^2 + g - 2)(g^2 + 2g + 2)]/[2(g^2 - 1)(g^2 - 4)]$$

$$A_3 = -2(1 + 3g + 6g^2 + 2g^3)/[g(g + 1)]$$

$$A_4 = -(g + 2)(1 + 5g + 6g^2 + 2g^3)/[g(g^2 - 1)] - 1/(g + 1)$$

$$\psi = [gA_3 - (4 + 2g)A_1]/[(4 + 2g)A_2 - g A_4]$$

$$(\alpha_G/2\pi) = c_G^{-1}[gA_4 - (4 + 2g)A_2]/(A_4A_1 - A_2A_3) \tag{21.2}$$

$$F_1 = (2/3)(1 - 3g^2/2 - g^3) - (\alpha_G/4\pi)[(2 + 4g + 4g^2)(g - 2) + \alpha_G\psi g^3]/[(g^2 - 1)(g - 2) + \\ + \alpha_G(2 + 4g + 4g^2)(g - 2) + \alpha_G\psi g^3] \tag{21.3}$$

using eq. 20.1.

21.2 SU(2) ElectroWeak "Fine Structure Constant" Eigenvalue Function

The U(1) QED eigenvalue function and eigenvalue was discussed in chapter 17. In this section we discuss and show plots of the SU(2) ElectroWeak sector eigenvalue function subject to the discussion in chapter 20 and the above. In particular we calculate the SU(2) eigenvalue function and eigenvalues using eqs. 21.2 and 20.1.

Due to the asymptotic freedom of non-abelian interactions we anticipate that the value of the exponential factor g will be positive at the "fine structure constant" eigenvalue of the SU(2) eigenvalue function which we take to have the form shown above due to the assumptions of chapter 20.

We define $F_{1su(2)}(\alpha)$ using eq. 21.3 above with $c_{GSU(2)}^{-1} = -0.003023589$ from eq. 20.1 using $C_{ad} = 2$ and $C_f = 2$.

In our discussion of QED we found a negative value for g and a consequent divergence in Z_3 as $\Lambda \rightarrow \infty$. We then found that $F_1(\alpha)$ did not yield the QED eigenvalue. We had to introduce $F_2(\alpha)$ by eliminating low order (in α) terms using intermediate renormalization. Those terms have logarithmic divergences. Their elimination did not create a problem since the QED renormalizations also diverge due to g < 0 at the eigenvalue point. The total divergence is generated by combining the divergent factors due to the $(p/\Lambda)^{2g}$ factor in Z_3 with the omitted divergent terms of F_1.

In the case of non-abelian interactions g > 0, which implies the renormalizations go to 1 as $\Lambda \rightarrow \infty$ (asymptotic freedom). Thus we will not truncate F_1 but will calculate using it (eq. 21.3) to find an approximation to the "fine structure constant" eigenvalues of non-abelian interactions. *As $\Lambda \rightarrow \infty$ non-abelian theories approach free field theories.*

The signature of the eigenvalue function essential singularity in the present case will be a divergence in F_1 for g > 0. It signals that we have an approximation to the expected essential singularity. See section 17.3.2.

The below plots show the overall form of $F_{1SU(2)}(\alpha)$ plotted vs. g and the divergent region of $F_{1SU(2)}(g)$ specifying the "fine structure constant" eigenvalue. Our approximate "fine structure constant" values in this case and the SU(3) and SU(4) cases are quite reasonable. We find the "essential singularity" point at

$$g = 0.54$$

where $\alpha_{SU(2)} = g_W^2/(4\pi)$

$$\alpha_{calculatedSU(2)}(g) = 0.0425$$

compared to the actual measured "fine structure constant" value

$$\alpha_{SU(2)} = 0.0305$$

displaying a fairly close match given the approximate nature of the calculations. The positive value of the exponential factor g above indicates that $Z_3 \rightarrow 1$ as $\Lambda \rightarrow \infty$ showing the SU(2) interaction is asymptotically free. (See eqs. 17.5 – 17.7.) It is interesting to note that the "fine structure constant" plot has a divergence at g = 0.58.

g = 0.54   Singularity at eigenvalue point

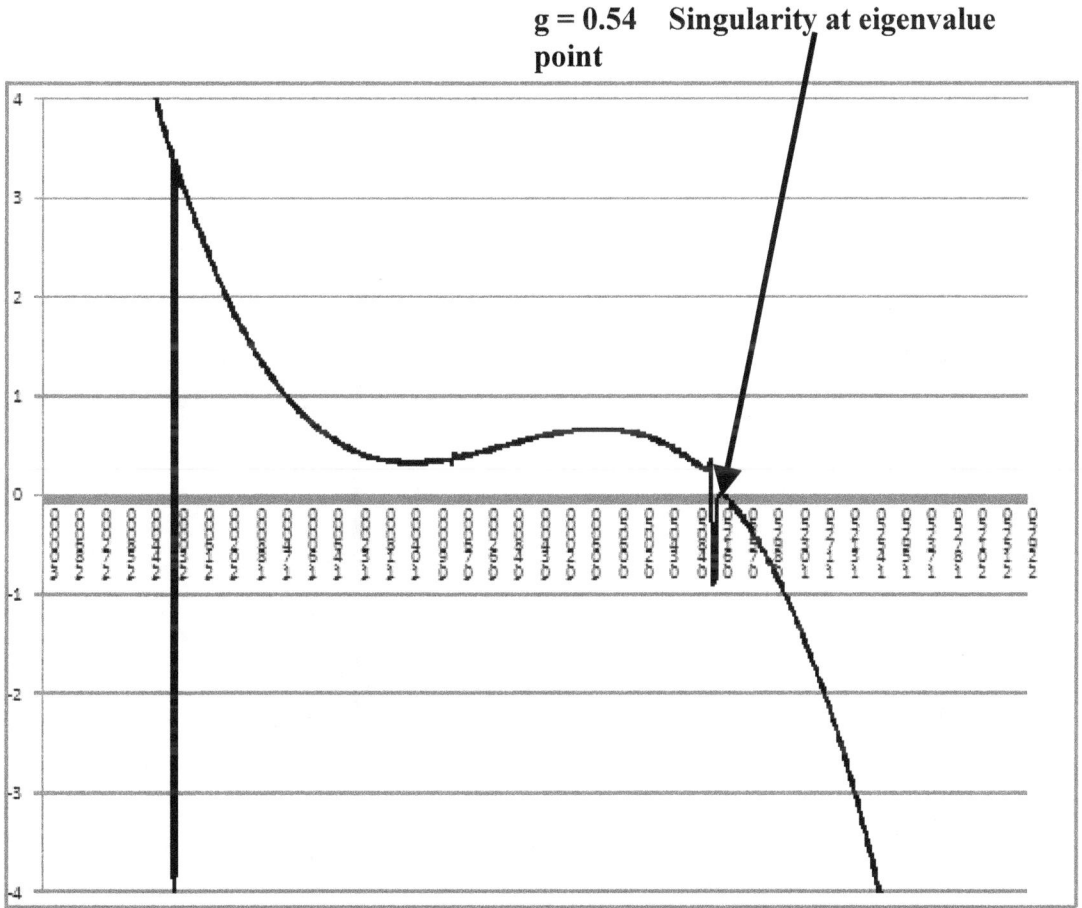

Figure 21.1. Plot of F$_{1su(2)}$ (vertical axis) as a function of g.

$$g = 0.54, \quad \alpha_{\text{calculatedSU(2)}}(g) = 0.0425$$

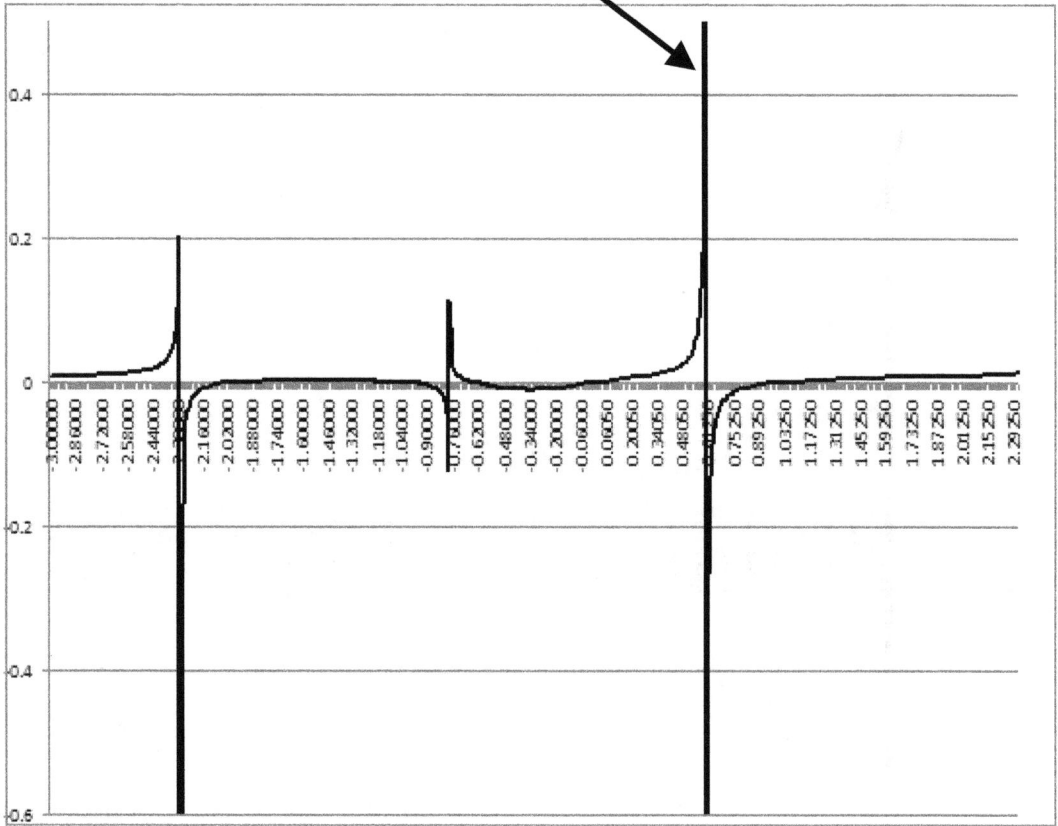

Figure 21.2. Plot of the "fine structure constant" $\alpha_{\text{calculatedSU(2)}}$ **(vertical axis) as a function of g.**

21.3 SU(3) Strong "Fine Structure Constant" Eigenvalue Function

The discussion of the SU(3) Strong interaction case is very much the same as the preceding SU(2) discussion. We define $F_{1su(3)}(\alpha)$ using eq. 21.3 above with

$$c_{GSU(3)}^{-1} = -0.004535384$$

from eq. 20.1 using $C_{ad} = 3$ and $C_f = 3$.

The signature of the eigenvalue function essential singularity in the present case will be a divergence in F_1 for $g > 0$. It signals that we have an approximation to the expected essential singularity. See section 17.3.2.

The below plots show the overall form of $F_{1SU(3)}(\alpha)$ plotted vs. g and the divergent region of $F_{1SU(3)}(g)$ specifying the "fine structure constant" eigenvalue. We find the "essential singularity" point at

$$g = 0.5605$$

where

$$\alpha_{calculatedSU(3)}(g) = 0.086$$

compared to the actual "measured" "fine structure constant" value (at $Q^2 = 2$ GeV)

$$\alpha_{SU(3)} = 0.118$$

again displaying a fairly close match given the approximate nature of the calculations.

The positive value of the exponential factor g above indicates that $Z_3 \to 1$ as $\Lambda \to \infty$ showing the SU(3) interaction is asymptotically free. (See eqs. 17.5 – 17.7.) It is interesting to note that the "fine structure constant" plot also has a divergence at g = 0.6105.

g = 0.5605   Singularity at eigenvalue point

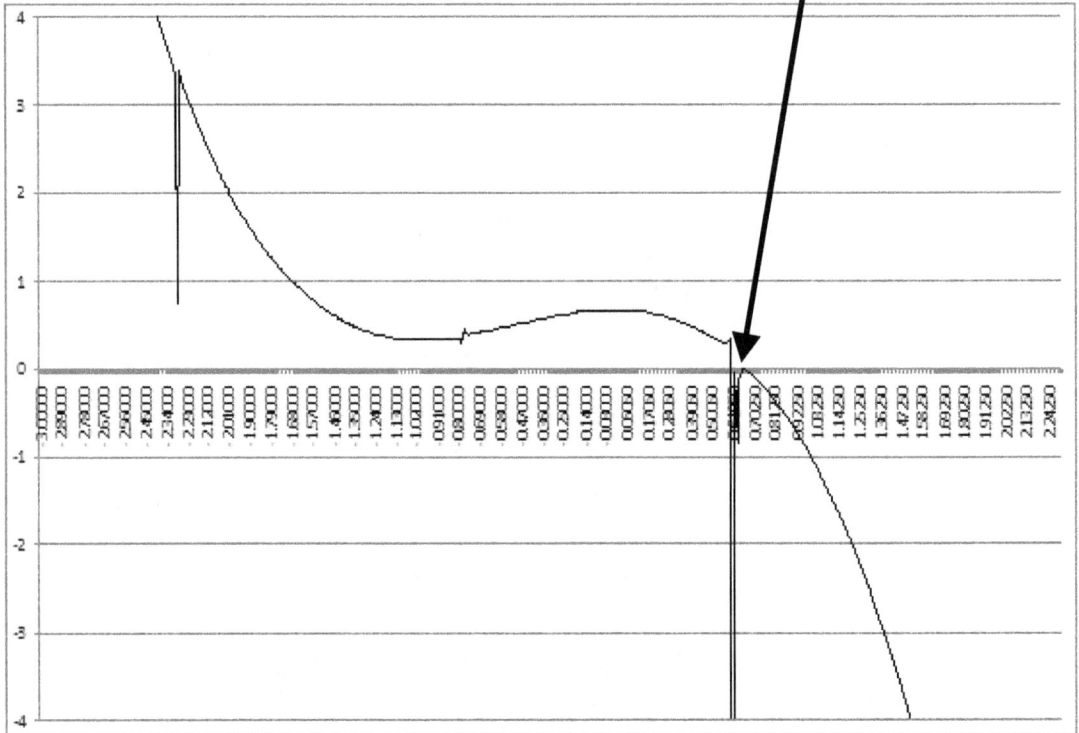

Figure 21.3. Plot of $F_{1su(3)}$ (vertical axis) as a function of g.

$$g = 0.5605, \quad \alpha_{\text{calculatedSU(3)}}(g) = 0.086$$

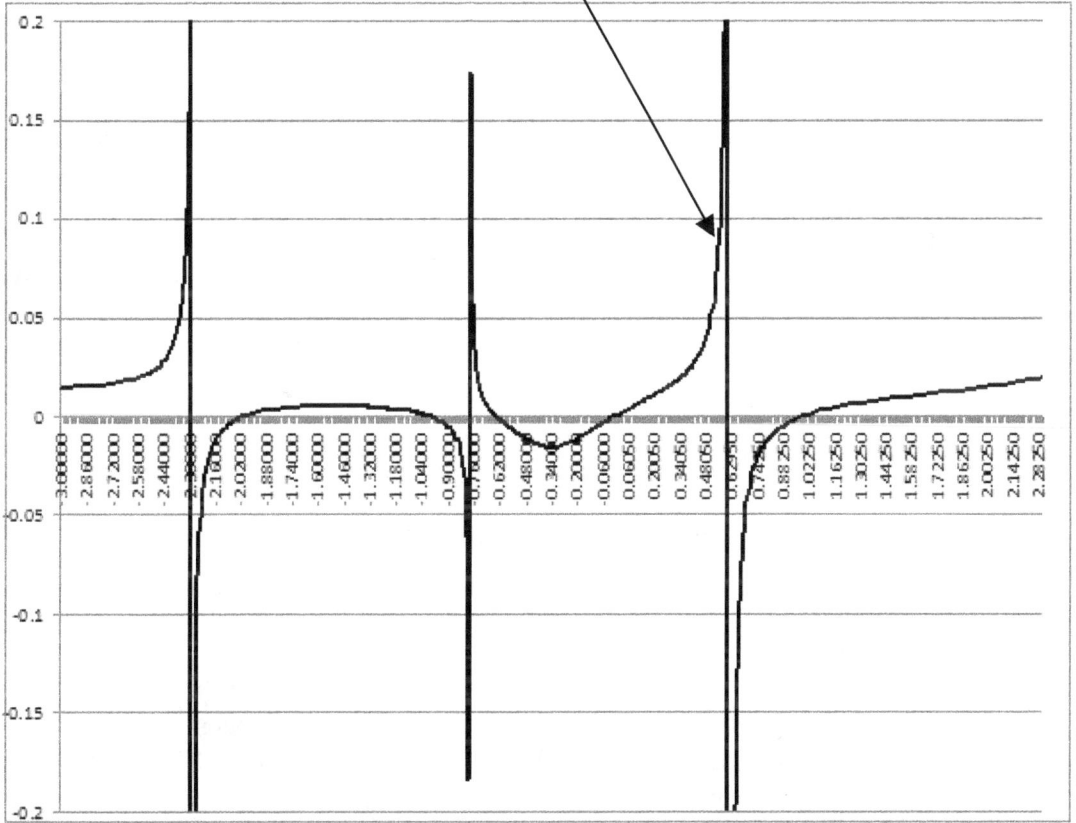

Figure 21.4. Plot of $\alpha_{\text{calculatedSU(3)}}$ **(vertical axis) as a function of g.**

21.4 SU(4) "Fine Structure Constant" Eigenvalue Function

The discussion of the SU(4) case of *other* interactions is very much the same as preceding discussions. We define $F_{1su(4)}(\alpha)$ using eq. 7.3 above with

$$c_{GSU(4)}^{-1} = -0.006047178$$

from eq. 20.1 using $C_{ad} = 4$ and $C_f = 4$.

The signature of the eigenvalue function essential singularity in the present case will be a divergence in F_1 for $g > 0$. It signals that we have an approximation to the expected essential singularity. See section 17.3.2.

The below plots show the overall form of $F_{1SU(4)}(\alpha)$ plotted vs. g, and the divergent region of $F_{1SU(4)}(g)$ specifying the "fine structure constant" eigenvalue. We find the "essential singularity" point at

$$g = 0.598$$

where

$$\alpha_{calculatedSU(4)}(g) = 0.384$$

compared to the conjectured[93] "fine structure constant" value

$$\alpha_{SU(4)} = 0.458$$

again displaying a fairly close match given the approximate nature of the calculations.

The positive value of the exponential factor g above indicates that $Z_3 \to 1$ as $\Lambda \to \infty$ showing the SU(4) interaction is asymptotically free. (See eqs. 17.5 – 17.7.) It is interesting to note that the "fine structure constant" plot has a divergence at $g = 0.618$.

---

[93] This value is based on the "doubling trend" seen in the three known coupling constants in chapter 19.

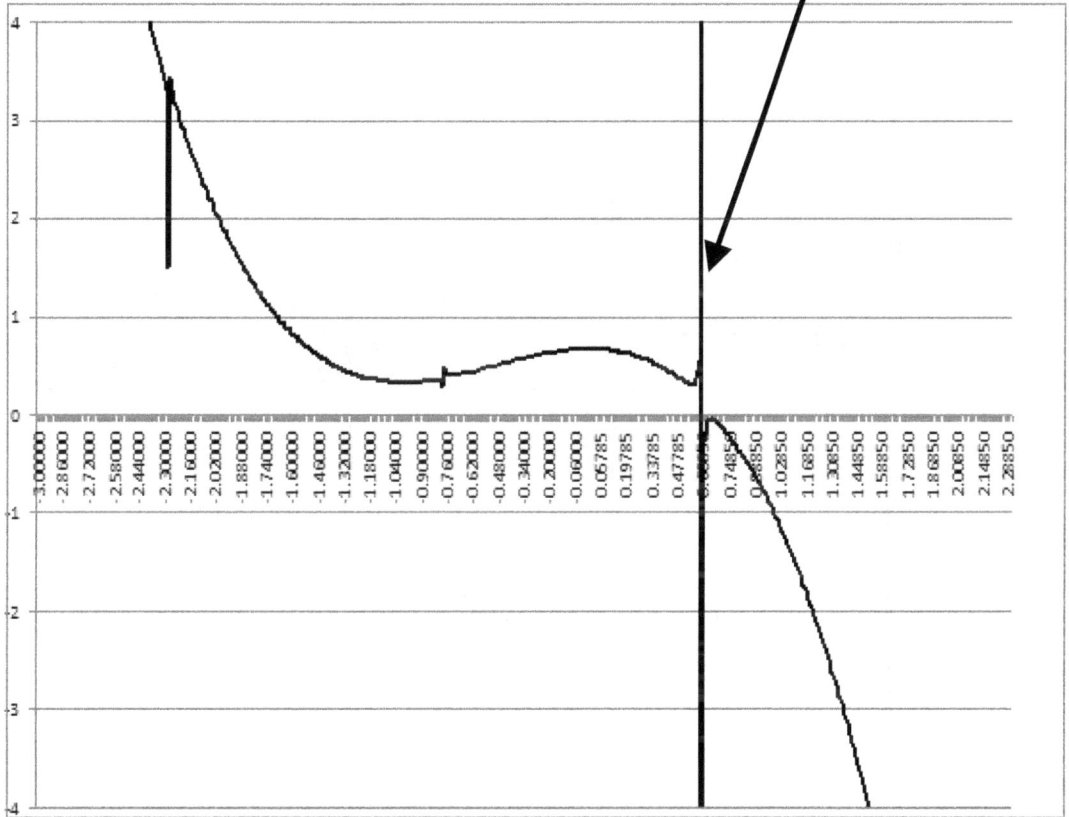

Figure 21.5. Plot of $F_{1su(4)}$ (vertical axis) as a function of g.

$$g = 0.598, \quad \alpha_{calculatedSU(4)}(g) = 0.384$$

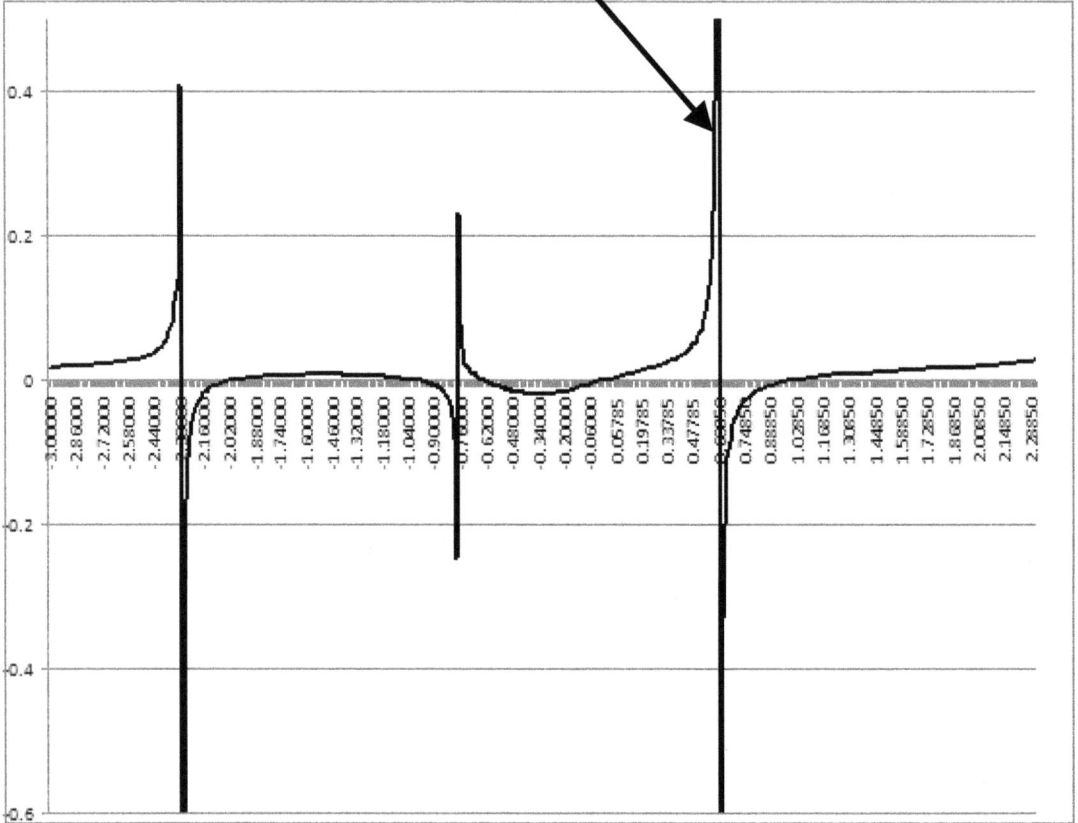

Figure 21.6. Plot of $\alpha_{calculatedSU(4)}$ (vertical axis) as a function of g.

22. "Fine Structure Constants" of the Unified SuperStandard Theory and Standard Model

In this chapter we summarize the values of the approximate values of the α_G determined in the previous chapter.

Group	*Known Coupling Constant*	*Known $g_G^2/(4\pi)$*	*Calculated $\alpha_G = g_G^2/(4\pi)$*	*Calculated[94] Exponent g_G*
QED, U(1)	~0.303	$\alpha = 0.0072973525693$	$\alpha = 0.0072973525693$	- 0.00058053691948
SU(2)	0.619	0.0305	0.0425	0.54
SU(3)	1.22	0.118	0.086	0.5605
SU(4)	2.4?[95]	0.458?	0.384	0.598

The relative closeness of the calculated values of "fine structure constants" to the experimentally known values is very encouraging—particularly in the case of the Electromagnetic fine structure constant α. It puts to rest other possible explanations for its value.

Our QED calculation of α has no free (adjustable) parameters unlike other attempts in the past. It also is soundly based on Quantum Field Theory. The calculation of the non-abelian coupling constants also has no free (adjustable) parameters.

Thus the coupling constant eigenfunctions depend only on inherent perturbation theory based on dynamics. Coupling constant values cannot be "tweaked" to their known values by adjusting input parameters.

---

[94] They appear in eqs. 17.5 – 17.8 above.
[95] This value is based on the "doubling trend" seen in the three known coupling constants above.

The ability of our 1973-4 calculation of the JBW eigenvalue function together with the new insights into understanding of the precise method to obtain its "fine structure constant" eigenvalues is also encouraging. It opens the possibility that the Unified SuperStandard Theory has within itself the mechanism for determining the constants appearing within it. It raises the hope that a similar self-determination mechanism may also exist within the theory to determine the masses appearing in the Higgs particles sector of the theory.

The result would be a self-contained all-encompassing fundamental theory—the Holy Grail of fundamental Physics.

23. Possible Exact Form of the Non-Abelian Eigenvalue Eigenfunctions

Figs. 21.4 - 21.6 of chapter 21 display a repetitive pattern that is similar to those of the trigonometric functions. In particular they show a similarity to the tangent function. Our approximate F_1 eigenvalue functions for non-abelian interactions, plotted in these figures, is, in each case, an approximation to the sum of one loop vacuum polarization diagrams that we calculated in our paper of 1974 (in Appendix 25-A).

In this chapter we further approximate F_1 for the purpose of studying non-abelian interaction running coupling constants.

Since the F_1 eigenvalue functions that we calculated in chapter 21 yielded good approximate values for the Weak and Strong coupling constants, it seems reasonable to believe that these eigenvalue function approximations are close to the real F_1 eigenvalue functions that would have been obtained in precise calculations summing one fermion loop Feynman diagrams.

It is also reasonable to believe that the true F_1 functions are less complicated than the approximate ones. Some past perturbation theory summations to all orders have shown a remarkable simplicity in the resulting summation. A particular example is the author's leading logarithm summation for the deep inelastic e-p structure functions.[96] In this paper a remarkable cancellation of diagrams, due to a Stirling Numbers of the Third Kind identity,[97] led to a simple compact result.

In the present case we will take the periodic pattern in the approximate F_1 functions to be tangent functions indications in the exact F_1 eigenfunctions. We view our approximate F_1 eigenfunctions as generated by a subset of the total one loop

---

[96] Stephen Blaha, Phys. Rev. D **3**, 510 (1971). This paper (the author's Ph.D. Thesis) showed that perturbation theory could not account for deep inelastic e-p scaling—an open question at the time.

[97] This type of Stirling number had not been encountered in perturbation theory before, or after, the author's paper.

contributions to the "vacuum polarization" of the non-abelian interaction coupling constants.

23.1 "Exact" Form of Non-Abelian Eigenvalue Functions

We suggest the correct form of the F_1 vacuum polarization eigenfunctions for the group G is

$$F_{G1}(g) = \tan[\pi(g + d_{Gf})/d_{Gd}] \tag{23.1}$$

and the coupling constant function is the absolute value[98]

$$\alpha_G(g) = |c_G \tan[\pi(g + d_{G\alpha})/d_{Gd}]| \tag{23.2}$$

For some value of g, $F_{G1}(g_0) \rightarrow \infty$ and $\alpha_G(g_0)$ is the coupling constant for interaction group G. We approximate the quantity d_{Gd} by $d_{Gd} = 1.29911 - 0.08929g$. Higher powers of g would be required to get a better fit—as we shall see in the following figures. We leave that issue to future work. The quantities d_{Gf} and $d_{G\alpha}$ are assumed to be constants. The graphs below show a good approximation of tangent fits to the approximate F_1 plots.

23.2 Similarity to the Madhava-Leibniz Representation of π

The form of the coupling constant eigenvalue function is comparatively simple compared to exact expressions found earlier. Remarkably it also suggests that α has a representation similar in character to the Madhava-Leibniz representation of $\pi = 3.14159...$:

$$\pi = 4\arctan(x) \tag{23.3}$$

for $x = 1$. Note the eigenvalue implied by eq. 23.2 is

---

[98] The absolute value is physically required to maintain the reality of coupling constants. Eq. 23.2 is an approximation that does not exclude imaginary coupling constants.

$$\alpha_G = |c_G \tan(x)| \tag{23.4}$$

(in absolute value) where

$$x = \pi(g_0 + d_{G\alpha})/d_{Gd} \tag{23.5}$$

for some g_0.

The displayed range of values of g will be g ε [-3, 3].

23.3 Comparison of Approximate Eigenfunctions and Eigenvalues to the Tangent Representation

In this section we display the graphs of the approximate Eigenfunctions and Eigenvalue functions, and the proposed tangent graphs for SU(2), SU(3) and SU(4).

23.3.1 SU(2)

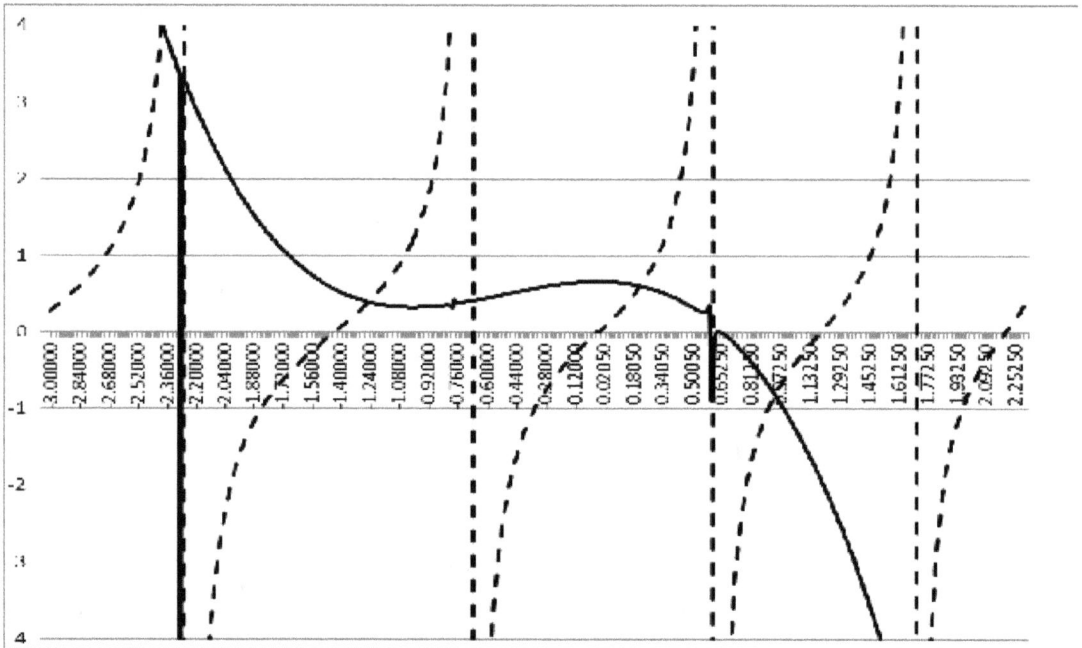

Figure 23.1 The SU(2) eigenvalue function F_1 is a solid line, while the tangent form of F_1 of eq. 23.1 is the broken line. The constants are d_{Gd} = 1.29911 - 0.08929g, and d_{Gf} = 0. F_1 is plotted vertically. The exponent g is plotted horizontally.

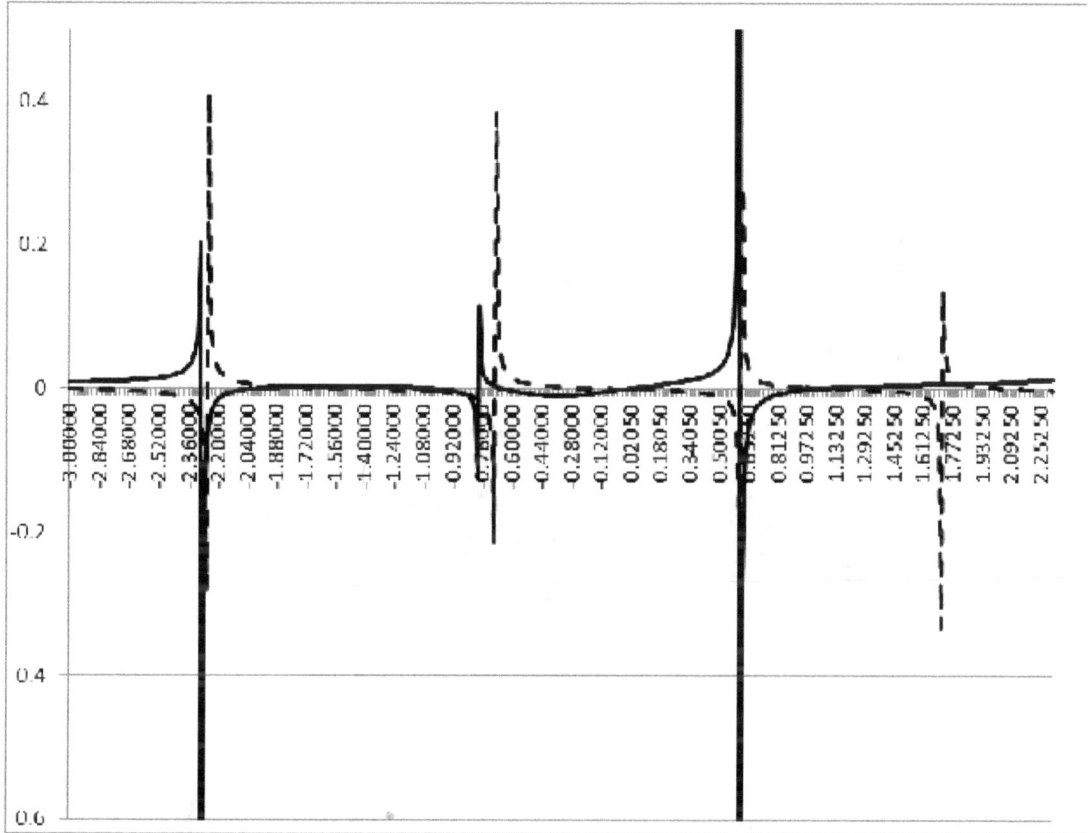

Figure 23.2 The SU(2) eigenvalue function α_G is a solid line, while the tangent form of α_G of eq. 23.2 is the broken line. The constants are d_{Gd} = 1.29911 - 0.08929g, and the α_G constant is $d_{G\alpha}$ = 0.00348417. α_G is plotted vertically. The exponent g is plotted horizontally.

23.3.2 SU(3)

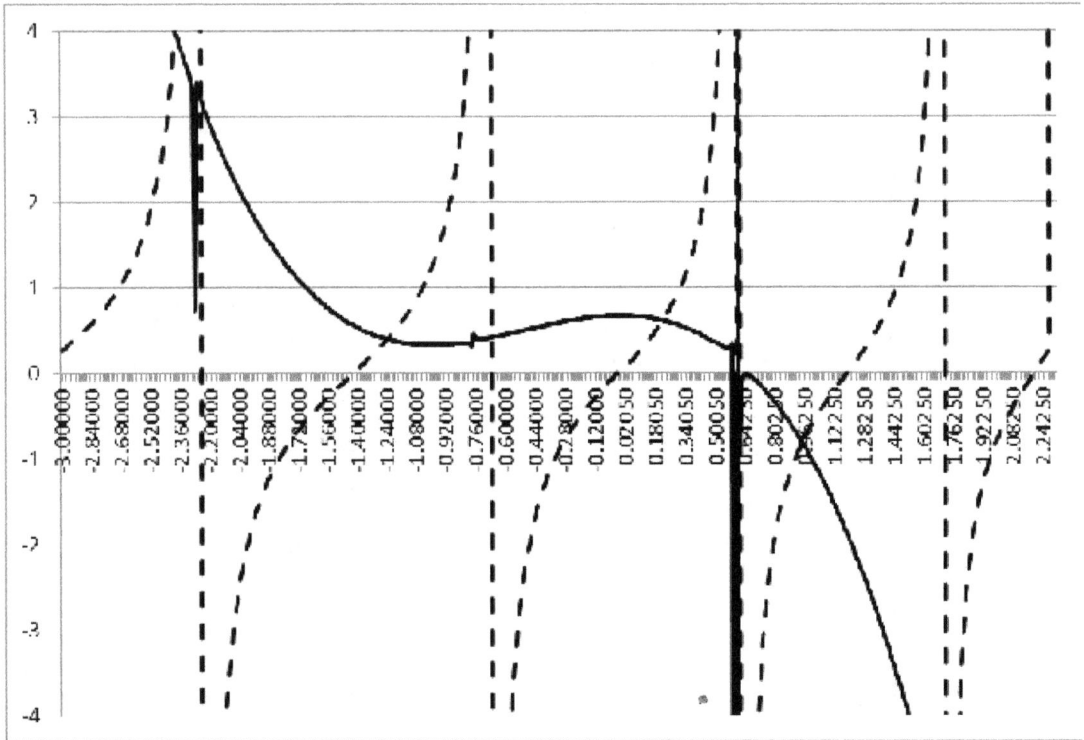

Figure 23.3 The SU(3) eigenvalue function F_1 is a solid line, while the tangent form of F_1 of eq. 23.1 is the broken line. The constants are d_{Gd} = 1.29911 - 0.08929g, and d_{Gf} = 0. F_1 is plotted vertically. The exponent g is plotted horizontally.

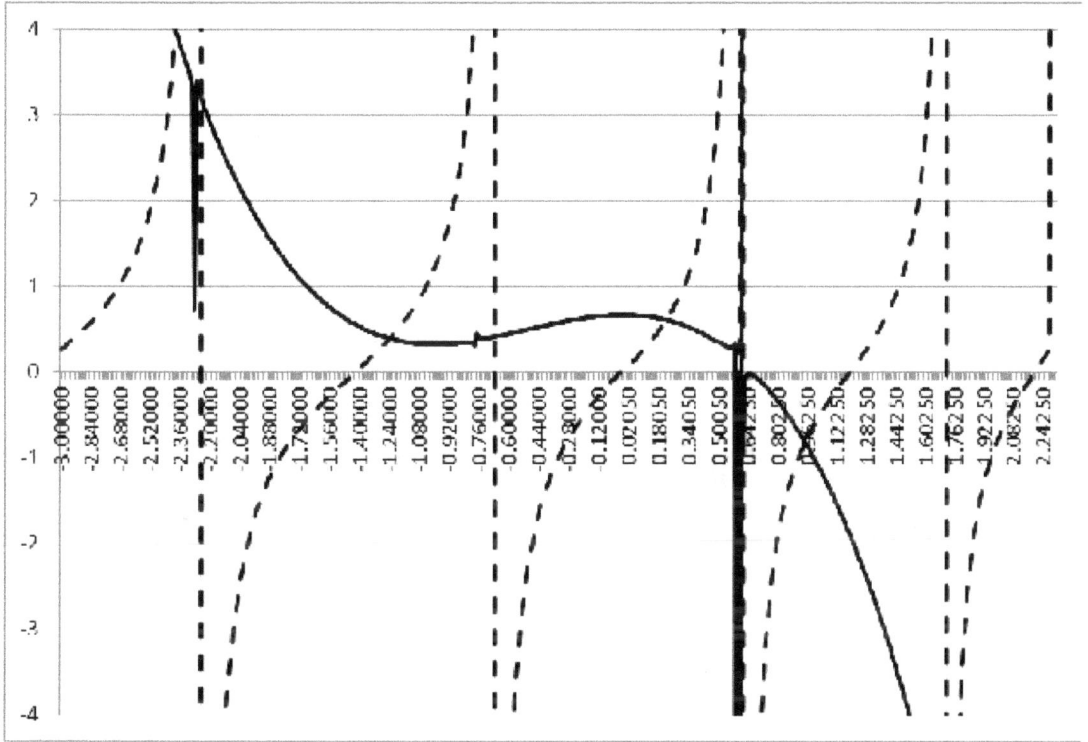

Figure 23.4 The SU(3) eigenvalue function α_G is a solid line, while the tangent form of α_G of eq. 23.2 is the broken line. The constants are d_{Gd} = 1.29911 - 0.08929g, and the α_G constant is $d_{G\alpha}$ = 0.00348417. α_G is plotted vertically. The exponent g is plotted horizontally.

23.3.3 SU(4)

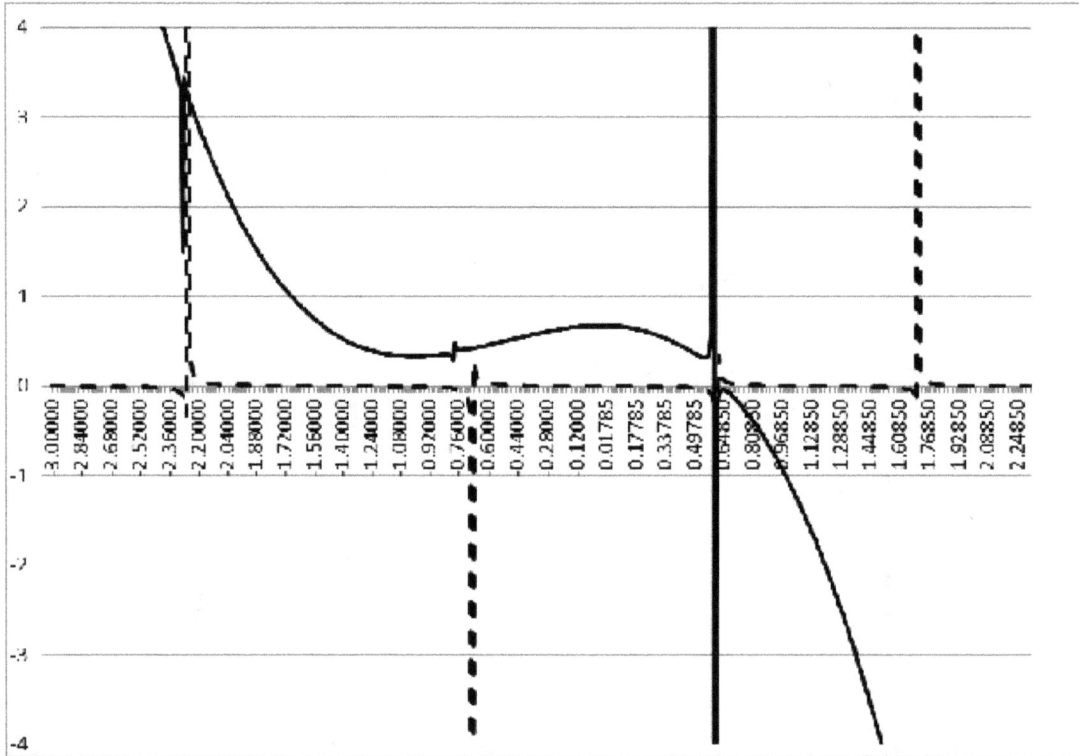

Figure 23.5 The SU(4) eigenvalue function F_1 is a solid line, while the tangent form of F_1 of eq. 23.1 is the broken line. The constants are d_{Gd} = 1.29911 - 0.08929g, and d_{Gf} = 0. F_1 is plotted vertically. The exponent g is plotted horizontally.

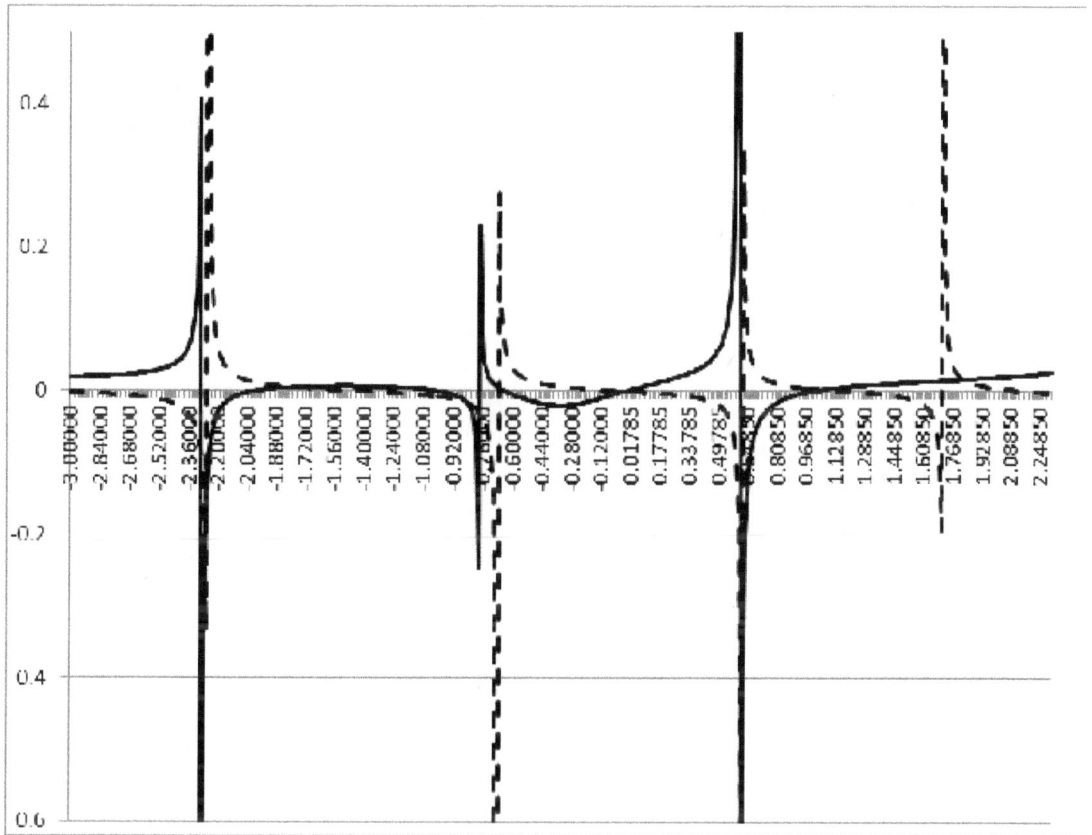

Figure 23.6 The SU(4) eigenvalue function $\alpha_G(g)$ is a solid line, while the tangent form of α_G of eq. 23.2 is the broken line. The constants are d_{Gd} = 1.29911 - 0.08929g, and the α_G constant is $d_{G\alpha}$ = 0.00348417. α_G is plotted vertically. The exponent g is plotted horizontally.

23.4 Comments

The quantities d_{Gd} and $d_{G\alpha}$ have the same values for SU(2), SU(3) and SU(4). The values are the same for QED as well. The group constants c_G differs from group to group. *The similarity in form for all non-abelian cases is due to our use of a "universal" form for eigenfunctions. Our justification is our success in the approximate calculation of all coupling constants.*

The similarity in the form, and the values of c_G and other constants in eq. 23.2 "explains" the difference in Standard Model coupling constant values.

24. Approximate Non-Abelian Running Coupling Constants with β Functions to All Orders in Perturbation Theory

In this chapter we consider the one fermion loop approximation for the β function of the Callan-Symanzik equation:

$$-\beta_1(\alpha) + m\partial(\alpha\pi^{(1)}{}_c)/\partial m \approx 0 \qquad (24.1)$$

where the vacuum polarization part is

$$\pi^{(1)}{}_c = F_1(\alpha)\ln(-q^2/m^2) + \dots \qquad (24.2)$$

resulting in the β function term

$$\beta_1(\alpha) = -2\alpha F_1(\alpha) \qquad (24.3)$$

The integration of β_1 yields

$$\ln(m_2/m_1) = \int_{\alpha_1}^{\alpha_2} d\alpha/\beta(\alpha) \qquad (24.4)$$

where

$$\alpha_i = \alpha(m_i) \qquad (24.5)$$

24.1 Approximate Form of the Eigenvalue Function $F_1(\alpha)$ to all Perturbative Orders

We first calculate the eigenvalue function $F_1(\alpha)$ and then proceed to calculate $\beta_1(\alpha)$. Then eq. 24.4 becomes soluble and an expression for the running coupling constant can be calculated.

For the sake of a transparent physical result[99] we assume $d_{Gd} = 1.5$, which is a reasonable approximation for g in the range [-3, 3] (the domain of physical interest).

Eq. 23.2 can be inverted to yield $g(\alpha)$ for a given group G:

$$g(\alpha) = (d_{Gd}/\pi)\, \arctan(\alpha_G/c_G) - d_{G\alpha} \qquad (24.6)$$

Substituting for g in eq. 23.1 using eq. 24.6 in gives

$$F_{G1}(\alpha) = [\alpha - c_G \tan(h_G)]/[1 + \alpha c_G \tan(h_G)] \qquad (24.7)$$

where

$$h_G = \pi(d_{G\alpha} - d_{Gf})/d_{Gd} \qquad (24.8)$$

Eq. 24.7 is a remarkably simple expression for the eigenvalue function in terms of α. It has the "eigenvalue zero" at[100]

$$\alpha_G = |c_G \tan(h_G)| \qquad (24.9)$$

We will see in chapter 11 that the QED[101] Fine Structure Constant α has the same form making it analogous to the Madhava-Leibniz formula for π discussed in section 23.2.

For massless QED, given the known value of α and using $\alpha = \tan(h)$ we find

---

[99] The expression used in our plots $d_{Gd} = 1.29911 - 0.08929g$ could have been used and would lead to a tractable calculation. The essence of the physics is well brought out by approximating $d_{Gd} = 1.5$.
[100] Eq. 10.9 uses an absolute value for α_G since coupling constants are real-valued.
[101] The use of intermediate renormalization and thus F_2 for QED does not change the character of the results. Only eq. 10.3 requires a superficial change.

$$h \cong 0.007297223 \tag{24.10}$$

and the QED quantities

$$h/\pi = (d_\alpha - d_f)/d_d \cong 0.002322780 \tag{24.11}$$

Using $d_\alpha = 0.00348417$ and $d_f = 0$ from chapter 25 we find

$$d_d = 1.499999784 \approx 1.5 \tag{24.12}$$

24.2 Approximate Form of Running Coupling Constant to all Perturbative Orders

Eqs. 24.3, 24.4 and 24.7 determine the approximate running coupling constant

$$\ln(m_2/m_1) = \int_{\alpha_1}^{\alpha_2} d\alpha/(-2\alpha F_{G1}(\alpha)) \tag{24.4}$$

The running coupling constant integration yields

$$(m_1/m_2)^2 = [(\alpha_2 - c_G \tan(h_G))/(\alpha_1 - c_G \tan(h_G))]^y (\alpha_2/\alpha_1)^z \tag{24.13}$$

where

$$y = c_G \tan(h_G) + (c_G \tan(h_G))^{-1} \tag{24.14}$$

$$z = - (c_G \tan(h_G))^{-1} \tag{24.15}$$

If

$$\alpha_i \gg |c_G \tan(h_G)|$$

for $i = 1, 2$, then eq. 24.13 becomes approximately

$$(m_1/m_2)^2 = [\alpha_2/\alpha_1]^{c_G \tan(h_G)} = [\alpha_2/\alpha_1]^{\alpha_G \, \text{signum}(c_G)} \tag{24.16}$$

where the exponent α_G is the constant coupling constant value in eq. 24.9 and signum(c_G) is the sign of c_G giving the proper asymptotic freedom of non-abelian couplings and the infinite bare charge of the QED (abelian) case.

Eq. 24.16 can be reexpressed as

$$\alpha(m_2) = \alpha_1(m_1)\, (m_1/m_2)^{2\text{signum}(c_G)/\alpha_G} \qquad (24.17)$$

24.3 Running Coupling Constant

The running coupling constant expression is applicable to QED, Weak SU(2), and Strong SU(3). Eqs. 24.13 and 24.17 show QED's running coupling constant gets stronger as the mass increases, while the non-abelian running coupling constants decrease as the mass increases yielding asymptotic freedom.

The growth/decline of the running coupling constant is a power law in this formulation while in other formulations, based on low order perturbation theory, it is logarithmic.[102] This difference impacts on GUT unified theories.

[102] H. Georgi, H. R. Quinn, and S. Weinberg, Phys. Rev. Lett. **33**, 451 (1974); W. E. Caswell, Phys. Rev. Lett. **33**, 244 (1974).

25. Approximate Massless QED Eigenvalue Function to All Orders in Perturbation Theory

The approximate calculation of F_2 in perturbation theory to all orders in chapter 15 and Appendix 25-A also resembles a tangent approximation with

$$d_d = d_{Gd} = 1.29911 - 0.08929*g$$
$$F_2(g) = \tan(\pi g/d_d)$$

and

$$\alpha(g) = \tan(\pi(g + 0.00348417)/d_d)$$

using the notation of chapter 9 and eqs. 24.10 - 24.12. Figs. 25.1 and 25.2 show the close approximation given by the tangent curves.

The general form of the expression for the fine structure constant α is

$$\alpha_G = |c_G \tan(x)| \tag{23.4}$$

where

$$x = \pi(g_0 + d_{G\alpha})/d_{Gd} \tag{23.5}$$

for some g_0.

We have shown the general form for α is

$$\alpha = \tan(h) \tag{25.1}$$

in close analogy to the Madhava-Leibniz representation of $\pi = 3.14159\ldots$.

We use
$$h = 0.007297223473 \tag{24.10}$$

in eq. 25.1 based on the α calculation of chapter 17, and the QED quantities

$$h/\pi = (d_\alpha - d_f)/d_d = 0.002322780 \qquad (24.11)$$

Thus the fine structure constant α is a feature of the dynamics of QED and is not a cosmological result, or a result of any other theory.

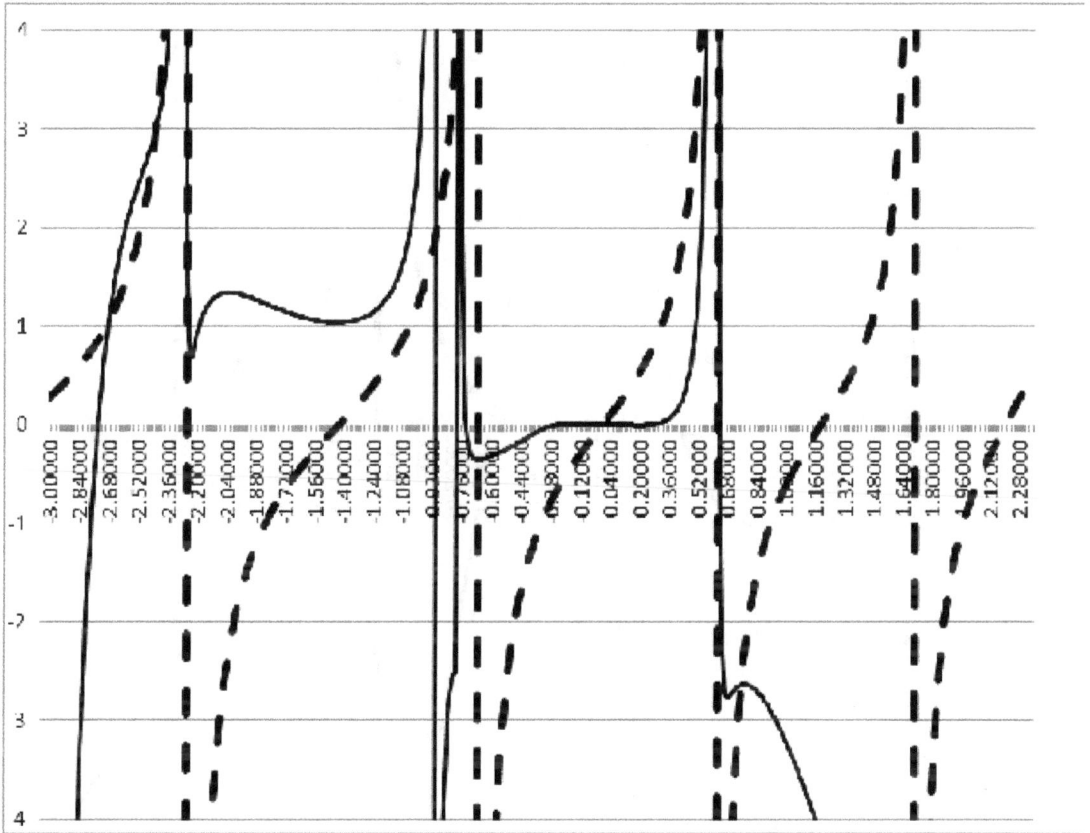

Figure 25.1 The massless QED eigenvalue function F$_2$ is the solid line, while the tangent form of F$_2$ of eq. 23.1 is the broken line. The constants are d$_{Gd}$ = 1.29911 - 0.08929g, and d$_{Gf}$ = 0. F$_2$ is plotted vertically. The exponent g is plotted horizontally.

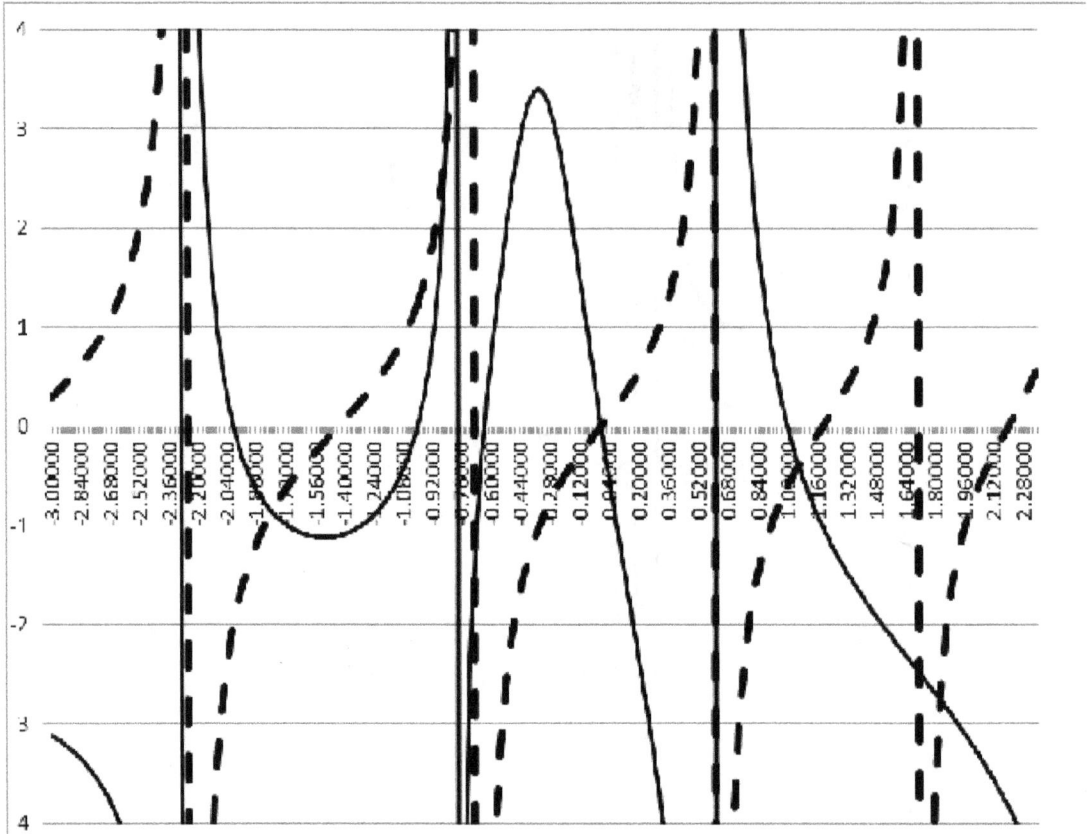

Figure 25.2 The QED fine structure eigenvalue function α(g) is a solid line, while the tangent form of α(g) of eq. 23.2 is the broken line. The constants are d_{Gd} = 1.29911 - 0.08929g, and the $α_G$ constant is $d_{Gα}$ = 0.00348417 for G = QED. α(g) is plotted vertically. The exponent g is plotted horizontally.

Appendix 25-A. "Approximate Calculation of the Eigenvalue Function in Massless Quantum Electrodynamics"

S. Blaha, Phys. Rev. **D9**, 2246 (1974).

The 3 + 1 Unified SuperStandard Theory of Our Universe: Details

In the chapters of this section we describe details of the Unified SuperStandard Theory of our universe. They show that the internal symmetries, and the form of the fermion and vector boson spectrums, of the Unified SuperStandard Theory directly follows from complex quaternion QUeST upon restricting complex quaternion coordinates to real coordinates.

26. Particle Cores, The Four Fermion Species, and Their Second Quantization

Summary

This chapter derives the form of the fermion spectrum from Complex Lorentz boosts. Under transformations of the Complex Lorentz group[103] from a state of rest, a spin ½ fermion can have one of four forms which we call *fermion species*.[104] These forms (species) are Dirac fermions and tachyon fermions with real-valued momenta; and complexon (Dirac-like) fermions and complexon tachyon fermions with real energies and complex-valued spatial momenta (subject to the condition that the spin, real part of the spatial momentum, and the imaginary part of the spatial momentum are orthogonal to each other.) *The energy of each boosted fundamental particle must be real-valued in the free field case since free fundamental particles do not decay.*

The number of distinct species is not changed by the extension of the set of coordinate transformations to Complex General Coordinate Transformations or to complex quaternion or octonion coordinates.

We map the four fermion species to charged leptons, neutrinos, up-type quarks, and down-type quarks respectively.

This derivation of four types (species) of fermions *directly* from Complex Lorentz group boosts from a rest state to a state with real-valued energy (thus giving stable fundamental fermions in the absence of interactions) is the only derivation of four fermion species that does not make any assumptions about internal symmetries.

Later the derivation will be brought to a deeper level by the introduction of fermion functionals that can combine with coordinate fourier expansions via an inner

---

[103] Complex Lorentz group boosts are require to obtain four species. It is one of our axioms and a requirement of proofs in axiomatic quantum field theory. See Streater (2000).

[104] Most of the material in this appendix was presented in the First Edition, Blaha (2012a), (2015a) and (2017b) as well as earlier books.

product to generate the fermion quantum fields described in this appendix. In addition we will define functionals for bosons that embody boson spin within a logical construct.

26.0 The Logic Core of Fundamental Fermions and Bosons

Previously we opened the possibility that fermions (and bosons) might have a core that embodies logic in the form of spin as well as bare masses in the case of fermions. In this book we expand our discussion to better describe the functionals space within which the core functionals of each type of fundamental particle reside. We will define functionals of various spins: 0, ½, 1, and 2. We will see that the core of spin ½ fermion functionals (that we call *qubes*) have four varieties that we will describe in this appendix.[105]Fermion functionals have a bare mass denoted m_0.

Bosons have cores as well that are boson functionals with integer spin. We call a boson core a *quba*[106] in analogy with the fermion functionals name of qubes. Boson functionals are massless. Bosons acquire masses through interactions.

The rationales for logic cores for particles is discussed in detail in chapters 3 and 8 of Blaha (2018e). We find them necessary to establish internal particle symmetries and interactions. We also will show that the formalism based on a space of all particle functionals can lead to an explanation of the 'spooky' action at a distance of Quantum Entanglement that has been the subject of much discussion.

26.0.1 The Logic Building Block of Fermions – Qube Cores

If we consider all possible 'things' that might constitute a fundamental building block for a fundamental fermion theory they are all, at best, *ad hoc* and raise questions of their necessity and whether they are composed of yet a more fundamental substructure.

There is only one choice of building block that avoids these issues – a logic unit or qubit. A qubit is a fundamental entity that is a complex form of computer bit. A bit (and thus a qubit) is known to have an energy, or equivalently a mass, and has no

---

[105] Each of the four varieties (which we call species) can be separated into left-handed and right-handed functionals.

[106] We use 'quba' simply because of its similarity to 'qube'. The leading 'b' signifies its bosonic use. We pronounce 'quba' as 'bub' with a silent 'e.' The word 'quba', itself, is the name of a Bantu language spoken by the Bubi people of Bioko Island in Equatorial Guinea.

constituents of a more primitive form.[107] We call a unit of logic that forms the core of a particle a *qube*.[108] It exists as the core of a particle. But, in itself, it has no *independent* material existence or space-time coordinates. A qube is a functional that acquires features such as coordinates, to become an elementary particle. We define a qube as a fermion field theory functional. (See chapters 3 and 8 of Blaha (2018e).) Later in this appendix we introduce physical features that will cloak qubes with properties and interactions making them into fundamental fermion quantum fields.

26.0.2 Mass of a Qube

Recent experiments have shown that a logical value of a qubit has an energy associated with it. One bit of information has about 3×10^{-21} joules of energy[109] or a rest mass, m_0, or about 0.02 eV using $m_0 = E/c^2$. This result was confirmed by E. Lutz et al.[110] who showed that there is a minimum amount of heat produced per bit of erased data. This minimal heat is called the *Landauer[111] limit*. The equivalent mass we will call the *Landauer mass* and denote it as m_0. We will assume that a fundamental Landauer mass exists in our discussions although the precise value of the mass will not be used since we may expect all physical particle masses to be renormalized to different values when interactions are taken into account.

We will assume all fermions contain a qube within them. (As stated above bosons do not have qubes within them.) We call their core a quba. A qube is assumed to have mass m_0. The masses of fermions are modified to their known values by interactions.

It is intriguing that the mass of the electron neutrino has been measured in a variety of experiments and found to be within an order of magnitude or so larger than

---

[107] A qube is a physical manifestation of a logical value. The relation of a qube to a logical value is analogous to the relation of a penciled point placed on paper to the concept of a point as a primitive in geometry.

[108] In the First Edition we called qubes iotas. However, since the name iota was previously used as a particle name many years ago it seemed reasonable to use a different name. We chose the name 'qube' for self-evident reasons. *'Qube' is pronounced 'cube.'*

[109] E. Muneyuki et al, *Nature Physics*, DOI: 10.1038/NPHYS1821.

[110] E. Lutz et al, Nature **483** (7388): 187–190,10.1038/nature10872, (2012).

[111] R. Landauer, "Irreversibility and heat generation in the computing process", IBM Journal of Research and Development **5** (3): 183–191, (1961)

our estimate of the Landauer mass (as we would expect since particles acquire a 'cloud of virtual particles' due to interactions.) This 'cloud' can be expected to increase its mass above the Landauer mass. Since neutrinos only have the weak interaction it is not surprising that the increase due to interactions should not be large. The Mainz Neutrino Mass Experiment, for example, estimates the electron neutrino mass to be less than 2 eV. The new Karlsruhe Tritium Neutrino Experiment (September, 2019) found an upper limit of less than 1.1 eV.

A number of astronomical studies have also generated estimates of neutrino masses. In July 2010 the 3-D MegaZ DR7 galaxy survey found a limit for the combined mass of the three neutrino varieties to be less than 0.28 eV.[112] A smaller upper bound for the sum of neutrino masses, 0.23 eV, was found in March 2013 by the Planck collaboration,[113] In February 2014 a new estimate of the sum was found to be 0.320 ± 0.081 eV due to discrepancies between the Planck's measurements of the Cosmic Microwave Background, and other predictions, combined with the assumption that neutrinos are the cause of weaker gravitational lensing than implied by massless neutrinos.[114]

Thus the experimentally measured values of neutrino masses are consistent with the qube Landauer mass estimate of 0.02 eV given above. We thus assume that *a fermion particle consists of a qube with a certain mass,[115] which is renormalized, together with other features. These features will emerge later in the derivation of the complete theory.[116]*

---

[112] S. Thomas et al, "Upper Bound of 0.28 eV on Neutrino Masses from the Largest Photometric Redshift Survey", Physical Review Letters **105**: 031301 (2010).

[113] Planck Collaboration, arXiv:1303.5076 (2013).

[114] R. A. Battye et al, "Evidence for Massive Neutrinos from Cosmic Microwave Background and Lensing Observations", Phys. Rev. Lett. **112**, 051303 (2014).

[115] Leibniz first proposed the idea of logic 'particles' which he called monads. Our definition of a logic 'particle' does not include (or exclude) the presence of a spiritual part which was part of the definition of Leibniz's monads.

[116] A recent experiment claims to separate the spin part (which we identify as a logical value later) of a molecule from the rest of the molecule.

We view Reality as ultimately a representation (or painting) of logic values evolving through interactions in time and space.[117]

26.0.3 Qube Spin

The spin of a qube is assumed to be spin ½. Qubes are solely a building block of fermions.

26.0.4 Qubes as Fermion Field Functionals

At this point qubes have an insubstantial appearance with only the attributes of mass and spin. In chapters 3 and 8 of Blaha (2018e) we suggested that they can be mathematically represented as fermion field functionals[118] and used to develop the structure of the fermion spectrum and The Unified SuperStandard Model.

We saw that the Standard Model interactions and features such as Quantum Entanglement *require* the use of a functional formalism for particle fields.

This new deeper formulation supports the theory presented in the First Edition (Blaha (2017f)). It adds a new level of depth that extends and clarifies the theory presented there.

We will begin here by defining a canonical functional approach to creating a simple Dirac fermion quantum field from a qube and a fourier quantum expression for the space-time part of a free fermion quantum field. We symbolize a qube for a fermion with no internal symmetry and spin ½ as f. We begin by defining a coordinate space Dirac fourier quantum expansion as

$$(s, x, t) = N(p)[b(p, s)u(p, s)e^{-ip \cdot x} + d^{\dagger}(p, s)v(p, s)e^{+ip \cdot x}]$$

where $N(p)$ is a normalization factor, u and v are functions of spin and momentum, and b and $d^{\dagger}$ are creation/annihilation operators.

---

[117] Those who might suggest matter is substantial, and logic values are not, should remember that matter would be completely insubstantial if there were no forces in nature. Neutrinos which are close to insubstantial would be completely insubstantial if there were no weak interactions.

[118] Functionals are a mathematical primitive of our theory. They have been used extensively by Feynman and others in quantum theories.

A Dirac quantum wave function can be defined as an inner product of a qube functional and a coordinate space fourier quantum expansion. For example

$$\psi(x) = (f, (s, x, t)) = \sum_{\pm s} \int d^3p N(p)[b(p, s)u(p, s)e^{-ip\cdot x} + d^\dagger(p, s)v(p, s)e^{+ip\cdot x}] \quad (26.1)$$

where we use a functional inner product formalism in the manner of Riesz $(1955)^{119}$ and others. A functional inner product yields a numeric value. In the present case, it yields a numeric (possibly quantum) function. In general an inner product of a functional f with a variable function g is expressed as

$$G(x) = (f, g(x))$$

For each value of x, G(x) has one numeric value modulo quantum smearing.

In this appendix we will implicitly use the above inner product expression for a fermion field for each of the four general types of fermions presented here:[120] a Dirac type of fermion, a tachyon type of fermion, a Dirac type of fermion with a complex 3-space momentum fourier expansion; and a tachyon type of fermion with a complex 3-space momentum fourier expansion.

Thus there are four differing functionals initially – one for each of the four types of fourier expansions. Internal symmetries, introduced later, will lead to a multi-dimensional space of functionals. We discuss these points in detail in chapter 3 of Blaha (2018e). *In this appendix we use conventional quantum field theory expressions deferring the functional space inner product representation discussions to subsequent chapters of Blaha (2018e).*

26.0.5 Boson Elementary Particles – Quba Cores

In defining qubes above we have considered only the fermion case. For reasons presented in chapter 3 of Blaha (2018e) there is also a need for boson core functionals called qubas. We can define a corresponding boson functional quba for each type of

---

[119] For example see pp. 61-2 of Blaha (2018e) where linear functionals and their inner products are defined.
[120] These types of fermion fields can be further subdivided into left-handed and right-handed fields.

boson. We will designate a boson functional as b_s where s specifies the spin which may be 0, 1, or 2. Every boson contains a boson functional core within it. For consistency we called a boson functional a *quba*. A quba has the spin of the elementary boson within which it resides. It has zero mass since bosons are typically massless prior to symmetry breaking effects. The functional content embodied in each type of elementary particle is summarized in Table 26.1.

PARTICLE TYPE	CORE	MASS	SPIN
Fermion	qube	m_0	½
Scalar Boson	quba	0	0
Vector Boson	quba	0	1
Graviton	quba	0	2

Table 26.1 Core functionals within the various types of fundamental elementary particles.

Having established the core concept for fundamental fermions we now determine the four basic species (types) of fermions with the implicit understanding that each fermion quantum field is the inner product of a core functional and a coordinate space fourier representation. The remainder of the appendix's derivations are based on the Complex Lorentz Group—as it was in the First Edition (Blaha (2017f)) and earlier books. [121]

26.1 Matrix Representation of Complex Lorentz Group L_C Boosts

The remainder of this appendix is based on the Complex Lorentz group which will be seen to have a primary role in defining the structure of the fermion spectrum.

We begin with Complex Lorentz Group (L_C) boosts because they will be crucial in the determination of the equations of motion of various types of spin ½ particles. An L_C boost can be expressed in the form

---

[121] The remainder of this appendix is the same as chapter 2 of the First Edition (Blaha (2017f)). A new section 2.10 has been added to depict the functional formulation of the quantum fields.

$$\Lambda_C(\mathbf{v_c}) = \exp[i\omega\hat{\mathbf{w}}\cdot\mathbf{K}] \tag{26.2}$$

where

$$\omega = (\omega_r^2 - \omega_i^2 + 2i\omega_r\omega_i\,\hat{\mathbf{u}}_r\cdot\hat{\mathbf{u}}_i)^{\frac{1}{2}} \tag{26.3}$$

and

$$\hat{\mathbf{w}} = (\omega_r\hat{\mathbf{u}}_r + i\omega_i\hat{\mathbf{u}}_i)/\omega \tag{26.4}$$

Since $\hat{\mathbf{u}}_r\cdot\hat{\mathbf{u}}_r = 1 = \hat{\mathbf{u}}_i\cdot\hat{\mathbf{u}}_i$

$$\hat{\mathbf{w}}\cdot\hat{\mathbf{w}} = 1 \tag{26.5}$$

and the complex relative velocity is

$$\mathbf{v_c} = \hat{\mathbf{w}}\,\tanh(\omega) \tag{26.6}$$

We now analytically continue to complex ω and complex unit vectors $\hat{\mathbf{w}}$. The resulting complex generalization will be the matrix form of proper L_C boosts:

$$\Lambda_C(\mathbf{v_c}) = \exp[i\omega\hat{\mathbf{w}}\cdot\mathbf{K}] \equiv \Lambda_C(\omega, \hat{\mathbf{w}})$$

$$
= \begin{bmatrix}
\cosh(\omega) & -\sinh(\omega)\hat{w}_x & -\sinh(\omega)\hat{w}_y & -\sinh(\omega)\hat{w}_z \\[2mm]
-\sinh(\omega)\hat{w}_x & 1+(\cosh(\omega)-1)\hat{w}_x^2 & (\cosh(\omega)-1)\hat{w}_x\hat{w}_y & (\cosh(\omega)-1)\hat{w}_x\hat{w}_z \\[2mm]
-\sinh(\omega)\hat{w}_y & (\cosh(\omega)-1)\hat{w}_x\hat{w}_y & 1+(\cosh(\omega)-1)\hat{w}_y^2 & (\cosh(\omega)-1)\hat{w}_y\hat{w} \\[2mm]
-\sinh(\omega)\hat{w}_z & (\cosh(\omega)-1)\hat{w}_x\hat{w}_z & (\cosh(\omega)-1)\hat{w}_y\hat{w}_z & 1+(\cosh(\omega)-1)\hat{w}_z^2
\end{bmatrix}
$$

$$\tag{26.7}$$

Since analytic continuations are unique, the above form for $\Lambda_C(\mathbf{v_c})$ is well-defined and unique. It spans the complete set of proper L_C boosts.

26.2 Left-handed and Right-handed Parts of L_C

We now describe the Left-handed and Right-handed parts[122] of L_C boosts.

---

[122] The designations Left-handed and Right-handed are chosen to reflect the Left-handed and Right-handed fermion fields that will be used to construct The Standard Model later. See Blaha (2007b) for more detail.

26.2.1 Left-handed Part of L_C

If we let

$$\hat{\mathbf{u}}_i = \hat{\mathbf{u}}_r \equiv \hat{\mathbf{u}} \qquad (26.8)$$

so that the vector $\hat{\mathbf{u}}_i$ is parallel to $\hat{\mathbf{u}}_r$, and

$$\omega_i = \pi/2 \qquad (26.9)$$

then $\Lambda_C(\mathbf{v_c})$ becomes a Left-handed L_C boost:

$$\Lambda_C(\mathbf{v_c}) = \Lambda_L(\omega_r, \mathbf{u}) \qquad (26.10)$$

26.2.2 Right-handed part of L_C

If we let

$$\hat{\mathbf{u}}_i = -\hat{\mathbf{u}}_r \equiv -\hat{\mathbf{u}} \qquad (26.11)$$

so that the vector $\hat{\mathbf{u}}_i$ is anti-parallel to $\hat{\mathbf{u}}_r$, and

$$\omega_i = -\pi/2 \qquad (26.12)$$

then $\Lambda_C(\mathbf{v_c})$ becomes a Right-handed L_C boost:

$$\Lambda_C(\mathbf{v_c}) = \Lambda_R(\omega_r, \mathbf{u}) \qquad (26.13)$$

as described in Blaha (2007b).

26.3 Difference between the Parts of L_C Reduced to Parallelism of $\hat{u}_r$ and $\hat{u}_i$

Since the Left-handed L_C part leads to the Standard Model's left-handed features, it seems that the parallel case $\hat{u}_i = \hat{u}_r \equiv \hat{u}$ is more favored by Nature.[123] To some extent this concept of parallel vectors $\hat{u}_i$ and $\hat{u}_r$, which leads to the Left-handed L_C, is more intuitively satisfying then the anti-parallel case that leads to the Right-handed L_C part. However, a deeper reason for Nature's choice remains to be found.

26.4 Free Spin ½ Particles – Leptons & Quarks

In this section we begin by developing dynamical equations for spin ½ particles based on the L_C parts. These spin ½ particles are conventional Dirac particles (Majorana particles are also allowed but not discussed), spin ½ tachyons, and "color" versions of both types totaling four species. We will identify leptons and quarks with these fields.

26.4.1 Introduction

Tachyons are particles that move faster than the speed of light. As we saw in earlier books tachyons exist inside Black Holes, and within current theories – particularly SuperString theories. There are also experimental indications that neutrinos are tachyons.

Attempts to create canonical tachyon quantum field theories began in the 1960's. These attempts were made within the framework of the Lorentz group and, consequently, were limited to spin 0 theories since there are no finite dimensional representations of the Lorentz group for negative m^2 except for the one-dimensional representation. None of these attempts, or attempts since then, succeeded in creating a canonically quantized spin 0 tachyon quantum field theory.[124]

In this section we will formulate a free spin ½ tachyon[125] Quantum Field Theory. We choose to develop a normal spin ½ theory first. Then we develop a free

---

[123] It is possible that parity violation might disappear at ultra-high energies. Then we would view the parity symmetric theory as broken to the left-handed Standard Model currently established by experiment with right-handed parts at higher energy.

[124] Except Blaha (2006).

[125] It fiffers significantly from tachyon theories such as those of G. Feinberg and E. C. Sudarshan.

spin ½ tachyon theory because, as we will see, spin ½ tachyon particles (quarks and leptons) play an extraordinary role in the Standard Model.

We will develop our spin ½ tachyon theory from the "ground up" by applying a Left-Handed L_C boost to the Dirac equation, and its Dirac spinor wave function, for a particle at rest. This procedure will give a tachyon spinor wave function, and the momentum space tachyon equation equivalent of the Dirac equation. Then we will obtain the coordinate space tachyon Dirac equation, define a lagrangian, and proceed to create a canonical quantum field theory for spin ½ tachyons.

The need for dynamical equations arises when we clothe each qube with coordinates to "make" a fermion. Having coordinates leads to describing the motion of particles. Dynamical equations specify the motion of particles. If they have a finite number of terms they can be derived from lagrangians. A lagrangian formalism yields a Hamiltonian (the energy) and the momentum.

26.4.2 First Step - Deriving the Conventional Dirac Equation

In this section we will review a method of obtaining the equation of motion of a particle using a free Dirac equation that is obtained by a Lorentz boost of a spinor wave function[126] of a particle at rest.

In the case of a Lorentz transformation the 4×4 matrix form of a Lorentz transformation of Dirac matrices is

$$S^{-1}(\Lambda(v))\gamma^{\nu}S(\Lambda(v)) = \Lambda^{\nu}{}_{\mu}(v)\gamma^{\mu} \qquad (26.14)$$

where $S(\Lambda(v))$ is

$$S(\Lambda(v)) = \exp(-i\omega\sigma_{0i}v_i/(2|\mathbf{v}|)) = \exp(-\omega\gamma^0\boldsymbol{\gamma}\cdot\mathbf{v}/(2|\mathbf{v}|))$$
$$= \cosh(\omega/2)I + \sinh(\omega/2)\gamma^0\boldsymbol{\gamma}\cdot\mathbf{p}/|\mathbf{p}| \qquad (26.15)$$

with $\omega = \text{arctanh}(|\mathbf{v}|)$, $\cosh(\omega/2) = [(E+m)/(2m)]^{\frac{1}{2}}$ and $\sinh(\omega/2) = |\mathbf{p}|[2m(E+m)]^{-\frac{1}{2}}$. Also

---

[126] The spinor wave function of a particle at rest is a 4-vector of the 4×4 matrix representation of 4-valued Asynchronous Logic.

$$S^{-1}(\Lambda(v)) = \gamma^0 S^\dagger(\Lambda(v))\gamma^0 = \exp(\omega\gamma^0\gamma\cdot\mathbf{v}/(2|\mathbf{v}|))$$
$$= \cosh(\omega/2)I - \sinh(\omega/2)\gamma^0\gamma\cdot\mathbf{p}/|\mathbf{p}| \qquad (26.16)$$

In constructing fermion dynamical equations *we shall assume that they are linear in derivatives* (although a quadratic form is possible.) We will use the sixteen 4×4 matrices that span the set of transformations of the four values of Asynchronous Logic. Since by theorem[127] all 4 × 4 γ matrices are equivalent up to a unitary transformation we can rotate any constant matrix into a multiple of γ^0 without loss of generality.

We begin by defining a generic positive energy plane wave solution of the Dirac equation for a normal fermion particle at rest with rest mass m, *which we take to be the qube bare mass in the absence of interactions,*[128] as

$$\psi(x) = e^{-imt}w(0) \qquad (26.17)$$

with w(0) a four component logic spinor column vector. *For a free particle at rest, the rest energy m = m_0, the qube mass.* The wave function satisfies the momentum space Dirac equation for a fermion at rest:

$$(m\gamma^0 - m)e^{-imt}w(0) = 0 \qquad (26.18)$$

Subsequently we will use a similar procedure to construct the free tachyonic Dirac equation.

If we now apply $S(\Lambda(v))$ we find

$$0 = S(\Lambda(v))(m\gamma^0 - m)e^{-imt}w(0) = [mS(\Lambda(v))\gamma^0 S^{-1}(\Lambda(v)) - m]S(\Lambda(v))w(0)$$

A straightforward evaluation shows

---

[127] R. H. Good, Rev. Mod. Phys., **27**, 187 (1955).

[128] As stated earlier the derivation proceeds in steps from the Complex Lorentz Group to free fermions and thence to interacting fermions. **We use the bare qube mass throughout the free fermion discussions in this appendix. m = m_0.**

$$mS(\Lambda(v))\gamma^0 S^{-1}(\Lambda(v)) = g_{\mu\nu}p^{\mu}\gamma^{\nu} = \not{p} \qquad (26.19)$$

where $p^0 = (p^2 + m^2)^{\frac{1}{2}}$, $\mathbf{p} = \gamma m\mathbf{v}$, and $p = |\mathbf{p}|$. In addition

$$S(\Lambda(v))w(0) = w(p) \qquad (26.20)$$

is a positive energy Dirac spinor. Therefore the Dirac equation for a fermion in motion in momentum space has the form:

$$(\not{p} - m)e^{-ip \cdot x}w(p) = 0 \qquad (26.21)$$

where the exponential factor, mt, is also boosted to p·x. Eq. 26.21 implies the well-known free, coordinate space Dirac equation:

$$(i\gamma^{\mu}\partial/\partial x^{\mu} - m)\psi(x) = 0 \qquad (26.22)$$

26.4.3 Derivation of the Tachyon Dirac Equation

The Left-handed boost has the form:

$$\Lambda_L(\omega, \mathbf{u}) = \Lambda(\omega + i\pi/2, \mathbf{u}) = \exp[i\omega_L \hat{\mathbf{u}} \cdot \mathbf{K}] \qquad (26.23)$$

where $\omega_L = \omega + i\pi/2$ and

$$\cosh(\omega_L) = i \sinh(\omega) = -\gamma = i\gamma_s \qquad (26.24)$$
$$\sinh(\omega_L) = i \cosh(\omega) = -\beta\gamma = i\beta\gamma_s$$

with, $\beta = v > 1$, $\gamma_s = (\beta^2 - 1)^{-\frac{1}{2}}$, and $\omega \geq 0$. Thus

$$\sinh(\omega) = \gamma_s \qquad (26.25)$$
$$\cosh(\omega) = \beta\gamma_s$$

The corresponding spinor transformation is:

$$S_L(\Lambda_L(\omega, \mathbf{u})) = \exp(-i\omega_L\sigma_{0i}v_i/(2|\mathbf{v}|)) = \exp(-\omega_L\gamma^0\boldsymbol{\gamma}\cdot\mathbf{v}/(2|\mathbf{v}|))$$

$$= \cosh(\omega_L/2)I + \sinh(\omega_L/2)\gamma^0\gamma\cdot\mathbf{p}/|\mathbf{p}| \qquad (26.26)$$

The inverse transformation is

$$S_L^{-1}(\Lambda_L(\omega, \mathbf{u})) = \gamma^2\gamma^0 K^{-1}S_L^\dagger K\gamma^0\gamma^2 = \gamma^2\gamma^0 S_L^{\,T}\gamma^0\gamma^2 = \exp(\omega_L\gamma^0\gamma\cdot\mathbf{v}/(2|\mathbf{v}|))$$
$$= \cosh(\omega_L/2)I - \sinh(\omega_L/2)\gamma^0\gamma\cdot\mathbf{p}/|\mathbf{p}| \qquad (26.27)$$

where the superscript T denotes the transpose and K is the complex conjugation operator (that also appears in the time-reversal operator). Note that S_L is not unitary just as the equivalent spinor Lorentz transformation $S(\Lambda(v))$ is not unitary.

We can now apply a left-handed superluminal transformation to the generic positive energy plane wave solution of the Dirac equation for a particle of mass m at rest. The result is

$$0 = S_L(\Lambda_L(\omega, \mathbf{u}))(m\gamma^0 - m)e^{-imt}w(0)$$
$$= [mS_L\gamma^0 S_L^{-1} - m]e^{-imt}S_L w(0)$$

where $S_L = S_L(\Lambda_L(\omega, \mathbf{u}))$. After some algebra

$$mS_L\gamma^0 S_L^{-1} = m[\cosh(\omega_L)\gamma^0 - \sinh(\omega_L)\gamma\cdot\mathbf{p}/|\mathbf{p}|]$$
$$= i\gamma^0 E - i\gamma\cdot\mathbf{p} = i\not{p} \qquad (26.28)$$

using the tachyon energy and momentum expressions

$$\mathbf{p} = m\mathbf{v}\gamma_s \qquad\qquad E = m\gamma_s \qquad (26.29)$$

Also

$$S_L w(0) = w_T(p) \qquad (26.30)$$

is a tachyon spinor. See section 26.12 for a discussion of tachyon spinors.

The momentum space tachyonic Dirac equation is

$$(i\not{p} - m)e^{ip\cdot x}w_T(p) = 0 \qquad (26.31)$$

where $p{\cdot}x = Et - \mathbf{p}{\cdot}\mathbf{x}$ after performing a corresponding left-handed superluminal coordinate transformation in the exponential factor. Thus a positive energy wave is transformed into a negative energy wave by the superluminal transformation.

If we apply $i\not{p}$ to it we find the tachyon mass condition is satisfied

$$-E^2 + \mathbf{p}^2 \ = \ m^2 \qquad (26.32)$$

Transforming back to coordinate space we obtain the *tachyon Dirac equation*:

$$(\gamma^\mu \partial/\partial x^\mu - m)\psi_T(x) = 0 \qquad (26.33)$$

The "missing" factor of i in the first term of eq. 26.33 requires the lagrangian to be different from the conventional Dirac lagrangian in order for the lagrangian to be real. The simplest, physically acceptable, free spin ½ tachyon lagrangian density is:

$$\mathcal{L}_T = \psi_T{}^S (\gamma^\mu \partial/\partial x^\mu - m)\psi_T(x) \qquad (26.34)$$

where

$$\psi_T{}^S = \psi_T{}^\dagger \, i\gamma^0\gamma^5 \qquad (26.35)$$

The corresponding action is

$$I = \smallint d^4x \mathcal{L}_T \qquad (26.36)$$

Appendix 3-B of Blaha (2007b) proves I is real. The Hamiltonian density is

$$\mathcal{H} = \pi_T\dot{\psi}_T - \mathcal{L} = i\psi_T{}^\dagger\gamma^5(\boldsymbol{\alpha}{\cdot}\nabla + \beta m)\psi_T = -i\psi_T{}^\dagger\gamma^5\dot{\psi}_T \qquad (26.37)$$

using the tachyon Dirac equation to obtain the last equality. The reader will note that the tachyon hamiltonian is hermitean by explicit calculation up to an irrelevant total spatial divergence.

26.4.3.1 Probability Conservation Law

The tachyon Dirac equation implies a probability conservation law:

$$\partial\rho_5/\partial t = \nabla\cdot\mathbf{j}_5 \qquad (26.38)$$

where

$$\rho_5 = \psi_T^{\dagger}\gamma^5\psi_T \qquad\qquad \mathbf{j}_5 = \psi_T^{\dagger}\gamma^5\boldsymbol{\alpha}\psi_T \qquad (26.39)$$

We are thus led to define the conserved axial charge Q_5

$$Q_5 = \int d^3x\ \psi_T^{\dagger}\gamma^5\psi_T \qquad (26.40)$$

26.4.3.2 Energy-Momentum Tensor

The tachyon energy-momentum tensor is

$$\mathfrak{T}_{T\mu\nu} = -\,g_{\mu\nu}\mathfrak{L}_T + \partial\mathfrak{L}_T/\partial(\partial\psi_T/\partial x_\mu)\ \partial\psi_T/\partial x^\nu \qquad (26.41)$$
$$= i\psi_T^{\dagger}\gamma^0\gamma^5\gamma_\mu\partial\psi_T/\partial x^\nu \qquad (26.42)$$

and thus the conserved energy and momentum are

$$P^0 = H = \int d^3x\ \mathfrak{T}_T^{00} = i\int d^3x\psi_T^{\dagger}\gamma^5(\boldsymbol{\alpha}\cdot\nabla + \beta m)\psi_T \qquad (26.43)$$

and

$$P^i = \int d^3x\ \mathfrak{T}_T^{0i} = -\,i\int d^3x\ \psi_T^{\dagger}\gamma^5\partial\psi_T/\partial x_i \qquad (26.44)$$

Both the energy and momentum differ significantly from the corresponding quantities for conventional Dirac fields.

26.4.4 Tachyon Canonical Quantization

Having defined a suitable tachyon lagrangian we can now proceed to its canonical quantization. The conjugate momentum can be calculated from the above lagrangian density:

$$\pi_{Ta} = \partial\mathfrak{L}_T/\partial\dot{\psi}_{Ta} \equiv \partial\mathfrak{L}_T/\partial(\partial\psi_{Ta}/\partial t) = -i(\psi_T^{\dagger}\gamma^5)_a \qquad (26.45)$$

The resulting non-zero, canonical anti-commutation relations are\

$$\{\pi_{T_a}(x), \psi_{Tb}(x')\} = i\, \delta_{ab}\, \delta^3(x - x')$$

or

$$\{\psi_T^{\dagger}{}_a(x), \psi_{Tb}(x')\} = -\, [\gamma^5]_{ab}\, \delta^3(x - x') \tag{26.46}$$

At this point we might attempt to complete the canonical quantization procedure in the conventional manner by fourier expanding the quantum field and specifying anti-commutation relations for the fourier component amplitudes. However the incompleteness of the set of plane waves, which are limited by the restriction $|p| \geq m$, causes the anti-commutator of the fields not to yield a $\delta^3(x - x')$. Thus the conventional approach fails to yield the required anti-commutation relations.[129]

Other approaches: 1) decompose the tachyon field into left-handed and right-handed parts and then second quantize each part; and 2) second quantize in light-front coordinates ($x^{\pm} = (x^0 \pm x^3)/\sqrt{2}$). These approaches also both fail.[130]

The only approach that does succeed[131] is to decompose the tachyon field into left-handed and right-handed parts and then second quantize in light-front coordinates. We follow that procedure in the following subsections.

26.4.4.1 Separation into Left-Handed and Right-Handed Fields

We will use a transformed set of Dirac matrices to develop our left-handed and right-handed tachyon formulations:

$$\gamma^0 = \begin{bmatrix} 0 & -I \\ -I & 0 \end{bmatrix} \qquad \gamma^i = \begin{bmatrix} 0 & \sigma_i \\ -\sigma_i & 0 \end{bmatrix} \qquad \gamma^5 = \begin{bmatrix} I & 0 \\ 0 & -I \end{bmatrix}$$

$$\tag{26.47}$$

which are obtained from the usual Dirac matrices by applying the unitary transformation $U = 2^{\frac{1}{2}}(I + \gamma^5\gamma^0)$. *I is the 4×4 identity matrix in eq. 26.47.* The γ^5 chirality

---

[129] See G. Feinberg, Phys. Rev. **159**, 1089 (1967) for example.
[130] See Blaha (2006) where these possibilities were considered and found to fail.
[131] Blaha (2006) discusses this case in detail.

operator's eigenvalues define handedness: +1 corresponds to right-handed; and −1 corresponds to left-handed:

$$\gamma^5\psi_L = -\psi_L \qquad \gamma^5\psi_R = \psi_R \qquad (26.48)$$

Consequently, we can define left-handed and right-handed tachyon fields with the projection operators:

$$\begin{aligned} C^\pm &= \tfrac{1}{2}(I \pm \gamma^5) \\ C^+ + C^- &= I \\ C^{\pm 2} &= C^\pm \\ C^+ C^- &= 0 \end{aligned} \qquad (26.49)$$

with the result

$$\begin{aligned} \psi_{TL} &= C^-\psi_T \\ \psi_{TR} &= C^+\psi_T \end{aligned} \qquad (26.50)$$

We can calculate the commutation relations of the left-handed and right-handed tachyon fields from eq. 26.46 by pre-multiplying and post-multiplying by $\tfrac{1}{2}(1 - \gamma^5)$ and $\tfrac{1}{2}(1 + \gamma^5)$. The results are:

$$\{\psi_{TLa}^\dagger(x), \psi_{TLb}(x')\} = \tfrac{1}{2}(1 - \gamma^5)_{ab}\,\delta^3(x - x') \qquad (26.51)$$

$$\{\psi_{TRa}^\dagger(x), \psi_{TRb}(x')\} = -\tfrac{1}{2}(1 + \gamma^5)_{ab}\,\delta^3(x - x') \qquad (26.52)$$

$$\{\psi_{TLa}^\dagger(x), \psi_{TRb}(x')\} = \{\psi_{TRa}^\dagger(x), \psi_{TLb}(x')\} = 0 \qquad (26.53)$$

The lagrangian density above decomposes into left-handed and right-handed parts:

$$\mathcal{L}_T = \psi_{TL}^\dagger\gamma^0 i\gamma^\mu\partial_\mu\psi_{TL} - \psi_{TR}^\dagger\gamma^0 i\gamma^\mu\partial_\mu\psi_{TR} - im[\psi_{TR}^\dagger\gamma^0\psi_{TL} - \psi_{TL}^\dagger\gamma^0\psi_{TR}] \quad (26.54)$$

26.4.4.2 Further Separation into + and – Light-Front Fields

There have been many studies of light-front (infinite momentum frame) physics in the past forty years.[132] Light-front coordinates *cannot* be obtained by a Lorentz transformation, or by a superluminal transformation, from a standard set of coordinate system variables even in a limiting sense. Instead they are a defined set of variables that have been used to develop quantum field theories that have been shown to be equivalent to quantum field theories based on conventional coordinates. In particular, light-front quantum field theories have been shown to yield fully Lorentz covariant S matrix elements that are the same as S matrix elements calculated in the conventional way.

Light-front variables can be defined by:

$$x^{\pm} = (x^0 \pm x^3)/\sqrt{2} \qquad (26.55)$$

$$\partial/\partial x^{\pm} \equiv \partial^{\mp} \equiv (\partial/\partial x^0 \pm \partial/\partial x^3)/\sqrt{2}$$

with the "transverse" coordinate variables, x^1 and x^2, unchanged.

The inner product of two 4-vectors has the form

$$x{\cdot}y = x^+ y^- + y^+ x^- - x^1 y^1 - x^2 y^2 \qquad (26.56)$$

and the light-front definition of Dirac matrices is:

$$\gamma^{\pm} = (\gamma^0 \pm \gamma^3)/\sqrt{2} \qquad (26.57)$$

with transverse matrices γ^1 and γ^2 defined as usual. Note the useful identity:

[132] L. Susskind, Phys. Rev. **165**, 1535 (1968); K. Bardakci and M. B. Halpern Phys. Rev. **176**, 1686 (1968), S. Weinberg, Phys. Rev. **150**, 1313 (1966); J. Kogut and D. Soper, Phys. Rev. **D1**, 2901 (1970); J. D. Bjorken, J. Kogut, and D. Soper, Phys. Rev. **D3**, 1382 (1971); R. A. Neville and F. Rohrlich, Nuov. Cim. **A1**, 625 (1971); F. Rohrlich, Acta Phys Austr. Suppl. **8**, 277 (1971); S-J Chang, R. Root, and T-M Yan, Phys. Rev. **D7**, 1133 (1973); S-J Chang, and T-M Yan, Phys. Rev. **D7**, 1147 (1973); T-M Yan, Phys. Rev. **D7**, 1761 (1973); T-M Yan, Phys. Rev. **D7**, 1780 (1973); C. Thorn, Phys. Rev. **D19**, 639 (1979); and references therein.

$$\gamma^{\pm 2} = 0$$

We define "+" and "–" tachyon fields with the projection operators:

$$R^{\pm} = \tfrac{1}{2}(I \pm \gamma^0 \gamma^3) \tag{26.58}$$

They are:

Left-handed, ± light-front fields: $\quad \psi_{TL}{}^{\pm} = R^{\pm} C^- \psi_T$

$$\tag{26.59}$$

Right-handed, ± light-front fields: $\quad \psi_{TR}{}^{\pm} = R^{\pm} C^+ \psi_T$

Now if we transform to light-front variables and fields as above we obtain the light-front free tachyon lagrangian:

$$
\begin{aligned}
\mathcal{L}_T = {}& 2^{\frac{1}{2}}\psi_{TL}{}^{+\dagger} i\partial^- \psi_{TL}{}^+ + 2^{\frac{1}{2}}\psi_{TL}{}^{-\dagger} i\partial^+ \psi_{TL}{}^- - \psi_{TL}{}^{+\dagger}\gamma^0 i\gamma^j \partial^j \psi_{TL}{}^- - \psi_{TL}{}^{-\dagger}\gamma^0 i\gamma^j \partial^j \psi_{TL}{}^+ - \\
& - 2^{\frac{1}{2}}\psi_{TR}{}^{+\dagger} i\partial^- \psi_{TR}{}^+ - 2^{\frac{1}{2}}\psi_{TR}{}^{-\dagger} i\partial^+ \psi_{TR}{}^- + \psi_{TR}{}^{+\dagger}\gamma^0 i\gamma^j \partial^j \psi_{TR}{}^- + \psi_{TR}{}^{-\dagger}\gamma^0 i\gamma^j \partial^j \psi_{TR}{}^+ - \\
& - im[\psi_{TR}{}^{+\dagger}\gamma^0\psi_{TL}{}^- - \psi_{TL}{}^{+\dagger}\gamma^0\psi_{TR}{}^- + \psi_{TR}{}^{-\dagger}\gamma^0\psi_{TL}{}^+ - \psi_{TL}{}^{-\dagger}\gamma^0\psi_{TR}{}^+]
\end{aligned}
\tag{26.60}
$$

with implied sums over j = 1,2. In contrast to the light-front tachyon lagrangian we note the corresponding light-front "normal" Dirac fermion lagrangian is

$$
\begin{aligned}
\mathcal{L}_{Dirac} = {}& 2^{\frac{1}{2}}\psi_L{}^{+\dagger} i\partial^- \psi_L{}^+ + 2^{\frac{1}{2}}\psi_L{}^{-\dagger} i\partial^+ \psi_L{}^- - \psi_L{}^{+\dagger}\gamma^0 i\gamma^j \partial^j \psi_L{}^- - \psi_L{}^{-\dagger}\gamma^0 i\gamma^j \partial^j \psi_L{}^+ - \\
& - 2^{\frac{1}{2}}\psi_R{}^{+\dagger} i\partial^- \psi_R{}^+ + 2^{\frac{1}{2}}\psi_R{}^{-\dagger} i\partial^+ \psi_R{}^- - \psi_R{}^{+\dagger}\gamma^0 i\gamma^j \partial^j \psi_R{}^- - \psi_R{}^{-\dagger}\gamma^0 i\gamma^j \partial^j \psi_R{}^+ - \\
& - im[\psi_R{}^{+\dagger}\gamma^0\psi_L{}^- + \psi_L{}^{+\dagger}\gamma^0\psi_R{}^- + \psi_R{}^{-\dagger}\gamma^0\psi_L{}^+ + \psi_L{}^{-\dagger}\gamma^0\psi_R{}^+]
\end{aligned}
\tag{26.61}
$$

The difference in signs between these lagrangians will turn out to be a crucial factor in the derivation of features of the Standard Model later.

Returning to the tachyon lagrangian eq. 26.60 we obtain equations of motion through the standard variational techniques:

$$2^{1/2}i\partial^{-}\psi_{TL}^{+} - \gamma^{0}i\gamma^{j}\partial^{j}\psi_{TL}^{-} + im\gamma^{0}\psi_{TR}^{-} = 0 \qquad (26.62)$$
$$2^{1/2}i\partial^{-}\psi_{TR}^{+} - \gamma^{0}i\gamma^{j}\partial^{j}\psi_{TR}^{-} + im\gamma^{0}\psi_{TL}^{-} = 0$$
$$2^{1/2}i\partial^{+}\psi_{TL}^{-} - \gamma^{0}i\gamma^{j}\partial^{j}\psi_{TL}^{+} + im\gamma^{0}\psi_{TR}^{+} = 0$$
$$2^{1/2}i\partial^{+}\psi_{TR}^{-} - \gamma^{0}i\gamma^{j}\partial^{j}\psi_{TR}^{+} + im\gamma^{0}\psi_{TL}^{+} = 0$$

Eqs. 26.62 show that ψ_{TL}^{-} and ψ_{TR}^{-} are dependent fields that are functions of ψ_{TL}^{+} and ψ_{TR}^{+} on the light-front where x^{+} equals a constant. They can be expressed in an integral form as well. (The independent fields ψ_{TL}^{+} and ψ_{TR}^{+} play a fundamental role in tachyon theory and are used to define "in" and "out" tachyon states in perturbation theory.)

The conjugate momenta are

$$\pi_{TL}^{+} = \partial\mathscr{L}/\partial(\partial^{-}\psi_{TL}^{+}) = 2^{1/2}i\psi_{TL}^{+\dagger} \qquad (26.63)$$
$$\pi_{TL}^{-} = \partial\mathscr{L}/\partial(\partial^{-}\psi_{TL}^{-}) = 0$$
$$\pi_{TR}^{+} = \partial\mathscr{L}/\partial(\partial^{-}\psi_{TR}^{+}) = -2^{1/2}i\psi_{TR}^{+\dagger} \qquad (26.64)$$
$$\pi_{TR}^{-} = \partial\mathscr{L}/\partial(\partial^{-}\psi_{TR}^{-}) = 0$$

Quantization on surfaces of constant x^{+} (light-front surfaces) has been shown to support satisfactory formulations of Quantum Electrodynamics and other quantum field theories. Thus x^{+} plays the role of the "time" variable in light-front quantized theories. So we will define canonical equal x^{+} anti-commutation relations for spin ½ tachyons.

The resulting canonical equal-light-front $(x^{+} = y^{+})$ anti-commutation relations of the independent fields are:

$$\{\psi_{TL}^{+\dagger}{}_{a}(x), \psi_{TL}^{+}{}_{b}(y)\} = 2^{-1}[C^{-}R^{+}]_{ab}\,\delta(x^{-} - y^{-})\delta^{2}(x - y) \qquad (26.65)$$
$$\{\psi_{TR}^{+\dagger}{}_{a}(x), \psi_{TR}^{+}{}_{b}(y)\} = -2^{-1}[C^{+}R^{+}]_{ab}\,\delta(x^{-} - y^{-})\delta^{2}(x - y) \qquad (26.66)$$
$$\{\psi_{TL}^{+}{}_{a}^{\dagger}(x), \psi_{TR}^{+}{}_{b}(y)\} = \{\psi_{TR}^{+}{}_{a}^{\dagger}(x), \psi_{TL}^{+}{}_{b}(y)\} = 0 \qquad (26.67)$$
$$\{\psi_{TL}^{+}{}_{a}(x), \psi_{TR}^{+}{}_{b}(y)\} = \{\psi_{TR}^{+}{}_{a}^{\dagger}(x), \psi_{TL}^{+\dagger}{}_{b}(y)\} = 0 \qquad (26.68)$$

where the factors of 2^{-1} are the result of the $2^{\frac{1}{2}}$ factor in eqs. 26.63 and 26.64, and the factor of $2^{-\frac{1}{2}}$ in the definition of x^{-} above.

If we compare eqs. 26.65 and 26.66 with the corresponding anti-commutation relations of *conventional Dirac* quantum fields:

$$\{\psi_{L}{}^{+\dagger}{}_{a}(x), \psi_{L}{}^{+}{}_{b}(y)\} = 2^{-1}[C^{-}R^{+}]_{ab} \, \delta(x^{-} - y^{-})\delta^{2}(x - y) \qquad (26.69)$$

$$\{\psi_{R}{}^{+\dagger}{}_{a}(x), \psi_{R}{}^{+}{}_{b}(y)\} = 2^{-1}[C^{+}R^{+}]_{ab} \, \delta(x^{-} - y^{-})\delta^{2}(x - y) \qquad (26.70)$$

we see that the right-handed tachyon anti-commutation relation has a minus sign relative to the corresponding right-handed conventional anti-commutation relation. The right-handed tachyon anti-commutation relation with its minus sign will require compensating minus signs in its creation and annihilation Fourier component operators' anti-commutation relations.

The sign differences between the lagrangian terms in eqs. 26.63 and 26.64 ultimately lead to parity violating features in the Standard Model lagrangian and thus resolve the long-standing question:

Why parity violation? Answer: Nature preferentially chooses the Left-handed part of the complex Lorentz group.. This choice is not a consequence of Ockham's Razor. But it does conform to Leibniz's Minimax Principle – a minor differentiation based on parity results in "maximal" physical consequences.

26.4.4.3 Left-Handed Tachyons

The free, "+" light-front, left-handed tachyon wave function Fourier expansion is:

$$\psi_{TL}{}^{+}(x) = \sum_{\pm s} \int d^{2}p dp^{+} N_{TL}{}^{+}(p)\theta(p^{+})[b_{TL}{}^{+}(p, s)u_{TL}{}^{+}(p, s)e^{-ip\cdot x} + d_{TL}{}^{+\dagger}(p, s)v_{TL}{}^{+}(p, s)e^{+ip\cdot x}]$$

$$(26.71)$$

and its hermitean conjugate is

$$\psi_{TL}{}^{+\dagger}(x) = \sum_{\pm s} \int d^{2}p dp^{+} N_{TL}{}^{+}(p)\theta(p^{+}) \, [b_{TL}{}^{+\dagger}(p, s)u_{TL}{}^{+\dagger}(p,s)e^{+ip\cdot x} + d_{TL}{}^{+}(p, s)v_{TL}{}^{+\dagger}(p, s)e^{-ip\cdot x}]$$

$$(26.72)$$

where † indicates hermitean conjugate, where

$$N_{TL}^{+}(p) = [2m|\mathbf{p}|/((2\pi)^3(p^+(p^+ - p^-) + p_{\perp}^{2}))]^{\frac{1}{2}} \qquad (26.73)$$

where the anti-commutation relations of the Fourier coefficient operators are

$$
\begin{aligned}
&\{b_{TL}^{+}(q,s),\, b_{TL}^{++}(p,s')\} = \delta_{ss'}\delta^2(\mathbf{q} - \mathbf{p})\delta(q^+ - p^+) \\
&\{d_{TL}^{+}(q,s),\, d_{TL}^{++}(p,s')\} = \delta_{ss'}\delta^2(\mathbf{q} - \mathbf{p})\delta(q^+ - p^+) \\
&\{b_{TL}^{+}(q,s),\, b_{TL}^{+}(p,s')\} = \{d_{TL}^{+}(q,s),\, d_{TL}^{+}(p,s')\} = 0 \\
&\{b_{TL}^{++}(q,s),\, b_{TL}^{++}(p,s')\} = \{d_{TL}^{++}(q,s),\, d_{TL}^{++}(p,s')\} = 0 \\
&\{b_{TL}^{+}(q,s),\, d_{TL}^{++}(p,s')\} = \{d_{TL}^{+}(q,s),\, b_{TL}^{++}(p,s')\} = 0 \\
&\{b_{TL}^{++}(q,s),\, d_{TL}^{++}(p,s')\} = \{d_{TL}^{+}(q,s),\, b_{TL}^{+}(p,s')\} = 0
\end{aligned}
\qquad (26.74)
$$

and where the spinors are

$$
\begin{aligned}
u_{TL}^{+}(p, s) &= C^- R^+ S_L(\Lambda_L(\mathbf{p}))w^1(0) \\
u_{TL}^{+}(p, -s) &= C^- R^+ S_L(\Lambda_L(\mathbf{p}))w^2(0) \\
v_{TL}^{+}(p, s) &= C^- R^+ S_L(\Lambda_L(\mathbf{p}))w^3(0) \\
v_{TL}^{+}(p, -s) &= C^- R^+ S_L(\Lambda_L(\mathbf{p}))w^4(0) \\
u_{TL}^{++}(p, s) &= w^{1T}(0)S_L^{\dagger}(\Lambda_L(\mathbf{p}))R^+C^- \\
u_{TL}^{++}(p, -s) &= w^{2T}(0)S_L^{\dagger}(\Lambda_L(\mathbf{p}))R^+C^- \\
v_{TL}^{++}(p, s) &= w^{3T}(0)S_L^{\dagger}(\Lambda_L(\mathbf{p}))R^+C^- \\
v_{TL}^{++}(p, -s) &= w^{4T}(0)S_L^{\dagger}(\Lambda_L(\mathbf{p}))R^+C^-
\end{aligned}
\qquad (26.75)
$$

where the superscript "T" indicates the transpose. (These spinors are described in section 26.12.)

The canonical left-handed, light-front anti-commutation relation results in:

$$\{\psi_{TL}^{+}{}_{a}(x),\, \psi_{TL}^{++}{}_{b}(y)\} = \sum_{\pm s,s'} \int d^2 p\, dp^+ \int d^2 p'\, dp'^+ \, N_{TL}^{+}(p)N_{TL}^{+}(p')\theta(p^+)\theta(p'^+) \cdot$$

$$\cdot[\{b_{TL}{}^{++}(p',s'), b_{TL}{}^{+}(p,s)\} u_{TL}{}^{+}{}_{a}(p,s) u_{TL}{}^{++}{}_{b}(p',s') e^{+ip'\cdot y - ip\cdot x} +$$
$$+ \{d_{TL}{}^{+}(p',s'), d_{TL}{}^{++}(p,s)\} v_{TL}{}^{+}{}_{a}(p,s) v_{TL}{}^{++}{}_{b}(p',s') e^{-ip'\cdot y + ip\cdot x}]$$

$$= \sum_{\pm s} \int d^2 p dp^+ N_{TL}{}^{+2}(p)\theta(p^+)[u_{TL}{}^{+}{}_{a}(p,s) u_{TL}{}^{+}{}^{\dagger}{}_{b}(p,s) e^{+ip\cdot(y-x)} +$$
$$+ v_{TL}{}^{+}{}_{a}(p,s) v_{TL}{}^{++}{}_{b}(p,s) e^{-ip\cdot(y-x)}]$$

$$= -i\int d^2 p dp^+ \theta(p^+) N_{TL}{}^{+2}(p)(2m|\mathbf{p}|)^{-1}\{[\,C^-R^+(i\not{p} - m)\gamma\cdot\mathbf{p}R^+C^-]_{ab} e^{+ip\cdot(y-x)} +$$
$$+ [C^-R^+(i\not{p} + m)\gamma\cdot\mathbf{p}R^+C^-]_{ab} e^{-ip\cdot(y-x)}\}$$

$$= -i\int d^2 p_\perp \int_0^\infty dp^+ N_{TL}{}^{+2}(p)\{[C^-R^+(ip^+(p^+ - p^-) + ip_\perp{}^2 - mp_\perp\cdot\gamma_\perp)C^-]_{ab} e^{+ip^+(y^- - x^-) - ip_\perp\cdot(y_\perp - x_\perp)} -$$
$$- [C^-R^+(-ip^+(p^+ - p^-) - ip_\perp{}^2 - mp_\perp\cdot\gamma_\perp)C^-]_{ab} e^{-ip^+(y^- - x^-) + ip_\perp\cdot(y_\perp - x_\perp)}\}/(2m|\mathbf{p}|)$$

$$= \int d^2 p_\perp \int_{-\infty}^\infty dp^+ N_{TL}{}^{+2}(p)[C^-R^+(p^+(p^+ - p^-) + p_\perp{}^2)]_{ab}\, e^{+ip^+(y^- - x^-) - ip_\perp\cdot(y_\perp - x_\perp)}/(2m|\mathbf{p}|)$$

upon letting $p^+ \to -p^+$ and $\mathbf{p}_\perp \to -\mathbf{p}_\perp$ in the second term after using $N_{TL}{}^{+2}(p)(p^+(p^+ - p^-) + p_\perp{}^2) = 1$. The result

$$= \tfrac{1}{2}\int d^2 p_\perp \int_{-\infty}^\infty dp^+ (2\pi)^{-3}[C^-R^+]_{ab} e^{+ip^+(y^- - x^-) - ip_\perp\cdot(y_\perp - x_\perp)}$$

$$= 2^{-1}[C^-R^+]_{ab}\,\delta(y^- - x^-)\delta^2(\mathbf{y} - \mathbf{x}) \tag{26.76}$$

Therefore we have left-handed, light-front quantized tachyons with canonical commutation relations and localized tachyons. As a result we have a canonical Tachyon Quantum Field Theory unlike previous efforts.

26.4.4.4 Right-Handed Tachyons

The case of right-handed tachyons is similar to the left-handed case with only two differences: a minus sign in the creation and annihilation operator anti-commutation relations, and the use of right-handed projection operators. The right-handed tachyon wave function light-front Fourier expansion is:

$$\psi_{TR}^{+}(x) = \sum_{\pm s} \int d^2pdp^+N_{TR}^{+}(p)\theta(p^+)[b_{TR}^{+}(p, s)u_{TR}^{+}(p, s)e^{-ip\cdot x} + d_{TR}^{++}(p, s)v_{TR}^{+}(p, s)e^{+ip\cdot x}]$$

(26.77)

and its hermitean conjugate is

$$\psi_{TR}^{++}(x) = \sum_{\pm s} \int d^2pdp^+N_{TR}^{+}(p)\theta(p^+) [b_{TR}^{++}(p, s)u_{TR}^{++}(p, s)e^{+ip\cdot x} + d_{TR}^{+}(p, s)v_{TR}^{++}(p, s)e^{-ip\cdot x}]$$

(26.78)

where $N_{TR}^{+}(p) = N_{TL}^{+}(p)$, where the anti-commutation relations of the Fourier coefficient operators are

$$\{b_{TR}^{+}(q,s), b_{TR}^{++}(p,s')\} = -\delta_{ss'}\delta^2(\mathbf{q} - \mathbf{p})\delta(q^+ - p^+) \qquad (26.79)$$
$$\{d_{TR}^{+}(q,s), d_{TR}^{++}(p,s')\} = -\delta_{ss'}\delta^2(\mathbf{q} - \mathbf{p})\delta(q^+ - p^+)$$
$$\{b_{TR}^{+}(q,s), b_{TR}^{+}(p,s')\} = \{d_{TR}^{+}(q,s), d_{TR}^{+}(p,s')\} = 0$$
$$\{b_{TR}^{++}(q,s), b_{TR}^{++}(p,s')\} = \{d_{TR}^{++}(q,s), d_{TR}^{++}(p,s')\} = 0$$
$$\{b_{TR}^{+}(q,s), d_{TR}^{++}(p,s')\} = \{d_{TR}^{+}(q,s), b_{TR}^{++}(p,s')\} = 0$$
$$\{b_{TR}^{++}(q,s), d_{TR}^{++}(p,s')\} = \{d_{TR}^{+}(q,s), b_{TR}^{+}(p,s')\} = 0$$

and where the spinors are

$$u_{TR}^{+}(p, s) = C^+R^+u_T(p,s) \qquad (26.80)$$
$$v_{TR}^{+}(p, s) = C^+R^+v_T(p,s) \qquad (26.81)$$

by section 26.12 (eq. 26.7).

The right-handed anti-commutation relation with the minus sign follows in particular because of the minus signs found earlier.

26.4.5 Interpretation of Tachyon Creation and Annihilation Operators

To properly discuss the physical interpretation of tachyon creation and annihilation operators we must first determine the Hamiltonian and momentum operators in terms of creation and annihilation operators.

The energy-momentum tensor density is the symmetrized version of

$$\mathfrak{I}^{\mu\nu} = \sum_i \partial \mathcal{L}/\partial(\partial\chi_i/\partial x_\mu) \, \partial\chi_i/\partial x_\nu - g^{\mu\nu}\mathcal{L} \tag{26.82}$$

where the sum over i is over the fields. The light-front hamiltonian is

$$H \equiv P^- = T^{+-} = \int dx^- d^2x \, \mathfrak{I}^{+-} \tag{26.83}$$

and the "momenta" are

$$P^+ = T^{++} = \int dx^- d^2x \, \mathfrak{I}^{++} \tag{26.84}$$

$$P^i = T^{+i} = \int dx^- d^2x \, \mathfrak{I}^{+i} \tag{26.85}$$

for i = 1,2.

The light-front, left-handed and right-handed tachyon lagrangian $\mathcal{L}_T$ and its equations of motion imply

$$H = i2^{-\frac{1}{2}}\int dx^- d^2x \; [\psi_{TL}^{++}\partial^-\psi_{TL}^+ - \partial^-\psi_{TL}^{++}\psi_{TL}^+ + \psi_{TL}^{-\dagger}\partial^+\psi_{TL}^- - \partial^+\psi_{TL}^{-\dagger}\psi_{TL}^- -$$
$$- \psi_{TR}^{++}\partial^-\psi_{TR}^+ + \partial^-\psi_{TR}^{++}\psi_{TR}^+ - \psi_{TR}^{-\dagger}\partial^+\psi_{TR}^- + \partial^+\psi_{TR}^{-\dagger}\psi_{TR}^- + \text{mass terms}] \tag{26.86}$$

After substituting for the various fields we find the *independent fields* (which create the in and out particle states) have the hamiltonian terms:

$$H = \sum_{\pm s} \int d^2 p dp^+ \, p^- [b_{TL}^{+\dagger}(p,s) b_{TL}^{+}(p,s) - d_{TL}^{+}(p,s) d_{TL}^{+\dagger}(p,s) - b_{TR}^{+\dagger}(p,s) b_{TR}^{+}(p,s) +$$
$$+ \, d_{TR}^{+}(p,s) d_{TR}^{+\dagger}(p,s)] \tag{26.87}$$

$$= \sum_{\pm s} \int d^2 p dp^+ \, p^- [b_{TL}^{+\dagger}(p,s) b_{TL}^{+}(p,s) + d_{TL}^{+\dagger}(p,s) d_{TL}^{+}(p,s) - b_{TR}^{+\dagger}(p,s) b_{TR}^{+}(p,s) -$$
$$- \, d_{TR}^{+\dagger}(p,s) d_{TR}^{+}(p,s)] \tag{26.88}$$

up to the usual infinite constants due to left-handed operator rearrangement and right-handed operator rearrangement that are discarded. Eq. 26.88 is the basis for our particle interpretation of tachyon creation and annihilation operators based on Dirac's hole theory. Dirac hole theory as applied in light-front coordinates assumes all negative p^- ("energy") states are filled.

26.4.5.1 Left-Handed Tachyon Creation and Annihilation Operators

1. We identify $b_{TL}^{+\dagger}(p,s)$ and $d_{TL}^{+}(p,s)$ as creation operators for left-handed tachyons. $b_{TL}^{+\dagger}(p,s)$ creates a positive p^- ("energy") state and $d_{TL}^{+}(p,s)$ creates a negative p^- ("energy") state.

2. $b_{TL}^{+}(p,s)$ and $d_{TL}^{+\dagger}(p,s)$ are the corresponding annihilation operators for left-handed tachyons. $b_{TL}^{+}(p,s)$ annihilates a positive p^- ("energy") state and $d_{TL}^{+\dagger}(p,s)$ annihilates a negative p^- ("energy") state.

3. We assume Dirac hole theory holds for the left-handed tachyon vacuum with all negative energy states filled. There is no tachyon energy gap as there is for Dirac fermions. There is also the problem that the left-handed tachyon vacuum is not invariant under ordinary Lorentz transformations or Superluminal transformations. *However if we confine ourselves to light-front coordinates for computations no ambiguity can result and the Lorentz covariant quantities that we calculate, such as the S matrix, are well-defined.*

4. Using tachyon hole theory we identify $b_{TL}{}^+(p,s)$ and $d_{TL}{}^{++}(p,s)$ as annihilation operators for left-handed tachyons. $b_{TL}{}^+(p,s)$ annihilates a positive p^- ("energy") state and $d_{TL}{}^{++}(p,s)$ annihilates a negative p^- ("energy") state – thus creating a hole in the tachyon sea that we view as the creation of a positive p^- ("energy"), left-handed antitachyon. $d_{TL}{}^+(p,s)$ annihilates a positive p^- ("energy"), left-handed antitachyon.

26.4.5.2 Right-Handed Tachyon Creation and Annihilation Operators

The anti-commutation relations of right-handed tachyon creation and annihilation operators and the right-handed Hamiltonian terms have the "wrong" sign compared to corresponding Dirac operators and left-handed tachyon operators. This situation is completely analogous to the situation of time-like photons in the covariant formulation of quantum Electrodynamics.[133] In the case of time-like photons it was possible to introduce an indefinite metric (Gupta-Bleuler formulation), and then to use the subsidary condition $\partial A^\nu / \partial x^\nu = 0$ to reduce the dynamics of QED to the transverse components. Thus the time-like photons were intermediate artifacts needed to have a manifestly covariant formulation while QED observables depended solely on the transverse components of the electromagnetic field.

In the present case of free tachyons, and in leptonic ElectroWeak Theory there is no evident "subsidary condition" to eliminate the right-handed tachyon fields. But since the only manner in which the right-handed leptonic tachyon fields[134] interact is through mass terms, which can be easily 'integrated out", right-handed leptonic tachyon fields are removed from the observable part of the leptonic ElectroWeak Theory by their "lack of interaction" with left-handed fields.

In the case of quark ElectroWeak Theory right-handed tachyon quark fields have charge ($-1/3$) and thus experience an electromagnetic interaction as well as a Z interaction. However, since quarks are totally confined, right-handed tachyon quarks will not be able to continuously emit photons or Z's due to energy conservation and

---

[133] Bogoliubov (1959) pp. 130-136.
[134] The tachyon fields are provisionally assumed to be neutrino fields in the leptonic sector, and d, s and b quarks in the quark sector.

their confinement to bound states of fixed positive energy. Earlier, when we consider complex Lorentz group boosts, we will suggest that quarks may not consist of Dirac particles or tachyons of the type considered up to this point in this appendix. Rather they may be variants on Dirac particles and tachyons satisfying different dynamical equations. However, the preceding comments on quarks would still apply.

Thus right-handed tachyons are analogous to time-like photons – necessary theoretically but prevented from causing a negative energy disaster by the forms of their interactions. We discuss this subject in more detail in the following appendices.

26.4.6 Tachyon Feynman Propagator

In this section we develop the light-front propagator for tachyons. We begin with a subsection describing the light-front propagators of Dirac fields.

26.4.6.1 Dirac Field Light-Front Propagators

The light-front Feynman propagator for the ψ^+ field of a Dirac fermion is

$$iS^+{}_F(x,y)\gamma^0 = \theta(x^+ - y^+)<0|\psi^+(x)\psi^{+\dagger}(y)|0> - \theta(y^+ - x^+)<0|\psi^{+\dagger}(y)\psi^+(x)|0> \quad (26.89)$$

and does not contain a non-covariant piece due to the projection operators:

$$iS^+{}_F(x,y) = \int d^2pdp^+\theta(p^+)[1/(2(2\pi)^3p^+)]\{\theta(x^+ - y^+)[R^+(\not{p} +m)R^-] \, e^{-ip\cdot(x-y)} +$$
$$+ \theta(y^+ - x^+)[R^+(-\not{p}+m)R^-]e^{+ip\cdot(x-y)}\}$$
$$= R^+iS_F(x,y)R^- \quad (26.90)$$

where $S_F(x,y)$ is the usual Feynman propagator.

The light-front Feynman propagator for a *left-handed* <u>Dirac</u> field ψ^+ is

$$iS^+{}_{LF}(x,y) = \int d^2pdp^+\theta(p^+)[1/(2(2\pi)^3p^+)]\{\theta(x^+ - y^+)[C^-R^+(\not{p}+m)R^-C^-]e^{-ip\cdot(x-y)} +$$
$$+ \theta(y^+ - x^+)[C^-R^+(-\not{p} + m)R^-C^-]e^{+ip\cdot(x-y)}\}$$

$$= C^-R^+iS_F(x,y)R^-C^- \quad (26.91)$$

26.4.6.2 Tachyon Field Light Front Propagators

Turning now to tachyons, the light-front Feynman propagator for the left-handed $\psi_{TL}{}^{+}$ *tachyon* field is (using the previous Fourier expansion of the left-handed tachyon field):

$$iS^{+}{}_{TLF}(x,y) = \theta(x^{+} - y^{+})<0|\psi_{TL}{}^{+}(x)\psi_{TL}{}^{++}(y)\gamma^{0}|0> - \theta(y^{+} - x^{+})<0|\psi_{TL}{}^{++}(y)\gamma^{0}\psi_{TL}{}^{+}(x)|0>$$
$$= -i\int d^{2}pdp^{+}\theta(p^{+})N_{TL}{}^{+2}(2m|\mathbf{p}|)^{-1}C^{-}R^{+}\{\theta(x^{+} - y^{+})[(i\not{p} - m)\gamma\cdot\mathbf{p}]e^{-ip\cdot(x-y)} +$$
$$+ \theta(y^{+} - x^{+})[(i\not{p} + m)\gamma\cdot\mathbf{p}]e^{+ip\cdot(x-y)}\}R^{+}C^{-}\gamma^{0}$$

If we define the on-shell momentum variable

$$p_{0}{}^{-} = (p_{0}{}^{1}p_{0}{}^{1} + p_{0}{}^{2}p_{0}{}^{2} - m^{2})/(2p_{0}{}^{+}), \; p_{0}{}^{+} = p^{+}, \; p_{0}{}^{j} = p^{j} \; (\text{for } j = 1, 2), \; p_{\perp 0}{}^{2} = p_{0}{}^{j}p_{0}{}^{j}$$

and

$$\not{p}_{0} = p_{0}\cdot\gamma$$

then the above equation can be rewritten as

$$S^{+}{}_{TLF}(x,y) = -C^{-}R^{+}\int d^{4}p[32\pi^{4}(p_{0}{}^{+}(p_{0}{}^{+} - p_{0}{}^{-}) + p_{0\perp}{}^{2})]^{-1}e^{-ip\cdot(x-y)}\{\theta(p^{+})(i\not{p}_{0} - m)\gamma\cdot\mathbf{p}_{0}]/[p^{-} -$$
$$- p_{0}{}^{-} + i\varepsilon] + \theta(-p^{+})(i\not{p}_{0} + m)\gamma\cdot\mathbf{p}_{0}]/[p^{-} + p_{0}{}^{-} - i\varepsilon]\}R^{+}C^{-}\gamma^{0}$$
$$= -\tfrac{1}{2} i\int d^{4}p(2\pi)^{-4}[C^{-}R^{+}(i\not{p} - m)\gamma\cdot\mathbf{p}R^{+}C^{-}\gamma^{0}]e^{-ip\cdot(x-y)}[(p^{2} + m^{2} + i\varepsilon)(p^{+}(p^{+} - p^{-}) + p_{\perp}{}^{2}))]^{-1}$$

and using $C^{-}R^{+}(i\not{p} - m)\gamma\cdot\mathbf{p}R^{+}C^{-} = i\, C^{-}R^{+}(p^{+}(p^{+} - p^{-}) + p_{\perp}{}^{2})$ we find

$$iS^{+}{}_{TLF}(x,y) = \tfrac{1}{2}C^{-}R^{+}\gamma^{0}\int d^{4}p(2\pi)^{-4} p^{+}e^{-ip\cdot(x-y)}/(p^{2} + m^{2} + i\varepsilon) \qquad (26.92)$$

Similarly the light-front Feynman propagator for the right-handed $\psi_{TR}{}^{+}$ tachyon field is

$$iS^{+}{}_{TRF}(x,y) = \theta(x^{+} - y^{+})<0|\psi_{TR}{}^{+}(x)\psi_{TR}{}^{++}(y)\gamma^{0}|0> - \theta(y^{+} - x^{+})<0|\psi_{TR}{}^{++}(y)\gamma^{0}\psi_{TR}{}^{+}(x)|0>$$

$$= -\tfrac{1}{2} C^+ R^+ \gamma^0 \int d^4 p (2\pi)^{-4} \, p^+ e^{-ip \cdot (x-y)} / (p^2 + m^2 + i\varepsilon) \qquad (26.93)$$

where the relative minus sign between eqs. 26.92 and 26.93 is due to the relative minus signs of the Fouier component operator anti-commutation relations.

Thus we find *tachyon* pole terms in the tachyon propagators as one would expect.

26.5 Complex Space and 3-Momentum & Real-Valued Energy Fermions (Quarks)

In this section we will use L_C boosts to develop a wider set of dynamical equations for free spin ½ fermions with real-valued energy and complex-valued 3-momentum.[135] We defined L_C boosts with

$$\Lambda_C(\mathbf{v_c}) = \exp[i\omega\hat{\mathbf{w}} \cdot \mathbf{K}] \qquad (26.94)$$

$$\omega = (\omega_r^2 - \omega_i^2 + 2i\omega_r\omega_i \, \hat{\mathbf{u}}_r \cdot \hat{\mathbf{u}}_i)^{\frac{1}{2}} \qquad (26.95)$$

$$\hat{\mathbf{w}} = (\omega_r\hat{\mathbf{u}}_r + i\omega_i\hat{\mathbf{u}}_i)/\omega \qquad (26.96)$$

$$\hat{\mathbf{w}} \cdot \hat{\mathbf{w}} = \hat{\mathbf{u}}_r \cdot \hat{\mathbf{u}}_r = \hat{\mathbf{u}}_i \cdot \hat{\mathbf{u}}_i = 1 \qquad (26.97)$$

$$\mathbf{v_c} = \hat{\mathbf{w}} \tanh(\omega) \qquad (26.98)$$

26.5.1 L_C Spinor "Normal" Lorentz Boosts & More Spin ½ Particle Types

Spinor boost transformations were used in previous sections to develop the dynamical equations for Dirac fields and tachyon fields. In this section we will use L_C spinor boosts to generate additional fermion field dynamical equations.

The form of the L_C spinor boost transformation corresponding to the coordinate transformation is:

$$S_C(\omega, \mathbf{v_c}) = \exp(-i\omega\sigma_{0k}\hat{w}_k/2) = \exp(-\omega\gamma^0\boldsymbol{\gamma} \cdot \hat{\mathbf{w}}/2)$$
$$= \cosh(\omega/2)I + \sinh(\omega/2)\gamma^0\boldsymbol{\gamma} \cdot \hat{\mathbf{w}} \qquad (26.99)$$

---

[135] The complexon theory that we develop and use for quark dynamics in the Standard Model is <u>not</u> required. Our Standard Model could use Dirac fermion dynamics for the up-type quarks and tachyon dynamics for down-type quarks. We choose to use complexon dynamics for all quark types because they have an internal SU(3)-like structure suggestive of color SU(3). More importantly, their spin dynamics is different and thus may resolve the differences between theory and experiment – particularly for the deep inelastic parton spin-dependent structure functions.

The inverse transformation is

$$S_C^{-1}(\omega, \mathbf{v_c}) = \gamma^2\gamma^0 K^{-1}S_C^\dagger K\gamma^0\gamma^2 = \gamma^2\gamma^0 S_C^{\ T}\gamma^0\gamma^2 = \exp(\omega\gamma^0\boldsymbol{\gamma}\cdot\hat{\mathbf{w}}/2)$$
$$= \cosh(\omega/2)I - \sinh(\omega/2)\gamma^0\boldsymbol{\gamma}\cdot\hat{\mathbf{w}} \qquad (26.100)$$

where the superscript T denotes the transpose and K is the complex conjugation operator (that also appears in the time-reversal operator). Note that S_C is not unitary just as in previous cases considered in this appendix.

We now redo the development of spin ½ dynamical equations of motion of earlier sections for this more general case of complex ω and $\hat{\mathbf{w}}$. Again we apply a boost to a Dirac equation for a positive energy plane wave particle of mass m at rest:

$$0 = S_C(\omega, \mathbf{v_c}))(m\gamma^0 - m)e^{-imt}w(0)$$
$$= [mS_C\gamma^0 S_C^{-1} - m]e^{-imt}S_C w(0) \qquad (26.101)$$

where $S_C = S_C(\omega, \hat{\mathbf{w}})$. After some algebra

$$mS_C\gamma^0 S_C^{-1} = m[\cosh(\omega)\gamma^0 - \sinh(\omega)\boldsymbol{\gamma}\cdot\hat{\mathbf{w}}] \qquad (26.102)$$

26.5.1.1 Case 1: Parallel Real and Imaginary Relative Vectors

If the real and imaginary relative vectors parts of $\hat{\mathbf{w}}$, namely $\hat{\mathbf{u}}_r$ and $\hat{\mathbf{u}}_i$, are parallel, then $\hat{\mathbf{u}}_r\cdot\hat{\mathbf{u}}_i = 1$ and

$$\omega = \omega_r + i\omega_i \qquad (26.103)$$

Eq. 26.102 can be re-expressed as

$$mS_C\gamma^0 S_C^{-1} = m[\cosh(\omega_r)\cos(\omega_i) + i\sinh(\omega_r)\sin(\omega_i)]\gamma^0 - m[\sinh(\omega_r)\cos(\omega_i) +$$
$$+ i\cosh(\omega_r)\sin(\omega_i)]\boldsymbol{\gamma}\cdot\hat{\mathbf{u}}_r \qquad (26.104)$$

or equivalently

$$mS_C\gamma^0 S_C^{-1} = \cos(\omega_i)\boldsymbol{\gamma}\cdot p_r + i\sin(\omega_i)\boldsymbol{\gamma}\cdot p_i \qquad (26.105)$$

where

$$p_r{}^0 = m\cosh(\omega_r) \qquad\qquad p_i{}^0 = m\sinh(\omega_r) \qquad (26.106)$$

and

$$\mathbf{p_r} = m\hat{\mathbf{u}}_r\sinh(\omega_r) \qquad\qquad \mathbf{p_i} = m\hat{\mathbf{u}}_r\cosh(\omega_r) \qquad (26.107)$$

If $\omega_i = 0$, then we recover the momentum space Dirac equation. If $\omega_i = \pi/2$, then we obtain the left-handed momentum space tachyon equation. Since the range of ω_i is [0, ∞> (due to the cut along the real ω-plane axis) eq. 26.105 corresponds to the results of the Left-Handed Lorentz boost part discussed earlier.

26.5.1.2 Case 2: Anti-Parallel Real and Imaginary Relative Vectors

If the real and imaginary relative vectors parts of $\hat{\mathbf{w}}$, $\hat{\mathbf{u}}_r$ and $\hat{\mathbf{u}}_i$, are anti-parallel $\hat{\mathbf{u}}_r = -\hat{\mathbf{u}}_i$, then $\hat{\mathbf{u}}_r \cdot \hat{\mathbf{u}}_i = -1$ and

$$\omega = \omega_r - i\omega_i \qquad (26.108)$$

We can then express eq. 26.105 as

$$mS_C\gamma^0 S_C{}^{-1} = m[\cosh(\omega_r)\cos(\omega_i) - i\sinh(\omega_r)\sin(\omega_i)]\gamma^0 - m[\sinh(\omega_r)\cos(\omega_i) - i\cosh(\omega_r)\sin(\omega_i)]\gamma\cdot\hat{\mathbf{u}}_r \qquad (26.109)$$

or

$$mS_C\gamma^0 S_C{}^{-1} = \cos(\omega_i)\gamma\cdot p_r - i\sin(\omega_i)\gamma\cdot p_i \qquad (26.110)$$

where

$$p_r{}^0 = m\cosh(\omega_r) \qquad\qquad p_i{}^0 = m\sinh(\omega_r) \qquad (26.111)$$

and

$$\mathbf{p_r} = m\hat{\mathbf{u}}_r\sinh(\omega_r) \qquad\qquad \mathbf{p_i} = m\hat{\mathbf{u}}_r\cosh(\omega_r) \qquad (26.112)$$

If $\omega_i = 0$, then we again recover the momentum space Dirac equation, If $\omega_i = \pi/2$, then we obtain the right-handed momentum space tachyon equation. (The range of ω_i is again [0, ∞>.)

Note: Since the matrix elements in the boost depend on $\gamma = (1 - \beta^2)^{-\frac{1}{2}}$ with a singularities at $\beta = \pm1$, which in turn corresponds to $\omega = \pm\infty$, there is a branch cut along the ω axis in the complex ω-plane. Therefore we point out again the product of three Left-handed transformations is not equivalent to a Right-handed transformation.

26.5.1.3 Case 3: Complexons: A New Type of Particle with Perpendicular Real and Imaginary 3-Momenta

   If the real and imaginary relative vectors parts of $\hat{\mathbf{w}}$, namely $\hat{\mathbf{u}}_r$ and $\hat{\mathbf{u}}_i$, are perpendicular, $\hat{\mathbf{u}}_r \cdot \hat{\mathbf{u}}_i = 0$, then

$$\omega = (\omega_r^2 - \omega_i^2)^{\frac{1}{2}} \qquad (26.113)$$

Thus ω is either pure real ($\omega_r \geq \omega_i$) or pure imaginary ($\omega_r < \omega_i$).

   The momentum space equation generated by the corresponding L_C spinor boost is

$$\{m \cosh(\omega)\gamma^0 - m \sinh(\omega)\gamma \cdot (\omega_r \hat{\mathbf{u}}_r + i\omega_i \hat{\mathbf{u}}_i)/\omega - m\} e^{-ip\cdot x} w_c(p) = 0 \qquad (26.114)$$

Defining the momentum 4-vector

$$p = (p^0, \mathbf{p}) \qquad (26.115)$$

where

$$p^0 = m \cosh(\omega) \qquad\qquad \mathbf{p} = \mathbf{p}_r + i\mathbf{p}_i \qquad (26.116)$$
$$\mathbf{p}_r = m\omega_r \hat{\mathbf{u}}_r \sinh(\omega)/\omega \qquad \mathbf{p}_i = m\omega_i \hat{\mathbf{u}}_i \sinh(\omega)/\omega \qquad (26.117)$$

and

$$\mathbf{p}_r \cdot \mathbf{p}_i = 0 \qquad (26.118)$$

then we obtain a positive energy Dirac-like equation with complex 3-momentum

or, explicitly,
$$[p\cdot\gamma - m]e^{-ip\cdot x} w_c(p) = 0$$
$$[p^0\gamma^0 - (\mathbf{p}_r + i\mathbf{p}_i)\cdot\gamma - m]e^{-ip\cdot x} w_c(p) = 0 \qquad (26.119)$$

with a complex 3-momentum $\mathbf{p}$ and the 4-momentum mass shell condition:

$$p^2 = p^{0\,2} - \mathbf{p}_r \cdot \mathbf{p}_r + \mathbf{p}_i \cdot \mathbf{p}_i = m^2 \qquad (26.120)$$

Note

$$|\mathbf{v}| = |\mathbf{p}|/p^0 = [(\mathbf{p}_r + i\mathbf{p}_i)\cdot(\mathbf{p}_r + i\mathbf{p}_i)]^{\frac{1}{2}}/p^0 = \tanh(\omega) \qquad (26.121)$$

and thus the Lorentz factor

$$\gamma = \cosh(\omega) \qquad (26.122)$$

Eq. 26.119 is the momentum space equivalent of the wave equation

$$[i\gamma^0 \partial/\partial t + i\gamma\cdot(\nabla_r + i\nabla_i) - m]\psi_C(t, \mathbf{x_r}, \mathbf{x_i}) = 0 \qquad (26.123)$$

where

$$x_c = (t, \mathbf{x_r} - i\mathbf{x_i}) \qquad (26.123a)$$

and where the grad operators ∇_r and ∇_i are with respect to $\mathbf{x_r}$ and $\mathbf{x_i}$ respectively. Since $\mathbf{\hat{u}_r}\cdot\mathbf{\hat{u}_i} = 0$, we see that there is a subsidiary condition on the wave function

$$\nabla_r\cdot\nabla_i\ \psi_C(t, \mathbf{x_r}, \mathbf{x_i}) = 0 \qquad (26.124)$$

We will call the particles satisfying eqs. 26.123 and 26.124 *complexons*. In addition eq. 26.118 implies the anti-commutation relation

$$\{\gamma\cdot\mathbf{p_r}, \gamma\cdot\mathbf{p_i}\} = 0 \qquad (26.125)$$

which in turn implies

$$\gamma\cdot\nabla_r\gamma\cdot\nabla_i\psi_C(t, \mathbf{x_r}, \mathbf{x_i}) = \gamma\cdot\nabla_i\gamma\cdot\nabla_r\psi_C(t, \mathbf{x_r}, \mathbf{x_i}) = 0 \qquad (26.126)$$

We note that eq. 26.125 is covariant under the real Lorentz group and eq. 26.126 can be easily put into covariant form since the difference of these 4-vectors squared is a real Lorentz group invariant: $[\gamma^0 \partial/\partial t + \gamma\cdot(\nabla_r + i\nabla_i)]^2 - [\gamma^0 \partial/\partial t + i\gamma\cdot(\nabla_r - i\nabla_i)]^2 = 4\nabla_r\cdot\nabla_i$.

Before considering a lagrangian formulation and the Fourier operator representation of $\psi_C(t, \mathbf{x_r}, \mathbf{x_i})$ we will define the spinors and associated real and imaginary spin operators.

The spinor generated from a spin up Dirac spinor at rest by a complex boost is

$$w_c(p) = S_C(p)w(0) = [\cosh(\omega/2)I + \sinh(\omega/2)\gamma^0\boldsymbol{\gamma}\cdot\hat{\mathbf{w}}]w(0) \qquad (26.127)$$

Following a procedure similar to section 26.12 (which the reader may wish to examine first) we define four spinors for Dirac particles at rest:

$$w^k(0) = \begin{bmatrix} \delta_{1k} \\ \delta_{2k} \\ \delta_{3k} \\ \delta_{4k} \end{bmatrix} \qquad (26.282)$$

where Kronecker deltas appear in the brackets. Then by applying eq. 26.127 to the spinors defined by eq. 26.282 we find the L_C spinors

$$S_C w^k(0) = w_{Cr}^k(p) + i w_{Ci}^k(p) \qquad (26.128)$$

where

$$
\begin{aligned}
S_{Cr} &= \cosh(\omega/2)I + (\omega_r/\omega)\sinh(\omega/2)\gamma^0\boldsymbol{\gamma}\cdot\hat{\mathbf{u}}_r \\
&= [(m + E)/(2m)]^{1/2}I + [m(m + E)]^{-1/2}\gamma^0\boldsymbol{\gamma}\cdot\mathbf{p}_r = aI + b\gamma^0\boldsymbol{\gamma}\cdot\mathbf{p}_r \qquad (26.129)
\end{aligned}
$$

Thus the "real" spinors $w_{Cr}^k(p)$ are the columns of

$$
S_{Cr} = \begin{array}{cccc}
\underline{w_{Cr}^1(p)} & \underline{w_{Cr}^2(p)} & \underline{w_{Cr}^3(p)} & \underline{w_{Cr}^4(p)} \\
\end{array}
$$

$$
S_{Cr} = \begin{bmatrix}
a & 0 & bp_{r\,z} & bp_{r-} \\
0 & a & bp_{r+} & -bp_{r\,z} \\
bp_{r\,z} & bp_{r-} & a & 0 \\
bp_{r+} & -bp_{r\,z} & 0 & a
\end{bmatrix} \qquad (26.130)
$$

where $p_{r\pm} = p_{r\,x} \pm i p_{r\,y}$. The "imaginary" spinors are the columns of

$$S_{Ci} = (\omega_i/\omega)\sinh(\omega/2)\gamma^0\boldsymbol{\gamma}\cdot\hat{\mathbf{u}}_i = [m(m + E)]^{-1/2}\gamma^0\boldsymbol{\gamma}\cdot\mathbf{p}_i = b\gamma^0\boldsymbol{\gamma}\cdot\mathbf{p}_i \qquad (26.131)$$

$$S_{Ci} = \begin{array}{cccc} \underline{w_{Ci}{}^1(p)} & \underline{w_{Ci}{}^2(p)} & \underline{w_{Ci}{}^3(p)} & \underline{w_{Ci}{}^4(p)} \\ \begin{bmatrix} 0 & 0 & bp_{i\,z} & bp_{i-} \\ 0 & 0 & bp_{i+} & -bp_{i\,z} \\ bp_{i\,z} & bp_{i-} & 0 & 0 \\ bp_{i+} & -bp_{i\,z} & 0 & 0 \end{bmatrix} \end{array} \tag{26.132}$$

where $p_{i\,\pm} = p_{i\,x} \pm ip_{i\,y}$.

Eqs. 26.127 through 26.132 imply that the wave function solution of eq. 26.123, subject to the subsidiary condition eq. 26.124, is[136, 137]

$$\psi_C(x_r, x_i) = \sum_{\pm s} \int d^3p_r d^3p_i \, N_C(p) \delta(\mathbf{p_r} \cdot \mathbf{p_i}/m^2) [b_C(p,s) u_C(p,\,s) e^{-i(p \cdot x + p^* \cdot x^*)/2} + \\ + \, d_C{}^\dagger(p,s) v_C(p,\,s) e^{+i(p \cdot x + p^* \cdot x^*)/2}] \tag{26.133}$$

where $\mathbf{p} = \mathbf{p_r} + i\mathbf{p_i}$ (eq. 3.95), $x = x_r - ix_i$, $p \cdot x = p^0 x^0 - \mathbf{p} \cdot \mathbf{x}$, and where we use

$$(p \cdot x + p^* \cdot x^*)/2 = p^0 x^0 - \mathbf{p_r} \cdot \mathbf{x_r} - \mathbf{p_i} \cdot \mathbf{x_i} \tag{26.134}$$

in the exponentials in order to avoid divergences that would appear in the calculation of the equal-time commutator, the Feynman propagator and other quantities of interest after second quantization. Note that

$$(\nabla_r + i\nabla_i)e^{-i(p \cdot x + p^* \cdot x^*)/2} = i(\mathbf{p_r} + i\mathbf{p_i})e^{-i(p \cdot x + p^* \cdot x^*)/2} \tag{26.135}$$

---

[136] Note that when $|\mathbf{p_i}| \geq |\mathbf{p_r}|$ (for imaginary $\omega = (\omega_r{}^2 - \omega_i{}^2)^{\frac{1}{2}}$) the 3-momentum becomes imaginary $\mathbf{p} \cdot \mathbf{p} < 0$. However, since we will be identifying confined quarks with this type of particle – much modified by a confining color quark interaction – the issue of an imaginary 3-momentum in the hypothetical free quark case becomes moot. We note the energy gap between positive and negative energy states disappears so $E = 0$ is possible. Thus real Lorentz transformations can mix positive and negative energy states. The solution is to do all calculations in the light-front frame as we do for tachyons. Then the mixing issue is resolved. In the present case we second quantize on the "time-front" for illustrative purposes.

[137] We scale $\mathbf{p_r} \cdot \mathbf{p_i}$ with m^2 in the delta function for convenience. All fermions have at least a minimal mass – the mass of the qube.

and

$$(\nabla_{\mathbf{r}} + i\nabla_{\mathbf{i}})e^{-ip^*\cdot x^*} = 0 \tag{26.136}$$

for all p.

The wave function's conjugate (the hermitean conjugate modified by letting $x_i \rightarrow -x_i$ in addition to hermitean conjugation) is

$$\psi_C{}^\dagger(x) = \psi_C{}^\dagger(x_{\mathbf{r}}, -x_{\mathbf{i}}) = \underset{\pm s}{\Sigma} \int d^3p_r d^3p_i\, \delta(\mathbf{p_r}\cdot\mathbf{p_i}/m^2)N_C(p^*)\cdot$$
$$\cdot[b_C{}^\dagger(p^*,s)u_C{}^\dagger(p^*,s)e^{+i(p\cdot x^* + p^*\cdot x)/2} + d_C(p^*,s)v_C{}^\dagger(p^*,s)e^{-i(p\cdot x^* + p^*\cdot x)/2}] \tag{26.137}$$

where $\mathbf{p} = \mathbf{p_r} + i\mathbf{p_i}$, $\mathbf{x} = \mathbf{x_r} - i\mathbf{x_i}$, $p\cdot x = p^0 x^0 - \mathbf{p}\cdot\mathbf{x}$, and † indicates hermitean hermitean conjugation.

The spinors are

$$u_C(p, s) = S_C(p)w^1(0)$$
$$u_C(p, -s) = S_C(p)w^2(0)$$
$$v_C(p, s) = S_C(p)w^3(0)$$
$$v_C(p, -s) = S_C(p)w^4(0) \tag{26.138}$$
$$u_C{}^\dagger(p^*, s) = w^{1T}(0)S_C{}^\dagger(p^*) = w^{1T}(0)S_C(p)$$
$$u_C{}^\dagger(p^*, -s) = w^{2T}(0)S_C{}^\dagger(p^*) = w^{2T}(0)S_C(p)$$
$$v_C{}^\dagger(p^*, s) = w^{3T}(0)S_C{}^\dagger(p^*) = w^{3T}(0)S_C(p)$$
$$v_C{}^\dagger(p^*, -s) = w^{4T}(0)S_C{}^\dagger(p^*) = w^{4T}(0)S_C(p)$$

with the superscript "T" indicating the transpose. Note that

$$S_C{}^\dagger(p^*) = [S_C(p^*)]^\dagger = S_C(p) \tag{26.139}$$

The normalization factor $N_C(p)$ is

$$N_C(p) = [2m/((2\pi)^6 p^0)]^{\frac{1}{2}} \tag{26.140}$$

Since $\mathbf{p_r} = \mathbf{p_i} = 0$ in the particle rest frame prior to the complex group boost, the boosted particle spin 4-vector s^μ satisfies

$$s^\mu p_r{}^\mu = s^\mu p_i{}^\mu = 0 \tag{26.141}$$

Note that s^μ is itself complex[138] and, if the spin points in the z-direction prior to the complex boost, then the boosted s^μ has the form

$$s^\mu = (-\sinh(\omega)\hat{w}_z, (0,0,1) + (\cosh(\omega) - 1)\hat{w}_z\hat{\mathbf{w}}) \tag{26.142}$$

with $\hat{\mathbf{w}}$ defined earlier: $\hat{\mathbf{w}} = (\omega_r\hat{\mathbf{u}}_r + i\omega_i\hat{\mathbf{u}}_i)/\omega = \mathbf{p}/(m\sinh(\omega))$.

26.5.1.4 A Global SU(3) Symmetry Revealed

Before proceeding to consider the second quantization of this case, we will consider a global SU(3) symmetry implicit in the previous equations. The defining property of the group SU(3) is that it preserves the invariance of inner products of complex 3-vectors of the form:

$$u^*\cdot v = u^1{}^*v^1 + u^2{}^*v^2 + u^3{}^*v^3 \tag{26.143}$$

If we examine the dynamical equation eq. 26.123 we see that the differential operator is invariant under an SU(3) transformation U (using $\nabla_c = (\nabla_c{}^*)^* = \mathbf{D}_c{}^*$)

$$[i\gamma^0\partial/\partial t + i\mathbf{D}_c{}^*\cdot\boldsymbol{\gamma} - m] = [i\gamma^0\partial/\partial t + i\mathbf{D}_c{}'{}^*\cdot\boldsymbol{\gamma}' - m] \tag{26.144}$$

where

$$\mathbf{D}_c{}^* = \nabla_c = \nabla_r + i\nabla_i$$

and

$$\gamma'^a = U^{ab}\gamma'^b$$
$$D_c'^{*a} = D_c'^{*b}U^{\dagger ab}$$

---

[138] This feature of partons, which is not present in ordinary Dirac particles, might be the source of the discrepancies between theory and experiment in deep inelstic parton spin physics which is based on conventional real parton spins.

where U is a global SU(3) transformation and $U^\dagger = U^{-1}$. By theorem[139] all 4×4 γ matrices such as γ' are equivalent up to a unitary transformation V. Thus $V^\dagger\gamma'V = \gamma$ and eq. 26.144 is equivalent to

$$[i\gamma^0\partial/\partial t + i\mathbf{D_c}*\cdot\boldsymbol{\gamma} - m] = [i\gamma^0\partial/\partial t + i\mathbf{D_c'}*\cdot\boldsymbol{\gamma} - m] \quad (26.145)$$

$$= [i\gamma^0\partial/\partial t + i\nabla_\mathbf{c}'\cdot\boldsymbol{\gamma} - m]$$

where $\nabla_{\mathbf{c'}_a} = U^{ab}\nabla_{\mathbf{cb}}$, This demonstrates that eq. 26.123 is invariant under an SU(3) transformation if

$$\psi_C(t, \mathbf{x_c}) = \psi_C(t, U\mathbf{x_c}) = \psi_C'(t, \mathbf{x_c'}) \quad (26.146)$$

where $\psi_C(t, \mathbf{x_c}) \equiv \psi_C(t, \mathbf{x_r}, \mathbf{x_i})$.

The subsidiary condition eq. 26.124 can be seen to transform as

$$\nabla_\mathbf{r}\cdot\nabla_\mathbf{i}\ \psi_C(t, \mathbf{x_c}) = \nabla_\mathbf{r}*\cdot\nabla_\mathbf{i}\ \psi_C(t, \mathbf{x_c}) = \nabla_\mathbf{r}'*\cdot\nabla_\mathbf{i}'\psi_C'(t, \mathbf{x_c'}) = 0 \quad (26.147)$$

under an SU(3) rotation. The invariance of the orthogonality condition is preserved.

The wave function (eq. 26.123) transforms in the following way under the SU(3) transformation U. If we define

$$q^{*\mu} = (q^0, \mathbf{q}*) = (p^0, \mathbf{p_r} + i\mathbf{p_i}) = (p^0, \mathbf{p}) = p^\mu \quad (26.148)$$

then eq. 26.133 can be rewritten in an invariant form under a SU(3) transformation:

$$\psi_C(x) = \sum_{\pm s} \int d^3q_r d^3q_i\ N_C(p^0)\delta(\mathbf{q_r}*\cdot\mathbf{q_i}/m^2)[b_C(q*,s)u_C(q*,s)e^{-i(q*\cdot x + q\cdot x*)/2} +$$
$$+ d_C^\dagger(q*,s)v_C(q*,s)e^{+i(q*\cdot x + q\cdot x*)/2}] \quad (26.149)$$

[139] R.H. Good, Rev. Mod. Phys., **27**, 187 (1955).

where $x = x_c$ subject to an examination of the transformation properties of the fourier coefficients and spinors. Note both terms in each exponential are separately invariant under global SU(3). (Note also $q_r^* = q_r$ since q_r is real.)

From the form of S_C above it is clear that an argument similar to that for the dynamical equations shows S_C is invariant under an SU(3) transformation and thus their spinors are also invariant under SU(3) transformations. The fourier coefficients, if second quantized in a direct generalization of the usual manner, have covariant anti-commutation relations under an SU(3) transformation. For example

$$\{b_C(q,s), b_C^\dagger(q'^*,s')\} = \delta_{ss'}\delta^3(q_r - q'_{r'})\delta^3(q_i - q'_{i'}) \qquad (26.150)$$

Under an SU(3) transformation, $z = Uq$ and $z' = Uq'$, the right side of eq. 26.150 transforms to

$$\delta^3(q_r - q'_{r'})\delta^3(q_i - q'_{i'}) \rightarrow \delta^3(z_r - z'_{r'})\delta^3(z_i - z'_{i'})/|\partial(q)/\partial(z)| = \delta^3(z_r - z'_{r'})\delta^3(z_i - z'_{i'}) \qquad (26.151)$$

where

$$|\partial(q)/\partial(z)| = |\partial(q_r^1,q_r^2,q_r^3,q_i^1, q_i^2, q_i^3)/\partial(z_r^1,z_r^2,z_r^3,z_i^1, z_i^2, z_i^3)| = 1 \quad (26.152)$$

is the Jacobian of the transformation U. Thus the fourier coefficients transform trivially under SU(3). For example,

$$b_C(q^*,s) \rightarrow b_C(z^*,s \qquad (26.153)$$

Since the integrand transforms as

$$\int d^3q_r d^3q_i \rightarrow \int d^3z_r d^3z_i \, |\partial(q)/\partial(z)| = \int d^3z_r d^3z_i \qquad (26.154)$$

the wave function $\psi_C(t, \mathbf{x})$ transforms as an SU(3) scalar up to an inessential unitary transformation V of γ matrices: $\psi_C(t, \mathbf{x}) \rightarrow V\psi_C(t, \mathbf{x})$.[140]

26.5.1.5 Global SU(3) Spin ½ Complexon Fields

Having uncovered an SU(3) symmetry in the scalar field equations of Case 3A the generalization of the scalar field equations to the **3** representation of SU(3) is direct:

$$\psi_C{}^a(x) = \sum_{\pm s} \int d^3p_r d^3p_i\, N_C(p)\delta(\mathbf{p_r \cdot p_i}/m^2)[b_C(p,a,s)u_C{}^a(p, s)e^{-i(p\cdot x + p^*\cdot x^*)/2} +$$
$$+ d_C{}^\dagger(p,a,s)v_C{}^a(p, s)e^{+i(p\cdot x + p^*\cdot x^*)/2}] \qquad (26.155)$$

where $x = x_c$ for a = 1,2, 3 with $u_C{}^a(p, s)$ and $v_C{}^a(p, s)$ being the product a spinor of type eq. 26.138 and a 3 element column vector c^a with b^{th} element

$$b^a(b) = \delta^{ab} \qquad (26.156)$$

Under a global SU(3) transformation U the **3** complexon wave functions transform as

$$\psi_C{}^{\prime a}(x) = U^{ab}\psi_C{}^b(x) \qquad (26.157)$$

In a subsequent discussion we will extend the global SU(3) symmetry described in these subsections to be color local SU(3) upon the introduction of the Yang-Mills color gluon interaction.

26.5.1.6 Lagrangian Formulation and Second Quantization of Complexons

In this subsection we will outline the canonical quantization of SU(3) singlet complexons with the quantum field equation

---

[140] The spinors $u_C(q^*,s)$ and $v_C(q^*,s)$ are unchanged up to a unitary transformation of the γ matrices $(V^\dagger\gamma'V = \gamma)$. Thus the term $(U\mathbf{w})^*\cdot\gamma' = \mathbf{w}^*\cdot V\gamma V^\dagger \equiv \mathbf{w}^*\cdot\gamma$ in the expressions for the $u_C(q^*,s)$ and $v_C(q^*,s)$ spinors.

$$[i\gamma^0 \partial/\partial t + i\gamma \cdot (\nabla_r + i\nabla_i) - m]\psi_C(t, \mathbf{x_r}, \mathbf{x_i}) = 0 \qquad (26.158)$$

and subsidiary condition

$$\nabla_r \cdot \nabla_i \, \psi_C(t, \mathbf{x_r}, \mathbf{x_i}) = 0 \qquad (26.159)$$

We begin with the Lagrangian density

$$\mathcal{L} = \bar{\psi}_C(i\gamma^\mu D_\mu - m)\psi_C(x) \qquad (26.160)$$

where $\bar{\psi}_C = \psi_C^\dagger \gamma^0$:

$$\psi_C^\dagger = [\psi_C(\mathbf{x_r}, \mathbf{x_i})]^\dagger \big|_{\mathbf{x_i} = -\mathbf{x_i}} \qquad (26.161)$$

$$D_0 = \partial/\partial x^0$$
$$D_k = \partial/\partial x^k + i \, \partial/\partial x_i^k \qquad (26.162)$$

with $x^k = x_r^k$ for $k = 1, 2, 3$. The invariant action (under real Lorentz transformations) is

$$I = \int d^7x \, \mathcal{L} \qquad (26.163)$$

It is easy to show that the action is real

$$I^* = I \qquad (26.164)$$

in a manner similar to the case considered in section 26.12 due to the form of $\psi_C^\dagger$ in eq. 26.161. (One has to change the integration over $\mathbf{x_i}$ to $-\mathbf{x_i}$ after taking the complex conjugate of I and performing manipulations similar to those in section 26.12.)

The conjugate momentum is

$$\pi_{Ca} = \partial\mathcal{L}/\partial\dot{\psi}_{Ca} \equiv \partial\mathcal{L}/\partial(\partial\psi_{Ca}/\partial x^0) = i\psi_{C\,a}^\dagger \qquad (26.165)$$

where a is a spinor index. It yields the non-zero anti-commutation relation

$$\{\psi_{C\,a}^\dagger(x), \psi_{Cb}(y)\} = \delta_{ab} \, \delta^3(x_r - y_r)\delta^3(x_i - y_i) \qquad (26.166)$$

where x and y are complex. However we will see that the constraint eq. 26.159 is required. So the correct anti-commutator turns out to be

$$\{\psi_{C_a}^{\dagger}(x), \psi_{Cb}(y)\} = -\delta_{ab}\delta'(\mathbf{\nabla_r}\cdot\mathbf{\nabla_i}/m^2)[\delta^3(x_r - y_r)\delta^3(x_i - y_i)] \qquad (26.167)$$

where all $\mathbf{\nabla_r}$ and $\mathbf{\nabla_i}$ are ∇ derivatives with respect to x, and where $\delta'(\mathbf{\nabla_r}\cdot\mathbf{\nabla_i})$ is the derivative of a delta function with the argument being differential operators such as those in eq. 26.159. The minus sign is due to the presence of a *derivative* of a delta-function and is not an issue.

The hamiltonian density is

$$\mathcal{H} = \pi_C\dot{\psi}_C - \mathcal{L} = \psi_C^{\dagger}(-i\mathbf{\alpha}\cdot\mathbf{D} + \beta m)\psi_C \qquad (26.168)$$

and the (unsymmetrized) energy-momentum tensor is

$$\mathcal{T}_{\mu\nu} = -g_{\mu\nu}\mathcal{L} + \partial\mathcal{L}/\partial(D^{\mu}\psi_C)D_{\nu}\psi_C \qquad (26.169)$$

The conserved energy and momentum are

$$P^0 = H = \int d^3x_r d^3x_i \,\mathcal{T}^{00} = \int d^3x_r d^3x_i \,\mathcal{H} \qquad (26.170)$$

and

$$P^i = \int d^3x_r d^3x_i \,\mathcal{T}^{0i} \qquad (26.171)$$

We now proceed to establish the canonical anti-commutation relations. First, the second quantization of the complexon field uses the above fourier coefficient anti-commutation relations (suitably rewritten):

$$\begin{aligned}
\{b_C(p,s), b_C^{\dagger}(p'^*,s')\} &= \delta_{ss'}\delta^3(\mathbf{p_r} - \mathbf{p'_{r'}})\delta^3(\mathbf{p_i} + \mathbf{p'_{i'}}) \\
\{d_C(p,s), d_C^{\dagger}(p'^*,s')\} &= \delta_{ss'}\,\delta^3(\mathbf{p_r} - \mathbf{p'_{r'}})\delta^3(\mathbf{p_i} + \mathbf{p'_{i'}}) \\
\{b_C(p,s), b_C(p'^*,s')\} &= \{d_C(p,s), d_C(p'^*,s')\} = 0 \\
\{b_C^{\dagger}(p,s), b_C^{\dagger}(p'^*,s')\} &= \{d_C^{\dagger}(p,s), d_C^{\dagger}(p'^*,s')\} = 0 \\
\{b_C(p,s), d_C^{\dagger}(p'^*,s')\} &= \{d_C(p,s), b_C^{\dagger}(p'^*,s')\} = 0
\end{aligned} \qquad (26.172)$$

$$\{b_C^\dagger(p,s), d_C^\dagger(p'^*,s')\} = \{d_C(p,s), b_C(p'^*,s')\} = 0$$

The delta-function arguments $\delta^3(\mathbf{p}_i + \mathbf{p}'_{i'})$ above have a positive sign in order to obtain $\delta^3(\mathbf{x}_i - \mathbf{y}_i)$ in the field anti-commutator eq. 26.167.

The spinors, eq. 26.138, satisfy

$$\sum_{\pm s} u_\alpha(p, s)\bar{u}_\beta(p^*, s) = (2m)^{-1}(\not{p} + m)_{\alpha\beta} \qquad (26.173)$$

$$\sum_{\pm s} v_\alpha(p, s)\bar{v}_\beta(p^*, s) = (2m)^{-1}(\not{p} - m)_{\alpha\beta}$$

remembering

$$\bar{u}_C(p^*,s) = w^{1T}(0)S_C(p)\gamma^0 = w^{1T}(0)[\cosh(\omega/2)I + \sinh(\omega/2)\gamma^0\gamma\cdot\hat{\mathbf{w}}]\gamma^0 \quad (26.174)$$

by eqs. 26.137 since $\hat{\mathbf{w}}^{**} = \hat{\mathbf{w}}$.

We will now evaluate the equal-time anti-commutation relation using eqs. 26.136 and 26.137:

$$\{\psi_{C\,a}^\dagger(x), \psi_{Cb}(y)\} = \sum_{\pm s,\, s'} \int d^3p_r d^3p_i\, d^3p'_r d^3p'_i\, \delta(\mathbf{p}_r\cdot\mathbf{p}_i/m^2)\delta(\mathbf{p}'_r\cdot\mathbf{p}'_i/m^2)\, N_C(p')N_C(p)\cdot$$

$$\cdot[\{b_C^\dagger(p^*,s)u_{Ca}^\dagger(p^*,s)e^{+i(p\cdot x^* + p^*\cdot x)/2}, b_C(p',s')u_{Cb}(p', s')e^{-i(p'\cdot y + p'^*\cdot y)/2}\}+$$

$$+ \{d_C(p^*,s)v_{Ca}^\dagger(p^*,s)e^{-i(p\cdot x^* + p^*\cdot x)/2}, d_C^\dagger(p',s')v_{Cb}(p', s')e^{+i(p'\cdot y + p'^*\cdot y)/2}\}]$$

$$= \int d^3p_r d^3p_i\, N_C^2(p)[\delta(\mathbf{p}_r\cdot\mathbf{p}_i/m^2)]^2[((\not{p} + m)\gamma^0)_{ba}e^{+i(p\cdot x^* + p^*\cdot x)/2 - i(p^*\cdot y + p\cdot y^*)/2} +$$

$$+((\not{p} - m)\gamma^0)_{ba}e^{-i(p\cdot x^* + p^*\cdot x)/2 + i(p^*\cdot y + p\cdot y^*)/2}]/(2m)$$

Next we use eq. 26.140 and the identity

$$[\delta(x - y)]^2 = -\tfrac{1}{2}\,\delta'(x - y) \equiv -\tfrac{1}{2}\, d\delta(x - y)/dx \qquad (26.175)$$

which can be derived from the step function identity $\theta(x - y) = [\theta(x - y)]^2$ to obtain

$$\{\psi_{C~a}^{\dagger}(x),\psi_{Cb}(y)\} = -\tfrac{1}{2}\!\int\! d^3p_r d^3p_i N_C^{~2}(p)\delta'(\mathbf{p_r}\!\cdot\!\mathbf{p_i}/m^2)[((\not p+m)\gamma^0)_{ba}e^{-ipr\cdot(xr-yr)+ipi\cdot(xi-yi)} +$$
$$+ ((\not p-m)\gamma^0)_{ba}e^{+ipr\cdot(xr-yr)-ipi\cdot(xi-yi)}]/(2m)$$

$$= -\tfrac{1}{2}\delta_{ba}\!\int\! d^3p_r d^3p_i N_C^{~2}(p)\delta'(\mathbf{p_r}\!\cdot\!\mathbf{p_i}/m^2)p^0 e^{-ipr\cdot(xr-yr)+ipi\cdot(xi-yi)}/m$$

$$= -\delta_{ab}\,\delta'(\nabla_r\!\cdot\!\nabla_i/m^2)[\delta^3(x_r-y_r)\delta^3(x_i-y_i)] \qquad (26.176)$$

The grad operators, ∇_r and ∇_i, are derivatives are with respect to real and imaginary x in the Dirac delta functions. The factor[141] $\delta'(\nabla_r\!\cdot\!\nabla_i)$ expresses the orthogonality constraint in coordinate space on the momenta. It is analogous to the transversality constraint on the electromagnetic vector potential commutator:

$$[\pi_A^{~j}(x), A_k(y)] = -i\,\delta^{tr}_{~jk}(x-y) \qquad (26.177)$$

$$\delta^{tr}_{~jk}(x-y) = (\delta_{jk} - \partial_j\partial_k/\nabla^2)\,\delta^3(x-y) \qquad (26.178)$$

where $\partial_k = \partial/\partial x_k$.

26.5.1.7 Complexon Feynman Propagator

The complexon Feynman propagator for ψ_C is[142]

$$iS_C(x,y) = \theta(x^0-y^0)\langle 0|\psi_C(x)\psi_C^{\dagger}(y)\gamma^0|0\rangle - \theta(y^0-x^0)\langle 0|\psi_C^{\dagger}(y)\gamma^0\psi_C(x)|0\rangle \quad (26.179)$$
$$= \int\! d^3p_r d^3p_i N_C^{~2}(p)[\delta(\mathbf{p_r}\!\cdot\!\mathbf{p_i}/m^2)]^2\{\theta(x^0-y^0)(\not p+m)e^{-i(p^*\cdot(x-y)+p\cdot(x^*-y^*))/2} -$$
$$- \theta(y^0-x^0)(\not p-m)e^{+i(p^*\cdot(x-y)+p\cdot(x^*-y^*))/2}\}/(2m)$$

---

[141] A derivative of a delta function containing grad operators.

[142] The reader, upon seeing the additional integrations $\int d^3p_i$ might suspect that they would ultimately lead to divergence issues in perturbation theory calculations. However the $\delta'(\mathbf{p_r}\!\cdot\!\mathbf{p_i}/m^2)$ term compensates in part for the additional integrations by four powers of momentum since $\delta'(\mathbf{p_r}\!\cdot\!\mathbf{p_i}/m^2) = (|\mathbf{p_r}||\mathbf{p_i}|/m^2)^{-2}\delta'(\cos\theta_{ri})$ where θ_{ri} is the angle between the momenta. As a result only 2 fermion and 3 fermion loop integrations would potentially have difficulties if one uses the conventional approach to perturbation theory. If one uses the approach of Blaha (2003) and (2005a) then there are no divergences.

$$= -(4\pi)^{-1}\int dp^0 d^3 p_r d^3 p_i (2\pi)^{-6}\delta'(\mathbf{p_r \cdot p_i}/m^2)(\not p + m)e^{-i(p^* \cdot (x-y) + p \cdot (x^* - y^*))/2}/(p^2 - m^2 + i\varepsilon)$$

$$= -\tfrac{1}{2}\int dp^0 d^3 p_r d^3 p_i \, \delta'(\mathbf{p_r \cdot p_i}/m^2)(\not p + m)(2\pi)^{-7}\exp[-ip^0(x^0 - y^0) +$$
$$+ i\mathbf{p_r \cdot (x_r - y_r)} - i\mathbf{p_i \cdot (x_i - y_i)}]/(p^2 - m^2 + i\varepsilon) \qquad (26.180)$$

The integral can be written in the form:

$$I = \int dp^0 d^3 p_r d^3 p_i \delta'(\mathbf{p_r \cdot p_i}/m^2)(\not p + m)\exp[-ip^0(x^0 - y^0) + i\mathbf{p_r \cdot (x_r - y_r)} - i\mathbf{p_i \cdot (x_i - y_i)}]/(p^2 - m^2 + i\varepsilon)$$
$$= \int d^4 p_r dM^2 \delta'(\nabla_r \cdot \nabla_i/m^2)(p^0\gamma^0 - (\mathbf{p_r} - \nabla_i)\cdot\gamma + m)\exp[-ip^0(x^0 - y^0) + i\mathbf{p_r \cdot (x_r - y_r)}]\cdot$$
$$\cdot J(\mathbf{x_i - y_i}, M^2)/(p_r^2 - M^2 + i\varepsilon) \qquad (26.181)$$

where $p_r^2 = p^{0\,2} - \mathbf{p_r \cdot p_r}$ and

$$J(\mathbf{x_i - y_i}, M^2) = (2\pi)^{-3}\int d^3 p_i \delta(M^2 + \mathbf{p_i}^2 - m^2)\exp[-i\mathbf{p_i \cdot (x_i - y_i)}] \qquad (26.182)$$
$$= (2\pi)^{-2}|\mathbf{x_i - y_i}|^{-1}\theta(m^2 - M^2)\sin((m^2 - M^2)^{1/2}|\mathbf{x_i - y_i}|)$$

The complexon Feynman propagator can be rearranged into the form of a spectral integral:

$$iS_C(x, y) = -\int dM\, (i\gamma^0\partial/\partial x^0 - i(\nabla_r - i\nabla_i)\cdot\gamma + m)\delta'(\nabla_r \cdot \nabla_i/m^2)J(\mathbf{x_i - y_i}, M^2)\Delta_F(x - y, M)$$
$$(26.183)$$

where

$$\Delta_F(x - y, M) = (2\pi)^{-4}\int d^4 p_r \exp[-ip^0(x^0 - y^0) + i\mathbf{p_r \cdot (x_r - y_r)}]/(p_r^2 - M^2 + i\varepsilon)$$
$$(26.184)$$

26.5.1.8 Case 4: Left-handed Tachyon Complexons

In this case $\hat{\mathbf{u}}_r \cdot \hat{\mathbf{u}}_i = 0$ again. However we add an imaginary term to ω to obtain a manifest Left-handed L_C boost[143]

$$\Lambda_{CL}(\mathbf{v_c}) = \exp[i(\omega + i\pi/2)\hat{\mathbf{w}}\cdot\mathbf{K}] \tag{26.185}$$

where ω remains

$$\omega = (\omega_r^2 - \omega_i^2)^{\frac{1}{2}} \tag{26.186}$$

and

$$\hat{\mathbf{w}} = (\omega_r \hat{\mathbf{u}}_r + i\omega_i \hat{\mathbf{u}}_i)/\omega \tag{26.187}$$
$$\hat{\mathbf{w}}\cdot\hat{\mathbf{w}} = \hat{\mathbf{u}}_r \cdot \hat{\mathbf{u}}_r = \hat{\mathbf{u}}_i \cdot \hat{\mathbf{u}}_i = 1 \tag{26.188}$$
$$\mathbf{v_c} = \hat{\mathbf{w}}\tanh(\omega + i\pi/2) = \hat{\mathbf{w}}\coth(\omega) \tag{26.189}$$

Letting $\omega_L = \omega + i\pi/2$ we find, as before,

$$\cosh(\omega_L) = i\sinh(\omega) = -\gamma = i\gamma_s \tag{26.190}$$
$$\sinh(\omega_L) = i\cosh(\omega) = -\beta\gamma = i\beta\gamma_s$$

with, $\beta = v_c = |\mathbf{v_c}| > 1$, $\gamma_s = (\beta^2 - 1)^{-\frac{1}{2}}$, and

$$\sinh(\omega) = \gamma_s \tag{26.191}$$
$$\cosh(\omega) = \beta\gamma_s$$

Thus we denote $\Lambda_{CL}(\mathbf{v_c})$ by

$$\Lambda_{CL}(\mathbf{v_c}) \equiv \Lambda_{CL}(\omega, \hat{\mathbf{w}}) \tag{26.192}$$

The corresponding spinor boost transformation is:

$$S_{CL}(\Lambda_{CL}(\omega, \hat{\mathbf{w}})) = \exp(-i\omega_L\sigma_{0i}\hat{w}_i/2) = \exp(-\omega_L\gamma^0\boldsymbol{\gamma}\cdot\hat{\mathbf{w}}/2)$$
$$= \cosh(\omega_L/2)I + \sinh(\omega_L/2)\gamma^0\boldsymbol{\gamma}\cdot\hat{\mathbf{w}} \tag{26.193}$$

The momentum space equation generated by $S_{CL}(\Lambda_{CL}(\omega, \hat{\mathbf{w}}))$ is

---

[143] The reader can readily verify the form is consistent that generated by an L_C boost transformation.

$$\{m \cosh(\omega_L)\gamma^0 - m \sinh(\omega_L)\gamma\cdot(\omega_r\hat{\mathbf{u}}_r + i\omega_i\hat{\mathbf{u}}_i)/\omega - m\}e^{+ip\cdot x}w_{cL}(p) = 0 \quad (26.194)$$

or

$$\{im \sinh(\omega)\gamma^0 - im \cosh(\omega)\gamma\cdot(\omega_r\hat{\mathbf{u}}_r + i\omega_i\hat{\mathbf{u}}_i)/\omega - m\}e^{+ip\cdot x}w_{cL}(p) = 0 \quad (26.195)$$

where p·x = Et − **p·x** after performing a corresponding left-handed superluminal coordinate transformation in the exponential factor. Thus the positive energy wave is transformed into a negative energy wave by the transformation.

The momentum 4-vector is defined by

$$p = (p^0, \mathbf{p}) \quad (26.196)$$

where

$$p^0 = m \sinh(\omega) \qquad \mathbf{p} = \mathbf{p}_r + i\mathbf{p}_i \quad (26.197)$$

with

$$\mathbf{p}_r = m\omega_r\hat{\mathbf{u}}_r \cosh(\omega)/\omega \quad \mathbf{p}_i = m\omega_i\hat{\mathbf{u}}_i \cosh(\omega)/\omega \quad (26.198)$$

and

$$\mathbf{p}_r\cdot\mathbf{p}_i = 0 \quad (26.199)$$

then eq. 26.195 becomes the complexon tachyon equation

$$[ip\cdot\gamma - m]e^{+ip\cdot x}w_{cL}(p) = 0 \quad (26.200)$$

with a complex 3-momentum **p** and the tachyon 4-momentum mass shell condition:[144]

$$p^2 = p^{0\,2} - \mathbf{p}_r^2 + \mathbf{p}_i^2 = -m^2 \quad (26.201)$$

Eq. 26.200 is the momentum space equivalent of the wave equation

$$[\gamma^0\partial/\partial t + \gamma\cdot(\nabla_r + i\nabla_i) - m]\psi_{CL}(t, \mathbf{x}_r, \mathbf{x}_i) = 0 \quad (26.202)$$

or

---

[144] Note that the presence of the $\mathbf{p}_i^2$ term does not change the tachyon requirement that $\mathbf{p}_r^2 \geq m^2$ as seen in the previous cases.

$$[\gamma\cdot\nabla - m]\psi_{CL}(t, \mathbf{x_r}, \mathbf{x_i}) = 0 \tag{26.203}$$

with the subsidiary condition on the wave function

$$\nabla_r\cdot\nabla_i\, \psi_{CL}(t, \mathbf{x_r}, \mathbf{x_i}) = 0 \tag{26.204}$$

also holds. We note that eq. 26.202 is covariant under the real Lorentz group and eq. 26.204 can be easily put into (real Lorentz group) covariant form.

Before considering a lagrangian formulation and the Fourier operator representation of $\psi_{CL}(t, \mathbf{x_r}, \mathbf{x_i})$ we will define the tachyon spinors, and its associated real and imaginary spin operators.

The spinor generated from a spin up Dirac spinor at rest by the L_C spinor boost eq. 26.193 is

$$w_{cL}(p) = S_{CL}w(0) = [\cosh(\omega_L/2)I + \sinh(\omega_L/2)\gamma^0\gamma\cdot\hat{\mathbf{w}}]w(0) \tag{26.205}$$

Following a procedure similar to section 26.12 (which the reader may wish to examine first) we define four spinors for Dirac particles at rest with eq. 26.282. Then by applying a boost to these rest spinors we find the L_C tachyon spinors:

$$S_{CL}w^k(0) = w_{cL}{}^k(p) \tag{26.206}$$

and from these tachyon spinors we generalize to tachyon spinors $u_{CL}(p, s)$ and $v_{CL}(p, s)$ in a manner similar to that of the previous case.

Eqs. 26.200 through 26.204 imply that the wave function solution of eq. 26.200, subject to the subsidiary condition eq. 26.204, has the form

$$\psi_{CL}(x) = \sum_{\substack{\pm s \\ p_r^2 \geq m^2}} \int d^3p_r d^3p_i\, N_{CL}(p)\delta(\mathbf{p_r}\cdot\mathbf{p_i}/m^2)[b_{CL}(p,s)u_{CL}(p, s)e^{-i(p\cdot x + p^*\cdot x^*)/2} +$$
$$+ d_{CL}{}^\dagger(p,s)v_{CL}(p, s)e^{+i(p\cdot x + p^*\cdot x^*)/2}] \tag{26.207}$$

where $\mathbf{p} = \mathbf{p_r} + i\mathbf{p_i}$, $\mathbf{x} = \mathbf{x_r} - i\mathbf{x_i}$, $p \cdot x = p^0 x^0 - \mathbf{p} \cdot \mathbf{x}$, and $b_{CL}(p, s)$ and $d_{CL}(p,s)$ are tachyon fourier coefficients.

26.5.1.9 Global SU(3) Symmetry

We can show that there is also a global SU(3) symmetry present here as shown in the previous case. The demonstration is similar to that of eqs. 26.143 – 26.156.

26.5.1.10 Light-Front Quantization of Tachyonic Complexons

Because of the momentum constraint $\mathbf{p_r}^2 \geq m^2$ the set of solutions of the form of eq. 26.207 is incomplete and the result of second quantization would not be an equal time anti-commutator expression consisting of derivatives of delta functions (eq. 26.176) but rather an analogue to previous unsuccessful attempts to create a second quantized tachyon theory.[145]

Therefore we will use light-front coordinates, and left and right handed field operators (as previously) to obtain a successful second quantization of this new type of tachyon.

The "missing" factor of i in the first term of eq. 26.203 requires the lagrangian to be different from the conventional Dirac lagrangian in order for the lagrangian to be real. The simplest, physically acceptable, free spin ½ tachyon lagrangian density for ψ_{CL} is:

$$\mathcal{L}_{CL} = \psi_{CL}{}^C(x)(\gamma \cdot \nabla - m)\psi_{CL}(x) \tag{26.208}$$

where

$$\psi_{CL}{}^C(x) = [\psi_{CL}(x)]^\dagger\big|_{\mathbf{x_i} = -\mathbf{x_i}} \, i\gamma^0\gamma^5 \tag{26.209}$$

is similar to eq. 26.161. In words, eq. 26.209 states: take the hermitean conjugate of $\psi_{CL}(x)$; change $\mathbf{x_i}$ to $-\mathbf{x_i}$; and then post-multiply by the indicated factors.

The free complexon invariant action (under real Lorentz transformations) is

---

[145] Such as G. Feinberg, Phys. Rev. **159**, 1089 (1967).

$$I = \int d^7x \mathcal{L}_{CL} \qquad (26.210)$$

The action can be shown to be real

$$I^* = I \qquad (26.211)$$

in a manner similar to the case considered in section 26.12. The tachyonic complexon's energy-momentum tensor is

$$\mathfrak{T}_{CL\mu\nu} = - g_{\mu\nu} \mathcal{L}_{CL} + \partial \mathcal{L}_{CL}/\partial(D^\mu \psi_{CL}) D_\nu \psi_{CL} \qquad (26.212)$$
$$= i \psi_{CL}{}^C \gamma^0 \gamma^5 \gamma_\mu D_\nu \psi_{CL}$$

where

$$D_0 = \partial/\partial x^0$$
$$D_k = \partial/\partial x_r{}^k + i\, \partial/\partial x_i{}^k \qquad (26.213)$$

and thus the conserved energy and momentum are

$$P^0 = H = \int d^3x_r d^3x_i\, \mathfrak{T}_{CL}{}^{00} = i\int d^3x_r d^3x_i \psi_{CL}{}^C \gamma^5 (\boldsymbol{\alpha}\cdot\mathbf{D} + \beta m)\psi_{CL}$$
$$\qquad (26.214)$$
$$P^k = \int d^3x_r d^3x_i\, \mathfrak{T}_{CL}{}^{0k} = - i\int d^3x_r d^3x_i\, \psi_{CL}{}^C \gamma^5 D^k \psi_{CL} \qquad (26.215)$$

Having defined a suitable tachyon lagrangian we can now proceed to its canonical quantization. The conjugate momentum can be calculated from the lagrangian density eq. 26.212:

$$\pi_{CLa} = \partial\mathcal{L}_{CL}/\partial\dot\psi_{CLa} \equiv \partial\mathcal{L}_{CL}/\partial(\partial\psi_{CLa}/\partial t) = -i([\psi_{CL}(x)]^\dagger|_{\mathbf{x}_i = -\mathbf{x}_i}\gamma^5)_a \qquad (26.216)$$

The resulting non-zero, canonical anti-commutation relations are

$$\{\pi_{CLa}(x),\ \psi_{CLb}(y)\} = i\, \delta_{ab}\, \delta^3(x_r - y_r)\delta^3(x_i - y_i)$$

based on locality in both real and imaginary coordinates:

$$\{\psi_{CL}{}_a^\dagger(x)\big|_{\mathbf{x_i} = -\mathbf{x_i}}, \ \psi_{Tb}(y)\} = - [\gamma^5]_{ab} \ \delta^3(x_r - y_r)\delta^3(x_i - y_i) \qquad (26.217)$$

At this point we might attempt to complete the canonical quantization procedure in the conventional manner by Fourier expanding the field and specifying anti-commutation relations for the fourier component amplitudes. However the incompleteness of the set of plane waves, which are limited by the restriction $\mathbf{p_r}^2 \geq m^2$, causes the equal time anti-commutator of the fields *not* to yield a δ-functions.

Therefore we turn to the previous successful approach to tachyon quantization[146] and decompose the tachyonic complexon field into left-handed and right-handed parts and then second quantize in light-front coordinates.

26.5.2 Separation into Left-Handed and Right-Handed Fields

As before we will use a transformed set of Dirac matrices to develop our left-handed and right-handed tachyon formulations. The γ^5 chirality operator's eigenvalues define handedness: +1 corresponds to right-handed; and −1 corresponds to left-handed:

$$\gamma^5 \psi_{CLL} = - \psi_{CLL} \qquad\qquad \gamma^5 \psi_{CLR} = \psi_{CLR} \qquad (26.218)$$

We define left-handed and right-handed tachyon fields with the projection operators:

$$\begin{aligned}
C^\pm &= \tfrac{1}{2}(I \pm \gamma^5) \\
C^+ + C^- &= I \\
C^{\pm 2} &= C^\pm \\
C^+ C^- &= 0
\end{aligned} \qquad (26.219)$$

with the result

$$\begin{aligned}
\psi_{CLL} &= C^- \psi_{CL} \\
\psi_{CLR} &= C^+ \psi_{CL}
\end{aligned} \qquad (26.220)$$

---

[146] Blaha (2006) discusses this case in detail.

We can calculate the commutation relations of the left-handed and right-handed tachyonic complexon fields from eq. 26.217 by pre-multiplying and post-multiplying by $\frac{1}{2}(1 - \gamma^5)$ and $\frac{1}{2}(1 + \gamma^5)$. The results are:

$$\{\psi_{CLLa}^{\dagger}(x)|_{\mathbf{x_i} = -\mathbf{x_i}}, \psi_{CLLb}(y)\} = C^-_{ab}\,\delta^6(x - y) \tag{26.221}$$

$$\{\psi_{CLRa}^{\dagger}(x)|_{\mathbf{x_i} = -\mathbf{x_i}}, \psi_{CLRb}(y)\} = -C^+_{ab}\,\delta^6(x - y) \tag{26.222}$$

$$\{\psi_{CLLa}^{\dagger}(x)|_{\mathbf{x_i} = -\mathbf{x_i}}, \psi_{CLRb}(y)\} = \{\psi_{CLRa}^{\dagger}(x)|_{\mathbf{x_i} = -\mathbf{x_i}}, \psi_{CLLb}(x')\} = 0 \tag{26.223}$$

where

$$\delta^6(x - y) = \delta^3(x_r - y_r)\delta^3(x_i - y_i) \tag{26.224}$$

The lagrangian density of eq. 26.208 decomposes into left-handed and right-handed parts: (The change $\mathbf{x_i}$ to $-\mathbf{x_i}$ will be understood in $\psi_{CLL}^{\dagger}(x)$ and $\psi_{CLR}^{\dagger}(x)$ in the following.)

$$\mathscr{L}_{CL} = \psi_{CLL}^{\dagger}\gamma^0 i\gamma^\mu \partial_\mu \psi_{CLL} - \psi_{CLR}^{\dagger}\gamma^0 i\gamma^\mu \partial_\mu \psi_{CLR} - im[\psi_{CLR}^{\dagger}\gamma^0 \psi_{CLL} - \psi_{CLL}^{\dagger}\gamma^0 \psi_{CLR}] \tag{26.225}$$

26.5.3 Further Separation into + and – Light-Front Complexon Fields

As previously, we now use light-front coordinates and quantization to obtain a successful second quantization of this form of tachyon field. Light-front variables, in the present case where we have to contend with complex 3-vectors, are defined by real coordinates and derivatives:

$$x^\pm = (x^0 \pm x_r^3)/\sqrt{2} \tag{26.226}$$
$$\partial/\partial x^\pm \equiv \partial^\mp \equiv (\partial/\partial x^0 \pm \partial/\partial x_r^3)/\sqrt{2}$$

with the "transverse" real coordinate variables, x_r^1 and x_r^2, and imaginary coordinate variables x_i^1, x_i^2, and x_i^3.

The inner product of two 4-vectors has the form

$$x \cdot y = x^+ y^- + y^+ x^- + i[y_i^3(x^+ - x^-) + x_i^3(y^+ - y^-)]/\sqrt{2} + x_i^3 y_i^3 - (\mathbf{x}_{r_\perp} - i\mathbf{x}_{i_\perp}) \cdot (\mathbf{y}_{r_\perp} - i\mathbf{y}_{i_\perp})$$
(26.227)

with

$$\begin{aligned}
\mathbf{x}_{r_\perp} &= (x_r^1, x_r^2) & \mathbf{x}_{i_\perp} &= (x_i^1, x_i^2) \\
\mathbf{y}_{r_\perp} &= (y_r^1, y_r^2) & \mathbf{y}_{i_\perp} &= (y_i^1, y_i^2)
\end{aligned}$$
(26.228)

where $x = (x^0, \mathbf{x} = \mathbf{x_r} - i\mathbf{x_i})$ and $y = (y^0, \mathbf{y} = \mathbf{y_r} - i\mathbf{y_i})$. Momenta are always defined as $p = (p^0, \mathbf{p} = \mathbf{p_r} + i\mathbf{p_i})$.

The light-front definition of Dirac matrices is:

$$\gamma^\pm = (\gamma^0 \pm \gamma^3)/\sqrt{2}$$
(26.229)

with transverse matrices γ^1 and γ^2 defined as usual. Note:

$$\gamma^{\pm 2} = 0$$

We define "+" and "–" tachyon fields with the projection operators:

$$R^\pm = \tfrac{1}{2}(I \pm \gamma^0 \gamma^3)$$
(26.230)

Left-handed, ± light-front fields: $\qquad \psi_{CLL}{}^\pm = R^\pm C^- \psi_{CL}$ (26.231)

Right-handed, ± light-front fields: $\qquad \psi_{CLR}{}^\pm = R^\pm C^+ \psi_{CL}$

Transforming to light-front variables and fields as above we obtain the light-front free tachyon lagrangian:

$$\begin{aligned}
\mathcal{L}_{CL} &= 2^{\frac{1}{2}}\psi_{CLL}{}^{++}i\partial^-\psi_{CLL}{}^+ + 2^{\frac{1}{2}}\psi_{CLL}{}^{-}i\partial^+\psi_{CLL}{}^- - \psi_{CLL}{}^{++}\gamma^0[i\boldsymbol{\gamma}_\perp\cdot\boldsymbol{\nabla}_{r_\perp} - \boldsymbol{\gamma}\cdot\boldsymbol{\nabla}_i]\psi_{CLL}{}^- - \\
&\quad - \psi_{CLL}{}^{-}\gamma^0[i\boldsymbol{\gamma}_\perp\cdot\boldsymbol{\nabla}_{r_\perp} - \boldsymbol{\gamma}\cdot\boldsymbol{\nabla}_i]\psi_{CLL}{}^+ - 2^{\frac{1}{2}}\psi_{CLR}{}^{++}i\partial^-\psi_{CLR}{}^+ - 2^{\frac{1}{2}}\psi_{CLR}{}^{-}i\partial^+\psi_{CLR}{}^- + \\
&\quad + \psi_{CLR}{}^{++}\gamma^0[i\boldsymbol{\gamma}_\perp\cdot\boldsymbol{\nabla}_{r_\perp} - \boldsymbol{\gamma}\cdot\boldsymbol{\nabla}_i]\psi_{CLR}{}^- + \psi_{CLR}{}^{-}\gamma^0[i\boldsymbol{\gamma}_\perp\cdot\boldsymbol{\nabla}_{r_\perp} - \boldsymbol{\gamma}\cdot\boldsymbol{\nabla}_i]\psi_{CLR}{}^+ -
\end{aligned}$$

$$-\,\mathrm{im}[\psi_{CLR}{}^{+\dagger}\gamma^0\psi_{CLL}{}^- - \psi_{CLL}{}^{+\dagger}\gamma^0\psi_{CLR}{}^- + \psi_{CLR}{}^{-\dagger}\gamma^0\psi_{CLL}{}^+ - \psi_{CLL}{}^{-\dagger}\gamma^0\psi_{CLR}{}^+]$$

$$(26.232)$$

(Note the similarity to the previous tachyon case.) Again the difference in signs between the left-handed and right-handed terms will be a crucial factor in the derivation of the left-handed features of the Standard Model.

Eq. 26.232 generates the equations of motion:

$$2^{\frac{1}{2}}i\partial^-\psi_{CLL}{}^+ - \gamma^0[i\boldsymbol{\gamma_\perp}\cdot\boldsymbol{\nabla_{r\perp}} - \boldsymbol{\gamma}\cdot\boldsymbol{\nabla_i}]\psi_{CLL}{}^- + im\gamma^0\psi_{CLR}{}^- = 0 \qquad (26.233)$$

$$2^{\frac{1}{2}}i\partial^-\psi_{CLR}{}^+ - \gamma^0[i\boldsymbol{\gamma_\perp}\cdot\boldsymbol{\nabla_{r\perp}} - \boldsymbol{\gamma}\cdot\boldsymbol{\nabla_i}]\psi_{CLR}{}^- + im\gamma^0\psi_{CLL}{}^- = 0$$

$$2^{\frac{1}{2}}i\partial^+\psi_{CLL}{}^- - \gamma^0[i\boldsymbol{\gamma_\perp}\cdot\boldsymbol{\nabla_{r\perp}} - \boldsymbol{\gamma}\cdot\boldsymbol{\nabla_i}]\psi_{CLL}{}^+ + im\gamma^0\psi_{CLR}{}^+ = 0$$

$$2^{\frac{1}{2}}i\partial^+\psi_{CLR}{}^- - \gamma^0[i\boldsymbol{\gamma_\perp}\cdot\boldsymbol{\nabla_{r\perp}} - \boldsymbol{\gamma}\cdot\boldsymbol{\nabla_i}]\psi_{CLR}{}^+ + im\gamma^0\psi_{CLL}{}^+ = 0$$

Eqs. 26.233 show that $\psi_{CLL}{}^-$ and $\psi_{CLR}{}^-$ are dependent fields that are functions of $\psi_{CLL}{}^+$ and $\psi_{CLR}{}^+$ on the light-front where x^+ equals a constant. They can be expressed in an integral form as well. (The independent fields $\psi_{CLL}{}^+$ and $\psi_{CLR}{}^+$ play a fundamental role in tachyonic complexon theory and are used to define "in" and "out" tachyon states in perturbation theory.)

The conjugate momenta implied by eq. 26.232 are

$$\pi_{CLL}{}^+ = \partial\mathscr{L}/\partial(\partial^-\psi_{CLL}{}^+) = 2^{\frac{1}{2}}i\psi_{CLL}{}^{+\dagger} \qquad (26.234)$$

$$\pi_{CLL}{}^- = \partial\mathscr{L}/\partial(\partial^-\psi_{CLL}{}^-) = 0$$

$$\pi_{CLR}{}^+ = \partial\mathscr{L}/\partial(\partial^-\psi_{CLR}{}^+) = -2^{\frac{1}{2}}i\psi_{CLR}{}^{+\dagger} \qquad (26.235)$$

$$\pi_{CLR}{}^- = \partial\mathscr{L}/\partial(\partial^-\psi_{CLR}{}^-) = 0$$

x^+ plays the role of the "time" variable in light-front quantized theories. So we define canonical equal x^+ anti-commutation relations for spin ½ tachyonic complexons also.

The canonical equal-light-front $(x^+ = y^+)$ anti-commutation relations of the independent fields would normally be:

$$\{\psi_{CLL}{}^{+\dagger}{}_a(x), \psi_{CLL}{}^+{}_b(y)\} = 2^{-1}[C^-R^+]_{ab}\delta(x^- - y^-)\delta^2(x_r - y_r)\delta^3(x_I - y_i)$$
(26.236)

$$\{\psi_{CLR}{}^{+\dagger}{}_a(x), \psi_{CLR}{}^+{}_b(y)\} = -2^{-1}[C^+R^+]_{ab}\,\delta(x^- - y^-)\delta^2(x_r - y_r)\delta^3(x_I - y_i)$$
(26.237)

$$\{\psi_{CLL}{}^+{}_a{}^\dagger(x), \psi_{CLR}{}^+{}_b(y)\} = \{\psi_{CLR}{}^+{}_a{}^\dagger(x), \psi_{CLL}{}^+{}_b(y)\} = 0$$
(26.238)

$$\{\psi_{CLL}{}^+{}_a(x), \psi_{CLR}{}^+{}_b(y)\} = \{\psi_{CLR}{}^+{}_a{}^\dagger(x), \psi_{CLL}{}^{+\dagger}{}_b(y)\} = 0$$
(26.239)

But as in the previous case they will be modified.

Again we see that the right-handed tachyon anti-commutation relation (eq. 26.237) has a minus sign relative to the corresponding conventional right-handed anti-commutation relation.

The sign differences between the left-handed and right-handed lagrangian terms ultimately lead to parity violating features in the Standard Model lagrangian.

26.5.3.1 Left-Handed Tachyonic Complexons

The free, "+" light-front, left-handed tachyonic complexon Fourier expansion is:

$$\psi_{CLL}{}^+(x_r, x_i) = \sum_{\pm s} \int d^2p_r dp^+ d^3p_i\, N_{CLL}{}^+(p)\theta(p^+)\delta((p_i{}^3(p^+ - p^-)/\surd 2 + \mathbf{p}_{r\perp}\!\cdot\!\mathbf{p}_{i\perp})/m^2)\cdot$$

$$\cdot[b_{CLL}{}^+(p, s)u_{CLL}{}^+(p, s)e^{-i(p\cdot x + p^*\cdot x^*)/2} + d_{CLL}{}^{+\dagger}(p, s)v_{CLL}{}^+(p, s)e^{+i(p\cdot x + p^*\cdot x^*)/2}]$$
(26.240)

Its hermitean conjugate is

$$\psi_{CLL}{}^{+\dagger}(x_r, x_i) = \sum_{\pm s} \int d^2p_r dp^+ d^3p_i\, N_{CLL}{}^+(p)\theta(p^+)\delta((p_i{}^3(p^+ - p^-)/\surd 2 + \mathbf{p}_{r\perp}\!\cdot\!\mathbf{p}_{i\perp})/m^2)\cdot$$

$$\cdot [b_{CLL}^{\dagger}(p^*,s)u_{CLL}^{\dagger}(p^*,s)e^{+i(p^*\cdot x + p\cdot x^*)/2} + d_{CLL}(p^*,s)v_{CLL}^{\dagger}(p^*,s)e^{-i(p^*\cdot x + p\cdot x^*)/2}]$$

$$(26.241)$$

where $p = p_r + ip_i$, $x = x_r - ix_i$, $p\cdot x = p^0 x^0 - \mathbf{p}\cdot\mathbf{x}$, and † indicates hermitean conjugate. The spinors are

$$u_{CLL}^{+}(p, s) = C^{-} R^{+} S_{CL} w^1(0)$$
$$u_{CLL}^{+}(p, -s) = C^{-} R^{+} S_{CL} w^2(0)$$
$$v_{CLL}^{+}(p, s) = C^{-} R^{+} S_{CL} w^3(0)$$
$$v_{CLL}^{+}(p, -s) = C^{-} R^{+} S_{CL} w^4(0)$$
$$u_{CLL}^{+\dagger}(p^*, s) = w^{1T}(0) S_{CL} R^{+} C^{-}$$
$$u_{CLL}^{+\dagger}(p^*, -s) = w^{2T}(0) S_{CL} R^{+} C^{-}$$
$$v_{CLL}^{+\dagger}(p^*, s) = w^{3T}(0) S_{CL} R^{+} C^{-}$$
$$v_{CLL}^{+\dagger}(p^*, -s) = w^{4T}(0) S_{CL} R^{+} C^{-}$$

$$(26.242)$$

where the superscript "T" indicates the transpose (These spinors are described in section 26.12.) and

$$N_{CLL}^{+}(p) = (2\pi)^{-3}(2m/p^+)^{\frac{1}{2}} \qquad (26.243)$$

The anti-commutation relations of the Fourier coefficient operators are

$$\{b_{CLL}(p,s), b_{CLL}^{\dagger}(p'^*,s')\} = 2^{-\frac{1}{2}}\delta_{ss'}\delta(p^+ - p'^+)\delta^2(\mathbf{p_r} - \mathbf{p'_{r'}})\delta^3(\mathbf{p_i} + \mathbf{p'_{i'}})$$
$$\{d_{CLL}(p,s), d_{CLL}^{\dagger}(p'^*,s')\} = 2^{-\frac{1}{2}}\delta_{ss'}\,\delta(p^+ - p'^+)\delta^2(\mathbf{p_r} - \mathbf{p'_{r'}})\delta^3(\mathbf{p_i} + \mathbf{p'_{i'}})$$
$$\{b_{CLL}(p,s), b_{CLL}(p'^*,s')\} = \{d_{CLL}(p,s), d_{CLL}(p'^*,s')\} = 0$$
$$\{b_{CLL}^{\dagger}(p,s), b_{CLL}^{\dagger}(p'^*,s')\} = \{d_{CLL}^{\dagger}(p,s), d_{CLL}^{\dagger}(p'^*,s')\} = 0 \qquad (26.244)$$
$$\{b_{CLL}(p,s), d_{CLL}^{\dagger}(p'^*,s')\} = \{d_{CLL}(p,s), b_{CLL}^{\dagger}(p'^*,s')\} = 0$$
$$\{b_{CLL}^{\dagger}(p,s), d_{CLL}^{\dagger}(p'^*,s')\} = \{d_{CLL}(p,s), b_{CLL}(p'^*,s')\} = 0$$

The delta-function arguments $\delta^3(\mathbf{p_i} + \mathbf{p'_{i'}})$ above have a positive sign in order to obtain $\delta^3(\mathbf{x_i} - \mathbf{y_i})$ in the field anti-commutators.

The spinors, eq. 26.242, satisfy

$$\sum_{\pm s} u_{CLL}{}^+{}_\alpha(p, s)\bar{u}_{CLL}{}^+{}_\beta(p*, s) = (2m)^{-1}[C^-R^+(i\not{p} + m)R^-C^+]_{\alpha\beta}$$

$$\sum_{\pm s} v_{CLL}{}^+{}_\alpha(p, s)\bar{v}_{CLL}{}^+{}_\beta(p*, s) = (2m)^{-1}[C^-R^+(i\not{p} - m)R^-C^+]_{\alpha\beta} \qquad (26.245)$$

where $\bar{u}_{CLL}{}^+ = u_{CLL}{}^{+\dagger}\gamma^0$ and $\bar{v}_{CLL}{}^+ = v_{CLL}{}^{+\dagger}\gamma^0$.

We now evaluate the canonical left-handed, light-front anti-commutation relation:

$$\{\psi_{CLL}{}^+{}_a(x), \psi_{CLL}{}^{+\dagger}{}_b(y)\} = \sum_{\pm s,s'} \int d^3p_i d^2p dp^+ \int d^3p_i' d^2p' dp'^+ N_{CLL}{}^+(p) \, N_{CLL}{}^+(p')\cdot$$

$$\cdot\theta(p^+)\theta(p'^+)\delta((p_i{}^3(p^+-p^-)/\sqrt{2} + \mathbf{p}_{r\perp}\cdot\mathbf{p}_{i\perp})/m^2) \, \delta((p_i'^3(p'^+ - p'^-)/\sqrt{2} + \mathbf{p}'_{r\perp}\cdot\mathbf{p}'_{i\perp})/m^2)\cdot$$

$$\cdot[\{b_{CLL}{}^{+\dagger}(p'*,s'),b_{CLL}{}^+(p,s)\}u_{CLL}{}^+{}_a(p,s)u_{CLL}{}^{+\dagger}{}_b(p'*,s')e^{+i(p'*\cdot y+p'\cdot y*)2 - i(p\cdot x+p*\cdot x*)/2} +$$

$$+\{d_{CLL}{}^+(p'*,s'),d_{CLL}{}^{+\dagger}(p,s)\}v_{CLL}{}^+{}_a(p,s)v_{CLL}{}^{+\dagger}{}_b(p'*,s')e^{-i(p'*\cdot y+p'\cdot y*)/2 + i(p\cdot x + p*\cdot x*)/2}]$$

$$= 2^{-1/2}\sum_{\pm s} \int d^3p_i d^2p_r dp^+ [N_{CLL}{}^+(p)]^2\theta(p^+)[\delta((p_i{}^3(p^+ - p^-)/\sqrt{2} + \mathbf{p}_{r\perp}\cdot\mathbf{p}_{i\perp})/m^2)]^2 \cdot$$

$$\cdot[u_{CLL}{}^+{}_a(p,s)u_{CLL}{}^{+\dagger}{}_b(p*,s)e^{+i(p*\cdot(y-x)+p\cdot(y*-x*))/2} + v_{CLL}{}^+{}_a(p,s)v_{CLL}{}^{+\dagger}{}_b(p*,s)e^{-i(p*\cdot(y-x)+p\cdot(y*-x*))/2}]$$

$$= -2^{-3/2}\int d^3p_i d^2p dp^+\theta(p^+)[N_{CLL}{}^+(p)]^2\delta'((p_i{}^3(p^+ - p^-)/\sqrt{2} + \mathbf{p}_{r\perp}\cdot\mathbf{p}_{i\perp})/m^2)(2m)^{-1}\cdot$$

$$\cdot\{[C^-R^+(i\not{p} + m)\gamma^0R^+C^-]_{ab}e^{+i(p*\cdot(y-x)+p\cdot(y*-x*))/2} +$$

$$+[C^-R^+(i\not{p} - m)\gamma^0R^+C^-]_{ab}e^{-i(p*\cdot(y-x)+p\cdot(y*-x*))/2}\}$$

$$= -(1/2)C^-R^+\delta_{ab} \int d^3p_i d^2p_\perp \int_0^\infty dp^+ \, \delta'((p_i{}^3(p^+ - p^-)/\sqrt{2} + \mathbf{p}_\perp\cdot\mathbf{p}_{i\perp})/m^2)(2\pi)^{-6}\cdot$$

$$\cdot\{e^{+i\{p^+(y^- - x^-) - \mathbf{p}_{r\perp}\cdot(\mathbf{y}_{r\perp} - \mathbf{x}_{r\perp}) + \mathbf{p}_i\cdot(\mathbf{y}_i - \mathbf{x}_i)\}} + e^{-i\{p^+(y^- - x^-) - \mathbf{p}_{r\perp}\cdot(\mathbf{y}_{r\perp} - \mathbf{x}_{r\perp}) + \mathbf{p}_i\cdot(\mathbf{y}_i - \mathbf{x}_i)\}}\}$$

$$= -C^-R^+\delta_{ab}(4\pi)^{-1}\int_0^\infty dp^+\delta'(\nabla_r\cdot\nabla_i/m^2)\delta^3(y_i-x_i) \, \delta^2(y_r-x_r)\{e^{+ip^+(y^- - x^-)}+e^{-ip^+(y^- - x^-)}\}$$

whereupon we revert back to the original form of the constraint: $\delta(\nabla_r \cdot \nabla_i / m^2)$

$$\{\psi_{CLL}{}^+{}_a(x), \psi_{CLL}{}^{+\dagger}{}_b(y)\} = -(1/2)C^-R^+\delta_{ab}\,\delta'(\nabla_r \cdot \nabla_i/m^2)\delta(y^- - x^-)\delta^2(y_r - x_r)\delta^3(y_i - x_i)$$

$$(26.246)$$

The result is the left-handed, light-front equivalent of the earlier non-tachyon result. Again the constraint is apparent in the anti-commutator. (The factor of 2 difference is due to light-front coordinate definitions.)

Therefore we have left-handed, light-front quantized tachyonic complexons with the equivalent of canonical anti-commutation relations, and with localized tachyonic complexons. As a result we have a canonical tachyonic complexon Quantum Field Theory.

26.5.3.2 Left-handed Case 4: Tachyonic Complexon Feynman Propagator

The light-front Feynman propagator for the left-handed $\psi_{CLL}{}^+$ *tachyonic complexon field is*

$$iS^+{}_{CLLF}(x,y) = \theta(x^+ - y^+)<0|\psi_{CLL}{}^+(x)\psi_{CLL}{}^{+\dagger}(y)\gamma^0|0> - \theta(y^+ - x^+)<0|\psi_{CLL}{}^{+\dagger}(y)\gamma^0\psi_{CLL}{}^+(x)|0>$$

$$(26.247)$$

$$= -\tfrac{1}{2}\int d^3p_i d^2p_r dp^+ \theta(p^+) N_{CLL}{}^{+2}\delta'((p_i{}^3(p^+ - p^-)/\sqrt{2} + \mathbf{p}_{r\perp}\cdot\mathbf{p}_{i\perp})/m^2)(2m)^{-1}C^-R^+ \cdot$$

$$\cdot\{\theta(x^+ - y^+)[(i\not{p} + m)\gamma^0]e^{+i(p^*\cdot(y-x)+p\cdot(y^*-x^*))/2} +$$

$$+ \theta(y^+ - x^+)[(i\not{p} - m)\gamma^0]e^{-i(p^*\cdot(y-x)+p\cdot(y^*-x^*))/2}\}R^+C^-\gamma^0$$

If we define the on-shell momentum variables

$$p_0{}^- = (p_{r0}{}^1 p_{r0}{}^1 + p_{r0}{}^2 p_{r0}{}^2 - \mathbf{p}_{i0}\cdot\mathbf{p}_{i0} - m^2)/(2p_0{}^+)$$
$$p_0{}^+ = p^+, \; p_{r0}{}^j = p_r{}^j \quad \text{(for } j = 1, 2),$$
$$\mathbf{p}_{i0} = \mathbf{p}_i, \; p_{r\perp 0}{}^2 = p_{r0}{}^j p_{r0}{}^j$$
$$\not{p}_0 = p_0 \cdot \gamma$$

with $p_0 = (p^0, \mathbf{p}_{r0} + i\mathbf{p}_{r0})$ then the above equation can be rewritten as

$$iS^+_{CLLF}(x,y) = -\tfrac{1}{2}C^-R^+\int d^4p d^3p_i N_{CLL}^{+2}\delta'((p_{i0}{}^3(p_0{}^+-p_0{}^-)/\sqrt{2}+\mathbf{p}_{r\perp 0}\cdot\mathbf{p}_{i\perp 0})/m^2)(4\pi m)^{-1}e^{+i(p^*\cdot(y-x)+p\cdot(y^*-x^*))/2}\cdot$$
$$\cdot\{\theta(p^+)(i\not{p}+m)\gamma^0]/[p^--p_0{}^-+i\varepsilon]+\theta(-p^+)(i\not{p}-m)\gamma^0]/[p^-+p_0{}^--i\varepsilon]\}R^+C\gamma^0$$

$$= -\tfrac{1}{2}\int d^4p_r d^3p_i\, N_{CLL}^{+2}\delta'((p_{i0}{}^3(p^+-p^-)/\sqrt{2}+\mathbf{p}_{r\perp}\cdot\mathbf{p}_{i\perp})/m^2)(p^+/4\pi m)\,e^{+i(p^*\cdot(y-x)+p\cdot(y^*-x^*))/2}\cdot$$
$$\cdot[C^-R^+(i\not{p}+m)\gamma^0 R^+C^-\gamma^0][(p^2+m^2+i\varepsilon)]^{-1}$$

with $p_r = (p^0, \mathbf{p}_r)$ and $p = (p^0, \mathbf{p}_r + i\mathbf{p}_r)$. Substituting for N_{CLL} and using $x\delta'(x) = -\delta(x)$ we obtain

$$= -\tfrac{1}{2}\int d^4p_r d^3p_i (2\pi)^{-7}\delta'(\mathbf{p}_r\cdot\mathbf{p}_i/m^2)\exp[ip^0(y^0-x^0)-i\mathbf{p}_r\cdot(\mathbf{y}_r-\mathbf{x}_r)+i\mathbf{p}_i\cdot(\mathbf{y}_i-\mathbf{x}_i)]\cdot$$
$$\cdot[C^-R^+(i\not{p}+m)R^-C^+]/(p^2+m^2+i\varepsilon)$$

since $C^-R^+(i\not{p}+m)\gamma^0 R^+C^-\gamma^0 = C^-R^+(i\not{p}+m)R^-C^+$. The integral can then be written:

$$iS^+_{CLLF}(x,y) = \int d^4p_r d^3p_i \delta'(\mathbf{p}_r\cdot\mathbf{p}_i/m^2)C^-R^+(i\not{p}+m)R^-C^+\cdot$$
$$\cdot\exp[-ip^0(x^0-y^0)+i\mathbf{p}_r\cdot(\mathbf{x}_r-\mathbf{y}_r)-i\mathbf{p}_i\cdot(\mathbf{x}_i-\mathbf{y}_i)]/(p^2+m^2+i\varepsilon)$$

$$= \int d^4p_r dM^2 \delta'(\nabla_r\cdot\nabla_i/m^2)C^-R^+(ip^0\gamma^0-(\nabla_r-i\nabla_i)\cdot\boldsymbol{\gamma}+m)R^-C^+\cdot$$
$$\cdot\exp[-ip^0(x^0-y^0)+i\mathbf{p}_r\cdot(\mathbf{x}_r-\mathbf{y}_r)]J_2(\mathbf{x}_i-\mathbf{y}_i,M^2)/(p_r{}^2+M^2+i\varepsilon)$$

where

$$J_2(\mathbf{x}_i-\mathbf{y}_i, M^2) = (2\pi)^{-3}\int d^3p_i\,\delta(M^2-\mathbf{p}_i{}^2-m^2)\exp[-i\mathbf{p}_i\cdot(\mathbf{x}_i-\mathbf{y}_i)] \quad (26.248)$$
$$= (2\pi)^{-2}|\mathbf{x}_i-\mathbf{y}_i|^{-1}\theta(M^2-m^2)\sin((M^2-m^2)^{\frac{1}{2}}|\mathbf{x}_i-\mathbf{y}_i|)$$

This tachyonic complexon Feynman propagator can be rearranged into the form of a spectral integral:

$$iS^+_{CLLF}(x, y) = -\int dM\, C^-R^+(\gamma^0\partial/\partial x^0+(\nabla_r-i\nabla_i)\cdot\boldsymbol{\gamma}-m)R^-C^+\delta'(\nabla_r\cdot\nabla_i/m^2)\cdot$$
$$\cdot J_2(\mathbf{x}_i-\mathbf{y}_i, M^2)\Delta_{FT}(x-y,M) \quad (26.249)$$

with ∇_r and ∇_i derivatives with respect to $\mathbf{x_r}$ and $\mathbf{x_i}$ and where

$$\Delta_{FT}(x - y, M) = (2\pi)^{-4}\int d^4p_r \exp[-ip^0(x^0 - y^0) + i\mathbf{p_r} \cdot (\mathbf{x_r} - \mathbf{y_r})]/(p_r^2 + M^2 + i\varepsilon) \tag{26.250}$$

26.5.3.3 Case 5: Right-Handed Tachyonic Complexons

The case of right-handed tachyonic complexons is similar to left-handed complexons with only one difference: a minus sign in the canonical right-handed equal-time commutation relations resulting in a minus sign in the creation and annihilation operator anti-commutation relations. The right-handed tachyonic complexon wave function light-front Fourier expansion is:

$$\psi_{CLR}{}^+(\mathbf{x_r}, \mathbf{x_i}) = \sum_{\pm s} \int d^2p_r dp^+ d^3p_i \, N_{CLR}{}^+(p)\theta(p^+)\delta((p_i{}^3(p^+ - p^-)/\sqrt{2} + \mathbf{p_{r\perp}} \cdot \mathbf{p_{i\perp}})/m^2) \cdot$$
$$\cdot [b_{CLR}{}^+(p, s)u_{CLR}{}^+(p, s)e^{-i(p \cdot x + p^* \cdot x^*)/2} + d_{CLR}{}^{+\dagger}(p, s)v_{CLR}{}^+(p, s)e^{+i(p \cdot x + p^* \cdot x^*)/2}] \tag{26.251}$$

where

$$N_{CLR}{}^+(p) = (2\pi)^{-3}(2m/p^+)^{\frac{1}{2}} \tag{26.252}$$

Its hermitean conjugate is

$$\psi_{CLR}{}^{+\dagger}(\mathbf{x_r}, \mathbf{x_i}) = \sum_{\pm s} \int d^2p_r dp^+ d^3p_i \, N_{CLR}{}^+(p)\theta(p^+)\delta((p_i{}^3(p^+ - p^-)/\sqrt{2} + \mathbf{p_{r\perp}} \cdot \mathbf{p_{i\perp}})/m^2) \cdot$$
$$\cdot [b_{CLR}{}^\dagger(p^*, s)u_{CLR}{}^+(p^*, s)e^{+i(p^* \cdot x + p \cdot x^*)/2} + d_{CLR}(p^*, s)v_{CLR}{}^+(p^*, s)e^{-i(p^* \cdot x + p \cdot x^*)/2}] \tag{26.253}$$

where $\mathbf{p} = \mathbf{p_r} + i\mathbf{p_i}$, $\mathbf{x} = \mathbf{x_r} - i\mathbf{x_i}$, $p \cdot x = p^0 x^0 - \mathbf{p} \cdot \mathbf{x}$, and † indicates hermitean conjugate. The right-handed spinors are

$$u_{CLR}{}^+(p, s) = C^+ R^+ S_{CR}w^1(0)$$
$$u_{CLR}{}^+(p, -s) = C^+ R^+ S_{CR}w^2(0)$$
$$v_{CLR}{}^+(p, s) = C^+ R^+ S_{CR}w^3(0)$$
$$v_{CLR}{}^+(p, -s) = C^+ R^+ S_{CR}w^4(0) \tag{26.254}$$

$$u_{CLR}^{++\dagger}(p^*, s) = w^{1T}(0)S_{CR}R^+C^+$$
$$u_{CLR}^{++\dagger}(p^*, -s) = w^{2T}(0)S_{CR}R^+C^+$$
$$v_{CLR}^{++\dagger}(p^*, s) = w^{3T}(0)S_{CR}R^+C^+$$
$$v_{CLR}^{++\dagger}(p^*, -s) = w^{4T}(0)S_{CR}R^+C^+$$

where the superscript "T" indicates the transpose. The anti-commutation relations of the Fourier coefficient operators are

$$\{b_{CLR}(p,s), b_{CLR}^\dagger(p'^*,s')\} = -2^{-\frac{1}{2}}\delta_{ss'}\delta(p^+ - p'^+)\delta^2(\mathbf{p_r} - \mathbf{p'_{r'}})\delta^3(\mathbf{p_i} + \mathbf{p'_{i'}})$$
$$\{d_{CLR}(p,s), d_{CLR}^\dagger(p'^*,s')\} = -2^{-\frac{1}{2}}\delta_{ss'}\,\delta(p^+ - p'^+)\delta^2(\mathbf{p_r} - \mathbf{p'_{r'}})\delta^3(\mathbf{p_i} + \mathbf{p'_{i'}})$$
$$\{b_{CLR}(p,s), b_{CLR}(p'^*,s')\} = \{d_{CLR}(p,s), d_{CLR}(p'^*,s')\} = 0$$
$$\{b_{CLR}^\dagger(p,s), b_{CLR}^\dagger(p'^*,s')\} = \{d_{CLR}^\dagger(p,s), d_{CLR}^\dagger(p'^*,s')\} = 0 \qquad (26.255)$$
$$\{b_{CLR}(p,s), d_{CLR}^\dagger(p'^*,s')\} = \{d_{CLR}(p,s), b_{CLR}^\dagger(p'^*,s')\} = 0$$
$$\{b_{CLR}^\dagger(p,s), d_{CLR}^\dagger(p'^*,s')\} = \{d_{CLR}(p,s), b_{CRR}(p'^*,s')\} = 0$$

The spinors satisfy

$$\sum_{\pm s} u_{CLR}^+{}_\alpha(p, s)\bar{u}_{CLR}^+{}_\beta(p^*, s) = (2m)^{-1}[C^+R^+(-i\not{p} + m)R^-C^-]_{\alpha\beta} \qquad (26.256)$$

$$\sum_{\pm s} v_{CLR}^+{}_\alpha(p, s)\bar{v}_{CLR}^+{}_\beta(p^*, s) = (2m)^{-1}[C^+R^+(-i\not{p} - m)R^-C^-]_{\alpha\beta}$$

where $\bar{u}_{CLR}^+ = u_{CLR}^{++\dagger}\gamma^0$ and $\bar{v}_{CLR}^+ = v_{CLR}^{++\dagger}\gamma^0$.

The right-handed anti-commutation relation with a minus sign follows in particular because of the minus signs in eqs. 26.255.

26.5.3.4 Right-handed Case 5: Tachyonic Complexon Feynman Propagator

The Feynman propagator for right-handed tachyonic complexons can be obtained from eqs. 26.249 and 26.250 by changing the parity projection operator and some numerator signs in the integral (basically $p \to -p$) resulting in

$$iS^+_{CLRF}(x, y) = \int dM \; C^+ R^+ (\gamma^0 \partial/\partial x^0 + (\nabla_r - i\nabla_i)\cdot\gamma - m)R^- C^- \delta'(\nabla_r \cdot \nabla_i/m^2) \cdot$$
$$\cdot J_2(x_i - y_i, M^2)\triangle_{FT}(x - y, M) \qquad (26.257)$$

with $\nabla_r + i\nabla_i$ derivatives with respect to x_r and x_i and where

$$\triangle_{FT}(x - y, M) = (2\pi)^{-4}\int d^4p_r \exp[-ip^0(x^0 - y^0) + ip_r\cdot(x_r - y_r)]/(p_r^2 + M^2 + i\varepsilon)$$
$$(26.258)$$

26.5.3.5 Other Cases? No

The four cases considered above are the only cases having symmetry under the real Lorentz group L and a single real energy (with a corresponding single real time parameter) that is independent of the direction of the boost thus preserving (real) spatial rotation invariance. The reality of the time variable survives the breakdown to conventional Lorentz invariance.

One might think that using the other type of spinor boost operator.

$$S_{CR}(\Lambda_{CR}(\omega, \hat{w})) = \exp(-i\omega_R\sigma_{0i}w_i/2) = \exp(-\omega_R\gamma^0\gamma\cdot\hat{w}/2) \quad (26.259)$$
$$= \cosh(\omega_R/2)I + \sinh(\omega_R/2)\gamma^0\gamma\cdot\hat{w}$$

where $\omega_R = \omega - i\pi/2$ might lead to more possible forms of spin ½ wave equations and particles. In fact it merely leads to the same particle types but with the role of the left-handed and right-handed fields reversed. The result would be a "right-handed" Standard Model contrary to experiment.

26.6 Spinor Boosts Generate 4 Species of Particles: Leptons and Quarks

In this appendix we have found four types of fermions using complex Lorentz boosts that correspond in a natural way with the four general *species* (types) of known fermions: charged leptons, neutrinos, up-type color quarks and down-type color quarks.[147]

26.6.1 Charged lepton fermions

The conventional Dirac equation and solutions.

26.6.2 Neutrinos

Simple tachyons with real energy and 3-momentum. Their free field equation is:

$$(\gamma^\mu \partial/\partial x^\mu - m)\psi_T(x) = 0 \tag{26.260}$$

and their left-handed ψ_{TL}^+ Feynman propagator is:

$$iS^+_{TLF}(x, y) = \tfrac{1}{2}C^-R^+\gamma^0 \int d^4p(2\pi)^{-4}\, p^+ e^{-ip\cdot(x-y)}/(p^2 + m^2 + i\varepsilon) \tag{26.261}$$

Similarly the light-front Feynman propagator for the right-handed ψ_{TR}^+ tachyon field is

$$iS^+_{TRF}(x,y) = -\tfrac{1}{2}C^+R^+\gamma^0 \int d^4p(2\pi)^{-4}\, p^+ e^{-ip\cdot(x-y)}/(p^2 + m^2 + i\varepsilon) \tag{26.262}$$

26.6.3 Up-type Color Quarks

Up-type quarks are assumed[148] to be fermions with complex 3-momenta - complexons, and an internal color SU(3) symmetry, that satisfy $p^2 = m^2$. Their field equation with a color SU(3) index, denoted a, inserted is

---

[147] We call each type of fermion a *species*. Each species has three known generations.

[148] The complexon theory that we develop and use for quark dynamics in the Standard Model is <u>not</u> required. Our Standard Model could use Dirac fermion dynamics for the up-type quarks and tachyon dynamics for down-type quarks. Then the (broken) Left-handed complex Lorentz boosts would have the basic space-time group rather than L_C. We choose to use complexon dynamics for quarks because they have an internal SU(3)-like structure suggestive of color SU(3). More importantly, their spin dynamics is different and thus may resolve the differences between theory and experiment for the deep inelastic parton spin-dependent structure functions.

$$[i\gamma^0\partial/\partial t + i\gamma\cdot(\nabla_r + i\nabla_i) - m]\psi_C^a(t, \mathbf{x_r}, \mathbf{x_i}) = 0 \qquad (26.263)$$

with the subsidiary condition

$$\nabla_r\cdot\nabla_i\ \psi_C^a(t, \mathbf{x_r}, \mathbf{x_i}) = 0 \qquad (26.264)$$

The free field solution is:

$$\psi_C^a(x) = \sum_{\pm s} \int d^3p_r d^3p_i\ N_C(p)\delta(\mathbf{p_r}\cdot\mathbf{p_i}/m^2)[b_C(p,a,s)u_C^a(p, s)e^{-i(p\cdot x + p^*\cdot x^*)/2} +$$
$$+ d_C^\dagger(p,a,s)v_C^a(p, s)e^{+i(p\cdot x + p^*\cdot x^*)/2}] \qquad (26.265)$$

The free Feynman propagator arranged into the form of a spectral integral is

$$iS_C^{ab}(x,y)= -\delta^{ab}\int dM\ (i\gamma^0\partial/\partial x^0 - i(\nabla_r-i\nabla_i)\cdot\gamma + m)\delta'(\nabla_r\cdot\nabla_i/m^2)J(\mathbf{x_i} - \mathbf{y_i}, M^2)\Delta_F(x - y, M) \qquad (26.266)$$

where

$$\Delta_F(x - y, M) = (2\pi)^{-4}\int d^4p_r\ \exp[-ip^0(x^0 - y^0) + i\mathbf{p_r}\cdot(\mathbf{x_r} - \mathbf{y_r})]/(p_r^2 - M^2 + i\varepsilon) \qquad (26.267)$$

and

$$J(\mathbf{x_i}, M^2) = (2\pi)^{-3}\int d^3p_i\ \delta(M^2 + \mathbf{p_i}^2 - m^2)\ \exp[-i\mathbf{p_i}\cdot(\mathbf{x_i} - \mathbf{y_i})] \qquad (26.268)$$
$$= (2\pi)^{-2}|\mathbf{x_i} - \mathbf{y_i}|^{-1}\theta(m^2 - M^2)\sin((m^2 - M^2)^{\frac{1}{2}}|\mathbf{x_i} - \mathbf{y_i}|)$$

26.6.4 Down-type Color Quarks

Tachyonic complexons with complex 3-momenta, and an internal global SU(3) symmetry, that have mass shell condition $p^2 = -m^2$. Their field equation with a color SU(3) index, denoted a, inserted is

$$[\gamma^0\partial/\partial t + \gamma\cdot(\nabla_r + i\nabla_i) - m]\psi_{CL}^a(t, \mathbf{x_r}, \mathbf{x_i}) = 0 \qquad (26.269)$$

with the subsidiary condition on the wave function

$$\nabla_r \cdot \nabla_i \, \psi_{CL}{}^a(t, \, \mathbf{x_r}, \, \mathbf{x_i}) = 0 \qquad (26.270)$$

Its free field left-handed solution is:

$$\psi_{CLL}{}^{+a}(\mathbf{x_r}, \mathbf{x_i}) = \sum_{\pm s} \int d^2 p_r dp^+ d^3 p_i \, N_{CLL}{}^+(p)\theta(p^+)\delta((p_i{}^3(p^+ - p^-)/\surd 2 + \mathbf{p_{r\perp}} \cdot \mathbf{p_{i\perp}})/m^2) \cdot$$
$$\cdot [b_{CLL}{}^+(p,a,s)u_{CLL}{}^a(p,a,s)e^{-i(p \cdot x \,+\, p^* \cdot x^*)/2} + d_{CLL}{}^{++}(p,a,s)v_{CLL}{}^{+a}(p,a,s)e^{+i(p \cdot x \,+\, p^* \cdot x^*)/2}]$$
$$(26.271)$$

and its right-handed solution is

$$\psi_{CLR}{}^{+a}(\mathbf{x_r}, \mathbf{x_i}) = \sum_{\pm s} \int d^2 p_r dp^+ d^3 p_i \, N_{CLR}{}^+(p)\theta(p^+)\delta((p_i{}^3(p^+ - p^-)/\surd 2 + \mathbf{p_{r\perp}} \cdot \mathbf{p_{i\perp}})/m^2) \cdot$$

$$\cdot [b_{CLR}{}^+(p,a,s)u_{CLR}{}^{+a}(p,a,s)e^{-i(p \cdot x + p^* \cdot x^*)/2} + d_{CLR}{}^{++}(p,a,s)v_{CLR}{}^{+a}(p,a,s)e^{+i(p \cdot x + p^* \cdot x^*)/2}]$$
$$(26.272)$$

The free left-handed Feynman propagator arranged into the form of a spectral integral is

$$iS^+{}_{CLLF}{}^{ab}(x,y) = -\delta^{ab}\int dM \, C^- R^+ (\gamma^0 \partial/\partial x^0 + (\nabla_r - i\nabla_i) \cdot \gamma - m)R^- C^+ \delta'(\nabla_r \cdot \nabla_i/m^2) \cdot$$
$$\cdot J_2(\mathbf{x_i} - \mathbf{y_i}, M^2)\Delta_{FT}(x - y, M) \qquad (26.273)$$

with ∇_r and ∇_i derivatives with respect to $\mathbf{x_r}$ and $\mathbf{x_i}$ and where

$$\Delta_{FT}(x - y, M) = (2\pi)^{-4}\int d^4 p_r \exp[-ip^0(x^0 - y^0) + i\mathbf{p_r} \cdot (\mathbf{x_r} - \mathbf{y_r})]/(p_r{}^2 + M^2 + i\varepsilon) \quad (26.274)$$

and

$$J_2(\mathbf{x_i}, M^2) = (2\pi)^{-3}\int d^3 p_i \delta(M^2 - \mathbf{p_i}^2 - m^2) \exp[-i\mathbf{p_i} \cdot (\mathbf{x_i} - \mathbf{y_i})] \qquad (26.275)$$

$$= (2\pi)^{-2}|\mathbf{x_i} - \mathbf{y_i}|^{-1}\theta(M^2 - m^2)\sin((M^2 - m^2)^{\frac{1}{2}}|\mathbf{x_i} - \mathbf{y_i}|)$$

The free right-handed Feynman propagator arranged into the form of a spectral integral is

$$iS^+_{CLRF}{}^{ab}(x, y) = \delta^{ab}\int dM\; C^+R^+(\gamma^0\partial/\partial x^0 + (\nabla_r - i\nabla_i)\cdot\gamma - m)R^-C^-\delta'(\nabla_r\cdot\nabla_i/m^2)\cdot$$
$$\cdot J_2(\mathbf{x_i} - \mathbf{y_i}, M^2)\triangle_{FT}(x - y, M) \qquad (26.276)$$

with ∇_r and ∇_i derivatives with respect to $\mathbf{x_r}$ and $\mathbf{x_i}$, and where

$$\triangle_{FT}(x - y, M) = (2\pi)^{-4}\int d^4 p_r \exp[-ip^0(x^0 - y^0) + i\mathbf{p_r}\cdot(\mathbf{x_r} - \mathbf{y_r})]/(p_r{}^2 + M^2 + i\varepsilon) \qquad (26.277)$$

26.7 First Step Towards The SuperStandard Model

Thus we have derived a set of four fermion species that corresponds to the known fermions of one fermion generation from the Complex Lorentz Group.[149] We derive the four fermion generation form of the model based on a U(4) group that we call the Generation group. We derive the Generation group from conservation laws for baryon and lepton number.

The overall pattern that begins to emerge from the developments in this appendix divides particles and interactions into two categories (as seen in Nature):

Particles with real 4-Momenta	Complexons (Complex 3-Momenta)
Leptons	Color quarks
SU(2)⊗U(1) Vector Bosons	Color SU(3) gluons
Higgs Particles	Possibly Higgs Particles

Basically the leptons, SU(2)⊗U(1) Vector Bosons and a set of Higgs particles appear to be primarily based on the Left-handed boosts. These particles have real energies and momenta although some are "normal" and some are tachyons.

---

[149] Complex Lorentz group boosts lead to tachyons.

Another category of particles, complexons, emerges from our study of L_C. These particles have real energies and complex 3-momenta. In perturbation theory the loop integrations of loops of these particles would consist of a 7-fold integration over energy and complex 3-momenta with corresponding 7-fold delta functions to enforce energy-momentum conservation. As pointed out earlier the complex 3-momenta of these types of fermions has an SU(3) symmetry that it is natural to generalize to local color SU(3). (The other category of fermions, leptons. lack global SU(3) symmetry.) Thus we see the beginnings of the structure of the SuperStandard Theory in this appendix on spin ½ particles.

26.8 Dirac-like Equations of Matter from 4-Valued Logic

In our derivation every truly fundamental particle of matter, whether quark or lepton, has spin ½. We have seen in chapter 10 of Blaha (2011c) that the basic algebra of Operator Logic eigenvalue operators, and that of its raising and lowering operators, is the same as the algebra of creation and annihilation operators for free spin ½ particles. Our goal is to build our theory on the scaffolding of Operator Logic. We view a fermion particle as a qube core which is dressed in spatial coordinates (and internal symmetries):

$$\text{Qube core} + \text{coordinates} \rightarrow \text{fermion particle} \qquad (26.278)$$

The creation and annihilation operators $b(p,s)$ and $d^{\dagger}(p,s)$ (and their hermitean conjugates $b^{\dagger}(p,s)$ and $d(p,s)$) are mathematically similar to the raising and lowering operators of Operator (Matrix) Logic. They satisfy the anticommutation relations

$$\{b(q,s), b^{\dagger}(p,s')\} = \delta_{ss'}\delta^3(\mathbf{q} - \mathbf{p}) \qquad (26.279)$$
$$\{d(q,s), d^{\dagger}(p,s')\} = \delta_{ss'}\delta^3(\mathbf{q} - \mathbf{p})$$

Thus we see spin ½ particle wave functions originating from the Dirac-like spinors, and raising and lowering operators of the spinor formulation of Operator Logic.

When particles interact, the quantum field theory interaction terms use fermion creation operators, $b(q,s)$ *and* $d^{\dagger}(q,s)$, *and annihilation operators,* $b^{\dagger}(p,s')$ *and* $d(q,s)$, *to implement the transformations between the Qubes of the interacting particles.*[150] *Thus the mathematics of the embedded Qubes' logic values is automatically implemented within quantum field theoretic calculations.*

An interesting point that emerges from this discussion is the nature of spin ½ particle states such as

$$|p, s> = b^{\dagger}(p, s)|0> \qquad (26.280)$$

This state is interpreted as a one particle state. It also has an analogous interpretation in Operator Logic as creating a one term universe of discourse – a construct which is in part linguistic and in part logic. Thus particles are embodiments of Logic values and particle interactions change the logic values of the initial particles to those of the emergent particles. All in all, our universe can be viewed as an extraordinarily intricate logic machine. Serendipitously we are now seeing the use of particles to create quantum computers, which, in a sense, is bringing us full circle. Particles are Logic; Logic machines emerge from particle interactions.

26.9 Why Second Quantization of Fields?

One might have argued that the fermion field types that we have found could be treated as ordinary c-number fields and not be second quantized. However, particles are discrete entities that can be enumerated with integers. Second quantization implements the discrete particle concept in the most direct way and thus by Leibiz's Principle as well as Ockham's Razor second quantization is the best solution to obtain particle discreteness.

Quantum Theory is required by the discreteness of particles.

26.10 Why lagrangians? For dynamic evolution

Lagrangians naturally emerge as the 'preferred' formalism for quantum field theory due to their intimate relation with the energy-momentum tensor (particularly the Hamiltonian) that provides the generators of time evolution and of spatial translation.

---

[150] See chapter 3.

26.11 Functional Expression for Each of the four Species of Fermions

We have derived the four species of fermions in this appendix. We have used a 'conventional' notation for quantum fields. In this section we will define these quantum fields as inner products of functionals and fourier coordinate expansions.

Dirac Quantum Field:

$$\psi(x) = (_1f, \text{ Dirac\_fourier\_expansion})$$

Tachyon Quantum Field:

$$\psi_T(x) = (_2f, \text{ Tachyon\_fourier\_expansion})$$

Complexon Quantum Field:

$$\psi_C(x) = (_3f, \text{ Complexon\_fourier\_expansion})$$

Complexon Tachyon Quantum Field:

$$\psi_{CT}(x) = (_4f, \text{ Tachyon\_Complexon\_fourier\_expansion})$$

The digit prefixes of $_kf$ for k = 1, 2, 3, 4 distinguish the functionals for each species.

In addition we can decompose the above quantum fields into left-handed and right-handed fields. The left-handed functional representations are:

Left Dirac Quantum Field:

$$\psi_L(x) = (_{1L}f, \text{ left-handed\_Dirac\_fourier\_expansion})$$

Left Tachyon Quantum Field:

$$\psi_{TL}(x) = (_{2L}f, \text{ left-handed\_Tachyon\_fourier\_expansion})$$

Left Complexon Quantum Field:

$$\psi_{CL}(x) = (_{3L}f, \text{left-handed\_Complexon\_fourier\_expansion})$$

Left Complexon Tachyon Quantum Field:

$$\psi_{CTL}(x) = (_{4L}f, \text{left-handed\_Tachyon\_Complexon\_fourier\_expansion})$$

The right-handed cases have analogous forms.

26.12 Leptonic Tachyon Spinors

The general form of the solutions of the free tachyon Dirac equation can be written

$$\psi_T^r(x) = e^{-i\chi_r p \cdot x} w^r(p) \tag{26.281}$$

where $\chi_r = +1$ for $r = 1, 2$ and $\chi_r = -1$ for $r = 3, 4$. Denoting the spinors $w^r(p) = w^r(0)$ for a particle is at rest in a frame ($E = m$) we see they can take the form

$$w^r(0) = \begin{bmatrix} \delta_{1r} \\ \delta_{2r} \\ \delta_{3r} \\ \delta_{4r} \end{bmatrix} \tag{26.282}$$

where Kronecker deltas appear in the brackets. From eq. 26.30 we find

$$S_L(\Lambda_L(\omega, \mathbf{u}))w^r(0) = w_T^r(p) \tag{26.283}$$

Using eq. 26.66 for $S_L(\Lambda_L(\omega, \mathbf{u}))$ and

$$\mathbf{p} = m\mathbf{v}\gamma_s \qquad\qquad E = m\gamma_s \tag{26.284}$$

we see that eq. 26.283 implies the columns of the resulting $S_L(\Lambda_L(\omega, \mathbf{u}))$ matrix are

$$S_L(\Lambda_L(\omega, \mathbf{u})) = \begin{bmatrix} \underline{w_T^3(p)} & \underline{w_T^4(p)} & \underline{w_T^1(p)} & \underline{w_T^2(p)} \\ \cosh(\omega_L/2) & 0 & \sinh(\omega_L/2)p_z/p & \sinh(\omega_L/2)p_-/p \\ 0 & \cosh(\omega_L/2) & \sinh(\omega_L/2)p_+/p & -\sinh(\omega_L/2)p_z/p \\ \sinh(\omega_L/2)p_z/p & \sinh(\omega_L/2)p_-/p & \cosh(\omega_L/2) & 0 \\ \sinh(\omega_L/2)p_+/p & -\sinh(\omega_L/2)p_z/p & 0 & \cosh(\omega_L/2) \end{bmatrix}$$

$$(26.285)$$

based on the superluminal transformation of positive energy states to negative energy states with $p_\pm = p_x \pm ip_y$ and where $p = |\mathbf{p}|$. It is easy to verify

$$(i\not{p} - \chi_r m)w_T^r(p) = 0 \qquad (26.286)$$

where $\chi_r = -1$ for $r = 1, 2$ and $\chi_r = +1$ for $r = 3, 4$.

The spinors that we defined earlier can be generalized in a manner similar to Dirac spinors. We will use a similar notation to the Dirac spinor notation:

$$\begin{aligned} u_T(p, s) &= w_T^1(p) \\ u_T(p, -s) &= w_T^2(p) \\ v_T(p, s) &= w_T^3(p) \\ v_T(p, -s) &= w_T^4(p) \end{aligned} \qquad (26.287)$$

We define "double dagger" spinors:

$$\begin{aligned} u_T^{\ddagger}(p, s) &= u_T^{\dagger}(p, s)i\boldsymbol{\gamma}\cdot\mathbf{p}/|\mathbf{p}| \\ u_T^{\ddagger}(p, -s) &= u_T^{\dagger}(p, -s)i\boldsymbol{\gamma}\cdot\mathbf{p}/|\mathbf{p}| \\ v_T^{\ddagger}(p, s) &= v_T^{\dagger}(p, s)i\boldsymbol{\gamma}\cdot\mathbf{p}/|\mathbf{p}| \\ v_T^{\ddagger}(p, -s) &= v_T^{\dagger}(p, -s)i\boldsymbol{\gamma}\cdot\mathbf{p}/|\mathbf{p}| \end{aligned} \qquad (26.288)$$

where † indicates hermitean conjugate, which appear in important spinor "completeness" sums:

$$\sum_{\pm s} u_{T\alpha}(p, s)u_T^{\ddagger}{}_\beta(p, s) = (2m)^{-1}(i\not{p} - m)_{\alpha\beta} \tag{26.289}$$

$$\sum_{\pm s} v_{T\alpha}(p, s)v_T^{\ddagger}{}_\beta(p, s) = (2m)^{-1}(i\not{p} + m)_{\alpha\beta} \tag{26.290}$$

or

$$\sum_{\pm s} u_{T\alpha}(p, s)u_T^{\dagger}{}_\beta(p, s) = -i(2m)^{-1}[(i\not{p} - m)\boldsymbol{\gamma}\cdot\mathbf{p}/|\mathbf{p}|]_{\alpha\beta} \tag{26.291}$$

$$\sum_{\pm s} v_{T\alpha}(p, s)v_T^{\dagger}{}_\beta(p, s) = -i(2m)^{-1}[(i\not{p} + m)\boldsymbol{\gamma}\cdot\mathbf{p}/|\mathbf{p}|]_{\alpha\beta} \tag{26.292}$$

Lastly we define light-front, left-handed tachyon spinors by

$$
\begin{aligned}
u_{TL}^{+}(p, s) &= C^- R^+ S_L(\Lambda_L(\omega, \mathbf{u}))w^1(0) \\
u_{TL}^{+}(p, -s) &= C^- R^+ S_L(\Lambda_L(\omega, \mathbf{u}))w^2(0) \\
v_{TL}^{+}(p, s) &= C^- R^+ S_L(\Lambda_L(\omega, \mathbf{u}))w^3(0) \\
v_{TL}^{+}(p, -s) &= C^- R^+ S_L(\Lambda_L(\omega, \mathbf{u}))w^4(0)
\end{aligned}
\tag{26.293}
$$

$$
\begin{aligned}
u_{TL}^{+\dagger}(p, s) &= w^{1T}(0)\, S_L^{\dagger}(\Lambda_L(\omega, \mathbf{u}))\, R^+ C^- \\
u_{TL}^{+\dagger}(p, -s) &= w^{2T}(0)\, S_L^{\dagger}(\Lambda_L(\omega, \mathbf{u}))R^+ C^- \\
v_{TL}^{+\dagger}(p, s) &= w^{3T}(0)\, S_L^{\dagger}(\Lambda_L(\omega, \mathbf{u}))R^+ C^- \\
v_{TL}^{+\dagger}(p, -s) &= w^{4T}(0)\, S_L^{\dagger}(\Lambda_L(\omega, \mathbf{u}))R^+ C^-
\end{aligned}
\tag{26.294}
$$

where the superscript "T" indicates the transpose and † indicates hermitean conjugate.

Appendix 26-A Experimental Evidence for Faster-Than-Light Particles & Physics

Among the key assumptions of our Unified SuperStandard Theory are 1) that the speed of light is the same in all inertial reference frames and 2) that some fundamental particles (neutrinos and down-type quarks) travel faster than the speed of light.

In this appendix we describe convincing evidence for faster than light physics.

Until 1907 physicists thought that there was no limit on the speed of a particle or lump of matter. In 1907 Einstein and Poincaré showed that there was an inherent limit on the speed of a massive object – the speed of light. For the past 100 odd years physicists have generally accepted the speed of light as the limiting speed for particles with mass. Several theoretical physicists in the 1960's (E. C. Sudarshan and Gerald Feinberg) investigated the possibility of faster than light particles. They found that faster than light particles were theoretically possible but their theories – particularly their quantum field theories – had numerous discrepancies from canonical quantum field theory. These differences were taken by many to indicate that faster than light particles (called tachyons) were not present in nature. This belief was further supported by the happenings at particle accelerators where it was impossible to accelerate normal charged particles such as protons faster than the speed of light.

In the past fifteen years this author[151] developed a satisfactory quantum field theory of faster than light particles and found that if neutrinos and down-type quarks were faster than light particles he could derive the form of The Standard Model of Elementary Particles in detail. This theoretical development seems to have stimulated experimental groups at the new Linear Hadron Collider (LHC) at the CERN laboratory in Switzerland and the Gran Sasso Laboratory in Italy to measure the speed of neutrinos emitted in LHC particle collisions. The results, described below, were mixed and one can fairly say they neither proved nor disproved that neutrinos were tachyons.

---

[151] See Blaha (2012b) and earlier books extending back nine years.

However there is other experimental data that strongly indicate that neutrinos are tachyons, and that quantum mechanics requires – not just faster than light behavior – but in some circumstances instantaneous effects at a distance – infinite speed of transmission!

In this appendix we will look at experimentally proven instantaneous Quantum Mechanical effects, at tritium decay experiments over the past 20 years that imply faster than light neutrinos, at neutrino speed measurements at the CERN LHC and Gran Sasso, at tachyonic particle behavior inside of Black Holes, and at the tachyonic behavior of Higgs particles, the "so-called God particle." *The cumulative result of these considerations is that faster than light particles, and physics, are a part of nature.*

26-A.1 Instantaneous Quantum Mechanical Effects

Quantum entanglement is a quantum phenomenon wherein parts of a physical system are in a certain quantum state but are separated by a space-like distance. If a change is made in part of a quantum entangled system then it is known theoretically, and experimentally, that other parts of the system change instantaneously.[152] Many experiments have shown that the change in other parts of a system is instantaneous and thus can be viewed as taking place at infinite speed – obviously beyond the speed of light.[153] The most recent experiment by Juan Yin et al[154] has shown directly that quantum mechanical effects travel faster than 10,000 times the speed of light. These experimental results are consistent with the instantaneous speed predicted by quantum mechanics. Thus faster than light behavior is implicit in quantum theory and is experimentally verified.

26-A.2 Tritium Decay Experiments Yielding Neutrinos

Fact: Particles with negative values for the square of their mass are tachyons – particles moving faster than light.

---

[152] Matson, John, "Quantum Teleportation Achieved Over Record Distances" *Nature* **13**, August 2012.

[153] Francis, Matthew, "Quantum Entanglement Shows that Reality Can't be Local", *Ars Technica*, 30 October 2012.

[154] Juan Yin et al, arXiv[quant-ph]: 1303.0614V1 (March 4, 2013).

A series of experiments by various groups over recent years imply that electron neutrinos produced in tritium decay have negative mass squared despite the best efforts of experimenters to obtain positive values for the neutrino mass squared.

Experiment	measured mass squared	Year
Mainz	$-1.6 \pm 2.5 \pm 2.1$	2000
Troitsk	$-1.0 \pm 3.0 \pm 2.1$	2000
Zürich	$-24 \pm 48 \pm 61$	1992
Tokyo INS	$-65 \pm 85 \pm 65$	1991
Los Alamos	$-147 \pm 68 \pm 41$	1991
Livermore	$-130 \pm 20 \pm 15$	1995
China	$-31 \pm 75 \pm 48$	1995
1998 Average	-27 ± 20	1998

Table 26-A.1 Electron neutrino mass squared values found in various tritium decay experiments. (Masses are in units of eV.) The average mass squared is negative suggesting electron neutrinos are tachyons.

Table 26-A.1 summarizes the measured electron mass squared in these experiments. These experiments strongly suggest that neutrinos have negative mass squared and are thus faster-than-light particles - tachyons. However their small masses indicate that they only exceed the speed of light by a small amount.

26-A.3 LHC/Gran Sasso Direct Measurements of Neutrino Speeds

Two groups performed experiments at Gran Sasso Laboratory in Italy. They detected neutrinos emitted in interactions at the CERN LHC in Switzerland. The LVD collaboration in an exhaustive study of neutrino velocities found that the question was still open according to their data. Their refereed Physical Review Letter Abstract stated:

We report the measurement of the time of flight of ν_μ on the CNGS baseline (732 km) with the Large Volume Detector (LVD) at the Gran Sasso Laboratory. The CERN-SPS accelerator has been operated from May 10th to May 24th 2012, with a tightly bunched-beam structure to allow the velocity of neutrinos to be accurately measured on an event-by-event basis. LVD has detected 48 neutrino events, associated with the beam, with a high absolute time accuracy. These events allow us to establish the following limit on the difference between the neutrino speed and the light velocity: $-3.8\times10^{-6} < (v_\nu - c)/c < 3.1\times10^{-6}$ (at 99% C.L.). This value is an order of magnitude lower than previous direct measurements.[155]

These results (involving at least 35 neutrino detections) slightly favor, and do not rule out, faster-than-light neutrinos. Another experiment at the same locations by the ATLAS group stated that they found neutrino velocities (Five neutrinos were measured.) were below c. This group has not published their results as yet. We conclude that the published data appears to support faster than light neutrinos – consistent with our theory of The Standard Model.

A new project is in the planning stages to measure neutrino beams at larger distances. The hope is that the masses of the various neutrinos will be determined by the experiment. If the neutrino mass squared values turn out to be negative then it will constitute additional proof that neutrinos are tachyons (confirming tritium decay data), and thus support this author's formulation of The Standard Model of Elementary Particles.

26-A.4 Tachyonic Behavior Within Black Holes

Inside a black hole (such as the Schwarzschild solution of General Relativity) the time coordinate effectively becomes a spatial coordinate and the radius coordinate effectively becomes a time coordinate. An in-falling particle has a constantly decreasing

---

[155] N. Yu. Agafonova et al. (LVD Collaboration), "Measurement of the Velocity of Neutrinos from the CNGS Beam with the Large Volume Detector" Phys. Rev. Lett. **109**, 070801 (15 August 2012).

radial distance from the center of the black hole just as time always increases outside a black hole.

As a result of the interchange of the roles of time and radius the velocity of a particle descending radially inside a Black Hole has a speed faster than light and is tachyonic.

26-A.5 Higgs Fields are Tachyons

Recently groups at the LHC CERN laboratory have announced the discovery of Higgs particles. The dynamic equations for Higgs bosons in The Standard Model have a negative mass squared. The mass squared must be negative or the Higgs Mechanism could not generate particle masses. Having negative mass terms implies that Higgs fields are tachyonic – faster than light particles. Their tachyonic nature is masked by a quartic self-interaction that generates a condensate and thereby the masses of other particles.

26-A.6 Conclusion: Faster-Than-Light Particles – Tachyons Exist in Nature

The bulk of the experimental and theoretical evidence presented in previous sections strongly favors the existence of faster-than-light particles such as neutrinos. Tachyonic neutrinos are an important part of our form of The Standard Model. This form of the theory also strongly suggests that quarks are tachyonic in parallel with tachyonic neutrinos in order to obtain the symmetries of The Standard Model.

27. Fermion Core Qube Space for Quarks and Leptons

27.1 Fundamental Fermion Structure

In the free fermion case we assume fundamental fermions – quarks and leptons – of all generations and both normal and Dark, have an 'inner' core consisting of a qube – a spin ½ functional with a spin value having an SU(2) qubit spin symmetry and a mass m_0 as we described briefly in earlier chapters. The qube within a fermion does not have a specific size or location within itself since it is merely a functional.

We cloaked each initially free fermion with coordinates using a functional inner product as we described in chapter 2 for a free Dirac particle. We symbolized a qube for a fermion with no internal symmetry and spin s as f. We began by defining a coordinate space Dirac fourier quantum expansion as

$$(s, x, t) = N(p)[b(p, s)u(p, s)e^{-ip\cdot x} + d^\dagger(p, s)v(p, s)e^{+ip\cdot x}]$$

where $N(p)$ is a normalization factor, u and v are functions of spin and momentum, and b and $d^\dagger$ are creation/annihilation operators. We defined a Dirac quantum wave function with the inner product of a qube functional and a coordinate space fourier quantum expansion:

$$\psi(\dot{x}) = (f, (s, x, t)) = \sum_{\pm s} \int d^3p N(p)[b(p, s)u(p, s)e^{-ip\cdot x} + d^\dagger(p, s)v(p, s)e^{+ip\cdot x}]$$

where we use a functional inner product formalism in the manner of Riesz (1955)[156] and others.

Earlier we defined fermion quantum fields for the various species.

---

[156] For example see pp. 61-2 of Riesz (1955) where linear functionals and their inner products are defined.

27.2 Why Factorize Fermion Quantum Fields into a Functional and an independent Fourier Expansion?

We identify two reasons for factorizing fundamental fermion quantum fields:

1. The internal symmetries of free fundamental fermions are only partially correlated with their fourier expansions: there is a spin/handedness relation and a species relationship based on the species of each fundamental fermion. The internal symmetries: color, generation number, layer number, and so on appear only as indices attached to creation and annihilation operators. While functionals are matched with corresponding fourier expansions by inserting indices, there is no dynamics, at the free fermion level, enforcing the specific matching of functionals and fourier expansions.

2. Quantum Entanglement between parts of a physical state separated by space-like distances suggest the correlated changes in the quantum numbers of a physical state with spatially separated parts are independent of spatial distance. Quantum number changes (in any quantum numbers) between spatially separated parts are instantaneous. This demonstrated phenomena strongly supports the factorization of the quantum numbers and the coordinates of quantum fields. We demonstrate this point in more detail below.

3. The close analogy between the subgroups of the Complex Lorentz group and the known symmetries of The Standard Model suggest that the internal symmetries of fermions represented by functionals have a somewhat similar group structure to space-time. We explore this topic below.

For these reasons we propose to consider fundamental fermion quantum fields as inner products of fermion functionals and space-time fourier expansions. In chapter 33 we will show that bosons also have core functionals that are part of a space containing both fermion and boson functionals. This space further supports the analogy between space-time and particle functional space.

27.3 Particle Functional Space

Because of a similarity between the Complex Lorentz group subgroups and the factors of The Standard Model groups we begin by assuming that particle functional space has a basic set of four functionals that correspond to space-time coordinates. Thus we assume that there is a set of four functional coordinates f^μ that can be transformed using elements, $\Lambda_L{}^\mu{}_\nu$, of an 'internal symmetry' Complex Lorentz group to a different set of functional coordinates.[157]

$$f^\mu = \Lambda_L{}^\mu{}_\nu(\omega, \mathbf{u})f^\nu \tag{27.1}$$

where ω and $\mathbf{u}$ are c-value parameters.[158] The indices μ and υ are not space-time indices but rather those of the functional space's Complex Lorentz group.

27.3.1 Coordinate Sets

We now consider one of the subgroups of the Complex Lorentz group and generate the set of all representations from it. (In chapters 28 - 31 we will consider all the Complex Lorentz subgroups in detail.)

For example we consider the set of all particle functional coordinate systems that can be generated by applying all SU(3) transformations within the Complex Lorentz group. We now map each such functional coordinate system, denoted a, to a *single* functional that we will denote $f_{\frac{1}{2}a}$ The set $\{f_{\frac{1}{2}a}\}$ has a functional for every possible SU(3) particle that can be generated. Thus we can combine these functionals with matching fourier expansions to generate the complete set of SU(3) particles.

We can repeat this procedure for each of the SU(2)⊗U(1) subgroups of the particle Complex Lorentz group.

---

[157] In the late 1960's the author considred the possibility of such a particle space based on the simple notion that an electron and an up color triplet bare a resemblance to space-time coordinates—as do a neutrino and a down color triplet. The general concept being that there is a space of particles as well as a space of coordinates. After all the specification of a coordinate location requires one or more particles. Without particles, coordinates become meaningless. Without coordinates, particle dynamics becomes trivial. The paucity of data and theoretical insights led the author to abandon pursuing this idea. Quantum Entanglement and a deeper knowledge of The Standard Model now makes the idea of interest.
[158] We omit the subscript 's' for the moment.

Thus we have sets of functionals for each Complex Lorentz subgroup. Each set of functionals furnishes a representation for the corresponding subgroup.

27.3.3 Combination of Functionals for Group Factors

Some fermions are functionals corresponding to the tensor product of two or more subgroups. For example a quark, q, is an SU(3)⊗SU(2)⊗U(1) particle. We define the quark functional as

$$f_{qab} = f_{SU(3)a}f_{SU(2)\otimes U(1)b} \qquad (27.2)$$

where a is an SU(3) index and b is an SU(2)⊗U(1) index.

Thus our particle functional group definition can handle the SU(3)⊗SU(2)⊗U(1) direct product of commuting subgroup factors.

27.4 Internal Standard Model Symmetry vs. Particle Functional Space Symmetry

The subgroups of the Complex Lorentz group do not commute. Although similar to the Standard Model Symmetry Groups, the Standard Model symmetry is a direct product of SU(3)⊗SU(2)⊗U(1) – commuting factors. The particle space definition and procedure enables us to have a commuting set of Standard Model symmetry factors. It also gives us an analogy between particle space and coordinate space.

27.5 An Explanation of Quantum Entanglement – No Spookiness

27.5.1 All The Ways of Understanding 'Spooky' Entanglement

A quantum entangled state may have its parts separate and yet due to the creation of the state, changing one part of the separated state – no matter what the distance or whether the distance is space-like or time-like – instantly changes the distant part of the state. This effect has been termed 'spooky action at a distance' by Einstein and Rosen and others since them.

The important question is how this effect takes place. One answer is Quantum Mechanics – but that could be viewed as a disguised form of *deus ex machina*. If one asks for a concrete mechanism that causes instantaneous response several conventional approaches present themselves:

1. There could be a universal, infinite velocity force that enforces an instantaneous response. This is clearly not possible since it would presumably occur between any two interacting particles – swamping the usual forces such as electromagnetism.

2. There could be an infinitely dense classical field throughout space that transfers a change instantaneously.

Thus from a traditional point of view Entanglement is spooky.

27.5.2 Particle Functional Space Eliminates 'Spookiness"

The new functional formulation of fundamental particles offers a solution – a solution that also provides strong support for the functional space concept. The key to the elimination of spookiness is to have a particle functional space **that has no coordinate distances**.[159] Thus all functionals can be viewed as existing at one 'point.' Then the parts of a state would have superimposed functionals that would instantly transform each other independent of our coordinate space. For example, a spin flip entangled effect would be

$$f_{a+\frac{1}{2}} \, f_{b-\frac{1}{2}} \rightarrow f_{a-\frac{1}{2}} \, f_{b+\frac{1}{2}}$$

Separating the particles' functional parts from the space-time fourier parts automatically makes spooky distance irrelevant.

---

[159] Particle functional Lorentz transformations would still be allowed. One could define distances and translations similar to those of the inhomogeneous Lorenz group. At this point in time, however, there is no physical justification for this extension.

Thus the phenomena of Quantum Entanglement can easily account for the absence of 'spooky' distance—providing strong support for the particle functional space concept. See chapter 8 for the elimination of spookiness in complex quaternion space.

Particle functional space consists of one point with commuting functional coordinates. All fundamental particles in the universe coexist at the one point of particle functional space. Particle functional space is not part of our coordinate space – but a separate space that combines with our space within each fundamental particle. Particle functionals can be transformed by a functional Complex Lorentz group.

28. Coordinate Reality Group Analogue of ElectroWeak Doublets

28.0 Introduction

In chapter 26 we established the four species of fermions based on Complex Lorentz group (L_C) boosts. In this chapter[160] we will introduce the coordinate space *analogue* of the ElectroWeak interactions based on Complex Lorentz group considerations. We begin by generalizing the free Dirac equation to a 2×2 matrix of Dirac-like equations that have a larger group covariance. This matrix equation is applied to a doublet consisting of a normal Dirac particle wave function and a tachyon wave function. We will identify these doublets as *analogues* of ElectroWeak lepton doublets.

Then we will consider a generalized 2×2 equation matrix (covariant under the L_C group) for doublets of *complexon* particles with complex 3-momenta consisting of an up-type complexon and a down-type tachyonic complexon. Because of an inherent SU(3) symmetry we will identify these doublets as analogues of quark ElectroWeak doublets. SU(3) symmetry leads us to identify each complexon quark in a doublet as a color SU(3) triplet analogue.

28.1 Transformations of Dirac and Tachyon Equations

A Left-handed boost of the Dirac equation transforms the Dirac equation into the spin ½ tachyon equation, and vice versa:

$$S_L(\Lambda_L(\omega, \mathbf{u}))\psi(x) \rightarrow \psi_T'(x') \tag{28.1a}$$
$$S_L(\Lambda_L(\omega, \mathbf{u}))\psi_T(x) \rightarrow \psi'(x')$$

---

[160] This chapter is extracted from Blaha (2007b).

Also, noting the appearance of a γ^5, we see

$$S_L(\Lambda_L(\omega, \mathbf{u}))(\gamma^\mu \partial/\partial x^\mu - m)S_L^{-1}(\Lambda_L(\omega, \mathbf{u})) = (i\gamma^\mu \partial/\partial x'^\mu - m) \qquad (28.1b)$$
$$S_L(\Lambda_L(\omega, \mathbf{u}))\gamma^5(i\gamma^\mu \partial/\partial x^\mu - m)\gamma^5 S_L^{-1}(\Lambda_L(\omega, \mathbf{u})) = (\gamma^\mu \partial/\partial x'^\mu - m)$$

where

$$x'^\mu = i\Lambda_L{}^\mu{}_\nu(\omega, \mathbf{u})x^\nu \qquad (28.1c)$$
$$\partial/\partial x'^\mu = -i\Lambda_L{}^\nu{}_\mu(\omega, \mathbf{u})\partial/\partial x^\nu$$

with

$$x' = E(\mathbf{v})x = i\Lambda_L(\mathbf{v})x$$

Eqs. 28.1a – 28.1c imply

$$S_L(\Lambda_L(\omega, \mathbf{u}))(\gamma^\mu \partial/\partial x^\mu - m)\psi_T(x) = (i\gamma^\mu \partial/\partial x'^\mu - m)S_L(\Lambda_L(\omega, \mathbf{u}))\psi_T(x)$$
$$= (i\gamma^\mu \partial/\partial x'^\mu - m)\psi'(x') \qquad (28.1d)$$

and

$$S_L(\Lambda_L(\omega, \mathbf{u}))\gamma^5(i\gamma^\mu \partial/\partial x^\mu - m)\psi(x) = (\gamma^\mu \partial/\partial x'^\mu - m)S_L(\Lambda_L(\omega, \mathbf{u}))\gamma^5\psi(x)$$
$$= (\gamma^\mu \partial/\partial x'^\mu - m)\psi_T'(x') \qquad (28.1e)$$

where

$$\psi'(x') = S_L(\Lambda_L(\omega, \mathbf{u}))\psi_T(x) \qquad (28.1f)$$

and

$$\psi_T'(x') = S_L(\Lambda_L(\omega, \mathbf{u}))\gamma^5\psi(x) \qquad (28.1g)$$

Note the Dirac equation is not a left-handed Complex Lorentz covariant.

28.2 Doublet Extended Dirac Equations

We will now consider the issue of generalizing the Dirac equation so that the extended equation is covariant under both Lorentz transformations and Left-handed Complex Lorentz transformations.

The only obvious method to obtain an extended Dirac equation that is covariant under Complex Lorentz transformations is to define an 8×8 matrix generalization. Let

$$
đ(x) \;=\; \begin{bmatrix} (\gamma^{\mu}\partial/\partial x^{\mu} - m) & 0 \\[2ex] 0 & (i\gamma^{\mu}\partial/\partial x^{\mu} - m) \end{bmatrix}
\tag{28.2}
$$

be an 8×8 matrix operator with the 4×4 matrix elements shown, and let

$$
\Psi(x) \;=\; \begin{bmatrix} \psi_T(x) \\[2ex] \psi(x) \end{bmatrix}
\tag{28.3}
$$

be an 8 component column vector composed of a Dirac field and a tachyon field. Then the extended free Dirac equation is

$$
đ(x)\Psi(x) = 0
\tag{28.4}
$$

We now define the 8×8 Left-handed Complex Lorentz transformation

$$
S_{L8}(v) \;=\; \begin{bmatrix} 0 & S_L(\Lambda_L(v))\gamma^5 \\[2ex] S_L(\Lambda_L(v)) & 0 \end{bmatrix}
\tag{28.5}
$$

with inverse transformation

$$
S_{L\,L8}{}^{-1}(\Lambda_L(v)) \;=\; \begin{bmatrix} 0 & S_L{}^{-1}(\Lambda_L(v)) \\[2ex] \gamma^5 S_L{}^{-1}(\Lambda_L(v)) & 0 \end{bmatrix}
\tag{28.6}
$$

Note: we use the notations $S_L(\Lambda_L(v))$ and $S_L(\Lambda_L(\omega, \mathbf{u}))$ interchangeably. Applying S_{L8} to eq. 28.4 yields

$$0 = S_{L8}(\Lambda_L(v))đ(x)\Psi(x) = đ(x')\Psi'(x') \tag{28.7}$$

where

$$\Psi'(x') = \begin{bmatrix} S_L\gamma^5\psi(x) \\ \\ S_L\psi_T(x) \end{bmatrix} = \begin{bmatrix} \psi_T'(x') \\ \\ \psi'(x') \end{bmatrix} \tag{28.8}$$

Thus the extended Dirac equation is covariant under generalized Left-handed Complex Lorentz transformations such as eqs. 28.5-28.6. Covariance requires the tachyon and the Dirac particles must have the same absolute value for the mass which is the qube mass in the free fermion case.

It is easy to show that the extended Dirac equation eq. 28.4 is also covariant under conventional Lorentz transformations in the 8×8 representation:

$$S_8(\Lambda(v)) = \begin{bmatrix} S(\Lambda(v)) & 0 \\ \\ 0 & S(\Lambda(v)) \end{bmatrix} \tag{28.9}$$

with inverse

$$S_8^{-1}(\Lambda(v)) = \begin{bmatrix} S^{-1}(\Lambda(v)) & 0 \\ \\ 0 & S^{-1}(\Lambda(v)) \end{bmatrix} \tag{28.10}$$

and non-diagonal Lorentz transformations:

$$S_{8A}(\Lambda(v)) = \begin{bmatrix} 0 & S(\Lambda(v)) \\ S(\Lambda(v)) & 0 \end{bmatrix} \qquad (28.11)$$

with inverse transformation

$$S_{8A}^{-1}(\Lambda(v)) = \begin{bmatrix} 0 & S^{-1}(\Lambda(v)) \\ S^{-1}(\Lambda(v)) & 0 \end{bmatrix} \qquad (28.12)$$

Under a conventional Lorentz transformation we find

$$0 = S_8(\Lambda(v))đ(x)\Psi(x) = đ(x')\Psi'(x') \qquad (28.13)$$

$$0 = S_{8A}(\Lambda(v))đ(x)\Psi(x) = đ(x')\Psi'(x')$$

The lagrangian density that corresponds to our 8-dimensional construction is

$$\mathcal{L}_8 = \overline{\Psi}(x)đ(x)\Psi(x) \qquad (28.14)$$

where

$$\overline{\Psi}(x) = \Psi^\dagger\Gamma^0 \qquad (28.15)$$

and

$$\Gamma^0 = \begin{bmatrix} i\gamma^0\gamma^5 & 0 \\ 0 & \gamma^0 \end{bmatrix} \qquad (28.16)$$

The action

$$I = \int d^4x \mathcal{L}_8 \qquad (28.17)$$

is invariant under Lorentz transformations S_8 and S_{8A}.

Then the Hamiltonian density for the 8-dimensional theory is

$$\mathcal{H}_8(x) = \begin{bmatrix} i\psi_T^\dagger \gamma^5 (\boldsymbol{\alpha}\cdot\nabla + \beta m)\psi_T & 0 \\ \\ 0 & \psi^\dagger(-i\boldsymbol{\alpha}\cdot\nabla + \beta m)\psi \end{bmatrix} \tag{28.18}$$

28.3 Non-Invariance of the Extended Free Action under a Left-handed Extended Lorentz Transformation

The action eq. 28.17 is not invariant under Left-handed Complex Lorentz transformations. The fundamental cause of this non-invariance is the three dimensional nature of space. In the case of Dirac particles one can define a Lorentz invariant action because time is one-dimensional. Thus one can use $\psi^\dagger\gamma^0 = \bar\psi$ to form the Dirac field lagrangian and action. A key factor in Lorentz invariance is the relation between the inverse and hermitean conjugate of the spinor boost operator

$$\gamma^0 S^{-1}\gamma^0 = S^\dagger \tag{28.19}$$

In the case of the tachyon lagrangian and action, Left-handed Complex Lorentz invariance is not possible because the tachyonic equivalent to eq. 28.19 is

$$S_L^{-1}(\Lambda(\mathbf{v}))\gamma\cdot\mathbf{p}/|\mathbf{p}| = i\gamma^0 S_L^\dagger(\Lambda(\mathbf{v})) \tag{28.20}$$

where $\mathbf{p} = m\gamma_s\mathbf{v}$. The appearance of $\gamma\cdot\mathbf{p}/|\mathbf{p}|$ in eq. 28.20 precludes the invariance of the free tachyon action.

We will now show the effect of a Left-handed Complex Lorentz transformation (eqs. 28.5 and 28.6) on the lagrangian density eq. 28.1c. The two non-zero parts of the lagrangian density $\mathcal{L}_8$ (eq. 28.14) are

$$\mathcal{L}_1 = \psi_T^{\dagger} i\gamma^0 \gamma^5 (\gamma^\mu \partial/\partial x^\mu - m)\psi_T(x) \qquad (28.21)$$

and

$$\mathcal{L}_2 = \psi^{\dagger} \gamma^0 (i\gamma^\mu \partial/\partial x^\mu - m)\psi(x) \qquad (28.22)$$

where † represents complex conjugation. The effect of the transformation, eqs. 28.5-28.6, on these terms is

$$
\begin{aligned}
\mathcal{L}_1' &= \psi_T^{\dagger} i\gamma^0 \gamma^5 S_L^{-1} S_L (\gamma^\mu \partial/\partial x^\mu - m) \, S_L^{-1} S_L \psi_T(x) \\
&= \psi_T^{\dagger} i\gamma^0 \gamma^5 S_L^{-1} (i\gamma^\mu \partial/\partial x'^\mu - m) S_L \psi_T(x) \\
&= -\psi_T^{\dagger} S_L^{\dagger} \gamma^5 (\boldsymbol{\gamma}\cdot\mathbf{p}/|\mathbf{p}|)(i\gamma^\mu \partial/\partial x'^\mu - m) S_L \psi_T(x) \\
&= \psi'^{\dagger}(x')(\boldsymbol{\gamma}\cdot\mathbf{p}/|\mathbf{p}|)\gamma^5 (i\gamma^\mu \partial/\partial x'^\mu - m)\psi'(x')
\end{aligned}
\qquad (28.23)
$$

and

$$
\begin{aligned}
\mathcal{L}_2' &= \psi^{\dagger} \gamma^0 \gamma^5 S_L^{-1} S_L \gamma^5 (i\gamma^\mu \partial/\partial x^\mu - m)\gamma^5 S_L^{-1} S_L \gamma^5 \psi(x) \\
&= \psi^{\dagger} \gamma^0 \gamma^5 S_L^{-1} (\gamma^\mu \partial/\partial x'^\mu - m) S_L \gamma^5 \psi(x) \\
&= i\psi^{\dagger} \gamma^5 S_L^{\dagger} (\boldsymbol{\gamma}\cdot\mathbf{p}/|\mathbf{p}|)(\gamma^\mu \partial/\partial x'^\mu - m) S_L \gamma^5 \psi(x) \\
&= i\psi_T'^{\dagger}(x')(\boldsymbol{\gamma}\cdot\mathbf{p}/|\mathbf{p}|)(\gamma^\mu \partial/\partial x'^\mu - m)\psi_T'(x')
\end{aligned}
\qquad (28.24)
$$

using eqs. 28.20, 28.1f and 28.1g, where $\psi'(x')$ is a solution of the Dirac equation obtained by Left-handed Complex Lorentz boosting (by $\mathbf{v} = \mathbf{p}/(\gamma m)$) of a tachyon field and where $\psi_T'(x')$ is a solution of the tachyon equation obtained by Left-handed Complex Lorentz boosting (by $\mathbf{v} = \mathbf{p}/(\gamma m)$) of a Dirac field. Eqs. 28.23 - 28.24 clearly show that $\mathcal{L}_8$ is *not* invariant under Left-handed Complex Lorentz transformations.

Consequently the action of eq. 28.17 is only invariant under inhomogeneous Lorentz transformations. *This state of affairs is actually an advantage when we derive internal symmetry features of the SuperStandard Theory because it will be seen to prevent any interplay between unbroken internal symmetry ElectroWeak SU(2) rotations and Left-handed Complex Lorentz transformations.*

28.4 The Diracian Dilemma – To what do Left-handed Extended Lorentz Boost Particles Correspond? Answer: Left-handed Leptons

The development of this 8-dimensional formalism, and in particular, the "bi-spinor" wave function consisting of a Dirac spinor and a tachyon spinor, raises the question, "Is there a particle interpretation for the "bi-spinor" wave function?" Dirac faced a similar issue in 1928-1930 with the negative energy states of the Dirac equation. He developed "hole theory" which eventually led to the interpretation of holes in the sea of filled negative energy states as *positrons*. We now face the same problem: with what pairs of particles do we identify the doublets consisting of a Dirac particle and a tachyon?

The obvious natural interpretation of these 8-spinors is ElectroWeak isodoublets such as:

$$\Psi_\ell(x) \;=\; \begin{bmatrix} \psi_{\ell T} \\ \\ \psi_\ell \end{bmatrix} \;\sim\; \begin{bmatrix} \nu \\ \\ e \end{bmatrix} \tag{28.25}$$

are for leptons to have "e" represent a charged lepton and ν represent a neutrino. With this interpretation we can introduce SU(2) gauge interactions and develop one-generation, leptonic Weak theory naturally.

28.5 To what do Complexons Correspond? Quarks

We have identified two of the four types of spin ½ fermions as leptons. The remaining two types of spin ½ fermions – complexons – ψ_C and ψ_{CT} seem to naturally correspond to quarks since their equations of motion and wave functions have a natural SU(3) symmetry as we pointed out earlier. We therefore associate an analogue color SU(3) symmetry with these two types of spin ½ complexons. The Electroweak doublet of quarks then is

$$\Psi_q^{\ a}(x) = \begin{bmatrix} \psi_C^{\ a} \\ \\ \psi_{CT}^{\ a} \end{bmatrix} \sim \begin{bmatrix} u^a \\ \\ d^a \end{bmatrix} \qquad (28.26)$$

where u is an "up" type quark and d is a "down" type quark.[161]

The rationale for constructing quark doublets is the same as in the leptonic case: We wish to define a generalization of the "Dirac-like" equations of motion that is covariant under L_C boosts.

28.6 Quark Doublets

We assume that quark doublets consist of a complexon[162] and a tachyonic complexon[163] and to this extent they mirror lepton doublets. In this section we will develop a generalized free complexon equation and describe its features.

---

[161] While the lepton situation is clear in the sense that charged leptons cannot be tachyons since their masses are known (Thus only tachyonic neutrinos are the only currently allowed possibility.), the quark situation is somewhat unclear. We have provisionally chosen the "down" type of quark (d, s, and b) as tachyonic. The association of bound states of these quarks such as the K^0 and B^0 systems which are known to have CP violation, and the CP violation engendered by tachyons, encourages this interpretation.

In addition, $W^\pm$ charge asymmetry in $p\bar{p}$ collisions indicate the d sea in a proton is greater than the u sea (K. Abe et al, PRL **74**, 850 (1995)) as does the asymmetry of Drell-Yan production in deep inelastic scattering on p and n targets (A. Baldit et al, Phys. Lett. **B332**, 244 (1994)). These results are to be expected since there is no mass gap for a d tachyon sea while there is a mass gap for a u Dirac particle sea. Complexon quarks may explicate the discrepancies between theory and experiment in the spin structure functions of the parton model for nucleons.

[162] **An "ordinary" complexon can "exceed the speed of light" just like a tachyonic complexon because a complexon has a complex valued velocity enabling it to evade the real-valued singularity at v = c.**

[163] The global SU(3) symmetry of complexons makes their identification with quarks reasonable. However, the complexon theory that we develop and use for quark dynamics in the Standard Model is <u>not</u> required. Our SuperStandard Theory could use Dirac fermion dynamics for the up-type quarks and tachyon dynamics for down-type quarks. Then the (broken) Left-handed Extended Lorentz group would be the basic space-time group rather than L_C. We choose to use complexon dynamics for quarks because they have an internal SU(3)-like structure suggestive of color SU(3). More importantly, their spin dynamics is different and thus may resolve the differences between theory and experiment for the deep inelastic parton spin-dependent structure functions. Nevertheless, quarks could be

28.6.1 Summary of L_C Boosts to Generate Spin ½ Equations

We begin by recapitulating L_C boost features for coordinates and spinors:[164]

$$\Lambda_C(\mathbf{v_c}) = \exp[i\omega\hat{\mathbf{w}}\cdot\mathbf{K}] \tag{26.61}$$

$$\omega = (\omega_r^2 - \omega_i^2 + 2i\omega_r\omega_i\,\hat{\mathbf{u}}_r\cdot\hat{\mathbf{u}}_i)^{½} \tag{26.62}$$

$$\hat{\mathbf{w}} = (\omega_r\hat{\mathbf{u}}_r + i\omega_i\hat{\mathbf{u}}_i)/\omega \tag{26.63}$$

$$\hat{\mathbf{w}}\cdot\hat{\mathbf{w}} = \hat{\mathbf{u}}_r\cdot\hat{\mathbf{u}}_r = \hat{\mathbf{u}}_i\cdot\hat{\mathbf{u}}_i = 1 \tag{26.64a}$$

$$\mathbf{v_c} = \hat{\mathbf{w}}\,\tanh(\omega) \tag{26.64b}$$

The corresponding L_C spinor boost for $m^2 > 0$ particles with complex 3-momenta from chapter 2 of Blaha (2018e) is

$$S_C(\omega, \mathbf{v_c}) = \exp(-i\omega\sigma_{0k}\hat{w}_k/2) = \exp(-\omega\gamma^0\boldsymbol{\gamma}\cdot\hat{\mathbf{w}}/2)$$
$$= \cosh(\omega/2)I + \sinh(\omega/2)\gamma^0\boldsymbol{\gamma}\cdot\hat{\mathbf{w}} \tag{2.99}$$

with inverse transformation

$$S_C^{-1}(\omega, \mathbf{v_c}) = \gamma^2\gamma^0 K^{-1} S_C^\dagger K\gamma^0\gamma^2 = \gamma^2\gamma^0 S_C^{\,T}\gamma^0\gamma^2 = \exp(\omega\gamma^0\boldsymbol{\gamma}\cdot\hat{\mathbf{w}}/2)$$
$$= \cosh(\omega/2)I - \sinh(\omega/2)\gamma^0\boldsymbol{\gamma}\cdot\hat{\mathbf{w}} \tag{2.100}$$

The Dirac-like complexon equation resulting from the boost is

$$[i\gamma^0\partial/\partial t + i\boldsymbol{\gamma}\cdot(\nabla_r + i\nabla_i) - m]\psi_C(t, \mathbf{x_r}, \mathbf{x_i}) = 0 \tag{2.123}$$

where $\mathbf{x} = \mathbf{x_r} - i\mathbf{x_i}$. The subsidiary condition is

$$\nabla_r\cdot\nabla_i\,\psi_C(t, \mathbf{x_r}, \mathbf{x_i}) = 0 \tag{2.123a}$$

---

similar to leptons in this regard and form a doublet of a Dirac fermion and an ordinary tachyon. Whether quarks are complexons or not is an experimental question!

[164] **The equation numbering of this subsection 28.6.1 follows that of Blaha (2007b).**

The L_C coordinate boost that leads to $m^2 < 0$ tachyonic complexons with complex 3-momenta is

$$\Lambda_{CL}(\mathbf{v_c}) \equiv \Lambda_{CL}(\omega, \hat{\mathbf{w}}) = \exp[i(\omega + i\pi/2)\hat{\mathbf{w}}\cdot\mathbf{K}] \qquad (2.185)$$

where

$$\omega = (\omega_r{}^2 - \omega_i{}^2)^{\frac{1}{2}}$$
$$\hat{\mathbf{w}} = (\omega_r\hat{\mathbf{u}}_r + i\omega_i\hat{\mathbf{u}}_i)/\omega$$
$$\hat{\mathbf{w}}\cdot\hat{\mathbf{w}} = \hat{\mathbf{u}}_r\cdot\hat{\mathbf{u}}_r = \hat{\mathbf{u}}_i\cdot\hat{\mathbf{u}}_i = 1$$
$$\mathbf{v_c} = \hat{\mathbf{w}}\tanh(\omega + i\pi/2) = \hat{\mathbf{w}}\coth(\omega)$$
$$\omega_L = \omega + i\pi/2$$

The L_C spinor boost for tachyonic complexons is

$$S_{CL}(\Lambda_{CL}(\omega, \hat{\mathbf{w}})) = \exp(-i\omega_L\sigma_{0i}\hat{w}_i/2) = \exp(-\omega_L\gamma^0\boldsymbol{\gamma}\cdot\hat{\mathbf{w}}/2)$$
$$= \cosh(\omega_L/2)I + \sinh(\omega_L/2)\gamma^0\boldsymbol{\gamma}\cdot\hat{\mathbf{w}} \qquad (2.193)$$

The resulting Dirac-like tachyonic complexon equation is

$$[\gamma^0\partial/\partial t + \boldsymbol{\gamma}\cdot(\boldsymbol{\nabla}_r + i\boldsymbol{\nabla}_i) - m]\psi_{CL}(t, \mathbf{x_r}, \mathbf{x_i}) = 0 \qquad (2.202)$$

with the subsidiary condition

$$\boldsymbol{\nabla}_r\cdot\boldsymbol{\nabla}_i\,\psi_{CL}(t, \mathbf{x_r}, \mathbf{x_i}) = 0 \qquad (2.204)$$

28.6.2 L_C Boosts between Complexons and Tachyonic Complexons

An L_C spinor boost of a complexon can change it into a tachyonic complexon and vice versa:

$$S_{CL}(\Lambda_{CL}(\omega, \hat{\mathbf{w}}))\psi_C(x) \rightarrow \psi_{CT}'(x')$$
$$(28.27)$$
$$S_{CL}(\Lambda_{CL}(\omega, \hat{\mathbf{w}}))\psi_{CT}(x) \rightarrow \psi_C'(x')$$

Similarly the differential operator used in the equations of motion can also be transformed.

$$S_{CL}(\Lambda_{CL}(\omega, \hat{\mathbf{w}}))(\gamma^\mu D_\mu - m)S_{CL}^{-1}(\Lambda_{CL}(\omega, \hat{\mathbf{w}})) = (i\gamma^\mu D'_\mu - m) \quad (28.28)$$
$$S_{CL}(\Lambda_{CL}(\omega, \hat{\mathbf{w}}))\gamma^5(i\gamma^\mu D_\mu - m)\gamma^5 S_{CL}^{-1}(\Lambda_{CL}(\omega, \hat{\mathbf{w}})) = (\gamma^\mu D'_\mu - m)$$

where

$$x'^\mu = i\Lambda_{CL}{}^\mu{}_\nu(\omega, \mathbf{u})x^\nu \quad (28.29)$$
$$D'_\mu = -i\Lambda_{CL}{}^\nu{}_\mu(\omega, \mathbf{u})D_\nu$$

or in matrix form

$$X' = E_{CL}(\omega, \hat{\mathbf{w}})X \equiv i\Lambda_{CL}(\omega, \hat{\mathbf{w}})X \quad (28.30)$$

Eqs. 28.27 – 28.29 imply

$$S_{CL}(\Lambda_{CL}(\omega, \hat{\mathbf{w}}))(\gamma^\mu D_\mu - m)\psi_{CT}(x) = (i\gamma^\mu D'_\mu - m)S_{CL}(\Lambda_{CL}(\omega, \hat{\mathbf{w}}))\psi_{CT}(x)$$
$$= (i\gamma^\mu D'_\mu - m)\psi_C'(x') \quad (28.31)$$

and

$$S_{CL}(\Lambda_{CL}(\omega,\hat{\mathbf{w}}))\gamma^5(i\gamma^\mu D_\mu - m)\psi_C(x) = (\gamma^\mu D'_\mu - m)S_{CL}(\Lambda_{CL}(\omega,\hat{\mathbf{w}}))\gamma^5\psi_C(x)$$
$$= (\gamma^\mu D'_\mu - m)\psi_{CT}'(x') \quad (28.32)$$

where

$$\psi_C'(x') = S_{CL}(\Lambda_{CL}(\omega, \hat{\mathbf{w}}))\psi_{CT}(x) \quad (28.33)$$

and

$$\psi_{CT}'(x') = S_{CL}(\Lambda_{CL}(\omega, \hat{\mathbf{w}}))\gamma^5\psi_C(x) \quad (28.34)$$

Thus neither complexon dynamical equation is L_C covariant.

28.6.3 Doublet Dynamical Equation for Complexons

We will now consider the issue of generalizing the complexon dynamical equations so that the generalized equation is covariant under both Lorentz transformations and L_C boosts.

The only obvious method to obtain a generalized equation that is covariant under L_C boosts is to define an 8×8 matrix generalization. Let

$$
đ_C(x) = \begin{bmatrix} (i\gamma^\mu D_\mu - m) & 0 \\ 0 & (\gamma^\mu D_\mu - m) \end{bmatrix} \tag{28.35}
$$

be an 8×8 matrix operator with the 4×4 matrix elements shown, and let

$$
\Psi_C(x) = \begin{bmatrix} \psi_C(x) \\ \psi_{CT}(x) \end{bmatrix} \tag{28.36}
$$

be an 8 component column vector composed of a complexon field and a tachyonic complexon field. Then the generalized complexon equation is

$$
đ_C(x)\Psi_C(x) = 0 \tag{28.37}
$$

We now define the 8×8 Left-handed L_C boost transformation

$$
S_{CL8} \equiv S_{CL8}(\Lambda_{CL}(\omega, \hat{\mathbf{w}})) = \begin{bmatrix} 0 & S_{CL}(\Lambda_{CL}(\omega, \hat{\mathbf{w}})) \\ S_{CL}(\Lambda_{CL}(\omega, \hat{\mathbf{w}}))\gamma^5 & 0 \end{bmatrix} \tag{28.38}
$$

with inverse transformation

$$
S_{CL8}^{-1} \equiv S_{CL8}^{-1}(\Lambda_{CL}(\omega, \hat{\mathbf{w}})) = \begin{bmatrix} 0 & \gamma^5 S_{CL}^{-1}(\Lambda_{CL}(\omega, \hat{\mathbf{w}})) \\ S_{CL}^{-1}(\Lambda_{CL}(\omega, \hat{\mathbf{w}})) & 0 \end{bmatrix} \tag{28.39}
$$

Applying S_{CL8} to eq. 28.37 yields

$$
0 = S_{CL8}đ_C(x)\Psi_C(x) = đ_C(x')\Psi_C'(x') \tag{28.40}
$$

where

$$\Psi_C'(x') = \begin{bmatrix} S_{CL8}\psi_{CT}(x) \\ \\ S_{CL8}\gamma^5\psi_C(x) \end{bmatrix} = \begin{bmatrix} \psi_C'(x') \\ \\ \psi_{CT}'(x') \end{bmatrix} \qquad (28.41)$$

Thus the generalized complexon equation is covariant under L_C boosts. Covariance requires the complexon, and the tachyonic complexon, must have the same absolute value for the mass.

It is easy to show that the generalized complexon equation is also covariant under conventional Lorentz transformations represented as 4×4 diagonal blocks in an 8×8 matrix representation. (The demonstration is analogous to eqs. 28.9 – 28.13.)

The lagrangian density that corresponds to our 8-dimensional construction is

$$\mathcal{L}_{C8} = \overline{\Psi}_C(x)đ_C(x)\Psi_C(x) \qquad (28.42)$$

where

$$\overline{\Psi}_C(x) = \Psi_C^{\dagger}|_{\mathbf{x_i} = -\mathbf{x_i}} \Gamma_C^0 \qquad (28.43)$$

and

$$\Gamma_C^0 = \begin{bmatrix} \gamma^0 & 0 \\ \\ 0 & i\gamma^0\gamma^5 \end{bmatrix} \qquad (28.44)$$

The action

$$I = \int d^4x\mathcal{L}_{C8} \qquad (28.45)$$

is invariant under Lorentz transformations S_8 and S_{8A} (eqs. 28.9 – 28.12).

The Hamiltonian density for the 8-dimensional theory is

$$\mathcal{H}_{C8}(x) = \begin{bmatrix} \psi_C^{\dagger}(-i\boldsymbol{\alpha}\cdot\nabla_C + \beta m)\psi_C & 0 \\ \\ 0 & i\psi_{CT}^{\dagger}\gamma^5(\boldsymbol{\alpha}\cdot\nabla_C + \beta m)\psi_{CT} \end{bmatrix} \tag{28.46}$$

where the spatial vector part of D^{μ} is complex

$$\nabla_C = \mathbf{D} \tag{28.47}$$

28.6.4 Non-Invariance of the Generalized Free Complexon Action under an L_C Boost

The action 28.45 is not invariant under L_C boosts. The reason is similar to that of section 28.3 for the "leptonic" type of particle: there is no simple relation between the hermitean conjugate of an L_C spinor boost and its inverse (a situation similar to eq. 28.19 for the Dirac boost case).

Consequently the action of eq. 28.45 is only invariant under inhomogeneous Lorentz transformations. *This state of affairs is again an advantage when we derive features of the Superstandard Model because it prevents any interplay between unbroken internal symmetry ElectroWeak SU(2) rotations and L_C transformations in the complexon (quark) sector.*

29. Coordinate Reality Group Analogue for ElectroWeak SU(2)⊗U(1) due to Real Superluminal Velocities

In the preceding chapter we developed a fermion doublet analogue framework for the ElectroWeak SU(2) interactions. We now turn to develop the ElectroWeak interactions analogue with an SU(2)⊗U(1) group structure, from the geometry of Complex Lorentz transformations.

In the discussions up to this point we have not considered imaginary (and more generally complex) coordinates resulting from a superluminal Lorentz transformation.[165] In this chapter we show that the coordinates generated from real-valued coordinates are complex-valued in general and require us to introduce another transformation that maps complex coordinates to real coordinates.[166] This transformation, which we will call a *Coordinate Reality Group* transformation, will be of significance because it has an SU(2)⊗U(1) group symmetry. It emerges when we consider superluminal transformations but is not required for ordinary sublight Lorentz transformations. This new SU(2)⊗U(1) symmetry is the *analogue* of the SU(2)⊗U(1) internal symmetry of the ElectroWeak sector of The Superstandard Model.

We introduce this new transformation by reconsidering the previous simple example wherein one coordinate system is traveling at a speed v in the x direction with respect to the "laboratory" system. See Fig. 29.1.

The (left-handed[167]) Lorentz transformation is given by eq. 29.1, and the coordinates in the two reference frames are related by eq. 29.2.

---

[165] Superluminal transformations are a subset of Complex Lorentz transformations.

[166] The complex coordinates resulting from a superluminal transformation are physically viewed as real-valued by an observer in the new coordinate system. The apparent complexity of the coordinates resulting from a superluminal transformation are an artifact of the transformation.

[167] The right-handed Lorentz transformation case is analogous.

$$\Lambda_L(\omega, \mathbf{u} = (1,0,0)) = \begin{bmatrix} i\gamma_s & -i\beta\gamma_s & 0 & 0 \\ -i\beta\gamma_s & i\gamma_s & 0 & 0 \\ 0 & 0 & 1 & 0 \\ 0 & 0 & 0 & 1 \end{bmatrix} \tag{29.1}$$

implementing the coordinate transformation:

$$X' = \Lambda_L(\omega, \mathbf{u} = (1,0,0))X$$

or

$$\begin{aligned} t' &= i\gamma_s(t - \beta x) \\ x' &= i\gamma_s(x - \beta t) \\ y' &= y \\ z' &= z \end{aligned} \tag{29.2}$$

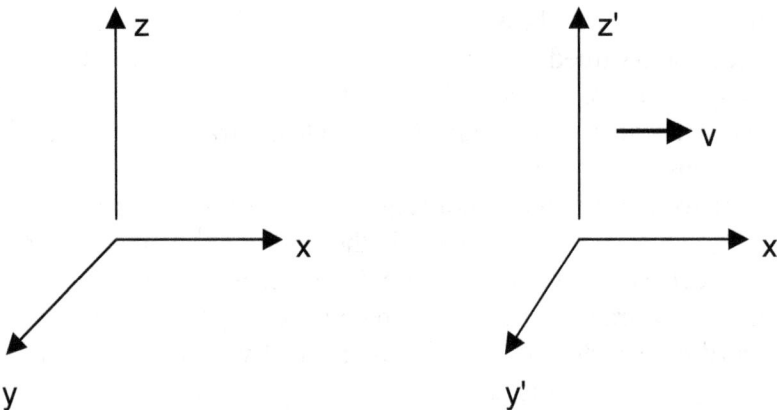

Figure 29.1. Depiction of two coordinate systems. The "primed" coordinate system is moving with velocity v in the positive x direction with respect to the "unprimed" coordinate system. We choose parallel axes for convenience.

We now define a Coordinate Reality group transformation $\Pi_L(\mathbf{u})$ that maps the real coordinates of the unprimed reference frame to real coordinates in the primed reference frame.

$$\Pi_L(\mathbf{u}) = \begin{bmatrix} -i & 0 & 0 & 0 \\ 0 & -i & 0 & 0 \\ 0 & 0 & 1 & 0 \\ 0 & 0 & 0 & 1 \end{bmatrix} \tag{29.3}$$

where $\mathbf{u}$ is the unit vector corresponding to the direction of $\mathbf{v}$ (the positive x direction in this example). Using $\Pi_L(\mathbf{u})$ we obtain an overall transformation from real coordinates to real coordinates in the case considered:

$$X'' = \Pi_L(\mathbf{u})\Lambda_L(\omega, \mathbf{u} = (1,0,0))X$$

or

$$\begin{aligned} t'' &= \gamma_s(t - \beta x) \\ x'' &= \gamma_s(x - \beta t) \\ y'' &= y \\ z'' &= z \end{aligned} \tag{29.4}$$

where $\gamma_s = (\beta^2 - 1)^{-\frac{1}{2}}$. An observer in the primed reference frame would consider his/her time to be real when measured on a clock, and distances along the x axis to be real when measured with a ruler. Thus eq. 29.4 makes good sense physically because in any reference frame, observers measure real distances and real times. For this reason we will call combined transformations of the type of eq. 29.4 – from real coordinates to real coordinates – *physical* superluminal transformations for real-valued velocities.

It is important to note that $\Pi_L(u)$ is position dependent in general for more complicated Λ_L transformations and so the Coordinate Reality group is a local group of the Yang-Mills type. This is clear from eqs. 29.1 and 29.2. We will see the Coordinate Reality group is the analogue of local (Yang-Mills) SU(2)⊗U(1) ElectroWeak symmetry.

This simple example generalizes to arbitrary relative real velocities **v**. First we note that the Lorentz transformation for a velocity **v** that is a rotation of the velocity in the x-direction (**v** = |**v**|R**u** where R is the relevant rotation matrix) has the form

$$\Lambda_L(\omega, \mathbf{v}) = \mathcal{R}(\mathbf{v}/v, \mathbf{u})\Lambda_L(\omega, \mathbf{u} = (1,0,0))\mathcal{R}^{-1}(\mathbf{v}/v, \mathbf{u}) \tag{29.5}$$

where $\mathcal{R}(\mathbf{v}/v, \mathbf{u})$ is a rotation from the velocity direction **u** to direction **v**/v.

The original transformation (eq. 29.2) can be written as

$$\Pi_L(\mathbf{u})\Lambda_L(\omega, \mathbf{u} = (1,0,0)) = \Pi_L(\mathbf{u})\mathcal{R}^{-1}(\mathbf{v}/v, \mathbf{u})\Lambda_L(\omega, \mathbf{v})\mathcal{R}(\mathbf{v}/v, \mathbf{u}) \tag{29.6}$$

Consequently the combined transformation for velocity **v** is

$$\mathcal{R}(\mathbf{v}/v, \mathbf{u})\Pi_L(\mathbf{u})\Lambda_L(\omega, \mathbf{u} = (1,0,0))\mathcal{R}^{-1}(\mathbf{v}/v, \mathbf{u})$$
$$= \mathcal{R}(\mathbf{v}/v, \mathbf{u})\Pi_L(\mathbf{u})\mathcal{R}^{-1}(\mathbf{v}/v, \mathbf{u})\Lambda_L(\omega, \mathbf{v})$$
$$= \Pi_L(\mathbf{v}/v)\Lambda_L(\omega, \mathbf{v}) \tag{29.7}$$

Thus for a Lorentz transformation $\Lambda_L(\omega, \mathbf{v})$ for velocity **v** we see that we can define a subsidiary transformation $\Pi_L(\mathbf{v}/v)$ of the form

$$\Pi_L(\mathbf{v}/v) = \mathcal{R}(\mathbf{v}/v, \mathbf{u})\Pi_L(\mathbf{u})\mathcal{R}^{-1}(\mathbf{v}/v, \mathbf{u}) \tag{29.8}$$

The general form of $\mathcal{R}(\mathbf{v}/v, \mathbf{u})$, is

$$\mathcal{R}(\mathbf{v}/v, \mathbf{u}) = \begin{bmatrix} 1 & 0 & 0 & 0 \\ 0 & & & \\ 0 & & \mathcal{R}_3(\mathbf{v}/v, \mathbf{u}) & \\ 0 & & & \end{bmatrix} \tag{29.9}$$

where $\mathcal{R}_3(\mathbf{v}/v, \mathbf{u})$ is a 3 × 3 rotation matrix that can be expressed in terms of the generators of the 3-dimensional rotation group as

$$\mathcal{R}_3(\mathbf{v}/v, \mathbf{u}) = \exp(i\boldsymbol{\theta}\cdot\mathbf{J}) \tag{29.10}$$

The rotation angles $\mathbf{\theta}$ are real numbers since we are rotating the real vector $\mathbf{u}$ to the real vector $\mathbf{v}/v$. Given the form of eq. 29.10 then we see that the form of $\Pi_L(\mathbf{v}//v)$ is

$$\Pi_L(\mathbf{v}/v) = \begin{bmatrix} -i & 0 & 0 & 0 \\ 0 & & & \\ 0 & & \mathcal{R}_3(\mathbf{v}/v, \mathbf{u})\Pi_{L3}(\mathbf{u})\mathcal{R}_3^{-1}(\mathbf{v}/v, \mathbf{u}) \\ 0 & & & \end{bmatrix} \qquad (29.11)$$

where

$$\Pi_{L3}(\mathbf{u}) = \begin{bmatrix} -i & 0 & 0 \\ 0 & 1 & 0 \\ 0 & 0 & 1 \end{bmatrix} \qquad (29.12)$$

If we consider the case of an infinitesimal rotation $\mathbf{\theta}$ to first order in $\mathbf{\theta}$

$$\mathcal{R}_3(\mathbf{v}/v, \mathbf{u}) \simeq I + i\mathbf{\theta}\cdot\mathbf{J} \qquad (29.13)$$

then

$$\Pi_{L3}(\mathbf{v}/v) = \mathcal{R}_3(\mathbf{v}/v, \mathbf{u})\Pi_{L3}(\mathbf{u})\mathcal{R}_3^{-1}(\mathbf{v}/v, \mathbf{u}) \simeq \Pi_{L3}(\mathbf{u}) + i\mathbf{\theta}\cdot\mathbf{J}\Pi_{L3}(\mathbf{u}) - i\Pi_{L3}(\mathbf{u})\mathbf{\theta}\cdot\mathbf{J}$$
$$\simeq \Pi_{L3}(\mathbf{u})[I + i\Pi_{L3}^{-1}(\mathbf{u})[\mathbf{\theta}\cdot\mathbf{J}, \Pi_{L3}(\mathbf{u})] \qquad (29.14)$$

where $\Pi_{L3}^{-1}(\mathbf{u})$ is the inverse of $\Pi_{L3}(\mathbf{u})$ and $[...]$ represents the commutator. Thus for arbitrary rotations eq. 29.14 implies

$$\Pi_{L3}(\mathbf{v}/v) = \mathcal{R}_3(\mathbf{v}/v, \mathbf{u})\Pi_{L3}(\mathbf{u})\mathcal{R}_3^{-1}(\mathbf{v}/v, \mathbf{u}) = \Pi_{L3}(\mathbf{u})\exp\{i\Pi_{L3}^{-1}(\mathbf{u})[\mathbf{\theta}\cdot\mathbf{J}, \Pi_{L3}(\mathbf{u})]\} \qquad (29.15)$$

We can find the general form of $\Pi_{L3}(\mathbf{v}/v)$ by considering the case of eq. 29.6 in more detail. The exponentiated matrix expression in 29.15 can be written

$$\Pi_{L3}^{-1}(\mathbf{u})[\mathbf{\theta}\cdot\mathbf{J}, \Pi_{L3}(\mathbf{u})] = \Pi_{L3}^{-1}(\mathbf{u})\mathbf{\theta}\cdot\mathbf{J}\Pi_{L3}(\mathbf{u}) - \mathbf{\theta}\cdot\mathbf{J} = \mathbf{\theta}\cdot\mathbf{Q} \qquad (29.16)$$

where

$$\mathbf{Q} = \Pi_{L3}^{-1}(\mathbf{u})\mathbf{J}\Pi_{L3}(\mathbf{u}) - \mathbf{J} = \mathbf{Q'} - \mathbf{J} \qquad (29.17)$$

The matrices Q_i can be evaluated using eq. 29.12 and the matrix representations of rotation generators J_i: which are equivalent in form to the SU(2) generators T_i:

$$J_1 = \begin{bmatrix} 0 & 0 & 0 \\ 0 & 0 & -i \\ 0 & i & 0 \end{bmatrix} = T_1 \qquad (29.18)$$

$$J_2 = \begin{bmatrix} 0 & 0 & i \\ 0 & 0 & 0 \\ -i & 0 & 0 \end{bmatrix} = T_2 \qquad (29.19)$$

$$J_3 = \begin{bmatrix} 0 & -i & 0 \\ i & 0 & 0 \\ 0 & 0 & 0 \end{bmatrix} = T_3 \qquad (29.20)$$

The rotation generators satisfy the commutation relations

$$[J_i, J_j] = i\epsilon_{ijk}J_k \qquad (29.21)$$

as do the SU(2) generators:

$$[T_i, T_j] = i\epsilon_{ijk}T_k \qquad (29.22)$$

We can calculate Q' and obtain

$$Q'_1 = \begin{bmatrix} 0 & 0 & 0 \\ 0 & 0 & -i \\ 0 & i & 0 \end{bmatrix} \qquad (29.23)$$

$$Q'_2 = \begin{bmatrix} 0 & 0 & -1 \\ 0 & 0 & 0 \\ -1 & 0 & 0 \end{bmatrix} \tag{29.24}$$

$$Q'_3 = \begin{bmatrix} 0 & 1 & 0 \\ 1 & 0 & 0 \\ 0 & 0 & 0 \end{bmatrix} \tag{29.25}$$

We note that each Q'_i is hermitean and the Q'_i satisfy the commutation relations:

$$[Q'_i, Q'_j] = i\epsilon_{ijk}Q'_k \tag{29.26}$$

Consequently the set of Q'_i are also equivalent to SU(2) generators. As a result the exponential factor

$$\Pi_{L3}(\mathbf{v}/v) = \Pi_{L3}(\mathbf{u})\exp\{i\mathbf{\theta}\cdot(\mathbf{Q}' - \mathbf{J})\} \tag{29.27}$$

is equivalent to a combination of SU(2) rotations not only in this case but in general for superluminal transformations. The factor $\Pi_{L3}(\mathbf{u})$ is not an SU(2) matrix since its determinant is not 1. However

$$\Pi'_{L3}(\mathbf{u}) = -i\Pi_{L3}(\mathbf{u}) \tag{29.28}$$

is an SU(2) matrix since

$$\Pi'_{L3}{}^{-1}(\mathbf{u}) = \Pi'_{L3}{}^{\dagger}(\mathbf{u}) \tag{29.29}$$
$$\det \Pi'_{L3}(\mathbf{u}) = 1 \tag{29.30}$$

and

$$\Pi'_{L3}(\mathbf{v}/v) = \Pi'_{L3}(\mathbf{u})\exp\{i\mathbf{\theta}\cdot(\mathbf{Q}' - \mathbf{J})\} \tag{29.31}$$

is similarly an SU(2) rotation.

Thus the general form of superluminal, *real* velocity, transformation from a real set of coordinates to a real set of coordinates is[168]

$$\Pi_L(\mathbf{v}/v)\Lambda_L(\omega, \mathbf{v}) \tag{29.32}$$

where

$$\Pi_L(\mathbf{v}/v) = \begin{bmatrix} -i & 0 & 0 & 0 \\ 0 & & & \\ 0 & & \Pi_{L3}(\mathbf{u})\exp\{i\,\boldsymbol{\theta}\cdot(\mathbf{Q'} - \mathbf{J})\} & \\ 0 & & & \end{bmatrix} \tag{29.33}$$

The Lorentz condition for real to real physical transformations generalizes to

$$\Lambda(\mathbf{v})^T \Pi_L(\mathbf{v}/v)^\dagger G\, \Pi_L(\mathbf{v})\Lambda(\mathbf{v}/v) = G \tag{29.34}$$

Since superluminal transformations $\Lambda_L(\omega, \mathbf{v})$ transform real coordinates to complex coordinates in general, we can generalize the form of a real-to-real superluminal transformation to

$$e^{i\varphi}\Pi_L(\mathbf{v'}/v')\Lambda_L(\omega, \mathbf{v}) \tag{29.35}$$

where φ is a constant phase and $\mathbf{v'}$ is an arbitrary velocity. This generalization will satisfy the generalized Lorentz condition

$$\Lambda(\mathbf{v})^T \Pi_L(\mathbf{v'}/v')^\dagger e^{-i\varphi}G\, e^{i\varphi}\Pi_L(\mathbf{v'}/v')\Lambda(\mathbf{v}) = G \tag{29.36}$$

but the transformation will, in general, yield a complex set of coordinates when applied to a set of real coordinates.

These considerations imply:

---

[168] The choice of the unit vector $\mathbf{u}$ and the angle vector $\boldsymbol{\theta}$ must be such that applying the transformation to a real set of coordinates yields a real set of coordinates.

1. Any observer in a coordinate system will treat a complex 4-dimensional coordinate system as if it were a real 4-dimensional coordinate system with complex-valued straight lines along each dimension (assuming rectangular coordinates).

2. The transformation $e^{i\varphi}\Pi'_{L3}(\mathbf{v}/v)$ is a SU(2)⊗U(1) transformation that takes complex 3-dimensional spatial coordinates to complex 3-dimensional spatial coordinates. In particular straight lines map to straight lines.

3. Physical observations in the observer's coordinate system are invariant under SU(2)⊗U(1) rotations of the spatial coordinates and the multiplication of the time component by an arbitrary phase.

4. The matrix

$$\Pi'_{L}(\mathbf{v}/v, \chi, \varphi) = \begin{bmatrix} e^{i\chi} & 0 & 0 & 0 \\ 0 & & & \\ 0 & & e^{i\varphi}\Pi'_{L3}(\mathbf{u})\exp\{i\,\boldsymbol{\theta}\cdot(\mathbf{Q'} - \mathbf{J})\} \\ 0 & & & \end{bmatrix} \qquad (29.37)$$

(where χ and φ are real numbers and $\mathbf{u}$ is a unit vector along any convenient coordinate axis) is a SU(2)⊗U(1) transformation that transforms complex 4-dimensional coordinates to complex 4-dimensional coordinates. Note, $\Pi_{L}(\mathbf{v}/v) = \Pi'_{L}(\mathbf{v}/v, 3\pi/2, \pi/2)$ is a special case of $\Pi'_{L}(\mathbf{v}/v, \chi, \varphi)$. Due to the manifest form of 29.37 we see

$$\Pi'_{L}{}^{\mu}{}_{\alpha}{}^{*}\Pi'_{L}{}^{\mu}{}_{\beta} = [\Pi'_{L}{}^{\dagger}\Pi'_{L}]_{\alpha\beta} = I_{\alpha\beta} \qquad (29.38)$$

(with an implied sum over μ) or, in matrix form,

$$\Pi'_{L}{}^{\dagger}\,\Pi'_{L} = I \qquad (29.39)$$

and also[169]

$$\Pi'_L{}^\dagger G \Pi'_L = G \tag{29.40}$$

5. Complex coordinate values of the type generated by superluminal transformations with real-valued velocities are transformable to real coordinates. The complex coordinates are thus physically equivalent to corresponding real coordinate values in the sense that an observer in that frame would automatically use the real coordinates so obtained since rulers and clocks always measure real spatial coordinates and times. *Therefore physical theory is invariant under global SU(2)⊗U(1) coordinate transformations since complex coordinates, so generated, can be rotated back to real coordinates.*

6. The complex coordinates of any point obtained through a superluminal transformation can be transformed to a real set of coordinates by the above SU(2)⊗U(1) transformation. This SU(2)⊗U(1) invariance is the analogue of the SU(2)⊗U(1) symmetry of the ElectroWeak interactions.

---

[169] Eq. D.40 is close to the defining condition for a Lorentz group element but the presence of complex conjugation rather than a transpose means Π'_L is outside the real and complex Lorentz groups.

30. Coordinate Reality Group Analogue for Dark ElectroWeak SU(2)⊗U(1)

In this chapter[170] we will consider superluminal transformations based on *complex-valued relative velocities*. The previous chapter considered the case of real-valued velocities. That case led to the Coordinate Reality Group analogue of ElectroWeak SU(2)⊗U(1).

We now consider Complex Lorentz transformations for complex-valued relative velocities. These transformations will require us to introduce another Coordinate Reality group with transformations that map complex coordinates to real coordinates. These transformations will be of significance because they lead to a hitherto unstated SU(2)⊗U(1) symmetry. We identify this SU(2)⊗U(1) symmetry as the analogue of the symmetry of the *Dark* ElectroWeak interactions of the Superstandard Model. Dark matter and interactions remain to be found experimentally but there may be some preliminary suggestive data from the CERN LHC.

We introduce these new Dark transformations by extending the previous simple example to Fig. 30.1 in which one coordinate system is traveling at a complex-valued velocity u in the x direction with respect to the "laboratory" system. In these new transformations the relative velocity is complex-valued and has two components: a real-valued component in the x direction and an imaginary-valued component in the y direction. $\mathbf{u} = u_x\mathbf{i} + iu_y\mathbf{j}$. In the complex case $\beta = \tanh(\omega_L)$ is real-valued by eqs. 26.20 – 26.22 where $\omega_L = \omega + i\pi/2$ and ω is real.

The (left-handed[171]) Lorentz transformation is given by eq. 26.23:

---

[170] Most of the material in this chapter appeared in Blaha (2011c) originally.
[171] The right-handed Lorentz transformation case is analogous.

$$\Lambda_L(\omega, \mathbf{u}) = \begin{bmatrix} i\gamma_s & -i\beta\gamma_s u_x & \beta\gamma_s u_y & 0 \\ -i\beta\gamma_s u_x & 1 + (i\gamma_s - 1)u_x^2 & i(i\gamma_s - 1)u_x u_y & 0 \\ \beta\gamma_s u_y & i(i\gamma_s - 1)u_x u_y & 1 - (i\gamma_s - 1)u_y^2 & 0 \\ 0 & 0 & 0 & 1 \end{bmatrix} \tag{30.1}$$

$$= \Lambda(\omega + i\pi/2, \mathbf{u})$$

implementing the coordinate transformation:

$$X' = \Lambda_L(\omega, \mathbf{u} = (u_x, iu_y, 0))X$$

or

$$t' = i\gamma_s(t - \beta u_x - i\beta u_y)$$
$$x' = -i\gamma_s\beta u_x t + i\gamma_s x + u_x x - u_x^2 x + i(i\gamma_s - 1)u_x u_y y \tag{30.2}$$
$$y' = \gamma_s\beta u_y t + i(i\gamma_s - 1)u_x u_y x + [1 - (i\gamma_s - 1)u_y^2]y$$
$$z' = z$$

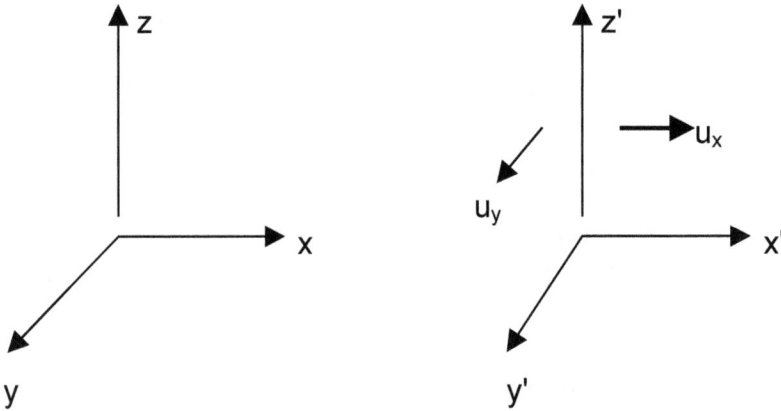

Figure 30.1. Depiction of two coordinate systems. The "primed" coordinate system is moving with velocity u = u_xi + iu_yj with respect to

the "unprimed" coordinate system. We choose parallel axes for convenience.

We now define a transformation that maps the real coordinates of the unprimed reference frame to real coordinates in the primed reference frame.

$$
\Pi_L(\mathbf{u}, \mathbf{X}) \;=\;
\begin{bmatrix}
e^{ia} & e^{ib} & e^{ic} & 0 \\
e^{id} & e^{ie} & e^{if} & 0 \\
e^{ig} & e^{ih} & e^{ij} & 0 \\
0 & 0 & 0 & 1
\end{bmatrix}
\tag{30.3}
$$

where $\mathbf{u}$ is the unit vector corresponding to the direction of the relative velocity.

It is important to note that $\Pi_L(\mathbf{u}, \mathbf{X})$ is position dependent and so the Coordinate Reality group is a local group. This is clear from eqs. 30.1 and 30.2. The Coordinate Reality Group in this case is $SU(2) \otimes U(1)$. It appears similar to the Reality group of Appendix B except for the crucial difference that the Coordinate Reality group here mixes time and spatial rows while the Coordinate Reality group of Appendix B only mixed spatial rows (See eqs. 29.35 and 29.37.).Again we have a local theory. Thus the combined Coordinate Reality group from this appendix and Appendix B is $SU(2) \otimes U(1) \otimes DSU(2) \otimes DU(1)$ where we prepend 'D' to the Dark Reality group parts. Elsewhere we defined complex boosts. We summarize the definition below:

$$
\Lambda_L(\omega, \mathbf{u}) = \Lambda_L(\mathbf{v_c}) = \exp[i(\omega + i\pi/2)\mathbf{u}\cdot\mathbf{K}]
\tag{30.4}
$$

where ω remains

$$
\omega = (\omega_r^2 - \omega_i^2)^{\frac{1}{2}}
$$

and

$$
\mathbf{u} = (\omega_r\mathbf{u_r} + i\omega_i\mathbf{u_i})/\omega
$$
$$
\mathbf{u}\cdot\mathbf{u} = \mathbf{u_r}\cdot\mathbf{u_r} = \mathbf{u_i}\cdot\mathbf{u_i} = 1
$$
$$
\mathbf{v_c} = \mathbf{u}\tanh(\omega + i\pi/2) = \mathbf{u}\coth(\omega)
$$

In the example, that we are considering, we set $\mathbf{u_x} = \omega_r\mathbf{u_r}$ and $\mathbf{u_y} = \omega_i\mathbf{u_i}$.

Using $\Pi_L(\mathbf{u}, \mathbf{X})$ we obtain an overall transformation from real coordinates to real coordinates:

$$X'' = \Pi_L(\mathbf{u}, \mathbf{X})\Lambda_L(\omega, \mathbf{u} = (u_x, u_y, 0))X \qquad (30.5)$$

with the coordinates of X and X" real-valued. An observer in the double primed reference frame would consider his/her time to be real when measured on a clock, and distances along the x and y axes to be real when measured with a ruler.

The velocity vectors: $u_x\mathbf{i}$ and $iu_y\mathbf{j}$ in our example define a plane in space. There are two types of rotations that are possible. 1) An angular rotation in the plane defined by the vectors. This is a U(1) transformation. 2) a spatial rotation of the plane that is an SU(2) rotation. Thus the joint rotations of **u** have an SU(2)⊗U(1) symmetry group. The Coordinate Reality group for 4-dimensions – has two SU(2)⊗U(1) factors that we denote SU(2)⊗U(1)⊗DSU(2)⊗DU(1). We see that the "newly found" group can be *assumed* to be the analogue of the Dark ElectroWeak symmetry group. We note that a further Coordinate Reality group part SU(3) will be introduced in the next chapter.

We consider Dark Matter and its interactions in greater detail elsewhere.

31. Coordinate Reality Group Analogue for Color SU(3)

31.1 Two Possible Approaches to Color SU(3)

There are two approaches to obtaining the Strong interaction and Color SU(3) symmetry:

A. Assume up-type and down-type quarks are in $\underline{3}$ representations of Color SU(3). This assumption sheds no light on a deeper origin of the Strong interaction and Color SU(3). It simply assumes the color SU(3) of the Strong interaction sector of the Standard Model. Thus our understanding is not deepened. A postulate corresponding to this assumption is:

Possible Postulate: Quarks are in the $\underline{3}$ representation of Color SU(3). The SU(3) symmetry is gauged with local Yang-Mills SU(3) fields called gluons that constitute the Strong interaction of the quark sector. Quarks are minimally coupled to the gluons in a gauge covariant fashion.

Comment: We will **not** use this postulate in our derivation.

B. In the preceding chapters the Internal Symmetry ElectroWeak interactions of the Unified Superstandard Model (modulo generations and their mixing) were shown to be an analogue to the Coordinate Reality Group associated with the Complex Lorentz group. The Reality group included $SU(2) \otimes U(1) \otimes SU(2) \otimes U(1)$.[172] Thus we found a significant geometrical

---

[172] We no longer prepend 'D' to the Dark $SU(2) \otimes U(1)$.

analogue for the form of the ElectroWeak interactions of normal and Dark matter.

We now establish a similar geometrical analogue of the internal symmetry Strong interaction and Color SU(3). *If we extend the parameters to be real functions of the space-time coordinates (i.e. local SU(3) transformations), then we obtain an analogue for color SU(3). A key factor in this interpretation is the global covariance of complexon equations of motion under global SU(3).*[173]

Comment: We follow this approach in our derivation.

31.2 A Global SU(3) Symmetry of Complexon Quarks

We will now consider a global SU(3) covariance. The defining property of the SU(3) group is that it preserves the invariance of inner products of complex 3-vectors of the form:

$$u^* \cdot v = u^1{}^* v^1 + u^2{}^* v^2 + u^3{}^* v^3 \tag{31.1}$$

If we examine the dynamical equation eq. 2.202 of Blaha (2018e) we see that the differential operator is covariant under a global SU(3) transformation U of the complex spatial 3-coordinates:

$$[i\gamma^0 \partial/\partial t + i\mathbf{D}_c^* \cdot \gamma - m] = [i\gamma^0 \partial/\partial t + i\mathbf{D}_c'^* \cdot \gamma' - m] \tag{31.2}$$

where

$$\mathbf{D}_c^* = \nabla_c = \nabla_r + i\nabla_i$$

and

$$\gamma^a = U^{ab}\gamma'^b \tag{31.3a}$$
$$D_c^{*a} = D_c'^{*b}U^{ab*} \tag{31.3b}$$

---

[173] This chapter was extracted from chapter 17 of Blaha (2011c) with some changes.

where $U^\dagger = U^{-1}$. We now exhibit the covariance of eq. 2.202. Since we can view the three spatial γ-matrices as SU(3) 3-vectors, we can express eq. 31.3 as the result of a SU(3) rotation V of the γ-matrices (on the spinor indices)

$$V\gamma^a V^{-1} = U^{ab}\gamma'^b \tag{31.4}$$

where V is a 4×4 reducible representation of SU(3), namely, $\underline{3}\oplus\underline{1}$. Since V commutes with γ^0 in the Pauli matrix representation of the γ matrices we see that V can have the form

$$V = \begin{bmatrix} A\exp(i\alpha_i\sigma_i) & 0 \\ 0 & B\exp(i\beta_i\sigma_i) \end{bmatrix}$$

where A, B, α_i and β_i are constants, and the zeroes represent 2×2 zero matrices. The inverse of V is $V^\dagger$. Thus eq. 31.4 becomes

$$V\gamma^a V^{-1} = \begin{bmatrix} 0 & AB^*\exp(i\alpha_i\sigma_i)\sigma_a\exp(-i\beta_i\sigma_i) \\ -A^*B\exp(i\beta_i\sigma_i)\sigma_a\exp(-i\alpha_i\sigma_i) & 0(-i\beta_i\sigma_i) \end{bmatrix}$$

We now note the generators of the global SU(3) symmetry under discussion have a 4×4 matrix reducible representation ($\underline{3}\oplus\underline{1}$). The generators of this reducible representation are F_i and F_0 (a diagonal matrix diag(0,0,0,0,0,0,0,0,0,1) with F_i being the Gell-Mann SU(3) generators for i = 1, 2, ..., 8.

Projection operators can be defined to project out the $\underline{3}$ representation piece P_3 and the $\underline{1}$ representations piece P_1, of the complexon spinor fields:

Thus the $\underline{3}$ complexon field is

$$\psi_{C3}(t, \mathbf{x_r}, \mathbf{x_i}) = P_3\psi_C(t, \mathbf{x_r}, \mathbf{x_i}) \tag{31.5}$$

while the $\underline{1}$ complexon field is

$$\psi_{C1}(t, \mathbf{x_r}, \mathbf{x_i}) = P_1\psi_C(t, \mathbf{x_r}, \mathbf{x_i}) \tag{31.6}$$

Since P_1 and P_3 do not commute with Lorentz transformations, a Lorentz transformation mixes ψ_{C1} and ψ_{C3}.[174] Since P_1 and P_3 do not commute with γ_5, left-handed and right-handed complexons would also be mixed by these projection operators. The matrix V has a $\underline{3}\oplus\underline{1}$ reducible representation.

In a manner similar to the covariance proof of the Dirac equation[175] we see that eq. 31.2 is covariant under SU(3) transformations:

$$V[i\gamma^0\partial/\partial t + i\gamma\cdot\mathbf{D_c}^* - m]V^{-1}V\psi_C(t, \mathbf{x_r}, \mathbf{x_i}) = 0$$

or

$$[i\gamma^{0\prime}\partial/\partial t' + i\mathbf{D_c'}^*\cdot\gamma' - m]V\psi_C(t, \mathbf{x_r}, \mathbf{x_i}) = 0 \tag{31.7}$$

(Note $\gamma^{0\prime} = V\gamma^0 V^{-1}$ and t' = t.) The SU(3) transformed wave function $\psi_C{}'(t, \mathbf{x'})$ is

$$\psi_C{}'(t', \mathbf{x'}) = V\psi_C(t, \mathbf{x}) = V\psi_C(t', U\mathbf{x'}) \tag{31.8}$$

Thus the complexon Dirac equation is covariant under global coordinate SU(3).

The subsidiary condition,

$$\nabla_r\cdot\nabla_i \, \psi_{Cu}(t, \mathbf{x_r}, \mathbf{x_i}) = 0 \tag{31.9}$$

---

[174] At this point it is worth noting that the construction of complexon fields, based on a boost from a particle rest state, guarantees that a reference frame exists in which any complexon particle has a single real time variable. Similarly a reference frame exists for a set of complexon particles (that is within a Lorentz of the center of momentum frame) with a single real time variable. The time variables of the individual complexon particles in the set are complex in general but are functions of the center of momentum real time variable. So there is only one real time variable for each complexon in the set although the time variable of an individual particle may be a complex function of the real-valued center of momentum time variable.

[175] For example see Bjorken (1964) pp. 18 – 20.

is also covariant under an SU(3) rotation:

$$\nabla_r'^* \cdot \nabla_i' \psi_C'(t, \mathbf{x}') = \nabla_r \cdot \nabla_i \, V\psi_C(t, \mathbf{x}) = V\nabla_r^* \cdot \nabla_i \, \psi_C(t, \mathbf{x}) = 0 \quad (31.10)$$

We now examine the transformation of the wave function eq. 31.8 under the SU(3) transformation U. If we define

$$q^{*\mu} = (q^0, \mathbf{q}^*) = (p^0, \mathbf{p_r} + i\mathbf{p_i}) = (p^0, \mathbf{p}) = p^\mu \quad (31.11)$$

then $\psi_C(t, \mathbf{x})$ will be seen to be covariant form under an SU(3) transformation:

$$\psi_C(t, x) = \sum_{\pm s} \int d^3 q_r d^3 q_i \, N_C(p^0) \delta(\mathbf{q_r}^* \cdot \mathbf{q_i}/m^2) [b_C(q^*,s) u_C(q^*,s) e^{-i(q^* \cdot x + q \cdot x^*)/2} +$$
$$+ \, d_C^\dagger(q^*,s) v_C(q^*,s) e^{+i(q^* \cdot x + q \cdot x^*)/2}] \quad (31.12)$$

Note both terms in each exponential are separately invariant under global SU(3). ($\mathbf{q_r}^* = \mathbf{q_r}$ since $\mathbf{q_r}$ is real.)

Eq. 31.8 implies that the spinors appearing in eq. 31.12 are covariant under SU(3) transformations

$$u_C'(q'^*,s') = Vu_C(q^*,s) \quad (31.13)$$
$$v_C'(q'^*,s') = Vv_C(q^*,s) \quad (31.14)$$

The fourier coefficients, if second quantized in a complex spatial coordinate generalization of the usual manner, also have covariant anti-commutation relations under an SU(3) transformation:

$$\{b_C(q,s), b_C^\dagger(q'^*,s')\} = \delta_{ss'} \delta^3(q_r - q'_{r'}) \delta^3(q_i - q'_{i'}) \quad (31.15)$$

Under an SU(3) transformation, z = Uq and z' = Uq', the right side of eq. 31.15 transforms to

$$\delta^3(q_r - q'_r)\delta^3(q_i - q'_i) \rightarrow \delta^3(z_r - z'_r)\delta^3(z_i - z'_i)/|\partial(q)/\partial(z)| = \delta^3(z_r - z'_r)\delta^3(z_i - z'_i)$$

$$(31.16)$$

where

$$|\partial(q)/\partial(z)| = |\partial(q_r^1, q_r^2, q_r^3, q_i^1, q_i^2, q_i^3)/\partial(z_r^1, z_r^2, z_r^3, z_i^1, z_i^2, z_i^3)| = 1 \qquad (31.17)$$

is the Jacobian of the transformation U. The fourier coefficients transform trivially under SU(3):

$$b_C(q^*, s) \rightarrow b_C(z^*, s) \qquad (31.18)$$

Since the integrand transforms as

$$\int d^3 q_r d^3 q_i \rightarrow \int d^3 z_r d^3 z_i \, |\partial(q)/\partial(z)| = \int d^3 z_r d^3 z_i \qquad (31.19)$$

we see that the wave function $\psi_C(t, \mathbf{x})$ transforms covariantly.

31.3 Local Color SU(3) and the Internal Symmetry Strong Interactions

In the previous section we showed that the equations of motion of free Dirac-like, complexon, up-type quarks are covariant under global SU(3) coordinate rotations.. The free, tachyon, complexon, down-type quark equations of motion are also easily seen to be covariant under this SU(3) subgroup. In this section we will show this covariance is the analogue of local Color SU(3) symmetry of quarks, and then we will introduce the Internal Symmetry Strong interaction analogue via minimal coupling to SU(3) Yang-Mills gluons in gauge covariant derivatives.

We now introduce a complexon field with a global SU(3) index a which takes values from 1 to 3 making the field a member of the $\underline{3}$ representation of global SU(3):

$$\psi_C^a(t, \mathbf{x}) \qquad (31.20)$$

Due to the SU(3) index the transformation property of $\psi_C^a(t, \mathbf{x})$ changes from eq. 31.8 to

$$\psi_C''^a(t, \mathbf{x}') = U^{ab}V\psi_C{}^b(t, \mathbf{x}) = U^{ab}V\psi_C{}^b(t, U\mathbf{x}') \qquad (31.21)$$

where U^{ab} is an SU(3) rotation of $\underline{3}$ representation "vectors" such as $\psi_C{}^b$ and $\mathbf{x}$. V is the corresponding rotation of the spinor indices of $\psi_C{}^b(t, \mathbf{x})$.

Note that the coordinate SU(3) rotation of the field factorizes into an SU(3) rotation of the three fields $\psi_C{}^b$ by U^{ab} and an SU(3) rotation of the four spinor components of each individual field $\psi_C{}^b$ by V.

This factorization enables us to consider a global SU(3) rotation of the $\psi_C{}^b$ fields while holding the coordinates fixed:

$$\psi_C'^a(t, \mathbf{x}) = U^{ab}\psi_C{}^b(t, \mathbf{x}) \qquad (31.22)$$

The equations of motion are covariant under this global transformation

$$0 = U^{ab}[i\gamma^0\partial/\partial t + i\gamma\cdot\mathbf{D_c}^* - m]\psi_C{}^b(t, \mathbf{x_r}, \mathbf{x_i})$$

$$= [i\gamma^0\partial/\partial t + i\gamma\cdot\mathbf{D_c}^* - m]\psi'_C{}^a(t, \mathbf{x_r}, \mathbf{x_i}) \qquad (31.23)$$

We now note the form of eq. 31.22 is the same as that of a *local* Yang-Mills rotation:

$$\psi_C'^a(t, \mathbf{x}) = \Theta^{ab}(t, \mathbf{x})\psi_C{}^b(t, \mathbf{x}) \qquad (31.24)$$

where $\mathbf{x} = \mathbf{x_r} + i\mathbf{x_i}$. Therefore if we introduce a local SU(3) Yang-Mills field $A_{Cv}(t, \mathbf{x_r}, \mathbf{x_i})$ and define a covariant derivative we can convert eq. 31.21 to the analogue case of Internal Symmetry local, color SU(3) if we do not perform the spinor rotation V.[176] The covariant derivative is

---

[176] This approach, by analogy, enables us to avoid the dilemmas associated with mixing coordinate and internal symmetries as described by Coleman, S., Phys. Rev. **138** B1262 (1965) and others in the case of SU(6) in the 1960's. Note that the spinor rotation V is expressed in terms of numerical matrices while, in the second quantized

$$\mathcal{D}_v = D_v - igA_{Cv} \tag{31.25}$$

where

$$A_{Cv} = A_C{}^a{}_v t^a \tag{31.26}$$

and where $D_v = D_{qv}$ is given by

$$D_0 = \partial/\partial x^0$$
$$D_k = \partial/\partial x_r{}^k + i \, \partial/\partial x_i{}^k \tag{31.27}$$

The SU(3) 3×3 matrix generators satisfy

$$[t^a, t^b] = if^{abc}t^c \tag{31.28}$$

We can represent $\Theta_{ab}(x)$ in the form:

$$\Theta_{ab}(x) = [\exp(-i\varphi_c(x)t^c)]_{ab} \tag{31.29}$$

where $\varphi_c(x)$ is a local parameter dependent on $x = (x^0, \mathbf{x} = \mathbf{x_r} + i\mathbf{x_i})$, and t^c is an SU(3) generator.

Applying a gauge transformation to the gauge covariant derivative of a complexon fermion field $\mathcal{D}_v\psi_C(x)$:

$$\Theta\mathcal{D}_v\psi_C(x) = \Theta D_v\psi_C(x) - ig\Theta A_{Cv}\Theta^{-1}\Theta\psi_C(x) \tag{31.30}$$
$$= D_v\psi_C'(x) - igA_C'{}_v\psi_C'(x) = (\mathcal{D}_v\psi_C(x))'$$

where

$$\psi_C'(x) = \Theta(x)\psi_C(x) \tag{31.31}$$

we find

$$A_C'{}_v = (-i/g)(D_v\Theta(x))\Theta^{-1}(x) + \Theta(x)A_{Cv}(x)\Theta^{-1}(x) \tag{31.32}$$

The reader will note that the form of eqs. 31.25 – 31.31 is identical to those associated with a conventional non-abelian gauge interaction with the replacement:

---

formulation, the U^{ab} rotation is expressed in terms of second quantized fields as well as numeric matrices. Thus the factorization is reflected in the form of the transformation.

$$\partial/\partial x^{\nu} \rightarrow D_{\nu} \qquad (31.33)$$

with D_{ν} given by eq. 31.27. Note that $\varphi_c(x)$, the local parameter in eq. 31.29 is dependent in general, on time, and the real and imaginary parts of the complex spatial 3-vector.

Introducing the SU(3) gauge covariant derivative transforms eq. 31.23 to

$$0 = [i\gamma^{\nu}\mathcal{D}_{\nu} - m]\psi_C^{\ a}(t, \mathbf{x_r}, \mathbf{x_i}) \qquad (31.34)$$

The preceding argumentation supports the following postulate:

Postulate: Quarks are in a $\underline{3}$ representation of an internal symmetry global SU(3) group.

We note the case of tachyon complexon quarks differs only in small details from the above discussion of Dirac-type complexon quarks.

31.4 Internal Symmetry Interactions Resulting by Analogy from Complex Space-Time Projected to Real Physical Space-Time

This chapter, and the preceding chapters, have shown that the Complex Lorentz group and the Coordinate Reality Group generate analogues of the familiar interactions of The Standard Model: SU(3)⊗SU(2)⊗U(1) plus an additional set of SU(2)⊗U(1) interactions that we take to be the interactions of Dark Matter.

32. Summary of Complex Lorentz Boosts of Fermions

32.1 Boosts at Real-Valued Velocities Greater Than the Speed of Light

In chapter 5 of Blaha (2017b) we showed that Complex Lorentz boosts for real-valued velocities with magnitude greater than the speed of light transform a rest frame coordinate system to a new complex-valued coordinate system. In general an SU(2)⊗U(1) transformation is needed to transform the target complex-valued coordinate system to a real-valued coordinate system for this type of boosts. We can symbolize these transformations with

Real-valued Coordinate System → $R_{SU(2)\otimes U(1)}\Lambda(|\mathbf{v}| > \mathbf{c})$ → Real-valued Coordinate System

where $\Lambda(|\mathbf{v}| > \mathbf{c})$ is a Complex Lorentz transformation with real-valued relative velocity $|\mathbf{v}| > c$, and $R_{SU(2)\otimes U(1)}$ is an SU(2)⊗U(1) transformation from complex-valued coordinates to real-valued coordinates.

The Complex Lorentz transformation $\Lambda(|\mathbf{v}| > \mathbf{c})$ when applied to a fermion at rest transforms it into a tachyon with real-valued energy. (See chapter 3 of Blaha (2017b).) The tachyons of this type are those of the neutral lepton species and of the Dark neutral lepton species.

32.2 Boosts at Complex-Valued Velocities

There are two types of boosts of this kind: those whose magnitude of velocity exceeds the speed of light yielding tachyons, and those whose magnitude of velocity is below the speed of light yielding a non-tachyon fermion. (See chapter 2, or chapters 6 and 7 of Blaha (2017b) for details. The forms of the tachyons of these varieties are described chapter 2.) The fermions generated by these boosts are identified as the up and down type normal and Dark quarks.

32.3 Mapping of the Boosted Fermions to Normal and Dark Fermion Species

Four types of fermion species, and the internal symmetries and interactions of particles, emerge from the Complex Lorentz group. We can summarize the map from our theory to the real world in the following way:

Normal Matter Fermions

Dirac fermions – Charged Leptons – Fields generated by Real Lorentz boosts – real v < c
Tachyon fermions – Neutrinos – Fields generated by Real Lorentz boosts – real v > c
Complexon fermions – Up-Type Quark Triplets – Fields generated by Lorentz boosts – complex v < c
Tachyon Complexon fermions – Down-Type Quark Triplets – From Lorentz boosts – complex v > c

Dark Matter Fermions

Dirac fermions – Dark Charged Leptons – Fields generated by Real Lorentz boosts – real v < c
Tachyon fermions – Dark Neutrinos – Fields generated by Real Lorentz boosts – real v > c
Complexon fermions – Dark Up-Type Quark Singlets – Fields generated by Lorentz boosts – complex v < c
Tachyon Complexon fermions – Dark Down-Type Quark Singlets – From Lorentz boosts – complex v > c

Normal Matter Gauge Bosons

$SU(2) \otimes U(1)$ - real space-time coordinates – not complexon coordinates
$SU(3)$ - complex space-time coordinates – complexon coordinates

Dark Matter Gauge Bosons

$SU(2) \otimes U(1)$ - real space-time coordinates – not complexon coordinates

33. New Particle Functional Space and Spin 0, 1, and 2 Boson Quba Functionals

33.1 Particle – Coordinate Duality

Elementary particles seem to be distinctly different from coordinates. However, there is a connection. One cannot measure a location's coordinates without the use of particles. This fact is evidenced by the number-phase uncertainty relation in Quantum Electrodynamics:

$$\Delta\varphi\Delta N \geq \hbar$$

where φ is the phase of a wave, N is the number of photons and $\hbar$ is Heisenberg's constant.

Earlier we discussed the particle functional space for fermions and showed how second quantized fermion fields could be the factorized into a fermion functional and a fourier coordinate expansion that were united by a functional inner product. We gave several important reasons why factorization of quantum fields made sense. The most important reason being that 'spooky action at a distance' quantum entanglement was demystified if the space of functionals consisted of a single point – eliminating spooky coordinate distance separation.

In this chapter we extend particle functional space to include boson functionals.[177] The reason is again evident from a consideration of quantum entanglement for states containing bosons. Factorizing free boson quantum fields enables us to separate boson functionals from spatial coordinates so that 'spooky action at a distance' is again precluded. An example would be a two photon state that upon

---

[177] Following up on the derivation of the particle functional part of complex quaternion and complex octonion dimensions described earlier.

separation retains quantum entanglement at a distance because the photon functionals, like all particle functionals, are concentrated at a single point.

Particle functional factorization is the only approach that eliminates 'spooky action at a distance.'

Based on a close analogy between Standard Model Reality Group coordinate symmetry: $SU(3) \otimes SU(2) \otimes U(1) \otimes SU(2) \otimes U(1)$ and particle functional space Lorentz group symmetry described in chapter 3 we derive the Internal Symmetry Reality Group of the extended Standard Model. The new Generation group, Layer groups, Species group, and Θ-Group are added to create the Unified SuperStandard Theory.

33.2 Boson Functional Symmetries

We now consider a quba functional for spin s, treat it as a type of coordinate[178] b_s^μ where s is the spin and $\mu = 0, 1, 2, 3$ is the particle space index, and apply particle functional Lorentz group transformations to a different set of functional coordinates.

$$b_s'^\mu = \Lambda_L{}^\mu{}_\nu(\omega, \mathbf{u})b_s^\nu \tag{33.1}$$

where $\Lambda_L{}^\mu{}_\nu$, of an 'internal symmetry' Complex Lorentz group for particle functional space, and where ω and $\mathbf{u}$ are c-number parameters with μ and υ functional space indices. The spin s is a label but does not affect functional Lorentz transformations.

We now consider one of the subgroups of the Complex Lorentz group and generate the set of all representations from it. (In chapters 4 – 7 we considered all the Complex Lorentz subgroups in detail.)

For example we consider the set of all particle functional coordinate systems that can be generated by applying all SU(3) transformations within the Complex Lorentz group. We now map each such functional coordinate system, denoted a, to a *single* functional that we will denote b_{sa} The set $\{b_{sa}\}$ has a functional for every possible SU(3) boson that can be generated. Thus we can combine these functionals with matching fourier expansions to generate the complete set of SU(3) boson functionals for spin s.

---

[178] As we did in quaternion and octonion spaces in earlier chapters.

We can repeat this procedure for each of the $SU(2) \otimes U(1)$ subgroups of the particle Complex Lorentz group.

Thus we have sets of functionals for each Complex Lorentz subgroup. Each set of functionals furnishes a representation for the corresponding subgroup.

First we apply all transformations of the $SU(3)$ subgroup to $b_{s\alpha}$ creating a set of coordinate systems denoted $b_{sSU(3)}$.[179] We then define an independent $SU(3)$ group whose sole purpose is to $SU(3)$ rotate the coordinate systems of $b_{sSU(3)}$. Next we follow the same procedure independently for each of $SU(2)$, $U(1)$, $SU(2)$, and $U(1)$ generating sets of coordinate systems for each. Then we again define $SU(2)$, $U(1)$, $SU(2)$, and $U(1)$ commuting groups to transform amongst each set.

We thus end up with five independent groups (not the Lorentz subgroups) that commute with each other and thus give us the Internal Symmetry group for the internal symmetry fields $SU(3) \otimes SU(2) \otimes U(1) \otimes SU(2) \otimes U(1)$.

The above process can be used for the vector bosons of the above groups and the additional groups defined in subsequent chapters.

The same procedure can be applied to Higgs bosons, Faddeev-Popov ghost fields, and gravitons.

33.3 Features of Core Boson Functionals - Qubas

We have briefly described the features of quba functionals. Their most important feature is that they are massless. Otherwise vector bosons would have a bare mass that would break gauge invariance.

An important consequence of the masslessness of qubas is that they have no tachyon equivalents. Note: the bare mass of qubes led to tachyons. The masslessness of qubas prevents Complex Lorentz boosts from generating tachyonic bosons.

---

[179] Each coordinate system in the set is one of the members of the set of items upon which the newly defined $SU(3)$ group operates. The individual coordinates of any coordinate system are not relevant to the newly defined $SU(3)$ group and the set upon which it operates.

UNIFIED SUPERSTANDARD THEORIES FOR QUATERNION UNIVERSES ... – *Stephen Blaha* **351**

33.4 Indices of Functionals

Quba functionals have a spin s and a set of indices that specify group elements. Their general form is

$$b_{s\xi} \qquad\qquad (33.2)$$

where ξ consists of a set of group indices for the Unified Standard Theory groups: SU(3),U(1), SU(2), Dark U(1) Dark SU(2), the Generation group, the Layer group, the Species group, and the Θ-group.[180] Similarly the qube functionals also have a set of internal symmetry indices

$$_i f_\xi \qquad\qquad (33.3)$$

where ξ is the set of internal symmetry indices, i identifies the species, and the fermion spin is ½. We can define a general form for each type of fermion coordinate fourier expansion as

$$_i(x)_\xi \qquad\qquad (33.4a)$$

where i specifies the fermion species; and the general form of boson coordinate expansion as

$$(x)_{s\xi}{}^\varsigma \qquad\qquad (33.4b)$$

where x represents the space-time coordinates, s the spin, ξ is the set of internal symmetry indices, and ς represents the *space-time coordinate space* indices: (NULL) for spin 0 bosons, one index α for vector bosons, and two indices, $\alpha\beta$, for gravitons:

$$(x)_{0\xi} \qquad\qquad (33.4c)$$
$$(x)_{1\xi}{}^\alpha \qquad\qquad (33.4d)$$

---

[180] The Dark U(1)⊗SU(2), the Generation group, the Layer group, and the Species group are derived in subsequent chapters.

$$(x)_{2\xi}{}^{\alpha\beta} \qquad\qquad (33.4e)$$

For example, earlier we defined a free Dirac-type species fermion quantum wave function with the inner product of a qube functional and a coordinate space fourier quantum expansion:

$$\psi(x) = ({}_1f, (s, x, t)) = \sum_{\pm s} \int d^3 p N(p)[b(p, s)u(p, s)e^{-ip\cdot x} + d^\dagger(p, s)v(p, s)e^{+ip\cdot x}]$$

where we use a functional inner product formalism in the manner of Riesz (1955)[181] and others. We can now represent this as

$$\psi_\xi(x) = ({}_1f_\xi, {}_1(x)_\xi) \qquad\qquad (33.5)$$

for the indices ξ where the Dirac fermion species is of type 1.

For scalar, vector and spin 2 bosons we use the notations

$$\varphi_\xi(x) = (b_{0\xi}, (x)_{0\xi}) \qquad\qquad (33.6)$$
$$\varphi_{1\xi}{}^\mu(x) = (b_{1\xi}, (x)_{1\xi}{}^\mu) \qquad\qquad (33.7)$$
$$\varphi_{2\xi}{}^{\mu\nu}(x) = (b_{2\xi}, (x)_{2\xi}{}^{\mu\nu}) \qquad\qquad (33.8)$$

Thus we have a compact notation for the particle functional factorized second quantized fields. The extensions of quantum field theory described later: Two-Tier Quantum Field Theory and PseudoQuantum Field Theory both can be easily expressed in this notation.
We note that qube and quba functionals have no space-time dependence.

33.5 Internal Symmetry Reality Group vs. Coordinate Reality Group

In this section[182] we will relate the Coordinate Reality Group that we developed earlier to an Internal Symmetry Reality group in the Particle Functional Space

---

[181] For example see pp. 61-2 of Riesz where linear functionals and their inner products are defined.
[182] This chapter outlines the detailed discussion presented in chapters 5 – 7 of Blaha (2017b) as well as earlier books such as Blaha (2015a).

introduced above. We begin by outlining the physical basis of the known Standard Model symmetry group SU(3)⊗SU(2)⊗U(1) and show that it should generalize to

$$SU(3)⊗SU(2)⊗U(1)⊗SU(2)⊗U(1)$$

with the extra SU(2)⊗U(1) factor, which we *postulate* describes Dark Matter ElectroWeak interactions.[183] Normal matter SU(2)⊗U(1) symmetry emerges from the consideration of boosts at real-valued velocities greater than the speed of light. The Dark Matter sector SU(2)⊗U(1) symmetry emerges from the consideration of boosts at complex-valued velocities less than or greater than the speed of light.

Before beginning the discussion of these cases we note that a local U(4) transformation can change any complex-valued coordinate system to a real-valued coordinate system in 4-dimensional flat space-time. However because of the peculiar nature of the Complex Lorentz group, and the defining relation of the Real and Complex Lorentz groups:

$$\Lambda(\mathbf{v}, \boldsymbol{\theta})^{T}G\Lambda(\mathbf{v}, \boldsymbol{\theta}) = G$$

where G is the flat space-time metric G = diag(1, −1, −1, −1), and where the superscript ^{T} specifies the transpose of the matrix; it is possible to *physically* specify the origin of each of the groups of the analogues of The Standard Model group: SU(2)⊗U(1), SU(3), and 'Dark' SU(2)⊗U(1). (See below.) These Lorentz subgroups do not commute with each other. We have shown in chapter 3 to create commuting equivalents of these groups.

33.5.1 Coordinate Reality Group

The Real Lorentz group transforms between coordinate systems that are in relative motion at a velocity below the speed of light. Consequently it transforms real-valued coordinate systems to real-valued coordinate systems. The Complex Lorentz

---

[183] The Dark Matter ElectroWeak interactions must be distinct from the normal matter ElectroWeak interactions or Dark Matter would have been found experimentally by now.

group includes the Real Lorentz group as a subgroup. It also includes transformations that transform real-valued coordinate systems into complex-valued coordinate systems. A subset of these transformations transform between coordinate systems at a velocity whose magnitude is below the speed of light. Another subset of transformations transform between coordinate systems at a relative velocity whose magnitude is above the speed of light.[184] These transformations correspond to faster than light motion and provide boosts discussed in chapter 26 that generate tachyonic fermions.

We showed in earlier books, such as Blaha (2017b), that faster than light physics is consistent and physically acceptable – even to the point of deriving faster-than-light Thermodynamics from the Maxwell-Boltzmann distribution including the law of increasing Entropy.

Chapters 28 through 31 derive the Coordinate Reality Group in detail based on the Complex Lorentz Group. *The elements of the Coordinate Reality Group are local transformations that map complex-valued coordinate systems generated by complex Lorentz transformations to real-valued coordinate systems.* We will see that the Coordinate Reality Group consists of the subgroups (SU(3), SU(2)⊗U(1), and SU(2)⊗U(1)) with 16 generators. The 16 generators are linear combinations of U(4) group generators. Local U(4) transformations can map any complex 4-dimensional space-time to real values.

The reader will notice that SU(3) and SU(2)⊗U(1) are the known symmetry groups of The Standard Model. However the Coordinate Reality Group *cannot* be an internal symmetry group without running into the difficulties of 'No Go' theorems that caused the demise of SU(6) in the 1960's.

So we were led to consider the possibility that the Standard Model internal symmetries are the result of an Internal Symmetry Reality Group[185] that is the analogue of the Coordinate Reality Group. This Internal Symmetry Reality group can be

---

[184] This was not noted by the rigorously mathematical derivation of Axiomatic Quantum Field Theory by Streater (2000) and others.

[185] In previous books we implicitly proceeded from the Coordinate Reality Group to the Internal Symmetry Reality Group. In this volume it becomes necessary to explicitly discuss the Internal Symmetry Reality Group. We see in this chapter that it follows from an Internal Symmetry Complex Lorentz Group, which, in turn, will be seen to be related to the 'Rotation of Interactions' Θ-Symmetry group described in previous books by the author and discussed later in this book.

constructed from Complex Lorentz group transformations of the Particle Functional Space.

We therefore postulate that the Standard Model internal symmetries are the consequence of an Internal Symmetry Reality Group. We will also postulate that the additional SU(2)⊗U(1) Coordinate Reality subgroup has a corresponding Internal Symmetry Group analogue for a Dark Matter SU(2)⊗U(1) ElectroWeak symmetry.[186] Thus the full Internal Symmetry Reality Group of the Unified SuperStandard Theory for the non-Dark sector is

$$SU(3) \otimes SU(2) \otimes U(1) \otimes SU(2) \otimes U(1)$$

33.5.2 Internal Symmetry Reality Group

Now following Decision Axiom Replication Principle: 'Nature Tends to Repeat Successful Strategies' we assume that the fundamental internal symmetry group of elementary particles arises from the Internal Symmetry Reality group originating in Complex Lorentz group transformations in Particle Functional Space. The Internal Symmetry Reality group elements transform complex-valued functional coordinate systems into real-valued functional coordinate systems.

We showed how to construct the commuting direct product Internal Symmetry Reality group with methods presented in chapter 27 and above. The choice of the Internal Symmetry Reality group[187] developed from Particle Functional Space considerations is the tensor product: $SU(3) \otimes SU(2) \otimes U(1) \otimes SU(2) \otimes U(1)$. We denote it as R and call it the Internal Symmetry Reality group:

$$R = SU(3) \otimes SU(2) \otimes U(1) \otimes SU(2) \otimes U(1)$$

---

[186] This postulate is based on the Decision Axiom Replication Principle, which, in brief, states that physical phenomena/models tend to be repeated in Nature. A good example is the appearance of harmonic oscillators in many areas of Physics. (See Appendix E.)

[187] In developing the Reality group for Particle Functional Space mapped into group elements, we note that this Reality group allows the set of mapped functional elements to be 'real-valued' and thus the quantum fields generated from these functionals via inner products to be 'real-valued' as well.

Each of the factors has an associated set of local Yang-Mills vector boson fields generated from functionals as described in section 33.2 and chapter 27. These fields embody the SuperStandard Theory particle interactions.

Again we see a parallel between coordinate space characteristics and particle functional space characteristics.

33.6 Particle Functional Space—A Space of Particle Functionals

The procedure we have developed puts particle functionals on a footing similar to coordinates. We are familiar with coordinate space with space-time coordinates that supports Lorentz group transformations between coordinate systems.

We now define functional space with Complex Lorentz group transformations between 4-dimensional functional coordinate systems. In the present case we see that functional space can be viewed as the direct sum of a spin ½ fermion functional subspace plus three boson functional subspaces for spin 0, 1, and 2 particles. These four subspaces support the known varieties of particles. The functional Complex Lorentz group does not transform between any of the four subspaces. Each subspace has its functionals generated within itself with no transformations yielding functionals in other subspaces.

We assume that there is no space-time distance between functionals in the particle functional space of fermion and boson functionals. We further assume that all particle functionals commute:

$$[_i f_\xi, {_i}'f_{\xi'}] = [_i f_\xi, b_{s'\xi'}] = [b_{s\xi}, b_{s'\xi'}] = 0 \qquad (33.9)$$

where ξ represents internal symmetry indices, i specifies the species, and s represents boson spin. Thus all functionals in the entire functional space may be viewed as located at a 'mathematical' point. The absence of coordinate space-time separation eliminates the 'spooky action at a distance' of Quantum Entanglement that has troubled generations of physicists.[188]

---

[188] The mechanism for the elimination of this problem reminds the author of Alexander the Great's solution of the problem of unraveling the Gordian knot – he reputedly simply sliced it open with his sword. The opening of the Gordian knot was doable only by the conquereor of the world. More modestly, we slice quantum fields into

Since particle functional space F is not located at any spatial (space-time) point we can view the universe as a direct product of F with space-time S, and specify the totality of space T with

$$T = S \otimes F$$

33.6.1 Fermion Functional Subspace

The coordinate systems of the spin ½ fermion functional subspace can be rotated into each other using the functional Complex Lorentz group. Since we map each coordinate system to a single functional (described earlier and in chapter 3) a Complex Lorentz group transformation causes a rotation of the set of functionals mapped from the set of functional coordinate systems. We can symbolize this process with the below analogue:[189]

Functional Coordinate System f Complex Lorentz Transformation

$$\Lambda \, {}_i f_\xi \rightarrow {}_i{}^{\cdot} f_\xi{}' \qquad (33.10)$$

Map of Individual Functionals f_f Corresponding to each Functional Coordinate System

$$\begin{aligned} {}_i f_\xi &\leftrightarrow {}_i f_{\xi f} \\ {}_i{}^{\cdot} f_\xi{}' &\leftrightarrow {}_i{}^{\cdot} f_{\xi f}{}' \end{aligned} \qquad (33.11)$$

Corresponding individual particle functional transformation

$$\Lambda_f \, {}_i f_f \rightarrow {}_i{}^{\cdot} f_f{}' \qquad (33.12)$$

---

functional and coordinate parts to eliminate spookiness and preserve the analogy between the Coordinate Reality group and the Internal Symmetry Reality group..

[189] In this example we omit inserting indices for the functional Complex Lorentz group transformations in the interests of clarity.

where the subscripted $_i f_f$ indicates a single functional mapped from a functional coordinate system $_i f$ (of species i) and Λ_f is the transformation corresponding to the Complex Lorentz group functional coordinate transformation Λ applied to $_i f$.[190]

A transformation of the form of eq. 33.12 would require a corresponding transformation of fermion fourier expansions to preserve the format of each inner product. Thus eq. 33.5 becomes upon transformation

$$\psi_{\xi\Lambda}(x) = (\Lambda_f \; _i f_\xi, \; \Lambda^\circ_f \; _i(x)_\xi) \tag{33.12a}$$

where Λ°_f is the corresponding transformation of the fourier coordinate expansion $_i(x)_\xi$.

By using the Internal Symmetry Reality group for fermion functional space we can force the particle fermion functionals $_i f_f$ to lead to real-valued inner products generating second quantized fields.

We note in passing that Complex Lorentz group transformations in fermion functional space can transform functionals from one species to a different species. (See section 26.11 for examples of functionals and quantum fields for various fermion species.)

33.6.2 Boson Functional Subspaces

The coordinate systems of each of the three integer spin boson functional subspaces can be separately rotated within each subspace using the functional Complex Lorentz group. Since we map each coordinate system to a single functional (described earlier and in chapter 3) a Complex Lorentz group transformation causes a rotation of the set of functionals of each spin mapped from the set of functional coordinate systems. We can symbolize this process with the analogues:[191]

---

[190] *We note that functional Lorentz transformations can map between functionals of differing species, which implies that the matching fourier expansions must map correspondingly via a matching transformation to preserve the symmetries of the inner products that generate quantum fields.* A a species index must be transformed synchronously with a functional Lorentz transformation that changes the species of a functional coordinate system.

[191] In the following examples we omit inserting indices for the functional Complex Lorentz group transformations in the interests of clarity.

Spin 0:
Functional Coordinate System $b_{0\xi}$ Complex Lorentz Transformation

$$\Lambda b_{0\xi} \rightarrow b_{0\xi}' \qquad (33.13)$$

Map of Individual Functionals $b_{0\xi f}$ Corresponding to each Functional Coordinate System

$$b_{0\xi} \leftrightarrow b_{0\xi f} \qquad (33.14)$$
$$b_{0\xi}' \leftrightarrow b_{0\xi f}'$$

Corresponding individual particle functional transformation

$$\Lambda_f b_{0\xi f} \rightarrow b_{0\xi f}' \qquad (33.15)$$

where the subscripted $b_{0\xi f}$ indicates a single functional mapped from a functional coordinate system $b_{0\xi}$ and Λ_f is the transformation corresponding to the Complex Lorentz group functional coordinate transformation Λ applied to $b_{0\xi}$.

Spin 1:
Functional Coordinate System $b_{1\xi}$ Complex Lorentz Transformation

$$\Lambda b_{1\xi} \rightarrow b_{1\xi}' \qquad (33.16)$$

Map of Individual Functionals $b_{1\xi f}$ Corresponding to each Functional Coordinate System

$$b_{1\xi} \leftrightarrow b_{1\xi f} \qquad (33.17)$$
$$b_{1\xi}' \leftrightarrow b_{1\xi f}'$$

Corresponding individual particle functional transformation

$$\Lambda_f b_{1\xi f} \rightarrow b_{1\xi f}' \tag{33.18}$$

where the subscripted $b_{1\xi f}$ indicates a single functional mapped from a functional coordinate system $b_{1\xi}$ and Λ_f is the transformation corresponding to the Complex Lorentz group functional coordinate transformation Λ applied to $b_{1\xi}$.

Spin 2:

Functional Coordinate System $b_{2\xi}$ Complex Lorentz Transformation

$$\Lambda b_{2\xi} \rightarrow b_{2\xi}' \tag{33.19}$$

Map of Individual Functionals $b_{2\xi f}$ Corresponding to each Functional Coordinate System

$$b_{2\xi} \leftrightarrow b_{2\xi f} \tag{33.20}$$
$$b_{2\xi}' \leftrightarrow b_{2\xi f}'$$

Corresponding individual particle functional transformation

$$\Lambda_f b_{2\xi f} \rightarrow b_{2\xi f}' \tag{33.21}$$

where the subscripted $b_{2\xi f}$ indicates a single functional mapped from a functional coordinate system $b_{2\xi}$ and Λ_f is the transformation corresponding to the Complex Lorentz group functional coordinate transformation Λ applied to $b_{2\xi}$.

The Internal Symmetry Reality group element functional rotations, denoted Λ, when combined with fourier coordinate expansions, via inner products, give the Internal Symmetry Reality group of The SuperStandard Theory.

33.7 Skeleton Functional Lagrangians

If we could imagine a 'snapshot' of the universe[192] at one instant of time we could presumably enumerate all the functionals of the universe's particles. Then succeeding snapshots would show an ebb and flow of functionals as time progresses. This thought brings us to the important issue of the transformations of particle functionals in particle interactions. The simplest statement that one could make about functional transformations is that they are created and annihilated according to the interaction terms of the skeletonized Unified SuperStandard Theory (excluding quadratic terms which do not transform functionals.)

We skeletonize a lagrangian density by deleting all quadratic terms and replacing all particle fields by their corresponding functionals.[193] For example the lagrangian

$$\mathcal{L} = \bar{\psi}_C(i\gamma^\mu D_\mu - m)\psi_C(x) + b(\bar{\psi}_C\psi_C(x))^2 \qquad (33.22)$$

becomes the skeleton lagrangian

$$\mathcal{L}_S = bf^4 \qquad (33.23)$$

where f is the fermion's functional.

Thus our skeletonized lagrangian formalism describes the transitions between functionals in an interaction. This formalism is made more concrete by considering Feynman diagrams for the interactions.

---

[192] We realize that such a snapshot is not possible since infinite velocity particles that could feed a camera this snapshot do not exist.

[193] In our construction of particle functional space we have not introduced complex conjugation of functionals for lack of a good reason. Complex conjugation takes place only in the fourier expansion part of a quantum field. Another issue is the appearance of lagrangian terms with factors that are divatives of fields. Since we do not do computations with skeleton lagrangians we can ignore the derivative in each such factor and simply substitute the functional. For example, $\varphi^3(\partial^\mu\varphi)^2$ becomes the quba expression b^5.

33.8 Functional Interactions and Feynman Diagrams

Feynman diagrams with their in and out ordering specify the transformations between functionals more completely. A simple example shows the interaction transformations of functionals. Consider the lagrangian term

$$(\bar{\psi}\psi(x))^2(\partial^\mu\varphi)^2$$

A corresponding Feynman diagram for it is

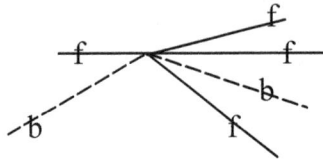

Figure 33.1.Functional Feynman diagram for the above interaction.

with qubes labeled f and qubas labeled b.

When internal symmetries are introduced then the skeletonized lagrangians and the corresponding Feynman diagram representations would be significantly more complicated.

33.9 Functional Space and Feynman Path Integrals

Functionals appear in Feynman Path Integrals and in Faddeev-Popov gauge fixing path integrals. We illustrate the use of functionals in the example:

$$Z(J) = N\int\prod_y dy \, \prod_\varphi d\varphi(y) \, \exp\{i\int d^4y[\mathscr{L}(\varphi(y) + J^\mu(y)\varphi(y)]\} \quad (33.24)$$

which in functional notational notation becomes

$$Z(J) = N\int\prod_y d(y) \, \prod_b db \, \exp\{i\int d^4y[\mathscr{L}(\varphi(y) + J^\mu(y)\varphi(y)]\} \quad (33.25)$$

where (y) represents the fourier expansion in the y coordinates, and with the implied inner product $\varphi(y) = (b, (y))$.

34. Factorization of Quantum Fields—Wave-Particles Realized Formally

Many years ago Dirac factored the Klein-Gordan equation and obtained the Dirac equation for spin ½ fermions. In this appendix we show that there is good reason to factor quantum mechanical wave functions and second quantized fields into an inner product of a particle functional and a corresponding fourier coordinate expansion. With this factorization, and the assumption that the space of all particle functionals, as well as the space of all fourier coordinate expansions, are both located at a point with the consequence that there is no distance measure in either space, we find a change in one of a pair of space-like separated parts of an initial state causes an instantaneous transformation of the other part (Einstein's spookiness). Because of the significance of this result we earlier implemented the factorization of wave functions and quantum fields as axioms.

These appendices describes more details of factorization for fermions and bosons. Much of it appears in Blaha (2018b).

34.1 Motivation for Quantum Field Factorization: Instantaneous Effects in Quantum Phenomena

Instantaneous quantum phenomena are apparent in many cases. For example:

1. Two particles placed in a definite spin state can separate to a space-like distance. If the z component of spin is flipped in one of the particles, the other particle instantaneously flips its spin in such a way as to conserve spin. This type of phenomena has been described as 'spooky' since it violates the law that no effect can travel at a rate faster than the speed of light.

2. Transitions between atomic levels take place instantaneously—in a zero time interval.

34.2 General Form of Factorization

Normally fermion and boson quantum fields are described by a wave function of the form

$$\chi(\mathbf{x}, t) \tag{34.1}$$

We can formally factorize quantum fields as an inner product of a functional f_k and a space-time fourier expansion denoted $(k, \mathbf{x}, t)$ (neglecting internal quantum numbers temporarily) wher k is the momentum.

$$\chi(\mathbf{x}, t) = (f_k, (k, \mathbf{x}, t)) \tag{34.2}$$

For a free two particle wave function (non-interacting) the wave function may be written as a product of inner products:

$$\chi(\mathbf{x}, t) = (f_{1k}, (k, \mathbf{x}, t)_1)\,(f_{2q}, (q, \mathbf{x}, t)_2) \tag{34.3}$$

where k and q are momenta.

34.3 Rationale for Factorization

The rationale for factorization lies in the nature of the functionals and coordinate fourier expansions that we use. For, we choose to create a space of particle functionals for fermions and bosons that consists of a single point with no distance measure (or alternately put, zero distance between all functionals.) We also choose to create a 'point' space of all coordinate fourier expansions for bosons and fermions, whose elements have all coordinate values, x.

For the moment we wish to note that the space of functionals consists of functionals for all fundamental particles in the universe (and Megaverse). We can describe transitions (interactions) in which functionals are transformed into other functionals. So the space of functionals has a dynamic aspect. Another important aspect of functional space is its universality—*all functionals of the Megaverse are present creating a type of link between all parts of the Cosmos.*

The space of coordinate fourier expansions consists of all possible expansions for particles in the coordinates of each respective universe and of the Megaverse. This space also has no distance measure.

The factorization that we propose, as exemplified by eqs. 34.2 and 34.3, enables instantaneous communication of a transition between two space-like separated parts of a state. A change in one part immediately causes a corresponding change in the other part because the changes take place in the functionals which are located at the same point in functional space.

In a certain sense we have divorced quantum phenomena from coordinate space by quantum field factorization.

34.4 Wave-Particle Duality Mathematically Realized

At the beginning of Quantum Theory in 1924, De Broglie postulated wave-particle duality and made progress in understanding quantum phenomena. Wave-particle duality was subsequently 'abandoned' in favor of the Quantum Mechanics of Heisenberg and Schrödinger.

The quantum field factorization that we propose provides a mathematical formulation of particles that separates the particle part (the functional) from the wave part (the coordinate fourier expansion.) Thus we have realized wave-particle duality for the purpose of understanding instantaneous effects of quantum entanglement.

However, in our formulation, quantum fields $\chi(\mathbf{x}, t)$ appear in the SuperStandard lagrangian. *Then* perturbation theory calculations of phenomena are made using free field fourier expansions, denoted $(\mathbf{x}, t)$, for all fermions and bosons. Thus we achieve a quantum theory (non-deterministic) and yet have wave-particle duality embedded within it (unlike De Broglie-Bohm theory.)

34.5 Visualization of a Particle

We can visualize an elementary particle located at point x as composed of two components: a functional f and a fourier expansion denoted by x. They follow from the above discussion and the derivation in Blaha (2018a):

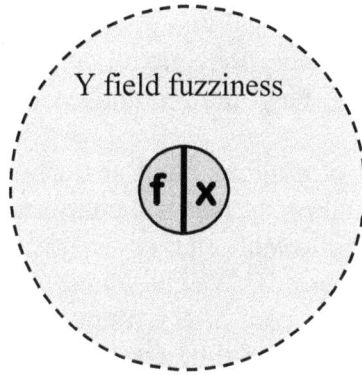

Figure 34.1 Symbolic view of a free particle having particle functional and fourier expansion parts. The 'smearing' of the particle by the Two-Tier Y field described in Blaha (2018a) and later in this volume is symbolically displayed.

The central disk represents the core of the particle which is located at x but 'smeared' by the cloud of Y particles of Two-Tier Quantization. Thus point particles do not exist in the full SuperStandard Theory and thus infinities are not encountered in the calculation of any diagram in perturbation theory.

The central disk (which is not necessarily truly centered) represents a Qube functional for fermions and a Quba functional for bosons. (See below.) As described earlier Qubes embody spin ½ and Qubas embody integer spin. Qubes have a bare mass m_0 and Qubas are massless. The bare mass of Qubes reflects the fact that fermions have spin and thus an intrinsic energy while Qubas can carry energy and thus spin while in motion but elementary bosons act only as 'delivery agents' for spin (and thus quantum values).

34.6 Factorization of Fermion Quantum Fields

If we consider all possible 'things' that might constitute a fundamental building block for a fundamental fermion theory they are all, at best, *ad hoc* and raise questions of their necessity and whether they are composed of a yet more fundamental substructure.

There is only one choice of building block that avoids these issues – a logic unit or qubit. A qubit is a fundamental entity that is a complex form of computer bit. A bit (and thus a qubit) is known to have an energy or equivalently a mass, and has no constituents of a more primitive form.[194] We call a unit of logic that forms the 'core' of a particle a *qube*.[195] It exists as the core of a fermion particle. But, in itself, it has no *independent* material existence or space-time coordinates. A qube is a functional that acquires features such as coordinates, to become an elementary particle. We define a qube as a fermion field theory functional. We now introduce physical features that will cloak qubes with properties and interactions making them into fundamental fermion quantum fields.

34.6.1 Dirac-like Equations of Matter from 4-Valued Logic

In our derivation every truly fundamental particle of matter, whether quark or lepton, has spin ½. We have seen in chapter 10 of Blaha (2011c) that the basic algebra of Operator Logic eigenvalue operators, and that of its raising and lowering operators, are the same as the algebra of creation and annihilation operators for free spin ½ particles. Our goal is to build our theory on the scaffolding of Operator Logic. We view a fermion particle as a qube core which is dressed in spatial coordinates (and internal symmetries):

$$\text{Qube core} + \text{coordinates} \rightarrow \text{fermion particle} \qquad (34.4)$$

---

[194] A qube is a physical manifestation of a logical value. The relation of a qube to a logical value is analogous to the relation of a penciled point placed on paper to the concept of a point as a primitive in geometry.

[195] In the First Edition we called qubes iotas. However, since the name iota was previously used as a particle name many years ago it seemed reasonable to use a different name. We chose the name 'qube' for self-evident reasons. *'Qube' is pronounced 'cube.'*

The creation and annihilation operators $b(p,s)$ and $d^\dagger(p,s)$ (and their hermitean conjugates $b^\dagger(p,s)$ and $d(p,s)$) are mathematically similar to the raising and lowering operators of Operator (Matrix) Logic. They satisfy the anticommutation relations

$$\{b(q,s), b^\dagger(p,s')\} = \delta_{ss'}\delta^3(\mathbf{q} - \mathbf{p}) \qquad (34.5)$$
$$\{d(q,s), d^\dagger(p,s')\} = \delta_{ss'}\delta^3(\mathbf{q} - \mathbf{p})$$

Thus we see spin ½ particle wave functions originating from the Dirac-like spinors, and raising and lowering operators of the spinor formulation of Operator Logic.

When particles interact the quantum field theory interaction terms use fermion creation operators, $b(q,s)$ *and* $d^\dagger(q,s)$, *and annihilation operators,* $b^\dagger(p,s')$ *and* $d(q,s)$, *to implement the transformations between the Qubes of the interacting particles.*[196] *Thus the mathematics of the embedded Qubes' logic values is automatically implemented within quantum field theoretic calculations.*

An interesting point that emerges from this discussion is the nature of spin ½ particle states such as

$$|p, s\rangle = b^\dagger(p, s)|0\rangle \qquad (34.6)$$

This state is interpreted as a one particle state. It also has an analogous interpretation in Operator Logic as creating a one term universe of discourse – a construct which is in part linguistic and in part logic. Thus particles are embodiments of Logic values and particle interactions change the logic values of the initial particles to those of the emergent particles. All in all, our universe can be viewed as an extraordinarily intricate logic machine. Serendipitously we are now seeing the use of particles to create quantum computers, which, in a sense, is bringing us full circle. Particles are Logic; Logic machines emerge from particle interactions.

---

[196] See chapter 27.

34.6.2 Functional Expression for Each of the four Species of Fermions

We have derived the four species of fermions in the book. We have used a 'conventional' notation for quantum fields. In this section we will define these quantum fields as inner products of functionals and fourier coordinate expansions.

Dirac Quantum Field:
$$\psi(x) = ({}_1f, \text{Dirac\_fourier\_expansion})$$

Tachyon Quantum Field:
$$\psi_T(x) = ({}_2f, \text{Tachyon\_fourier\_expansion})$$

Complexon Quantum Field:
$$\psi_C(x) = ({}_3f, \text{Complexon\_fourier\_expansion})$$

Complexon Tachyon Quantum Field:
$$\psi_{CT}(x) = ({}_4f, \text{Tachyon\_Complexon\_fourier\_expansion})$$

The digit prefixes of ${}_kf$ for k = 1, 2, 3, 4 distinguish the functionals for each species.

In addition we can decompose the above quantum fields into left-handed and right-handed fields. The left-handed functional representations are:

Left Dirac Quantum Field:
$$\psi_L(x) = ({}_{1L}f, \text{left-handed\_Dirac\_fourier\_expansion})$$

Left Tachyon Quantum Field:
$$\psi_{TL}(x) = ({}_{2L}f, \text{left-handed\_Tachyon\_fourier\_expansion})$$

Left Complexon Quantum Field:
$$\psi_{CL}(x) = ({}_{3L}f, \text{left-handed\_Complexon\_fourier\_expansion})$$

34.7 Factorization of Boson Quantum Fields

Blaha (2017f) opened the possibility that fermions (and bosons) might have a core that embodies logic in the form of spin as well as bare masses in the case of fermions. In Blaha (2018a) we expanded our discussion to better describe the functionals' space, within which the core functionals of each type of fundamental particle reside.

We defined functionals of various elementary boson spins: 0, ½, 1, and 2. Bosons have cores as well that are boson functionals with integer spin. We called a boson core a *quba*[197] in analogy with the fermion functionals name of qubes. Boson functionals are massless. Bosons acquire masses through interactions. The rationales for logic cores for particles was discussed in detail earlier.

34.7.1 Quba Cores of Fundamental Bosons

We define a corresponding boson functional quba for each type of elementary boson. We will designate a boson functional as b_s where s specifies the spin which may be 0, 1, or 2. Every boson contains a boson functional core within it. A quba has the spin of the elementary boson within which it resides. It has zero mass since bosons are typically massless prior to symmetry breaking effects. The space of Quba functionals is described earlier.

34.8 Types of Functionals for Fermions and Bosons

The functional content embodied in each type of elementary particle is summarized in Table 34.1.

---

[197] We use 'quba' simply because of its similarity to 'qube'. The leading 'b' signifies its bosonic use. We pronounce 'quba' as 'bub' with a silent 'e.' The word 'quba', itself, is the name of a Bantu language spoken by the Bubi people of Bioko Island in Equatorial Guinea.

PARTICLE TYPE	CORE	MASS	SPIN
Fermion	qube	m_0	½
Scalar Boson	quba	0	0
Vector Boson	quba	0	1
Graviton	quba	0	2

Table 34.1 Core functionals within the various types of fundamental elementary particles.

35. Functional Space of Particles and Internal Symmetry Groups

35.1 Particle Functional Subspace

The functionals of elementary particles form a space[198] that includes all the free field fermion and boson functionals of our universe and any other universe that might exist (the Megaverse). All fundamental fermions and bosons have a corresponding particle functional. Fermion particle functionals f... are labeled with momentum k, internal symmetry quantum numbers denoted λ, and possibly other subscripts ζ such as handedness, and PseudoQuantum[199] field type (1 or 2). Boson particle functionals b... are labeled with momentum k, spin s, internal symmetry quantum numbers denoted λ, and possibly other subscripts ζ such as handedness, and PseudoQuantum field type (1 or 2):

$$f_{k\lambda\zeta}$$
$$b_{ks\lambda\zeta} \tag{35.1}$$

35.2 Functional Coordinates and Internal Symmetry Groups

Functional Space also includes generic quadruplets of functionals that we can treat as functional coordinates of a 4-space of functionals as we did in the Second Edition. These functional coordinates can be transformed by Complex Lorentz transformations to be coordinates of a different coordinate system.

Further, one can create sets of coordinate systems that can be mapped to representations of the subgroups of the Complex Lorentz group: SU(3), SU(2), U(1), SU(2), and U(1). These groups, defined by maps of sets of representations, can combine to form a functional Reality group with commuting factors for Internal Symmetries: R =

---

[198] Much of this appendix appears in the Blaha (2018a).
[199] Discussed later.

$SU(3) \otimes SU(2) \otimes U(1) \otimes SU(2) \otimes U(1)$. In addition functional coordinates can be transformed by U(4) transformations to be coordinates of sets of coordinate systems that can be mapped to U(4) representations of a functional Generation group, the Layer groups, and the Species group. Thereby we define the set of commuting functional Internal Symmetry groups of Fig. 35.1 which comprise the groups of the Unified SuperStandard Theory: $[SU(3) \otimes SU(2) \otimes U(1) \otimes SU(2) \otimes U(1) \otimes U(4) \otimes U(4)]^4 \otimes U(4)$

Operators of the functional groups that we define, transform fundamental particles among each other. Thus our Functional Space contains both functional Internal Symmetry groups and the fundamental particle functionals upon which they operate. The functional Internal Symmetry groups can be dressed in space-time coordinates to yield the Internal Symmetry groups of the Unified SuperStandard Theory. The fundamental particle functionasl can be 'dressed' in space-time coordinates using functional inner products as described earlier and in the Second Edition.

35.3 Functional Transformations that Lead to Functional Internal Symmetry Groups

The procedure, described in section 3.3, for generating functional Internal Symmetry groups from sets of transformations of a functional 4-vector is somewhat involved. In this section we will diagram the procedure for creating a functional group U_f of the form such as

$$[SU(3) \otimes SU(2) \otimes U(1) \otimes SU(2) \otimes U(1) \otimes U(4) \otimes U(4)]^4 \otimes U(4) \qquad (35.2)$$

from a generic functional 4-vector f_μ.

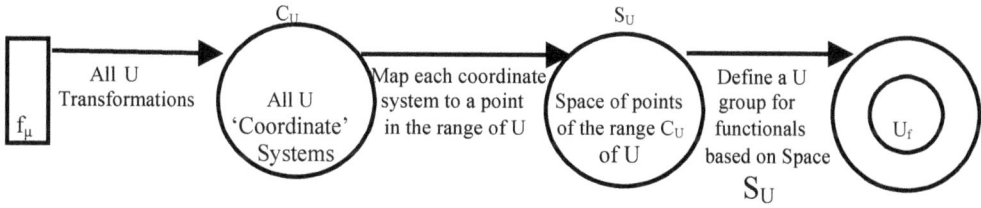

Figure 35.1. Schematic for the procedure for generating a functional Internal Symmmetry group U_f from sets of U transformations of a functional 4-vector f_μ.

Applying this procedure for each factor of eq. 35.2 gives a set of groups that commute with each other. From these functional groups we obtain the symmetry groups of the Unified SuperStandard Theory. Also the symmetry groups, that we obtain, have no difficulties with No Go theorems such as the Coleman-Mandula Theorem.

An alternate approach, based on the above procedure, is to define each group factor of eq. 35.2 in functional space in such a way that the defined functional implementations of the group factors commute, and then to use them to define fermion and boson particle functionals of the form of eq. 35.1 as eigenstates of subsets of factors.

35.4 Subspaces of Particle Functional Space (a Universe)

Particle subspaces of the particle functional space can be categorized by spin: spin ½ functionals, spin 0 functionals, spin 1 functionals, and spin 2 functionals. Each subspace has elements for all physical values of k, s, λ, and ζ.

35.5 Distance in Particle Functional Space

There is no distance in particle functional space—or, otherwise stated, all distances are zero and the space consists of a single mathematical point.

35.6 Products of Functionals – Functional Algebra

Due to the Feynman propagators that appear for all particles in Dyson-Wick perturbation theory calculations[200], which contain quantum fields (the inner products of functionals and fourier expansions) the product of two functionals for different particles is not defined. One can assume

$$f_{k\lambda\zeta}f_{k\lambda\zeta} = 1$$
$$b_{ks\lambda\zeta}b_{ks\lambda\zeta} = 1 \qquad (35.2)$$

or not. We will treat particle functionals as commuting c-number quantities:

$$[_if_\xi , _{i'}f_{\xi'}] = [_if_\xi , b_{s'\xi'}] = [b_{s\xi} , b_{s'\xi'}] = 0$$

as we did in the Second Edition.

35.7 Transformations between Functionals

The Unified SuperStandard Theory lagrangian has numerous interaction terms. Each interaction term defines a transformation between particle functionals in perturbation theory diagrams. Chapter 39 discusses particle transformations in the perturbation theory of quantum field theories. Chapter 37 shows how a lagrangian can be put in a form that gives functional transitions. Chapter 38 shows that these transformations can be put in the form of language grammar rules.

35.8 Functional Internal Symmetry Group Transformations

The λ subscript contains the indices of various fundamental representations of internal symmetry groups: $[SU(3) \otimes SU(2) \otimes U(1) \otimes SU(2) \otimes U(1) \otimes U(4) \otimes U(4)]^4$ where the last two U(4) factors are for the Generation groups and the Layer groups.[201]

---

[200] F. J. Dyson, Phys. Rev. **82**, 428 (1951); G. C. Wick, Phys. Rev., **80**, 268 (1950); See also chapter 6 of Blaha (2007b), and chapter 17 of Bjorken (1965).

[201] The U(4) Species group, resulting from Complex General Relativity (broken) is assumed to have a singlet representation for all particles $\underline{1}$ since it appears to be the only reasonable choice for a symmetry originating in General Relativity. All particles should have a unique interaction arising from Relativity and should not be gravity quadruplets.

Fig. 35.2 shows the internal symmetry groups of Particle Functional Space that 'rotate' particle functionals.

Since Particle Functional Space does not have coordinate space dependence the internal symmetry groups are independent of coordinate space and evade 'No Go' theorems of the 1960's. (The dummy 'k' index is merely to indicate that an integration over momentum takes place in functional inner products.)

35.9 Analogy Between Coordinate and Particle Functional Spaces

Earlier we discussed the relation of Particle Functional Space Internal Symmetries and the coordinate space Reality group. We saw an analogous relationship between the Reality group of coordinate space and of corresponding groups in Particle Functional Space. We *have found a new infinite, pointlike space—Particle Functional Coordinate Space (that are not space-time coordinates.)*

Functional Space contains the functionals of all fermion and boson particles, and the functionals of all fields of the groups of the Unified SuperStandard Theory. See Fig. 35.2

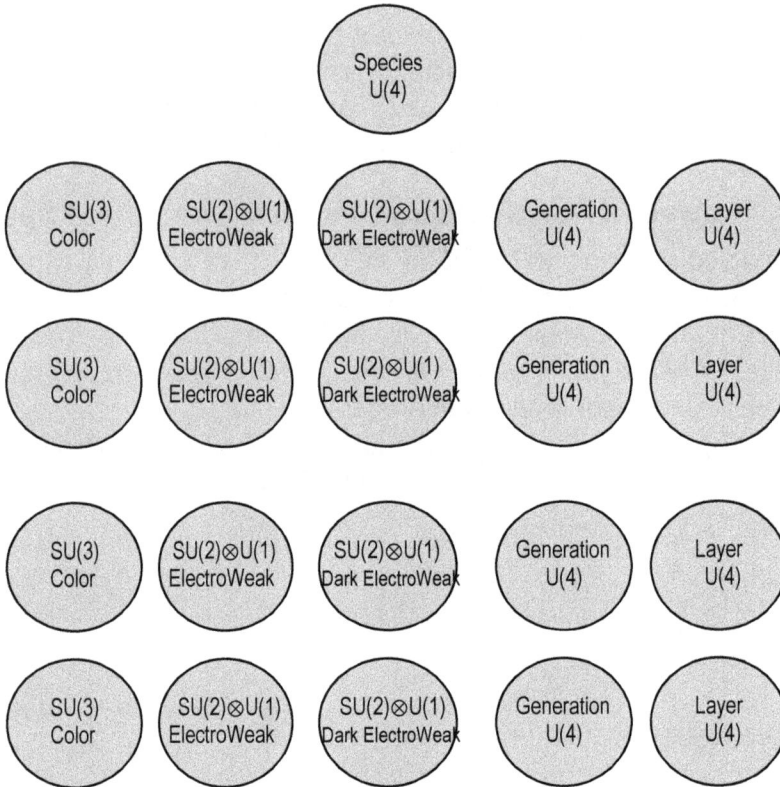

Figure 35.2. The set of four layers of internal symmetry groups in Particle Functional Space corresponding to the four layers of spin ½ fermions and the four layers of vector bosons. In addition there is the Species group.

36. The Wave Space of Free Fourier Expansions

The set of *free*[202] fourier wave function expansions for all fundamental fermion and boson fields forms a space[203] that we call *Wave Space*. The fourier wave function expansions contain creation and annihilation operators. Therefore Wave Space is a q-number space with commuting wave function expansions and non-commuting fourier wave function expansions as the case may be.

Wave Space supports Lorentz transformations of the space-time coordinates and momenta, and Internal Symmetry transformations as well.

36.1 Subspaces of Wave Space (Universe)

Wave Space can be divided into subspaces according to their spin and internal quantum numbers. The spin subspaces are for spin 0, ½, 1, and 2. Within spin subspaces there are subspaces corresponding to combinations of the internal symmetry group factors of eq. 35.2.

36.2 Internal Quantum Numbers of Wave Fourier Expansions

We can label fourier wave function expansions as

$$(s, \mathbf{x}, t)_{k\lambda\zeta} \tag{36.1}$$

---

[202] We take a perturbation theory view of the Unified SuperStandard Model, and other quantum field theories. The free fields obtained from functional inner products are used in perturbation theory to calculate quantities and processes of interest. Thus our restriction to free fourier expansions does not impede the determinations of perturbation theory results.

[203] Much of this chapter appears in the Second Edition, Volume 2.

where the labels are: momentum k, internal symmetry quantum numbers denoted by λ, and possibly other subscripts ζ such as handedness. For example, a fermion fourier expansion has the general form

$$(s, x, t)_{k\lambda\zeta} = N_{\lambda\zeta}(k)[b_{\lambda\zeta}(k, s)u_{\lambda\zeta}(k, s)e^{-ik\cdot x} + d_{\lambda\zeta}^{\dagger}(k, s)v_{\lambda\zeta}(k, s)e^{+ik\cdot x}]$$

(36.2)

where $s = \frac{1}{2}$, and a free field defined by the functional inner product

$$\psi_{\lambda\zeta}(x) = (f_{k\lambda\zeta}, (s, x, t)_{k\lambda\zeta}) = \sum_{\pm s} \int d^3k N_{\lambda\zeta}(k)[b_{\lambda\zeta}(k, s)u_{\lambda\zeta}(k, s)e^{-ik\cdot x} + d_{\lambda\zeta}^{\dagger}(k, s)v_{\lambda\zeta}(k, s)e^{+ik\cdot x}]$$

(36.3)

36.3 Distance in Wave Space

We assume the space-time distance between fourier wave function expansions to be *zero* in keeping with the zero distance between particle functionals in functional space. This assumption is solidly based on the instantaneity of transformations of parts of entangled states. (No spookiness!) Separating the parts of a quantum state S into space-like separated parts S_1 and S_2.we find a change in one part causes an instantaneous change in the other part:

$$<x|S> \rightarrow <x_1|S_1><x_2||S_2>$$

(36.4)

irrespective of distance since the implicit functionals and fourier expansions within each has no space-time separation from each other.

36.4 Space-Time Transformations Among Wave Fourier Expansions

Space-time Lorentz and Poincaré transformations of the coordinates and momenta can be applied to Wave Fourier Expansions.

36.5 Internal Symmetry Transformations of Wave Space Fourier Expansions

Internal Symmetry transformations can be applied to Wave Space Fourier Expansions using the indices within λ and ζ.

37. Particle Functional Lagrangians

37.1 Skeleton Functional Lagrangians

If we could imagine a 'snapshot' of the universe[204] at one instant of time we could presumably enumerate all the functionals of the universe's particles. Then succeeding snapshots would show an ebb and flow of functionals as time progresses. This thought brings us to the important issue of the transformations of particle functionals in particle interactions. The simplest statement that one could make about functional transformations is that they are created and annihilated according to the interaction terms of the skeletonized Unified SuperStandard Theory lagrangian (excluding quadratic terms which do not transform functionals.)

We skeletonize a lagrangian density by deleting all quadratic terms and replacing all particle fields by their corresponding functionals.[205] For example the lagrangian

$$\mathcal{L} = \bar{\psi}_C(i\gamma^\mu D_\mu - m)\psi_C(x) + b(\bar{\psi}_C\psi_C(x))^2 \qquad (37.1)$$

becomes the skeleton lagrangian

$$\mathcal{L}_S = bf^4 \qquad (37.2)$$

where f is the fermion's functional.

---

[204] We realize that such a snapshot is not possible since infinite velocity particles that could feed a camera a snapshot do not exist.

[205] In our construction of particle functional space we have not introduced complex conjugation of quantum functionals for lack of a good reason. Complex conjugation takes place only in the fourier expansion part of a quantum field. Another issue is the appearance of lagrangian terms with factors that are derivatives of fields. Since we do not do computations with skeleton lagrangians we can ignore the derivative in each such factor and simply substitute the functional. For example, $\varphi^3(\partial^\mu\varphi)^2$ becomes the quba expression b^5.

Thus our skeletonized lagrangian formalism describes the transitions between functionals in an interaction. This formalism is made more concrete by considering Feynman diagrams for the interactions.

A skeletonized lagrangian defines a Particle Functional Theory Language. See Blaha (2005b), (2005c) and (2009)

37.2 Functional-Lagrangian and Feynman Diagrams

Feynman diagrams with their "in and out" ordering specify the transformations between functionals more completely. A simple example shows the interaction transformations of functionals. Consider the lagrangian term

$$(\bar{\psi}\psi(x))^2(\partial^\mu\varphi)^2$$

The corresponding Feynman diagram appears in Fig. 37.1.with qube functionals labeled f and quba functionals labeled b. In the above example we have not introduced internal symmetries. When internal symmetries are introduced then the skeletonized lagrangians and the corresponding Feynman diagram representations are significantly more complex.

37.3 Production Rules for Functional-Lagrangians

Each term in a Functional-Lagrangian represents a set of language production rules. See chapter 38 for an example of production rules so generated. The mapping to language production rules illustrates the language interpretation of the Unified SuperStandard Theory.

37.4 Functional Interactions and Feynman Diagrams

Feynman diagrams with their in and out ordering specify the transformations between functionals more completely. A simple example shows the interaction transformations of functionals. Consider the lagrangian term

$$(\bar{\psi}\psi(x))^2(\partial^\mu\varphi)^2$$

A corresponding Feynman diagram for it is

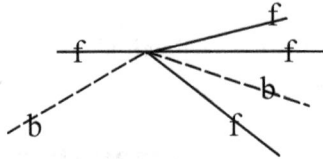

Figure 37.1.Functional Feynman diagram for the above interaction.

with qubes labeled f and qubas labeled b.
    When internal symmetries are introduced, then the skeletonized lagrangians and the corresponding Feynman diagram representations are significantly more complicated.

37.5 Functional Space and Feynman Path Integrals

    Functionals appear in Feynman Path Integrals and in Faddeev-Popov gauge fixing path integrals. We illustrate the use of functionals in the example:

$$Z(J) = N\int \prod_y dy \; \prod_\varphi d\varphi(y) \; \exp\{i\int d^4y[\mathcal{L}(\varphi(y) + J^\mu(y)\varphi(y)]\} \quad (37.3)$$

which in a functional notational notation becomes

$$Z(J) = N\int \prod_y d(y) \; \prod_b db \; \exp\{i\int d^4y[\mathcal{L}(\varphi(y) + J^\mu(y)\varphi(y)]\} \quad (37.4)$$

where (y) represents the fourier expansion in the y coordinates, and with the implied inner product $\varphi(y) = (b, (y))$.

38. Computational Language Interpretation of Particle Functional Transformations

In this chapter[206] we will discuss a language interpretation of particle functional transformations based on a Chomsky-like language and grammar. We will see that particle functional transformations can be viewed as grammar production rules with the net result that the evolution of the universe (Megaverse!) can be viewed as the evolution of an enormous Word consisting of a very large but finite number of (terminal) symbols.

38.1 Languages and Grammars

In chapter 3 of Blaha (2005b) we described a linguistic interpretation of particle interactions. In this interpretation particles play the role of symbols (terminal symbols and nonterminal symbols[207]) in an alphabet (of a finite number of symbols) for a Chomsky-like language. Chomsky defines four types of language ranging from type 0 through type 3. Particle theories, when viewed in terms of their perturbation expansions, can be viewed as a generalization of a type 0 language. A type 0 language (also called an unrestricted rewriting system) allows any grammar production rule of the form

$$X \to Y$$

where X and Y are sets of particles (strings).

---

[206] Most of this chapter appeared in Blaha (2005b) and other books by the author. For more than a hundred years mathematicians and physicists have been describing Physics and Mathematics as being a language in colloquial, layman's terms. Our books show that elementary particle physics, such as our SuperStandard Model, is precisely a type 0 Chomsky language, in which production rules are generated from lagrangian interaction terms. Specific examples are presented in Blaha (2005b) and (2005c).

[207] From a particle view a terminal symbol is a particle that appears in input or output states (strings) of a perturbation theory diagram. A nonterminal symbol is a particle that appears in an intermediate state of a perturbation theory diagram. In the theories that we have considered particles are both terminal and nonterminal symbols.

A production rule is a specification of a transformation of a string of symbols (set of particles) to another string of symbols (set of particles). In the case of quantum theories production rules are inherently quantum probabilistic.

A grammar is specified by a quadruple of items symbolized by the expression

$$<N, T, S, P>$$

where N is a set of nonterminal symbols, T is a set of terminal symbols, S is a special terminal symbol called the head or start symbol, and P is a finite set of production rules. In quantum theories such as the SuperStandard Theory N and T coincide. The start symbol S corresponds to the bare vacuum. Chomsky's definition of a language is the set of all strings of terminal symbols that can be generated from the start symbol using the production rules. *We extend the definition of a particle language to the set of all finite strings of particles (symbols) whether or not they can be generated from the start symbol.*[208]

The set of production rules is finite in Chomsky's definition of language. In the context of quantum field theories we note that a lagrangian of the form of a finite polynomial expression is equivalent to a finite set of production rules.

38.2 Example of Production Rules

In this section we will consider a simple example of production rules with the alphabet:

> Start Symbol: S
> Nonterminal symbols: A, B
> Terminal symbols: x, y

We choose the production rules:

$$S \rightarrow AB \qquad \text{Rule I}$$

---

[208] If one considers the fact that all particles originate either directly or indirectly from the Big Bang (the Start symbol), then the Chomsky definition of type zero languages applies where all strings originate in the Start symbol of the Big Bang.

$$A \rightarrow y \qquad \text{Rule II}$$
$$A \rightarrow Ay \qquad \text{Rule III}$$
$$B \rightarrow x \qquad \text{Rule IV}$$
$$B \rightarrow Bx \qquad \text{Rule V}$$

An example: Generating a string ('particles') yyxxx from the head symbol S using the above production rules:

$$S \rightarrow AB$$
$$AB \rightarrow AyB$$
$$AyB \rightarrow yyB$$
$$yyB \rightarrow yyBx$$
$$yyBx \rightarrow yyBxx$$
$$yyBxx \rightarrow yyxxx$$

38.3 Example of the Production Rules for a Lagrangian Interaction Term

Earlier we noted that a particle lagrangian with a finite number of terms polynomial in particle fields terms would always have a corresponding finite set of production rules. In this section we consider the example of a lagrangian electromagnetic interaction term for electrons and positrons:

$$\overline{e}\gamma \cdot Ae$$

This lagrangian term yields the production rules:

$$e \rightarrow eA$$
$$e \rightarrow Ae$$
$$eA \rightarrow e$$

$$Ae \rightarrow e$$
$$p \rightarrow pA$$
$$p \rightarrow Ap$$
$$pA \rightarrow p$$
$$Ap \rightarrow p$$
$$ep \rightarrow A$$
$$pe \rightarrow A$$
$$A \rightarrow ep$$
$$A \rightarrow pe$$

where e represents an electron, p represents a positron, and A represents the electromagnetic field. Blaha (2005b) presents sequences of transitions using the above production rules and their corresponding Feynman-like diagrams.

Blaha (2005b) also presents other examples such as the ElectroWeak Interaction:

$$\nu_e W^- e$$

where ν_e is an electron type neutrino, and W^- is a negative Weak W vector boson.

38.4   Particle Functional Transformations Identified as Production Rules

Earlier we showed how to define a lagrangian for functionals that provided transformation rules for functionals. In this section we consider the example of a functional lagrangian electromagnetic interaction term for electron and positron functionals:

$$f_e \gamma \cdot b_A f_e$$

This functional lagrangian term yields the functional production rules:

$$f_e \rightarrow f_e b_A$$
$$f_e \rightarrow b_A f_e$$
$$f_e b_A \rightarrow f_e$$
$$b_A f_e \rightarrow f_e$$
$$f_p \rightarrow f_p b_A$$
$$f_p p \rightarrow b_A f_p$$
$$f_p b_A \rightarrow f_p$$
$$b_A f_p \rightarrow f_p$$
$$f_e f_p \rightarrow b_A$$
$$f_p f_e \rightarrow b_A$$
$$b_A \rightarrow f_e f_p$$
$$b_A \rightarrow f_p f_e$$

where f_e represents an electron functional, f_p represents a positron functional, and b_A represents the electromagnetic field functional.

39. Quantum Field Theory With Wave-Particle Functional Formalism

Quantum field theory calculations are almost always performed in perturbation theory. Perturbation theory expansions[209] use vacuum expectation values of time ordered products of pairs of quantum fields. Since quantum fields, in our functional wave-particle formulation, are the result of inner products of functionals (particle cores) and waves (fourier wave expansions), the form of quantum field vacuum expectation values is the same as usually found.

Therefore our new deeper level of our understanding of particle structure does not change perturbation theory. However it does account for the instantaneity of effects in separated parts of a quantum entangled process.

---

[209]See chapter 6 of Blaha (2007b), and chapter 17 of Bjorken (1965) for formulations of perturbation theory as well. .

40. Two-Tier Coordinates

Originally Two-Tier coordinates were developed by this author to remove infinities that appear in perturbation theory calculations. We showed that the quantum smeared coordinates of Two-Tier Quantum Field Theory succeeded in removing all ultra-violet infinities in perturbation theory including the fermion triangle infinities. Remarkably the high precision, low energy[210] predictions of QED remained true in Two-Tier QED and thus remained consistent with experiment to a hitherto unsurpassed level of accuracy. 'Low' energy predictions in other quantum field theories also remained unchanged. At high energies, Two-Tier perturbation theory results are finite and consequently all ultra-violet infinities, to any order in perturbation theory, in *any number of space-time dimensions* were eliminated.

In addition to removing perturbation theory infinities Two-Tier coordinates enable us to define finite theories of Quantum Gravity and 'non-renormalizable' quantum field theories based on polynomial lagrangians, to tame vacuum fluctuations, to eliminate infinities associated with the Big Bang, and possibly to generate the explosive growth of the universe in its role as Dark Energy.[211]

Two-Tier Quantum Field Theory is established on the most fundamental level.

40.1 Two-Tier Features in 4-Dimensional Space-Time

Two-Tier Quantum Field Theory,[212] which was based on a new method[213] in the Calculus of Variations, uses two sets of fields to introduce quantum coordinates. We shall consider this technique for the specific case of a massless vector field $V^i(y)$ analogous to the electromagnetic field.

---

[210] Relative to a mass scale that was perhaps of the order of the Planck mass.

[211] See Blaha (2017b) and earlier books for details. This section is basically a summary of some features.

[212] See Blaha (2005a), and Blaha (2002), for discussions of this new method to eliminate infinities in quantum field theory calculations.

[213] Appendix D describes our method for the composition of extrema in some detail.

In 4-dimensional space-time the massless vector field has the form $Y^\mu(y)$ where the index μ ranges from 0 through 3. The X^μ coordinate system, where it appears, has a c-number real part and a q-number imaginary part. Thus particle fields which are normally defined on four-dimensional real space-time will now be defined on a complex four-dimensional space-time where four imaginary dimensions will appear as *Quantum Dimensions* embodied in a vector quantum field $Y^\mu(y)$:

$$X^\mu(y) = y^\mu + i\ Y^\mu(y)/M_c^2$$

where M_c is an extremely large mass of the order of the Planck mass or perhaps much larger.

The $Y^\mu(y)$ field is a function of the subspace y coordinates. The real part of the space-time dimensions will be taken to be the space of real-valued y coordinates.[214]

The imaginary part of space-time coordinates is the a massless $Y^\mu(y)$ vector quantum field that is suppressed further by a very large mass scale – perhaps of the order of the Planck mass – that reduces the imaginary Quantum Dimensions to the infinitesimal except at large momenta. The effects of Quantum Dimensions only become appreciable in quantum field theory at energies of the order of M_c. At these energies exponential Gaussian factors in each particle (and ghost) propagator are generated by the Quantum Dimensions and serve to make perturbation theory calculations ultra-violet finite – including calculations in Quantum Gravity.

The formalism introduces a new form of interaction that does not have the form of the simple polynomial interactions that have hitherto dominated quantum field theories. This form of interaction takes place via the composition of quantum fields and can be called a *Dimensional Interaction* or an *Interdimensional Interaction* since it affects particle behavior through Quantum Dimensions.

The basic ansatz of the Two-Tier formalism is to replace every appearance of a coordinate x in a quantum field with the variable

---

[214] In a deeper theory the real part might also be a quantum field that undergoes a condensation to generate c-number coordinates. We will not consider this possibility in this book.

$$x^\mu \to X^\mu = (y^0, \mathbf{y} + \mathbf{Y}(y^0, \mathbf{y})/M_c^2)$$

where $\mathbf{Y}(y^0, \mathbf{y})$ is the spatial part of a free massless vector field with features that are identical to the free QED field in the Radiation gauge.

Then one finds that the momentum space free field Feynman propagators $G(k)$ of all particles acquires a Gaussian factor $\exp(h(k))$:

$$G(k) \to G(k)\, \exp(h(k))$$

so that all perturbation theory diagrams are finite. The result is finite perturbative results for all calculations to any order in perturbation theory. Blaha (2005a) shows that Two-Tier theories are finite, Poincare covariant, and unitary. (See Blaha (2005a), chapter 5, for a complete discussion.)

40.2 Simple Two-Tier X^μ Formalism

In this subsection we will describe the basic Two-Tier formalism. Taking the lagrangian described in Blaha (2005a):[215]

$$\mathscr{L}(y) = \mathscr{L}_F\,(X^\mu(y))J + \mathscr{L}_C(X^\mu(y), \partial X^\mu(y)/\partial y^\nu, y) \qquad (40.1)$$

where

$$X^\mu(y) = y^\mu + i\, Y^\mu(y)/M_c^2 \qquad (40.2)$$

with M_c being a large mass scale, $Y_\mu(y)$ a vector quantum field, and where J is the absolute value of the Jacobian of the transformation from X to y coordinates:

$$J = |\partial(X)/\partial(y)|$$

The lagrangian term $\mathscr{L}_C$ is

$$\mathscr{L}_C = +\tfrac{1}{4}\, M_c^4 F^{\mu\nu} F_{\mu\nu}$$

with

---

[215] Eq. 7.1. See Appendix D for more detail.

$$F_{\mu\nu} = \partial X_\mu/\partial y^\nu - \partial X_\nu/\partial y^\mu \qquad (40.3)$$
$$\equiv i\,(\partial Y_\mu/\partial y^\nu - \partial Y_\nu/\partial y^\mu)/M_c^2$$

The lagrangian term $\mathscr{L}_F(X^\mu(y))$ contains the terms for scalar, fermion and other gauge terms in general. The sign in $\mathscr{L}_C$ is not negative – contrary to the conventional electromagnetic Lagrangian. The reason for this difference is that the quantum field part of X^μ is imaginary. Thus $\mathscr{L}_C$ ends up having the correct sign after taking account of the factor of i in the field strength $F_{\mu\nu}$.

Defining

$$F_{Y\mu\nu} = (\partial Y_\mu/\partial y^\nu - \partial Y_\nu/\partial y^\mu)$$

we see the Lagrangian assumes the form of the conventional electromagnetic Lagrangian:

$$\mathscr{L}_C = -\tfrac{1}{4}\, F_Y^{\mu\nu} F_{Y\mu\nu}$$

The action of this theory has the form

$$I = \int d^4y\, \mathscr{L}(y)$$

40.3 Y^μ Gauge

The gauge invariance of the Lagrangian allows us to choose a convenient gauge. The gauge invariance of the full Lagrangian

$$\mathscr{L}_s = L_F(\phi(X),\, \partial\phi/\partial X^\mu)\, J + \mathscr{L}_C(X^\mu(y),\, \partial X^\mu(y)/\partial y^\nu)$$

is based on the standard gauge invariance of $\mathscr{L}_C$, and the gauge invariance of $J\mathscr{L}_F$ in the form of translational invariance

$$X^\mu(y) \rightarrow X^\mu(y) + \delta X^\mu(y)$$

for the special case of a translation of X with the form of a gauge transformation:

$$\delta X^{\mu}(y) = \partial \Lambda(y)/\partial y_{\mu}$$

In this case we find

$$\int d^4 y \, \Lambda(y) \, \partial \, [\, J \partial/\partial X^{\mu} \, \mathcal{T}_{F\mu\nu} \,]/\partial y_{\nu} = 0 \qquad (40.4)$$

after a partial integration and so we have the differential conservation law:

$$\partial \, [\, J \partial \mathcal{T}_{F\mu\nu}/\partial X^{\mu}]/\partial y_{\nu} = 0$$

since $\Lambda(y)$ is arbitrary. This conservation law is trivially obeyed:

$$\partial \mathcal{T}_{F\mu\nu}/\partial X^{\mu} = 0 \qquad (40.5)$$

Thus translational invariance in the $\mathcal{L}_F$ sector together with standard gauge invariance in the $\mathcal{L}_C$ sector automatically guarantees Y field gauge invariance of the total Lagrangian. We use the separate invariance of each term of

$$L = \int d^4 y \, [\mathcal{L}_F \, J + \, \mathcal{L}_C \,] = \int d^4 X \, \mathcal{L}_F + \int d^4 y \, \mathcal{L}_C = L_F + L_C$$

under a constant translation $X^{\mu} \rightarrow X^{\mu} + \delta X^{\mu}$ where δX^{μ} is constant. Then we consider a position dependent translation/gauge transformation, which taken together with the above equation, establishes the invariance under the position dependent translation/gauge transformation.

An alternate approach that leads to the same result is to start with the particle part of the Lagrangian $\mathcal{L}_F$ rewritten to be invariant under general coordinate transformations, as it must, when we generalize to include General Relativity. Since position dependent translations are a form of general coordinate transformation the full theory must be invariant under position dependent translations due to invariance under general coordinate transformations.

Having established invariance under gauge transformations we now choose to use the most convenient gauge – the radiation gauge[216]:

$$\partial Y^i / \partial y^i = 0 \qquad (40.6)$$

where i = 1, 2, 3, which, in the absence of external sources, allows us to set

$$Y^0 = 0$$

since Y^0 does not have a canonically conjugate momentum. A conventional treatment leads to the equal time commutation relations:

$$[Y^\mu(\mathbf{y}, y^0), Y^\nu(\mathbf{y'}, y^0)] = [\pi^\mu(\mathbf{y}, y^0), \pi^\nu(\mathbf{y'}, y^0)] = 0 \qquad (40.7)$$

$$[\pi^j(\mathbf{y}, y^0), Y_k(\mathbf{y'}, y^0)] = -i\, \delta^{tr}_{jk}(\mathbf{y} - \mathbf{y'})$$

(Note the locations of the j indexes above introduce a minus sign.) where

$$\pi^k = \partial \mathcal{L}_c / \partial Y_k'$$
$$\pi^0 = 0$$

$$\delta^{tr}_{jk}(\mathbf{y} - \mathbf{y'}) = \int d^3k\, e^{i\,\mathbf{k}\cdot(\mathbf{y} - \mathbf{y'})}(\delta_{jk} - k_j k_k / \mathbf{k}^2)/(2\pi)^3$$

$$Y_k' = \partial Y_k / \partial y^0$$

The Radiation gauge reveals the two degrees of freedom that are present in the vector potential. The Fourier expansion of the vector potential is:

---

[216] It is also possible to quantize using an indefinite metric that preserves manifest Lorentz covariance as was done by Gupta and Bleuler for the electromagnetic field. We will use the Gupta-Bleuler approach later to establish covariance under special relativity later. Now we opt for manifest positivity and use the radiation gauge.

$$Y^i(y) = \int d^3k \, N_0(k) \sum_{\lambda=1}^{2} \varepsilon^i(k, \lambda)[a(k,\lambda) \, e^{-ik\cdot y} + a^\dagger(k,\lambda) \, e^{ik\cdot y}] \qquad (40.8)$$

where

$$N_0(k) = [(2\pi)^3 2\omega_k]^{-\frac{1}{2}}$$

and (since m = 0)

$$\omega_k = (\mathbf{k}^2)^{\frac{1}{2}} = k^0$$

with $\vec{\varepsilon}(k, \lambda)$ being the polarization unit vectors for $\lambda = 1,2$ and $k^\mu k_\mu = 0$.

The further development of this theory is described in Part 3 of Blaha (2005a).

40.4 Scalar Field Quantization Using X^μ

We will begin by considering the case of a scalar quantum field theory. We assume a real underlying y subspace. Since X^μ is a set of coordinates, we choose to define a scalar field ϕ as a function of X^μ, which, in turn, is a function of the y^ν coordinates. We will provisionally second quantize ϕ treating X^μ as c-number coordinates using a conventional approach.[217]

We assume a Lagrangian, with the momentum conjugate to ϕ:

$$\pi_\phi = \partial L_F / \partial \phi' \equiv \partial L_F / \partial(\partial\phi/\partial X^0) \qquad (40.9)$$

Following the canonical quantization procedure, π and ϕ become hermitian operators with equal time ($X^0 = X^{0\prime}$) commutation rules:

$$[\phi(X), \phi(X')] = [\pi_\phi(X), \pi_\phi(X')] = 0 \qquad (40.10)$$

---

[217] Some texts are: Bogoliubov, N. N., Shirkov, D. V., *Introduction to the Theory of Quantized Fields* (Wiley-Interscience Publishers Inc., New York, 1959); Bjorken, J. D., Drell, S. D., *Relativistic Quantum Fields* (McGraw-Hill, New York, 1965); Huang, K., *Quarks, Leptons & Gauge Fields Second Edition* (World Scientific, River Edge, NJ, 1992); Kaku, M., *Quantum Field Theory* (Oxford University Press, New York, 1993); Weinberg, S., *The Quantum Theory of Fields* (Cambridge University Press, New York, 1995).

$$[\pi_\phi(X), \phi(X')] = -i\,\delta^3(\mathbf{X} - \mathbf{X}')$$

The standard Fourier expansion of the solution to the Klein-Gordon equation is:

$$\phi(X) = \int d^3p \; N_m(p) \; [a(p) \; e^{-ip\cdot X} + a^\dagger(p) \; e^{ip\cdot X}]$$

where

$$N_m(p) = [(2\pi)^3 2\omega_p]^{-\frac{1}{2}}$$

and

$$\omega_p = (\mathbf{p}^2 + m^2)^{\frac{1}{2}}$$

The commutation relations of the Fourier coefficient operators are:

$$[a(p), a^\dagger(p')] = \delta^3(\mathbf{p} - \mathbf{p}')$$
$$[a^\dagger(p), a^\dagger(p')] = [a(p), a(p')] = 0$$

The reader will recognize the quantization procedure is formally identical to the standard canonical quantization procedure of a free scalar quantum field.

In the case of spin ½, spin 1 and spin 2 fields the standard quantization procedure *in terms of the X coordinate system* can also be followed in a way similar to the procedure in standard texts.

40.5 Scalar Feynman Propagators

The momentum space free field Feynman propagators G...(k) of all particles and ghosts in all Two-Tier Quantum Field Theories acquires a Gaussian factor exp(h(k)):

$$G...(k) \rightarrow G...(k) \; \exp(h(k))$$

so that all perturbation theory diagrams are finite. The result is a finite perturbative result in all calculations to any order in perturbation theory. Blaha (2005a) shows that Two-Tier theories are finite, Poincare covariant, and unitary.

An example of the Two-Tier effect on propagators is the case of the Two-Tier photon propagator[218] is:

$$iD_F^{TT}(y_1 - y_2)_{\mu\nu} = -i \int \frac{d^4p\, e^{-ip \cdot z}\, g_{\mu\nu}\, R(\mathbf{p}, z)}{(2\pi)^4\, (p^2 + i\varepsilon)} \tag{40.11}$$

(since the imaginary parts can be taken to be zero: $y_{1i}^{\mu} - y_{2i}^{\mu} = 0$) where

$$z^{\mu} = y_{1r}^{\mu} - y_{2r}^{\mu}$$

$$R(\mathbf{p}, z) = \exp[-p^i p^j \Delta_{Tij}(z)/M_c^4]$$

$$= \exp\{ -\mathbf{p}^2[A(v) + B(v)\cos^2\theta] / [4\pi^2 M_c^4 |\mathbf{z}|^2}$$

with i, j = 1, 2, 3, and with $\Delta_{Tij}(z)$ being the commutator of the positive frequency part $Y^+_k(y)$ and the negative frequency part $Y^-_k(y)$ of $Y_k(y)$:

$$\Delta_{Tij}(z) = [Y^+_j(y_{1r}), Y^-_k(y_{2r})] = \int d^3k\, e^{ik \cdot (y_{1r} - y_{2r})} (\delta_{jk} - k_j k_k/\mathbf{k}^2)/[(2\pi)^3 2\omega_k] \tag{40.12}$$

and

$$v = |z^0|/|\mathbf{z}|$$
$$A(v) = (1 - v^2)^{-1} + .5v\, \ln[(v - 1)/(v + 1)]$$
$$B(v) = v^2(1 - v^2)^{-1} - 1.5v\, \ln[(v - 1)/(v + 1)]$$
$$\mathbf{p \cdot z} = |\mathbf{p}|\, |\mathbf{z}|\, \cos\theta$$

---

[218] Blaha (2005a).

with $|\mathbf{p}|$ denoting the length of a spatial vector $\mathbf{p}$, $|\mathbf{z}|$ denoting the length of a spatial vector $\mathbf{z}$, and with $|z^0|$ being the absolute value of z^0.

The gaussian factors $R(\mathbf{p}, z)$ which appear in all Two-Tier propagators damp the large momentum behavior of all perturbation theory integrals producing a completely finite perturbation theory and yet give the usual results of perturbation theory at energies that are small compared to the mass scale M_c.

40.6 String-like Substructure of the Theory

Two-tier Quantum field Theory endows each particle with an extended structure that resembles the extended structure seen in boson string and Superstring theories. For example, Bailin (1994) use the operator[219]

$$V_\Lambda(k) = \int d^2\sigma \sqrt{-h}\, W_\Lambda(\tau, \sigma)\, e^{-ik\cdot X}$$

where X^μ is a quantized fourier expansion of the string fields (see eq. 7.22 of Bailin (1994)).

We note our X^μ coordinate-field has two transverse degrees of freedom due to gauge invariance, which also invites comparison to the boson string. A point of difference is that we have a well-defined quantum field theoretic formulation in conventional space-time that has the Standard Model as its "large distance" behavior thus introducing a note of reality that is not apparent in Superstring theories. We see that the interacting quantum field theories based on this approach also have good, finite, short distance behavior just as string theories.

The scalar, and other particles', Feynman propagators can be viewed as describing the propagation of a particle cloaked (accompanied) by a cloud of Y particles (which generates the $R(\mathbf{p}, y_1 - y_2)$ factor in the above propagator). If we examine the fourier transform of $R(p, z)$ we see:

$$(2\pi)^4 R(\mathbf{p}, q) = \int d^4z\, e^{iq\cdot z}\, R(\mathbf{p}, z) = \int d^4z\, e^{iq\cdot z}\, \exp[-p^i p^j \Delta_{Tij}(z)/M_c^4] \qquad (40.13)$$

---

[219] D. Bailin and A. Love, *Supersymmetric Gauge Field Theory and String Theory* (Institute of Physics Publishing, Philadelphia, PA, 1994) page 272.

and we find

$$R(\mathbf{p},q) = \sum_{n=0}^{\infty} [i(2\pi M_c)^4]^{-n} (n!)^{-1} \prod_{j=1}^{n} [\int d^4 k_j \, \theta(k_j^0)(\mathbf{p}^2 - (\mathbf{p}\cdot\mathbf{k}_j)^2/\mathbf{k}_j^2)/(k_j^2 + i\varepsilon)] \, \delta^4(q - \sum k_r)$$

which can be interpreted as a "cloud" of Y particles dressing the "bare" particle propagator. (The apparent divergences for $R(p, q)$ are an artifact of the expansion and the subsequent fourier transformation. They are not present in the $R(\mathbf{p}, y_1 - y_2)$ factor in the propagator. See Fig. 40.1 for the Feynman diagram of the Two-Tier 'cloaked' propagator as compared to the normal scalar particle Feynman propagator. The Two-Tier Feynman propagator is basically a conventional scalar propagator that is modified by coherent Y particle emission.[220]

---

[220] T. W. B. Kibble, Phys. Rev. **173**, 1527 (1968) and references therein. In particular see p. 1532 of Kibble's paper.

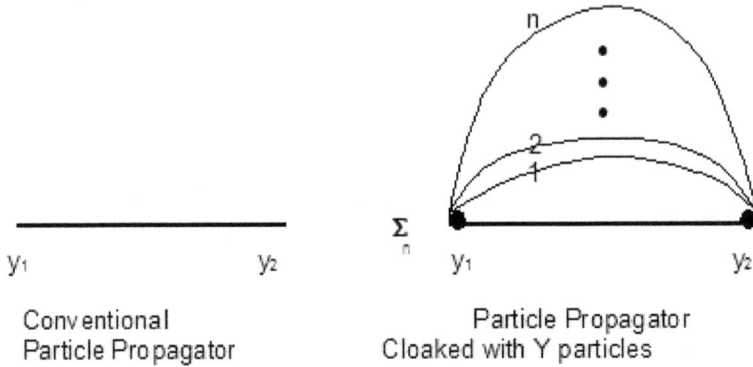

Figure 40.1. Feynman diagram for conventional and the nth diagram of a cloaked Two-Tier propagator.

We note that $R(p, q)$ satisfies the convolution theorem:

$$\int d^4k \, R(\mathbf{p}, k) \, R(\mathbf{p}, q-k) = [R(\mathbf{p}, q)]^2$$

or

$$(2\pi)^4 \int d^4z \, e^{iq \cdot z} R(\mathbf{p}, z) \, R(\mathbf{p}, z) = [\int d^4z \, e^{iq \cdot z} R(\mathbf{p}, z)]^2 \qquad (40.14)$$

The proof follows from the Binomial theorem.

40.7 Two-Tier Complexon Quantum Fields

In the case of the Complexon Standard Model we will need two variables $X_r{}^\mu$ and $X_i{}^\mu$ since we have complex spatial 3-coordinates. We define them similarly to the previous case:

$$X_r{}^\mu(y_r) = y_r{}^\mu + i \, Y_r{}^\mu(y_r)/M_c{}^2$$
$$X_i{}^\mu(y_i) = y_i{}^\mu + i \, Y_i{}^\mu(y_i)/M_c{}^2$$

where we choose the same mass scale for both the "real" and "imaginary" variables. The Two-Tier, single generation, version of the Complexon Standard Model then has an action of the form

$$I_{CSMtt} = \int dy^0 d^3y_r d^3y_i \left(\mathcal{L}_{CSM}(X_r^\mu(y_r), \mathbf{X}_i^k(y_i))J_2\right)\Big|_{y_{i0} = 0,\ Y_{r0} = Y_{i0} = 0} + \qquad (40.15)$$

$$+ \int dy_r^0 d^3y_r\, \mathcal{L}_C(X_r^\mu(y_r), \partial X_r^\mu(y_r)/\partial y_r^\nu, y_r) +$$

$$+ \int dy_i^0 d^3y_i\, \mathcal{L}_C(X_i^\mu(y_i), \partial X_i^\mu(y_i)/\partial y_i^\nu, y_i)$$

where the replacements

$$x^\mu \equiv x_r^\mu \ \rightarrow\ X_r^\mu(y_r)$$

$$x_i^k \ \rightarrow\ X_i^k(y_i)$$

for $\mu = 0, 1, 2, 3$ and $k = 1, 2, 3$ are made, followed by defining $y_r^0 = y^0$ and making a Complex Lorentz transformation to a frame where $y_i^0 = 0$. J_2 is the absolute value of the Jacobian of the transformation from (X_r, X_i) to (y_r, y_i) coordinates:

$$J_2 = |\partial(X_r, X_i)/\partial(y_r, y_i)|$$

We also choose gauges where $Y_r^0 = Y_i^0 = 0$. These types of transformations and gauge choices are discussed in detail in Blaha (2005a). The lagrangian terms $\mathcal{L}_C(X_r^\mu(y_r), \partial X_r^\mu(y_r)/\partial y_r^\nu, y_r)$ and $\mathcal{L}_C(X_i^\mu(y_i), \partial X_i^\mu(y_i)/\partial y_i^\nu, y_i)$ have the same form:

$$\mathcal{L}_C = +\tfrac{1}{4}\, M_c^4 F^{\mu\nu} F_{\mu\nu} \qquad (40.16)$$

with

$$F_{\mu\nu} = \partial X_\mu/\partial y^\nu - \partial X_\nu/\partial y^\mu$$

$$\equiv i \, (\partial Y_\mu/\partial y^\nu - \partial Y_\nu/\partial y^\mu)/M_c^2$$

or defining

$$F_{Y\mu\nu} = (\partial Y_\mu/\partial y^\nu - \partial Y_\nu/\partial y^\mu)$$

we see each lagrangian assumes the form of the conventional electromagnetic Lagrangian:

$$\mathcal{L}_C = -\tfrac{1}{4} \, F_Y{}^{\mu\nu} F_{Y\mu\nu}$$

The lagrangian is supplemented with the following condition on all complexon fields $\Phi_{...}$:

$$(\partial/\partial X_r{}^k(y_r)) \, (\partial/\partial X_i{}^k(y_i)) \Phi... = 0 \tag{40.17}$$

summed over $k = 1, 2, 3$. Non-complexon fields $\Omega...$ in our left-handed formulation satisfy the subsidiary condition:

$$\{(\partial/\partial X_r{}^k(y_r))(\partial/\partial X_i{}^k(y_i)) - [(\partial/\partial X_r{}^k(y_r))^2(\partial/\partial X_i{}^m(y_i))^2]^{1/2}\}\Omega... = 0 \tag{40.18}$$

summed over $k = 1, 2, 3$ and over $m = 1, 2, 3$ separately in each of the two terms.

40.8 Complexon Feynman Propagator

In the case of complexons, the Two-Tier Feynman propagator differs from the non-complexon case by having an integration over imaginary spatial 3-momenta, a derivative of a delta function embodying the orthogonality of the real and imaginary 3-momenta, and two factors of $R(\mathbf{p}, z)$: one factor being $R(\mathbf{p_r}, z_r)$ and the other factor being $R(\mathbf{p_i}, z_i)$ (where the time components $z_r^0 = z^0$ and $z_i^0 = 0$ since there is only one real time coordinate[221]) thus providing large momentum convergence for both real and imaginary 3-momentum integrations.

---

[221] We can arrange for $z_i^0 = 0$ by making a Complex Lorentz transformation to an inertial frame where z is real.

For a normal scalar particle the Feynman propagator is:

$$i\Delta_{CTF}(x - y) = \theta(x^+ - y^+)<0|\phi_{CT}(x)\,\phi_{CT}(y)|0> + \theta(y^+ - x^+)<0|\phi_{CT}(y)\phi_{CT}(x)|0>$$

$$= i\int d^4p_r d^3p_i (2\pi)^{-7}\delta'(\mathbf{p_r \cdot p_i}/m^2)e^{-ip^+(x^- - y^-) - ip^-(x^+ - y^+) + ip_\perp \cdot (x_\perp - y_\perp) - ip_i \cdot (x_i - y_i)}/(p^2 + m^2 + i\varepsilon)$$

$$(40.19)$$

in conventional quantum field theory.

In the case of Two-Tier quantum field a scalar *complexon* particle has the Feynman propagator

$$i\Delta_{CTFtt}(x - y) = i\int d^4p_r d^3p_i (2\pi)^{-7}\delta'(\mathbf{p_r \cdot p_i}/m^2)\ R(\mathbf{p_r}, z_r)R(\mathbf{p_i}, z_i) \cdot \qquad (40.20)$$

$$\cdot e^{-ip^+(x^- - y^-) - ip^-(x^+ - y^+) + ip_\perp \cdot (x_\perp - y_\perp) - ip_i \cdot (x_i - y_i)}/(p^2 - m^2 + i\varepsilon)$$

where the time components $z_r^{\,0} = z^0$ and $z_i^{\,0} = 0$ since there is only one time coordinate, where $R(\mathbf{p}, z)$ is given in the previous subsection, and where $p^2 = p^{0\,2} - p_r^{\,2} + p_i^{\,2}$.

Propagators for other types of particles are similarly modified in the Two-Tier formalism (See Blaha 2005a).

40.9 Vacuum Fluctuations

While the expectation value of a *conventional* free scalar field $\phi_{conv}(x)$ is zero in a conventional quantum field theory:

$$<0|\phi_{conv}(x)|0> = 0 \qquad (40.21)$$

the vacuum fluctuations of *conventional* scalar quantum field theory are quadratically divergent:

$$<0|\phi_{conv}(x)\phi_{conv}(x)|0> = \int d^3p/[(2\pi)^3 2\omega_p] \qquad (40.22)$$

In "Two-Tier" quantum field theory we find the vacuum expectation value of a free field is zero *and the expectation value of the square of the field is also zero:*

$$\langle 0|\phi(X)\phi(X)|0\rangle = \int d^3p \; e^{-p^i p^j \Delta_{Tij}(0)/Mc^4}/[(2\pi)^3 2\omega_p] = 0$$

since the exponential factor in the integral is $-\infty$. The exponent contains

$$\Delta_{Tij}(z) = \int d^3k \; e^{-ik\cdot z} \, (\delta_{ij} - k_i k_j/\mathbf{k}^2)/[(2\pi)^3 2\omega_k] \qquad (40.23)$$

where "T" is for "Two-Tier". Thus *vacuum fluctuations are zero in Two-Tier quantum field theory*. Correspondingly, we will see that renormalization constants are finite in the Two-Tier versions of QED, Electroweak Theory, the Standard Model and Quantum Gravity. See Blaha (2017b) and references therein for more details.

40.10 Time Intervals in General Relativity

Wigner[222] has studied the measurement of time intervals in General Relativity and sees a problem in the measurement of extremely short intervals. According to Wigner, the measurement of a time interval in a region of space requires the measurement of the length of time required for an event to happen. The measurement requires an accurate clock. But the accuracy of the clock is limited by the energy-time uncertainty relation:

$$\Delta E \Delta t \geq \hbar \qquad (40.24)$$

Thus the uncertainty in the clock's time measurement is related to the uncertainty in the clock's energy which is, in turn, related to the uncertainty in the clock's mass:

$$\Delta E = (\Delta m)c^2$$

To obtain "infinite" accuracy the uncertainty (fluctuations) in the clock's mass must be infinite and thus the clock's mass must be infinite. Infinite fluctuations in the clock's mass will produce corresponding infinite fluctuations in the gravitational field.

$$\Delta h \propto \Delta E \qquad \text{(in conventional General Relativity)}$$

---

[222] E. P. Wigner, Rev. Mod. Phys. **29**, 255 (1957); J. Math. Phys. **2**, 207 (1961).

As a result the notion of space-time and time intervals (which depend on the geometry through General Relativity) become uncertain. Thus, according to Wigner, and others, the concept of time intervals and space-time points becomes questionable.

The Two-Tier version of Quantum Gravity offers a way out of this dilemma. The gravitational force becomes stronger as one goes to shorter distances (higher energies) down to a distance (up to an energy) whose scale is set by M_c. At shorter distances (higher energies) the gravitational force becomes weaker and declines to zero at zero distance. Thus at very high energy the gravitational field fluctuations (Δh) are at worst inversely proportional to the energy (and probably decline by a higher power of inverse energy.) (The same considerations would apply if one chooses to consider fluctuations in the Riemann-Christoffel symbols.)

$$\Delta h < c_1/E < c_1/(\Delta E) \quad \text{(in Two-Tier Quantum Gravity)} \quad (40.25)$$

where c_1 is a constant. Thus Wigner's conclusion does not hold in the Two-Tier version of Quantum Gravity as gravitational fluctuations actually become smaller at energies above a critical energy whose scale is set by M_c.

In fact, combining the above equations we see

$$c_1 \Delta t/\Delta h \geq \hbar$$

at sufficiently high energy. Therefore the time uncertainty Δt, and the gravitational field fluctuations Δh, can both decrease while maintaining the energy-time uncertainty relation. *Thus the notion of a space-time point "is saved" in Two-Tier quantum gravity.*

40.11 Vacuum Fluctuations in the Gravitation Fields

While the expectation value of the free graviton field $h_{\mu\nu conv}(x)$ (weak field approximation) is zero in a conventional quantum field theoretic approach:

$$<0|h_{\mu\nu conv}(x)|0> = 0 \quad (40.26)$$

the vacuum fluctuations of the *conventional* quantum graviton field is quadratically divergent since

$$<0|h_{\mu\nu\text{conv}}(x)h_{\alpha\beta\text{conv}}(x)|0> = \int d^3p \; b'_{\mu\nu\alpha\beta}(p)/[(2\pi)^3 \; 2\omega_p] = \infty \qquad (40.27)$$

where $b'_{\mu\nu\alpha\beta}(p)$ is a rational function of the momentum p.
    In "Two-Tier" quantum field theory we find

$$<0|h_{\mu\nu}(X)h_{\alpha\beta}(X)|0> = \int d^3p \; b'_{\mu\nu\alpha\beta}(p) \; e^{-p^i p^j \Delta_{Tij}(0)}/[(2\pi)^3 2\omega_p] = 0 \qquad (40.28)$$

since the exponential factor in the integrand is $-\infty$. The exponent contains

$$\Delta_{Tij}(z) = \int d^3k \; e^{-ik \cdot z}(\delta_{ij} - k_i k_j/\mathbf{k}^2)/[(2\pi)^3 2\omega_k]$$

Thus the vacuum fluctuations of $h_{\mu\nu}$ are zero in "Two-Tier" quantum field theory and, correspondingly, the weak field Two-Tier quantization of Quantum Gravity is consistently finite (and weak in perturbation theory calculations.)

40.12 Two-Tier Features in D-Dimensional Space-Time (such as the Megaverse)

    Since a field, quantized in D-dimensional conventional coordinates (D > 4), would lead to divergences in perturbation theory calculations, we can use D-dimensional Two-Tier coordinates to avoid divergences in perturbation theory:

$$Y^i(y) = y^i + i \; Y_u^i(y)/M_u^{D/2} \qquad (40.29)$$

where $Y_u^i(y)$ for i = 1, ..., D is a D-dimensional free gauge field and M_u is a mass of the order of the Planck mass or greater. The $Y_u^i(y)$ term adds a quantum field to the D coordinates making them a set of quantum coordinates. Quantum coordinate derivatives are defined by

$$\partial_i = \partial/\partial Y^i(y) = \partial/\partial(y^i - Y_u^i(y)/M_u^{D/2}) \qquad (40.30)$$

The use of these coordinates to quantize particle fields leads to a completely finite perturbation theory. We applied them in Blaha (2017b) to create a finite fundamental theory of mater. We applied them to fields in the Megaverse[223] to achieve a finite theory of Megaverse dynamics for elementary particles and universe particles.

The second quantization of a vector gauge field $V^i(y)$ is analogous to the second quantization of the electromagnetic field. The lagrangian density terms for the free $V^i(Y(y))$ fields is

$$\mathscr{L}_{Vu} = -\tfrac{1}{4} F_{Vu}^{ij}(Y(y))F_{Vuij}(Y(y)) \qquad (40.31)$$

The lagrangian is

$$L_{Vu} = \int d^D y \, \mathscr{L}_{Vu}(Y(y))$$

with

$$F_{Vuij} = \partial V_i(Y(y))/\partial Y^j(y) - \partial V_j(Y(y))/\partial Y^i(y)$$

where the values of i and j range from 1 to D in this section.

The equal time commutation relations, using the Dth coordinate as the time coordinate, are specified in the usual way:

$$[V^i(Y(\mathbf{y}, y^0)), V^j(Y(\mathbf{y}', y^0))] = [\pi^i(Y(\mathbf{y}, y^0)), \pi^j(Y(\mathbf{y}', y^0))] = 0$$
$$[\pi_j(Y(\mathbf{y}, y^0)), V_k(Y(\mathbf{y}', y^0))] = -i \, \delta^{(D-1)tr}{}_{jk}(Y(\mathbf{y},0) - Y(\mathbf{y}',0))$$

where

$$\pi_u^k = \partial \mathscr{L}_{Vu}(V(Y(y)))/\partial V_k'(Y(y))$$
$$\pi_u^D = 0$$

for k = 1, ... , (D – 1), and

---

[223] Blaha (2017c).

$$\delta^{(D-1)tr}_{jk}(\mathbf{y}-\mathbf{y}') = \int d^{(D-1)}k \; e^{i\,\mathbf{k}\cdot(Y(\mathbf{y},0)-Y(\mathbf{y}',0))}\,(\delta_{jk}-k_jk_k/\mathbf{k}^2)/(2\pi)^{D-1} \quad (40.32)$$
$$V_k'(Y(y)) = \partial V_k(Y(y))/\partial y^{1D}$$

for j, k = 1, 2, ... , (D – 1).

   If we choose the Radiation gauge for $V_k(Y(y))$:

$$V^D(Y(y)) = 0$$
$$\partial V^j(Y(y))/\partial Y^j(y) = 0 \quad (40.33)$$

for j = 1, 2, ... , (D – 1) then (D – 2) degrees of freedom (polarizations) are present in the vector potential.[224] The Fourier expansion of the vector potential $V^i(Y(y))$ is:

$$V^i(Y(y)) = \int d^{(D-1)}k \; N_{0v}(k) \sum_{\lambda=1}^{D-2} \varepsilon^i(k,\,\lambda)[a_V(k,\lambda):e^{-ik\cdot Y(y)}: + a_V{}^\dagger(k,\lambda):e^{ik\cdot Y(y)}:] \quad (40.34)$$

for i = 1, ... , (D – 2) where

$$N_{0v}(k) = [(2\pi)^{(D-1)}2\omega_k]^{-\frac{1}{2}}$$

and (since the field is massless)

$$k^D = \omega_k = (\mathbf{k}^2)^{\frac{1}{2}}$$

where k^D is the energy, and where the $\varepsilon^i(k,\,\lambda)$ are the polarization unit vectors for λ = 1, ... , (D – 2) and $k^\mu k_\mu = k^{D\,2} - \mathbf{k}^2 = 0$.

   The commutation relations of the Fourier coefficient operators are:

$$[a_V(k,\lambda),\, a_V{}^\dagger(k',\lambda')] = \delta_{\lambda\lambda'}\delta^{D-1}(\mathbf{k}-\mathbf{k}')$$
$$[a_V{}^\dagger(k,\lambda),\, a_V{}^\dagger(k',\lambda')] = [a_V(k,\lambda),\, a_V(k',\lambda')] = 0$$

and the polarization vectors satisfy

$$\sum_{\lambda=1}^{D-2} \varepsilon_i(k,\,\lambda)\varepsilon_j(k,\,\lambda) = (\delta_{ij}-k_ik_j/\mathbf{k}^2)$$

---

[224] Note we use the Radiation gauge for $Y^\mu(y)$ also.

The V^μ Feynman propagator is

$$iD_F^{trTT}(y_1 - y_2)_{jk} = <0|T(V_j(Y(y_1))V_k(Y(y_2)))|0>$$ (40.35)

$$= -ig_{jk} \int \frac{d^D k \, e^{-ik \cdot (y_1 - y_2)} \, R(\mathbf{k}, y_1 - y_2)}{(2\pi)^D \, (k^2 + i\varepsilon)}$$

where g_{jk} is the D-dimensional Lorentz metric and where $R(\mathbf{k}, y_1 - y_2)$ is given by

$$R(\mathbf{k}, y_1 - y_2) = \exp[-k^i k^j \Delta_{Tij}(y_1 - y_2)/M_u^D]$$
$$= \exp\{-k^2[A(v) + B(v)\cos^2\theta] / [(2\pi)^{D-2} M_u^4 z^2]\}$$

where k^2 is *the sum of the squares of the D − 1 spatial components* with

$$z^\mu = y_1{}^\mu - y_2{}^\mu$$
$$z = |\mathbf{z}| = |\mathbf{y}_1 - \mathbf{y}_2|$$
$$k = |\mathbf{k}|$$
$$v = |z^0|/z$$
$$A(v) = (1 - v^2)^{-1} + .5v \ln[(v-1)/(v+1)]$$
$$B(v) = v^2(1 - v^2)^{-1} - 1.5v \ln[(v-1)/(v+1)]$$

$$\mathbf{k} \cdot \mathbf{z} = kz \cos\theta$$

and $|\mathbf{k}|$ denoting the length of a spatial $(D-1)$-vector $\mathbf{k}$ while $|z^0|$ is the absolute value of $z^0 \equiv z^D$.

As the above equations indicate, the Gaussian damping factor $R(k, z)$ for *all* large spatial momentum k^j is the same for both the positive and negative frequency parts of the (Two Tier) V Feynman propagator. We are assuming the spatial momentum is real-valued in this discussion. It is also important to note that $R(k, z)$ does not depend on $k^0 = k^D$ (in the V and Y_u Radiation gauges) and thus the integration over k^0 proceeds in the usual way to produce time-ordered positive and negative frequency parts.

The Gaussian exponential factor in *all* spatial coordinates causes the Feynman propagator to be finite and, together with the Gaussian factor in universe particle propagators, causes all perturbation theory calculations when interactions are introduced to be finite as we have seen in Blaha (2017b).

For small momentum much less than M_u then $R(\mathbf{k}, y_1 - y_2) \rightarrow 1$ and the Feynman propagator is the "normal" propagator of conventional D-dimensional quantum field theory. For large momentum the corresponding potential approaches r^{D-3} in contrast to the electromagnetic Coulomb potential r^{-1}. The V potential is highly non-singular at large energies.

Thus using Two-Tier Quantum Field Theory we can perform perturbation theory calculations that always yield a finite result.[225] This is not true if conventional Quantum Field is used.[226]

---

[225] In particular, the fermion triangle divergence (anomaly) does not occur in our Two Tier Quantum Field Theory of the fermion sector. Thus there is no requirement for axion-like particles in the Megaverse (or in universes) although the possible existence of this type of particle is not ruled out.

[226] Blaha (2005a) provides a complete discussion of Two-Tier Quantum Field Theory.

41. PseudoQuantum Field Theory

PseudoQuantum Field Theory (and its Quantum Mechanics analogue CQ Mechanics[227]) originate in the need to second quantize in unusual coordinate systems, and in curved space-time coordinate systems. The papers in Appendices I and J provide a detailed introduction to PseudoQuantum Field Theory to which the reader is referred.

In this subsection we point out its advantages in a variety of field theory contexts that are relevant for the Unified SuperStandard Theory. The advantages of PseudoQuantum Field Theory are:

1. Quantization in any coordinate system in flat or curved space-times with an invariant definition of asymptotic particle states. An n particle asymptotic state in one coordinate system is a unitarily equivalent n particle asymptotic state in any other coordinate system. Therefore particle number is invariant under change of coordinate system. This is important for the Unified SuperStandard Theory in curved space-times. It is also important for quantization in higher dimensional Euclidean spaces such as the Megaverse. The method was developed in the late 1970's by the author to provide a quantization procedure which supports a unique particle interpretation of states in arbitrary non-static space-times where no global timelike coordinate (Killing vector) exists. PseudoQuantum Field Theory which we developed in a series of books[228] also can be formulated in the Megaverse. Thus we can use it in the Megaverse to

---

[227] See Appendix E for details, which contains Blaha (2016f). CQ Mechanics encompasses both classical mechanics and quantum mechanics, and provides a method of rotating between them. It has applications to transitions between Quantum/Semi-Classical Entanglement, and Quantum/Classical Path Integrals, and Quantum/Classical Chaos.

[228] See Blaha (2017b) for the discussion of the PseudoQuantum field theory formalism for Higgs particles in our Extended Standard Model. See chapter 20 of Blaha (2017b), and earlier books, for a more detailed view than that presented here.

implement the Higgs Mechanism to generate particle masses and symmetry breaking.

2. PseudoQuantum Field Theory enables one to define Higgs particle dynamics in such a way that a non-zero vacuum expectation value cleanly separates from the quantum field part of the Higgs fields. This technique can be used in symmetry breaking mechanisms, mass generation, and possible generation of coupling constants as vacuum expectation values.

3. It supports the canonical definition of higher derivative field theories through the use of the Ostrogradski bootstrap. See Appendix B where a fourth order theory of the Strong interaction is defined that has color confinement and a linear r potential. The potential part of this theory was used by the Cornell group to calculate the Charmonium spectrum. (See Blaha (2017b) for details.)

An associated advantage of using PseudoQuantum Field Theory is that it provides for retarded propagators and an Arrow of Time.

41.1 General Case of PseudoQuantization in Differing Coordinate Systems

Papers in Appendix A describe the PseudoQuantization procedure that relates second quantizations in differing coordinate systems. We can epitomize the general concept in the following short example.

Consider the case of a scalar particle in D space-time dimensions that we second quantize in coordinate system denoted 1 with coordinates x based on a timelike Killing vector

$$\varphi(x) = \sum_{\alpha} [\chi_\alpha(x)A_\alpha + \chi_\alpha^*(x)A_\alpha^\dagger] \qquad (41.1)$$

where the $\chi_\alpha(x)$ are positive frequency with respect to a definition of positive frequency within a universe – following the notation of Appendix A.

Consider now the second quantization of the particle field in a second coordinate system denoted 2 with coordinates y based on a different timelike Megaverse Killing vector

$$\varphi(y) = \sum_{\beta} [\psi_\beta(y)b_\beta + \psi_\beta^*(y)b_\beta^\dagger] \tag{41.2}$$

where the $\psi_\beta(y)$ are positive frequency with respect to 2's definition of positive frequency.

Comparing above definitions we see the difference in the definition of the coordinates used in the field expansions as well as the implicit difference in the definitions of positive frequency. To relate the quantizations to each other, we must use the relation between the x and y coordinates:

$$y_i = f_i(x)$$

or, in vector form,

$$y = f(x)$$

for i = 1, 2, ... , D. Thus

$$\varphi(f(x)) = \sum_{\beta} [\psi_\beta(f(x))b_\beta + \psi_\beta^*(f(x))b_\beta^\dagger] \tag{41.3}$$

Inverting the above equations to obtain the relation of the fourier coefficient operators we see:

$$A_\alpha = \sum_{\beta} [C_{\alpha\beta} b_\beta + C'_{\alpha\beta} b_\beta^\dagger]$$

where $C_{\alpha\beta}$ and $C'_{\alpha\beta}$ are c-number functions of α and β:

$$C_{\alpha\beta} = (\chi_\alpha(x), \varphi(f(x)))$$
$$C'_{\alpha\beta} = (\chi_\alpha^*(x), \varphi(f(x))) \tag{41.4}$$

The above equations imply an N particle state in one coordinate system will appear as a superposition of states of various numbers of particles in the other coordinate system IF the standard quantum field theory formulation is used.

TO REMEDY this situation – which we take to be unphysical – we must reformulate quantum field theory using the PseudoQuantum formulation presented Appendix A. The scalar particle case is discussed in Appendix A between eqs. 6 – 31, to which the reader is referred.

The conclusions of that section, and the sections following it, in Appendix A are:

1. One can define corresponding unitarily equivalent particle states in two quantizations with invariant particle numbers.

2. The fourier coefficient operators of the two quantizations are related by Bogoliubov transformations and are unitarily equivalent.

3. The group of the local Bogoliubov transformations is an infinite tensor product of $SU_{1,1}$ groups.

4. The vacuums of the particles are invariant under Bogoliubov transformations that relate the Megaverse and the universe quantizations.

5. Unitarily equivalent perturbation theories of both quantizations can be defined.

We now consider the case of Two-Tier PseudoQuantization, and then turn to various applications of PseudoQuantization.

41.2 Two-Tier PseudoQuantum Field Theory

The combination of the Two-Tier procedure with the PseudoQuantization procedure leads to a somewhat more complicated situation. In principle, both are required for a Unified SuperStandard Theory in any coordinate system in flat or curved space-times in any number of dimensions. However their direct combination is both complicated and unphysical.

The main purpose of PseudoQuantization is to have particle number invariance under a change of coordinate system. Two-Tier Field Theory 'cloaks' each particle in infinite 'clouds' of Y^μ quanta as Fig. 40.1 illustrates. We define PseudoQuantization as implementing particle number invariance for 'bare' particles without their clouds of Y^μ quanta. Thus an asymptotic particle state of n particles (neglecting its Y^μ quanta cloud) remains a unitarily equivalent n particle state (neglecting its Y^μ quanta cloud) under a change of coordinate system.

To implement this concept we first define quantizations of a particle in coordinate systems without Two-tier quanta. We then 'dress' the quantizations by replacing the coordinates y^μ in each coordinate system with the corresponding Two-Tier coordinates:

$$y^\mu \rightarrow X^\mu(y) = y^\mu + i\, Y^\mu(y)/M_c^{\,2} \tag{41.5}$$

It appears the most convenient gauge in each coordinate system is the Lorentz gauge:

$$\partial Y^\mu/\partial y^\mu = 0 \tag{41.6}$$

We now briefly consider the case of a scalar particle PseudoQuantization. This case is considered in more detail in Appendix A. Following Appendix A we must introduce two fields $\varphi_1(y)$ and $\varphi_2(y)$ with the free fields' lagrangian

$$\mathscr{L}(y) = \partial^\mu\varphi_1\partial_\mu\varphi_2 - \tfrac{1}{2}\,\partial^\mu\varphi_1\partial_\mu\varphi_1 - m^2\varphi_1\varphi_2 + \tfrac{1}{2}\,m^2\,\varphi_1^{\,2} \tag{41.7}$$

in a coordinate system with coordinates y. Then following the steps indicated in paper 1 of Appendix A from eq. 7 onward we arrive at a PseudoQuantum formulation in the coordinate system with coordinates y that is unitarily equivalent to that of a different coordinate system defined a similar manner.

From eq. 43 onwards we can replace the c-number coordinates x and y with Two-Tier coordinates of the form

$$X^\mu(y) = y^\mu + i\, Y^\mu(y)/M_c^2$$

and proceed to calculate propagators and perturbation theory diagrams.

Thus we have a straight-forward procedure to unite the PseudoQuantum formalism with Two-Tier coordinates to obtain finite perturbation theory results with unitary equivalence to quantization in other coordinate systems in both flat and curved space-times.

The use of two fields per particle of PseudoQuantum field theory will be seen to part of the applications consider in the remainder of this subsection. We will put aside the consideration of quantizations in other coordinate systems in what follows to keep the presentation as simple as possible.

41.3 PseudoQuantum Higgs Scalar Particle Field Theory in D-dimensional Space-Time

41.3.1 The Enigma of Higgs Particles and the Higgs Mechanism

In our previous work on the Standard Model, and its generalization to The Unified SuperStandard Theory described in a series of books entitled *Physics is Logic* ..., we showed that the fermion spectrum results from Complex Special Relativity, the gauge interactions result from the Reality group, the fermion generations result from the Generation group, and the Theory of Everything results from a combination with Complex General Relativity. The Higgs particles and the Higgs Mechanism were inserted to generate particle masses and symmetry breaking effects.

Whence comes Higgs particles? A more fundamental cause has not been suggested until our analysis, which is presented here. So the Higgs sector appeared to be an expedient mechanism to insert much needed symmetry breaking and masses into the theory.

There are a number of peculiarities in the implementation of the Higgs Mechanism:

1. First, it is selective in the sense that some gauge fields have associated Higgs particles and utilize the Higgs Mechanism, and some gauge fields do not have associated Higgs particles. In particular, the ElectroWeak gauge fields, the Generation group gauge fields, the Layer group fields, and the complex gravity Species gauge fields have associated Higgs particles. The strong interaction (gluon) gauge fields do not.

2. The Higgs potentials have a quadratic mass term of the "wrong" sign plus a quartic interaction term, which together, generate non-zero vacuum expectation values. They obviously accomplish their goal. But the source of these potentials, and why they have the same form, is unknown. One expects a fundamental principle should be operative here.

3. One can imagine creating a Higgs microscope at some super-accelerator. Using this microscope in the presence of a (classical) condensate could enable the Uncertainty Principle to be violated. This possibility, in the case of a microscope using electromagnetic fields, was the source of a heuristic argument for the need to quantize the electromagnetic field.[229]

4. The formulation of the Higgs Mechanism uses classical fields under the assumption that a path integral formulation justifies their use. While this may be true, the path integral formulation relies on implicit, unstated boundary conditions that obscure the physics of the quantum field theoretic nature of the mechanism. A direct quantum field theoretic study of the Higgs Mechanism is needed and would further elucidate its character.

5. Scalar fields have a cloud hanging over them that spin ½ fields do not. A spin ½ particle cannot transition to negative energy because there is a filled sea of negative energy particles. No additional particles can fall into the sea due to the Pauli Exclusion Principle that forbids two fermions with the same

---

[229] Heitler (1954) p. 86 provides a good discussion of the need to quantize the electromagnetic field.

4-momentum and quantum numbers. In the case of scalar particles the Pauli Exclusion Principle does not apply and so a *filled* negative energy sea of scalar particles is not possible and positive energy scalar particles can transition to negative energy without hindrance. This problem has been "resolved" by an appropriate definition of the scalar particle vacuum to exclude transitions to negative energy. But the rationale for the definition is lacking. Dirac was asked about this issue many years ago. He said he had a solution to the problem. However he did not present it – in keeping with his well-known taciturn nature. So the issue remains an open question.

For the above reasons we will show that a more satisfactory method of achieving the goals of mass generation and symmetry breaking exists.[230] This method relies on a larger Fock space that enables the appearance of a vacuum expectation value for Higgs particles to be understood within a truly quantum framework. More importantly, this method is a consequence the PseudoQuantization procedure described above that enables unitarily equivalent quantizations in different coordinate systems. So a profound fundamental justification for our Higgs boson formulation exists. One major consequence of this approach is the appearance of a local Arrow of Time – a concept that has been a subject of interest for over one hundred years. Another consequence is a rationale for ElectroWeak Higgs bosons and for their absence for the strong (gluon) interaction.

41.3.2 PseudoQuantization of Scalar Particles

We now consider the PseudoQuantization[231] of a scalar particle field that will become a Higgs particle with a non-zero vacuum expectation value.[232] We begin by

---

[230] In the Extended Standard Model of Blaha (2015a) we have shown that the basic particles have a mass, the Landauer mass, so that the theory is symmetry violating from the very start. We have also shown that our Two-Tier formalism for quantum field theories always yields finite results in perturbation theory calculations – making the renormalization approach of t'Hooft and others, which relied on initially massless gauge fields, unnecessary.

[231] PseudoQuantization in a D-dimensional space-time is described in Blaha (2017c). This discussion is relevant to PseudoQuantization in the Megaverse, or in other universes.

defining two fields that correspond to the scalar particle: $\varphi_1(x)$ and $\varphi_2(x)$.[233] These fields will be assumed to have the equal time commutators

$$[\varphi_i(x), \pi_j(y)] = i(1 - \delta_{ij})\delta^3(\mathbf{x} - \mathbf{y}) \tag{41.8}$$
$$[\varphi_i(x), \varphi_j(y)] = 0$$
$$[\pi_i(x), \pi_j(y)] = 0$$

where δ_{ij} is the Kronecker δ and where $\pi_i(x)$ is the canonically conjugate momentum to $\varphi_i(x)$. The fields $\varphi_1(x)$ and $\pi_1(y)$ will be observable classical fields. The fields $\varphi_2(x)$ and $\pi_2(y)$ will not be observables so that $\varphi_1(x)$ and $\pi_1(y)$ can both be sharp on the set of physical states.

We now specify the lagrangian density for a scalar Klein-Gordon particle:

$$\mathcal{L} = \partial\varphi_1/\partial x_\mu \partial\varphi_2/\partial x^\mu \tag{41.9}$$

with hamiltonian density

$$\mathcal{H} = \pi_1\,\pi_2 + \partial\varphi_1/\partial x_i \partial\varphi_2/\partial x^i$$

where i labels spatial coordinates, and $\pi_1 = \partial\varphi_2/\partial t$ and $\pi_2 = \partial\varphi_1/\partial t$. The lagrangian $\mathcal{L}$ is without a potential or mass term.

The lagrangian and hamiltonian for a massive scalar particle in this formalism are

$$\mathcal{L} = \partial\varphi_1/\partial x_\mu \partial\varphi_2/\partial x^\mu - m^2\,\varphi_1\varphi_2 \tag{41.10}$$

with hamiltonian density

$$\mathcal{H} = \pi_1\,\pi_2 + \partial\varphi_1/\partial x_i \partial\varphi_2/\partial x^i + m^2\,\varphi_1\varphi_2$$

---

[232] Much of this section appears in Blaha (2016c), and earlier books, as well as in S. Blaha, Phys. Rev. **D17**, 994 (1978). The case of fermion PseudoQuantization is also discussed in Appendix A – S. Blaha, Il Nuovo Cimento **49A**, 35 (1979).

[233] The subscripts on the fields are not gauge symmetry indices but simply identifiers distinguishing the fields from each other.

The fields can be fourier expanded in terms of creation and annihilation operators:

$$\varphi_i(\mathbf{x}, t) = \int d^3k \, [a_i(k)f_k(x) + a_i^\dagger(k)f_k^*(x)] \qquad (41.11)$$

for i = 1, 2 where

$$f_k(x) = e^{-ik\cdot x} /(2\omega_k(2\pi)^3)^{\frac{1}{2}}$$

with $\omega_k = |\mathbf{k}|$.

The creation and annihilation operators satisfy the commutation relations:

$$[a_i(k), a_j^\dagger(k')] = (1 - \delta_{ij})\delta^3(\mathbf{k} - \mathbf{k}')$$
$$[a_i(k), a_j(k')] = 0$$
$$[a_i^\dagger(k), a_j^\dagger(k')] = 0$$

for i, j = 1, 2.

In this formulation the defining properties of a physical state are:

$$\varphi_1(x)|\Phi, \Pi> = \Phi(x)|\Phi, \Pi>$$
$$\pi_1(x)|\Phi, \Pi> = \Pi(x)|\Phi, \Pi> \qquad (41.12)$$

where $\Phi(x)$ and $\Pi(x)$ are sharp on the states and thus classical fields with

$$\Phi(\mathbf{x}, t) = \int d^3k \, [\alpha(k)f_k(x) + \alpha^*(k)f_k^*(x)] \qquad (41.13)$$

and correspondingly for $\Pi(x)$.

41.3.3 Vacuum States for Scalar (Higgs) Particles with Non-Zero Vacuum Expectation Values

When we implement the mass mechanism, Φ is constant. We can define a set of states

$$a_1(k)|\alpha> = \alpha(k)|\alpha>$$

$$a_1^\dagger(k)|\alpha> = \alpha^*(k)|\alpha>$$

and correspondingly a set of coherent states

$$|\alpha> = C\exp\left\{\int d^3k \, [\alpha(k)a_2^\dagger(k) + \alpha^*(k)a_2(k)]\right\}|0> \qquad (41.14)$$

where C is a normalization constant and where the vacuum state |0> satisfies

$$a_1(k)|0> = a_1^\dagger(k)|0> = 0 \qquad (41.15)$$

$$a_2(k)|0> \neq 0 \qquad\qquad a_2^\dagger(k)|0> \neq 0$$

The dual vacuum state satisfies

$$<0|a_2(k) = <0|a_2^\dagger(k) = 0$$

$$<0|a_1(k) \neq 0 \qquad\qquad <0|a_1^\dagger(k) \neq 0$$

With this coherent state formalism, which gives purely classical fields and yet also has quantum fields through the use of φ_2 and its creation and annihilation operators, we now have the machinery to define a mass mechanism without the introduction of a potential whose origin can only be described as dubious.

For we can define a coherent state for some k as

$$|\Phi, \Pi> = C\exp\{[(2\pi)^3\omega_k/2]^{1/2}\Phi[a_2^\dagger(k) + a_2(k)]\}|0> \qquad (41.16)$$

where C is a normalization constant, which yields a non-zero vacuum expectation value:

$$\varphi_1(x)|\Phi, \Pi> = \Phi| \, \Phi, \Pi> \qquad (41.17)$$

where Φ is a constant. Evaluating a fermion interaction term we find a mass term emerges[234]

$$\psi\,(\varphi_1 + \varphi_2)\psi \;\rightarrow\; \overline{\psi}(\Phi + \varphi_2)\psi \qquad (41.18)$$

It generates a mass for an interaction with a gauge field of the form

$$A^\mu(\varphi_1 + \varphi_2)^2 A_\mu \;\rightarrow\; A^\mu(\Phi + \varphi_2)^2 A_\mu \qquad (41.19)$$

It also yields a quantum field theoretic interaction that would result in the production of ElectroWeak particles from these scalar fields. The production of Higgs particles that decay into ElectroWeak gauge particles has recently been found at CERN.

The present formalism provides a clean way to separate the vacuum expectation value of a scalar particle from its quantum field part in contrast to the Higgs Mechanism where one has to separate a Higgs field into parts manually.

41.3.4 Interpretation of Negative Energy Scalar Particle States

As we noted earlier, scalar particle physics has the problem of no barrier to the decay of positive energy states t̅o̅ negative energy states due to the absence of a Pauli Exclusion Principle for bosons. The PseudoQuantization procedure that we developed in 1978 and describe here allows negative energy states as one would physically expect and raises the possibility of disastrous particle decays to negative energy. The above equations show that negative energy states are possible in this theory.

However they also show that combined positive and negative energy boson states can be interpreted as classical field states. In addition, the ability of any number of boson particles to have the same 4-momentum and quantum numbers shows that a *macroscopic* classical scalar field state can be constructed.

Thus we can view states containing negative energy particles as classical field states and thus solve[235] the issue of interpreting negative energy particle states – a more

---

[234] When matrix elements with a "vacuum state" are taken.

satisfactory approach than the standard quantization procedure does – with due respect to Professor Dirac.

We note that macroscopic many particle fermion states can only have one particle in any mode unlike bosons. Therefore we cannot use this formalism to create macroscopic classical fermion field states.[236] And the filled Dirac sea of negative energy fermions precludes the transition of a positive energy Dirac fermion to a negative energy state. *Thus there is a certain complementarity between fermions that cannot become classical fields but have a filled sea precluding decays to negative energy states, and bosons that can become classical fields but support decays to negative energy states.*

41.3.5 Contrast with Conventional Second Quantization of Scalar Particles

The PseudoQuantization procedure followed here uses different boundary conditions than the usual scalar particle quantization procedure. The essence of the difference is embodied in a comparison of the definition of the vacuum above and the definition of the conventional second quantized field vacuum:

$$a|0> = 0 \qquad \text{Conventional Approach}$$
$$a^\dagger|0> \neq 0$$

In the conventional approach the creation of negative energy boson states is eliminated *ab initio* whereas in our approach it is allowed in order to support classical field states with non-zero vacuum expectation values that are a form of classical field. While one cannot discredit the conventional choice for conventional scalar fields, one can see that our approach yields a physically more important result – particularly for Higgs fields – because it leads to an Arrow of Time *locally* – an important feature of physical phenomena that has been a subject of much discussion and dispute. One can say that the conventional approach sweeps the issue "under the rug" rather than seeking a deeper

---

[235] Also a boson that has no interactions cannot transition from to a positive energy state to a negative energy state due to conservation of energy.

[236] However we can create PseudoQuantum fermion states. See S. Blaha, Phys. Rev. **D17**, 994 (1978) (reproduced in Appendix I) and references therein to earlier papers by the author.

justification – differing from Dirac's implied notion that the issue merited attention. We will discuss the "Arrow of Time" within the framework of our PseudoQuantization approach later.

41.3.6 Why Inertial Reference Frames are Special

The great physicists of the early 20th century raised numerous questions about Special Relativity after Einstein and Poincarè's discovery. Prominent among them was the question of why inertial reference frames are of especial importance in Special Relativity, and afterwards in General Relativity.

It appears that our formulation of the mass generation mechanism sheds significant light on the reason for the special prominence of inertial frames. Earlier we considered the case of a massless PseudoQuantized scalar. We now consider massive scalars since experiments at CERN have apparently discovered a Higgs particle with a 125 GeV/c mass. The above equations describe a massive scalar particle. If the scalar is massive, then the "vacuum" state that yields a non-zero expectation value must change to

$$|\Phi, \Pi> = C\exp\{(2\pi)^3 m/2]^{\frac{1}{2}}[a_2^\dagger(\mathbf{0},m) + a_2(\mathbf{0},m)]\}|0> \qquad (41.20)$$

to have operators for a particle of mass m in its rest frame. Then, having established this preferred frame for a Higgs particle, in The Unified SuperStandard Theory, and requiring that invariant intervals

$$ds^2 = dt^2 - d\mathbf{x}^2 \quad \text{(in rectangular coordinates)}$$

are unchanged by a (complex or real) Lorentz transformation, we find that inertial reference frames are singled out as "special" in the sense that they are the only accessible reference frames that can be generated by a Lorentz boost/transformation from the Higgs particle rest frame. *The Higgs particle vacuum state singles out the class of inertial reference frames.*

Thus Higgs particles play a central role in establishing the basis of physical reality.

41.3.7 PseudoQuantization Reveals More Physical Consequences than the Higgs Mechanism of Scalar Particles

Earlier we pointed out that our PseudoQuantization theory of Higgs particles reveals more physical consequences than the conventional approach, which implements the Higgs Mechanism by simply using a potential term that has a minimum at a non-zero vacuum expectation value. This section shows the major results of a properly implemented mechanism. We find a better explanation of the negative energy state problem of boson field theories. We find a local arrow of time that explains the direction of time that we, and all of nature, experiences. We find the reason why inertial reference frames have a special physical significance – a result long sought by physicists.

In addition we will see in chapter 11 that real gauge fields should have an associated Higgs particle, while necessarily complex gauge fields (the Strong interaction gauge field in The Unified SuperStandard Theory) do not have an associated gauge field. These results correspond to experimental reality.

41.3.8 The T Invariance Issues of Our PseudoQuantized Scalar Particle Theory

The PseudoQuantized scalar particle hamiltonian equations are invariant under time reversal $t \rightarrow t' = -t$. The 'new' vacuum states defined above break the time reversal invariance of the theory resulting in retarded particle propagators.

The hamiltonian equations

$$[H, \varphi_1(\mathbf{x}, t)] = -i\partial\varphi_1/\partial t \qquad (41.21)$$
$$[H, \varphi_2(\mathbf{x}, t)] = -i\partial\varphi_2/\partial t$$

are invariant under time reversal. If we define a time reversal operator transformation U then the time reversed equations are

$$[UHU^{-1}, \varphi_1(\mathbf{x}, -t)] = +i\partial\varphi_1(\mathbf{x}, -t)/\partial(-t)$$
$$[UHU^{-1}, \varphi_2(\mathbf{x}, -t)] = +i\partial\varphi_2(\mathbf{x}, -t)/\partial(-t)$$

The operator U, which is unitary, transforms H into **–H**. This operation is legal because the hamiltonian – in this case – is not positive definite and admits negative energy states.[237] Thus

$$[H, \varphi_1(\mathbf{x}, -t)] = -i\partial\varphi_1(\mathbf{x}, -t)/\partial (-t)$$
$$[H, \varphi_2(\mathbf{x}, -t)] = -i\partial\varphi_2(\mathbf{x}, -t)/\partial (-t)$$

and the time reversal invariance of the equations of motion is established for this case.

Time reversal invariance is broken by our choice of vacuum states. This choice is necessary to obtain classical field states as we showed earlier. A demonstration of the time reversal symmetry breaking is presented later where we show theory has retarded propagators for particle propagation to and from asymptotic states.

Within the interaction region the particle propagators are the sum of retarded and advanced parts that combine to yield principle value propagators – not Feynman propagators. Many years ago Feynman and Wheeler championed principle value propagators for electrodynamics to obtain an action-at-a distance theory of Quantum Electrodynamics. While their theory, and ours, differ from the standard quantum field theory approach there is no reason to view them as faulty, or having serious physical defects. The only question is whether nature chooses conventional quantum field theory or PseudoQuantized quantum field theory. In our case the need for a classical scalar particle non-zero vacuum expectation value strongly motivates our choice of pseudoquantized Higgs particles.

41.3.9 Retarded Propagators for Our Quantized Higgs Particles

In the previous section we pointed out that our PseudoQuantization Higgs theory has an arrow of time due to is boundary conditions as expressed by its definition of the vacuum state and its dual. In this section we will show that the theory uses retarded propagators for propagation to and from the interaction region to asymptotic in-states and out-states. Within an interaction region the theory uses half-retarded – half-

---

[237] Unlike the usual case of second quantized Klein-Gordon quantum field theory.

advanced propagators. We discuss aspects of the perturbation theory and propagators of our scalar particles in this chapter.

First we note that in-states at $t = -\infty$ are composed of superpositions of $a_2(k)$ and $a_2^\dagger(k)$ creation and annihilation operators:

$$a_2(k)|0> \neq 0 \qquad\qquad a_2^\dagger(k)|0> \neq 0$$

while the out-states composed of superpositions of $a_1(k)$ and $a_1^\dagger(k)$ creation and annihilation operators:

$$<0|a_1(k) \neq 0 \qquad\qquad <0|a_1^\dagger(k) \neq 0$$

Consequently when in-state particles (x_1) propagate into the interaction region (x_2) the relevant propagators are retarded propagators with the form

$$
\begin{aligned}
G_{in}(x_2, x_1) &= <0|T(\varphi_{1\,in}(x_2), \varphi_{2\,in}(x_1))|0> \qquad (41.22)\\
&= \theta(x_{20} - x_{10})<0|[\varphi_{1\,in}(x_2), \varphi_{2\,in}(x_1)]\,|0>
\end{aligned}
$$

This is a manifestly retarded propagator. The choice of vacuums clearly results in a time asymmetry giving a retarded propagation reflecting the familiar Arrow of Time.

A similar situation prevails for propagation to out-states (x_3) from the interaction (x_2) region:

$$
\begin{aligned}
G_{out}(x_3, x_2) &= <0|T(\varphi_{1\,out}(x_3), \varphi_{2\,out}(x_2))|0> \qquad (41.23)\\
&= \theta(x_{30} - x_{20})<0|[\varphi_{1\,out}(x_3), \varphi_{2\,out}(x_2)]\,|0>
\end{aligned}
$$

Within the interaction region the Higgs particles have principle value propagators.

Thus we find PseudoQuantized Higgs particles embody a local Arrow of Time. The locality of the Arrow of Time is embodied in all the particles that interact with the Higgs particle. Since the mass of *every* particle – bosons and fermions – has a Higgs contribution, and thus *every* particle interacts with the Higgs particles, the Arrow of

Time permeates The Unified SuperStandard Theory as well as the more familiar Standard Model known from experiment.

41.3.10 The Local Arrow of Time

In the *Physics is Logic* series of monographs we saw that complex coordinates led to the form of the fermion spectrum, that the mapping of complex coordinates to real-valued coordinates yielded the Reality group and The Unified SuperStandard Theory gauge interactions, that Complex General Relativity led to Higgs particles that were directly united with elementary particle masses and gave us the equality of inertial mass and gravitational mass. Later we will see the reduction of complex gauge fields to real gauge fields explains the appearance of Higgs fields in The Unified SuperStandard Theory.

The PseudoQuantization procedure leads to retarded Higgs field propagators and thence to a *local* arrow of time. Many arguments have been put forward over the past hundred plus years for the Arrow of Time. Many arguments based on Statistical Mechanics, Entropy, and Boltzmann's statistical atomic theory have suggested the Arrow of Time is a global statistical consequence. This view seems to contradict the results of elementary particle experiments where a *local* Arrow of Time is evident.

Our rationale for the Arrow of Time begins with retarded Higgs fields. Then we note that Higgs field quantum interactions appear for all fermions and gauge particles. Thus all particle interactions are imbued with an Arrow of Time. Particles united to form macroscopic matter inherit their combined Arrows of Time producing the global Arrow of Time we experience.

Thus our PseudoQuantization approach offers a more satisfactory solution of the origin of the Arrow of Time.

It is remarkable that complex quantities – coordinates and fields – through the Higgs phenomena that we have considered, lead to the equality of inertial mass and gravitational mass, and an Arrow of Time. This unity of mass and time phenomena may reflect the deeper fact that we can have no practical Arrow of Time if all particles were massless, for particle dynamics at light speed would then be pointless. This view has been expressed by DeWitt, Unruh, and others who have pointed out that, physically, time is meaningful and measurable only if masses exist; the larger the mass, the more accurate the time measurement in principle.[238]

41.3.11 Space-Time Dependent Particle Masses

It is possible that the ultimate Unified SuperStandard Theory has masses that evolve with time and may also be spatially varying – different values in different parts of the universe. Presently there is no decisive evidence for this possibility although astrophysical studies continue. In this section we will describe the mechanism for space-time dependent masses.

Consider a classical field (time and spatially varying):

$$\Phi(\mathbf{x}, t) = \int d^3k \, [\alpha(k)f_k(x) + \alpha^*(k)f_k{}^*(x)] \qquad (41.24)$$

If we define the coherent vacuum state

$$|\alpha> = C \exp\left\{\int d^3k \, [\alpha(k)a_2{}^\dagger(k) + \alpha^*(k)a_2(k)]\right\}|0> \qquad (41.25)$$

then

$$\varphi_1(x)|\Phi, \Pi> = \Phi(x)|\Phi, \Pi>$$
$$\pi_1(x)|\Phi, \Pi> = \Pi(x)|\Phi, \Pi>$$

where

$$\varphi_i(\mathbf{x}, t) = \int d^3k \, [a_i(k)f_k(x) + a_i{}^\dagger(k)f_k{}^*(x)] \qquad (41.26)$$

for i = 1, 2 and where

$$f_k(x) = e^{-ik\cdot x} / (2\omega_k(2\pi)^3)^{\frac{1}{2}}$$

with ω_k equal to the energy.

---

[238] No mass, no clock; no clock, no physical time. See Blaha (2015a) pp. 368-371 for a discussion including comments by DeWitt and Unruh.

41.3.12 Inertial Mass Equals Gravitational Mass

From the days of Newton through Einstein[239] to the present the equality of gravitational mass and inertial mass has been a topic of interest. Mach, who played an important role, in this ongoing discussion, thought distant masses in the universe were the source of the equality. However the origin of the equality, which has been shown experimentally to very high accuracy, remained uncertain until the *Physics is Logic* series of books, in which we showed the interconnection of the Unified SuperStandard Theory and Complex Gravitation via Higgs generated masses that united gravitational and inertial mass.

In Blaha (2016h) we showed that a Complex General Relativity transformation can be factored into the product of a complex-valued transformation and a real-valued General Coordinate transformation. The set of complex valued transformations form a U(4) group that we called the General Coordinate Reality group. Later we will define the Internal Symmetry Species Group as the corresponding analogue. The Species Group has gauge fields that undergo spontaneous symmetry breaking and generate contributions to all fermion masses.

Since fermion field masses are now sums of ElectroWeak Higgs contributions, Generation group Higgs contributions, Layer group Higgs contributions, and Species group contributions, and since the gravitational Higgs fields appear in all fermion masses, the equality of inertial and gravitational mass is proven. The gravitational Higgs particles' equations depend, in part, on the gravitational field by Blaha (2016h) and so set the mass scale of gravitational mass, and thereby of all Higgs mass contributions. They set the scale of inertial masses equal to the scale of gravitational masses. **Since an expression cannot mix mass scales, the gravitational mass scale must be the same as the inertial mass scale. Inertial Mass equals gravitational mass.**

We have established the equality of inertial and gravitational mass at the short distance quantum level. In our view, this explanation is far more satisfying than basing the equality on a combination of large distance phenomena and quantum phenomena.

---

[239] For example, Einstein and Grossman in 1913 stated, "The theory herein described originates in the conviction that the proportionality between the inertial and gravitational mass of a body is an exact law of nature that must be expressed as a foundation principle of theoretical physics."

As Einstein and Weyl have pointed out, all fundamental physics phenomena should be based on a local theory. Complex Gravity as we have constructed it, combined with the Unified SuperStandard Theory, furnishes a completely local basic Theory of Everything.

The equation above contains a coherent state $|\alpha\rangle$ for a time and spatially varying mass. The above equations can be generalized to the case of multiple space-time varying masses.[240]

$$|\Phi_1, \Phi_2, \ldots, \Phi_n; \Pi_1, \Pi_2, \ldots, \Pi_n\rangle = C \prod_{i=1}^{n} \exp\left\{\int d^3k \left[\alpha_i(k)a_{2i}^{\dagger}(k) + \alpha_i^{*}(k)a_{2i}(k)\right]\right\}|0\rangle \quad (41.27)$$

Then all n mass vacuum expectation values are space-time dependent:

$$\varphi_{1i}(x) \mid \Phi_1, \Phi_2, \ldots, \Phi_n; \Pi_1, \Pi_2, \ldots, \Pi_n\rangle = \Phi_i(x) \mid \Phi_1, \Phi_2, \ldots, \Phi_n; \Pi_1, \Pi_2, \ldots, \Pi_n\rangle$$
$$(41.28)$$

Thus our formalism can accommodate space-time varying masses should they be found in the Cosmos.

41.3.13 Benefits of the PseudoQuantization Method

In this book, and in earlier work, we showed that a more physically satisfactory method for avoiding the negative energy state problem exists. This method relies on the use of a larger Fock space in which negative energy states (or partially negative energy states) are interpreted as states containing classical fields or a mix of classical fields and individual boson particles. This approach resolves the negative energy boson issue and provides a common framework for boson particles and classical boson fields.

One consequence of the PseudoQuantization method is that it enables the appearance of a vacuum expectation value for Higgs particles (a constant classical field) to be understood within a truly quantum framework. Another major consequence of this approach is the appearance of a *local* Arrow of Time due to the Higgs mass generation

---

[240] The "vacuum" state $|0\rangle$ also implicitly has factors for the vacuum expectation values used for fields that give masses to fermions and vector bosons as described in Blaha (2016h).

mechanism – a concept that has been a subject of interest for over one hundred years. A macroscopic arrow of time is often described as a statistical result. But our approach yields an arrow of time at the single particle level.

The conventional approach to boson field quantization sweeps these issues "under the rug" rather than seeking a deeper justification. It differs from Dirac's implied notion that the issue merited attention.

Another important consequence of the PseudoQuantization method is that it singles out inertial reference frames when applied to the case of Higgs particles.

Yet another more subtle consequence of boson PseudoQuantization is that it provides a rationale/explanation for the presence of ElectroWeak Higgs bosons, *and for their absence for the strong (gluon) interactions. The question of why there are no strong interaction Higgs bosons has not been previously considered to the best of this author's knowledge.*

42. Higgs Particles, Gauge Fields, and Higgs Mechanism Generated Coupling Constants

Higgs particles appear in many contexts in the Unified SuperStandard Theory. In this subsection we consider a possible origin of Higgs particles in complex-valued gauge fields that explains why there are no Strong Interaction gauge field Higgs particles. We also show it is possible to define Higgs particles that generate gauge field coupling constants. Using this mechanism we show that the known coupling constants appear to correspond to Higgs values of the same order of magnitude suggesting that we are close to a form of unification. We also show the Higgs Mechanism for fermion particle masses may explain the equality of inertial and gravitational mass – a topic of continuing interest for many years.

42.1 The Genesis of Scalar (Higgs?) Particle Fields from Complex Gauge Fields

In the past[241] we showed that scalar particles can be 'extracted' from all spin 1 gauge fields except color SU(3).

Since our Unified SuperStandard Theory is ultimately based on the Complex General Coordinate Transformations and the Complex Lorentz group (thus complex-valued coordinate systems), it appears reasonable to assume all spin 1 gauge fields to initially be similarly complex-valued. Most of these gauge fields can be rotated to real values. However we shall see that color SU(3) gauge fields are *necessarily* complex-valued. All other gauge fields can be rotated to real values. The price of rotation is the introduction of scalar fields. Some of these fields may be Higgs particle fields and generate gauge boson masses (symmetry breaking) and fermion masses.

---

[241] Blaha (2015c) and (2016c).

Thus we can view scalar particles including Higgs particles as inherently associated with most gauge fields. From the viewpoint of our derivation from basic axioms, the origin of Higgs bosons in complex-valued gauge fields gives a 'tighter' derivation of the overall theory. Higgs fields are an inherently a part of the theory.

42.2 The Difference between the Strong Gauge Field and the Other Gauge Fields in the SuperStandard Theory

In our Unified SuperStandard Theory the only gauge field without an associated Higgs particle is the strong interaction gluon gauge field. *We view this exception as a particularly important clue as to the nature of the relation between gauge fields and Higgs particles.*

How does the strong interaction gauge field differ from all other gauge fields in the Unified SuperStandard Theory? An examination of the gauge fields dynamic equations (and other lagrangian terms) of our Unified SuperStandard Theory reveals that all gauge field dynamic equation kinetic terms *except those of the strong interaction gauge field* have the form:

$$\partial/\partial x_\mu\, F^a_{\mu\nu} + gf^{abc}A^{b\mu}\,F^c_{\mu\nu} = j^a_\nu \tag{42.1}$$

where

$$F^a_{\mu\nu} = \partial/\partial x^\nu A^a_\mu - \partial/\partial x^\mu A^a_\nu + gf^{abc}A^b_\mu A^c_\nu \tag{42.2}$$

where the coordinates x^ν *are real-valued,*[242] where a, b, c are structure constant indices, where g is a coupling constant, and where j^a_ν is the corresponding current. The gauge field A^a_μ is real for all normal and Dark ElectroWeak gauge fields, Generation group gauge fields, and Layer group gauge fields. Thus the above equations are real-valued.

The strong interaction gauge field[243] in our Unified SuperStandard Theory differs from the other gauge fields by being *necessarily* complex[244] due to the complex 3-space complexon derivatives that appear in the corresponding dynamic equations:

---

[242] Before the introduction of the Two-Tier formalism.
[243] This field is called a complexon gauge field in Blaha (2017b), (2015a) and earlier books.

$$D^\mu F_C{}^a{}_{\mu\nu} + gf^{abc} A_C{}^{b\mu} F_C{}^c{}_{\mu\nu} = j^a{}_\nu \qquad (42.3)$$

with

$$F_C{}^a{}_{\mu\nu} = D_\nu A_C{}^a{}_\mu - D_\mu A_C{}^a{}_\nu + gf^{abc} A_C{}^b{}_\mu A_C{}^c{}_\nu \qquad (42.4)$$

where

$$D_k = \partial/\partial x_r{}^k + i \, \partial/\partial x_i{}^k$$

$$D_0 = \partial/\partial x^0$$

for k = 1, 2, 3 where $A_C{}^a{}_\mu$ is the complexon color Strong interaction gauge field. The complexon spatial coordinates have the form $x_r{}^k + i \, x_i$. The time coordinate is real-valued. These equations are eqs. 10.16 and 5.162 of Blaha (2015a) for complexon gauge fields,[245] the carriers of the strong interaction in the Unified SuperStandard Theory.

This difference enables us to differentiate the strong gauge field from all other gauge fields in The Unified SuperStandard Theory. Thereby we can develop a unified formalism for the non-strong gauge fields and their corresponding Higgs particles.

The necessarily complex nature of the color SU(3) field is the reason that the Strong Interaction gauge fields do not acquire a mass via the Higgs Mechanism. As shown below, the necessary complexity of Strong Interaction gauge fields precludes the generation of Higgs fields from Strong Yang-Mills gauge fields.

42.3 Generation of Higgs Fields from Non-Abelian Gauge Fields

In the prior section we considered the difference between the strong gauge field and the other gauge fields of The Unified SuperStandard Theory. Unlike strong gauge fields the other gauge fields (ElectroWeak and so on) could be real or complex. In a manner similar to what we did in the preceding *Physics is Logic* books (and earlier books) we can assume gauge fields are initially complex, and then transform them to

---

[244] One cannot cleanly separate the real and imaginary parts of its dynamic equations.

[245] In The Unified SuperStandard Model we also identify quark species particles as having complex 3-momentum. We call them complexon fermions.

real-valued fields using a phase transformation that introduces scalar fields, some of which we will take to be Higgs fields.

We define a complex phase transformation for a gauge field $A^{b\mu}$ with

$$A'^{a\mu}(x) = \Phi(x)^a{}_b A^{b\mu}(x) \tag{42.5}$$

where $\Phi(x) = \mathrm{diag}(\exp[i\varphi_1(x)], \exp[i\varphi_2(x)], \ldots , \exp[i\varphi_n(x)])$, and n is the number of symmetry components of $A^{b\mu}$. Inserting $A'^{a\mu}(x)$ above we find:

$$\partial/\partial x_\mu \, F'^a{}_{\mu\nu} + gf^{abc} A'^{b\mu} \, F'^c{}_{\mu\nu} = j^a{}_\nu \tag{42.6}$$

where

$$F'^a{}_{\mu\nu} = \partial/\partial x^\nu \{\exp[i\varphi_a(x)] A^a{}_\mu\} - \partial/\partial x^\mu \{\exp[i\varphi_a(x)] A^a{}_\nu\} + gf^{abc} \exp[i\varphi_b(x)] A^b{}_\mu \exp[i\varphi_c(x)] A^c{}_\nu \tag{42.7}$$

If we now assume that $\varphi_a(x)$ is small for all a then

$$\exp[i\varphi_a(x)] \simeq 1 + i\varphi_a(x)$$

to first order. Substituting above, and keeping terms to leading order yields the real part:

$$\partial/\partial x_\mu \, F^a{}_{\mu\nu} + gf^{abc} A^{b\mu} \, F^c{}_{\mu\nu} = j^a{}_\nu \tag{42.8}$$

where $F^a{}_{\mu\nu}$ is given above, and the imaginary part is:

$$\partial/\partial x_\mu \, F_i{}^a{}_{\mu\nu} + gf^{abc} A^{b\mu} \, F_i{}^c{}_{\mu\nu} = 0 \tag{42.9}$$

to leading order where

$$F_i{}^a{}_{\mu\nu} = \partial/\partial x^\nu \, \varphi_a(x) A^a{}_\mu - \partial/\partial x^\mu \, \varphi_a(x) A^a{}_\nu$$

Then we find

$$A^a_\nu \square \varphi_a(x) - A^a_\mu \, \partial/\partial x_\mu \partial/\partial x^\nu \, \varphi_a(x) - gf^{abc} A^{b\mu} \, [A^c_\mu \, \partial/\partial x^\nu \, \varphi_a(x) - A^c_\nu \, \partial/\partial x^\mu \, \varphi_a(x)] = 0$$
$$(42.10)$$

in the Lorentz gauge, with no sum over a. This equation is a form of Klein-Gordon equation having interaction terms with the gauge field. If the gauge field is weak then only the first two terms are important.

Note that only derivatives of $\varphi_a(x)$ appear above. Consequently shifts of the $\varphi_a(x)$ field by a constant still yield solutions. This feature makes $\varphi_a(x)$ a candidate to be a Higgs particle field.

Note also that complexon gauge fields cannot have such a phase change, with a subdivision into real and imaginary dynamic equations, due to the complexity of the spatial coordinates. This difference appears to be the reason why the strong interaction gauge field does not have an associated Higgs particle.

The $\varphi_a(x)$ particles can be made into Higgs particles by adding an appropriate potential:

$$V = A \, \varphi_a^2(x) + B \, \varphi_a^4(x) \qquad (42.11)$$

where A and B are constants. Approximating with its first two terms and inserting the potential term (with an A^a_ν factor) we find the Higgs-like equation:

$$A^a_\nu \square \varphi_a(x) - A^a_\mu \, \partial/\partial x_\mu \partial/\partial x^\nu \, \varphi_a(x) + A^a_\nu \partial V/\partial \varphi_a = 0 \qquad (42.12)$$

$\varphi_a(x)$ has a minimum at the minimum of the potential in the corresponding lagrangian. The second and third terms constitute the interaction. Neglecting these terms we see that we obtain the free, massless, field Klein-Gordon equation

$$\square \varphi_a(x) = 0 \qquad (42.13)$$

The pairing of Higgs particles with real-valued gauge fields is thus established.[246] The non-existence of a matching Higgs field for the strong interaction is due to the inherently complex nature of the strong interaction (complexon) gauge field in the Unified SuperStandard Theory also follows.

The derivation presented here is analogous to the derivation of Higgs fields in Complex General Relativity – also a gauge theory – in *Physics is Logic Part II*.

One of the remarkable aspects of The Unified SuperStandard Theory is its ability to directly prove qualitative properties of elementary particles: four fermion species, Parity violation, the distinction between leptons and quarks, the match of the SuperStandard Theory's (broken) symmetries with the internal symmetry Reality group, and now the existence of Higgs gauge fields in all interaction sectors except for the strong interactions. We take these successes to be indicators of the correctness of The Unified SuperStandard Theory.

42.4 General Higgs Formulation of Gauge and Fermion Particle Masses

We have seen seven of the interactions present in our Unified SuperStandard Theory. Four more interactions will be presented later. One of them is the Species gauge field $A_S{}^\mu$ generated from the Reality group of complex General Relativistic transformations. There are two more interactions associated with General Relativistic transformations. We shall discuss the Higgs Mechanism associated with the spin 1 gauge field interactions that appear in fermion covariant derivatives. They can be put in a vector form:[247]

$$\mathbf{A}_I{}^\mu = (g_1\mathbf{A}_{SU(3)}{}^\mu(x_C), g_2\mathbf{W}^\mu(x), g_3\mathbf{A}_E{}^\mu(x), g_4\mathbf{W}_D{}^\mu(x), g_5\mathbf{A}_{DE}{}^\mu(x), g_6\mathbf{A}_{DSU(3)}{}^\mu(x_C), g_7\mathbf{U}^\mu(x), g_8\mathbf{V}^\mu(x))$$
(42.14)

where each element is a vector of the gauge fields in the group of the gauge field and the respective coupling constants are labeled $g_1, g_2, \ldots, g_8$. The subscript 'D' labels

---

[246] Some of the Higgs fields so generated may not have vacuum expectation values and so may only play a role in interactions.
[247] Later we will reformulate this discussion in terms of PseudoQuantum field theory.

Dark matter interactions. 'W' labels Weak fields, 'E' labels Electromagnetic fields. 'U' labels U(4) Generation group fields., and 'V' labels one of the U(4) Layer groups fields.

$[SU(3){\otimes}SU(2){\otimes}U(1){\otimes}SU(2){\otimes}U(1){\otimes}SU(3){\otimes}U(4){\otimes}U(4)]^4$ is the interactions' symmetry[248] since, as we shall see, each of the four fermion layers due to the Layer groups discussed later, has a separate set of interactions.[249] In each layer the number of fields for the interactions is 8. 3, 1, 3, 1, 8, 16, and 16 – totaling 56 fields.

Similarly we define a 7-vector of 48 generators

$$\mathbf{T}_I = (\mathbf{T}_{SU(3)}, \boldsymbol{\tau}_{SU(2)}, \mathbf{I}_{U(1)}, \boldsymbol{\tau}_{DSU(2)}, \mathbf{I}_{DU(1)}, \mathbf{T}_{DSU(3)}, \mathbf{G}_{U(4)}, \mathbf{G}_{LU(4)}) \qquad (42.15)$$

Then the total gauge fields interaction term within a covariant derivative corresponding to the interactions, the General Relativistic U(4) Species interaction $A_S{}^\mu$, and the spinor interaction B^μ can be expressed as

$$\mathbf{A}_I{}^\mu{}_k\mathbf{T}_{Ik} + A_S{}^\mu{}_k\mathbf{G}_{Sk} + g_B B^\mu \qquad (42.16)$$

summed separately over layers k for each interaction. The remaining additional interactions are real-valued gravitational connections that we will describe later. The covariant derivative of a fermion field (neglecting General Relativistic terms) for each layer is

$$\{\partial^\mu + i\,[\mathbf{A}_{Itot}{}^\mu + A_S{}^\mu + g_B B^\mu]\}\gamma_\mu\psi = 0 \qquad (42.17)$$

where

$$\mathbf{A}_{Itot}{}^\mu = \mathbf{A}_I{}^\mu{}_k\mathbf{T}_{Ik}$$

Note the complexon nature of the SU(3) gauge field makes us use the covariant derivative

---

[248] Excepting the Species group and the Θ-Symmetry group.
[249] Otherwise the various layers would have interactions between them which would have appeared in experiments. Then the upper three layers would not be Dark.

$$\{\partial /\partial x_{C\mu} + i\, [\mathbf{A}_{Itot}{}^{\mu} + A_S{}^{\mu} + g_B B^{\mu}]\}\gamma^{\mu}\psi = 0 \qquad (42.18)$$

for the SU(3) quark dynamic equations where the other gauge fields are functions of x_r = Re x_C.

We now consider the combined effects of the interactions, $\mathbf{A}_I{}^{\mu}$, on generating gauge boson masses (symmetry breaking) and fermion masses.[250] We begin by defining a composite Higgs field for all interactions:

$$\eta = \prod_{k=1}^{8} \eta_{kTSLg}$$

where k labels the group, T labels the type of matter, S labels the species, L labels the layer and g labels the generation. We now consider

$$D^{\mu}\eta = \{\partial^{\mu} + i\,[\mathbf{A}_{Itot}{}^{\mu} + g_B B^{\mu}]\}\eta \qquad (42.19)$$

Letting η be a product of real fields $\rho..$, whose elements are composed of zeroes and non-zero real fields

$$\eta = \prod_{k=1}^{8} \rho_{kTSLg} \qquad (42.20)$$

Then we find that

$$(D_{\mu}\eta)^{\dagger} D^{\mu}\eta = \sum_{kTSLg} \partial_{\mu}\rho_{kTSLg}\, \partial^{\mu}\rho_{kTSLg} + \sum_{kTSL} g_k{}^2 \beta_{kTSL} U_{kTSL}{}^2 \qquad (42.21)$$

where β_{kTSL} is a sum of terms, each of which is quadratic in the vacuum expectation value of a Higgs field. The second term above yields the masses of the gauge fields of non-zero mass.

---

[250] We developed the Higgs Mechanism in detail in Blaha (2017b) and earlier books for fermions and bosons for all seven interactions above plus the Reality group $A_S{}^{\mu}$ of Complex General Relativistic transformations, the Generation group, and the Layer groups.

The lagrangian terms that generate fermion masses have the form

$$\mathscr{L}_{FermionMasses} = \sum_{kTSLgh} \bar{\psi}_{L\,kTSLg}\,\rho_{0kTSL}m_{kTSLgh}\psi_{R\,kTSLh} \qquad (42.22)$$

where the sums over g and h are over the generations of a specific T, S, and L of a group labeled k, and ρ_{0kTSL} is the vacuum expectation value of a Higgs particle. The initial 'L' and 'R' subscripts represent Left and Right.

In addition to the mass contributions of the interactions the mass contributions of the Species group, denoted $m_{G...}$, must eventually also be taken into account (later).

We note that the total fermion mass matrix can be diagonalized using a matrix A_{TSL}:

$$m_{TSLphys} = A_{L\,TSL}\sum_{k} \rho_{0kTSL}m_{kTSL}A_{R\,TSL}{}^{-1} \qquad (42.23)$$

where $m_{TSLphys}$ is the diagonal mass matrix for the generations specified by T, S, and L.

Thus the lagrangian fermion mass terms for physical fermions become

$$\mathscr{L}_{FermionMasses} = \sum_{TSL} \bar{\psi}_{LphysTSL}\,m_{TSLphys}\psi_{RphysTSL} + c.c. \qquad (42.24)$$

with diagonal mass matrices $m_{TSLphys}$.

42.5 The Mixing Pattern in the Fermion Periodic Table

The preceding discussions describe the pattern of mixing resulting from the ElectroWeak, Generation group, and Layer group. Fig. 42.1 pictorially presents an example of the mixing pattern within the Periodic Table of Fermions.

42.6 The Full Unified SuperStandard Theory Fermion Mass Matrices

The above equations lead to the total mass matrices *for the four layers* listed below. The masses for each particle in the various layers are different although we use the same symbol for each type of mass. We include the mass contributions, m_{Gi}, from the Species group with gauge fields $A_S{}^\mu$.

   The below list is for the 'lowest' generation of the lowest layer. The other generations and the other three layers have a similar pattern. The masses listed in the list are symbolic and are not of the same value for each particle type. We denote them generally as: m_{Wi} for the Weak group contribution, m_{Li} for the Layer group contribution, m_{Geni} for the Generation group contribution, m_{Gi} for the (Gravitational) Species group contribution with gauge fields $\mathbf{A_S}{}^\mu$.[251]

Charged Lepton Species Total Mass Matrix
$$m_{etot} = m_{We} + m_{Le} + m_{Ge}$$

Neutral Lepton Species Mass Matrix
$$m_{\upsilon tot} = m_{W\upsilon} + m_{Lv} + m_{G\upsilon}$$

Up-Type Quark Species Mass Matrix (for each color)
$$m_{utot} = m_{Wu} + m_{Lu} + m_{Gen-u} + m_{Gu}$$

Down-Type Quark Species Mass Matrix (for each color)
$$m_{dtot} = m_{Wd} + m_{Ld} + m_{Gen-d} + m_{Gd}$$

Dark Charged Lepton Species Total Mass Matrix
$$m_{Detot} = m_{DWe} + m_{DLe} + m_{Ge}$$

Dark Neutral Lepton Species Mass Matrix
$$m_{D\upsilon tot} = m_{DW\upsilon} + m_{DL\upsilon} + m_{G\upsilon}$$

Dark Up-Type Quark Species Mass Matrix
$$m_{Dutot} = m_{DWu} + m_{DLu} + m_{DGen-u} + m_{Gu}$$

---

[251] The Species group is discussed in a subsequent chapter.

Dark Down-Type Quark Species Mass Matrix

$$m_{Ddtot} = m_{DWd} + m_{DLd} + m_{DGen-d} + m_{Gd}$$

The (gravitational) Species group contribution to each fermion mass, m_{Gi} for each fermion type i, sets the scale for all fermion masses (and secondarily of massive gauge bosons' masses) yielding the "principle" of Newton, Einstein and others that **inertial mass equals gravitational mass**.

The generation group contributions, in the spontaneous breakdown that we described, appear only in quark and Dark quark mass matrices possibly providing a reason why quark masses are so much larger than lepton masses. See Blaha (2017b).

42.7 Higgs Mechanism for Coupling Constants

Particle masses are attributed to the operation of the Higgs Mechanism. Coupling constants are also constants that appear in the Unified SuperStandard Theory. In this section[252] we define a Higgs Mechanism that yields the values of coupling constants as vacuum expectation values of Higgs particles.

---

[252] This chapter is largely extracted from Blaha (2015d).

The Fermion Periodic Table
NORMAL FERMIONS | DARK FERMIONS

Layer 4
Generation mixing in the
generations of each species for each
species separately for each layer.

Four layer Mixing
for each generation
of each species

...

Layer 3

Layer 2

Layer 1 – Our Layer

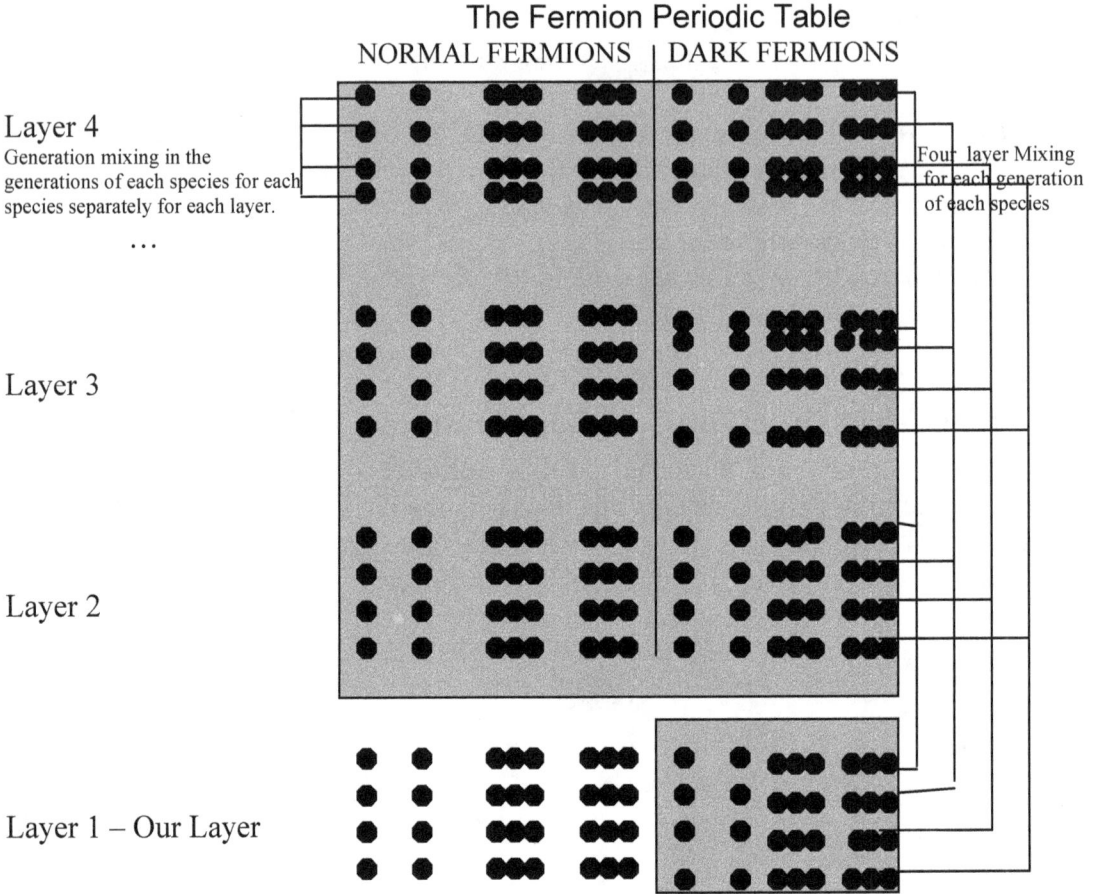

Figure 42.1. Fermion particle spectrum and partial example of pattern of mass mixing of the Generation group and of the Layer group. Dark sector parts of the periodic table are shaded gray. Unshaded parts are the known fermions with an additional, as yet not found, 4th generation shown. The lines on the left side (only shown for one layer) display the Generation mixing within each layer's species. The Generation mixing applies within each layer using a separate Generation group for each layer. The lines on

the right side show Layer group mixing with the mixing amongst all four layers for each of the four generations individually. There are four Layer groups. There are 256 fundamental fermions.

42.7.1 The Interaction Coupling Constants and their PseudoQuantum Field Vacuum Expectation Values

Unified SuperStandard Theory coupling constants are:[253]

- The Strong interaction coupling constant field g_S.
- The Weak SU(2) coupling constant g_W.
- The Electromagnetic U(1) coupling constant g_E.
- The Dark Weak SU(2) coupling constant g_{DW}.
- The Dark Electromagnetic U(1) coupling constant g_{DE}.
- The Dark Strong SU(3) coupling constant g_{DS}.
- The U(4) Generation group coupling constant g_G.
- The U(4) Layer group coupling constant g_V.
- The U(4) Species group coupling constant g_S.
- The complex gravitational coupling constant $g_{GR} = \kappa^{-1} = (4\pi G)^{-\frac{1}{2}}$.

Based on the discussions of Blaha (2015d) we can define Higgs vacuum expectation values for these coupling constants using a mass factor to obtain the correct coupling constant dimensions. We use the PseudoQuantum formalism of section 10.3.3 where we set the constants as follows:

- The Strong interaction coupling constant field $\Phi_1 = m_1 g_S$.
- The Weak SU(2) coupling constant $\Phi_2 = m_2 g_W$.

---

[253] The following groups and their coupling constants are duplicated four-fold for the four layers. The Strong interaction coupling constant field g_S, The Weak SU(2) coupling constant g_W, The Electromagnetic U(1) coupling constant g_E, The Dark Weak SU(2) coupling constant g_{DW}, The Dark Electromagnetic U(1) coupling constant g_{DE}, The U(4) Generation group coupling constant g_G, and The U(4) Layer group coupling constant g_V. We will only consider the known first layer here The following group constants appear once – the same for all four layers – the U(4) Species group, and Gravitational coupling constant. The interactions of these groups will be discussed later.

- The Electromagnetic U(1) coupling constant $\Phi_3 = m_3 g_E$.
- The Dark Weak SU(2) coupling constant $\Phi_4 = m_4 g_{DW}$.
- The Dark Electromagnetic U(1) coupling constant $\Phi_5 = m_5 g_{DE}$.
- The U(4) Generation group coupling constant $\Phi_6 = m_6 g_G$.
- The U(4) Layer group coupling constant $\Phi_7 = m_7 g_V$.
- The U(4) Species group coupling constant $\Phi_8 = m_8 g_S$.
- The complex gravitational coupling constant $\Phi_{10} = m_{10} g_{GR} = \kappa^{-1} = (4\pi G)^{-\frac{1}{2}}$.

The ten masses, m_1, m_2, ... , m_{10} may be equal or they may have different values. It is also possible they all may be equal to κ^{-1}, which would yield

- The Strong interaction coupling constant field $\Phi_1 = \kappa^{-1} g_S$.
- The Weak SU(2) coupling constant $\Phi_2 = \kappa^{-1} g_W$.
- The Electromagnetic U(1) coupling constant $\Phi_3 = \kappa^{-1} g_E$.
- The Dark Weak SU(2) coupling constant $\Phi_4 = \kappa^{-1} g_{DW}$.
- The Dark Electromagnetic U(1) coupling constant $\Phi_5 = \kappa^{-1} g_{DE}$.
- The U(4) Generation group coupling constant $\Phi_6 = \kappa^{-1} g_G$.
- The U(4) Layer group coupling constant $\Phi_7 = \kappa^{-1} g_V$.
- The U(4) Species group coupling constant $\Phi_8 = \kappa^{-1} g_S$.
- The complex gravitational coupling constant $\Phi_{10} = \kappa^{-1} g_{GR} = \kappa^{-1} = (4\pi G)^{-\frac{1}{2}}$.

Then scaling the above vacuum expectation values by κ^{-1} would give:[254]

- The strong interaction coupling constant[255] vacuum expectation value $\Phi_1' = g_S = 1.22$
- The Weak SU(2) coupling constant vacuum expectation value $\Phi_2' = g_W = 0.619$.
- The Electromagnetic U(1) coupling constant vacuum expectation value $\Phi_3' = e = g_E = 0.303$.
- The Dark Weak SU(2) coupling constant vacuum expectation value $\Phi_4' = \Phi_2'$. (?)
- The Dark Electromagnetic U(1) coupling constant vacuum expectation value $\Phi_5' = = \Phi_3'$. (?)

---

[254] All coupling constant values are based on data extracted from K. A. Olive et al (Particle Data Group), Chinese Physics **C38**, 090001 (2014).
[255] Based on the running coupling constant value $\alpha_s (M_Z^2) = 0.1193 \pm 0.0016$.

- The U(4) Generation group coupling constant $\Phi_6' = g_G$. (?)
- The U(4) Layer group coupling constant $\Phi_7' = g_V$. (?)
- The U(4) Species group coupling constant $\Phi_8' = g_S$. (?)
- The complex gravitational coupling constant $\Phi_{10}' = 1$.

The known *scaled* vacuum expectation values,[256] which are in fact the scaled Higgsian coupling constants, have a comparable range of values[257] as opposed to the range of values for the unscaled constants which range from the ultra-small gravitational vacuum expectation value to values, perhaps, within a few orders of magnitude of unity.

Given the range of known values above, it appears reasonable to conjecture that the unknown values would also be of the order of unity.

The known coupling constant values above are of comparable value, which suggests that our Unified SuperStandard Theory, at current energies, may be close to the GUT level at which properly scaled coupling constants are equal.

42.7.2 Unified SuperStandard Theory Lagrangian Coupling Constants

We begin with the Unified SuperStandard Theory lagrangian density $\mathcal{L}_{TE}$ with coupling constants explicitly displayed[258]

$$\mathcal{L}_{TE} = \mathcal{L}_{TE}(g_S, g_W, g_E, g_{DW}, g_{DE}, g_G, g_{DS}, g_V, g_S, g_{GR}) \qquad (42.25)$$

and fields and space-time coordinates not displayed.

In terms of vacuum expectation values as discussed earlier we see we can write[259]

---

[256] The closeness of all the values to one is suggestive: The value $\alpha = 1$ (or $e = (4\pi)^{\frac{1}{2}} = 3.54$) was the value found in our calculation in the Johnson, Baker, Willey model of QED. Perhaps a larger calculation along the lines of our paper in massless ElectroWeak theory might yield scaled coupling constant values near unity.

[257] The weakness of the Weak interactions is primarily due to the large masses of the Z and W vector bosons – not the values of their coupling constants g and g'.

[258] The 11th coupling constant g_Θ and the Θ symmetry group are discussed later.

[259] The "vacuum" state |0> above also has factors for the vacuum expectation values used for fields that give masses to fermions and vector bosons as described in Blaha (2015b).

$$\mathscr{L}_{TE} = \mathscr{L}_{TE}(\Phi_1/m_1, \Phi_2/m_2, \dots, \Phi_{11}/m_{11}) \qquad (42.26)$$

where

$$| \Phi_1, \Phi_2, \dots, \Phi_{11}; \Pi_1, \Pi_2, \dots, \Pi_{11}> = C \prod_{i=1}^{11} \{\exp[[(2\pi)^3 m_i/2]^{\frac{1}{2}} \Phi_i[a_{i2}^\dagger(\mathbf{0}, m_i) + a_{i2}(\mathbf{0}, m_i)]]\} |0>$$

Assuming all $m_i = \kappa^{-1}$ we obtain

$$| \Phi_1, \Phi_2, \dots, \Phi_{11}; \Pi_1, \Pi_2, \dots, \Pi_{11}> = C \prod_{i=1}^{11} \{\exp[[(2\pi/\kappa)^3/2]^{\frac{1}{2}} \Phi_i'[a_{i2}^\dagger(\mathbf{0}, \kappa^{-1}) + a_{i2}(\mathbf{0}, \kappa^{-1})]]\} |0>$$

$$(42.27)$$

Then we can write

$$\mathscr{L}_{TE} = \mathscr{L}_{TE}(\Phi_1', \Phi_2', \dots, \Phi_{11}') \qquad (42.28)$$

Setting $m_i = g_{CG} = \kappa^{-1}$ = the Planck mass, simplifies the above expressions and *supports the belief that we are close to the unification of all interactions.* However having particles of such large mass makes them undetectable by accelerators. It also seems too large from the viewpoint of physical intuition. Consequently the above may be the correct expressions with masses perhaps in the TeV range.

We finally note

$$\phi_{1i}| \Phi_1, \Phi_2, \dots, \Phi_{11}; \Pi_1, \Pi_2, \dots, \Pi_{11}> = \Phi_{1i}| \Phi_1, \Phi_2, \dots, \Phi_{11}; \Pi_1, \Pi_2, \dots, \Pi_{11}>$$

$$(42.29)$$

42.7.3 Big Bang Vacuum

At the origin of the universe – the Big Bang – there was a vacuum state in principle. In our earlier books[260] we showed that the universe existed in an ultra-small, but finite, region for an infinitesimal time before it began an explosive inflationary

---

[260] Blaha (2015a) and Blaha (2004).

expansion to become the familiar universe. In this time period there were no infinities – a finite temperature and so on.

Thus it is reasonable to assume one of two possibilities for the above ten coupling constants: 1) they have remained unchanged since the beginning, or 2) they have changed with time.

In this section we note, that if our scaling with the Planck mass κ^{-1} in preceding discussions is correct, then it is reasonable to assume that the vacuum state in the beginning is that defined above with |0> including factors setting fermion and vector boson masses as described in Blaha (2015b).

42.7.4 Evolving/Space-Time Dependent Coupling Constants

It is possible that the Unified SuperStandard Theory coupling constants evolve with time and may also be spatially varying – different constants in different parts of the universe. Presently there is no decisive evidence for either possibility. In this section we will describe the mechanism to support either or both possibilities.

Consider a classical field (time and spatially varying):

$$\Phi(\mathbf{x}, t) = \int d^3k \, [\alpha(k)f_k(x) + \alpha^*(k)f_k^*(x)] \qquad (42.30)$$

If we define the coherent vacuum state

$$|\alpha> = C \exp\left\{\int d^3k \, [\alpha(k)a_2^\dagger(k) + \alpha^*(k)a_2(k)]\right\}|0> \qquad (42.31)$$

then

$$\varphi_1(x)|\Phi, \Pi> = \Phi(x)|\Phi, \Pi>$$

$$\pi_1(x)|\Phi, \Pi> = \Pi(x)|\Phi, \Pi>$$

where

$$\varphi_i(\mathbf{x}, t) = \int d^3k \, [a_i(k)f_k(x) + a_i^\dagger(k)f_k^*(x)] \qquad (42.32)$$

for i = 1, 2 with

$$f_k(x) = e^{-ik \cdot x} / (2\omega_k (2\pi)^3)^{\frac{1}{2}}$$

where $\omega_k = |\mathbf{k}|$.

The coherent vacuum state $|\alpha\rangle$ has a time and spatially varying vacuum expectation value (classical) field. The above equations can be generalized to the case of the eleven coupling constant vacuum expectation values:[261]

$$|\Phi_1, \Phi_2, \dots, \Phi_{11}; \Pi_1, \Pi_2, \dots, \Pi_{11}\rangle = C \prod_{i=1}^{11} \exp\left\{\int d^3k \left[\alpha_i(k)a_{2i}^\dagger(k) + \alpha_i^*(k)a_{2i}(k)\right]\right\}|0\rangle$$

(42.33)

Then all eleven coupling constant vacuum expectation values are space-time dependent:

$$\varphi_{1i}(x) | \Phi_1, \Phi_2, \dots, \Phi_{11}; \Pi_1, \Pi_2, \dots, \Pi_{11}\rangle = \Phi_i(x) | \Phi_1, \Phi_2, \dots, \Phi_{11}; \Pi_1, \Pi_2, \dots, \Pi_{11}\rangle$$

(42.34)

and the Unified SuperStandard Theory lagrangian becomes

$$\mathcal{L}_{TE} = \mathcal{L}_{TE}(\Phi_1(x), \Phi_2(x), \dots, \Phi_{11}(x))$$

(42.35)

for matrix elements between the vacuum defined by $|\alpha\rangle$ and its conjugate,

Thus our formalism can accommodate space-time varying coupling constants should they be found in the Cosmos.

42.7.5 A Unified SuperStandard Theory Lagrangian Without Any Constants

The preceding sections put coupling constants within the same framework as particle masses completing the process of eliminating constants from The Unified SuperStandard Theory lagrangian. Instead the vacuum contains the values of all coupling constants and particle masses. In one sense this new formulation is a tradeoff. The values of all constants are shifted to the vacuum. However the shift has some advantages technically. One advantage is the ability to have space-time dependent coupling constants as shown above. It would also be straightforward to make masses,

---

[261] The "vacuum" state $|0\rangle$ also has factors for the vacuum expectation values used for fields that give masses to fermions and vector bosons as described in Blaha (2015b).

and mixing angles, space-time dependent. The possible space-time dependence of coupling constants and particle masses has been an active area of experimental interest for many years although cosmological data seems to indicate these quantities have not changed significantly since the universe began.

The question of changes in lagrangian coupling constant physical values is of great philosophical importance since it appears that the existence of life, as we know it, depends sensitively on their values. This dependence has been embodied in the Anthropomorphic Principle and studied by a number of physicists and philosophers.

Since our formulation allows space-time varying physical constants the question of the Anthropomorphic Principle attains new importance. As we saw in the case of the theory of Black Holes, which was a theory without evidence for over forty years before Black Holes were discovered, Nature seems to provide phenomena that have been shown to be theoretically possible. Many other cases of this sort have also occurred – the most recent example at the time of this writing is Weyl fermions.

Lastly, our approach opens the possibility of a study of all the many constants in The Unified SuperStandard Theory lagrangian *on the same footing* rather than in the piecemeal fashion used up to the present. It replaces the scattered hodge-podge of constants in the lagrangian with a centralized location for all constants in the vacuum state permitting a direct study of their interconnection. *The study of the vacuum now becomes of central importance.*

42.7.6 The Form is Determined But Not the Constants

The derivation of The Unified SuperStandard Theory here, and in the Blaha (2015) books and earlier work, was based on Asynchronous Logic (to support parallel physical processes spread in space and time); on complex space-time coordinates, the Complex Lorentz group and complex General Coordinate transformations; the Reality group to map complex coordinates to the real-valued coordinates that we observe; the Generations group that yields the four fermion generations and particle number interactions such as the baryon number and lepton number interactions, and the Species group for complex general coordinate transformations.

This firm basis in fundamental considerations enables us to forge a path to The Unified SuperStandard Theory, which included the known features of The Standard

Model. We thus were able to avoid the many possible variants and extensions of The Standard Model that have been considered in the Physics literature over the past thirty years.

Two remarkable features of The Unified SuperStandard Theory derivation were:

1. A precise fixing of the form of the Unified SuperStandard Theory.
2. The absence of any constraints on the values of its coupling constants or masses.

Particle masses were fixed by either the original Higgs Mechanism or by our new mechanism that was based on an extension of Quantum Field Theory to include classical fields such as the vacuum expectation values that cropped up in the original Higgs Mechanism and were handled "by hand." (See Blaha (2015c).)

Thus, up to this point, we have a Unified SuperStandard Theory (known) except for a basis for the values of the coupling constants that appear in the theory. The coupling constants have a wide range of values. A fundamental basis for their values has been wanting.

Status of the Derivation Now

Up to this point in our derivation of the Unified SuperStandard Theory, starting from chapter 26, we have relied directly on the Complex Lorentz group for the derivation of the form of the fermion spectrum with four fermion species (three up-type and down-type quark subspecies) in one generation,[262] and for the derivation of the form of the Internal Symmetry Reality group as the analogue of the subgroups of the Coordinate Reality group of the Complex Lorentz group. The Internal Symmetry Reality group is identical to that of the Standard Model with the addition of a Dark SU(2)⊗U(1) symmetry and Dark fermions (and Higgs fields to generate particle masses.) We also developed the formalisms for Two-Tier Field Theory and PseudoQuantum Field Theory to remedy shortcomings in the conventional formulation of Quantum Field Theory.

Now we turn to derive two additional symmetry groups based on the form of the one generation Standard Model lagrangian with SU(3)⊗SU(2)⊗U(1)⊗SU(2)⊗U(1)⊗ SU(3) symmetry group. An examination of this lagrangian with its dynamic terms and interaction terms corresponding to this symmetry group reveals that there is a set of conserved particle numbers. (The most prominent of these conserved particle numbers is Baryon number. Baryon number conservation violation has been fruitlessly sought in many high energy experiments. To high accuracy, Baryon Number is conserved.) Since these particle numbers are conserved in the one generation Standard Model we find that they imply additional group symmetries that we call the U(4) Generation group, and the Layer groups consisting of a set of four U(4) groups.

Thus we use the Complex Lorentz group to derive the interactions and species of fermions: neutrino, charged lepton, up-type quarks and down-type quarks (with four Dark species equivalents) in the one generation case. We then use the structure of the

---

[262] Generations and layers will be described next.

corresponding lagrangian to derive the additional Generation and Layers groups[263] with their additional interactions and levels of fermions. Since these groups are all U(4) groups we find four generations of fermions in four layers. (The fourth fermion generation remains to be found. We expect the fourth generation fermions – excepting neutrinos – to be very massive.)

---

[263] These group symmetries are badly broken.

43. Interactions Based on Particle Numbers

The particle interactions following directly from Complex General Relativity are

$$SU(2) \otimes U(1) \otimes SU(3) \otimes SU(2) \otimes U(1) \otimes SU(3) \otimes U(4)$$

where the U(4) factor is for the Species Group.

They have a SU(14) covering group that contains this direct product of groups. Unlike other attempts to develop a formulation of the Standard Model (or generalizations) the Extended Standard Model is directly based on a foundation consisting of Complex General Relativity and Quantum Field Theory.

To those who might prefer to base a theory on real General Relativity we note that proofs in Quantum Field Theory *require* the Complex Lorentz Group.[264] Thus the Complex Lorentz group is unavoidable for a properly (and rigorously) formulated Quantum Field Theory. Since the formulation of the Complex Lorentz Group in flat space-time can only be as the limit of Complex General Relativity, the choice of a foundation of Complex General Relativity is required.

Earlier we also specified a foundation of Quantum Field Theory. Our motivation was that it provided the simplest possible formulation of particles capable of engaging in creation and annihilation.

Since particles are countable and thus have discrete particle numbers Quantum Field Theory brings particle numbers, and particle number laws such as particle conservation laws, into consideration.

Appendix F shows that Complex Lorentz boosts generate four types of fermion particles that we call *particle species*. We map these four species to charged leptons (such as electrons), neutral leptons (such as neutrinos), up-type quarks (such as the u quark), and down-type quarks (such as the d quark).

---

[264] Streater (2000).

43.1 Basis of the Generation Group

We define a particle number operator for quark-type particles that we call Baryon Number, and a particle number operator for lepton-type particles that we call Lepton Number.

By analogy, we assume that there are four species of Dark matter: charged Dark leptons, neutral Dark leptons, Dark up-type quarks, and Dark down-type quarks.[265] Thus we are led to Dark particle numbers: Dark Baryon Number, and Dark Lepton Number.

These four fermion particle number operators are assumed to be conserved in particle interactions modulo any experimentally found violations of their conservation.

Regardless of whether their conservation is violated or not we may use these four numbers, denoted B, L, B_D, and L_D respectively, as "diagonal" operators within a U(4) group. We will call this group the Generation Group (chapter 5) and relate it to the generations of the eight (normal and Dark) species of fermions. On this basis we assume there are four generations of each species with one generation (of large masses) as yet not found. (The gauge vector bosons of the Generation Group also have large masses.) If the conservation of the fermion particle numbers is broken then we will view it as a consequence of Generation Group symmetry breaking.

43.2 Basis of the Layer Group

The set of particle number operators can be further refined if we take account of the fourfold fermion generations. In further refining the set of particle number operators we temporarily neglect interactions that would violate conservation laws for the set.

We therefore subdivide the above particle number set into four particle numbers per generation. For the ith generation we define

L_{iB} – The Baryon particle number for the ith generation
L_{iDB} – The Dark Baryon particle number for the ith generation
L_{iL} – The Lepton particle number for the ith generation

---

[265] Later we specify Dark quarks to be Color singlets to prevent interactions between Dark and normal quarks. Such interactions have not been found experimentally. Normal quarks will be Color triplets.

L_{iDL} – The Dark Lepton particle number for the ith generation

for each generation i = 1, 2, 3, 4. Individual fermions have positive $L_{ia} = +1$ values and anti-fermions have negative $L_{ia} = -1$ values for species a = 1, 2, 3, 4 (with the three color subspecies of quarks treated as part of one species.)

At this point we have four particle number operators for each generation. We wish to define a group framework for each set of particle numbers. The simplest way is to assume that each generation consists of four layers with particles in each generation being in a U(4) fundamental representation.[266] Then each generation has a U(4) group with the generation's four number operators as its diagonal operators. We call this group the Layer Group of the generation. With four generations we obtain four U(4) Layer groups.

The consequence of this expansion of particle numbers and groups is that the set of fermions increases fourfold. We now have four layers, with each having four generations, Experimentally we know of three generations of fermions—the lowest generations of the lowest level. The remaining generation and three levels of fermions is of much higher mass and yet to be found.

We can denote the 16 number operators of the $[U(4)]^4$ total Layer Group by

L_{ijB} – The layer j Baryon particle number for the ith generation
L_{ijDB} – The layer j Dark Baryon particle number for the ith generation
L_{ijL} – The layer j Lepton particle number for the ith generation
L_{ijDL} – The layer j Dark Lepton particle number for the ith generation

See chapter 45 for a detailed discussion of the Layer Group. We note in passing that the conservation of these number operators is badly broken. Yet the underlying group structure remains.

The definition of four layers of fermions raises the question of their vector gauge field interactions. Since we see no evidence of the higher layers in particle interactions it seems reasonable to assume that each layer has its own Extended

---

[266] See Fig. 45.2 for a depiction of the "splitting" of fermions: first into generations, then into layers.

Standard Model set of interactions and its own Generation Group. Consequently the total symmetry group (although broken) is[267]

$$[SU(2)\otimes U(1)\otimes SU(3)\otimes SU(2)\otimes U(1))\otimes SU(3)\otimes U(4)]^4\otimes U(4)\otimes [U(4)]^4$$
$$= [SU(2)\otimes U(1)\otimes SU(3)\otimes SU(2)\otimes U(1)\otimes SU(3)]^4\otimes U(4)]^9 \qquad (43.1)$$

---

[267] The Species group is not multiplied fourfold since all fermions in all layers experience the same gravitation and Species group.

44. Generation Group

We now turn to the derivation of the Generation Group and, in the following chapter, the Layer Groups, which are based on conserved (and almost conserved) particle numbers such as Baryon Number Conservation.

In this chapter we consider the extension of the one generation SuperStandard theory to four generations based on a new U(4) symmetry group that we call the Generation Group.

It is based on four conservation laws for Baryon, Lepton, Dark Baryon, and Dark Lepton Numbers. The conservation laws are manifest in the fermion terms of the one generation version of The Standard Model which all have the form:

$$\mathcal{L} \sim \overline{\psi}_\alpha \gamma^\mu D_{\mu\alpha\beta} \, \psi_\beta$$

where the covariant derivative terms preserve the four conservation laws. Symmetry breaking is brought in at a later point through Higgs fields terms. There is an evident $SU(3) \otimes SU(2) \otimes U(1) \otimes SU(2) \otimes U(1)$ symmetry in the sum of these lagrangian terms. The next relevant question is whether the particle number conservation laws have associated gauge fields that provide interactions. This question can be partially answered in our current state of experimental knowledge.

In Blaha (2017b), and earlier books, we showed that the existence of a long range baryonic force[268] supports baryon number conservation in a manner similar to electric charge conservation due to the electromagnetic force. Lepton number conservation suggests a very weak long range force as well. By analogy we postulate two similar Dark particle number conservation laws and forces.

---

[268] See Blaha (2017b) and earlier books for a discussion of evidence for a baryonic force and conservation law. We found that gravity experiments suggested a possible baryonic potential with the (order of magnitude) coupling constant $\alpha_B = \beta^2/4\pi \simeq .118 \, Gm_H^2$ where m_H is the mass of the hydrogen atom. Section 44.1 provides details.

Having four conserved (or almost conserved) particle numbers with their attendant forces (interactions) leads naturally to a U(4) symmetry with four 'diagonal' generators. We call the U(4) group the Generation Group.

If Baryon Number and Lepton Number are both conserved quantities then any linear combination of them is also conserved. Therefore

$$B' = aB + bL$$

is also conserved.

If we consider the Dark Matter sector of the Unified SuperStandard Theory it is reasonable to assume that *Dark Baryon Number B_D and Dark Lepton Number L_D are conserved* also (although there is no experimental evidence available as yet to confirm (or deny) these assumptions.)

Thus we have four conserved particle Numbers. Linear combinations of these numbers are also conserved:

$$B' = aB + bL + cB_D + dL_D$$
$$L' = eB + fL + gB_D + hL_D$$
$$B_D' = iB + jL + kB_D + lL_D$$
$$L_D' = mB + nL + oB_D + pL_D$$

The set of 4×4 matrices form an U(4) group if we wish to perform these transformations within lagrangians of the type of the Unified SuperStandard Theory. The choice of U(4) rather than SU(4) is required since there are four independent particle numbers. U(4) has four diagonal matrices in its algebra while SU(4) only has three diagonal matrices. U(4) preserves the independence of the four independent particle Numbers. It also allows complex rotations.

The new U(4) Generation Group symmetry leads immediately to four fermion generations. We add an index to each fermion field ranging from 1 through four, add U(4) gauge fields to the covariant derivative,[269] and thus Yang-Mills local gauge field terms to the lagrangian.

---

[269] See subsection 43.2.8.4.

Then we perform the following tasks:[270]

- Define the Two-Tier Lepton and Quark Sectors

- Introduce Symmetry Breaking via the Higgs Mechanism for Fermions and U(4) Gauge Fields

Thus we now have four fermion generations, a broken Generation Group symmetry, and masses for fermions and Generation Group gauge fields.

44.1 Baryon Number Conservation and a Possible Baryonic Force

We have considered baryon number conservation and a possible baryonic force in Blaha (2014a) and (2014b). The primary forces involved in the interactions and collisions of baryons include the forces of The Standard Model, the force of gravity, and possibly a fifth force which we take to be the baryonic force, a much discussed force that depends on the baryon numbers of objects experiencing it. The Gravitation and baryonic forces (neglecting other SuperStandard Theory interactions) between two clumps of baryonic matter containing baryons and other particles: clump1 being of mass m_1 and baryon number n_1, and clump2 being of mass m_2 and baryon number n_2 is

$$F = -Gm_1m_2/r^2 + (\beta^2/4\pi)n_1n_2/r^2 \qquad (44.1)$$

where G is the gravitational constant, β is the baryonic coupling constant and r is the distance between the 'widely' separated clumps. Experimentally a baryonic force between baryons has not been detected with any degree of certainty. Sakurai (1964) discusses early efforts to determine the force of gravity in detail. Eőtvős experiments on the ratio of the observed gravitational mass to the inertial mass showed that that the gravity force is constant to within one part in 100,000,000 as far back as 1922

---

[270] These tasks are described in detail in chapters 12 and 13 in Blaha (2017b) as well as earlier books.

indicating the baryonic force, if it exists, as we believe it does, is extremely weak compared to the gravitational force. Eőtvős et al[271] found

$$(\beta^2/4\pi)/(Gm_p^{\,2}) < 10^{-5}$$

where m_p is the proton mass.

Since then, the experiment has been redone with improved accuracy by Dicke and collaborators.[272] They have improved the accuracy to one part in 100 billion. A further analysis showed a very small discrepancy that suggested the ratio, while small, was non-zero, implying the equivalence principle might not be exact and that the discrepancy changed with the material used in the experiment – just what one might expect if a very small baryonic force was present – often called the "fifth force." At present the existence and amount of the discrepancy is unclear. *Nevertheless, we will assume a fifth force – Baryonic force. We will also assume there are three additional forces corresponding to the three other conserved particle numbers. These forces are not relevant to these considerations.*

The conservation of baryon number has been repeatedly investigated by experimenters and found to be true to extremely high accuracy. For decades theorists have suggested that a baryon conservation law[273] follows from the existence of a gauge field in a manner much like electric charge conservation follows from the properties of the electromagnetic gauge field.

44.1.1 Estimate of the Baryonic Coupling Constant

The baryonic force, and coupling constant, if it exists, is known to be very small in comparison to gravity and the other known forces. However, measurements of the gravitational constant G are significantly different.[274,275] The reason(s) for these

---

[271] Eőtvős, R. V., Pekár, D., Fekete, E., Ann. d. Physik **68**, 11 (1922).

[272] P. G. Roll, R. Krotkov, R. H. Dicke, Annals of Physics, 26, 442, 1964.

[273] See Gell-Mann, M. and Levy, M. *Nuovo Cimento* 16, 705 (1960) for a proof and Sakurai (1964) for a discussion of the relation of the baryonic gauge field to gravity experimentally.

[274] T. Quinn et al, Phys. Rev. Lett. **111**, 101102 (2013).

[275] P. J. Mohr, B.N. Taylor, and D. B. Newell, Rev. Mod. Phys. 84, 1527 (2012).

discrepancies is not known. We will assume that both the 2010 and 2013 measurements of G are experimentally correct but disagree because of the baryonic force term in eq. 44.1 that would create a difference in effective G values if the experiments used different masses and thus baryon numbers. Quinn et al found a value for the gravitational constant of $G_1 = 6.67545 \times 10^{-11}$ $m^3 kg^{-1} s^{-2}$. The combined 2010 CODATA value for the gravitational constant was $G_2 = 6.67384 \times 10^{-11}$ $m^3 kg^{-1} s^{-2}$. Both values are subject to estimated uncertainties.

Suppose these values are correct and due to a difference in the chemical composition (metals) of the test masses used in the experiment. Quinn et all use 1.2 kg test masses composed of Cu-0.7% Te free machining alloy. The CODATA value being a composite of many experiments does not have an effective equivalent test mass value or composition specified.[276] Suppose the test mass value is $N_1{}^2 m_1{}^2 + N_{1e}{}^2 m_e{}^2$ for the G_1 result giving

$$-(N_1{}^2 m_1{}^2 + N_{1e}{}^2 m_e{}^2)G_1 = [-G(m_1{}^2 N_1{}^2 + N_{1e}{}^2 m_e{}^2) + (\beta^2/4\pi)N_1{}^2] \qquad (44.2)$$

where G is the *real* value of the gravitational constant. The total test mass is $(m_1{}^2 N_1{}^2 + N_{1e}{}^2 m_e{}^2)$ with N_1 baryons of average mass m in each test mass and N_{1e} leptons of average mass m_e.

Suppose further the test mass value is $N_2{}^2 m_2{}^2 + N_{2e}{}^2 m_e{}^2$ for the G_2 result giving

$$-(N_2{}^2 m_2{}^2 + N_{2e}{}^2 m_e{}^2)G_2 = [-G(m_2{}^2 N_2{}^2 + N_{2e}{}^2 m_e{}^2) + (\beta^2/4\pi)N_2{}^2] \qquad (44.3)$$

where G is again the *real* value of the gravitational constant. The total test mass is $(m_2{}^2 N_2{}^2 + N_{2e}{}^2 m_e{}^2)$ with N_2 baryons of average mass m_2 in each test mass and N_{2e} leptons of average mass m_e. Since the test masses are electrically neutral and there are approximately equal numbers of protons and neutrons in a test mass it follows approximately that

$$N_{1e} = \tfrac{1}{2}N_1 \quad \text{and} \quad N_{2e} = \tfrac{1}{2}N_2 \qquad (44.4)$$

---

[276] The Eötvös' experiment used a 0.1 gm test mass of $RaBr_2$. R. v. Eötvös, D. Pekár, E. Fekete, Annalen der Physik (Leipzig) 68, 11, 1922.

Subtracting eq. 44.2 from eq. 44.3 after some algebra[277] we find

$$\Delta G = -G_2 + G_1 = (\beta^2/4\pi)/(m_2^2 + m_e^2/2) - (\beta^2/4\pi)/(m_1^2 + m_e^2/2)$$
$$\simeq (\beta^2/4\pi)(1/m_2^2 - 1/m_1^2) \tag{44.5}$$

The masses m_1 and m_2 can differ. For example, if m_H is mass of the hydrogen atom, then $m^{-1} = 1.0m_H^{-1}$ for hydrogen, for carbon $m^{-1} = 1.00782m_H^{-1}$, for copper $m^{-1} = 1.00895m_H^{-1}$, and for lead $m^{-1} = 1.00794m_H^{-1}$.[278] Thus using the Quinn et al results, and CODATA results, and assuming copper and lead test masses, we find the order of magnitude *estimate*:

$$\alpha_B = \beta^2/4\pi \simeq \Delta G/[(1.00895^2 - 1.00794^2)\, m_H^2]$$
$$\simeq \Delta G/G\, G\, m_H^2/.002037$$
$$\simeq (0.000241/0.002037)Gm_H^2$$
$$\simeq 0.118\, Gm_H^2 \tag{44.6}$$

indicating a very weak baryonic force consistent with our general view of the universe. The baryon fine structure constant is minute in comparison to the electromagnetic fine structure constant $\alpha \simeq 1/137$.

Due to our assumptions in the calculation of α_B, which makes it merely an order of magnitude estimate at best, we suggest that an experimental group measure G with differing test masses in the same apparatus to obtain a better value for α_B.

44.2 Four Generation Unified SuperStandard Theory

Previously we derived the form of a one generation Unified SuperStandard Theory that included the known parts of the Standard Model (excepting the Higgs sector) and an $SU(2)\otimes U(1)$ part for Dark Matter.

---

[277] The reduction of the calculation to algebra reminds the author of Nobelist Hans Bethe's remark that he only felt he understood a physical phenomenon when he could reduce it to algebra. This was quite evident when the author collaborated with Professor Bethe on a study of pion condensation in neutron stars some years ago.

[278] "One Hundred Years of the Eötvös Experiment", l. Bod, E. Fischbach, G. Marx and Maria Náray-Ziegler, August, 1990.

In this section we generalize to the four generation SuperStandard Theory that results.[279] Covariant derivatives acquire another interaction term with 16 U(4) fields $U_i{}^\mu$. In addition we add another index to each fermion field specifying its generation. Lastly a set of initially massless gauge field dynamics terms is added to the Unified SuperStandard Theory lagrangian to specify U(4) gauge field evolution and interactions.

44.2.1 Two-Tier Lepton Sector

We begin with the definition of a quadruplet of leptons – a pair of doublets, one normal and one Dark, instead of a single doublet. We define left and right lepton quadruplets with[280]

$$\Psi_{L,Ra}(X) \;=\; \begin{bmatrix} \psi_{DL,Ra}(X) \\ \psi_{NL,Ra}(X) \end{bmatrix} \tag{44.7}$$

where a is a generation index ranging from 1 to 4, $\psi_{NL,R}(X)$ is a "normal" ElectroWeak-like lepton doublet, and where $\psi_{DL,R}(X)$ is a similar Dark ElectroWeak-like lepton doublet consisting of a Dark electron-like fermion and a Dark neutrino-like fermion.

We define covariant derivative terms which we express in matrix form are

$$D_{L,R}(X) \;=\; \begin{bmatrix} \gamma^\mu D_{DL,R\mu} & 0 \\ 0 & \gamma^\mu D_{NL,R\mu} \end{bmatrix} \tag{44.8}$$

---

[279] It is based on the three principles based on Ockham's Razor ("The simplest choice is often the best."): 1) The only connecting interaction is a weak interaction, 2) The form of ElectroWeak theory remains unchanged, and 3) Dark Matter parallels normal matter in its general characteristics: four generations, SU(3) singlets, an SU(2)⊗U(1) symmetry analogous to ElectroWeak symmetry, SU(2)⊗U(1) dark lepton and dark quark doublets.
[280] The X's are Two-Tier coordinates.

where the normal matter left-handed covariant derivative is

$$D_{NL\mu} = \partial/\partial X^\mu - \tfrac{1}{2}ig\boldsymbol{\sigma}\cdot\mathbf{W}_\mu + \tfrac{1}{2}ig'B_\mu - \tfrac{1}{2}ig_G\mathbf{G}\cdot\mathbf{U}_\mu \qquad (44.9)$$

where g_G is an ultra-weak generational coupling constant, $\mathbf{G}\cdot\mathbf{U}_\mu$ is the sum of the inner product of 16 U(4) generators G_i and gauge fields $U_i(X)$, and where the Dark matter left-handed covariant derivative is

$$D_{DL\mu} = \partial/\partial X^\mu - \tfrac{1}{2}ig_D\boldsymbol{\sigma}\cdot\mathbf{W}'_\mu + \tfrac{1}{2}ig_D'B'_\mu - \tfrac{1}{2}ig_G\mathbf{G}\cdot\mathbf{U}_\mu \qquad (44.10)$$

with $\boldsymbol{\sigma}$ a vector composed of the Pauli matrices. The right-handed covariant derivatives have a simpler form. The normal matter right-handed covariant derivative is

$$D_{NR\mu} = \partial/\partial X^\mu + \tfrac{1}{2}ig'B_\mu - \tfrac{1}{2}ig_G\mathbf{G}\cdot\mathbf{U}_\mu \qquad (44.11)$$

and the Dark matter right-handed covariant derivative is

$$D_{DR\mu} = \partial/\partial X^\mu + \tfrac{1}{2}ig_D'B'_\mu - \tfrac{1}{2}ig_G\mathbf{G}\cdot\mathbf{U}_\mu \qquad (44.12)$$

The normal and Dark electroweak fields above are functions of a Two-Tier X. The Faddeev-Popov mechanism operative for these types of fields is described in appendix 19-A of Blaha (2011c).

44.2.2 Quark Sector

In the *quark* sector we define left and right quark quadruplets with

$$\Psi_{qL,Ra}(X_c) = \begin{bmatrix} \psi_{DqL,Ra}(X_c) \\ \psi_{NqL,Ra}(X_c) \end{bmatrix} \qquad (44.13)$$

where $\psi_{NqL,Ra}(X_c)$ is a "normal" ElectroWeak-like quark doublet, and where $\psi_{DqL,Ra}(X_c)$ is a Dark ElectroWeak-like quark doublet consisting of a SU(3) singlet Dark up-quark of unit Dark charge and a SU(3) singlet Dark down-quark of zero Dark charge in the ath generation.

The covariant derivative terms are contained in $D_q(X_c)$ which we express in matrix form as

$$D_{qL,R}(X_c) = \begin{bmatrix} \gamma^\mu D_{qDL,R\mu}(X_c) & 0 \\ 0 & \gamma^\mu D_{qNL,R\mu}(X_c) \end{bmatrix} \tag{44.14}$$

where the normal quark matter left-handed covariant derivative is

$$D_{qNL\mu} = \partial/\partial X_c{}^\mu - \tfrac{1}{2}ig\boldsymbol{\sigma}\cdot\mathbf{W}_\mu - ig'B_\mu/6 - \tfrac{1}{2}ig_G\mathbf{G}\cdot\mathbf{U}_\mu + ig_C\boldsymbol{\tau}\cdot A_{C\mu} \tag{44.15}$$

and where the Dark quark left-handed covariant derivative is

$$D_{qDL\mu} = \partial/\partial X_c{}^\mu - \tfrac{1}{2}ig_D\boldsymbol{\sigma}\cdot\mathbf{W'}_\mu + \tfrac{1}{2}ig_D'B'_\mu - \tfrac{1}{2}ig_G\mathbf{G}\cdot\mathbf{U}_\mu \tag{44.16}$$

since Dark quarks are SU(3) singlets with unit or zero Dark charge. The right-handed quark covariant derivatives have a simpler form. The normal quark right-handed covariant derivative is

$$D_{qNR\mu} = \partial/\partial X_c{}^\mu + \tfrac{1}{2}ig'B_\mu/3 - \tfrac{1}{2}ig_G\mathbf{G}\cdot\mathbf{U}_\mu + ig_C\boldsymbol{\tau}\cdot A_{C\mu} \tag{44.17}$$

and the Dark quark right-handed covariant derivative is

$$D_{qDR\mu} = \partial/\partial X_c{}^\mu + \tfrac{1}{2}ig_D'B'_\mu - \tfrac{1}{2}ig_G\mathbf{G}\cdot\mathbf{U}_\mu \tag{44.18}$$

The normal and Dark gauge boson fields are functions of $X_c. = (X_{r_\mu}(y_r), X_{i_\mu}(y_i))$. The Faddeev-Popov mechanism is operative for gauge boson fields and is described later in this book and in Appendix 19-A of Blaha (2011c).[281] The *complexon* quark SuperStandard Theory ElectroWeak Sector covariant derivatives in quadruplet matrix form are

$$D_{qL,R}(X_c) = \begin{bmatrix} \gamma^\mu D_{qDL,R\mu} & 0 \\ 0 & \gamma^\mu D_{qNL,R\mu} \end{bmatrix} \qquad (44.19)$$

The remaining parts of the complexon Standard Model are described in chapter 23 of Blaha (2011) and summarized below. The addition of singlet Dark quark Higgs terms is also required.

The lagrangian density and action is

$$\mathcal{L}_{CSM} = \Psi_{L}{}^{a\dagger}\gamma^0 i\gamma^\mu D_{L\mu}\Psi_L{}^a - \Psi_R{}^{a\dagger}\gamma^0 i\gamma^\mu D_{R\mu}\Psi_{3R}{}^a + \Psi_{CL}{}^{a\dagger}\gamma^0 i\gamma^\mu \mathcal{D}_{qL\mu}\Psi_{CL}{}^a +$$
$$+ \Psi_{CR}{}^{a\dagger}\gamma^0 i\gamma^\mu \mathcal{D}_{qR\mu}\Psi_{CR}{}^a - \mathcal{L}_{BareMasses} + \mathcal{L}_{Gauge} + \mathcal{L}_{Mass} + \mathcal{L}_{Ufields}$$

$$(44.20)$$

where a is the generation index. $\mathcal{L}_{BareMasses}$ contains the fermion bare mass terms. Also,

$$\mathcal{L}_{Gauge} = \mathcal{L}_{GaugeEW} + \mathcal{L}_{GaugeC} + \mathcal{L}_{GaugeEWD} \qquad (44.21)$$

with

$$\mathcal{L}_{GaugeEW} = -\tfrac{1}{4}\, F_W{}^{a\mu\nu}F_W{}^a{}_{\mu\nu} - \tfrac{1}{4}\, F_B{}^{\mu\nu}F_{B\mu\nu} + \mathcal{L}_{EW}{}^{ghost} \qquad (44.22)$$

$$\mathcal{L}_{GaugeEWD} = -\tfrac{1}{4}\, F'_W{}^{a\mu\nu}F'_W{}^a{}_{\mu\nu} - \tfrac{1}{4}\, F_B{}^{\mu\nu}F_{B'\mu\nu} + \mathcal{L}_{W'}{}^{ghost} \qquad (44.23)$$

and

$$\mathcal{L}_{GaugeC} = \mathcal{L}_{CCG} + \mathcal{L}_C{}^{ghost} + \mathcal{L}_{CC}{}^{ghost} \qquad (44.24)$$

---

[281] Those who might be concerned about the propagator term $<W_i(X), W_j(X_c)>$ and similar propagators where one field is a function of X and the other field is a function of X_c should note that such terms are to very good approximation equal to $<W_i(X), W_j(X)>$ for energies much less than M_c (which could be as large as the Planck energy.)

$$\mathcal{L}_{Ufields} = -\tfrac{1}{4}\, F_U{}^{a\mu v} F_{U\mu vv} + \mathcal{L}_U{}^{ghost} + \mathcal{L}_U{}^{UHiggs} \tag{44.25}$$

where $\mathcal{L}_U{}^{UHiggs}$ is discussed later. The ElectroWeak gauge bosons W_μ^a, B_μ and B'_μ field tensors are:

$$F_W{}^a{}_{\mu v} = \partial W^a{}_\mu/\partial X^v - \partial W^a{}_v/\partial X^\mu + g_2 f^{abc} W^b{}_\mu W^c{}_v \tag{44.26}$$

$$F_{B\mu v} = \partial B_\mu/\partial X^v - \partial B_v/\partial X^\mu \tag{44.27}$$

and the Dark ElectroWeak gauge bosons $W'{}_\mu^a$ and B'_μ field tensors are:

$$F_{B'{}_{\mu v}} = \partial B'{}_\mu/\partial X^v - \partial B'{}_v/\partial X^\mu$$

$$F'_W{}^a{}_{\mu v} = \partial W'^a{}_\mu/\partial X^v - \partial W'^a{}_v/\partial X^\mu + g_2 f^{abc} W'^b{}_\mu W'^c{}_v \tag{44.28}$$

The U fields' tensor is:

$$F_U{}^a{}_{\mu v} = \partial U^a{}_\mu/\partial X^v - \partial U^a{}_v/\partial X^\mu + g_G f_4{}^{abc} U^b{}_\mu U^c{}_v \tag{44.29}$$

where $f_4{}^{abc}$ are the U(4) algebra commutator constants.

$\mathcal{L}_{EW}{}^{ghost}$ contains the Faddeev-Popov ghost terms for the ElectroWeak W_μ^a gauge bosons. The complexon color gluon lagrangian $\mathcal{L}_{CCG}$ is defined by

$$\mathcal{L}_{CCG} = -\tfrac{1}{4}\, F_{CC}{}^{a\mu v}(X) F_{CC}{}^a{}_{\mu v}(X) \tag{44.30}$$

where

$$F_{CC}{}^a{}_{\mu v} = \partial/\partial X_c{}^v\, A_C{}^a{}_\mu - \partial/\partial X_c{}^\mu\, A_C{}^a{}_v + g f_{su(3)}{}^{abc} A_C{}^b{}_\mu A_C{}^c{}_v \tag{44.31}$$

where $A_C{}^a{}_v$ is the color gluon gauge field, g is the color coupling constant, and the f $_{su(3)}{}^{abc}$ are the SU(3) structure constants.

In addition $\mathcal{L}_C{}^{ghost}$ is the color SU(3) Faddeev-Popov ghost terms which we discuss later.[282] The mass sector $\mathcal{L}_{Mass}$ is presumably based on the Higgs Mechanism which creates the fermion and ElectroWeak vector boson masses, and generation mixing.

The lagrangian is supplemented with the following condition on all complexon fields $\Phi_{...}$:[283]

$$\nabla_r \cdot \nabla_i \Phi \ldots = 0 \qquad (44.32)$$

The subsidiary condition:

$$[\nabla_r \cdot \nabla_i - (\nabla_r{}^2 \nabla_i{}^2)^{\frac{1}{2}}]\Omega \ldots = 0 \qquad (44.33)$$

would guarantee a particle's real momentum is parallel to its imaginary momentum. The Faddeev-Popov Method can be used to implement the eq. 44.32 constraint as we show later.

44.3 U(4) Gauge Symmetry Breaking and Long Range Forces

Above we showed that there was good experimental evidence for a conserved Baryon Number B and we proceeded to develop a simple U(1) gauge theory that would imply Baryon Number conservation in a manner analogous to QED's implying electric charge conservation. In section 44.2 we used a new symmetry group local U(4) to generalize the one generation Unified SuperStandard Theory to a four generation Unified SuperStandard Theory based on four conserved particle numbers: B, L, B_D, and L_D.[284]

We now assume in our construction that the four generation Unified SuperStandard Theory has a local U(4) symmetry that is broken by mass terms generated by the Higgs Mechanism.

---

[282] Faddeev-Popov gauge fixing is described below, and in appendix 19-A of Blaha (2011c), for the complexon Lorentz gauge. $\mathcal{L}_{CC}{}^{ghost}$ is the complexon color SU(3) constraint ghost terms defined through the Faddeev-Popov mechanism.

[283] These conditions implement the orthogonality of the real and imaginary parts of complexon 3-momentum.

[284] Charge, although a conserved number, is a part of the ElectroWeak sector, account of which has already been taken.

Further, we will assume that the Higgs breakdown yields two massless (long range) fields which we associate with Baryon Number B and Dark Baryon Number B_D. The remaining fields acquire masses and generate short range forces.

We use the following U(4) diagonal matrices:

$$G_1 = \text{diag}(1, 1, 1, 1) \qquad\qquad (44.34)$$
$$G_2 = \text{diag}(0, 1, 0, 0)$$
$$G_3 = \text{diag}(0, 0, 1, 0)$$
$$G_4 = \text{diag}(0, 0, 0, 1)$$

The U(4) algebra has 16 hermitean matrices that satisfy

$$G_i^\dagger = G_i \qquad\qquad (44.35)$$

The particle numbers can be expressed in terms of the diagonal generators as

$$B = G_1 - G_2 - G_3 - G_4 \qquad\qquad (44.36)$$
$$B_D = G_2$$
$$L = G_3$$
$$L_D = G_4$$

The covariant derivatives have the general form:

$$D_{...\,\mu} = \partial / \partial X^\mu + ... - \tfrac{1}{2} i g_G \mathbf{G} \cdot \mathbf{U}_\mu \qquad\qquad (44.37)$$

where the ellipses indicate the other details of the particular covariant derivative. We now wish to express the four gauge fields $U_i(X)$ for $i = 1, 2, 3, 4$ corresponding to the diagonal generators in terms of the fields of the four particle number gauge fields: B_μ, L_μ, $B_{D\mu}$, and $L_{D\mu}$.

$$U_{i\mu} = A_{ik} N_{k\mu} \qquad\qquad (44.38)$$

where A_{ik} are the elements of a matrix of constants and

$$N_\mu = \begin{bmatrix} B_\mu(X) \\ L_\mu(X) \\ B_{D\mu}(X) \\ L_{D\mu}(X) \end{bmatrix} \tag{44.39}$$

is a column vector consisting of the gauge fields corresponding to each of the conserved particle numbers.

The matrix A must have non-zero determinant so that eq. 44.38 can be inverted to express the particle number fields in terms of the four $U_i(X)$ gauge fields:

$$N_\mu = A^{-1}U_\mu \tag{44.40}$$

resulting in

$$B_\mu(X) = U_{1\mu} \tag{44.41}$$
$$L_\mu(X) = U_{1\mu} + U_{2\mu}$$
$$B_{D\mu}(X) = U_{1\mu} + U_{3\mu}$$
$$L_{D\mu}(X) = U_{1\mu} + U_{4\mu}$$

Then

$$D_{\ldots\mu} = \partial/\partial X^\mu + \ldots - \tfrac{1}{2}ig_G[\sum_{i=5}^{16} \mathbf{G_i}U_{i\mu} + BB_\mu(X) + LL_\mu(X) + B_DB_{D\mu}(X) + L_DL_{D\mu}(X)] \tag{44.42}$$

where the particle numbers, which are analogous to the charges Q and Q' in ElectroWeak theory, are B, L, B_D, and L_D. They are expressed in terms of U(4) generators by eqs. 44.36.

45. Layer Group

This chapter[285] describes the four Internal Symmetry Layer Groups. The transformations of the Complex Lorentz group transformations lead to the internal symmetry Reality group: $R = SU(3) \otimes SU(2) \otimes U(1) \otimes SU(2) \otimes U(1)$, which is the symmetry group of the 'original' Standard Model plus Dark $SU(2) \otimes U(1)$. The Internal Symmetry $U(4)$ Generation group was shown to follow from the four particle number conservation laws,[286] and increased the Standard Model group to $SU(3) \otimes SU(2) \otimes U(1) \otimes SU(2) \otimes U(1) \otimes U(4)$.

45.1 New Conserved Particle Numbers

The four generations of enhanced form of the model lead to a new set of particle conservation laws.

If we examine the form of the $SU(3) \otimes SU(2) \otimes U(1) \otimes SU(2) \otimes U(1) \otimes SU(3) \otimes U(4)$ free fermion lagrangian terms of the Unified SuperStandard Theory

$$\mathcal{L} \sim \bar{\psi}_\alpha \gamma^\mu D_{\mu\alpha\beta} \psi_\beta$$

we find that the $SU(3) \otimes SU(2) \otimes U(1) \otimes SU(2) \otimes U(1) \otimes SU(3) \otimes U(4)$ symmetry of the terms not only yields the B, L, B_D, and L_D particle conservation laws but it also yields four quartets of conservation laws – one quartet for each of the four generations – totaling 16 conserved particle numbers.

The four 'conserved' layer numbers per generation are:

L_{iB} – The Baryon layer particle number for the ith generation

L_{iDB} – The Dark Baryon particle layer number for the ith generation

---

[285] Much of this chapter appears in Blaha (2017c).
[286] Baryon number, Lepton number, and their Dark analogues.

L_{iL} – The Lepton layer particle number for the ith generation

L_{iDL} – The Dark Lepton particle layer number for the ith generation

for each generation i = 1, 2, 3, 4. Fermions have positive L_{ia} = +1 values and anti-fermions have negative L_{ia} = −1 values for species a = 1, 2, 3, 4 (with the three color subspecies treated as part of one species.)

45.2 Four Layers of Fermions

These four sets of particle numbers lead to four U(4) Layer groups. The rationale for the new Layer groups is similar to that of the Generation group which was based on the four particle numbers B, L, B_D, and L_D. The Generation group was based on U(4) rotations related to the four number operators B, L, B_D, and L_D in the one generation SuperStandard Theory.

Similarly we base four Layer groups on the four quartets of conserved layer numbers. Each generation has a Layer group. Consequently, for each fermion species, *we assume the fermion in each generation acquires a Layer index and 'becomes' a set of four fermions in a Layer U(4) 4 representation.* Thus the set of four generations becomes four layers of four generations as shown in Figs. 45.1 and 45.2.

45.3 Layer Group Rotations

A U(4) rotation of the ith generation transforms the four i^{th} generation fermions of the four layers amongst each other (Fig. 45.3). The vertical rotations for generation k can be symbolized by:

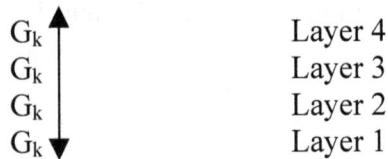

G_k ↑	Layer 4
G_k	Layer 3
G_k	Layer 2
G_k ↓	Layer 1

We assume layer 1 is the known layer of fermions.

Figure 45.1. The four layers of fermions. Each layer (oval) has four generations of fermions. Layer 1 is the layer that we have found experimentally. The 4th generation of layer 1, the Dark part of layer 1, and the remaining three more massive layers constitute Dark matter.

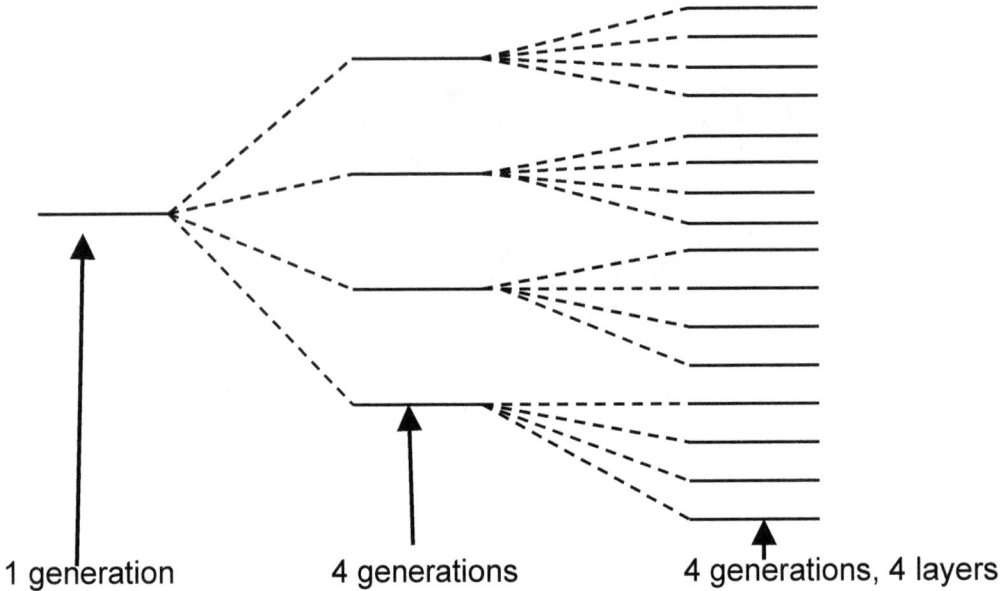

Figure 45.2. The 'splitting of a single generation fermion into four generations and then into four layers.

45.4 Layer Groups

We assume that each of the four Layer groups has associated gauge fields and particle interactions. Each generation in the four layers has a U(4) Layer symmetry group mixing the fermions in its generation across all four layers.(Fig. 45.3) Since the coefficients in Layer group transformations can be local functions, the new Layer groups are implemented with Yang-Mills fields.

It is important to note that the Layer particle numbers are independent of the baryon and lepton particle numbers that form the basis of the Generation group, and so the physics embodied in the Generation group is not the same as the physics of the Layer groups

Layer numbers are conserved under strong and electromagnetic interactions but broken by the Electromagnetic and Weak interactions as well as their Dark counterparts.

Further the gauge fields for SU(3)⊗SU(2)⊗U(1)⊗SU(2)⊗U(1)⊗SU(3)⊗U(4) must now be different for each layer. Thus interactions of these types between fermions in different layers is prevented. As a result the SuperStandard Theory symmetry now is

$$[SU(3) \otimes SU(2) \otimes U(1) \otimes SU(2) \otimes U(1) \otimes SU(3) \otimes U(4) \otimes U(4)]^4$$

where, for compactness, we place the four Layer groups within the quartic expression.

45.5 Steps to Introduce the Layer Groups in the SuperStandard Theory

To implement this new Layer symmetry we expand the SuperStandard Theory lagrangian with the following steps:

1, All covariant derivatives must expand to incude four U(4) Layer gauge fields terms V^μ_i. for i = 1, 2, 3, 4. The new terms are for four layers of SU(3)⊗SU(2)⊗U(1)⊗SU(2)⊗U(1)⊗SU(3)⊗U(4) fields. Each gauge field then has an additional index specifying its layer. The rationale for the choice of these sets of groups to be duplicated is that they, and only they, all play a necessary role in determining the structure of the Periodic Table of Fermions.

2. The new Layer groups are the same as in the Unified SuperStandard Theory of Blaha (2017b). Each fermion particle field has an index labeling the layer of the particle making four layers of four generations of fermions.

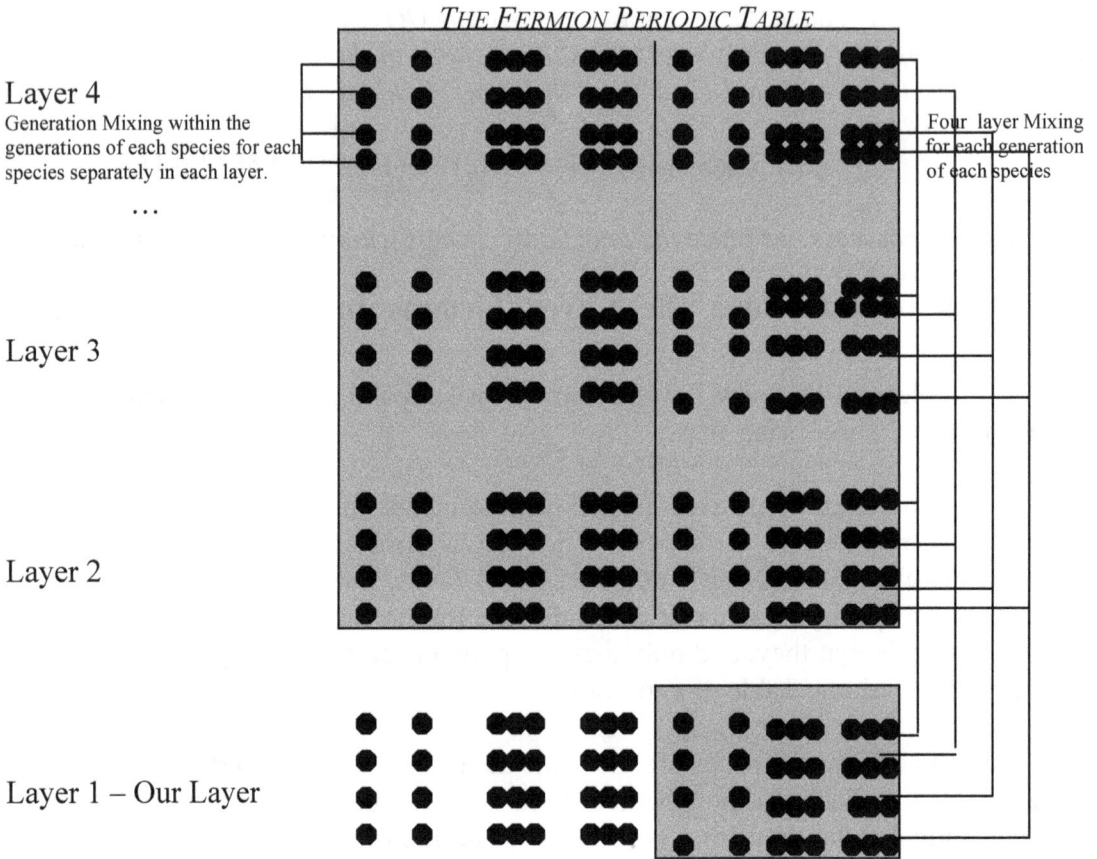

Figure 45.3. Partial example of pattern of particle transformations of the Generation group and of the Layer groups.. Current Dark matter parts of the periodic table are gray. Light parts are the known fermions with an additional, as yet not found, 4th generation shown. The lines on the left side show an example of the Generation mixing within one species. The

Generation mixing applies to each species in each layer. The lines on the right side show the Layer mixing generation by generation.among all four layers for each generation individually.

3. Each layer should have its own set of Higgs particles (modulo mixing) contributing to fermion masses. A layer index number must be added to each Higgs field. One expects that the masses of fermions should be substantially larger for the three 'upper' layers beyond our layer.[287] Otherwise we would have found particles from these upper layers.

The form of the "periodic tables" of fermion and vector bosons that results appears in Fig. 45.4. The anticipated[288] large breakdown of Layer groups symmetries causes the fermion masses of each generation of each species to be significantly different. *This difference may explain the large increase in masses in each species as one goes up from the lowest mass fermion in the species (such as the e mass in the charged lepton species) to the highest (known) mass in the species (such as the τ mass in the charged lepton species.) It also explains why simple models of the fermion generations mass spectrums do not work.*

---

[287] The interplayed mixing of the particles in each generation between different layers may partly explain the vast increases seen in fermion masses as one goes from generation to generation in each species. The Higgs particles in different layers are different and a possible source of the growth of fermion masses.

[288] Experimentally the top three layers of fermions have not been found. Therefore we expect that they have extremely large masses.

46. Equipartition Principle for Fermions and Gauge Fields

We now[289] will suggest a rationale for the dominant abundance of Dark mass-energy:

46.1 Equipartition Principle for Particle Degrees of Freedom

In a closed system at equilibrium the thermal energy of a system is equally partitioned (distributed) among its degrees of freedom. This Equipartition Principle is well known. The application of this principle to the beginning of the universe *when all particles were massless* and all symmetries were unbroken suggests that the distribution of mass-energy should be the same for all degrees of freedom at that time. Thus there should be approximately equal numbers and energies of 192 fermions and 192 vector bosons with the same fraction of the total thermal energy.

We now estimate the relative proportion of Normal and Dark matter in the universe at its beginning based on this Equipartition Principle.

46.2 Proportion of Dark Mass-Energy in the Universe

First we note that 8 of the 12 fermion species (counting quarks of each color as a separate species) in layer 1 – the layer with which we are familiar are Normal matter fermions. (Our discussion is based on our Unified SuperStandard Theory.) Four of the 12 first layer species are Dark.

The other three fermion layers are all Dark from our point of view since they have not been detected. Thus we find that 40 of the 48 species are Dark yielding a *percentage of Dark Matter equal to 40/48 = 83.33%*. (The same counting could have been done by counting individual fermions with the same results.)

---

[289] Some of the material in this section appeared in Blaha (2016a).

Recent studies of the proportion of Dark Matter in the universe have yielded two estimates: 84.5% by Aghanim et al in Astronomy and Astrophysics 1303;5062 and 81.5% from a NASA fit to various models.

Thus our estimate based on our fermion Equipartition Principle is midway between these experimental estimates.

Two possibilities emerge with respect to the present proportion of Dark Matter:

1. The percentage has not changed from the Beginning and the approximate estimates are slightly off. The lack of change could be due to the extremely small decay rates of the fermions in the higher layers.

2. The percentage of matter in the upper layers has decreased due to decay and so the current proportion may be somewhat below 83.33%.

46.3 Proportion of Dark Mass-Energy in the Universe

We know of 12 of the 192 vector bosons in the SuperStandard Theory and 24 fermions. Thus we find a total of 348 out of 384 particles are Dark yielding a Dark mass-energy of 91% of the universes mass-energy at the beginning of the universe.

The sum of Dark energy in the universe currently has been estimated to be 68% of the total energy and the energy of Dark Matter is estimated to be 26.8%. The total is 95% - a value that compares favorably with our above approximate estimate of 91%. The difference in these values can be attributed to various factors. One distinct possibility is the decay of Dark mass-energy from higher layers to the known layer in the 13.8 billion years since the Big Bang.

47. Higgs Symmetry Breaking

There are five broken internal symmetries in the Unified SuperStandard Theory: ElectroWeak symmetry breaking, Generation group symmetry breaking, Layer groups symmetry breaking, and Species group symmetry breaking. In this chapter we will consider ElectroWeak symmetry breaking, Generation group symmetry breaking, and Layer groups symmetry breaking. Each group has a different set of gauge fields[290] in each of the four layers of fermions and vector bosons. While the parameters of these groups differ from layer to layer *the overall form of the symmetry breaking for each layer can be assumed to be the same.*

When we consider the Species group later we will examine its symmetry breaking. *This group has the same gauge fields in all layers – a necessary assumption given their nature.*

47.1 ElectroWeak Symmetry Breaking using the Higgs Mechanism

In this section, and the following sections, we consider the case of symmetry breaking in one layer. Symmetry breaking in other layers are assumed to have a similar pattern.

47.1.1 Higgs Mechanism for ElectroWeak Gauge Field Masses

We require that there is one massless field, the electromagnetic field coupled to electric charge and three massive fields that receive masses via the Higgs Mechanism which breaks $SU(2) \otimes U(1)$ symmetry[291]. The Dark sector is assumed to be analogous to the normal particle sector in this respect since it has a $SU(2) \otimes U(1)$ symmetry in the

---

[290] This would appear to be a necessary assumption since the higher layers have not been found experimentally although ultra-high masses and/or ultra-low coupling constants might explain the lack of experimental evidence for higher layers at present.

[291] We will use the standard Higgs Mechanism formulation because of its familiarity rather than the PseudoQuantum formulation that we developed in an earlier chapter.

Unified SuperStandard Theory. The massive fields become the short range Weak interactions. These assumptions in the construction of the Unified SuperStandard Theory conform to the Leibniz Minimax Principle.

We assume that a doublet Higgs field exists with two components:

$$\eta = \begin{bmatrix} \varphi_+ \\ \varphi_0 \end{bmatrix} \tag{47.1}$$

with conjugate Higgs doublet

$$\eta' = \begin{bmatrix} \varphi_0 \\ -\varphi_- \end{bmatrix} \tag{47.2}$$

The Higgs sector lagrangian has the form:

$$\mathcal{L}_{EW}{}^{Higgs} = (\partial\eta^\dagger/\partial X^\mu)(\partial\eta/\partial X^\mu) - \lambda(\eta^\dagger\eta - \rho^2)^2 + \mathcal{L}_{EW}{}^{Higgs}{}_{EWMasses} \tag{47.3}$$

where the symmetry breaking follows from the choice of unitary gauge

$$\eta = \begin{bmatrix} 0 \\ \rho \end{bmatrix} \tag{47.4}$$

where ρ is a real field. Then the covariant derivative of η is

$$D_{...\mu}\eta = \{\partial/\partial X^\mu + ... + ig t\cdot W_\mu + + ig' t_0 W_{0\mu}\} \begin{bmatrix} 0 \\ \rho \end{bmatrix} \tag{47.5}$$

where g and g' are coupling constants, t_0.is a ½ the identity matrix, and the **t** matrices are ½ the vector of Pauli matrices. The ellipses indicate additional indices and additional terms respectively.

Then

$$D_{...\mu}\eta = \begin{bmatrix} \tfrac{1}{2}ig\rho(W_{1\mu} - iW_{2\mu}) \\ \partial\rho/\partial X^\mu - \tfrac{1}{2}ig\rho(\cos\theta_W)^{-1}Z_\mu \end{bmatrix} \tag{47.6}$$

where θ_W is the Weinberg angle and

$$W_3{}^\mu = Z^\mu\cos\theta_W + A^\mu\sin\theta_W$$
$$W_0{}^\mu = -Z^\mu\sin\theta_W + A^\mu\cos\theta_W \tag{47.7}$$

From eq. 47.6 we find the corresponding Higgs field kinetic terms in the lagrangian are

$$(D_{...\mu}\eta)^\dagger D_{...}{}^\mu\eta = \partial\rho/\partial X^\mu\partial\rho/\partial X_\mu + g^2\rho^2[W_1{}^\mu W_{1\mu} + W_2{}^\mu W_{2\mu}]/4 + g^2\rho^2 Z^\mu Z_\mu/(2\cos\theta_W)^2 \tag{47.8}$$

with the $W_1{}^\mu$ and $W_2{}^\mu$ and Z^μ gauge bosons acquiring masses and the electromagnetic field A^μ massless.

47.1.2 ElectroWeak Higgs Mechanism Generation of Fermion Masses

We now consider the ElectroWeak Higgs Mechanism for the eight species of fermions (four species of "normal" matter and four species of Dark Matter). We shall consider the mass terms for the four normal species which is the same as that of the four Dark species except for the values in the various species mass matrices. Therefore we define the initial 4-vector for the generations of the normal species by

$$\Psi_s = \begin{bmatrix} \psi_{11} \\ \psi_{12} \\ \psi_{13} \\ \psi_{14} \\ \cdots \\ \psi_{41} \\ \psi_{42} \\ \psi_{43} \\ \psi_{44} \end{bmatrix} \tag{47.9}$$

where ψ_{ki} is the generation index for the ith generation of the kth species. ψ_{k1} is the wave function for the 1st generation, ψ_{k4} is the 4th generation member of the kth species, and we omit other indices in the interests of clarity. The normal fermion species are ordered: charged lepton (k = 1), up-type quark, neutral lepton, and down-type quark (k = 4). Other indices of these wave functions are suppressed in the interests of clarity. A 4th generation fermion of any species is yet to be found experimentally.

We assume that a "double" doublet Higgs field exists with four components:

$$\eta = \begin{bmatrix} \varphi_{1+} \\ \varphi_{10} \\ \varphi_{2+} \\ \rho_{20} \end{bmatrix} \tag{47.10}$$

with conjugate Higgs doublet

$$\eta' = \begin{bmatrix} \varphi_{10} \\ -\varphi_{1-} \\ \varphi_{20} \\ -\rho_{2-} \end{bmatrix} \tag{47.11}$$

The Higgs sector lagrangian has the form:

$$\mathcal{L}_{EW}{}^{Higgs} = (\partial\eta^{\dagger}/\partial X^{\mu})(\partial\eta/\partial X^{\mu}) - \lambda(\eta^{\dagger}\eta - \rho^2)^2 + \mathcal{L}_{EW}{}^{Higgs}{}_{FermionMasses} \tag{47.12}$$

where the symmetry breaking follows from the choice of unitary gauge

$$\eta = \begin{bmatrix} 0 \\ \rho_1 \\ 0 \\ \rho_2 \end{bmatrix} \qquad \eta' = \begin{bmatrix} \rho_1 \\ 0 \\ \rho_2 \\ 0 \end{bmatrix} \tag{47.13}$$

where η is a real field quadruplet.

The lagrangian density mass term for the four normal fermion species is

$$\mathcal{L}_{EW}{}^{Higgs}{}_{FermionMasses} = \Sigma_{\alpha,\beta} \{\overline{\psi}_{kL\alpha}\eta m_{k\alpha\beta}\,\psi_{kR\beta} + \overline{\psi}_{kL\alpha}\,\eta' m'_{k\alpha\beta}\psi_{kR\beta}\} + c.c. \tag{47.14}$$

where $m_{k\alpha\beta}$ and $m'_{k\alpha\beta}$ are complex constant matrices, where $\alpha, \beta = 1, \dots , 4$, and where the second term is the double conjugation doublet used to produce a total mass term invariant under weak hypercharge. The total fermion lagrangian mass terms are

$$\mathcal{L}^{Higgs}{}_{FermionMasses} = \mathcal{L}_U{}^{UHiggs}{}_{FermionMasses} + \mathcal{L}_{EW}{}^{Higgs}{}_{FermionMasses}$$

plus the qube mass which is negligible except perhaps for neutrinos. $\mathcal{L}_{EW}{}^{Higgs}{}_{FermionMasses}$ is the contribution of ElectroWeak Higgs Mechanism to the fermion masses. Using the vacuum expectation value of η in eq. 47.13 we find

$$\mathcal{L}_{EW}{}^{Higgs}{}_{FermionMasses} = \Sigma_{\alpha,\beta} \{ \bar{\Psi}_{2L\alpha} \rho_1 m_{2\alpha\beta} \Psi_{2R\beta} + \bar{\Psi}_{4L\alpha} \rho_2 m_{4\alpha\beta} \Psi_{4R\beta} +$$
$$+ \bar{\Psi}_{1L\alpha} \rho_1 m'_{1\alpha\beta} \Psi_{1R\beta} + \bar{\Psi}_{3L\alpha} \rho_2 m'_{3\alpha\beta} \Psi_{3R\beta} \} + c.c$$
$$(47.15)$$

giving mass terms for all four species. **There is an implicit color summation over the color quarks in each generation and quark species.**

The four mass matrices m_1 , ... , m_4 are all complex, constant mass matrices. They can be brought to diagonal form D_k with non-negative values by U(4) matrices A_k and B_k:

$$A_k m_k B_k^{-1} = D_k \qquad (47.16)$$

or

$$m_k = A_k^{-1} D_k B_k \qquad (47.17)$$

for k= 1, ..., 4.

We now note, that although, D_k has non-negative real values, down-type quarks are all tachyonic and up-type quarks are all non-tachyonic, and neutral leptons are all tachyonic and charged leptons are all Dirac non-tachyonic leptons, due to their lagrangian kinetic terms.

We further note that $m_k^\dagger m_k$ is hermitean, and A_k and B_k are members of U(4) as is D_k for k = 1, 2, 3, 4, with the result that all matrices m_k are members of the U(4) group.

We can use these U(4) transformations A_k and B_k to define the sixteen "physical" fermion fields:

$$\Psi_{2L\alpha} \rho_1 m_{2\alpha\beta} \Psi_{2R\beta} + \Psi_{4L\alpha} \rho_2 m_{4\alpha\beta} \Psi_{4R\beta} + \Psi_{1L\alpha} \rho_1 m_{1\alpha\beta} \Psi_{1R\beta} + \Psi_{3L\alpha} \rho_2 m_{3\alpha\beta} \Psi_{3R\beta} \qquad (47.18)$$

$$= (\overline{\psi}_{2L} A_2^{-1})_\alpha \rho_1 D_{2\alpha\beta} (B_2 \psi_{2R})_\beta + (\overline{\psi}_{4L} A_4^{-1})_\alpha \rho_2 D_{4\alpha\beta} (B_4 \psi_{4R})_\beta +$$

$$+ (\overline{\psi}_{1L} A_1^{-1})_\alpha \rho_1 D_{1\alpha\beta} (B_1 \psi_{1R})_\beta + (\overline{\psi}_{3L} A_3^{-1})_\alpha \rho_2 D_{3\alpha\beta} (B_3 \psi_{3R})_\beta$$

$$= \overline{\psi}_{2L phys \alpha} \rho_1 D_{2\alpha\beta} \psi_{2R phys \beta} + \overline{\psi}_{4L phs \alpha} \rho_2 D_{4\alpha\beta} \psi_{4R phys \beta} + \overline{\psi}_{1L phys \alpha} \rho_1 D_{1\alpha\beta} \psi_{1R phys \beta} +$$

$$+ \overline{\psi}_{3L phs \alpha} \rho_2 D_{3\alpha\beta} \psi_{3R phys \beta}$$

Species:

| Up-type quarks | down-type quarks | charged leptons | neutral leptons |

The preceding discussion with changes in the values of constants and constant matrices holds for Dark Matter also where the Dark quarks and leptons acquire ElectroWeak mass terms. The Dark Matter species ElectroWeak mass terms, with the subscript D signifying Dark Matter, are[292]

$$\overline{\psi}_{D2L phys \alpha} \rho_{D1} D_{D2\alpha\beta} \psi_{D2R phys \beta} + \overline{\psi}_{D4L phs \alpha} \rho_{D2} D_{D4\alpha\beta} \psi_{D4R phys \beta} + \overline{\psi}_{D1L phys \alpha} \rho_{D1} D_{D1\alpha\beta} \psi_{D1R phys \beta} + \overline{\psi}_{D3L phs \alpha} \rho_{D2} D_{D3\alpha\beta} \psi_{D3R phys \beta}$$

$$(47.19)$$

Dark
Species: Up-type quarks   down-type quarks   charged leptons   neutral leptons

47.1.3 ElectroWeak Higgs Mechanism for Gauge Fields Including Dark Gauge Fields

In section 47.1.2 we introduced a double doublet (a quadruplet) to derive the symmetry breaking fermion mass spectrum of the four normal fermion families and saw that a similar derivation should be operative for Dark fermions by Ockham's Razor – it is the simplest possible approach to handling the broken SU(2)⊗U(1) symmetry that the Dark ElectroWeak sector has in the Unified SuperStandard Theory.

Now we generalize the SU(2)⊗U(1) ElectroWeak symmetry breaking via the Higgs Mechanism that gives mass to three SU(2)⊗U(1) gauge bosons to include Dark SU(2)⊗U(1) symmetry breaking via the Higgs Mechanism that will give mass to three

---

[292] Dark quarks are Color SU(3) singlets by *assumption* in the Unified SuperStandard Model..

Dark SU(2)⊗U(1) gauge bosons and also yield a massless gauge boson analogous to the electromagnetic field. This assumption is consistent both with the Leibniz Minimax Principle and Ockham's Razor having both simplicity within the context of the Unified SuperStandard Theory and achieving a maximal effect with a minimal extension of the formalism. We will therefore be constructing an SU(2)⊗U(1)⊗SU(2)⊗U(1) Higgs Mechanism symmetry breaking derivation using the double doublets of section 47.1.2.

We assume that a "double" doublet Higgs field exists with four components:

$$\eta = \begin{bmatrix} \varphi_{1+} \\ \varphi_{10} \\ \varphi_{2+} \\ \rho_{20} \end{bmatrix} \tag{47.20}$$

with conjugate Higgs doublet

$$\eta' = \begin{bmatrix} \varphi_{10} \\ -\varphi_{1-} \\ \varphi_{20} \\ -\rho_{2-} \end{bmatrix} \tag{47.21}$$

The Higgs sector lagrangian has the form:

$$\mathcal{L}_{EW}^{Higgs} = (\partial\eta^\dagger/\partial X^\mu)(\partial\eta/\partial X^\mu) - \lambda(\eta^\dagger\eta - \rho^2)^2 + \mathcal{L}_{EW}^{Higgs}{}_{EWMasses} \tag{47.22}$$

where the symmetry breaking follows from the choice of unitary gauge (similar in form to eq. 47.13):

$$\eta = \begin{bmatrix} 0 \\ \rho_1 \\ 0 \\ \rho_2 \end{bmatrix} \qquad \eta' = \begin{bmatrix} \rho_1 \\ 0 \\ \rho_2 \\ 0 \end{bmatrix} \tag{47.23}$$

where η is a real field quadruplet.

The covariant derivative of η in the unitary gauge is

$$
D_{...\mu}\,\eta = \begin{bmatrix} \{\partial/\partial X^{\mu} + ... + igt\cdot\mathbf{W}_{\mu} + + ig't_0 W_{0\mu}\} & 0 \\ \\ 0 & \{\partial/\partial X^{\mu} + ... + ig_D t\cdot\mathbf{W}_{D\mu} + + ig_D't_0 W_{D0\mu}\} \end{bmatrix} \begin{bmatrix} 0 \\ \rho_1 \\ 0 \\ \rho_2 \end{bmatrix}
$$

$$(47.24)$$

where g and g' are coupling constants with g_D and g_D' their Dark equivalents, t_0.is a ½ the identity matrix, the **t** matrices are ½ the vector of Pauli matrices, and the zeros and derivative terms are all 2×2 submatrices.The ellipses indicate additional indices and additional terms respectively.

Then

$$
D_{...\mu}\,\eta = \begin{bmatrix} \tfrac{1}{2}ig\rho_1(W_{1\mu} - iW_{2\mu}) \\ \partial\rho_1/\partial X^{\mu} - \tfrac{1}{2}ig\rho_1(\cos\theta_W)^{-1}Z_{\mu} \\ \tfrac{1}{2}ig_D\rho_2(W_{D1\mu} - iW_{D2\mu}) \\ \partial\rho_2/\partial X^{\mu} - \tfrac{1}{2}ig_D\rho_2(\cos\theta_{WD})^{-1}Z_{D\mu} \end{bmatrix}
$$

$$(47.25)$$

where θ_{WD} is the Dark Weinberg angle and

$$
\begin{aligned}
W_3{}^{\mu} &= Z^{\mu}\cos\theta_W + A^{\mu}\sin\theta_W \\
W_0{}^{\mu} &= -Z^{\mu}\sin\theta_W + A^{\mu}\cos\theta_W \\
W_{D3}{}^{\mu} &= Z_D{}^{\mu}\cos\theta_{WD} + A_D{}^{\mu}\sin\theta_{WD} \\
W_{D0}{}^{\mu} &= -Z_D{}^{\mu}\sin\theta_{WD} + A_D{}^{\mu}\cos\theta_{WD}
\end{aligned}
$$

$$(47.26)$$

From eq. 47.25 we find the corresponding Higgs field kinetic terms in the lagrangian are

$$(D_{...\mu}\eta)^{\dagger}D_{...}^{\mu}\eta = \partial\rho_1/\partial X^{\mu}\partial\rho_1/\partial X_{\mu} + g^2\rho_1^2[W_1^{\mu}W_{1\mu} + W_2^{\mu}W_{2\mu}]/4 +$$
$$+ g^2\rho_1^2 Z^{\mu}Z_{\mu}/(2\cos\theta_W)^2 + \partial\rho_2/\partial X^{\mu}\partial\rho_2/\partial X_{\mu} +$$
$$+ g_D^2\rho_2^2[W_{D1}^{\mu}W_{D1\mu} + W_{D2}^{\mu}W_{D2\mu}]/4 + g_D^2\rho_2^2 Z_D^{\mu}Z_{D\mu}/(2\cos\theta_{WD})^2 \quad (47.27)$$

with the W_1^{μ} and W_2^{μ} and Z^{μ} gauge bosons acquiring masses and the electromagnetic field A^{μ} massless, and with the Dark W_{D1}^{μ} and W_{D2}^{μ} and Z_D^{μ} gauge bosons acquiring masses and the Dark electromagnetic field A_D^{μ} massless.

Thus the SU(2)⊗U(1)⊗SU(2)⊗U(1) Higgs Mechanism symmetry breaking yields the desired results. Note it is consistent with a massless Dark electromagnetic field.

47.1.4 Higgs Mechanism for Tachyons

The Higgs mechanism is currently the favored mechanism for spontaneous symmetry breaking and to give masses to fermions and bosons. The nature of the free tachyon lagrangian terms,

$$\mathcal{L}_{free} = \psi_T^{\dagger}i\gamma^0\gamma^5(\gamma^{\mu}\partial/\partial x^{\mu} + m_0)\psi_T(x) \quad (47.28)$$

where m_0 is a possible bare mass, requires a Higgs sector that contributes to a tachyon mass through spontaneous symmetry breaking having the general form:

$$\mathcal{L}_{Higgs} = \frac{1}{2}\partial\phi/\partial x^{\mu}\partial\phi/\partial x_{\mu} - \psi_T^{\dagger}i\gamma^0\gamma^5\psi_T\phi - V(\phi) \quad (47.29)$$

where

$$V(\phi) = g^2(\phi^2 - (\delta m)^2)^2 \quad (47.30)$$

Note the quadratic term in ϕ – the "mass" term has the negative sign of a tachyon – again showing that tachyons are a feature of modern physics. In the present case the quartic term "stabilizes" the tachyon field which then can be shifted to the minimum of the potential.

The spontaneous symmetry breaking resulting from the potential $V(\phi)$ causes the mass of the tachyon to change to

$$m = m_0 \pm \delta m \tag{47.31}$$

A choice of vacuum state corresponding to the positive sign in eq. 47.31 causes the tachyon mass to increase. A choice of the vacuum state corresponding to the negative sign causes the tachyon mass to decrease, and could cause m to become negative. However this event would not make the tachyon into a normal particle. Rather it would essentially transform the tachyon into its antiparticle.

47.2 Generation Group Symmetry Breaking using the Higgs Mechanism

47.2.1 Generation U(4) Gauge Symmetry Breaking and Long Range Forces

Earlier we showed that there was good experimental evidence for a conserved Baryon Number B and we proceeded to develop a simple U(1) gauge theory that would imply Baryon Number conservation in a manner analogous to QED's implying electric charge conservation. We used a new local U(4) symmetry group to generalize the one generation SuperStandard Theory to a four generation SuperStandard Theory based on four conserved particle numbers: B, L, B_D, and L_D.[293]

We now assume that the four generation SuperStandard Theory has a local U(4) symmetry that is broken by mass terms generated by the Higgs Mechanism .

Further, we will assume that the Higgs breakdown yields two massless (long range) fields which we associate with Baryon Number B and Dark Baryon Number B_D. The remaining fields acquire masses and generate short range forces.

We use the following U(4) diagonal matrices:

---

[293] Charge, although a conserved number, is a part of the ElectroWeak sector, account of which has already been taken.

$$G_1 = \text{diag}(1, 1, 1, 1) \tag{47.32}$$
$$G_2 = \text{diag}(0, 1, 0, 0)$$
$$G_3 = \text{diag}(0, 0, 1, 0)$$
$$G_4 = \text{diag}(0, 0, 0, 1)$$

The U(4) algebra has 16 hermitean matrices that satisfy

$$G_i^{\dagger} = G_i \tag{47.33}$$

The particle numbers can be expressed in terms of the diagonal generators as

$$B = G_1 - G_2 - G_3 - G_4 \tag{47.34}$$
$$B_D = G_2$$
$$L = G_3$$
$$L_D = G_4$$

The covariant derivatives have the general form *for both normal and Dark U gauge fields*:

$$D_{...\mu} = \partial / \partial X^{\mu} + ... - \tfrac{1}{2} i g_G \mathbf{G} \cdot \mathbf{U}_{\mu} \tag{47.35}$$

where the ellipses indicate the other details of the particular covariant derivative. We now wish to express the four gauge fields $U_i(X)$ for $i = 1, 2, 3, 4$ corresponding to the diagonal generators in terms of the fields of the four particle number gauge fields: B_{μ}, L_{μ}, $B_{D\mu}$, and $L_{D\mu}$.

$$U_{i\mu} = A_{ik} N_{k\mu} \tag{47.36}$$

where A_{ik} are the elements of a matrix of constants and

$$N_\mu = \begin{bmatrix} B_\mu(X) \\ L_\mu(X) \\ B_{D\mu}(X) \\ L_{D\mu}(X) \end{bmatrix} \tag{47.37}$$

is a column vector consisting of the gauge fields corresponding to each of the conserved particle numbers.

The matrix A must have non-zero determinant so that eq. 47.36 can be inverted to express the particle number fields in terms of the four $U_i(X)$ gauge fields:

$$N_\mu = A^{-1}U_\mu \tag{47.38}$$

Resulting in

$$\begin{aligned} B_\mu(X) &= U_{1\mu} \\ L_\mu(X) &= U_{1\mu} + U_{2\mu} \\ B_{D\mu}(X) &= U_{1\mu} + U_{3\mu} \\ L_{D\mu}(X) &= U_{1\mu} + U_{4\mu} \end{aligned} \tag{47.39}$$

Then

$$D_{\dots\mu} = \partial/\partial X^\mu + \dots - \tfrac{1}{2}ig_G\left[\sum_{i=5}^{16} G_iU_{i\mu} + BB_\mu(X) + LL_\mu(X) + B_DB_{D\mu}(X) + L_DL_{D\mu}(X)\right]$$

$$\tag{47.40}$$

where the particle numbers, which are analogous to the charges Q and Q' in ElectroWeak theory, are B, L, B_D, and L_D. They are expressed in terms of U(4) generators by eqs. 47.34.

47.2.2 Higgs Mass Mechanism for U(4) Generation Gauge Fields

We now require that there are two massless fields, one coupled to Baryon number and one coupled to Dark Baryon number. The Dark sector is assumed to be analogous to the normal particle sector in this respect. There are fourteen remaining fields that acquire masses and longitudinal components. These fields become short range, ultra-weak generational forces. The masses they acquire through the Higgs Mechanism are presumably very large as these gauge particles have not been found experimentally.[294]

We assume that a scalar Higgs field exists which is a U(4) vector with four components corresponding to the fermion generations. It is an SU(2)⊗U(1)⊗SU(3) ElectroWeak scalar. Its lagrangian density is

$$\mathcal{L}_U^{\text{UHiggs}} = (\partial \eta^\dagger / \partial X^\mu)(\partial \eta / \partial X^\mu) - \lambda (\eta^\dagger \eta - \rho^2)^2 + \mathcal{L}_U^{\text{UHiggs}}{}_{\text{FermionMasses}}$$

where $\mathcal{L}_U^{\text{UHiggs}}{}_{\text{FermionMasses}}$ are the fermion masses produced by the U Higgs Mechanism and where we choose a unitary gauge in which

$$\eta = \begin{bmatrix} 0 \\ \rho_1 \\ 0 \\ \rho_2 \end{bmatrix} \qquad (47.41)$$

where ρ_1 and ρ_2 are real fields. Then the covariant derivative of η is

$$D_{...\mu}\eta = \{\partial / \partial X^\mu + ... - \tfrac{1}{2}ig_G[\Sigma G_i U_{i\mu} + BB_\mu(X) + LL_\mu(X) + B_D B_{D\mu}(X) + L_D L_{D\mu}(X)]\} \begin{bmatrix} 0 \\ \rho_1 \\ 0 \\ \rho_2 \end{bmatrix}$$

$$(47.42)$$

The sum over i is from 5 through 16, and $[G_i]_{jk}$ is the jk element of G_i. Then

---

[294] Section 16.4 discusses this topic in more detail.

$$
D_{...\mu} \eta = \begin{bmatrix} -\frac{1}{2}ig_G\{\rho_1\Sigma[\mathbf{G_i}]_{12}U_{i\mu} + \rho_2\Sigma[\mathbf{G_i}]_{14}U_{i\mu}\} \\ \partial\rho_1/\partial X^\mu - \frac{1}{2}ig_G\rho_1 L_\mu - \frac{1}{2}ig_G\{\rho_1\Sigma[\mathbf{G_i}]_{22}U_{i\mu} + \rho_2\Sigma[\mathbf{G_i}]_{24}U_{i\mu}\} \\ -\frac{1}{2}ig_G\{\rho_1\Sigma[\mathbf{G_i}]_{32}U_{i\mu} + \rho_2\Sigma[\mathbf{G_i}]_{34}U_{i\mu}\} \\ \partial\rho_2/\partial X^\mu - \frac{1}{2}ig_G\rho_2 L_{D\mu} - \frac{1}{2}ig_G\{\rho_1\Sigma[\mathbf{G_i}]_{42}U_{i\mu} + \rho_2\Sigma[\mathbf{G_i}]_{44}U_{i\mu}\} \end{bmatrix} \quad (47.43)
$$

$$
= \begin{bmatrix} -\frac{1}{2}ig_G\Sigma\{\rho_1[\mathbf{G_i}]_{12} + \rho_2[\mathbf{G_i}]_{14}\}U_{i\mu} \\ \partial\rho_1/\partial X^\mu - \frac{1}{2}ig_G\rho_1 L_\mu - \frac{1}{2}ig_G\rho_2\Sigma[\mathbf{G_i}]_{24}U_{i\mu} \\ -\frac{1}{2}ig_G\Sigma\{\rho_1[\mathbf{G_i}]_{32} + \rho_2[\mathbf{G_i}]_{34}\}U_{i\mu} \\ \partial\rho_2/\partial X^\mu - \frac{1}{2}ig_G\rho_2 L_{D\mu} - \frac{1}{2}ig_G\rho_1\Sigma[\mathbf{G_i}]_{42}U_{i\mu} \end{bmatrix} \quad (47.44)
$$

since the generators $\mathbf{G_i}$ have zeroes along their diagonals for $i = 5, \dots, 16$.

From eq. 47.43 we find the corresponding Higgs field kinetic terms in the lagrangian are

$$
(D_{...\mu}\eta)^\dagger D_{...}{}^\mu \eta = \partial\rho_1/\partial X^\mu\partial\rho_1/\partial X_\mu + \partial\rho_2/\partial X^\mu \partial\rho_2/\partial X_\mu + g_G^2\rho_1^2 L_\mu L^\mu/4 + g_G^2\rho_2^2 L_{D\mu} L_D{}^\mu/4 + \dots
$$
$$(47.45)$$

Note there are differing mass squared terms for the Lepton ($g_G^2\rho_1^2/4$) and Dark Lepton ($g_G^2\rho_2^2/4$) gauge fields making them short range fields with the likelihood of very large masses much beyond ElectroWeak gauge field masses, and with an ultraweak coupling constant g_G as suggested by the "experimental" coupling for the Baryonic force.

The Baryonic and Dark Baryonic gauge fields are massless and thus long range although their coupling constant appears to be ultraweak – much below the gravitational coupling constant G.

We now turn to calculating the remaining terms in eq. 47.45 that determine the masses of the remaining 14 gauge fields. We begin by assigning matrix elements for the remaining hermitean U(4) generators:

$$
[G_5]_{ik} = \delta_{i1}\delta_{k2} + \delta_{i2}\delta_{k1} \quad (47.46)
$$

$$[G_6]_{ik} = -i\delta_{i1}\delta_{k2} + i\delta_{i2}\delta_{k1}$$
$$[G_7]_{ik} = \delta_{i1}\delta_{k3} + \delta_{i3}\delta_{k1}$$
$$[G_8]_{ik} = -i\delta_{i1}\delta_{k3} + i\delta_{i3}\delta_{k1}$$
$$[G_9]_{ik} = \delta_{i1}\delta_{k4} + \delta_{i4}\delta_{k1}$$
$$[G_{10}]_{ik} = -i\delta_{i1}\delta_{k4} + i\delta_{i4}\delta_{k1}$$
$$[G_{11}]_{ik} = \delta_{i2}\delta_{k3} + \delta_{i3}\delta_{k2}$$
$$[G_{12}]_{ik} = -i\delta_{i2}\delta_{k3} + i\delta_{i3}\delta_{k2}$$
$$[G_{13}]_{ik} = \delta_{i2}\delta_{k4} + \delta_{i4}\delta_{k2}$$
$$[G_{14}]_{ik} = -i\delta_{i2}\delta_{k4} + i\delta_{i4}\delta_{k2}$$
$$[G_{15}]_{ik} = \delta_{i3}\delta_{k4} + \delta_{i4}\delta_{k3}$$
$$[G_{16}]_{ik} = -i\delta_{i3}\delta_{k4} + i\delta_{i4}\delta_{k3}$$

Then completing eq. 47.45 using eq. 47.44 we find

$$(D_{...\mu}\eta)^\dagger D_{...}{}^\mu\eta = \partial\rho_1/\partial X^\mu \partial\rho_1/\partial X_\mu + \partial\rho_2/\partial X^\mu \, \partial\rho_2/\partial X_\mu + g_G{}^2\rho_1{}^2 L_\mu L^\mu/4 + g_G{}^2\rho_2{}^2 L_{D\mu} L_D{}^\mu/4 +$$
$$+ (g_G/2)^2\rho_1{}^2(U_5{}^2 + U_6{}^2) + (g_G/2)^2\rho_2{}^2(U_9{}^2 + U_{10}{}^2) + (g_G/2)^2\rho_1{}^2(U_{11}{}^2 +$$
$$+ U_{12}{}^2) + + (g_G/2)^2(\rho_1{}^2 + \rho_2{}^2)(U_{13}{}^2 + U_{14}{}^2) + + (g_G/2)^2\rho_2{}^2(U_{15}{}^2 + U_{16}{}^2)$$
$$(47.47)$$

up to total divergences which generate surface terms which we discard and assuming that all fields satisfy the gauge condition

$$\partial U_i{}^\mu/\partial X^\mu = 0 \qquad (47.48)$$

Note that there are no mass terms for $U_7(X)$ and $U_8(X)$ as well as $B_\mu(X)$ and $B_{D\mu}(X)$ due to our choice of unitary gauge eq. 47.41. Consequently there are four massless long range fields and 12 gauge fields that acquire masses of three different values: $(g_G/2)\rho_{10}$, $(g_G/2)\rho_{20}$, and $(g_G/2)(\rho_{10}{}^2 + \rho_{20}{}^2)^{1/2}$ where ρ_{10} and ρ_{20} are the vacuum expectation values of ρ_1 and ρ_2 respectively. The fields $U_7(X)$ and $U_8(X)$ are not "diagonal" and thus appear in the fermion sector as terms connecting fermions in different generations within the four species of normal fermions and within the four species of Dark fermions.[295] Therefore they do not change the values of any of the four types of particle numbers.

Based on an earlier estimate (eq. 44.6) the ultraweak value of the coupling constant is

$$g_G = (4\pi\alpha_B)^{1/2} \approx 1.218 \, (Gm_H^2)^{1/2} \tag{47.49}$$

The gauge field $B_\mu(X)$ is now part of a quadruplet of long range fields. It is the massless, long range field discussed in chapter 44. The two non-diagonal long range forces, being between different generations of a species and having an ultra-weak coupling constant are not of great consequence because of the short lifetime of the higher generations of a species. Therefore, despite their long range, they have only the "shortest" time to exert an inter-generation force before a higher generation particle decays.

Since we expect the other massive fields to have very large masses (and thus very large Higgs field vacuum expectation values), and ultraweak coupling, they are not likely to be experimentally found for the foreseeable future.

47.2.3 Impact of the Generation U(4) Higgs Mechanism on Fermion Generation Masses

The fermion masses of the charged lepton, and the up-type quark, and down-type quark species' generations all show a rapid increase of mass with the generation. For example the u quark mass is a few MeV while the t quark (third generation) has a mass of about 170 GeV/c. The ratio of these masses is about 170,000. While one can account for this great difference by the judicious choices of Higgs' parameter values, when one considers the Generation group, and its associated numerical quantities: ultraweak coupling, very large U particle masses – perhaps of the order of hundreds or thousands of GeV/c, and the corresponding very large Higgs particle vacuum expectation values in this U gauge field sector[296] then the differences in fermion masses within a species become more understandable and natural from a Leibniz Principle perspective.

---

[295] Neutral lepton, charged lepton, up-type quark and down-type quark plus the four corresponding Dark species..
[296] They are not the Higgs particles of the SU(2)⊗U(1) ElectroWeak sector.

Thus the popular view that the ElectroWeak gauge field symmetry breaking is solely via ElectroWeak Higgs fields is not part of our SuperStandard Theory unless the U(4) sector is removed. In our model there are two sets of contributions to fermion symmetry breaking: ElectroWeak Higgs particles symmetry breaking, and Generation group U(4) Higgs particles symmetry breaking. The Generation group causes each species to break into four generations.

In the conventional Standard Model the breakup of species into generations is inserted "by hand." It is not a consequence of the existence of $SU(2)\otimes U(1)$ symmetry or symmetry breaking. In our approach the U(4) Generation group causes the appearance of generations. We base the existence of the Generation group[297] on the four conserved particle numbers. Leibniz' Principle and Ockham's Razor then lead to the above construction/derivation.

47.2.4 Generation Group Higgs Mechanism for Fermion Masses

We now consider the Generation group Higgs Mechanism for the eight species of fermions (four species of "normal" matter[298] and four species of Dark Matter). We shall consider the mass terms for the four normal species, which is the same in form as that of the four Dark species except for the values in the various species mass matrices. Therefore we define the initial 4-vector for the generations of the normal species by

---

[297] In earlier books we suggested the fermion generations might be the result of a wormhole to another 4-dimensional universe. The new approach is simpler and more consistent with known facts – thus more consistent with the Leibniz Minimax Principle.

[298] Not taking account of the three color quark species of normal matter yet.

$$\Psi_s = \begin{bmatrix} \psi_{11} \\ \psi_{12} \\ \psi_{13} \\ \psi_{14} \\ \ldots \\ \psi_{41} \\ \psi_{42} \\ \psi_{43} \\ \psi_{44} \end{bmatrix} \qquad (47.50)$$

where ψ_{ki} is the generation index for the ith generation of the kth species. ψ_{k1} is the wave function for the 1st generation, ψ_{k4} is the 4th generation member of the kth species, and we omit other indices in the interests of clarity. The normal fermion species order here are: charged lepton ($k = 1$), up-type quark, neutral lepton, and down-type quark ($k = 4$). Other indices of these wave functions are suppressed in the interests of clarity. A 4th generation fermion of any species is yet to be found experimentally. The lagrangian density mass terms for the four normal fermion species are

$$\mathcal{L}_U{}^{UHiggs}{}_{FermionMasses} = \Sigma_{k,\alpha,\beta}\ \bar{\psi}_{kL\alpha}\ \eta_k m_{k\alpha\beta}\psi_{kR\beta} + c.c. \qquad (47.51)$$

where $m_{k\alpha\beta}$ is complex constant matrix, where k labels species, and where $\alpha, \beta = 1, \ldots , 4$. The total of fermion lagrangian mass terms is

$$\mathcal{L}^{Higgs}{}_{FermionMasses} = \mathcal{L}_U{}^{UHiggs}{}_{FermionMasses} + \mathcal{L}_{EW}{}^{Higgs}{}_{FermionMasses} \qquad (47.52)$$

where $\mathcal{L}_{EW}{}^{Higgs}$ is the contribution of ElectroWeak Higgs Mechanism to the fermion masses (discussed above). Using the vacuum expectation value of η in eq. 47.37 we find

$$\mathcal{L}_U^{\text{UHiggs}}{}_{\text{FermionMasses}} = \Sigma_{\alpha,\beta} \{ \overline{\Psi}_{2L\alpha}\, \rho_1 m_{2\alpha\beta} \Psi_{2R\beta} + \overline{\Psi}_{4L\alpha}\, \rho_2 m_{4\alpha\beta} \Psi_{4R\beta} \} + \text{c.c.}$$

$$(47.53)$$

giving mass terms tor the up-type and down-type quark species but not to lepton species. There is an implicit color summation over the color quarks in each generation and quark species. *Qualitatively eq. 47.53 could be viewed as corresponding to the experimentally known largeness of quark masses relative to lepton masses in each generation of normal matter.*

The mass matrices $m_2 = [m_{2\alpha\beta}]$ and $m_4 = [m_{4\alpha\beta}]$ are both complex, constant mass matrices. They can be brought to diagonal form with non-negative values by $U(4)$ matrices A_k and B_k:

$$A_2 m_2 B_2^{-1} = D_2 \tag{47.54}$$
$$A_4 m_4 B_4^{-1} = D_4$$

or

$$m_2 = A_2^{-1} D_2 B_2 \tag{47.55}$$
$$m_4 = A_4^{-1} D_4 B_4$$

We now note, that although, both D_2 and D_4 have non-negative real values, down-type quarks are all tachyonic and up-type quarks are all non-tachyonic due to their lagrangian kinetic terms as seen earlier.

We further note that $m_2^\dagger m_2$ and $m_4^\dagger m_4$ are hermitean, and A_k and B_k are members of $U(4)$ as is D_k for $k = 2,4$, with the result that m_2 and m_4 are also both members of the $U(4)$ group. Thus

$$m_2^{-1} = m_2^\dagger \tag{47.56}$$
$$m_4^{-1} = m_4^\dagger$$

We can express the mass matrices in terms of $U(4)$ generators

$$m_2 = \Sigma G_i m_{2i} \tag{47.57}$$
$$m_4 = \Sigma G_i m_{4i}$$

$$m_2^{-1} = m_2^{\dagger} = \Sigma G_i m_{2i}^* \tag{47.58}$$
$$m_4^{-1} = m_4^{\dagger} = \Sigma G_i m_{4i}^*$$

since the matrices G_i are all hermitean, where $\{m_{2i}\}$ and $\{m_{4i}\}$ are each a set of sixteen complex constants.

While we do not as yet know the 4th generation fermions or their masses, the third generation quarks have masses that are far greater than the 1st and 2nd generation quarks or their sum suggesting that the trace of m_2 and m_4.is dominated by the 4th generation mass of the two quark species with a similar situation holding for the two Dark quark species. Therefore if we take the trace of m_2 and m_4 then it seems probable based on the trend of the generations that the 4th generation mass dominates the trace:

$$D_{24} \approx \text{tr } D_2 \tag{47.59}$$
$$D_{44} \approx \text{tr } D_4$$

We can use these A_k and B_k U(4) transformations to define the eight "physical" (up to further ElectroWeak Higgs Mechanism effects) up-type and down-type quark generations fields:

$$\bar{\psi}_{2L\alpha} \, \rho_1 m_{2\alpha\beta} \psi_{2R\beta} + \bar{\psi}_{4L\alpha} \, \rho_2 m_{4\alpha\beta} \psi_{4R\beta} =$$
$$= (\bar{\psi}_{2L} A_2^{-1})_\alpha \rho_1 D_{2\alpha\beta} (B_2 \psi_{2R})_\beta + (\bar{\psi}_{4L} A_4^{-1})_\alpha \rho_2 D_{4\alpha\beta} (B_4 \psi_{4R})_\beta$$
$$= \bar{\psi}_{2L\text{phys}\alpha} \, \rho_1 D_{2\alpha\beta} \psi_{2R\text{phys}\beta} + \bar{\psi}_{4L\text{phs}\alpha} \, \rho_2 D_{4\alpha\beta} \psi_{4R\text{phys}\beta} \tag{47.60}$$

Species: up-type quarks   down-type quarks

The preceding discussion with changes in the values of constants and constant matrices holds for Dark Matter also where the Dark quarks acquire mass terms but the Dark leptons do not. The Dark Matter species mass terms, with the subscript D signifying Dark Matter, are

$$= \bar{\psi}_{D2L\text{phys}\alpha}\, \rho_{D1} D_{D2\alpha\beta} \psi_{D2R\text{phys}\beta} + \bar{\psi}_{D4L\text{phs}\alpha}\, \rho_{D2} D_{D4\alpha\beta} \psi_{D4R\text{phys}\beta} \qquad (47.61)$$

Dark Species: up-type quarks   down-type quarks

47.3 Layer Groups Higgs Symmetry Breaking

The Layer groups symmetry is also broken. *In this section we use the Higgs Mechanism to implement symmetry breaking for the lowest generation of the four layers for Layer group '1'.*[299] *The symmetry breaking for each of the other three generations of the four layers is assumed to have a similar form.*

47.3.1 Layer Group Higgs Mechanism Contributions to Layer Gauge Field Masses

In this section we will determine a Layer group's Higgs contributions to gauge field masses. (The fermion mass contributions from the various Higgs interactions are discussed in section 47.3.2. We will see that all[300] layers have Layer group Higgs contributions to the layer's fermion masses.) There are four Layer groups – one for each of the four fermion generations. We will consider the case of the generation '1' Layer group below.[301] The cases of other layers is assumed to be similar.

We begin by assuming that four scalar Higgs fields η_a exist, which are each U(4) Layer group 4-vectors. The η_a are SU(2)⊗U(1)⊗SU(3) ElectroWeak and Strong Interaction scalars. The lagrangian density terms for the ath η_a are[302]

---

[299] We omit the designation of the Layer group and gauge fields. The presentation is only for generation '1' in this section (whose species contain e, ν_e, u and d type fermions.)

[300] All Layers have Layer group Higgs contributions to avoid massless Layer group gauge fields. *Since transitions between fermions layers have not been found we assume all Layer groups gauge fields are very massive and/or the coupling constants are ultra-small.*

[301] We omit an index number for the Layer group and other quantities in the interest of simplicity.

[302] Again we use the standard formulation of the Higgs Mechanism because of its familiarity.

$$\mathscr{L}_V^{\text{Higgs}} = (\partial \eta_a^\dagger / \partial X^\mu)(\partial \eta_a / \partial X^\mu) - \lambda(\eta_a^\dagger \eta_a - \rho_a^2)^2 + \mathscr{L}_V^{\text{Higgs}}{}_{\text{FermionMasses}}$$

(47.62)

where $\mathscr{L}_V^{\text{Higgs}}{}_{\text{FermionMasses}}$ are the fermion masses produced by the Layer Higgs Mechanism and where we set the η Layer 4-vector with Higgs field components to

$$\eta_a = \begin{bmatrix} \rho_{1a} \\ \rho_{2a} \\ \rho_{3a} \\ \rho_{4a} \end{bmatrix}$$

(47.63)

where ρ_{1a}, ρ_{2a}, ρ_{3a} and ρ_{4a} are real fields.[303] Then the covariant derivative of η_a is

$$D_{\dots\mu}\eta_a = \{\partial / \partial X^\mu + \dots - \tfrac{1}{2}ig_V[\Sigma_i \mathbf{G}_{\text{Li}}\mathbf{V}_{i\mu} + \mathbf{G}_{\text{L1}}\mathbf{V}_{1\mu} + \mathbf{G}_{\text{L2}}\mathbf{V}_{2\mu} + \mathbf{G}_{\text{L3}}\mathbf{V}_{3\mu} + \mathbf{G}_{\text{L4}}\mathbf{V}_{4\mu}]\} \begin{bmatrix} \rho_{1a} \\ \rho_{2a} \\ \rho_{3a} \\ \rho_{4a} \end{bmatrix}$$

(47.64)

The sum over i is from 5 through 16 (non-diagonal matrices), and $[\mathbf{G}_{\text{Li}}]_{jk}$ is the jk[th] element of $\mathbf{G}_{\text{Li}}$. Then

$$D_{\dots\mu}\eta_a = \begin{bmatrix} \partial\rho_{1a}/\partial X^\mu - \tfrac{1}{2}ig_V\{\rho_{1a}\mathbf{G}_{\text{L1}}\mathbf{V}_{1\mu} + \rho_{2a}\Sigma[\mathbf{G}_{\text{Li}}]_{11}\mathbf{V}_{i\mu} + \rho_{2a}\Sigma[\mathbf{G}_{\text{Li}}]_{12}\mathbf{V}_{i\mu} + \rho_{3a}\Sigma[\mathbf{G}_{\text{Li}}]_{13}\mathbf{V}_{i\mu} + \rho_{4a}\Sigma[\mathbf{G}_{\text{Li}}]_{14}\mathbf{V}_{i\mu}\} \\ \partial\rho_{2a}/\partial X^\mu - \tfrac{1}{2}ig_V\{\rho_{2a}\mathbf{G}_{\text{L2}}\mathbf{V}_{2\mu} + \rho_{1a}\Sigma[\mathbf{G}_{\text{Li}}]_{21}\mathbf{V}_{i\mu} + \rho_{2a}\Sigma[\mathbf{G}_{\text{Li}}]_{22}\mathbf{V}_{i\mu} + \rho_{3a}\Sigma[\mathbf{G}_{\text{Li}}]_{23}\mathbf{V}_{i\mu} + \rho_{4a}\Sigma[\mathbf{G}_{\text{Li}}]_{24}\mathbf{V}_{i\mu}\} \\ \partial\rho_{3a}/\partial X^\mu - \tfrac{1}{2}ig_V\{\rho_{3a}\mathbf{G}_{\text{L3}}\mathbf{V}_{3\mu} + \rho_{1a}\Sigma[\mathbf{G}_{\text{Li}}]_{31}\mathbf{V}_{i\mu} + \rho_{2a}\Sigma[\mathbf{G}_{\text{Li}}]_{32}\mathbf{V}_{i\mu} + \rho_{3a}\Sigma[\mathbf{G}_{\text{Li}}]_{33}\mathbf{V}_{i\mu} + \rho_{4a}\Sigma[\mathbf{G}_{\text{Li}}]_{34}\mathbf{V}_{i\mu}\} \\ \partial\rho_{4a}/\partial X^\mu - \tfrac{1}{2}ig_V\{\rho_{4a}\mathbf{G}_{\text{L4}}\mathbf{V}_{4\mu} + \rho_{1a}\Sigma[\mathbf{G}_{\text{Li}}]_{41}\mathbf{V}_{i\mu} + \rho_{2a}\Sigma[\mathbf{G}_{\text{Li}}]_{42}\mathbf{V}_{i\mu} + \rho_{3a}\Sigma[\mathbf{G}_{\text{Li}}]_{43}\mathbf{V}_{i\mu} + \rho_{4a}\Sigma[\mathbf{G}_{\text{Li}}]_{44}\mathbf{V}_{i\mu}\} \end{bmatrix}$$

(47.65)

---

[303] Each field ρ_{ia} can be expressed as a pseudoquantum field: $\rho_{ia} = \varphi_{1ia} + \varphi_{2ia}$ where φ_{1ia} has the vacuum expectation value ρ_{i0a} for i = 1, ... , 4. Thus our pseudoquantum field theory version is implemented easily.

$$= \begin{bmatrix} \partial\rho_{1a}/\partial X^{\mu} - \tfrac{1}{2}ig_V\{\rho_{1a}\mathbf{G}_{L1}\mathbf{V}_{1\mu} + \rho_{2a}\Sigma[\mathbf{G}_{Li}]_{12}\mathbf{V}_{i\mu} + \rho_{3a}\Sigma[\mathbf{G}_{Li}]_{13}\mathbf{V}_{i\mu} + \rho_{4a}\Sigma[\mathbf{G}_{Li}]_{14}\mathbf{V}_{i\mu}\} \\ \partial\rho_{2a}/\partial X^{\mu} - \tfrac{1}{2}ig_V\{\rho_{2a}\mathbf{G}_{L2}\mathbf{V}_{2\mu} + \rho_{1a}\Sigma[\mathbf{G}_{Li}]_{21}\mathbf{V}_{i\mu} + \rho_{3a}\Sigma[\mathbf{G}_{Li}]_{23}\mathbf{V}_{i\mu} + \rho_{4a}\Sigma[\mathbf{G}_{Li}]_{24}\mathbf{V}_{i\mu}\} \\ \partial\rho_{3a}/\partial X^{\mu} - \tfrac{1}{2}ig_V\{\rho_{3a}\mathbf{G}_{L3}\mathbf{V}_{3\mu} + \rho_{1a}\Sigma[\mathbf{G}_{Li}]_{31}\mathbf{V}_{i\mu} + \rho_{2a}\Sigma[\mathbf{G}_{Li}]_{32}\mathbf{V}_{i\mu} + \rho_{4a}\Sigma[\mathbf{G}_{Li}]_{34}\mathbf{V}_{i\mu}\} \\ \partial\rho_{4a}/\partial X^{\mu} - \tfrac{1}{2}ig_V\{\rho_{4a}\mathbf{G}_{L4}\mathbf{V}_{4\mu} + \rho_{1a}\Sigma[\mathbf{G}_{Li}]_{41}\mathbf{V}_{i\mu} + \rho_{2a}\Sigma[\mathbf{G}_{Li}]_{42}\mathbf{V}_{i\mu} + \rho_{3a}\Sigma[\mathbf{G}_{Li}]_{43}\mathbf{V}_{i\mu}\} \end{bmatrix} \quad (47.66)$$

since the generators $\mathbf{G}_i$ have zeroes along their diagonals for i = 5, ... , 16.

From eq. 47.66 we find the corresponding a^{th} Higgs field kinetic terms in the lagrangian are

$$(D_{...\mu}\eta_a)^{\dagger} D_{...}{}^{\mu}\eta_a = \partial\rho_{1a}/\partial X^{\mu}\, \partial\rho_1/\partial X_{\mu} + \partial\rho_{2a}/\partial X^{\mu}\, \partial\rho_2/\partial X_{\mu} + \partial\rho_{3a}/\partial X^{\mu}\, \partial\rho_3/\partial X_{\mu} +$$
$$+ \partial\rho_{4a}/\partial X^{\mu}\, \partial\rho_4/\partial X_{\mu} + g_V^2\rho_{1a}^2\mathbf{V}_{1\mu}\mathbf{V}_1{}^{\mu}/4 + g_V^2\rho_{2a}^2\mathbf{V}_{2\mu}\mathbf{V}_2{}^{\mu}/4 +$$
$$+ g_V^2\rho_{3a}^2\,\mathbf{V}_{3\mu}\mathbf{V}_3{}^{\mu}/4 + g_V^2\rho_{4a}^2\,\mathbf{V}_{4\mu}\mathbf{V}_4{}^{\mu}/4 + ...$$

$$(47.67)$$

We now turn to calculating the remaining terms in eq. 47.67 that determine the masses of the remaining 14 gauge fields. We begin by assigning matrix elements for the remaining hermitean U(4) generators:

$$[\mathbf{G}_{L5}]_{ik} = \delta_{i1}\delta_{k2} + \delta_{i2}\delta_{k1} \quad\quad (47.68)$$
$$[\mathbf{G}_{L6}]_{ik} = -i\delta_{i1}\delta_{k2} + i\delta_{i2}\delta_{k1}$$
$$[\mathbf{G}_{L7}]_{ik} = \delta_{i1}\delta_{k3} + \delta_{i3}\delta_{k1}$$
$$[\mathbf{G}_{L8}]_{ik} = -i\delta_{i1}\delta_{k3} + i\delta_{i3}\delta_{k1}$$
$$[\mathbf{G}_{L9}]_{ik} = \delta_{i1}\delta_{k4} + \delta_{i4}\delta_{k1}$$
$$[\mathbf{G}_{L10}]_{ik} = -i\delta_{i1}\delta_{k4} + i\delta_{i4}\delta_{k1}$$
$$[\mathbf{G}_{L11}]_{ik} = \delta_{i2}\delta_{k3} + \delta_{i3}\delta_{k2}$$
$$[\mathbf{G}_{L12}]_{ik} = -i\delta_{i2}\delta_{k3} + i\delta_{i3}\delta_{k2}$$
$$[\mathbf{G}_{L13}]_{ik} = \delta_{i2}\delta_{k4} + \delta_{i4}\delta_{k2}$$
$$[\mathbf{G}_{L14}]_{ik} = -i\delta_{i2}\delta_{k4} + i\delta_{i4}\delta_{k2}$$
$$[\mathbf{G}_{L15}]_{ik} = \delta_{i3}\delta_{k4} + \delta_{i4}\delta_{k3}$$
$$[\mathbf{G}_{L16}]_{ik} = -i\delta_{i3}\delta_{k4} + i\delta_{i4}\delta_{k3}$$

Then completing eq. 47.67 using eq. 47.66 we find

$$(D_{...\mu}\eta_a)^\dagger D_{...}^{\ \mu}\eta_a = \partial\rho_{1a}/\partial X_\mu \, \partial\rho_1/\partial X_\mu + \partial\rho_{2a}/\partial X_\mu \, \partial\rho_{2a}/\partial X_\mu + \partial\rho_{3a}/\partial X_\mu \, \partial\rho_{3a}/\partial X_\mu + \partial\rho_{4a}/\partial X_\mu \, \partial\rho_{4a}/\partial X_\mu +$$
$$+ g_V^2\rho_{1a}^2\mathbf{V}_{1\mu}\mathbf{V}_1^\mu/4 + g_V^2\rho_{2a}^2 V_2^2/4 + g_V^2\rho_{3a}^2 V_3^2/4 + g_V^2\rho_{4a}^2 V_4^2/4 +$$
$$+ (g_V/2)^2(\rho_{1a}^2 + \rho_{2a}^2)(V_5^2 + V_6^2) + (g_V/2)^2(\rho_{1a}^2 + \rho_{3a}^2)(V_7^2 + V_8^2) +$$
$$+ (g_V/2)^2(\rho_{1a}^2 + \rho_{4a}^2)(V_9^2 + V_{10}^2) + (g_V/2)^2(\rho_{2a}^2 + \rho_{3a}^2)(V_{11}^2 + V_{12}^2) +$$
$$+ (g_V/2)^2(\rho_{2a}^2 + \rho_{4a}^2)(V_{13}^2 + V_{14}^2) + (g_V/2)^2(\rho_{3a}^2 + \rho_{4a}^2)(V_{15}^2 + V_{16}^2)$$

$$(47.69)$$

up to total divergences, which generate surface terms which we discard, and also assuming that all fields satisfy the gauge condition

$$\partial V_{ia}^\mu/\partial X^\mu = 0 \qquad\qquad (47.70)$$

Eq.47.69 shows all Layer groups gauge fields have masses. The combination of an ultra-weak coupling constant and very large gauge field masses results in extremely weak interactions between the fields in each layer, which leads to almost independent layers of normal and Dark fermions. Thus the Darkness! They result in very rare decays between layers, and very weak interactions between fermions in different layers. The higher layers with presumably much more massive fermions are thus well "insulated" from our layer. Thus they are Dark to us as well.

We estimate Layer groups gauge field masses to be very large – of the order of many TeV or they would have been detected at CERN by now. Their detection must await the construction of much more powerful accelerators. *The "non-diagonal" Layer gauge fields are the means by which we may hope to eventually find fermions of the higher layers.*

47.3.2 Layer Group Higgs Mechanism Contributions to Fermion Masses

The fermion masses of the charged lepton, and the up-type quark, and down-type quark species' generations all show a rapid increase of mass with the generation. For example the u quark mass is a few MeV while the t quark (third generation) has a mass of about 170 GeV/c. The ratio of these masses is about 170,000. While one can account for this great difference by the judicious choices of Higgs' parameter values, when one considers the Layer groups and their associated numerical quantities: ultra-weak coupling, very large Layer gauge field masses – perhaps of the order of hundreds or thousands of GeV/c, then a large difference in particle masses between layers is understandable and natural.

The form of the fermion lagrangian mass terms for the charged lepton species, generation '1' Layer group L_1 due to Higgs symmetry breaking is

$$\mathcal{L}^{Higgs}_{FermionMasses,\, l = '1',e} = \Sigma_{\delta,\gamma} \bar{\Psi}_{L1'e\delta}\, \eta_{'1e} m_{'1'e\delta\gamma} \Psi_{R'1'e\gamma} +$$
$$+ \Sigma_{\delta,\gamma} \bar{\Psi}_{DL'1'e\delta}\, \eta_{D'1'e} m_{D'1'e\delta\gamma} \Psi_{DR'1'e\gamma}$$

(47.71)

The indices δ and γ label *layer* rows and columns. The fields labeled η (with subscripts) are Higgs fields that have non-zero vacuum expectation values. Replacing 'e' with another species designator gives the terms for another species in the normal or Dark sectors. Replacing '1' with another generation number '2', '3', or '4' specifies a different generation corresponding to a different one of the other three Layer groups.

47.4 The Total Fermion Mass Matrix

The total fermion mass matrix lagrangian terms are

$$\mathcal{L}^{Higgs}_{FermionMasses} = \mathcal{L}_{EWFermionMasses} + \mathcal{L}_{UFermionMasses} + \mathcal{L}_{VtotFermionMasses} + \mathcal{L}_{\Theta\text{-}SymmetryFermionMasses} +$$
$$+ \mathcal{L}_{GravSpeciesFermionMasses} + \text{c.c.}$$

(47.72)

where $\mathcal{L}_{EW}^{Higgs}$ is the contribution of ElectroWeak Higgs Mechanism to the fermion masses, $\mathcal{L}_{UFermionMasses}$ is the Generation group contribution, $\mathcal{L}_{VtotFermionMasses}$ is the sum of the four Layer groups contributions, and $\mathcal{L}_{GravSpeciesFermionMasses}$ is the Gravitational Species group (discussed later) contribution.

The mass matrices in eq. 47.72 are complex, constant mass matrices that can be totaled and brought to diagonal form with non-negative values by U(4) matrices. (They cannot be separately diagonalized.) The resulting diagonalized mass matrices are the mass matrices of the physical fermions.

47.5 Four Generation CKM Matrix

The four generation generalization for a single layer of the CKM matrix[304]

$$C_4 = A_2 B_4^{-1} \tag{47.73}$$

is a constant 4×4 U(4) Generation group matrix. Redefining field phases it can be reduced to an SU(4) matrix. It appears in the charge-changing quark current

$$J^\mu_{charged} = \overline{\psi}_{4Lphs}\gamma^\mu C_4\psi_{2Lphys} \tag{47.74}$$

Thus a specific Generation group matrix determines the mixing between the generations. This matrix may be expected to generate CP violation. C_4 can be constructed as a product of SU(4) factors with nine arbitrary parameters.

47.6 Four Layer CKM-like Matrix

The four layer 16×16 CKM-like matrices would have been four blocks of 4×4 CKM-like matrices along the diagonal for each of the leptons, quarks, Dark leptons and Dark quarks *IF* the Layer groups did not mix the four layers of fermions. However since the Layer groups do mix the four layers of fermions, the four[305] 16×16 CKM-like matrices C_{4i} (for i = leptonic, quark, Dark leptonic and Dark quark) are not block diagonal.[306] Each CKM-like matrix is a 16×16 non-block-diagonal, constant, complex matrix.

The charged ElectroWeak currents have the form

$$J^\mu_i = \overline{\psi}_{physa}\gamma^\mu[C_i]_{ab}\psi_{physb} \tag{47.75}$$

with sums over the indices a and b. They may be expected to generate CP violation.

---

[304] M. Kobayashi and K. Maskawa, Prog. Theo. Phys. **49**, 652 (1975) and references therein.

[305] For leptons, quarks, Dark leptons and Dark quarks.

[306] Thus attempts to guess the form of the CKM-like matrices based on simple group theory hypotheses are unlikely to be successful. In view of the contributions of five contributions to the total fermion mass matrix (eq. 15.72), and the need to diagonalize the total mass matrix, partial diagonalizations (for example, based solely on ElectroWeak symmetry breaking) are not physically meaningful.

48. Faddeev-Popov Method for Gauge Fields

48.1 Faddeev-Popov Method for 'Normal' Yang-Mills Gauge Fields

The Faddeev-Popov Method for normal Yang-Mills fields (those with real-valued parameters) has been described in numerous textbooks. For example, the reader may read Huang (1992) or Kaku (1993).

48.2 Pure Gauge Complexon Path Integral Formulation and Faddeev-Popov Method

The path integral formalism for complexon non-abelian, pure, Yang-Mills fields differs significantly from the conventional gauge field path integral formalism. The path integral for a complexon gauge field can be written symbolically as:

$$Z(J^\mu) = N \int DA_C \Delta_{FP}(A_C) \delta(F(A_C)) \Delta_C(A_C) \delta(F_C(A_C)) \exp\{i \int d^7 y [\mathscr{L} + J^\mu(y) A_{C\mu}(y)]\}$$
(48.1)

where $\delta(F(A_C))$ specifies the gauge, $\Delta_{FP}(A_C)$ is its Faddeev-Popov determinant; and $\delta(F_C(A_C))$ specifies the complexon condition (eqs. 2.123a and 2.124) with $\Delta_C(A_C)$ the Faddeev-Popov determinant for the complexon condition. In both cases the Faddeev-Popov determinant can be calculated in the standard way.[307]

First we consider the gauge fixing delta function. Note that it can be written as a delta function in the gauge times a determinant:

$$\delta(F(A_C{}^\omega)) = \delta(\omega - \omega_0) |\det \delta F(A_{C\mu}{}^\omega(x))/\delta \omega(x)|^{-1}\big|_{F(A_C)=0}$$
(48.2)

where ω_0 is a reference gauge, where

---

[307] See for example Huang (1992).

$$Ac_{\mu}^{a\,\omega}(x) = Ac_{\mu}^{a}(x) - g^{-1}D_{\mu}\omega^{a} + f^{abc}\omega^{b}(x)Ac_{\mu}^{c}(x) \tag{48.3}$$
$$= Ac_{\mu}^{a}(x) + \delta Ac_{\mu}^{a\,\omega}(x)$$

and

$$Re\; Fc_{\mu k}^{a\,\omega} = Re\; Fc_{\mu k}^{a} + f^{abc}\omega^{b}(x)Fc_{\mu k}^{c} \tag{48.4}$$
$$= Re\; Fc_{\mu k}^{a} + \delta(Re\; Fc_{\mu k}^{a\,\omega})$$

under an infinitesimal gauge transformation, and where

$$\Delta_{FP}(Ac) = \; |det\; \delta F(Ac_{\mu}^{\omega}(x))/\delta\omega(x)||_{F(Ac)\,=\,0,\;\omega\,=\,0} \tag{48.5}$$

We will choose the complexon Lorentz gauge to evaluate the Faddeev-Popov determinant:

$$F^{a}(Ac) = D_{\mu}Ac^{a\mu}(x) = 0 \tag{48.6}$$

We find

$$F^{a}(Ac_{\mu}^{\omega}(x)) = D^{\mu}(Ac_{\mu}^{a}(x) - g^{-1}D_{\mu}\omega^{a}(x) + f^{abc}\omega^{b}(x)Ac_{\mu}^{c}(x))$$
$$= -g^{-1}D^{\mu}D_{\mu}\omega^{a}(x) + f^{abc}Ac_{\mu}^{c}(x)D^{\mu}\omega^{b}(x) \tag{48.7}$$

Thus

$$\delta F^{a}(Ac_{\mu}^{\omega}(x))/\delta\omega^{b}(x) = -g^{-1}\delta^{ab}D^{\mu}D_{\mu} + f^{abc}Ac^{c\mu}(x)D_{\mu} \tag{48.8}$$

and

$$\Delta_{FP}(Ac) = \; |det\; (g^{-1}\delta^{ab}D^{\mu}D_{\mu} - f^{abc}Ac^{c\mu}(x)D_{\mu})| \tag{48.9}$$

where | ... | represent absolute value.

We can rewrite the Faddeev-Popov determinant as a path integral over anti-commuting c-number fields with a ghost Lagrangian term:

$$\Delta_{FP}(Ac) = \; \int D\chi^{*}D\chi\; exp[\; i\int d^{7}x\; \mathscr{L}^{ghost}(x)] \tag{48.10}$$

where

$$\mathscr{L}^{ghost}(x) = \chi^{a*}(x)[\delta^{ab}D^{\mu}D_{\mu} - gf^{abc}Ac^{c\mu}(x)D_{\mu}]\chi^{b}(x) \tag{48.11}$$

48.3 Faddeev-Popov Application to the Complexon Condition

The complexon condition can also be implemented within the path integral formalism using the Faddeev-Popov Mechanism. Using the identity

$$1 = \int DA_C \Delta_C(A_C)\delta(F_C(A_C)) \tag{48.12}$$

we see that an infinitesimal gauge transformation yields eqs. 48.3 and 48.4. This enables us to relate $\Delta_C(A)$ to the determinant

$$\delta(F_C(A_C{}^{\omega})) = |\det \delta F_C(A_{C_\mu}{}^{\omega}(x))/\delta\omega(x)|^{-1}|_{F_C(A_C)=0,\, \omega\,=\,0} \tag{48.13}$$

and

$$\Delta_C(A_C) = |\det \delta F_C(A_{C_\mu}{}^{\omega}(x))/\delta\omega(x)||_{F_C(A_C)\,=\,0,\, \omega\,=\,0} \tag{48.14}$$

From eq. 48.31 we see

$$F_C(A_{C_\mu}(x)) = A_C{}^{a\mu}[\partial^2 A_C{}^a{}_\mu/\partial x_r{}^k \partial x_i{}^k - \partial^2 A_C{}^a{}_k/\partial x_r{}^\mu \partial x_i{}^k + gf^{abc}\partial(A_C{}^b{}_\mu A_C{}^c{}_k)/\partial x_i{}^k] \tag{48.15}$$

with

$$F_C{}^a(A_{C_\mu}) = 0 \tag{48.16}$$

Inserting eq. 48.3 and 48.4 we find

$$[\delta F_C(A_{C_\mu}{}^{\omega}(x))/\delta\omega^a(x)]|_{F_C(A_C)\,=\,0,\, \omega\,=\,0} =$$
$$= \delta[\delta A_C{}^{b\mu\omega}\mathrm{Re}\,\partial F_C{}^b{}_{\mu}k/\partial x_{ik} + A_C{}^{b\mu}\partial\delta(\mathrm{Re}\,F_C{}^b{}_{\mu}k{}^{\omega})/\partial x_{ik}]/\delta\omega^a(x)|_{\omega\,=\,0}$$
$$= -g^{-1}(\mathrm{Re}\,\partial F_C{}^a{}_{\mu}k/\partial x_{ik})D^\mu - f^{abc}A_C{}^{b\mu}(\mathrm{Re}\,F_C{}^c{}_{\mu}k)\partial/\partial x_{ik} \tag{48.17}$$

Thus

$$\Delta_C(A_C) = |\det (g^{-1}(\mathrm{Re}\,\partial F_C{}^a{}_{\mu}k/\partial x_{ik})D^\mu + f^{abc}A_C{}^{b\mu}(\mathrm{Re}\,F_C{}^c{}_{\mu}k)\partial/\partial x_{ik}| \tag{48.18}$$

where | ... | represent absolute value.

We can rewrite this Faddeev-Popov determinant as a path integral over anti-commuting c-number fields with a ghost Lagrangian:

$$\Delta_C(A_C) = \lim_{r \to \infty} \int D\chi_C{}^* D\chi_C \exp[ir^{-2} \int d^7x \, \mathscr{L}_C{}^{ghost}(x)] \qquad (48.19)$$

where r is a constant that is taken to the limit ∞, and where

$$\mathscr{L}_C{}^{ghost}(x) = \chi_C{}^*(x)\{D^\mu D_\mu + r^2 t^a[(Re \, \partial F_C{}^a{}_{\mu k}/\partial x_{ik})D^\mu + gf^{abc} A_C{}^{b\mu}(Re \, F_C{}^c{}_{\mu k})\partial/\partial x_{ik}]\}\chi_C(x) \qquad (48.20)$$

where t^a is a 3×3 matrix of the <u>3</u> representation of color SU(3) and $\chi_C(x)$ is a three row field in the <u>3</u> representation. *The introduction of $D^\mu D_\mu$ is based on consistency with the complexon formalism. It is needed to establish a perturbative expansion of the path integral. Its effect vanishes in the limit $r \to \infty$ reducing ghost loops of this type to point interactions.* The reader will note that second order and third order derivative terms appear in the interaction in $\mathscr{L}_C{}^{ghost}(x^\mu)$ and raise the issue of non-renormalizable divergences. If one uses the Two-Tier approach to quantum field theory developed by Blaha (2005a) then all potential divergences disappear. *The Two-Tier formulation of the pure, complexon, Yang-Mills theory that we are discussing is finite.*

The complete pure complexon, Yang-Mills path integral is

$$Z(J^\mu) = N \int DA_C D\chi^* D\chi D\chi_C{}^* D\chi_C \Delta_{FP}(A_C)\delta(F(A_C))\Delta_C(A_C)\delta(F_C(A_C)) \cdot$$
$$\cdot \exp\{i\int d^7y \, [\mathscr{L} + J^\mu A_{C\mu}]\} \qquad (48.21)$$

where

$$\mathscr{L} = \mathscr{L}_{CG} + \mathscr{L}^{ghost} + \mathscr{L}_C{}^{ghost} \qquad (48.22)$$

with the lagrangian terms specified by eqs. 48.11, and 48.20.

49. Unified SuperStandard Theory Multi-Quark Particles, and Dark Matter 'Chemistry'

49.1 New Multi-quark Particles such as Penta-Quark Particles

The CERN LHC has found evidence for penta-quark particles – particles consisting of five quarks.[308] This discovery is a step beyond the tetraquark (four quark) particle named $Z_C(3900)$ found, and announced,[309] in 2013. The Z_C particle could be viewed as two D-mesons bound together by the strong color force in a color singlet "hadron molecule."

Similarly, the pentaquark could be viewed as a particle consisting of a two quark meson and a three quark baryon formed into a type of molecule by the strong interaction.

49.2 Multi-Quark Molecules – Quark Chemistry

These recent discoveries are the initial steps to a spectrum of multi-quark particles consisting of

$$2m + 3b$$

quarks where m is the number of "meson-like" constituents and b is the number of "baryon-like" constituents. The pentaquark has $m = b = 1$ constituents.

We are now confronted by a new type of "molecule" where the binding force is not electromagnetic but, instead, the strong color force. The tight binding of these "molecules" by the strong force makes them effectively into particles. But they are

---

[308] Announced July 15, 2015. See arXiv.org/abs/1507.03414.

[309] June 20, 1013. Discovery by the Belle Collaboration, High Energy Accelerator Research Organization, in Tsukuba, Japan; and BESIII Collaboration at the Beijing Electron Positron Collider in China. Phys. Rev. Lett. **110**, 252001, 2013 and **110**, 252002, 2013 .

perhaps more comparable to the binding of protons and neutrons into atomic nuclei by the nuclear force.

The development of a quark molecular chemistry would, if molecules were produced in quantity, lead to quark materials with extremely important physical characteristics – quark matter – with both new and important "chemical" properties. However, quark molecules have an extraordinarily short lifetime and so the creation of a number of quark molecules, and their fabrication into materials, does not appear to be feasible. Nevertheless, their theoretical and experimental study might serve to drive conventional chemistry and materials science forward.

49.2.1 Quarkium Particles Numbering Scheme

If multi-quark particles are found to be numerous with some being 'quasi-stable' in the sense that their decays are much longer than expected, then it might be sensible to have a labeling standard for these particles. For u and d quark particles with, perhaps, small admixtures of strange quarks one could define them as Q(m, b) specifying their meson and baryon parts.

49.2.2 Quarkium Islands of Stability?

If the set of multi-quark particles becomes as numerous as atomic nuclei then it is possible that an analogous 'island of stability' may be found with longer-lived particles just as were found for atomic nuclei.

49.3 Large Multi-Quark "Molecules" and Eventually Quark Stars

One can envision the eventual creation of very "large" multi-quark particles/molecules. Taken to the limit of extremely large "molecules" of extraordinary size it appears possible to develop a dynamics of quark stars where the dominant force is not the nuclear force as we usually know it and electromagnetism, but rather the strong force. Astronomy has yet to discover an unambiguous quark star.

Multi-quark molecules, like atomic nuclei, can be expected to have at least two types of atomic-like models: 1) a shell model with shells of meson-like and baryon-like constituents, and 2) a liquid drop model with constituents floating within the molecule.

The development of these models remains for the future since the qualitative and quantitative features of the inter-constituent strong force is totally unknown at present.

49.4 Dark Matter Features: Particle Spectrum and Basic Chemistry of Dark Matter, Dark Matter Bodies

In chapter 39 of Blaha (2012a) we initially described Dark Matter particles, Dark atoms, the Dark Periodic Table, and Dark basic chemistry – all based on the additional SU(2)⊗U(1) symmetry in the SuperStandard Theory that more or less mirrors normal ElectroWeak symmetry. The major differences were 1) that Dark quarks are assumed to be SU(3) singlets as suggested by the known weakness of Dark Matter interactions with normal matter and 2) that *physical* hadron charges – both electric charge and Dark electric charge – are quantized with whole number values. (Dark quarks are the emasculated "hadrons" of this theory since they do not experience the Strong Interaction.) This chapter contains material from chapter 39 of Blaha (2012a) for completeness as well as some additional thoughts.

Another important issue is the effect of gravitation on Dark Matter—Can Dark matter aggregate under the force of gravity to form galactic clusters, galaxies, and smaller objects such as Dark suns and perhaps Dark planets? We discuss these possibilities below.

This chapter is based on the assumption of four generations.[310] Previous chapters present a strong theoretical case for four generations of fermions.

49.5 Fundamental Dark Particles

We now consider the Dark particles that are associated with our new Dark SU(2)⊗U(1) symmetry. Since Dark Matter only interacts weakly with known matter, Dark particles must be color singlets. Thus there will be 12 (in the three generation case) or 16 (in the four generation case) Dark particles (plus their antiparticles.) Recent experiments suggest Dark particles have extremely large masses of the order of 8.6

---

[310] This chapter is based on the assumption of one layer of four generations. The introduction of the Layer group and four layers of four fermion generations dramatically increases the number of Dark fermion states.

GeV/c or larger. Dark quarks are color singlets with complex 3-momenta in our Unified SuperStandard Theory.

"Periodic Table" of 'Known' Fermions

NORMAL ELECTROWEAK FERMION DOUBLETS

Generation Quarks	Leptons		Color Triplet	
	Real-Valued 3-Momenta		Complex-Valued 3-Momenta	
1	e	ν_e	$u_1\ u_2\ u_3$	$d_1\ d_2\ d_3$
2	μ	ν_μ	$c_1\ c_2\ c_3$	$s_1\ s_2\ s_3$
3	τ	ν_τ	$t_1\ t_2\ t_3$	$b_1\ b_2\ b_3$
4^{311}	υ	ν_υ	$V_1\ V_2\ V_3$	$W_1\ W_2\ W_3$

DARK SU(2)⊗U(1) FERMION DOUBLETS

Generation Dark Quarks	Dark Leptons		Color Singlet	
	Real-Valued 3-Momenta		Complex-Valued 3-Momenta	
1	e_D	ν_{eD}	u_D	d_D
2	μ_D	$\nu_{\mu D}$	c_D	s_D
3	τ_D	$\nu_{\tau D}$	t_D	b_D
4?	υ_D	$\nu_{\upsilon D}$	V_D	W_D

Figure 49.1 "Periodic Table" of four generations of normal and Dark fundamental fermions assuming only one layer of fermions.

---

[311] Recently it appears that a new heavier neutrino has been found, that we denote ν_υ, where υ is the lower case Greek letter Ypsilon. We suggest that there is also a corresponding charged heavy electron-like particle, that we denote υ, to be found.

In addition to Dark fermions there will be four $SU(2) \otimes U(1)$ Dark gauge bosons – also with large masses:

Dark Particle Gauge Bosons

U(1): $W'_0{}^\mu$	SU(2): $W'_1{}^\mu$	$W'_2{}^\mu$	$W'_3{}^\mu$

49.6 Dark Particle Chemistry

The chemistry of Dark Matter will be different from the chemistry of known matter due to the absence of the color interaction. Dark particles cannot combine through a strong interaction to form a hadronic spectrum, or atomic nuclei, such as we see in normal matter. Thus Dark particle atoms are like hydrogen atoms, and consist of a Dark quark particle bound to a Dark lepton particle by the Dark electric force assuming Dark charge is quantized and has equal integer absolute values for Dark quarks and leptons. More complex Dark molecules are also possible as we will see later.

Suppose all Dark charged particles have charge $\pm 1 e_D$ (e_D is the Dark unit of charge which may possibly be equal to e, the electromagnetic charge), and each Dark doublet has a Dark particle with unit Dark charge and a neutral Dark particle. Then we would expect that (in the case of four generations) Dark Matter would consist of:

1. Lepton-like fundamental particles: Four Dark charged and four Dark charge neutral particles and their anti-particles.

2. Quark-like fundamental particles: Four Dark charged and four Dark charge neutral particles and their anti-particles.

3. There would be a total of eight neutral Dark quarks and leptons.

4. Atoms are composed of oppositely charged Dark particles of different types. There are 16 Dark "atoms" of the form leptoDark particle - Darkquark particle, $e_D u_D$, $e_D c_D$, $e_D t_D$, $\mu_D u_D$, $\mu_D c_D$, $\mu_D t_D$, $\tau_D u_D$, $\tau_D c_D$, $\tau_D t_D$, $e_D v_D$, $\mu_D v_D$, $\tau_D v_D$, $\omega_D v_D$, $\omega_D u_D$, $\omega_D c_D$, and $\omega_D t_D$ plus their anti-matter equivalents. There are 12 "quasi-stable" particle-

anti-particle combinations: six leptoDark - antileptoDark particle combinations, and six Darkquark - antiDarkquark particle combinations. (There is no attractive nuclear force.) All of these atoms are bound by the Dark electric force.

5. Simple molecules of the type of Figs. 49.2 – 49.4 below, and so on, based on Dark dipole interactions, Dark van der Waals forces and other Dark electromagnetic interactions are possible.

After a sufficiently long time, collisions would lead perhaps to the dominance of Dark particles and the "disappearance" of anti-Dark particles if the number of Dark particles is overwhelmingly dominant in a fashion similar to normal matter.[312] (The other possibility is not excluded.) The Dark Periodic Table is presented in Fig. 49.2.

Periodic Table of Simple Dark Particle Atoms
(Assuming Dark anti-particles are Annihilated)

$e_D u_D$	$\mu_D u_D$	$\tau_D u_D$	$\upsilon_D u_D$	d_D
$e_D c_D$	$\mu_D c_D$	$\tau_D c_D$	$\upsilon_D c_D$	s_D
$e_D t_D$	$\mu_D t_D$	$\tau_D t_D$	$\upsilon_D t_D$	b_D
$e_D v_D$	$\mu_D v_D$	$\tau_D v_D$	$\upsilon_D v_D$	w_D

Figure 49.2 "Periodic Table" of Simple Dark atoms assuming only one layer of fermions.

There also are similar antiparticle atoms. Bound states are assumed bound, perhaps into hydrogen-like atoms, through a Dark electromagnetic force. The last column consists of Dark charge neutral quarks. Antiparticle atoms of these states might also exist or be created through the Dark ElectroWeak interactions. One could extend the table with a

---

[312] The study of electron-positron production through Weak Interaction with Dark Matter by M. Aguilar et al, Phys. Rev. Letters **110**, 141102 (2013) does not seem to clarify this issue.

column of Dark neutrinos. The decays, and mixing between generations, remains to be determined in the unknown Dark Higgs sector.

The periodic table (Fig. 49.2) that we constructed is based on an analogy with the features of normal matter: quarks are much heavier than leptons and leptons "revolve" around the Dark quark nuclei. If this view is correct then one can conceive of a chemistry of Dark Matter with molecules bound by Dark electromagnetic forces. Pair bonding of Dark leptons would be possible and so one could conceive of a fairly complex Dark chemistry bounded by the fact that a Dark particle atom has only one Dark lepton. Thus there would only be 64 bound pairs of atoms[313] similar to H_2.

$$e_D$$
$$u_D \qquad u_D$$
$$e_D$$

Figure 49.3. Example of two Dark atoms binding in a manner similar to the binding of hydrogen molecules in H_2. Considering the possible combinations of quarks and leptons there are 64 bound varieties of this type. Chains of these molecules are also possible in principle.

Dark particles can be combined into chains and more complex molecules. A simple chain of Dark particles appears in Fig. 49.4.

$$e_D \qquad\qquad u_D$$
$$u_D \qquad u_D \quad e_D \qquad e_D$$
$$e_D \qquad\qquad u_D$$

Figure 49.4. Example of a simple Dark Matter chain.

---

[313] Plus anti-particle equivalents.

$$e_D \qquad\qquad u_D \qquad\qquad u_D \qquad\qquad e_D$$

$$u_D \qquad u_D \qquad e_D \qquad e_D \qquad e_D \qquad e_D \qquad u_D \qquad u_D$$

$$e_D \qquad\qquad u_D \qquad\qquad u_D \qquad\qquad e_D$$

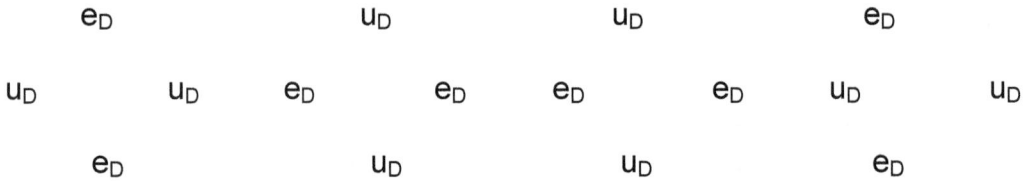

Figure 49.5. A segment of two strands interacting with each other. Other combinations are possible using other Dark charged fermions. This diagram is suggestive of a Dark form of DNA. Dark Life?

More complex chains as well as two and three dimensional bound aggregates are also possible. These considerations suggest something approaching the complexity of simple life forms may be possible.

Dark dipole effects could lead to the (weaker) binding of larger assemblages of Dark atoms. For example, if the masses of the leptonic and quark Dark particles are not too dissimilar, crystalline Dark Matter appears possible.

We thus arrive at a Dark Matter sector with much less variety than normal matter. It may, or may not, preclude the existence of Dark life forms, and possibly of Dark solids composed of Dark Matter. Solid Dark planets may exist. The detection of planet-like and star-like Dark Matter objects within the Dark Matter "cloud" surrounding a galaxy would be of great importance.

49.7 Gravitation and Dark Matter Bodies: Galactic Clusters, Galaxies

Dark Matter is thought to equal approximately 83% of the total matter in the universe. It is known to form halos around galaxies and to form filaments between galaxies in galactic clusters. It appears to influence the orbits of stars in galaxies possibly causing the rotational speed of stars at large radial distances from galactic centers to be roughly constant and independent of the radius.

When two galaxies collide it appears that the Dark Matter drags between the galaxies as they separate. Thus Dark Matter has forces as we suggested earlier.

We see these effect in the Dark Matter halos around galaxies. Galaxies are permeated with Dark Matter. Our galaxy, for example, appears to have ten times as much Dark Matter as normal matter.[314] Galaxies that are composed of predominantly Dark matter have been found.

49.8 Dark Stars, Dark Planets, Dark Globular Clusters—Or Just Darkened?

However we have not seen (as yet?) clumps of Dark Matter within our galaxy similar in size to stars and planets or globular clusters. Nor do we see the earth, or other Solar System planets, having a discrepancy in mass due to Dark Matter clumps within these bodies. Nor is there a Dark planet in our Solar System.

Yet there is no apparent reason why Dark Matter should not clump within planets (and our sun) causing a discrepancy in solar system dynamics and ostensibly the force of gravity. The ten to one dominance of Dark Matter in our galaxy creates a striking discrepancy with the apparent absence of Dark Matter gravitational effects in the Solar System.

49.8.1 Open Questions on Dark Matter in Our Solar System

1. Why does not the earth's mass have a significant contribution from the Dark Matter within the earth since our galaxy has 10 times more Dark Matter than normal matter? Does not Dark Matter clump somewhat around planet size objects?

2. Same question as 1 but applied to the sun. Where is the effect of the Dark Matter within the sun?

3. Why is there no Dark Jupiter in our Solar System with normal moons circling it?

---

[314] Therefore one would expect more gravitational phenomena reflecting the presence of Dark Matter locally in our galaxy than its effect on galactic rotation. Since Dark Matter hardly interacts with normal matter it would intersperse with normal matter throughout the galaxy forming a hidden part of the galaxy – present but almost completely not detectable. The hidden part would occupy the same space as normal galactic matter rather like an evanescent ghost in a horror movie.

4. Why are not the planetary dynamics of the Solar System affected by the presence of Dark Matter?

49.8.2 Open Questions on Dark Stars and Dark Globular Clusters of Stars

There are many stars in the galaxy composed of normal matter apparently. However the ten to one ratio of Dark Matter to normal matter in our galaxy should be reflected in stellar dynamics which is a supposedly well understood field. Stellar dynamics depends significantly on the mass of stars.

1. Why does not stellar dynamics take account of the Dark Matter contribution to the mass of stars?

2. When a star contracts generating gravitational energy why does not the contraction of the Dark Matter clumped within the star impact gravitationally on the contraction of the normal matter?

3. Why is not the evolution and structure of globular clusters of stars affected by their presumably large Dark Matter content?

4. Will we find a Dark star circled by some large normal matter planets?

5. Can we detect a Dark cluster by its gravitational effects?

49.8.3 Comments on these Questions

The only currently apparent resolution of these questions is a lack of significant Dark chemical interactions/reactions between Dark Matter atoms that may prevent the formation of Dark solids and liquids.[315] Dark nuclear decays similar to the normal nuclear reactions that power stars may also not exist. Thus no Dark stars. But Dark aggregates of the mass of the sun or more should be possible.

---

[315] Thus Dark atoms may not have Dark dipole forces or other multipole forces that would induce the formation of Dark Matter liquids and solids.

Dark Matter in galaxies appears to be rather like a distributed gas within a gravitational well formed by the combined masses of a galaxy. It also forms a filament between galaxies.[316] Within a galaxy Dark Matter appears dispersed in a somewhat uniform way with only very minor gravitational clumping by clusters, stars and planets.

---

[316] A galaxy composed overwhelmingly of Dark Matter has recently been found.

50. Complex General Relativity Reformulated

We have seen that Complex Special Relativity is the basis of flat space-time phenomena. Flat space-time coordinates are complex-valued in general. The real-valued coordinates that we experience in everyday life are the result of our measuring instruments: clocks and rulers. Real-valued coordinates are generated from complex-valued coordinates by Reality group transformations.

If flat space-time is governed by Complex Special Relativity then it is clear that curved 'space-time' is governed by Complex General Relativity. Here again there is a Reality group   the General Relativistic Reality group – a U(4) group – that maps complex-valued General Relativity coordinates to real-valued curved coordinates.[317] There is a corresponding U(4) Internal Symmetry Reality group that we call the *Species group*. This group rotates fermions.

We can isolate the General Relativistic Reality group by factoring complex General Relativistic coordinate transformations into parts that consist of a real-valued General Coordinate transformation and complex-valued coordinate transformations. It will be apparent that the General Relativistic Reality group emerges in this discussion. The Species group is distinct from the Internal Symmetry Reality group of the SuperStandard Theory: R = SU(3)⊗SU(2)⊗U(1)⊗SU(2)⊗U(1)⊗SU(3). We begin by defining the tetrad notation.

50.1 Tetrad (Vierbein) Formalism

The *vierbein* formalism begins with the Equivalence Principle that allows us to define an inertial coordinate system in the neighborhood of any point Z in space-time. We will use the notation $\varsigma^{\alpha}(Z)$ to denote the inertial coordinates at Z. We define a tetrad or vierbein as

---

[317] Much of this chapter appears in Blaha (2016h) and (2017a).

$$v^{\alpha}{}_{\mu}(x) = (\partial\varsigma^{\alpha}(x)/\partial x^{\mu})_{x=Z} \tag{50.1}$$

and, in a neighborhood of Z, we can invert the relation between ς and x to define an inverse

$$w^{\mu}{}_{\alpha}(x) = (\partial x^{\mu}(\varsigma)/\partial\varsigma^{\alpha})_{x=X} \tag{50.2}$$

such that

$$w^{\mu}{}_{\alpha}(x)v^{\alpha}{}_{\nu}(x) = \delta^{\mu}{}_{\nu}$$
$$w^{\mu}{}_{\beta}(x)v^{\alpha}{}_{\mu}(x) = \delta^{\alpha}{}_{\beta} \tag{50.3}$$

In real General Relativity all *tetrads* are real-valued. In Complex General Relativity a *tetrad* $v^{\alpha}{}_{\mu}(x)$ is complex-valued.

The metric at a curved space-time point X is defined in terms of *tetrads* as

$$g_{\rho\sigma}(x) = \eta_{\alpha\beta}\, v^{\alpha}{}_{\rho}(x)v^{\beta}{}_{\sigma}(x) \tag{50.4}$$
$$g^{\rho\sigma}(x) = \eta^{\alpha\beta}\, w^{\rho}{}_{\alpha}(x)w^{\sigma}{}_{\beta}(x)$$

The inverse of a *tetrad* transformation can also be expressed as

$$w_{\beta}{}^{\nu}(x) = v_{\beta}{}^{\nu}(x) = \eta_{\beta\alpha}g^{\nu\mu}(x)v^{\alpha}{}_{\mu}(x)$$

Then a *tetrad* and its inverse satisfy

$$v^{\alpha}{}_{\mu}(x)v_{\beta}{}^{\mu}(x) = \delta^{\alpha}{}_{\beta} \tag{50.5}$$

and

$$v^{\alpha}{}_{\mu}(x)v_{\alpha}{}^{\nu}(x) = \delta^{\nu}{}_{\mu}$$

There are two general types of space-time transformations that can be performed on a tetrad.

1. A complex-valued (possibly real-valued) General Relativistic coordinate transformation:

$$v'^{\alpha}{}_{\mu}(x) = \partial x^{\nu}/\partial x'^{\mu} \; v^{\alpha}{}_{\nu}(x)$$

2. A complex-valued, local *Lorentzian transformation*

$$v'^{\beta}{}_{\mu}(x) = \Lambda(x)^{\beta}{}_{\alpha} \, v^{\alpha}{}_{\mu}(x)$$

where $\Lambda(x)^{\beta}{}_{\alpha}$ is an element of a subset of the local Complex Lorentz Group.

The local Lorentzian transformations $\Lambda(x)^{\beta}{}_{\alpha}$ consist of local Lorentz transformations that are real-valued, and complex-valued Lorentz transformations. Both types of transformations satisfy the orthogonality condition:

$$\eta_{\alpha\beta}\Lambda^{\alpha}{}_{\rho}(x)\Lambda^{\beta}{}_{\sigma}(x) = \eta_{\rho\sigma} \tag{50.6}$$

Thus the *tetrad* partakes of both local (position dependent) General Relativistic transformations and local Lorentzian transformations.

50.2 Complex General Relativistic Transformations

The General Relativistic Reality group interaction emerges from complex General Relativistic transformations. We can separate elements of the set of all complex General Coordinate transformations into a product of two factors: a real-valued General Coordinate transformation and a complex-valued General Coordinate transformation. The set of complex factors can be further factored into those that satisfy

$$\Lambda(\omega, \mathbf{u})^{\mathrm{T}}G\Lambda(\omega, \mathbf{u}) = G \tag{50.7}$$

and those that do not. We then see that the set of those that do not satisfy the above equation form a curved space representation of the U(4) group under 'multiplication' of transformations.

The elements of the set of real and complex General Coordinate transformations whose flat complex space-time limit satisfy the above equation form the elements of the Complex Lorentz group.[318]

We thus find the set of all 4-dimensional complex, curved space General coordinate transformations can be visualized as in Fig. 50.1. The next section describes the interplay of the three parts displayed in Fig. 50.1.

50.3 Structure of Complex General Coordinate Transformations

Complex General Coordinate transformations can be uniquely factored into products of two terms, which will later be further factored into three factors. They have the form

$$\partial x''^{\nu}(x)/\partial x^{\mu} = U(x'')^{\nu}{}_{\beta}\, \partial x'^{\beta}(x)/\partial x^{\mu} \tag{50.8}$$

where

$$x''^{\nu}(x) = U(x'')^{\nu}{}_{\beta} x'^{\beta}$$
$$x'^{\mu}(x) = U^{-1\mu}{}_{b}(x'')\, x''^{b}$$

where $U(x')^{\nu}{}_{\beta}$ is complex and where $\partial x'^{\beta}(x)/\partial x^{\mu}$ is a purely real General Coordinate transformation.

We define

$$U(x'')^{\mu}{}_{\nu} = w^{\mu}{}_{a}(x'')\big[\exp(i\textstyle\sum_{k} g_{k}\Phi_{k}(x'')\tau_{k})\big]^{a}{}_{b}\, v^{b}{}_{\nu}(x'') \tag{50.9}$$

$$U^{-1}(x'')^{\mu}{}_{\nu} = w^{\mu}{}_{a}(x'')\big[\exp(-i\textstyle\sum_{k} g_{k}\Phi_{k}(x'')\tau_{k})\big]^{a}{}_{b}\, v^{b}{}_{\nu}(x'')$$

where the constants g_{k} are real, and Φ_{k} and τ_{k} are hermitean. The uniqueness of the factorization follows from the Reality group (and U(4)) property that any complex 4-vector can be uniquely mapped to any specified real 4-vector.

---

[318] It is this part of curved space-time General Relativity that becomes the flat space-time Complex Lorentz group, which leads to the SU(3)⊗SU(2)⊗U(1)⊗SU(2)⊗U(1)⊗SU(3) Standard Model Reality group.

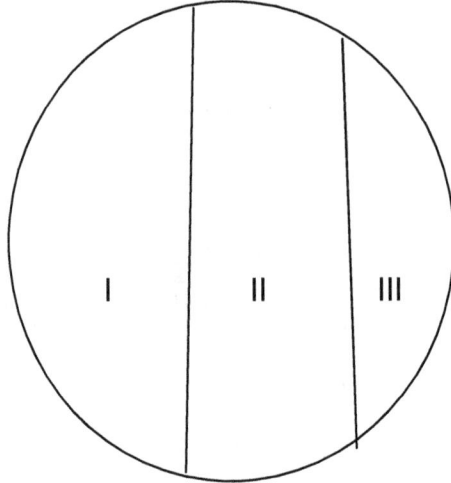

Figure 50.1. A visualization of the set of General Coordinate transformations separated into real-valued General coordinate transformations (part I), complex transformations that satisfy $\Lambda(\omega, u)^T G \Lambda(\omega, u) = G$ (part II), and complex transformations that do not satisfy $\Lambda(\omega, u)^T G \Lambda(\omega, u) = G$ (part III). Part I and part II combine in the limit of flat space-time to form the Complex Lorentz group. Parts II and III elements form a U(4) group that we call the General Relativistic Reality group.

Given the factorization above it becomes possible to separate the affine connection correspondingly.

50.4 Complex Affine Connection – General Relativistic Reality Group

The structure of a complex general coordinate transformation enables us to calculate its affine connection for later use in determining the covariant derivative, and the dynamic equations. First the transformation to the real-valued x' coordinates from inertial coordinates is

$$\Gamma^{\sigma}{}_{\lambda\mu}(x') = \partial x'^{\sigma}/\partial\varsigma^{\rho}\ \partial^2\varsigma^{\rho}/\partial x'^{\lambda}\partial x'^{\mu} \tag{50.10}$$

Next the Reality group transformation has the affine connection

$$\Gamma^{\sigma}_{\lambda\mu}(x'') = \partial x''^{\sigma}/\partial \varsigma^{\rho} \; \partial^2 \varsigma^{\rho}/\partial x''^{\lambda}\partial x''^{\mu}$$

which can be re-expressed as

$$\Gamma^{\sigma}_{\lambda\mu}(x'') = \partial x''^{\sigma}/\partial x'^{\beta} \; \partial x'^{\beta}(\varsigma)/\partial \varsigma^{\rho} \; \partial/\partial x''^{\mu}[\partial \varsigma^{\rho}/\partial x'^{\alpha} \; \partial x'^{\alpha}/\partial x''^{\lambda}]$$
$$= \partial x''^{\sigma}/\partial x'^{\beta} \; \partial x'^{\alpha}/\partial x''^{\lambda} \; \partial x'^{\gamma}/\partial x''^{\mu} \; \Gamma^{\beta}_{\alpha\gamma}(x') + \partial x''^{\sigma}/\partial x'^{\beta} \; \partial^2 x'^{\beta}/\partial x''^{\lambda}\partial x''^{\mu}$$

$$(50.11)$$

Next substituting the General Relativistic Reality group transformation

$$x''^{\nu}(x) = U(x'')^{\nu}_{\beta} x'^{\beta}$$
$$x'^{\mu}(x) = U^{-1}(x'')^{\mu}_{\beta} x''^{\beta}$$

together with

$$\partial x''^{\sigma}/\partial x'^{\beta} = \partial[U(x'')^{\sigma}_{\alpha} x'^{\alpha}]/\partial x'^{\beta} = U(x'')^{\sigma}_{\beta} + x'^{\alpha} \, \partial U(x'')^{\sigma}_{\alpha}/\partial x'^{\beta}$$

$$\partial x'^{\sigma}/\partial x''^{\beta} = \partial[U^{-1}(x'')^{\sigma}_{\alpha} x''^{\alpha}]/\partial x''^{\beta} = U^{-1}(x'')^{\sigma}_{\beta} + x''^{\alpha} \, \partial U^{-1}(x'')^{\sigma}_{\alpha}/\partial x''^{\beta}$$

we find the second term above is the Reality fields affine connection

$$\Gamma_R^{\sigma}_{\lambda\mu}(x'') = \partial[U(x'')^{\sigma}_{\alpha} x'^{\alpha}]/\partial x'^{\beta} \; \partial\{\partial[U^{-1}(x'')^{\beta}_{\alpha} x''^{\alpha}]/\partial x''^{\lambda}\}/\partial x''^{\mu}$$

and so we find the affine connections are approximately additive. Thus approximately

$$\Gamma^{\sigma}_{\lambda\mu}(x'') = \Gamma_{GR}^{\sigma}_{\lambda\mu}(x') + \Gamma_R^{\sigma}_{\lambda\mu}(x'')$$

if $x''^{\sigma} \simeq x'^{\sigma}$.

A complex transformation of types II and III in Fig. 50.1 has the form:

$$U(x'')^{\mu}_{\nu} = w^{\mu}_a(x'')[\exp(i \sum_k \Phi_k(x'')\tau_k)]^a_b \, l^b_{\nu}(x'')$$
$$U^{-1}(x'')^{\mu}_{\nu} = w^{\mu}_a(x'')[\exp(-i\sum_k \Phi_k(x'')\tau_k)]^a_b \, l^b_{\nu}(x'')$$

where τ_k is a U(4) generator matrix. Its infinitesimal transformation is approximately

$$U(x'')^\nu_\beta \approx \delta^\nu_\beta + i \sum_k \Phi_k(x'')[\tau_k]^\nu_\beta \qquad (50.12)$$

$$U^{-1}(x'')^\nu_\beta \approx \delta^\nu_\beta - i \sum_k \Phi_k(x'')[\tau_k]^\nu_\beta$$

using the *vierbein* flat space-time limits

$$w^\mu_a(x'') \approx \delta^\mu_a$$
$$l^b_\nu(x'') \approx \delta^b_\nu$$

where

$$\Phi_k(x) = \int^x dy_\lambda A_{Rk}{}^\lambda(y) \qquad (50.13)$$

Then

$$\Gamma_R{}^\sigma{}_{\lambda\mu} = -\tfrac{1}{2}i\{\sum_k A_{Rk}(x'')_\mu[\tau_k]^\sigma{}_\lambda + \sum_k A_{Rk}(x'')_\lambda[\tau_k]^\sigma{}_\mu\} \qquad (50.14)$$
$$= A_R{}^\sigma{}_{\mu\lambda} + A_R{}^\sigma{}_{\lambda\mu}$$

(summed over k) with the matrix $A_R{}^\sigma{}_{\mu\lambda}$ given by

$$A_R{}^\sigma{}_{\mu\lambda} = -\tfrac{1}{2}i\sum_k A_{Rk\mu}[\tau_k]^\sigma{}_\lambda \qquad (50.15)$$

with $A_R{}^\sigma{}_{\mu\lambda}$ transformable to matrix row and column numbers

$$A_{R_{flat}}{}^{\mu a}{}_b = A_{R_{flatk}}{}^\mu[\tau_k]^\sigma{}_\lambda \delta_\sigma{}^a \delta^\lambda{}_b$$

using the flat space-time vierbein values, and so $A_{R_{flat}}{}^a{}_{\mu b}$ may be written in matrix form as

$$A_{R_{flat\mu}} = -\tfrac{1}{2}i\sum_k A_{R_{flatk\mu}}\tau_k \qquad (50.16)$$

In the flat space-time limit the $A_{Rk}{}^{\lambda}(y)$ becomes the Coordinate Species group U(4) gauge fields $A_{R_{flatk}}{}^{\lambda}(y)$.

The relevant *quadratic* $A_R{}^{\sigma}{}_{\mu\lambda}$ terms from eq. 22.22 below that are needed to find the dynamic equation for the gauge fields $A_{R_{flat}}{}^i{}_{\mu}$ are contained in

$$\mathcal{L}_A = \mathrm{Tr}\ \sqrt{g}[M\partial_\nu R^1{}_{\sigma\mu}\partial^\nu R^{2\sigma\mu} + aR^1{}_{\sigma\mu}R^{2\sigma\mu} + bg^{\sigma\mu}(R^1{}_{\sigma\mu} + R^2{}_{\sigma\mu}) + 1/4(g_{\mu\nu} + g^2{}_{\mu\nu})T^{\mu\nu}] \tag{50.17}$$

We can let

$$R^i{}_{\sigma\mu} = R^{i\beta}{}_{\sigma\beta\mu} \equiv \partial_\mu(A_R{}^{i\beta}{}_{\sigma\beta} + A_R{}^{i\beta}{}_{\beta\sigma}) - \partial_\beta(A_R{}^{i\beta}{}_{\sigma\mu} + A_R{}^{i\beta}{}_{\mu\sigma}) \tag{50.18}$$

for i = 1, 2. In the flat space-time limit we chose the Landau gauge

$$\partial_\mu A_{R_{flat}}{}^{i\mu a}{}_b = 0 \tag{50.19}$$

As a result

$$R^i{}_{\sigma\mu} \equiv \partial_\mu(A_R{}^{i\beta}{}_{\sigma\beta} + A_R{}^{i\beta}{}_{\beta\sigma}) \tag{50.20}$$

Using

$$A_{R_{flat}}{}^{i\mu\sigma}{}_\lambda = A_{R_{flat}}{}^{i\mu}[\tau_k]^a{}_b\delta^\sigma{}_a\delta_\lambda{}^b \tag{50.21}$$
$$A_{R_{flat}}{}^i{}_\mu = -\tfrac{1}{2}i\sum_k A_{R_{flat}}{}^i{}_{k\mu}$$

and taking the trace in eq. 50.17 we obtain

$$\mathcal{L}_A = \mathrm{Tr}\ \sqrt{g}[8M\partial_\nu\partial_\mu A_{R_{flat}}{}^1{}_\sigma\partial^\nu\partial^\mu A_{R_{flat}}{}^{2\sigma} + 8a\partial_\mu A_{R_{flat}}{}^1{}_\sigma\partial^\mu A_{R_{flat}}{}^{2\sigma} + 1/4(g_{\mu\nu} + g^2{}_{\mu\nu})T^{\mu\nu}] \tag{50.22}$$

in the flat space-time limit. Eq. 50.17 needs to take account of the complex nature of $(g_{\mu\nu} + g^2{}_{\mu\nu})$ until transformed by the infinitesimal form of the complex Reality transformation:

$$(g_{\beta\alpha} + g^2{}_{\beta\alpha})' \rightarrow U\,(x'')_{\beta}{}^{\mu}(g_{\mu\nu} + g^2{}_{\mu\nu})U^{-1}(x'')^{\nu}{}_{\alpha}$$

$$= (\delta_{\beta}{}^{\mu} + i\sum_k \Phi_k(x'')[\tau_k]_{\beta}{}^{\mu})(g_{\mu\nu} + g^2{}_{\mu\nu})(\delta^{\nu}{}_{\alpha} - i\sum_k\Phi_k(x'')[\tau_k]^{\nu}{}_{\alpha})$$

$$\cong (g_{\beta\alpha} + g^2{}_{\beta\alpha}) + i\{\sum_k \Phi_k(x'')[\tau_k]_{\beta}{}^{\mu} - i\sum_k \Phi_k(x'')[\tau_k]^{\nu}{}_{\alpha})(g_{\mu\nu} + g^2{}_{\mu\nu})$$

$$\cong (g_{\beta\alpha} + g^2{}_{\beta\alpha}) + i\sum_k \Phi_k(x'')\{[\tau_k]_{\beta\alpha} - [\tau_k]_{\beta\alpha}\} \qquad (50.23)$$

Approximating $\Phi_k(x'')$ with an infinitesimal line we find

$$\sum_k\Phi_k(x) \cong \delta x_\lambda A_{R\mathrm{flat}}{}^{1\lambda}(x) \qquad (50.24)$$

by eq. 50.13. Thus

$$1/4(g_{\mu\nu} + g^2{}_{\mu\nu})T^{\mu\nu} \cong \tfrac14\,[(g_{\beta\alpha} + g^2{}_{\beta\alpha}) + i\,\delta x_\lambda A_{R\mathrm{flat}}{}^{1\lambda}(x)\{[\tau_k]_{\beta\alpha} - [\tau_k]_{\beta\alpha}\}]T^{\alpha\beta}$$

Applying the canonical Euler-Lagrange method we obtain the dynamical equations (using integration by parts to handle higher order derivative terms):[319]

$$\Box^2 A_{R\mathrm{flat}}{}^1{}_{\sigma} + (a/M)\Box A_{R\mathrm{flat}}{}^1{}_{\sigma} + i\delta x_{\sigma}\{[\tau_k]_{\beta\alpha} - [\tau_k]_{\beta\alpha}\}T^{\alpha\beta}/(32M) = 0 \qquad (50.25)$$

$$\Box^2 A_{R\mathrm{flat}}{}^2{}_{\sigma} + (a/M)\Box A_{R\mathrm{flat}}{}^2{}_{\sigma} = 0 \qquad (50.26)$$

Since the $A_{R\mathrm{flat}}$ gauge field is gravitational in nature it exists, as eq. 50.25 shows, as a type of gravitational interaction whose source is the energy-momentum tensor. Following the standard derivation of the gravitational potential we find the Coulomb interaction of $A_{R\mathrm{flat}}{}^{10}$.

50.5 Species Interaction Gravity Potential

Assuming that we are dealing with non-relativistic matter we can calculate the gravity potential contribution from section 54.4 below:[320]

---

[319] It is possible that the Reality transformation also depends on $A_{R\mathrm{flat}}{}^2{}_{\sigma}$. Then eq. 50.26 would have an energy-momentum tensor term as well. Consequently there would be an additional interaction of the same form as in eq. 50.27 below.

$$V_{GA1}(\mathbf{x}) = -\,(1/32M) \int d^3k \, \exp(i\mathbf{k}\cdot\mathbf{x})V_{GA1}(\mathbf{k})/(2\pi)^3 \qquad (50.27)$$

where

$$V_{GA1}(\mathbf{k}) = (\mathbf{k}^4 + (a/M)\mathbf{k}^2)^{-1} \qquad (50.28)$$

The eq. 50.28 can be separated into two terms:

$$V_{GA1}(\mathbf{k}) = (M/a)[1/k^2 - 1/(\mathbf{k}^2 + a/M)] \qquad (50.29)$$

which yield

$$V_{GA1}(\mathbf{r}) = -[1/(96\pi a)][1/r - e^{-m_A r}/r] \qquad (50.30)$$

Later in section 55.3 and 55.5 we will see

$$M \sim 1.46\times10^{179} \text{ GeV}^{-2} \qquad (50.31)$$

$$m_A = (a/M)^{\frac{1}{2}} = 2.49\times10^{-90} \text{ GeV} \qquad (50.32)$$

Since $a \cong 1$ the coupling constant

$$1/(96\pi a) \cong 0.0033 \qquad (50.33)$$

In comparison the electromagnetic fine structure constant is

$$\alpha \cong 0.0073$$

---

[320] **In absence of Higgs breaking. Later we add Higgs Species breaking terms that further increases the masses of the Species gauge fields. The addition of Higgs mass contributions to Species gauge fields does not conflict with the mass term found here but merely adds further contributions to Species gauge field masses. These contributions change these masses but would not appear to significantly change their order of magnitude.**

Thus the Species A_R coupling constant is approximately ½ of the fine structure constant.

50.6 Influence of Gravitational Gauge Field on Gravitation

The gravitational gauge field potential in eq. 50.30 has a relatively large coupling constant that makes it competitive with the known force of gravity at large distances of the scale of galactic distances. This force, which is negligible at short distances of the order of planetary distances, may be part of the MoND phenomena that affects the motion of stars.

50.7 PseudoQuantization of Affine Connections

Having obtained the form of the general affine connection we now PseudoQuantize them for later use in our unification program. We define

$$R^{1\beta}{}_{\sigma\nu\mu} = \partial_\mu H^{1\beta}{}_{\sigma\nu} - \partial_\nu H^{1\beta}{}_{\sigma\mu} + H^{1\gamma}{}_{\nu\sigma}H^{1\beta}{}_{\gamma\mu} - H^{1\gamma}{}_{\mu\sigma}H^{1\beta}{}_{\gamma\nu}$$
$$R^{2\beta}{}_{\sigma\nu\mu\rho} = \partial_\mu H^{2\beta}{}_{\sigma\nu} - \partial_\nu H^{2\beta}{}_{\sigma\mu} + H^{2\gamma}{}_{\nu\sigma}H^{2\beta}{}_{\gamma\mu} - H^{2\gamma}{}_{\mu\sigma}H^{2\beta}{}_{\gamma\nu} +$$
$$+ H^{1\gamma}{}_{\nu\sigma}H^{2\beta}{}_{\gamma\mu} - H^{1\gamma}{}_{\mu\sigma}H^{2\beta}{}_{\gamma\nu} + H^{2\gamma}{}_{\nu\sigma}H^{1\beta}{}_{\gamma\mu} - H^{2\gamma}{}_{\mu\sigma}H^{1\beta}{}_{\gamma\nu}$$

$$(50.34)$$

where

$$H^\sigma{}_{\nu\mu} = \Gamma_{GR}{}^\sigma{}_{\nu\mu} + \Gamma_{GR}{}^{2\sigma}{}_{\nu\mu} + \Gamma_R{}^{1\sigma}{}_{\nu\mu} + \Gamma_R{}^{2\sigma}{}_{\nu\mu} \qquad (50.35)$$

and where $\Gamma_{GR}{}^\sigma{}_{\nu\mu}$ and $\Gamma_{GR}{}^{2\sigma}{}_{\nu\mu}$ are affine connections for real-valued General Relativity, and $\Gamma_R{}^{1\sigma}{}_{\nu\mu}$ and $\Gamma_R{}^{2\sigma}{}_{\nu\mu}$ are affine connections for a complex-valued set of transformations embodying a U(4) gauge group that combine with real-valued General Relativistic transformations to yield Complex General Relativistic transformations.

The affine connection is most often viewed as a derived quantity—part of the derivation of the curvature tensor in General Relativity. It is typically derived from manipulations of the metric $g_{\mu\nu}$. However, the affine connection can also be viewed as a set of independent fields that become related to the metric via dynamic equations.

Some years ago A. Einstein and H. Weyl[321] pointed out that the metric and the affine connection should be treated as independent quantities and subject to independent arbitrary infinitesimal variations:

> "In contrast to Einstein's original "metric" conception in terms of the $g_{\nu\mu}$ there was later developed, by Eddington, by Einstein himself, and recently by Schrödinger, an affine field theory operating with the components $\Gamma^{\sigma}_{\nu\mu}$ of an affine connection. But in 1925 Einstein also advocated a "mixed" formulation by means of a lagrangian in which both the $g_{\nu\mu}$ and the $\Gamma^{\sigma}_{\nu\mu}$ are taken as basic field quantities and submitted to independent arbitrary infinitesimal variations.[322] In certain respects this seems to be the most natural procedure."

Following this approach we have introduced the above affine connections for use in the construction of our unification of particle interactions.

---

[321] H. Weyl, Phys. Rev. **77**, 699 (1950).
[322] A. Einstein, Sitzungsber., Preuss. Akad. Der Wissensch. (1925), p. 414.

51. Species Group U(4) Gauge Fields

From the discussion in sections 50.4 - 50.5 we see the flat space-time limit of $A_{Rk}{}^{\lambda}(y)$ is a local U(4) coordinate space gauge field. There are, *by assumption*,[323] a corresponding internal symmetry gauge fields $A_{Sk}{}^{\lambda}(y)$ – the Internal Symmetry U(4) Species Group gauge fields. The mathematical features of this field is quite similar to the U(4) Generation group fields. The interaction that appears in covariant derivatives is $g_8 A_S{}^{\mu}(x) = g_8 A_{Sk}{}^{\mu}(x) \mathbf{G}_{Sk}$ where the $\mathbf{G}_{Sk}$ are U(4) generator matrices and k is summed from 1, ... , 16.

Below we will see that the effect of the Internal Symmetry Species Group is to U(4) rotate the four components of each fermion's field. Since it preserves the species of each fermion we call this group the *Species Group*. It performs a U(4) rotation of the spinor representation of each fermion

We will see that the Higgs Mechanism breakdown of the Species Group endows each fermion with a mass contribution that breaks the scale invariance of the Unified SuperStandard Theory.

Since the Species Group Higgs Mechanism breaking gives each fermion a 'gravity generated' mass, and since this mass sets the mass scale for each fermion, we conclude later that the principle of the equality of inertial and gravitational mass is a direct consequence. ***Inertial mass equals gravitational mass.***

---

[323] In this discussion we *assume* that the Coordinate Species Reality Group with gauge fields $A_{Rk}{}^{\lambda}(y)$ has a corresponding Internal Symmetry Reality group that we call the Internal Symmetry Species Group. This assumption parallels the assumptions for the $SU(3) \otimes SU(2) \otimes U(1) \otimes SU(2) \otimes U(1)$ Internal Symmetry Reality Group presented in previous chapters.

51.1 Species Group Covariance

A Species Group transformation on a Dirac equation must be covariant. Consider the Dirac equation lagrangian term under an Internal Symmetry Species Group transformation:

$$\bar{\psi}(x)[i\gamma_\mu(\partial/\partial x_\mu - ig_8 A_{Sk}{}^\mu(x)G_{Sk}) - m]\psi(x) = 0 \tag{51.1}$$

summed over k. If we perform a Species group transformation U on lagrangian terms:

$$\bar{\psi}(x)[i\gamma_\mu(\partial/\partial x_\mu - ig_8 A_{Sk}{}^\mu(x)G_{Sk}) - m]U^{-1}U\psi(x)$$

or

$$\bar{\psi}(x)U^{-1}U[iU^{-1}U\gamma_\mu U^{-1}U(\partial/\partial x_\mu - ig_8 A_{Sk}{}^\mu(x)G_{Sk}) - m]U^{-1}U\psi(x)$$

we find

$$\bar{\psi}'(x)[i\gamma_\mu'U(\partial/\partial x_\mu - ig_8 A_{Sk}{}^\mu(x)G_{Sk})U^{-1} - m]\psi'(x)$$

where

$$\gamma_\mu'(x) = U\gamma_\mu U^{-1}$$

is locally equivalent to a Dirac matrix by Good's Theorem.[324] If we set

$$A'_S{}^\mu(x) = -(i/g_8)U[\partial U^{-1}/\partial x^\mu] + UA_S{}^\mu(x)U^{-1}$$

then the transformed lagrangian terms are

$$\bar{\psi}'(x)[i\gamma_\mu'(x)(\partial/\partial x_\mu - ig_8 A'_{Sk}{}^\mu(x)G_{Sk}) - m]\psi'(x) \tag{51.2}$$

They have the same form as the original terms above and thus the expression is covariant. We note the indices of the matrices G_{Sk} are spinor indices and so $G_{Sk}\gamma_\mu$ has an implicit spinor matrix summation. But the symmetry group is U(4).

---

[324] R. H. Good, Jr., Rev. Mod. Phys., **27**, 187 (1955).

The coordinate dependence of $\gamma_\mu'(x)$ introduces locality into the Dirac matrix. This locality might be viewed with concern except that an inverse Species group transformation exists that removes the locality. Thus the physical impact of this 'new' locality is eliminated.

51.2 Spontaneous Symmetry Breaking of the General Relativity U(4) Reality Group – The Species Group

We begin the discussion of the Internal Symmetry Species Group symmetry breaking[325] by defining a Higgs field η which is a Species group 4-vector

$$\eta = \begin{bmatrix} \rho_1 \\ \rho_2 \\ \rho_3 \\ \rho_4 \end{bmatrix} \tag{51.3}$$

where ρ_1, ρ_2, ρ_3 and ρ_4 are real fields.[326] Then the covariant derivative of η (taking account only of the Species group) is

$$D_{...\mu}\eta = \{\partial/\partial X^\mu + ... - \tfrac{1}{2}ig_8\Sigma\, A_{Sk}{}^\mu(x)G_{Sk}\} \begin{bmatrix} \rho_1 \\ \rho_2 \\ \rho_3 \\ \rho_4 \end{bmatrix}$$

$$\tag{51.3}$$

Following steps similar to eqs. 47.4 through 47.17 for the Generation Group symmetry breaking we find with ρ_i being the vacuum expectation value of the Higgs field:

---

[325] Since the Species gauge fields have been shown to have a mass it might seem redundant to introduce Higgs symmetry breaking as well. However the Higgs breaking introduces the further benefit of giving a mass term to each particle – thus establishing the equality of gravitational mass and inertial mass as we discuss in section 20.4.

[326] Each field ρ_i can be expressed as a PseudoQuantum field: $\rho_i = \varphi_{1i} + \varphi_{2i}$ where φ_{1i} has the vacuum expectation value $\rho_{i0.\ for\ i\ =\ 1,\ ...\ ,\ 4}$. Thus our PseudoQuantum field theory version is implemented easily.

$$(D_{...\mu}\eta)^{\dagger} D_{...}{}^{\mu}\eta = \partial\rho_1/\partial X^{\mu} \, \partial\rho_1/\partial X_{\mu} + \partial\rho_2/\partial X^{\mu} \, \partial\rho_2/\partial X_{\mu} + \partial\rho_3/\partial X^{\mu} \, \partial\rho_3/\partial X_{\mu} +$$
$$+ \partial\rho_4/\partial X^{\mu} \, \partial\rho_4/\partial X_{\mu} +$$
$$+ \tfrac{1}{4} g_8^2 \{ \rho_1^2 A_{S1}^2 + \rho_2^2 A_{S2}^2 + \rho_3^2 A_{S3}^2 + \rho_4^2 A_{S4}^2 +$$
$$+ (\rho_1^2 + \rho_2^2)(V_5^2 + V_6^2) + \tfrac{1}{4}(\rho_1^2 + \rho_3^2)(V_7^2 + V_8^2) +$$
$$+ (\rho_1^2 + \rho_4^2)(V_9^2 + V_{10}^2) + \tfrac{1}{4}(\rho_2^2 + \rho_3^2)(V_{11}^2 + V_{12}^2) +$$
$$+ (\rho_2^2 + \rho_4^2)(V_{13}^2 + V_{14}^2) + \tfrac{1}{4}(\rho_3^2 + \rho_4^2)(V_{15}^2 + V_{16}^2) \}$$

$$(51.4)$$

up to total divergences, which generate surface terms which we discard. We also assume that all fields satisfy the gauge condition

$$\partial A_{Si}{}^{\mu}/\partial X^{\mu} = 0 \qquad (51.5)$$

Eq. 51.4 shows all Species Group gauge fields have masses. Thus Species Group symmetry is completely broken. The combination of an ultra-weak coupling constant and very large gauge field masses results in extremely Species interactions.

We assume Species group gauge field masses to be very large – of the order of the Planck mass in view of its origin in Complex General Relativity.

51.3 Species Group Higgs Mechanism Contributions to Fermion Masses

The symmetry breaking of the Species Group results in a contribution to each fermion mass of all types, species, generations, and layers. The Species Group contributions to normal and Dark fermion mass terms are

$$\mathcal{L}^{Higgs}_{FermionMassesSpecies} = \Sigma_{s,g,l} \bar{\psi}_{sglL} \rho_s \, m_{sgl} \psi_{sglR} + \Sigma_{s,g,l} \bar{\psi}_{DsglL} \rho_s \, m_{Dsgl} \psi_{DsglR} + c.c.$$

$$(51.6)$$

The η field expectation value has components labeled ρ_s.[327] The mass matrices m_{sgl} and m_{Dsgl} are the complex constant Species mass matrix contributions for normal and Dark species.

51.4 Species Group Higgs Masses Shows Inertial Mass Equals Gravitational Mass

In Blaha (2016h) we showed that a Complex General Relativity transformation can be factored into the product of a complex-valued transformation and a real-valued General Coordinate transformation. The set of complex valued transformations form a U(4) group that we called the General Coordinate Reality group. The analogous Internal Symmetry Species Group has gauge fields that undergo spontaneous symmetry breaking and generate contributions to all fermion masses.

Since fermion field masses are now sums of ElectroWeak Higgs contributions, Generation group Higgs contributions, Layer group Higgs contributions, and General Coordinate Species Group contributions, and since the Species Group Higgs fields appear in all fermion masses, the equality of inertial and gravitational mass is proven. The Species Group Higgs particles' equations set the mass scale of gravitational mass, and thereby of all Higgs mass contributions. The scale of inertial masses equals to the scale of gravitational masses. **Since an expression cannot mix mass scales, the gravitational mass scale must be the same as the inertial mass scale.**

Inertial mass equals gravitational mass.

We have established the equality of inertial and gravitational mass at the short distance quantum level. In our view, this explanation is far more satisfying than basing the equality on a combination of large distance phenomena and quantum phenomena. As Einstein and Weyl have pointed out, all fundamental physics phenomena should be based on a local theory.

---

[327] The Higgs fields η... in our PseudoQuantum formulation are η... $= \varphi_{1...}(x) + \varphi_{2...}(x)$ as described earlier.

We have mapped Complex General Relativistic transformations consisting of U(4) transformations and Real General Relativistic transformations into transformations consisting of Internal Symmetry Species Group factors and Real General Relativistic transformations factors. The Higgs Mechanism breakdown of the Species Group has the important consequence that it prevents Species Group transformations that rotate between fermions and anti-fermions.

52. Derivation Status at this Point

At this point in our derivation of the Unified SuperStandard Theory we have completed the following:

1. Derived the dimension of space-time based on the requirement that Physical phenomena must support parallel processes.
2. We showed how fermion and boson quantum fields should each be factored into the inner product of a functional and a fourier coordinate expansion.
3. Assuming Complex Lorentz group coordinate symmetry we have derived the existence of exactly four fermion species and four Dark fermion species. Using the Reality group associated with the Complex Lorentz group as a guide we have postulated the existence of the Standard Model symmetry group SU(3)⊗SU(2)⊗U(1)⊗SU(2)⊗U(1)⊗SU(3) with the additional SU(2)⊗U(1) factors for Dark ElectroWeak symmetry.
4. We introduced a space of functionals, established a Lorentz group for this space and then developed a parallel to the Coordinate Space Reality group. This parallel group is the Internal Symmetry SU(3)⊗SU(2)⊗U(1)⊗SU(2)⊗U(1)⊗SU(3) group of the extended Standard Model. We extended this group to the Generation and Layers groups subsequently to form the SuperStandard Theory.
5. Assuming that the Unified SuperStandard Theory can be described by a lagrangian we have derived conserved particle numbers such as Baryon Number. The four conserved particle numbers for baryons, leptons, Dark baryons and Dark leptons suggest a U(4) symmetry that we call the Generation Group. This group's $\underline{4}$ representation gives a four generation fermion spectrum.
6. The enhanced free fermion lagrangian, now having four generations of fermions, has four new conserved fermion numbers for each of the four generations. The four conserved particle numbers for each generation are for baryons, leptons, Dark baryons and Dark leptons of each generation. These numbers suggest four U(4) symmetries – one U(4) symmetry for each generation. We call these symmetries Layer Groups symmetries. We then add indices to fermion fields of each generation to create a U(4) $\underline{4}$ representation. Thus we have four sets of four generations. We call each set of four generations a layer.

7. We assume each Layer has its own set of SU(3)⊗SU(2)⊗U(1)⊗SU(2)⊗U(1)⊗U(4) where the U(4) factor is for the Generation group. We assume the gauge interactions of each layer are different since transitions between layers due to these interactions have not been found experimentally. Introducing a U(4) Layer group factor for each of the generations we can specify the Unified SuperStandard Theory symmetry in a compact form as

$$[SU(3)\otimes SU(2)\otimes U(1)\otimes SU(2)\otimes U(1)\otimes SU(3)\otimes U(4)\otimes U(4)]^4$$

at this point in the derivation.

8. Turning to Complex General Relativity we find that it factors into a U(4) group, which becomes the Internal Symmetry Species Group, combined with Real General Relativity.

9. Thus the Unified SuperStandard Theory symmetry is

$$[SU(3)\otimes SU(2)\otimes U(1)\otimes SU(2)\otimes U(1)\otimes SU(3)\otimes U(4)\otimes U(4)]^4\otimes U(4)$$

plus Real General Relativity.

10. We also showed how to eliminate divergences in perturbation theory calculations using the Two-Tier formulation of Quantum Field Theory. The Two-Tier theory essentially smears space-time points at ultra-small distances eliminating divergences.

11. In addition we described PseudoQuantum Field Theory which enables quantization in any coordinate system, offers a cleaner formulation of symmetry breaking and supports canonical formulations of higher derivative field theories.

12. The Higgs Mechanism for symmetry breaking and particle mass generation for all the above interactions were also described.

The appearance of the integer 4 throughout the Unified SuperStandard Theory can be traced back to the axiom that physical processes can take place in parallel, which determined the dimension of space-time in our universe to be greater than or equal to 4.

53. The Riemann-Christoffel Curvature Tensor and Vector Boson Part of Unified SuperStandard Theory Lagrangian

In view of our goal of defining a unified theory of elementary particles and General Relativity we begin by defining a Riemann-Christoffel curvature tensor which we will use to construct a lagrangian for the theory.

53.1 The Covariant Derivative

The covariant derivative[328] which appears in fermion and gravitation equations uses the vector boson 10-vector:

$$^a\mathbf{A_I}^\mu(x) = (^ag_1{}^a\mathbf{A}_{SU(3)}{}^\mu(x_C),\ ^ag_2{}^a\mathbf{W}^\mu(x)\ ,\ ^ag_3{}^a\mathbf{A_E}^\mu(x),\ ^ag_4{}^a\mathbf{W_D}^\mu(x),\ ^ag_5{}^a\mathbf{A_{DE}}^\mu(x),\ ^ag_6{}^a\mathbf{U}^\mu(x),\ ^ag_7{}^a\mathbf{V}^\mu(x),$$
$$^ag_8{}^a\mathbf{A}_{DSU(3)}{}^\mu(x_C),\ ^ag_9{}^a\mathbf{U_D}^\mu(x),\ ^ag_{10}{}^a\mathbf{V_D}^\mu(x)) \tag{53.1}$$

where a labels the layer, a = 1, 2, 3, 4. We label the respective coupling constants in each layer a as $^ag_1$, $^ag_2$, …, $^a g_{10}$. In the equation above: the subscript 'D' labels Dark matter interactions, 'W' labels Weak fields, 'E' labels Electromagnetic fields, U labels U(4) Generation group fields, 'V' labels U(4) Layer group fields, U_D labels Dark U(4) Generation group fields, and V_D labels Dark U(4) Layer group fields. A_S labels the U(4) Species Group fields.

We define the sum over a and the components of the vector $^a\mathbf{A_I}^\mu(x)$ labeled with i = 1, 2 for each of the paired PseudoQuantum fields, by

$$\mathbf{C_I}^\mu(x) = \Sigma_{a,i}\ ^a\mathbf{A}_{Ii}{}^\mu(x) + \Sigma_i\ (g_{11}\mathbf{A_S}^{i\mu}(x) \tag{53.2}$$

---

[328] This section has equations obtained from Blaha (2017d) and (2018e).

We begin by considering the case of one layer of vector bosons below omitting the ^{a} superscript. **The generalization to four layers is straightforward.**

Using the above definitions the *PseudoQuantum* covariant derivative of a 4-vector Z_μ is

$$D_\nu Z_\mu = (\partial_\nu + iF_\nu)Z_\mu - H^\sigma{}_{\nu\mu}Z_\sigma \qquad (53.3)$$
$$= [g^\sigma{}_\mu \partial_\nu + ig^\sigma{}_\mu F_\nu - H^\sigma{}_{\nu\mu}]Z_\sigma$$
$$= [g^\sigma{}_\mu \partial_\nu + iD^\sigma{}_{\mu\nu}]Z_\sigma$$

where[329]

$$F^\mu = C_I{}^{1\mu}(x) + C_I{}^{2\mu}(x) + A_R{}^{1\mu} + A_R{}^{2\mu} + B^{1\mu} + B^{2\mu} \qquad (53.4)$$

and

$$H^\sigma{}_{\nu\mu} = \Gamma_{GR}{}^\sigma{}_{\nu\mu} + \Gamma_{GR}{}^{2\sigma}{}_{\nu\mu} \qquad (53.5)$$
$$D^\sigma{}_{\mu\nu} = g^\sigma{}_\mu F_\nu + iH^\sigma{}_{\nu\mu}$$

where we have abstracted the complex part of the complex affine connection into the U(4) gauge field $A_S{}^\mu$. $H^\sigma{}_{\nu\mu}$ is the real-valued part of the complex affine connection.

We define the full vector gauge field covariant derivative to be

$$D^\mu = \partial^\mu - i(C_I{}^{1\mu}(x) + C_I{}^{2\mu}(x)) \times \qquad (53.5a)$$

Commutation relations of the vector fields in F_μ are implicit when the covariant derivative is applied to vectors and tensors such as Z_σ. This is indicated by the $\times$ symbol above.

53.2 The Curvature Tensor

The curvature tensor applied to a 4-vector Z_β is[330]

$$R'^\beta{}_{\sigma\nu\mu}Z_\beta = g^\alpha{}_\mu(\partial_\nu + iF_\nu)g^\beta{}_\sigma(\partial_\alpha + iF_\alpha)Z_\beta - H^\alpha{}_{\mu\nu}g^\beta{}_\sigma(\partial_\alpha + iF_\alpha)Z_\beta + \qquad (53.6)$$

---

[329] We will omit the insertion of the spinor coupling constants of the spinor connection $B^{1\mu}$ and $B^{2\mu}$ in eq. 11.2 in the interests of simplifying expressions.

[330] **With an implicit summation over layers understood.**

$$+ H^{\alpha}{}_{\mu\nu}H^{\beta}{}_{\sigma\alpha}Z_{\beta} - g^{\alpha}{}_{\mu}(\partial_{\nu} + iF_{\nu})H^{\beta}{}_{\sigma\alpha}Z_{\beta} - H^{\gamma}{}_{\nu\sigma}\{g^{\alpha}{}_{\gamma}(\partial_{\mu} + iF_{\mu})Z_{\alpha} - H^{\alpha}{}_{\gamma\mu}Z_{\alpha}\} -$$
$$- \{\mu \leftrightarrow \nu\}$$

$$= ig^{\beta}{}_{\sigma}(\partial_{\nu}F_{\mu} - \partial_{\mu}F_{\nu} - i[F_{\nu}, F_{\mu}])Z_{\beta} + (\partial_{\mu}H^{\beta}{}_{\sigma\nu} - \partial_{\nu}H^{\beta}{}_{\sigma\mu} + H^{\gamma}{}_{\nu\sigma}H^{\beta}{}_{\gamma\mu} - H^{\gamma}{}_{\mu\sigma}H^{\beta}{}_{\gamma\nu})Z_{\beta}$$

$$= ig^{\beta}{}_{\sigma}(F_{E}{}^{1}{}_{\nu\mu} + F_{E}{}^{2}{}_{\nu\mu} + F_{W}{}^{1}{}_{\nu\mu} + F_{W}{}^{2}{}_{\nu\mu} + F_{DE}{}^{1}{}_{\nu\mu} + F_{DE}{}^{2}{}_{\nu\mu} + F_{DW}{}^{1}{}_{\nu\mu} + F_{DW}{}^{2}{}_{\nu\mu} + F_{SU(3)}{}^{1}{}_{\nu\mu} + F_{SU(3)}{}^{2}{}_{\nu\mu} +$$
$$+ F_{DSU(3)}{}^{1}{}_{\nu\mu} + F_{DSU(3)}{}^{2}{}_{\nu\mu} + F_{U}{}^{1}{}_{\nu\mu} + F_{U}{}^{2}{}_{\nu\mu} + F_{V}{}^{1}{}_{\nu\mu} + F_{V}{}^{2}{}_{\nu\mu} + F_{UD}{}^{1}{}_{\nu\mu} + F_{UD}{}^{2}{}_{\nu\mu} + F_{VD}{}^{1}{}_{\nu\mu} + F_{VD}{}^{2}{}_{\nu\mu} +$$
$$+ F_{S}{}^{1}{}_{\nu\mu} + F_{S}{}^{2}{}_{\nu\mu} + F_{\Theta}{}^{1}{}_{\nu\mu} + F_{\Theta}{}^{2}{}_{\nu\mu} + F_{B}{}^{1}{}_{\nu\mu} + F_{B}{}^{2}{}_{\nu\mu} +$$
$$+ F_{A}{}^{1}{}_{\nu\mu} + F_{A}{}^{2}{}_{\nu\mu})Z_{\beta} + (\partial_{\mu}H^{\beta}{}_{\sigma\nu} - \partial_{\nu}H^{\beta}{}_{\sigma\mu} + H^{\gamma}{}_{\nu\sigma}H^{\beta}{}_{\gamma\mu} - H^{\gamma}{}_{\mu\sigma}H^{\beta}{}_{\gamma\nu})Z_{\beta}$$

$$= R'_{E}{}^{\beta}{}_{\sigma\nu\mu}Z_{\beta} + R'_{SU(2)}{}^{\beta}{}_{\sigma\nu\mu}Z_{\beta} + R'_{DE}{}^{\beta}{}_{\sigma\nu\mu}Z_{\beta} + R'_{DSU(2)}{}^{\beta}{}_{\sigma\nu\mu}Z_{\beta} + R'_{SU(3)}{}^{\beta}{}_{\sigma\nu\mu}Z_{\beta} +$$
$$+ R'_{DSU(3)}{}^{\beta}{}_{\sigma\nu\mu}Z_{\beta} + R'_{U}{}^{\beta}{}_{\sigma\nu\mu}Z_{\beta} + R'_{V}{}^{\beta}{}_{\sigma\nu\mu}Z_{\beta} + R'_{UD}{}^{\beta}{}_{\sigma\nu\mu}Z_{\beta} + R'_{VD}{}^{\beta}{}_{\sigma\nu\mu}Z_{\beta} + R'_{S}{}^{\beta}{}_{\sigma\nu}Z_{\beta} +$$
$$+ R'_{A}{}^{\beta}{}_{\sigma\nu}Z_{\beta} + R'_{B}{}^{\beta}{}_{\sigma\nu}Z_{\beta} + R'_{G}{}^{\beta}{}_{\sigma\nu\mu}Z_{\beta}$$

where all $F_{...}{}^{1}{}_{\nu\mu}$ and $F_{...}{}^{2}{}_{\nu\mu}$ terms have summations over the four layers (see below) except the terms[331] $F_{S}{}^{1}{}_{\nu\mu} + F_{S}{}^{2}{}_{\nu\mu} + F_{\Theta}{}^{1}{}_{\nu\mu} + F_{\Theta}{}^{2}{}_{\nu\mu} + F_{A}{}^{1}{}_{\nu\mu} + F_{A}{}^{2}{}_{\nu\mu} + F_{B}{}^{1}{}_{\nu\mu} + F_{B}{}^{2}{}_{\nu\mu}$, and where[332]

$$R'_{SU(3)}{}^{\beta}{}_{\sigma\nu\mu} = ig^{\beta}{}_{\sigma}(F_{SU(3)}{}^{1}{}_{\nu\mu} + F_{SU(3)}{}^{2}{}_{\nu\mu}) \qquad (53.7)$$
$$R'_{SU(2)}{}^{\beta}{}_{\sigma\nu\mu} = ig^{\beta}{}_{\sigma}(F_{W}{}^{1}{}_{\nu\mu} + F_{W}{}^{2}{}_{\nu\mu})$$
$$R'_{E}{}^{\beta}{}_{\sigma\nu\mu} = ig^{\beta}{}_{\sigma}(F_{E}{}^{1}{}_{\nu\mu} + F_{E}{}^{2}{}_{\nu\mu})$$
$$R'_{U}{}^{\beta}{}_{\sigma\nu\mu} = ig^{\beta}{}_{\sigma}(F_{U}{}^{1}{}_{\nu\mu} + F_{U}{}^{2}{}_{\nu\mu})$$
$$R'_{V}{}^{\beta}{}_{\sigma\nu\mu} = ig^{\beta}{}_{\sigma}(F_{V}{}^{1}{}_{\nu\mu} + F_{V}{}^{2}{}_{\nu\mu})$$
$$R'_{DSU(3)}{}^{\beta}{}_{\sigma\nu\mu} = ig^{\beta}{}_{\sigma}(F_{DSU(3)}{}^{1}{}_{\nu\mu} + F_{DSU(3)}{}^{2}{}_{\nu\mu})$$
$$R'_{DSU(2)}{}^{\beta}{}_{\sigma\nu\mu} = ig^{\beta}{}_{\sigma}(F_{DW}{}^{1}{}_{\nu\mu} + F_{DW}{}^{2}{}_{\nu\mu})$$
$$R'_{DE}{}^{\beta}{}_{\sigma\nu\mu} = ig^{\beta}{}_{\sigma}(F_{DE}{}^{1}{}_{\nu\mu} + F_{DE}{}^{2}{}_{\nu\mu})$$
$$R'_{UD}{}^{\beta}{}_{\sigma\nu\mu} = ig^{\beta}{}_{\sigma}(F_{UD}{}^{1}{}_{\nu\mu} + F_{UD}{}^{2}{}_{\nu\mu})$$
$$R'_{VD}{}^{\beta}{}_{\sigma\nu\mu} = ig^{\beta}{}_{\sigma}(F_{VD}{}^{1}{}_{\nu\mu} + F_{VD}{}^{2}{}_{\nu\mu})$$

---

[331] The U(4) General Relativity Reality Group fields $A_{R}{}^{\beta}$ have $F_{A}{}^{1}{}_{\nu\mu} + F_{A}{}^{2}{}_{\nu\mu}$.

[332] The B field is the General Relativistic spinor connection. Its effects are miniscule in physical situations except for extreme cases that have not as yet been encountered experimentally.

$$R'_S{}^\beta{}_{\sigma\nu\mu} = ig^\beta{}_\sigma(F_S{}^1{}_{\nu\mu} + F_S{}^2{}_{\nu_\mu})$$
$$R'_B{}^\beta{}_{\sigma\nu\mu} = ig^\beta{}_\sigma(F_B{}^1{}_{\nu\mu} + F_B{}^2{}_{\nu_\mu})$$
$$R'_A{}^\beta{}_{\sigma\nu\mu} = ig^\beta{}_\sigma(F_A{}^1{}_{\nu\mu} + F_A{}^2{}_{\nu_\mu})$$

and

$$R'_G{}^\beta{}_{\sigma\nu\mu} = \partial_\mu H^{1\beta}{}_{\sigma\nu} - \partial_\nu H^{1\beta}{}_{\sigma\mu} + H^{1\gamma}{}_{\nu\sigma}H^{1\beta}{}_{\gamma\mu} - H^{1\gamma}{}_{\mu\sigma}H^{1\beta}{}_{\gamma\nu} + \partial_\mu H^{2\beta}{}_{\sigma\nu} - \partial_\nu H^{2\beta}{}_{\sigma\mu} + \qquad (53.8)$$
$$+ H^{2\gamma}{}_{\nu\sigma}H^{2\beta}{}_{\gamma\mu} - H^{2\gamma}{}_{\mu\sigma}H^{2\beta}{}_{\gamma\nu} + H^{1\gamma}{}_{\nu\sigma}H^{2\beta}{}_{\gamma\mu} - H^{1\gamma}{}_{\mu\sigma}H^{2\beta}{}_{\gamma\nu} + H^{2\gamma}{}_{\nu\sigma}H^{1\beta}{}_{\gamma\mu} - \Gamma^{2\gamma}{}_{\mu\sigma}\Gamma^\beta{}_{\gamma\nu}$$
$$= R^{1\beta}{}_{\sigma\nu\mu} + R^{2\beta}{}_{\sigma\nu\mu}$$

with

$$H^\beta{}_{\sigma\nu\mu} = \partial_\mu H^\beta{}_{\sigma\nu} - \partial_\nu H^\beta{}_{\sigma\mu} + H^\gamma{}_{\nu\sigma}H^\beta{}_{\gamma\mu} - H^\gamma{}_{\mu\sigma}H^\beta{}_{\gamma\nu} \qquad (53.9)$$
$$R^{1\beta}{}_{\sigma\nu\mu} = \partial_\mu H^{1\beta}{}_{\sigma\nu} - \partial_\nu H^{1\beta}{}_{\sigma\mu} + H^{1\gamma}{}_{\nu\sigma}H^{1\beta}{}_{\gamma\mu} - H^{1\gamma}{}_{\mu\sigma}H^{1\beta}{}_{\gamma\nu}$$
$$R^{2\beta}{}_{\sigma\nu\mu p} = \partial_\mu H^{2\beta}{}_{\sigma\nu} - \partial_\nu H^{2\beta}{}_{\sigma\mu} + H^{2\gamma}{}_{\nu\sigma}H^{2\beta}{}_{\gamma\mu} - H^{2\gamma}{}_{\mu\sigma}H^{2\beta}{}_{\gamma\nu} +$$
$$+ H^{1\gamma}{}_{\nu\sigma}H^{2\beta}{}_{\gamma\mu} - H^{1\gamma}{}_{\mu\sigma}H^{2\beta}{}_{\gamma\nu} + H^{2\gamma}{}_{\nu\sigma}H^{1\beta}{}_{\gamma\mu} - H^{2\gamma}{}_{\mu\sigma}H^{1\beta}{}_{\gamma\nu}$$

and

$$H^{1\sigma}{}_{\nu\mu} = \Gamma_{GR}{}^\sigma{}_{\nu\mu} \qquad (53.10)$$
$$H^{2\sigma}{}_{\nu\mu} = \Gamma_{GR}{}^{2\sigma}{}_{\nu\mu}$$

and with summations over four layers indicated by Σ (Layer numbers on fields are not shown to avoid clutter.) As a result we have

$$F_{SU(3)}{}^1{}_{\varkappa\mu} = \Sigma \ \{\partial A_{SU(3)}{}^1{}_\mu/\partial x^\varkappa - \partial A_{SU(3)}{}^1{}_\varkappa/\partial x^\mu + ig_1[A_{SU(3)}{}^1{}_\varkappa, A_{SU(3)}{}^1{}_\mu]\ \} \quad (53.11)$$
$$F_W{}^1{}_{\varkappa\mu} = \Sigma \ \{\partial W^1{}_\mu/\partial x^\varkappa - \partial W^1{}_\varkappa/\partial x^\mu + ig_2[W^1{}_\varkappa, W^1{}_\mu]\ \}$$
$$F_E{}^1{}_{\varkappa\mu} = \Sigma \ \{\partial A_E{}^1{}_\mu/\partial x^\varkappa - \partial A_E{}^1{}_\varkappa/\partial x^\mu\ \}$$
$$F_{DW}{}^1{}_{\varkappa\mu} = \Sigma \ \{\partial W_D{}^1{}_\mu/\partial x^\varkappa - \partial W_D{}^1{}_\varkappa/\partial x^\mu + ig_4[W_D{}^1{}_\varkappa, W_D{}^1{}_\mu]\ \}$$
$$F_{DE}{}^1{}_{\varkappa\mu} = \Sigma \ \{\partial A_{DE}{}^1{}_\mu/\partial x^\varkappa - \partial A_{DE}{}^1{}_\varkappa/\partial x^\mu\}$$
$$F_{DSU(3)}{}^1{}_{\varkappa\mu} = \Sigma \ \{\partial A_{DSU(3)}{}^1{}_\mu/\partial x^\varkappa - \partial A_{DSU(3)}{}^1{}_\varkappa/\partial x^\mu + ig_8[A_{DSU(3)}{}^1{}_\varkappa, A_{DSU(3)}{}^1{}_\mu]\ \}$$
$$F_U{}^1{}_{\varkappa\mu} = \Sigma \ \{\partial U^1{}_\mu/\partial x^\varkappa - \partial U^1{}_\varkappa/\partial x^\mu + ig_6[U^1{}_\varkappa, U^1{}_\mu]\ \}$$
$$F_V{}^1{}_{\varkappa\mu} = \Sigma \ \{\partial V^1{}_\mu/\partial x^\varkappa - \partial V^1{}_\varkappa/\partial x^\mu + ig_7[V^1{}_\varkappa, V^1{}_\mu]\ \}$$

$$F_{U_D}{}^1_{\kappa\mu} \equiv F_{DU}{}^1_{\kappa\mu} = \Sigma\ \{\partial U_D{}^1_\mu/\partial x^\kappa - \partial U_D{}^1_\kappa/\partial x^\mu + ig_9[U_D{}^1_\kappa, U_D{}^1_\mu]\ \}$$

$$F_{V_D}{}^1_{\kappa\mu} \equiv F_{DV}{}^1_{\kappa\mu} = \Sigma\ \{\partial V_D{}^1_\mu/\partial x^\varkappa - \partial V_D{}^1_\varkappa/\partial x^\mu + ig_{10}[V_D{}^1_\varkappa, V_D{}^1_\mu]\ \}$$

$$F_S{}^1_{\kappa\mu} = \partial A_S{}^1_\mu/\partial x^\varkappa - \partial A_S{}^1_\varkappa/\partial x^\mu + ig_{11}[A_S{}^1_\varkappa, A_S{}^1_\mu]$$

$$F_B{}^1_{\kappa\mu} = \partial B^1_\mu/\partial x^\varkappa - \partial B^1_\varkappa/\partial x^\mu + i[B^1_\varkappa, B^1_\mu]$$

$$F_A{}^1_{\kappa\mu} = \partial A_R{}^1_\mu/\partial x^\varkappa - \partial A_R{}^1_\varkappa/\partial x^\mu + i[A_R{}^1_\varkappa, A_R{}^1_\mu]$$

$$F_{SU(3)}{}^2_{\varkappa\mu} = \Sigma\ \{\partial A_{SU(3)}{}^2_\mu/\partial x^\varkappa - \partial A_{SU(3)}{}^2_\varkappa/\partial x^\mu + ig_1[A_{SU(3)}{}^2_\varkappa, A_{SU(3)}{}^2_\mu] + \tag{53.12}$$
$$+ ig_1[A_{SU(3)}{}^1_\varkappa, A_{SU(3)}{}^2_\mu] + ig_1[A_{SU(3)}{}^2_\varkappa, A_{SU(3)}{}^1_\mu]\}$$

$$F_W{}^2_{\varkappa\mu} = \Sigma\ \{\partial W^2_\mu/\partial x^\varkappa - \partial W^2_\varkappa/\partial x^\mu + ig_2[W^2_\varkappa, W^2_\mu] + ig_2[W^1_\varkappa, W^2_\mu] + ig_2[W^2_\varkappa, W^1_\mu]\ \}$$

$$F_E{}^2_{\varkappa\mu} = \Sigma\ \{\partial A_E{}^2_\mu/\partial x^\varkappa - \partial A_E{}^2_\varkappa/\partial x^\mu\}$$

$$F_{DSU(3)}{}^2_{\varkappa\mu} = \Sigma\ \{\partial A_{DSU(3)}{}^2_\mu/\partial x^\varkappa - \partial A_{DSU(3)}{}^2_\varkappa/\partial x^\mu + ig_8[A_{DSU(3)}{}^2_\varkappa, A_{DSU(3)}{}^2_\mu] +$$
$$+ ig_8[A_{DSU(3)}{}^1_\varkappa, A_{DSU(3)}{}^2_\mu] + ig_8[A_{DSU(3)}{}^2_\varkappa, A_{DSU(3)}{}^1_\mu]\}$$

$$F_{DW}{}^2_{\varkappa\mu} = \Sigma\ \{\partial W_D{}^2_\mu/\partial x^\varkappa - \partial W_D{}^2_\varkappa/\partial x^\mu + ig_4[W_D{}^2_\varkappa, W_D{}^2_\mu] + ig_4[W_D{}^1_\varkappa, W_D{}^2_\mu] +$$
$$+ ig_4[W_D{}^2_\varkappa, W_D{}^1_\mu]\}$$

$$F_{DE}{}^2_{\varkappa\mu} = \Sigma\ \{\partial A_{DE}{}^2_\mu/\partial x^\varkappa - \partial A_{DE}{}^2_\varkappa/\partial x^\mu\}$$

$$F_U{}^2_{\varkappa\mu} = \Sigma\ \{\partial U^2_\mu/\partial x^\varkappa - \partial U^2_\varkappa/\partial x^\mu + ig_6[U^2_\varkappa, U^2_\mu] + ig_6[U^1_\varkappa, U^2_\mu] + ig_6[U^2_\varkappa, U^1_\mu]\}$$

$$F_V{}^2_{\varkappa\mu} = \Sigma\ \{\partial V^2_\mu/\partial x^\varkappa - \partial V^2_\varkappa/\partial x^\mu + ig_7[V^2_\varkappa, V^2_\mu] + ig_7[V^1_\varkappa, V^2_\mu] + ig_7[V^2_\varkappa, V^1_\mu]\}$$

$$F_{U_D}{}^2_{\varkappa\mu} = \Sigma\ \{\partial U_D{}^2_\mu/\partial x^\varkappa - \partial U_D{}^2_\varkappa/\partial x^\mu + ig_9[U_D{}^2_\varkappa, U_D{}^2_\mu] + ig_9[U_D{}^1_\varkappa, U_D{}^2_\mu] +$$
$$+ ig_6[U_D{}^2_\varkappa, U_D{}^1_\mu]\}$$
$$\equiv F_{DU}{}^2_{\varkappa\mu}$$

$$F_{V_D}{}^2{}_{\varkappa\mu} = \Sigma \ \{\partial V_D{}^2{}_\mu/\partial x^\varkappa - \partial V_D{}^2{}_\varkappa/\partial x^\mu + ig_7[V_D{}^2{}_\varkappa, V_D{}^2{}_\mu] + ig_7[V_D{}^1{}_\varkappa, V_D{}^2{}_\mu] +$$
$$+ \ ig_7[V_D{}^2{}_\varkappa, V_D{}^1{}_\mu]\}$$
$$\equiv F_{DV}{}^2{}_{\varkappa\mu}$$

$$F_S{}^2{}_{\varkappa\mu} = \partial A_S{}^2{}_\mu/\partial x^\varkappa - \partial A_S{}^2{}_\varkappa/\partial x^\mu + ig_8[A_S{}^2{}_\varkappa, A_S{}^2{}_\mu] + ig_{11}[A_S{}^1{}_\varkappa, A_S{}^2{}_\mu] + ig_{11}[A_S{}^2{}_\varkappa, A_S{}^1{}_\mu]$$

$$F_B{}^2{}_{\varkappa\mu} = \partial B{}^2{}_\mu/\partial x^\varkappa - \partial B{}^2{}_\varkappa/\partial x^\mu + i[B{}^2{}_\mu, B{}^2{}_\varkappa] + i[B{}^1{}_\mu, B{}^2{}_\varkappa] + i[B{}^2{}_\mu, B{}^1{}_\varkappa]$$

$$F_A{}^2{}_{\varkappa\mu} = \partial A_R{}^2{}_\mu/\partial x^\varkappa - \partial A_R{}^2{}_\varkappa/\partial x^\mu + i[A_R{}^2{}_\mu, A_R{}^2{}_\varkappa] + i[A_R{}^1{}_\mu, A_R{}^2{}_\varkappa] + i[A_R{}^2{}_\mu, A_R{}^1{}_\varkappa]$$

Note that $R'{}^\beta{}_{\sigma\nu\mu}$ factorizes into a

$$[U(1) \otimes SU(2) \otimes U(1) \otimes SU(2) \otimes SU(3) \otimes SU(3) \otimes U(4) \otimes U(4)]^4 \otimes U(4)$$

part and a Riemann-Christoffel Gravitational curvature tensor part. For later use in defining a lagrangian we define

$$R'{}^\beta{}_{\sigma\nu\mu} = R'_E{}^{1\beta}{}_{\sigma\nu\mu} + R'_E{}^{2\beta}{}_{\sigma\nu\mu} + R'_{SU(2)}{}^{1\beta}{}_{\sigma\nu\mu} + R'_{SU(2)}{}^{2\beta}{}_{\sigma\nu\mu} + R'_{DE}{}^{1\beta}{}_{\sigma\nu\mu} + R'_{DE}{}^{2\beta}{}_{\sigma\nu\mu} + \quad (53.13)$$
$$+ R'_{DSU(2)}{}^{1\beta}{}_{\sigma\nu\mu} + R'_{DSU(2)}{}^{2\beta}{}_{\sigma\nu\mu} + R'_{SU(3)}{}^{1\beta}{}_{\sigma\nu\mu} + R'_{SU(3)}{}^{2\beta}{}_{\sigma\nu\mu} + R'_{DSU(3)}{}^{1\beta}{}_{\sigma\nu\mu} +$$
$$+ R'_{DSU(3)}{}^{2\beta}{}_{\sigma\nu\mu} + R'_U{}^{1\beta}{}_{\sigma\nu\mu} + R'_U{}^{2\beta}{}_{\sigma\nu\mu} + R'_V{}^{1\beta}{}_{\sigma\nu\mu} + R'_V{}^{2\beta}{}_{\sigma\nu\mu} + R'_{DU}{}^{1\beta}{}_{\sigma\nu\mu} +$$
$$+ R'_{DU}{}^{2\beta}{}_{\sigma\nu\mu} + R'_{DV}{}^{1\beta}{}_{\sigma\nu\mu} + R'_{DV}{}^{2\beta}{}_{\sigma\nu\mu} + R'_S{}^{1\beta}{}_{\sigma\nu\mu} + R'_S{}^{2\beta}{}_{\sigma\nu\mu} +$$
$$+ R'_B{}^{1\beta}{}_{\sigma\nu\mu} + R'_B{}^{2\beta}{}_{\sigma\nu\mu} + R'_A{}^{1\beta}{}_{\sigma\nu\mu} + R'_A{}^{2\beta}{}_{\sigma\nu\mu} + R{}^{1\beta}{}_{\sigma\nu\mu} + R{}^{2\beta}{}_{\sigma\nu\mu}$$

where

$$R'_E{}^{1\beta}{}_{\sigma\nu\mu} = ig^\beta{}_\sigma F_E{}^1{}_{\nu\mu} \qquad\qquad (53.14)$$
$$R'_E{}^{2\beta}{}_{\sigma\nu\mu} = ig^\beta{}_\sigma F_{DE}{}^2{}_{\nu\mu}$$
$$R'_{DE}{}^{1\beta}{}_{\sigma\nu\mu} = ig^\beta{}_\sigma F_E{}^1{}_{\nu\mu}$$
$$R'_{DE}{}^{2\beta}{}_{\sigma\nu\mu} = ig^\beta{}_\sigma F_{DE}{}^2{}_{\nu\mu}$$
$$R'_{SU(2)}{}^{1\beta}{}_{\sigma\nu\mu} = ig^\beta{}_\sigma F_W{}^1{}_{\nu\mu}$$
$$R'_{SU(2)}{}^{2\beta}{}_{\sigma\nu\mu} = ig^\beta{}_\sigma F_{DW}{}^2{}_{\nu\mu}$$
$$R'_{DSU(2)}{}^{1\beta}{}_{\sigma\nu\mu} = ig^\beta{}_\sigma F_W{}^1{}_{\nu\mu}$$

$$R'_{DSU(2)}{}^{2\beta}{}_{\sigma\nu\mu} = ig^{\beta}{}_{\sigma}F_{DW}{}^{2}{}_{\nu\mu}$$

$$R'_{SU(3)}{}^{1\beta}{}_{\sigma\nu\mu} = ig^{\beta}{}_{\sigma}F_{SU(3)}{}^{1}{}_{\nu\mu}$$

$$R'_{SU(3)}{}^{2\beta}{}_{\sigma\nu\mu} = ig^{\beta}{}_{\sigma}F_{SU(3)}{}^{2}{}_{\nu\mu}$$

$$R'_{DSU(3)}{}^{1\beta}{}_{\sigma\nu\mu} = ig^{\beta}{}_{\sigma}F_{DSU(3)}{}^{1}{}_{\nu\mu}$$

$$R'_{DSU(3)}{}^{2\beta}{}_{\sigma\nu\mu} = ig^{\beta}{}_{\sigma}F_{DSU(3)}{}^{2}{}_{\nu\mu}$$

$$R'_{U}{}^{1\beta}{}_{\sigma\nu\mu} = ig^{\beta}{}_{\sigma}F_{U}{}^{1}{}_{\nu\mu}$$

$$R'_{U}{}^{2\beta}{}_{\sigma\nu\mu} = ig^{\beta}{}_{\sigma}F_{U}{}^{2}{}_{\nu\mu}$$

$$R'_{V}{}^{1\beta}{}_{\sigma\nu\mu} = ig^{\beta}{}_{\sigma}F_{V}{}^{1}{}_{\nu\mu}$$

$$R'_{V}{}^{2\beta}{}_{\sigma\nu\mu} = ig^{\beta}{}_{\sigma}F_{V}{}^{2}{}_{\nu\mu}$$

$$R'_{DU}{}^{1\beta}{}_{\sigma\nu\mu} = ig^{\beta}{}_{\sigma}F_{DU}{}^{1}{}_{\nu\mu}$$

$$R'_{DU}{}^{2\beta}{}_{\sigma\nu\mu} = ig^{\beta}{}_{\sigma}F_{DU}{}^{2}{}_{\nu\mu}$$

$$R'_{DV}{}^{1\beta}{}_{\sigma\nu\mu} = ig^{\beta}{}_{\sigma}F_{DV}{}^{1}{}_{\nu\mu}$$

$$R'_{DV}{}^{2\beta}{}_{\sigma\nu\mu} = ig^{\beta}{}_{\sigma}F_{DV}{}^{2}{}_{\nu\mu}$$

$$R'_{S}{}^{1\beta}{}_{\sigma\nu\mu} = ig^{\beta}{}_{\sigma}F_{S}{}^{1}{}_{\nu\mu}$$

$$R'_{S}{}^{2\beta}{}_{\sigma\nu\mu} = ig^{\beta}{}_{\sigma}F_{S}{}^{2}{}_{\nu\mu}$$

$$R'_{B}{}^{1\beta}{}_{\sigma\nu\mu} = ig^{\beta}{}_{\sigma}B^{1}{}_{\nu\mu}$$

$$R'_{B}{}^{2\beta}{}_{\sigma\nu\mu} = ig^{\beta}{}_{\sigma}B^{2}{}_{\nu\mu}$$

$$R'_{A}{}^{1\beta}{}_{\sigma\nu\mu} = ig^{\beta}{}_{\sigma}A_{R}{}^{1}{}_{\nu\mu}$$

$$R'_{A}{}^{2\beta}{}_{\sigma\nu\mu} = ig^{\beta}{}_{\sigma}A_{R}{}^{2}{}_{\nu\mu}$$

The total Ricci tensor is

$$R'_{\sigma\mu} = R'^{\beta}{}_{\sigma\beta\mu} \tag{53.15}$$

$$\begin{aligned}
&= iF_{E}{}^{1}{}_{\sigma\mu} + iF_{E}{}^{2}{}_{\sigma\mu} + iF_{W}{}^{1}{}_{\sigma\mu} + iF_{W}{}^{2}{}_{\sigma\mu} + iF_{DE}{}^{1}{}_{\sigma\mu} + iF_{DE}{}^{2}{}_{\sigma\mu} + iF_{DW}{}^{1}{}_{\sigma\mu} + iF_{DW}{}^{2}{}_{\sigma\mu} + iF_{SU(3)}{}^{1}{}_{\sigma\mu} + iF_{SU(3)}{}^{2}{}_{\sigma\mu} + \\
&\quad + iF_{DSU(3)}{}^{1}{}_{\sigma\mu} + iF_{DSU(3)}{}^{2}{}_{\sigma\mu} + iF_{U}{}^{1}{}_{\sigma\mu} + iF_{U}{}^{2}{}_{\sigma\mu} + iF_{V}{}^{1}{}_{\sigma\mu} + iF_{V}{}^{2}{}_{\sigma\mu} + iF_{DU}{}^{1}{}_{\sigma\mu} + iF_{DU}{}^{2}{}_{\sigma\mu} + iF_{DV}{}^{1}{}_{\sigma\mu} + \\
&\quad + iF_{DV}{}^{2}{}_{\sigma\mu} + iF_{S}{}^{1}{}_{\sigma\mu} + iF_{S}{}^{2}{}_{\sigma\mu} + iF_{B}{}^{1}{}_{\sigma\mu} + iF_{B}{}^{2}{}_{\sigma\mu} + \\
&\quad + iF_{A}{}^{1}{}_{\sigma\mu} + iF_{A}{}^{2}{}_{\sigma\mu} + \partial_{\mu}H^{1\beta}{}_{\sigma\beta} - \partial_{\beta}H^{1\beta}{}_{\sigma\mu} + H^{1\gamma}{}_{\beta\sigma}H^{1\beta}{}_{\gamma\mu} - H^{1\gamma}{}_{\mu\sigma}H^{1\beta}{}_{\gamma\beta} + \\
&\quad + \partial_{\mu}H^{2\beta}{}_{\sigma\beta} - \partial_{\beta}H^{2\beta}{}_{\sigma\mu} + H^{2\gamma}{}_{\beta\sigma}H^{2\beta}{}_{\gamma\mu} - H^{2\gamma}{}_{\mu\sigma}H^{2\beta}{}_{\gamma\beta} + H^{1\gamma}{}_{\beta\sigma}H^{2\beta}{}_{\gamma\mu} - H^{1\gamma}{}_{\mu\sigma}H^{2\beta}{}_{\gamma\beta} +
\end{aligned}$$

$$+ H^{2\gamma}_{\beta\sigma} H^{1\beta}_{\gamma\mu} - H^{2\gamma}_{\mu\sigma} H^{1\beta}_{\gamma\beta}$$

$$= R'^1_{E\,\sigma\mu} + R'^2_{E\,\sigma\mu} + R'^1_{SU(2)\,\sigma\mu} + R'^2_{SU(2)\,\sigma\mu} + R'^1_{DE\,\sigma\mu} + R'^2_{DE\,\sigma\mu} + R'^1_{DSU(2)\,\sigma\mu} + R'^2_{DSU(2)\,\sigma\mu} +$$
$$+ R'^1_{SU(3)\,\sigma\mu} + R'^2_{SU(3)\,\sigma\mu} + R'^1_{U\,\sigma\mu} + R'^2_{U\,\sigma\mu} + R'^1_{V\,\sigma\mu} + R'^2_{V\,\sigma\mu} + R'^1_{DSU(3)\,\sigma\mu} + R'^2_{DSU(3)\,\sigma\mu} +$$
$$+ R'^1_{DU\,\sigma\mu} + R'^2_{DU\,\sigma\mu} + R'^1_{DV\,\sigma\mu} + R'^2_{DV\,\sigma\mu} + R'^1_{S\,\sigma\mu} + R'^2_{S\,\sigma\mu} +$$
$$+ R'^{1\beta}_{A\,\sigma\beta\mu} + R'^{2\beta}_{A\,\sigma\beta\mu} + R'^{1\beta}_{B\,\sigma\beta\mu} + R'^{2\beta}_{B\,\sigma\beta\mu} + R^1_{\sigma\mu} + R^2_{\sigma\mu}$$
$$= R'^1_{\sigma\mu} + R'^2_{\sigma\mu}$$

where

$$R'^1_{\sigma\mu} = R'^1_{E\,\sigma\mu} + R'^1_{SU(2)\,\sigma\mu} + R'^1_{DE\,\sigma\mu} + R'^1_{DSU(2)\,\sigma\mu} + R'^1_{SU(3)\,\sigma\mu} + R'^1_{U\,\sigma\mu} + R'^1_{V\,\sigma\mu} + \quad (53.16)$$
$$+ R'^1_{DSU(3)\,\sigma\mu} + R'^1_{DU\,\sigma\mu} + R'^1_{DV\,\sigma\mu} + R'^1_{S\,\sigma\mu} + R'^{1\beta}_{A\,\sigma\beta\mu} + R'^{1\beta}_{B\,\sigma\beta\mu} + R^1_{\sigma\mu}$$

$$R'^2_{\sigma\mu} = R'^2_{E\,\sigma\mu} + R'^2_{SU(2)\,\sigma\mu} + R'^2_{DE\,\sigma\mu} + R'^2_{DSU(2)\,\sigma\mu} + R'^2_{SU(3)\,\sigma\mu} + R'^2_{U\,\sigma\mu} + R'^2_{V\,\sigma\mu} +$$
$$+ R'^2_{DSU(3)\,\sigma\mu} + R'^2_{DU\,\sigma\mu} + R'^2_{DV\,\sigma\mu} + R'^2_{S\,\sigma\mu} + R'^{2\beta}_{A\,\sigma\beta\mu} + R'^{2\beta}_{B\,\sigma\beta\mu} + R^2_{\sigma\mu}$$

$$(53.17)$$

with

$$R'^1_{E\,\sigma\mu} = iF^1_{E\,\sigma\mu} \quad\quad (53.18)$$
$$R'^2_{E\,\sigma\mu} = iF^2_{E\,\sigma\mu}$$
$$R'^1_{SU(2)\,\sigma\mu} = iF^1_{W\,\sigma\mu}$$
$$R'^2_{SU(2)\,\sigma\mu} = iF^2_{W\,\sigma\mu}$$
$$R'^1_{DE\,\sigma\mu} = iF^1_{DE\,\sigma\mu}$$
$$R'^2_{DE\,\sigma\mu} = iF^2_{DE\,\sigma\mu}$$
$$R'^1_{DSU(2)\,\sigma\mu} = iF^1_{DW\,\sigma\mu}$$
$$R'^2_{DSU(2)\,\sigma\mu} = iF^2_{DW\,\sigma\mu}$$
$$R'^1_{SU(3)\,\sigma\mu} = iF^1_{SU(3)\,\sigma\mu}$$
$$R'^2_{SU(3)\,\sigma\mu} = iF^2_{SU(3)\,\sigma\mu}$$
$$R'^1_{U\,\sigma\mu} = iF^1_{U\,\sigma\mu}$$
$$R'^2_{U\,\sigma\mu} = iF^2_{U\,\sigma\mu}$$
$$R'^1_{V\,\sigma\mu} = iF^1_{V\,\sigma\mu}$$
$$R'^2_{V\,\sigma\mu} = iF^2_{V\,\sigma\mu}$$

$$R'_{DSU(3)}{}^{1}{}_{\sigma\mu} = iF_{DSU(3)}{}^{1}{}_{\sigma\mu}$$
$$R'_{DSU(3)}{}^{2}{}_{\sigma\mu} = iF_{DSU(3)}{}^{2}{}_{\sigma\mu}$$
$$R'_{DU}{}^{1}{}_{\sigma\mu} = iF_{DU}{}^{1}{}_{\sigma\mu}$$
$$R'_{DU}{}^{2}{}_{\sigma\mu} = iF_{DU}{}^{2}{}_{\sigma\mu}$$
$$R'_{DV}{}^{1}{}_{\sigma\mu} = iF_{DV}{}^{1}{}_{\sigma\mu}$$
$$R'_{DV}{}^{2}{}_{\sigma\mu} = iF_{DV}{}^{2}{}_{\sigma\mu}$$
$$R'_{S}{}^{1}{}_{\sigma\mu} = iF_{S}{}^{1}{}_{\sigma\mu}$$
$$R'_{S}{}^{2}{}_{\sigma\mu} = iF_{S}{}^{2}{}_{\sigma\mu}$$
$$R'_{A}{}^{1}{}_{\sigma\mu} = iF_{B}{}^{1}{}_{\sigma\mu}$$
$$R'_{A}{}^{2}{}_{\sigma\mu} = iF_{B}{}^{2}{}_{\sigma\mu}$$
$$R'_{B}{}^{1}{}_{\sigma\mu} = iF_{B}{}^{1}{}_{\sigma\mu}$$
$$R'_{B}{}^{2}{}_{\sigma\mu} = iF_{B}{}^{2}{}_{\sigma\mu}$$

with the further definition of $R''^{1}{}_{\sigma\mu}$ and $R''^{2}{}_{\sigma\mu}$:

$$R''^{1}{}_{\sigma\mu} = R'_{SU(3)}{}^{1}{}_{\sigma\mu} + R'_{DSU(3)}{}^{1}{}_{\sigma\mu} + R^{1}{}_{\sigma\mu} \tag{53.19}$$
$$R''^{2}{}_{\sigma\mu} = R'_{SU(3)}{}^{2}{}_{\sigma\mu} + R'_{DSU(3)}{}^{2}{}_{\sigma\mu} + R^{2}{}_{\sigma\mu}$$

$R'^{1}{}_{\sigma\mu}$ is the Ricci tensor. An additional Ricci-like tensor is

$$H_{\sigma_{\mu}} = H^{\beta}{}_{\sigma\beta\mu} \tag{53.20}$$

The curvature scalar is

$$R' = g^{\sigma\mu}R'_{\sigma\mu} = + \partial^{\sigma}H^{1\beta}{}_{\sigma\beta} - \partial_{\beta}H^{1\beta}{}_{\sigma}{}^{\sigma} + H^{1\gamma}{}_{\beta\sigma}H^{1\beta}{}_{\gamma}{}^{\sigma} - H^{1\gamma}{}_{\mu\sigma}H^{1\beta}{}_{\gamma\beta} + \partial^{\sigma}H^{2\beta}{}_{\sigma\beta} - \partial_{\beta}H^{2\beta}{}_{\sigma}{}^{\sigma} +$$
$$+ H^{2\gamma}{}_{\beta\sigma}H^{2\beta}{}_{\gamma}{}^{\sigma} - H^{2\gamma\sigma}{}_{\sigma}H^{2\beta}{}_{\gamma\beta} + H^{1\gamma}{}_{\beta\sigma}H^{2\beta}{}_{\gamma}{}^{\sigma} - H^{1\gamma\sigma}{}_{\sigma}H^{2\beta}{}_{\gamma\beta} + H^{2\gamma}{}_{\beta\sigma}H^{1\beta}{}_{\gamma}{}^{\sigma} -$$
$$- H^{2\gamma\sigma}{}_{\sigma}H^{1\beta}{}_{\gamma\beta} \tag{53.21}$$

$$= g^{\sigma\mu}(R^{1\beta}{}_{\sigma\beta\mu} + R^{2\beta}{}_{\sigma\beta\mu})$$

53.3 Vector Boson and Graviton Lagrangian Terms

We choose the vector boson and gravitational part of the lagrangian[333] of the Unified SuperStandard Theory (with the Higgs sector and the Faddeev-Popov terms gauge sector not displayed here) to be:

$$\mathcal{L} = \text{Tr } \sqrt{g}[MD_v R''^1_{\sigma\mu} D^v R''^{2\sigma\mu} + aR'^1_{\sigma\mu} R'^{2\sigma\mu} + bR' + cg^{\sigma\mu} g^2_{\sigma\mu} + c'g^{2\sigma\mu} g^2_{\sigma\mu} - dA_{SU(3)}{}^2_{\mu} A_{SU(3)}{}^{2\mu}] \tag{53.22}$$

with a sum over layers understood, where D_v is given by eq. 53.5a, where M, a, b, c, c', and d are constants, and $R''^i_{\sigma\mu}$ for i = 1, 2 determined above.[334]

This higher derivative lagrangian maintains the locality of the theory but does entail a modest modification in the derivation of the Euler-Lagrange equations of motion. It also requires the use of principal value propagators rather than ordinary

---

[333] The rationale for this choice is 1) to obtain the known Stanard Model interactions, 2) to obtain a canonical formulation for this higher derivative theory, and 3) to introduce higher derivative terms that yield quark confinement and a theory of gravity that accounts for known deviations from Newtonian gravity such as described by MoND. See Blaha (2019g) and (2018e) and earlier books for details on these points.

[334] One may ask why $R''^1_{\sigma\mu}$ and $R''^2_{\sigma\mu}$ appear in the first term of the lagrangian, and not other interaction terms. We believe the primary reason is: "The extended vierbein $l^{\mu ai}(x)$ can be viewed as located at a point in a higher dimensional complex-valued space.

$$l^{\mu ai}(x) = (\partial \xi_X{}^{ai}(x)/\partial x_\mu)_{X=h(x)}$$

where $\xi_X{}^{ai}$ is a set of locally inertial coordinates located at point X, and x = h(x) is a 4-dimensional point in a tangent subspace of the higher dimensional space:

$$X = h(x)$$

The relation between complex 4-dimensional coordinates x and the higher dimensional coordinates X is an embedding of a 4-dimensional surface within the higher dimensional complex space when account is taken of the range of possible x values. We have considered such embeddings in Blaha (2015a), and in earlier books, and developed a theory of a higher dimensional complex-valued space (the *Megaverse*) that contains our universe and probably many other universes." Thus SU(3) and Gravitation have a special role in our particle dynamics based on geometry. The second reason is the common feature of color SU(3) and real-valued General Relativity is that they are the only interactions that do not participate in 'rotations of interactions' as described earlier and in chapter 31 of Blaha (2017b). The third, practical reason is the experimental reality that the Strong Interaction and Gravitation are known to have 'anomalous' features that will be seen to be remedied by these insertions while the other interactions are 'conventional.'

Feynman propagators for gluon and graviton interactions. Thus the Strong Interaction sector, and the Gravitation sector are Action-at-a-Distance theories that are similar in spirit to Wheeler-Feynman Electrodynamics. The two U(1) Electromagnetic sectors, the Generation group U(4) gauge field sector, the Layer group U(4) gauge field sector, the two SU2) Weak sectors, the U(4) A_s gauge field sector, the spinor connection sector, and the Θ-interaction sector may, or may not, be Action-at-a-Distance fields. They are not constrained to be Action-at-a-Distance by the present considerations.

Since we wish to apply our theory cosmologically, and within hadrons, where the gravitational spinor connections are negligible due to the smallness of the gravitational constant G and the 'smallness' of Gravitational B spinor connection effects on the cosmological scale, we set $F^1_{\nu\mu} = F^2_{\nu\mu} = 0$ and find[335]

$$\mathcal{L} = \text{Tr} \ \sqrt{g}[MD_\nu(R''^1_{SU(3)\sigma\mu} + R''^1_{DSU(3)\sigma\mu} + R^1_{\sigma\mu})D^\nu(R'_{SU(3)}{}^{2\sigma\mu} + R'_{DSU(3)}{}^{2\sigma\mu} + R^{2\sigma\mu}) +$$
$$+ aR''^1_{\sigma\mu}R'^{2\sigma\mu} + bR' + cg^{\sigma\mu}g^2_{\sigma\mu} + c'g^{2\sigma\mu}g^2_{\sigma\mu} - dA_{SU(3)}{}^2{}_\mu A_{SU(3)}{}^{2\mu}]$$

$$(53.23)$$

Since there are no strong interaction fields in 'empty' space and gravity is negligible within hadrons,[336] we can drop the interaction terms between the Strong interaction and the Gravity interaction. However, we cannot drop the interaction terms amongst Electromagnetism, the Weak interaction, the Strong Interaction, the Generation groups U(4) interactions, the U(4) Layer groups interactions, and the U(4) Species group interaction – within, and between, hadrons. The interaction terms between Electromagnetism and Gravitation are important cosmologically.

The above lagrangian terms can therefore be expressed as:[337]

$$\mathcal{L} = \mathcal{L}_E + \mathcal{L}_{SU(2)} + \mathcal{L}_{DE} + \mathcal{L}_{DSU(2)} + \mathcal{L}_{SU(3)} + \mathcal{L}_{DSU(3)} + \mathcal{L}_U + \mathcal{L}_V + \mathcal{L}_{DU} + \mathcal{L}_{DV} + \mathcal{L}_S +$$
$$+ \mathcal{L}_G + \mathcal{L}_{int}$$

$$(53.24)$$

---

[335] The constants have the dimensions: M has the dimension of inverse mass squared, b has dimension mass squared, a is dimensionless, c and c' have dimension mass, and d has dimension mass squared.

[336] We show gravity weakens at very short distances using our Two-Tier Quantum Field Theory formalism. See Appendix A, and Blaha (2003) and (2005a) among other books by the author.

[337] We only consider the gauge field lagrangian terms.

where taking traces of $\mathcal{L}$s terms is understood, and with coupling constants not displayed below to avoid clutter,

$$\mathcal{L}_E = \text{Tr } \sqrt{g}\{M\{[\partial_\nu + i(A_E{}^1{}_\nu + A_E{}^2{}_\nu)]F^1{}_{E\sigma\mu}[\partial^\nu + i(A_E{}^{1\nu} + A_E{}^{2\nu})]F^2{}_E{}^{\sigma\mu}\} + aF_E{}^1{}_{\sigma\mu}F_E{}^{2\sigma\mu}\}$$

(53.25)

$$\mathcal{L}_{SU(2)} = \text{Tr } \sqrt{g}[aF_W{}^1{}_{\sigma\mu}F_W{}^{2\sigma\mu}]$$

$$\mathcal{L}_{DE} = \text{Tr } \sqrt{g}\{M\{[\partial_\nu + i(A_{DE}{}^1{}_\nu + A_{DE}{}^2{}_\nu)]F^1{}_{DE\sigma\mu}[\partial^\nu + i(A_{DE}{}^{1\nu} + A_{DE}{}^{2\nu})]F_{DE}{}^{2\sigma\mu}\} + aF_{DE}{}^1{}_{\sigma\mu}F_{DE}{}^{2\sigma\mu}\}$$

$$\mathcal{L}_{DSU(2)} = \text{Tr } \sqrt{g}[aF_W{}^1{}_{\sigma\mu}F_W{}^{2\sigma\mu}]$$

$$\mathcal{L}_{SU(3)} = \text{Tr } \sqrt{g}\{M[\partial_\nu + i(A_{SU(3)}{}^1{}_\nu + A_{SU(3)}{}^2{}_\nu)]F_{SU(3)}{}^1{}_{\sigma\mu}[\partial^\nu + i(A_{SU(3)}{}^{1\nu} + A_{SU(3)}{}^{2\nu})]F_{SU(3)}{}^{2\sigma\mu} + aF_{SU(3)}{}^1{}_{\sigma\mu}F_{SU(3)}{}^{2\sigma\mu} - dA_{SU(3)}{}^2{}_\mu A_{SU(3)}{}^{2\mu}\}$$

$$\mathcal{L}_{DSU(3)} = \text{Tr } \sqrt{g}\{M[\partial_\nu + i(A_{DSU(3)}{}^1{}_\nu + A_{DSU(3)}{}^2{}_\nu)]F_{DSU(3)}{}^1{}_{\sigma\mu}[\partial^\nu + i(A_{DSU(3)}{}^{1\nu} + A_{DSU(3)}{}^{2\nu})]F_{DSU(3)}{}^{2\sigma\mu} + aF_{DSU(3)}{}^1{}_{\sigma\mu}F_{DSU(3)}{}^{2\sigma\mu} - dA_{DSU(3)}{}^2{}_\mu A_{DSU(3)}{}^{2\mu}\}$$

$$\mathcal{L}_U = \text{Tr } \sqrt{g}[aF_U{}^1{}_{\sigma\mu}F_U{}^{2\sigma\mu}]$$

$$\mathcal{L}_V = \text{Tr } \sqrt{g}[aF_V{}^1{}_{\sigma\mu}F_V{}^{2\sigma\mu}]$$

$$\mathcal{L}_{DU} = \text{Tr } \sqrt{g}[aF_{DU}{}^1{}_{\sigma\mu}F_{DU}{}^{2\sigma\mu}]$$

$$\mathcal{L}_{DV} = \text{Tr } \sqrt{g}[aF_{DV}{}^1{}_{\sigma\mu}F_{DV}{}^{2\sigma\mu}]$$

$$\mathcal{L}_S = \text{Tr } \sqrt{g}[aF_S{}^1{}_{\sigma\mu}F_S{}^{2\sigma\mu}]$$

$$\mathcal{L}_G = \text{Tr } \sqrt{g}[MD_\nu R^1{}_{\sigma\mu}D^\nu R^{2\sigma\mu} + aR^1{}_{\sigma\mu}R^{2\sigma\mu} + bg^{\sigma\mu}(R^{1\beta}{}_{\sigma\beta\mu} + R^{2\beta}{}_{\sigma\beta\mu}) + cg^{\sigma\mu}g^2{}_{\sigma\mu} + c'g^{2\sigma\mu}g^2{}_{\sigma\mu}]$$
$$= \text{Tr } \sqrt{g}[MD_\nu R^1{}_{\sigma\mu}D^\nu R^{2\sigma\mu} + aR^1{}_{\sigma\mu}R^{2\sigma\mu} + bH + cg^{\sigma\mu}g^2{}_{\sigma\mu} + c'g^{2\sigma\mu}g^2{}_{\sigma\mu}]$$

$$\mathcal{L}_{int} = \mathcal{L} - (\mathcal{L}_E + \mathcal{L}_{SU(2)} + \mathcal{L}_{DE} + \mathcal{L}_{DSU(2)} + \mathcal{L}_{SU(3)} + \mathcal{L}_U + \mathcal{L}_V + \mathcal{L}_{DSU(3)} + \mathcal{L}_{DU} + \mathcal{L}_{DV} + \mathcal{L}_S + \\ + \mathcal{L}_G) \tag{53.26}$$

with appropriate sums over layers and gravitational B spinor connection terms omitted. Thus $\mathcal{L}_{SU(3)}$, $\mathcal{L}_{SU(2)}$, $\mathcal{L}_E$, $\mathcal{L}_{DE,}$ $\mathcal{L}_{DSU(2)}$, $\mathcal{L}_{U,}$, $\mathcal{L}_V$, $\mathcal{L}_S$, and parts of $\mathcal{L}_{int}$ are the dominant interactions within hadrons, and $\mathcal{L}_G$, $\mathcal{L}_E$ and parts of $\mathcal{L}_{int}$ are the dominant interactions in space within the framework of this discussion.

The $D_\nu R^1{}_{\sigma\mu}$ and $D^\nu R^{2\sigma\mu}$ terms have the form:

$$D_\nu R^i{}_{\sigma\mu} = + \partial_\nu R^i{}_{\sigma\mu} - H^{1\beta}{}_{\sigma\nu} R^i{}_{\beta\mu} - H^{1\beta}{}_{\nu\mu} R^i{}_{\sigma\beta} \tag{53.27}$$

for $i = 1, 2$ while covariant derivatives for internal symmetries are given by eq. 53.5a.

Blaha (2019g) and (2018e) discusses further details of the lagrangian obtained from the Riemann-Christoffel tensor, and its gravitation and Strong Interaction terms. The Dark Strong interaction terms generated by the first term in eq. 53.23 has an effect that parallels the Strong interaction case and leads to the confinement of Dark quarks.[338]

53.4 New Vector Boson Interactions

The above lagrangian can be broken up into pieces in the following manner:

$$\mathcal{L}_E = Tr \sqrt{g}\{M\{[\partial_\nu + i(A_E^1{}_\nu + A_E^2{}_\nu)]F^1{}_{E\sigma\mu}[\partial^\nu + i(A_E^{1\nu} + A_E^{2\nu})]F^2{}_E{}^{\sigma\mu}\} + aF_E^1{}_{\sigma\mu}F_E^{2\sigma\mu}\} \tag{53.28}$$

$$\mathcal{L}_{SU(2)} = Tr \sqrt{g}[aF_W^1{}_{\sigma\mu}F_W^{2\sigma\mu}]$$
$$\mathcal{L}_{DE} = Tr \sqrt{g}\{M\{[\partial_\nu + i(A_{DE}^1{}_\nu + A_{DE}^2{}_\nu)]F^1{}_{DE\sigma\mu}[\partial^\nu + i(A_{DE}^{1\nu} + A_{DE}^{2\nu})]F_{DE}^{2\sigma\mu}\} + \\ + aF_{DE}^1{}_{\sigma\mu}F_{DE}^{2\sigma\mu}\}$$
$$\mathcal{L}_{DSU(2)} = Tr \sqrt{g}[aF_W^1{}_{\sigma\mu}F_W^{2\sigma\mu}]$$
$$\mathcal{L}_{SU(3)} = Tr \sqrt{g}\{M[\partial_\nu + i(A_{SU(3)}^1{}_\nu + A_{SU(3)}^2{}_\nu)]F_{SU(3)}^1{}_{\sigma\mu}[\partial^\nu + i(A_{SU(3)}^{1\nu} + \\ + A_{SU(3)}^{2\nu})]F_{SU(3)}^{2\sigma\mu} + aF_{SU(3)}^1{}_{\sigma\mu}F_{SU(3)}^{2\sigma\mu} - dA_{SU(3)}^2{}_\mu A_{SU(3)}^{2\mu}\} \tag{53.29}$$

---

[338] See Blaha (2019g) and (2018e).

$$\mathcal{L}_U = \text{Tr } \sqrt{g}[aF_U^1{}_{\sigma\mu}F_U^{2\sigma\mu}]$$

$$\mathcal{L}_V = \text{Tr } \sqrt{g}[aF_V^1{}_{\sigma\mu}F_V^{2\sigma\mu}]$$

$$\mathcal{L}_S = \text{Tr } \sqrt{g}[aF_S^1{}_{\sigma\mu}F_S^{2\sigma\mu}]$$

$$\mathcal{L}_G = \text{Tr } \sqrt{g}[MD_\nu R^1{}_{\sigma\mu}D^\nu R^{2\sigma\mu} + aR^1{}_{\sigma\mu}R^{2\sigma\mu} + bg^{\sigma\mu}(R^{1\beta}{}_{\sigma\beta\mu} + R^{2\beta}{}_{\sigma\beta\mu}) + cg^{\sigma\mu}g^2{}_{\sigma\mu} + c'g^{2\sigma\mu}g^2{}_{\sigma\mu}]$$

$$= \text{Tr } \sqrt{g}[MD_\nu R^1{}_{\sigma\mu}D^\nu R^{2\sigma\mu} + aR^1{}_{\sigma\mu}R^{2\sigma\mu} + bH + cg^{\sigma\mu}g^2{}_{\sigma\mu} + c'g^{2\sigma\mu}g^2{}_{\sigma\mu}]$$

$$\mathcal{L}_{int} = \mathcal{L} - (\mathcal{L}_E + \mathcal{L}_{SU(2)} + \mathcal{L}_{DE} + \mathcal{L}_{DSU(2)} + \mathcal{L}_{SU(3)} + \mathcal{L}_U + \mathcal{L}_V + \mathcal{L}_S + \mathcal{L}_G) \quad (53.30)$$

again with appropriate sums over layers and with coupling constants not displayed to avoid clutter. Thus $\mathcal{L}_{SU(3)}$, $\mathcal{L}_{SU(2)}$, $\mathcal{L}_E$, $\mathcal{L}_{DE}$, $\mathcal{L}_{DSU(2)}$, $\mathcal{L}_U$, $\mathcal{L}_V$, $\mathcal{L}_S$, and parts of $\mathcal{L}_{int}$ are the dominant interactions within hadrons, and $\mathcal{L}_G$, $\mathcal{L}_E$ and parts of $\mathcal{L}_{int}$ are the dominant interactions in space within the framework of this discussion. The terms of $\mathcal{L}_{int}$ have 'new' interactions between gauge fields that are described in some detail in Blaha (2017b) and other books. These interactions are not in the conventional Standard Model. They lead to modifications of gravity, the Strong Interactions, spin dynamics and so on.

54. Gravitational Potential on the Three Distance Scales

This chapter[339] and the next chapter describe some of the possible results of the unified lagrangian terms in eq. 53.22 and 53.30. In this chapter we put together results on the gravitational potential that were previously found in Blaha (2016g), (2016h) and (2017a). We note that a new experimental study of 33,000 galaxies indicate that the gravitational potential at inter-galactic distances deviates significantly from the Newtonian potential G/r. In Blaha (2017a) we showed that such a deviation occurs in our theory. This experimental result was not known at the time of its writing.[340,341]

In this chapter we see that the theory has higher derivative dynamic equations but, unlike the Strong Interaction sector (next chapter) which yields color (quark-gluon) confinement, the gravitation dynamic equations do not have confinement. They do yield a modified form of gravity at various intermediate distances of the order of the average galactic radius, and beyond.

The modification of gravity implied by our theory is consistent with the need for Dark Matter described in our Theory of Everything books (Blaha (2015a) and (2016c)).[342] It is also consistent with a MoND theory predictions[343] with the addition of

---

[339] Most of this chapter appears in Blaha (2017a) and earlier books by the author.

[340] The gravitational potential was found to be greater than G/r at inter-galactic distances in a survey of 33,000 galaxies by M. Brouwer and colleagues at the Leiden Observatory (The Netherlands) in an announcement on December 18, 2016.

[341] There are other higher derivative theories – some with two metrics, and some with a metric plus vector plus scalar field formulation. The present work is based on a unification of Strong and Gravity sectors and a totally different formalism. Some significant references are: M. Milgrom, Phys. Rev. **D80**, 123536 (2009); C. Skordis et al, Phys. Rev. Lett. **96**, 011301 (2006); R. H. Sanders, Astrophysical Journal **480**, 492 (1997); and references therein; J. D. Bekenstein, Phys. Rev. **D70**, 083509 (2004) and references therein; J-P. Bruneton, Phys. Rev. **D76**, 124012 (2007) with a higher derivative gravity and metric, vector, scalar fields. See also references within these articles.

[342] I. Ferreras et al, Phys. Rev. Lett., **96**, 011301 (2006) shows the need for both Dark Matter and MOND based on studies of astrophysical data.

sixth order derivatives in the Gravitation sector lagrangian in a manner consistent with the higher order terms appearing in the Strong Interaction sector of the unified theory.

Thus our approach to MoND does not require a major change in Mechanics, quantum theory, or General Relativity (modulo higher order derivatives). The consistency between the need for higher order derivatives in both the Strong Interaction and Gravitation sectors is encouraging.

The developments in this chapter are a generalization with higher order derivative terms of the unified theory presented in chapter 6 of Blaha (2016d).

54.1 Gravitation Sector Dynamic Equations

Our gravity sector has two metric fields, $g_{\mu\nu}$ and $g^2{}_{\mu\nu}$ derived from the unified formalism described earlier. Some of the relevant gravitation equations found in sections 53.2 and 53.3 are:

$$H^{\sigma}{}_{\nu\mu} = \Gamma^{\sigma}{}_{\nu\mu} + \Gamma^{2\sigma}{}_{\nu\mu}$$

$$H^{\beta}{}_{\sigma\nu\mu} = \partial_{\mu}H^{\beta}{}_{\sigma\nu} - \partial_{\nu}H^{\beta}{}_{\sigma\mu} + H^{\gamma}{}_{\nu\sigma}H^{\beta}{}_{\gamma\mu} - H^{\gamma}{}_{\mu\sigma}H^{\beta}{}_{\gamma\nu}$$

$$H_{\sigma\mu} = H^{\beta}{}_{\sigma\beta\mu}$$

$$H = g^{\sigma\mu}H_{\sigma\mu}$$
$$\mathcal{H} = R'^{1} + R'^{2}$$

We use the gravitational sector lagrangian:

$$\mathcal{L}_{G} = \sqrt{g}[MD_{\nu}R'^{1}{}_{G\sigma\mu}D^{\nu}R'^{2}{}_{G}{}^{\sigma\mu} + aR'^{1}{}_{G\sigma\mu}R'^{2}{}_{G}{}^{\sigma\mu} + bg^{\sigma\mu}(R'^{1}{}_{G}{}^{\beta}{}_{\sigma\beta\mu} + R'^{2}{}_{G}{}^{\beta}{}_{\sigma\beta\mu}) + cg^{\sigma\mu}g^2{}_{\sigma\mu} +$$
$$+ eg^{2\sigma\mu}g^2{}_{\sigma\mu}]$$

$$\mathcal{L}_{G} = \sqrt{g}[MD_{\nu}R'^{1}{}_{G\sigma\mu}D^{\nu}R'^{2}{}_{G}{}^{\sigma\mu} + aR'^{1}{}_{G\sigma\mu}R'^{2}{}_{G}{}^{\sigma\mu} + bH + cg^{\sigma\mu}g^2{}_{\sigma\mu} + eg^{2\sigma\mu}g^2{}_{\sigma\mu}]$$

where

---

[343] Our gravity theory has aspects that are very similar to the MoND theories described in A. Balakin et al, Phys. Rev. **D70**, 064027 (2004); H-S Zhao et al, Phys. Rev. **D82**, 103001 (2010); and references therein. However our approach is very different.

$$D_v Z_\mu = (\partial_v + iF_v) Z_\mu - H^\sigma{}_{v\mu} Z_\sigma$$

$$= [g^\sigma{}_\mu \partial_v + ig^\sigma{}_\mu F_v - H^\sigma{}_{v\mu}] Z_\sigma$$
$$= [g^\sigma{}_\mu \partial_v + iD^\sigma{}_{\mu v}] Z_\sigma$$

where M, a, b, c, and e are constants determined later. We determine a[344]

$$a = 1/(2f) = 7.47 \times 10^{181} \tag{54.0}$$

The remaining values are determined later in this chapter in section 54.8.

The lagrangian dynamic equations that result are difficult. We consequently will examine the weak gravitation limiting case where we can approximate the two metrics with

$$g_{\mu v} \cong \eta_{\mu v} + h_{\mu v} \tag{54.1}$$
$$g^2{}_{\mu v} \cong \eta_{\mu v} + h^2{}_{\mu v} \tag{54.2}$$

where

$$|h_{\mu v}| \ll 1 \tag{54.3}$$
$$|h^2{}_{\mu v}| \ll 1$$

Using the relations

$$\partial_\mu h^\mu{}_v = \tfrac{1}{2} \partial_v h^\mu{}_\mu \tag{54.4}$$
$$\partial_\mu h^{2\mu}{}_v = \tfrac{1}{2} \partial_v h^{2\mu}{}_\mu \tag{54.5}$$

and neglecting higher order terms in $h_{\mu v}$ and $h^2{}_{\mu v}$ we find

$$R'^1{}_{\mu v} = \tfrac{1}{2}[\Box h_{\mu v} - \partial_\mu \partial_\lambda h^\lambda{}_v - \partial_v \partial_\lambda h^\lambda{}_\mu + \partial_v \partial_\mu h^\lambda{}_\lambda]$$
$$\cong \tfrac{1}{2} \Box h_{\mu v} \tag{54.6}$$
$$R'^1 \cong \tfrac{1}{2} \Box h_\mu{}^\mu$$

$$R'^2{}_{\mu v} = \tfrac{1}{2}[\Box h^2{}_{\mu v} - \partial_\mu \partial_\lambda h^{2\lambda}{}_v - \partial_v \partial_\lambda h^{2\lambda}{}_\mu + \partial_v \partial_\mu h^{2\lambda}{}_\lambda]$$
$$\cong \tfrac{1}{2} \Box h^2{}_{\mu v} \tag{54.7}$$

---

[344] The value of a is obtained from the Charmonium calculation.

$$R'^2 \cong \tfrac{1}{2}\Box h^2{}_\mu{}^\mu$$

with

$$D_v = \partial_v$$

upon neglecting higher order terms.

Substituting we find the *effective quadratic* part of the lagrangian (in $h_{\mu\nu}$ and $h^2{}_{\mu\nu}$) is

$$\mathscr{L}_G = \sqrt{g}[M\partial_\nu R'^1{}_{G\sigma\mu}\partial^\nu R'^2{}_G{}^{\sigma\mu} + a\Box h_{\sigma\mu}\Box h^{2\sigma\mu}/4 + \tfrac{1}{2}b(\partial_\alpha h^{\sigma\mu}\partial^\alpha h_{\sigma\mu} + \partial_\alpha h^{2\sigma\mu}\partial^\alpha h^2{}_{\sigma\mu}) +$$
$$+ c(4 + \eta^{\sigma\mu}h^2{}_{\sigma\mu} + h^{\sigma\mu}\eta_{\sigma\mu} + h^{\sigma\mu}h^2{}_{\sigma\mu}) +$$
$$+ e(2\eta^{\sigma\mu}h^2{}_{\sigma\mu} + h^{2\sigma\mu}h^2{}_{\sigma\mu}) + 1/4(h_{\mu\nu} + h^2{}_{\mu\nu})T^{\mu\nu}] \qquad (54.8)$$

Using partial integrations, we find the standard technique for determining the equations of motion from a lagrangian using independent variations with respect to $h_{\mu\nu}$ and $h^2{}_{\mu\nu}$ yields

$$-M\Box^3 h^{2\mu\nu}/4 + a\Box^2 h^{2\mu\nu}/4 + \tfrac{1}{2}b\Box h^{\mu\nu} + c(\eta^{\mu\nu} + h^{2\mu\nu}) + 1/4T^{\mu\nu} = 0 \qquad (54.9)$$

$$-M\Box^3 h^{\mu\nu}/4 + a\Box^2 h^{\mu\nu}/4 + \tfrac{1}{2}b\Box h^{2\mu\nu} + c(\eta^{\mu\nu} + h^{\mu\nu}) + 2e(\eta^{\mu\nu} + h^{2\mu\nu}) + 1/4T^{\mu\nu} = 0 \qquad (54.10)$$

The term $c\eta^{\mu\nu}$ can be viewed as part of the total energy-momentum tensor $T'^{\mu\nu}$:

$$T'^{\mu\nu} = T^{\mu\nu} + 2c\eta^{\mu\nu} \qquad (54.10a)$$

It plays a role similar to the Cosmological Constant. Subtracting the equations we find

$$M(\Box^3 h^{\mu\nu}/4 - \Box^3 h^{2\mu\nu}/4) + a\Box^2 h^{2\mu\nu}/4 - a\Box^2 h^{\mu\nu}/4 + \tfrac{1}{2}b\Box h^{\mu\nu} - \tfrac{1}{2}b\Box h^{2\mu\nu} + c(h^{2\mu\nu} - h^{\mu\nu}) - 2e(\eta^{\mu\nu} + h^{2\mu\nu}) = 0 \qquad (54.11)$$

and thus

$$[-M\Box^3 + a\Box^2 - \tfrac{1}{2}b\Box + (4c - 8e)]h^{2\mu\nu} = 8e\eta^{\mu\nu} - M\Box^3 h^{\mu\nu} + a\Box^2 h^{\mu\nu} - \tfrac{1}{2}b\Box h^{\mu\nu} + 4ch^{\mu\nu} \qquad (54.12)$$

Therefore we determine the metric equation for $h^{2\mu\nu}$ to be

$$h^{2\mu\nu} = [-M\Box^3 + a\Box^2 - \tfrac{1}{2}b\Box + (4c - 8e)]^{-1}[8e\eta^{\mu\nu} - M\Box^3 h^{\mu\nu} + a\Box^2 h^{\mu\nu} - \tfrac{1}{2}b\Box h^{\mu\nu} + 4ch^{\mu\nu}]$$

(54.13)

Substituting in eq. 54.9 we obtain

$$[-M\Box^3/4 + a\Box^2/4 + c][-M\Box^3 + a\Box^2 - \tfrac{1}{2}b\Box + (4c - 8e)]^{-1}[8e\eta^{\mu\nu} - M\Box^3 + a\Box^2 h^{\mu\nu} - \tfrac{1}{2}b\Box h^{\mu\nu} + 4ch^{\mu\nu}] + \\ + \tfrac{1}{2}b\Box h^{\mu\nu} + 1/4T^{\mu\nu} = 0$$

(54.14)

We now redefine the energy-momentum tensor with the result eq. 54.14 becomes

$$\{[-M\Box^3/4 + a\Box^2/4 + c][-M\Box^3 + a\Box^2 - \tfrac{1}{2}b\Box + (4c - 8e)]^{-1}[-M\Box^3 + a\Box^2 - \tfrac{1}{2}b\Box + 4c] + \\ + \tfrac{1}{2}b\Box\}h^{\mu\nu} = -1/4T'^{\mu\nu}$$

(54.15)

54.2 Real General Relativity Gravity Potential

Assuming that we are dealing with non-relativistic matter we can calculate the gravity potential contribution from eq. 54.15

$$V_{G1}(\mathbf{x}) = -\int d^3k \, \exp(i\mathbf{k}\cdot\mathbf{x})V_{G1}(\mathbf{k})/(2\pi)^3$$

(54.16)

where

$$V_{G1}(\mathbf{k}) = \{[Mk^6/4 + ak^4/4 + c][Mk^6 + ak^4 + \tfrac{1}{2}bk^2 + (4c - 8e)]^{-1}[Mk^6 + ak^4 + \tfrac{1}{2}bk^2 + 4c] - \\ - \tfrac{1}{2}bk^2\}^{-1} \\ = [Mk^6 + ak^4 + \tfrac{1}{2}bk^2 + (4c - 8e)]/\{[Mk^6/4 + ak^4/4 - \tfrac{1}{2}bk^2 + c][Mk^6 + ak^4 + \tfrac{1}{2}bk^2 + \\ + 4c] + 4ebk^2\}$$

(54.17)

Similarly eq. 54.13 implies the other contribution to the total gravity potential is

$$V_{G2}(\mathbf{x}) = -\int d^3k \, \exp(i\mathbf{k}\cdot\mathbf{x})V_{G2}(\mathbf{k})/(2\pi)^3$$

(54.18)

where

$$V_{G2}(\mathbf{k}) = \{[Mk^6 + ak^4 + \tfrac{1}{2}bk^2 + 4c]^{-1}[Mk^6 + a\,k^4 + \tfrac{1}{2}bk^2 + 4c - 8e]\}V_{G1}(\mathbf{k}) \quad (54.19)$$

The total gravity potential energy *for real-valued General Relativity* is thus

$$V^{tot}_{RG}(\mathbf{x}) = V_{G1}(\mathbf{x}) + V_{G2}(\mathbf{x}) \tag{54.20}$$

54.3 Real-valued General Relativistic Gravity Solution

Eqs. 54.16 - 54.17 can generate a massless graviton, which seems a requirement based on cosmological considerations, and also generate a pair of massive gravitons of very low mass. The massive gravitons generate a MoND-like potential that might explicate the anomalous gravitation effects seen at distances of the order of galactic dimensions.

If we set the "Cosmological Constants" $c = e = 0$, then eq. 54.17 becomes

$$V_{G1}(\mathbf{k}) = 4/\{Mk^2[k^4 + ak^2/M - 2b/M]\} \tag{54.21}$$

The denominator of eq. 54.21 can be factored into the form

$$V_{G1}(\mathbf{k}) = 4/\{Mk^2[k^2 + \tfrac{1}{2}m_A^2 + \tfrac{1}{2}(m_A^4 + 8bm_A^2/a)^{\frac{1}{2}}][k^2 + \tfrac{1}{2}m_A^2 - \tfrac{1}{2}(m_A^4 + 8bm_A^2/a)^{\frac{1}{2}}]\}$$
$$(54.22)$$

where

$$m_A = (a/M)^{\frac{1}{2}}$$

Assuming $m_A^2 \ll 8b/a$, or $a^2/8 < 1/8 \ll bM = M(2\pi G)^{-1}$, which is reasonable since a is approximately one and M is enormous, we see

$$V_{G1}(\mathbf{k}) \cong 4/\{Mk^2[k^2 + (2bm_A^2/a)^{\frac{1}{2}}][k^2 - (2bm_A^2/a)^{\frac{1}{2}}]\} \qquad (54.23)$$
$$= (1/b)\{-2/k^2 + 1/[k^2 + (2bm_A^2/a)^{\frac{1}{2}}] + 1/[k^2 - (2bm_A^2/a)^{\frac{1}{2}}]\}$$

up to negligible terms.

From eq. 54.19 we see $V_{G2}(\mathbf{k}) = V_{G1}(\mathbf{k})$ if $c = e = 0$. Thus the total gravitational potential due to eq. 54.8 *for **real-valued** General Relativity* (in momentum space) is

$$V^{tot}_{RG}(\mathbf{k}) = \tfrac{1}{2}(V_{G1}(\mathbf{k}) + V_{G2}(\mathbf{k})) \tag{54.24}$$
$$= (1/b)\{-2/\mathbf{k}^2 + 1/[\mathbf{k}^2 + (2bm_A^2/a)^{\frac{1}{2}}] + 1/[\mathbf{k}^2 - (2bm_A^2/a)^{\frac{1}{2}}]\}$$
$$\cong \pi G\{-2/\mathbf{k}^2 + 1/[\mathbf{k}^2 + (2bm_A^2/a)^{\frac{1}{2}}] + 1/[\mathbf{k}^2 - (2bm_A^2/a)^{\frac{1}{2}}]\}$$

with b set to

$$b = (2\pi G)^{-1} = 2.364 \times 10^{55} \text{ eV}^2 \tag{54.24a}$$

introducing the connection of b to G, Newton's gravitational constant and using J.1d.
The coordinate space potential is[345,346]

$$V^{tot}_{RG}(\mathbf{r}) = -G/r + a_1 G e^{-m_G r}/r + a_2 G\cos(m_G r)/r \tag{54.25}$$

where

$$m_G^2 = (2bm_A^2/a)^{\frac{1}{2}} = (2b/M)^{\frac{1}{2}} \tag{54.25a}$$
$$a_1 = \tfrac{1}{2}$$
$$a_2 = \tfrac{1}{2}$$

where m_G is an extremely small mass.

54.4 U(4) General Relativistic Reality Group Gravity Contribution

The Reality group of Complex General Relativity generates a gravitational interaction. Its role is to rotate 4-vectors of coordinates to complex-values.

Assuming that we are dealing with non-relativistic matter we can calculate the gravity potential contribution from the General Relativistic Reality Group:

$$V_{GA1}(\mathbf{x}) = -1/(32M) \int d^3k \exp(i\mathbf{k}\cdot\mathbf{x})V_{GA1}(\mathbf{k})/(2\pi)^3 \tag{54.26}$$

where

$$V_{GA1}(\mathbf{k}) = (\mathbf{k}^4 + (a/M)\mathbf{k}^2)^{-1} \tag{54.27}$$

---

[345] Since the theory has higher order drivatives that could lead to unitarity problems Feynman propagators must be taken in Principal order. Since potentials are a part of Feynman propagators the potentials real value must be used.

[346] The third term in eq. 44.25 is an oscillating Yukawa term that, because of the smallness of m_G, is slowly varying towards the end of a galaxy and thus could be well within observational error bounds. It appears that the real part of the third term is the contribution to the total gravitational potential using Principal value propagators.

up to a U(4) matrix factor with matrix elements of order 1.

The eq. 54.29 denominator can be separated into two terms:

$$V_{GAl}(\mathbf{k}) = (M/a)[1/k^2 - 1/(\mathbf{k}^2 + a/M)] \tag{54.28}$$

which upon integration yield

$$V_{GAl}(\mathbf{r}) = -[1/(128\pi a)][1/r - e^{-m_A r}/r] \tag{54.29}$$

We use:[347]

$$M \sim 1.61 \times 10^{163} \text{ eV}^{-2} \tag{54.30}$$

and thus

$$m_A = (a/M)^{\frac{1}{2}} = 1.94 \times 10^{-101} \text{ eV} \tag{54.31}$$

since $a = 7.47 \times 10^{181}$ by eq. 54.0. The coupling constant is

$$\alpha_R = 1/(128\pi a) \cong 3.3 \times 10^{-185} \tag{54.32}$$

In comparison the electromagnetic fine structure constant is

$$\alpha_{QED} \cong 0.0073$$

Thus the General Relativistic Reality Group fields (A_R) coupling constant is negligible in comparison.

54.5 Total Gravity

We now turn to combine the General Relativistic Reality group gravity contribution with the gravitational potential of real-valued General Relativity using:

$$V^{tot}_{RG}(\mathbf{r}) = -G/r + a_1 Ge^{-m_G r}/r + a_2 G\cos(m_G r)/r \tag{54.25}$$

---

[347] M is determined later in section 44.9.

The total combined gravitational potential of Complex General Relativity (modulo Higgs corrections) [348]

$$V_{TOT}(\mathbf{r}) = -G/r - a_1 G e^{-m_G r}/r + a_2 G \cos(m_G r)/r - [1/(128\pi a)][1/r - e^{-m_A r}/r] \quad (54.33)$$

with the mass constant calculated from eq. 54.25a above:

$$m_G = (2b/M)^{\frac{1}{4}} = 1.31 \times 10^{-27} \text{ eV} \quad (54.34)$$

The *other* graviton mass m_A for the Species gravity potential part is given by

$$m_A = (a/M)^{\frac{1}{2}} \cong 1.94 \times 10^{-101} \text{ eV} \quad (54.35)$$

The input constants are[349]

$$b = (2\pi G)^{-1} = 2.364 \times 10^{55} \text{ eV}^2 \quad (54.36)$$

$$M \sim 1.61 \times 10^{163} \text{ eV}^{-2} \quad (54.37)$$

We now examine $V_{TOT}(\mathbf{r})$ at short distances (within the solar system), distances of tens of thousands of light years (intra-galactic distances), and distances between galaxies (hundreds of thousands to millions of light years and beyond).

54.6 Influence of Gravitational Gauge Field on Gravitation

The gravitational gauge field potential $V_{TOT}(\mathbf{r})$ in eq. 54.35 has a relatively large coupling constant that makes it competitive with the known force of gravity at large distances of the scale of galactic distances. This force, which is negligible at short distances of the order of planetary distances, may be part of the MoND phenomena that affects the motion of stars.

---

[348] The third term in eq. 44.25 is an oscillating Yukawa term that, because of the smallness of m_G, is slowly varying towards the end of a galaxy and thus could be well within observational error bounds. It appears that the real part of the third term is the contribution to the total gravitational potential using Principal value propagators.

[349] M is determined later in section 54.9.

54.7 Intra-Solar System Distance Scale

Since m_G and m_A are extremely small the gravitational potential at distances of perhaps up to at least several light years is

$$V_{TOT}(\mathbf{r}) \cong -G/r \qquad (54.38)$$

to well within feasible experimental limits since the factor Reality Group contribution is negligible due to the extremely small values of m_G and m_A causing the terms

$$- a_1 Ge^{-m_G r}/r + a_2 G\cos(m_G r)/r - [1/(128\pi a)][1/r - e^{-m_A r}/r] \cong 0 \qquad (54.39)$$

in eq. 54.33 to be zero to very good approximation.

54.8 Galactic Distance Scale

At distances of several tens of thousands of light years up to perhaps 100,000's of light years we find the m_G terms in eq. 54.33 to be the major cause of the deviations from Newton's law since

$$m_G \gg m_A \qquad (54.40)$$

by eqs. 54.34 and 54.25. Thus

$$\begin{aligned} V_{TOT}(\mathbf{r}) &\cong -G/r - 0.5Ge^{-m_G r}/r + 0.5G\cos(m_G r)/r \\ &\cong -G/r + 0.5Gm_G - 0.5Gm_G^2 r + Gm_G^3 r^2/12 \end{aligned} \qquad (54.41)$$

for $r \approx m_G^{-1}$. The resultant force is

$$\mathbf{F} = -\nabla V_{TOT}(\mathbf{r})(\mathbf{r}) \sim -Gr/r^3 + 0.5Gm_G^2\mathbf{r}/r - Gm_G^3 r\, \mathbf{r}/6 + ... \qquad (54.42)$$

54.9 Value of M

We determine M by requiring it to influence galactic motions at radii of the order of 100,000 light years (the radius of the Andromeda galaxy).[350]

Transforming all distances to electron volts (eV) using

$$1 \text{ eV}^{-1} = 1.2398 \times 10^{-6} \text{ m} \qquad (54.43)$$

we find 100,000 ly (light years) $\equiv 7.63 \times 10^{26}$ eV^{-1} implying one of the graviton masses is approximately (order of magnitude):

$$m_G \approx 1.31 \times 10^{-27} \text{ eV} \qquad (54.44)$$

if the gravitation potential (force) terms are to modify gravity at galactic distances of the order of a hundred thousand light years.

Solving for M from

$$m_G = (2b/M)^{\frac{1}{4}} \qquad (54.36)$$

we find

$$M = 2b/m_G^{\ 4} = 1.61 \times 10^{163} \text{ eV}^{-2} \qquad (54.45)$$

Thus a *zero mass graviton determines gravity at short and ultra-long distances. A graviton of mass m_G affects gravity at intermediate distances.*

54.10 Intergalactic Distance Scale

We can estimate the gravitational potential of $V_{TOT}(\mathbf{r})$ in eq. 54.33 for large distances $r \approx m_A^{-1}$ of the order of many hundreds of thousands of light years, and beyond. We find

$$V_{TOT}(\mathbf{r}) \cong -G/r + [1/(128\pi a)](m_A^2 r/2 - m_A) \qquad (54.46)$$

for $r \approx m_A^{-1}$. The resultant force is

---

[350] All coupling constant values are based on data from K. A. Olive et al (Particle Data Group), Chinese Physics **C38**, 090001 (2014).

$$\mathbf{F} = -\nabla V_{TOT}(\mathbf{r})(\mathbf{r}) \sim -Gr/r^3 - [1/(256\pi M)]\mathbf{r}/r \qquad (54.47)$$

to good approximation since m_A sets a distance scale of the order of tens of thousands of light years causing the oscillating term to 'wash out,' and causing the $a_1 Ge^{-m_G r}/r$ term to be negligible.

Consequently we find a deeper potential, and thus a larger attractive gravitational force at inter-galactic distances, in agreement with the 33,000 galaxy survey of M. Brouwer and colleagues.

54.11 Qualitative Agreement With Gravitational Data at all Known Distances

Our results are to be compared to the MoND force of A. Balakin et al, Phys. Rev. **D70**, 064027 (2004):[351]

$$F = -\lambda Gm[M/r^2 - |\Pi_c| r/c^2]$$

and the vector form suggested by H-S Zhao et al, Phys. Rev. **D82**, 103001 (2010):

$$\partial \Phi/\partial \mathbf{r} = Gmr/r^3 + (Gm)^{\frac{1}{2}}\mathbf{r}/r^2$$

Recent studies of 153 galaxies confirm the MoND discrepancy from Newtonian gravitation.[352]

The resemblance of our results in eqs. 54.42 and 54.47 to MOND estimates is clear.

We chose the value of M such that the gravity potential at distances of the order of a 100,000 light years (the radius of the Andromeda galaxy) is increased by the Yukawa like terms due to the small value of m_G.

We conclude our unified theory agrees with known gravitational data at the three distance scales.

---

[351] The constant c above is the speed of light, and M is the mass (not the M used in our lagrangian equations.)
[352] S. S. McGaugh, F. Lelli, and J. M. Schombert, arXiv: 1609.0591 (2016).

54.12 Coherence of Large Scale Structures in the Universe

We have chosen M to be such that m_G is of the order of the inverse radius of a galaxy. If we chose m_G^{-1} to be of the order of the distances associated with the recently observed[353] coherence of large scale structures, such as galaxies separated by distances of the order of 20 million light years, then M would be of the order of 25.8×10^{191} eV^{-2} and could account for large scale coherence through gravitational effects.

Then m_A would be many tens of billions of light years (parsecs) accounting, perhaps, for the alignments of supermassive black holes inside quasars over distances of billions of parsecs.[354]

54.13 Unity with the Strong Interaction

The value of a and M are also relevant to our Strong Interaction sector. In chapter 13 we show that they are consistent with chatmonium data.

54.14 Three Gravitons

The sixth order graviton dynamic equations lead to three gravitons: one with mass zero and the other two being m_G and m_A. The three gravitons generate variations from Newtonian gravity on different distance scales giving three regions of gravity.

---

[353] See J. H. Lee *et al*, The Astrophysical Journal 884, Number 2 (2019) for a study of large scale structures in the universe.
[354] D. Hutsemékers *et al*, Astronomy and Astrophysics Manuscript Number aa24631 (Septrmber, 2014) arXiv:1409.6098 (2014).

55. The Strong Interaction Sector

In this chapter[355] we describe some of the implications of the Strong Interaction sector of our theory. We will show that a linear quark potential and quark confinement follows, and then determine the parameter a used earlier.

55.1 Strong Interaction Lagrangian Terms

The flat space-time Strong Interaction gauge field lagrangian terms is

$$\mathcal{L}_{SU(3)} = \text{Tr } \sqrt{g}\{M[\partial_\nu + ig_1(A_{SU(3)}{}^1{}_\nu + A_{SU(3)}{}^2{}_\nu)]F_{SU(3)}{}^1{}_{\sigma\mu}[\partial^\nu + i\,g_1(A_{SU(3)}{}^{1\nu} + \\ + A_{SU(3)}{}^{2\nu})]F_{SU(3)}{}^{2\sigma\mu} + aF_{SU(3)}{}^1{}_{\sigma\mu}F_{SU(3)}{}^{2\sigma\mu} - dA_{SU(3)}{}^2{}_\mu A_{SU(3)}{}^{2\mu}\}$$

Dropping the subscript $_{SU(3)}$ for added clarity and adding color fermion terms we obtain[356]

$$\mathcal{L}_{SU(3)} = \text{Tr }\{MD_\nu F^1{}_{\sigma\mu}D^\nu F^{2\sigma\mu} + aF^1{}_{\sigma\mu}F^{2\sigma\mu} - dA^2{}_\mu A^{2\mu}\} + \bar{\psi}[i\slashed{\nabla} + g_1(\slashed{A}^1 + \slashed{A}^2) - m]\psi \tag{55.1}$$

where (for j = 1, 2)

$$D_\nu F^j{}_{\sigma\mu} = \partial_\nu F^j{}_{\sigma\mu} + ig_1[A^1{}_\nu, F^j{}_{\sigma\mu}] + ig_1[A^2{}_\nu, F^j{}_{\sigma\mu}] \tag{55.2}$$

We should start with eq. 55.1. However, with a view towards perturbation theory which appears reasonable in view of the smallness of the strong interaction coupling constant $f^2/4\pi = 0.024$ seen below, we will abstract a quadratic expression in the fields from eq.

---

[355] Most of this chapter appears in Blaha (2016h) and earlier books by the author.
[356] We note the constant a, that appears in this chapter is NOT the Charmonium constant denoted $a_{cornell}$.

55.1 and then proceed to develop gluon propagators and the strong interaction potential. The 'free' Strong Interaction lagrangian that we use is

$$\mathcal{L}_{SU(3)F} = Tr\{MD_{Fv}F_F{}^{1a}{}_{\sigma\mu}D_F{}^{v}F_F{}^{2a\sigma\mu} + aF_F{}^{1a}{}_{\sigma\mu}F_F{}^{2a\sigma\mu} - dA^{2a}{}_{\mu}A^{2a\mu}\} + \\ + \bar{\psi}[i\nabla + f(A^1 + A^2) - m]\psi \tag{55.3}$$

where

$$F^{1a}{}_{\mu\varkappa} = \partial A^{1a}{}_{\mu}/\partial x^{\varkappa} - \partial A^{1a}{}_{\varkappa}/\partial x^{\mu} \tag{55.4}$$
$$F^{2a}{}_{\mu\varkappa} = \partial A^{2a}{}_{\mu}/\partial x^{\varkappa} - \partial A^{2a}{}_{\varkappa}/\partial x^{\mu}$$

and

$$D_{Fv} = \partial_v \tag{55.5}$$

The conjugate momenta to $A^{1a}{}_{\mu}$ and $A^{2a}{}_{\mu}$ are respectively

$$\pi^{1a}{}_{\mu} = \partial\mathcal{L}_{SU(3)F}/(\partial A^{1a}{}_{\mu}/\partial t) = aF_F{}^{2a\mu t} \tag{55.6}$$
$$\pi^{2a}{}_{\mu} = \partial\mathcal{L}_{SU(3)F}/(\partial A^{2a}{}_{\mu}/\partial t) = aF_F{}^{1a\mu t}$$

The non-zero, equal time commutation relations are

$$[\pi^{ia}{}_{\mu}(\mathbf{x}, t), A^{jb}{}_{v}(\mathbf{y}, t)] = i(1 - \delta^{ij})\delta^{ab}\delta^{G(\mu v)}(\mathbf{x} - \mathbf{y}) \tag{55.7}$$

where i and j label the fields, and $G(\mu v)$ indicates the gauge[357] G and the associated index expressions, with

$$\delta^{G(\mu v)}(\mathbf{x} - \mathbf{y}) = \int d^4k \, \exp(-ik \cdot x) b^G{}_{\mu v}(k)/(2\pi)^4 \tag{55.8}$$

where $b^G{}_{\mu v}(k)$ is a polynomial in k with a δ-function factor restricting the integration over k.

---

[357] Not the gravitational coupling constant.

55.1.1 Dynamical Equations

After performing partial integrations on the $MD_{Fv}F_F{}^1{}_{\sigma\mu}D_F{}^vF_F{}^{2\sigma\mu}$ term (and discarding surface terms at 'infinity') the Euler-Lagrange dynamical equations (in the Landau gauge) due to independent variations with respect to $A^{1a}{}_\mu$ is

$$2M\square^2 A^{2a}{}_\mu - 2a\square A^{2a}{}_\mu = -g_1\bar\psi\, T^a\gamma_\mu\psi \qquad (55.9)$$

and, with respect to $A^{2a}{}_\mu$, is

$$2M\square^2 A^{1a}{}_\mu - 2a\square A^{1a}{}_\mu - 2dA^{2a}{}_\mu = -g_1\,\bar\psi\, T^a\gamma_\mu\psi \qquad (55.10)$$

where T^a is an SU(3) generator. Subtracting the equations we find

$$2M\square^2 A^{1a}{}_\mu - 2a\square A^{1a}{}_\mu - 2M\square^2 A^{2a}{}_\mu + 2a\square A^{2a}{}_\mu - 2dA^{2a}{}_\mu = 0$$

or

$$A^{2a}{}_\mu = [2M\square^2 - 2a\square + 2d]^{-1}[2M\square^2 A^{1a}{}_\mu - 2a\square A^{1a}{}_\mu] \qquad (55.11)$$

with the result

$$\{2M\square^2 - 2a\square - 2d[2M\square^2 - 2a\square + 2d]^{-1}[2M\square^2 - 2a\square]\}A^{1a}{}_\mu = -g_1\bar\psi\, T^a\gamma_\mu\psi$$

or

$$\{2M\square^2 - 2a\square - 2d[2M\square^2 - 2a\square + 2d]^{-1}[2M\square^2 - 2a\square]\}A^{1a}{}_\mu = -g_1\bar\psi\, T^a\gamma_\mu\psi$$

$$\{2M\square^2 - 2a\square - 2d + 4d^2[2M\square^2 - 2a\square + 2d]^{-1}\}A^{1a}{}_\mu = -g_1\bar\psi\, T^a\gamma_\mu\psi \qquad (55.12)$$

Eq. 55.12 leads to the Principal Value (Feynman) propagator:

$$D^{11}{}_{\mu v}(x-y) = P -i<0|T(A^1{}_\mu(x),\, A^1{}_v(y)|0>$$
$$= P \int d^4k\, \exp(-ik\cdot x)b_{\mu v}(k)D_1(k)/(2\pi)^4 \qquad (55.13)$$

where $b_{\mu\nu}(k)$ is a Landau gauge polynomial in k, and

$$D_1(k) = \{2Mk^4 - 2ak^2 - 2d + 4d^2[2Mk^4 - 2ak^2 + 2d]^{-1}\}^{-1}$$
$$= [2Mk^4 - 2ak^2 + 2d](2Mk^4 - 2ak^2)^{-2}$$

Thus

$$D^{11}{}_{\mu\nu}(x-y) = P \int d^4k \, \exp(-ik\cdot x) b_{\mu\nu}(k)[2Mk^4 - 2ak^2 + 2d]/[(2\pi)^4(2Mk^4 - 2ak^2)^2]$$
$$= P \int d^4k \, \exp(-ik\cdot x) b_{\mu\nu}(k)[2Mk^4 - 2ak^2 + 2d]/[(2\pi)^4 k^4 (2Mk^2 - 2a)^2] \quad (55.14)$$

indicating a linear potential r term as well as terms of lower powers in r and Yukawa-like terms with a mass of $(a/M)^{\frac{1}{2}}$. We will describe the resulting effective Strong Interaction potential in more detail later.

Eq. 55.9 leads to the other propagator:

$$D^{12}{}_{\mu\nu}(x-y) = P - i<0|T(A^1{}_\mu(x), A^2{}_\nu(y)|0>$$
$$= P \int d^4k \, \exp(-ik\cdot x) b_{\mu\nu}(k) D_2(k)/(2\pi)^4 \quad (55.15)$$

where

$$D_2(k) = [2Mk^4 - 2ak^2]^{-1} \quad (55.16)$$

Thus

$$D^{12}{}_{\mu\nu}(x-y) = P \int d^4k \, \exp(-ik\cdot x) b_{\mu\nu}(k)/[(2\pi)^4 k^2 (2Mk^2 - 2a)] \quad (55.17)$$

indicating a 1/r potential term plus a Yukawa term with a mass of $(a/M)^{\frac{1}{2}}$.

Due to the form of the interaction with quarks the total effective gluon interaction between quarks is

$$D^{tot}{}_{\mu\nu}(x-y) = D^{11}{}_{\mu\nu}(x-y) + 2D^{12}{}_{\mu\nu}(x-y)$$
$$= P \int d^4k \, \exp(-ik\cdot x) b_{\mu\nu}(k)\{[3Mk^4 - 3ak^2 + 2d]/[(2\pi)^4 k^4 (2Mk^2 - 2a)^2]\}$$
$$(55.18)$$

55.2 Strong Interaction Potential

Eq. 55.18 leads to the form of the total Strong Interaction potential. We note that the $\mu = \nu = 0$ part of the Feynman propagator for transverse gluons has the form:

$$D^{tot}_{00}(x - y) = \ldots - \int d^4k\, V_{SI}(\mathbf{k})\, \exp(-ik\cdot(x - y))/(2\pi)^4 = \ldots + V_{SI}(\mathbf{x} - \mathbf{y})\delta(x_0 - y_0)$$
$$(55.19)$$

where

$$V_{SI}(x) = - \int d^3k\, \exp(ik\cdot x)V_{SI}(k)/(2\pi)^3$$

with

$$V_{SI}(\mathbf{k}) = [3M\mathbf{k}^4 + 3a\mathbf{k}^2 + 2d]/[\mathbf{k}^4(2M\mathbf{k}^2 + 2a)^2] \qquad (55.20)$$
$$= (2M)^{-2}\{2d(M/a)^2/\,\mathbf{k}^4 + [3a(M/a)^2 - 4d(M/a)^3]/\mathbf{k}^2$$
$$+ 2d(M/a)^2/(\mathbf{k}^2 + a/M)^2 + [-3a(M/a)^2 + 4d(M/a)^3]/(\mathbf{k}^2 + a/M)\}$$

Letting

$$m_{SI} = (a/M)^{\frac{1}{2}} \equiv m_A \qquad (55.21)$$

we find

$$V_{SI}(\mathbf{k}) = (2a)^{-2}\{2d/\mathbf{k}^4 + [3a - 4dm_{SI}^{-2}]/\mathbf{k}^2 + 2d/(\mathbf{k}^2 + m_{SI}^2)^2 + [4dm_{SI}^{-2} - 3a]/(\mathbf{k}^2 + m_{SI}^2)\}$$
$$(55.22)$$

The constant, a, is dimensionless and of order 1. The constant M has the dimension of inverse mass squared. We anticipate M will be extremely large resulting in a very small gluon mass m_{SI}.

There are massless gluon terms that generate color confinement reducing the impact of the massive gluon terms to a negligible effect outside hadronic regions.

We also note that the value of the inverse of the large distance graviton mass is of the order of the average galactic radius (the average galactic radius is large) and thus generate a Modified Newtonian potential (MoND).

Substituting eq. 55.26 we obtain a sum of massless and Yukawa-like potentials. A Yukawa potential has the expansion:

$$V_Y(\mathbf{r}) = \int d^3k\, \exp(i\mathbf{k}\cdot\mathbf{r})/[(2\pi)^3(\mathbf{k}^2 + m^2)] = \exp(-mr)/[4\pi r] \qquad (55.23)$$

Thus we obtain

$$V_{SI}(\mathbf{r}) = -(2a)^{-2}\{2d(dV_Y(\mathbf{r})/dm^2)|_{m=0} + [3a - 4dm_{SI}^{-2}]/(4\pi r) - 2d(dV_Y(\mathbf{r})/dm^2|_{m=m_{SI}}) + [4dm_{SI}^{-2} - 3a]V_Y(\mathbf{r})|_{m=m_{SI}}\}$$

$$(55.24)$$

with the form

$$V_{SI}(\mathbf{r}) = \alpha_1 r + \alpha_2/r + \alpha_3 e^{-m_{SI}r}/(4\pi m_{SI}) + \alpha_4 e^{-m_{SI}r}/(4\pi r) \qquad (55.25)$$

where the constants α_i are:

$\alpha_1 = -(2a)^{-2}d/(8\pi)$ (up to an infrared divergent constant) $\qquad\qquad (55.26)$
$\alpha_2 = (2a)^{-2}[3a - 4dm_{SI}^{-2}]/(4\pi)$
$\alpha_3 = -(2a)^{-2}d$
$\alpha_4 = (2a)^{-2}[4dm_{SI}^{-2} - 3a]$

Thus we find the form of the potential of eq. 55.24 is $1/r$ and linear terms plus Yukawa-like terms with very small mass m_{SI} – perhaps near zero. As a result the relevant form of the effective potential is

$$V(r) = -2g^2/r + g^2\lambda^2 r \qquad (55.27)$$

where

$$-2g^2 = \alpha_2 = (2a)^{-2}[3a - 4dm_{SI}^{-2}]/(4\pi) \qquad (55.28)$$
$$g^2\lambda^2 = \alpha_1 = (2a)^{-2}d/(16\pi)$$

55.3 Charmonium and the Strong Interaction

In 1974 a bound state of a charmed and an anti-charmed quark was discovered by two experimental groups. Since charmed quarks are quite massive theoretical attempts were made to understand the charmed quark bound states within the framework of non-relativistic quantum mechanics. The "Cornell group" developed a fairly satisfactory[358] charmed quark bound state spectrum in 1974-5 using a

---

[358] As did a Harvard group.

combination of a linear and inverse 1/r potential as the strong interaction. In a recent fit[359] they gave the potential energy:

$$V(r) = -\kappa/r + r/a_{cornell}^2 \qquad (55.29)$$

where $\kappa = 0.61$, $a_{cornell} = 2.38$ GeV^{-1} and the charmed quark mass was 1.84 GeV. Based on eq. 55.29 we find

$$g^2 = \kappa/2$$
$$\lambda = (2/(\kappa a_{cornell}^2))^{\frac{1}{2}}$$

giving

$$g = 0.552$$
$$\lambda = 0.761 \text{ GeV}^2$$

The constants m_G and b, and thus M, are known from chapter 12. The mass $m_A = m_{SI}$ is given by eq. 55.21. Solving for a and d:

$$-2g^2 = (2a)^{-2}(3a - 4dm_{SI}^{-2})/(4\pi) = (3a - 4dm_{SI}^{-2})/(16\pi a^2) = -\kappa = -0.61 \qquad (55.30)$$

$$g^2\lambda^2 = d/(64\pi a^2) = 1/a_{cornell}^2 = 1/2.38^2 \text{ GeV}^2 = 0.177 \text{ GeV}^2$$

we find

$$a = [16(0.177 \times 10^{18})\pi M - 3/16]/(0.61\pi)$$
$$= 7.47 \times 10^{181}$$
$$d = 1.99 \times 10^{383} \text{ eV}^2$$

using

$$M \sim 1.61 \times 10^{163} \text{ eV}^{-2} \qquad (12.37)$$

Consequently the masses of chapter 12 are

---

[359] E. J. Eichten, K. Lane, and C. Quigg, arXiv:hep-ph/ 0206018 (2002). See this paper for references to earlier work by the "Cornell group" and the "Harvard group" as well as papers by other researchers. Their results appear in the Particle Data Group (PDG) tables.

$$m_G = (2b/M)^{\frac{1}{4}} = 1.31 \times 10^{-27} \text{ eV} \tag{12.34}$$

and

$$m_A = m_{SI} = (a/M)^{\frac{1}{2}} \cong 1.94 \times 10^{-101} \text{ eV} \tag{55.31}$$

55.4 The Origin of the Linear Potential

The linear potential appears to have originated in a suggestion of Feynman in the Spring of 1974. This author proposed[360] a non-Abelian gauge quantum field theory, which yielded a linear potential. These papers, which had 4th order dynamic equations for the gauge fields, showed how to avoid the problems previously associated with higher derivative theories by using principal-value gauge field propagators that were similar in concept to the action-at-a-distance propagators used by Feynman and Wheeler in the late 1940's to formulate action-at-a-distance Quantum Electrodynamics.

Thus a non-Abelian quantum field theory of the strong interaction yielding a linear potential was created. In parallel with this development, Kenneth Wilson (later a Nobelist) was developing lattice gauge theory. Because lattice lines focus the field of gauge boson, lattice gauge theory exhibited a linear potential as well between quarks. Thus it offered an alternative to our gauge theory. However, the linearity of the lattice potential was "built-in" by the lattice theory formulation and thus was an artifact of the lattice formulation. This approach, and other proposed approaches, all share the problem that the linear potential that they produce cannot be proven to truly be a consequence. Rather the linear potential is the "likely result."

On the other hand our higher dimensional theory produces the linear potential if the standard rules of quantum field theory are followed with the proviso that gauge field propagators are principal-value propagators.

This author had several discussions with Professor Wilson in late 1974 in the author's office and while walking to lunch at the Cornell Faculty Club. Professor Wilson proposed possible flaws in the author's theory on almost a daily basis. The author was able to show these suggested flaws were not flaws but physically acceptable.

---

[360] S. Blaha, Phys. Rev. D**10**, 4268 (July, 1974) and Phys. Rev. D**11**, 2921 (December, 1974). These papers appeared before the charmonium calculations of the Cornell and Harvard groups in 1975.

The final discussion with Wilson ended with Wilson stating words to the effect, "Your theory may be a correct phenomenological approximation to my theory of the strong interaction and quark confinement. But my theory is the correct one. Your theory is only a phenomenology." In the forty plus years since this concluding discussion no one has proved that the conventional strong interaction theory truly has a linear potential and quark confinement although some approximations suggest it does.

In the absence of a demonstration of a linear potential in the standard strong interaction model we suggest our theory is a viable alternative. Since the linear potential appears to fairly successfully describe much of the charmonium spectrum we feel our theory with its explicit derivation of a linear potential is worthy of interest – especially because it is in agreement with experiment as far as we know. *An experimentally completely correct phenomenology is a theory*

55.5 Numeric Constants of the Gravitational and Strong Sectors

The numeric constants appearing in

$$\mathcal{L}_G = \sqrt{g}[MD_\nu R'^1{}_{G\sigma\mu}D^\nu R'^2{}_G{}^{\sigma\mu} + aR'^1{}_{G\sigma\mu}R'^2{}_G{}^{\sigma\mu} + bH + cg^{\sigma\mu}g^2{}_{\sigma\mu} + eg^{2\sigma\mu}g^2{}_{\sigma\mu}]$$

and

$$\mathcal{L}_{SU(3)} = \text{Tr}\,\sqrt{g}\{M[\partial_\nu + ig_1(A_{SU(3)}{}^1{}_\nu + A_{SU(3)}{}^2{}_\nu)]F_{SU(3)}{}^1{}_{\sigma\mu}[\partial^\nu + i\,g_1(A_{SU(3)}{}^{1\nu} +$$
$$+ A_{SU(3)}{}^{2\nu})]F_{SU(3)}{}^{2\sigma\mu} + aF_{SU(3)}{}^1{}_{\sigma\mu}F_{SU(3)}{}^{2\sigma\mu} - dA_{SU(3)}{}^2{}_\mu A_{SU(3)}{}^{2\mu}\}$$

are

$$M = 1.61 \times 10^{163} \text{ eV}^{-2}$$
$$a = 7.47 \times 10^{181}$$
$$b = 2.364 \times 10^{55} \text{ eV}^2$$
$$d = 1.99 \times 10^{383} \text{ eV}^2$$
$$c = e = 0$$
$$g_1 = 0.552$$

They have remarkably large values. Yet they conspire to yield small constants in the gravity and Strong potentials.

Appendix 55-A. Comparison of S. Blaha, Phys. Rev. D11, 2921 (1974) and the Present Work

The eq. 55.1 lagrangian is approximately the same as eq. 17 of S. Blaha, Phys. Rev. D11, 2921 (1974). We now discuss the relation of the 1974 paper to the present work except for additional terms $MD_\nu F^1_{\sigma\mu} D^\nu F^{2\sigma\mu}$ and $[A^2_{\varkappa}, A^2_{\mu}]$; and the following changes in parameters:

$$a = -\tfrac{1}{2} \qquad\qquad d = \tfrac{1}{2}\lambda^2$$

Since that paper essentially contains a complete description of our Strong Interaction theory (modulo the additional terms) we refer the reader to it and to its predecessor paper referenced therein. There are a few additional changes required to bring the 1974-5 papers into agreement with our current theory:

1. Eq. 30 of the above referenced paper must be modified to

$$F^2_{\varkappa\mu} = \partial A^2_{\mu}/\partial x^\varkappa - \partial A^2_{\varkappa}/\partial x^\mu + \mathbf{ig_1[A^2_{\varkappa}, A^2_{\mu}]} + ig_1[A^1_{\varkappa}, A^2_{\mu}] + ig_1[A^2_{\varkappa}, A^1_{\mu}]$$

$$(55\text{-A.1})$$

with the addition of the 'bolded' term if$[A^2_{\varkappa}, A^2_{\mu}]$. There is also a trivial change of notation of coupling constant from 'g_1' to 'f'.

2. Eqs. 6 and 18 should have the interaction term expanded to

$$g_1 A^1 \quad\rightarrow\quad g_1(A^1 + A^2) \quad```$$

$$(55\text{-A.2})$$

and similarly in eq. 20. Eqs. 38 – 41 directly show that the additional interaction term leads to a gluon propagator[361] $<A^1 + A^2, A^1 + A^2> = 2<A^1, A^2> + <A^1, A^1>$, and introduces a 1/r term in the potential part of the gluon propagator.

As a result the effective gluon propagator in the theory, **if the $Mf^2D_vF^1_{\sigma\mu}D^vF^{2\sigma\mu}$ term is neglected**, combines eqs. 38 and 39 to give the *short-distance*[362] gluon propagator between quarks:

$$g_{\mu\nu}\delta_{ab}P[\lambda^2/k^4 - 1/k^2] \qquad (55\text{-A.3})$$

up to a constant factor.[363]

These changes *explicitly* lead to a Strong Interaction potential of the form

$$V(r) = -2g^2/r + g^2\lambda^2 r \qquad (55\text{-A.4})$$

Naturally one can expect perturbative corrections to eq. 55.6 in higher order in f. However the apparent relative smallness of f suggests eq. 55-A.4 is a good approximation to the short-distance, inter-quark interaction.

55-A.1 Canonical Equal Time Commutation Relations

The Euler-Lagrange equations of motion, eqs. 27 – 31 in the author's 1975 paper, are modified most significantly by the $Mf^2D_vF^1_{\sigma\mu}D^vF^{2\sigma\mu}$ lagrangian term in eq. 55.1. In order to use the canonical method to obtain the contributions to the equations of motion of this higher derivative term we use integration by parts and discard surface terms for eq. 55.1, as is usually done in quantum field theory.

---

[361] Eqs. 40-41 in the above referenced paper.

[362] We see that the $Mf^2D_vF^1_{\sigma\mu}D^vF^{2\sigma\mu}$ term will affect the short-distance behavior of the inter-quark interaction. The equivalent term in the gravitation sector influences the long-distance form of the gravity potential and leads to a MoND-like behavior.

[363] This propagator is taken in Principal value to avoid potential unitarity problems. This topic is described in detail in earlier papers and books by the author.

56. Other Effects of New Interactions Between Bosons

56.1 Missing Nucleon Spin Puzzle

The estimates of nucleon spin that are obtained from parton analyses of deep inelastic electron – nucleon interactions are woefully short of the spin expected in quark models of nucleons. The missing spin has been attributed to a number of causes. However the Missing Spin Puzzle remains.

From eq. 53.28 - 53.30 it is clear that there are important new interaction terms between the Electromagnetic and Strong interaction fields. After taking traces we find

$$\mathcal{L}_{\text{intEM-S}} = -\text{Tr } iM\{(A_E{}^1{}_\nu + A_E{}^2{}_\nu)\, F_{SU(3)}{}^1{}_{\sigma\mu} D^\nu F_{SU(3)}{}^{2\sigma\mu} + iD_\nu F_{SU(3)}{}^1{}_{\sigma\mu}(A_E{}^{1\nu} + A_E{}^{2\nu})F_{SU(3)}{}^{2\sigma\mu}\}$$

$$(56.1)$$

$\mathcal{L}_{\text{intEM-S}}$ generates a combined photon-gluon vertex insertion in gluon interactions between quarks within a hadron. Figs. 56.1 and 56.2 show two simple possible vertex insertions in a gluon line.

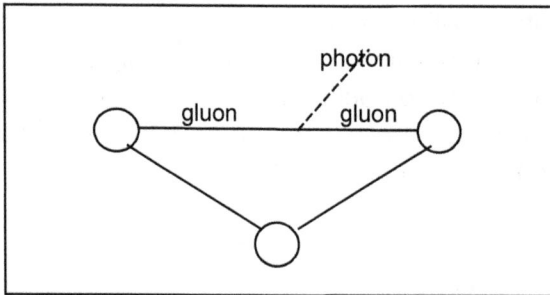

Figure 56.1 A single 'outgoing' photon vertex insertion in a gluon line. Only single gluon lines between the three quarks are displayed. This

gluon-gluon-photon interaction is possible because the photon field is an SU(1) field. Thus the diagram is one of the possibilities embedded in eq. 56.1.

The gluon line, by itself, has $1/k^4$ and $1/k^2$ momentum space propagator terms. *The insertion of the vertex in Fig.56.1 generated by $\mathscr{L}_{intEM\text{-}S}$ yields a combined momentum factor of $k^3(k^4k^4)^{-1} = k^{-5}$ which would make it (summed over all gluon lines) a significant contribution to the proton spin determination in deep inelastic electron-nucleon scattering.[364] The insertion of the vertex in Fig. 56.2 generated by $\mathscr{L}_{intEM\text{-}S}$ yields a combined momentum factor of $k^2(k^4k^4)^{-1} = k^{-6}$ which may have a less significant effect.*

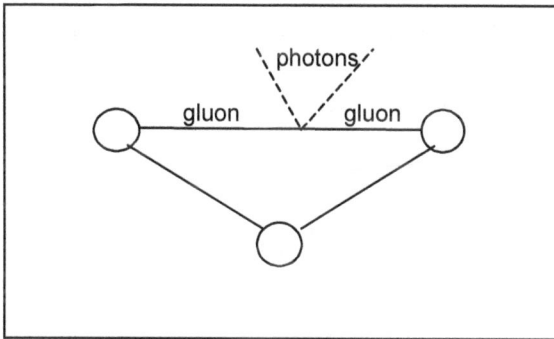

Figure 56.2 An 'outgoing' two photon vertex insertion in a gluon line. Only single gluon lines between the three quarks are displayed.

Thus our unified theory may help solve the nucleon spin puzzle. The interactions in Figs. 56.1 and 56.2 introduce a new direct connection between photons and spin one gluons. Thus their contributions to the summations of proton spin interactions in parton

---

[364] See C. A. Aidala, S. D. Bass, D. Hasch, and G. K. Mallot, arXiv: 1209.2803v2 (2013) and references therein for a review of the 'missing' nucleon spin puzzle.

models may account for the 'missing' two-thirds of proton spin. Our unified theory has a new gluon-photon interaction that is not found in the conventional Standard Model.

These considerations apply also to the Dark fermion sector, and also apply layer by layer to the four normal and Dark layers of fermions.

56.2 Vector Meson Dominance (VMD) Due to New Gluon – Electromagnetic Interactions

Vector Meson Dominance (VMD) is a model describing the hadronic part of the physical photon consisting of a purely electromagnetic part and a hadronic part.[365] The hadronic part contains of light vector mesons: ρ, ω, and φ. Consequently physical photon-hadron interactions have a hadronic part as well as a conventional electromagnetic photon part. Features of the interaction include a greater intensity than a purely photon part and a similarity of the physical photon interactions between protons and neutrons despite the difference in their charge structures. The physical photon thus appears to be a superposition of a purely electromagnetic photon and a vector meson.

The boson lagrangian term

$$\mathcal{L}_M = \text{Tr} \sqrt{g} M D_\nu R''^1_{\sigma\mu} D^\nu R''^{2\sigma\mu} \tag{56.2}$$

has a part with photon-gluon interactions. One of these terms has the form

$$AAGG \tag{56.3}$$

where A represents a photon and G represents a gluon, with indices and derivatives not shown. This lagrangian term (and other similar terms in eq. 56.3) has the Feynman diagram:

---

[365] J. J. Sakurai, Ann. Phs. 11 (1960).

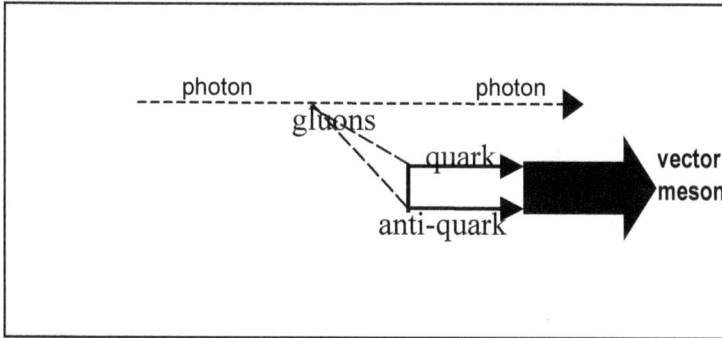

Figure 56.3 An energetic purely electromagnetic photon emitting a gluon pair that generate a quark – anti-quark pair that then combine to form a vector meson of the VMD type. The physical photon thus combines a pure electromagnetic part with a vector meson.

Thus VMD is part of the boson lagrangian terms in eq. 56.3.

These considerations apply also to the physical Dark photon consisting of a Dark electromagnetic photon part and a Dark hadronic part consisting of a Dark vector meson part. They also apply layer by layer to the four normal and Dark layer vector bosons.

56.2.1 Gluon – SU(2) Weak Vector Boson Interaction – A New Form of Vector Meson Dominance (VMD)

The preceding discussion of this section 56.2 also applies to SU(2) Weak Interaction vector bosons, denoted U, if one simply substitutes each of the three SU(2) Weak Interaction vector bosons above. The terms of the form

$$UUGG \tag{56.4}$$

where U represents one of the SU(2) Weak vector bosons and G represents a gluon, with indices and derivatives not shown. The types of Feynman diagrams that result are analogous to Fig. 56.4 with the photon lines replaced with U vector boson lines. Since gluons do not have electric charge the produced vector mesons are neutral.

One concludes that charged and uncharged SU(2) Weak vector bosons are each accompanied by a neutral vector meson yielding a form of VMD in the Weak Interaction sector. This applies to 'normal' and Dark Weak vector bosons in each of the four layers separately.

56.3 Layer Field – Gluon Interactions

The lagrangian in the four layer extension of eq. 53.23 indicates that there are direct interactions between gluons and the four[366] Layer group gauge fields, V_a, such as

$$V_a{}^{1i}{}_\nu V_a{}^{2iv} \partial_\alpha A_{bSU(3)\mu}{}^1 \partial^\alpha A_{bSU(3)}{}^{2\mu} \tag{56.5}$$

where a labels the Layer group for generation a, and b labels the specific layer[367] with other indices and derivatives not shown. These interactions result in Feynman diagrams that modify the Strong Interactions between quarks to include intermediate transitions between quarks in different layers. Fig. 56.5 shows the simplest forms of this interaction between two quarks.

---

[366] There is a separate Layer group field, labeled with subscript a, for each of the four generations comprising each layer.
[367] With implied summations over a nd b.

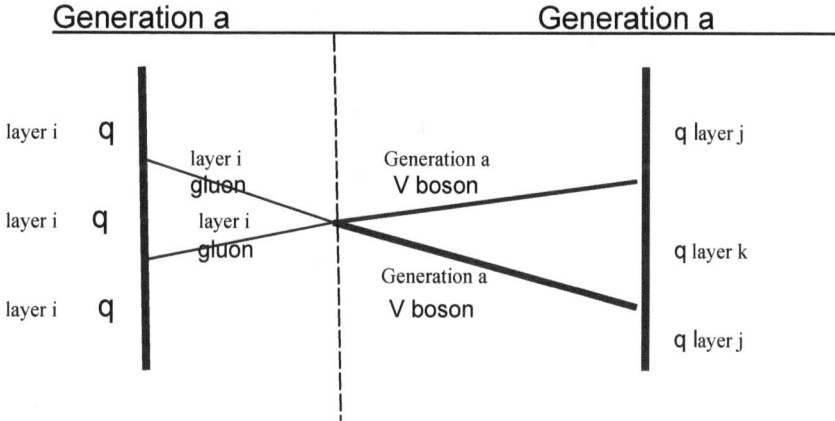

Figure 56.4 Gluon-Layer V gauge particle interaction between two quarks (one of layer i and one of layer j) with each of generation a (for one of the four generations appearing in each layer.) The layer k quark is generated by the V boson and then transformed back to layer j. The result is an interaction between quarks in different layers. Since only layer 1 is experimentally known this interaction must be quite weak and/or the masses of quarks in higher layers must be very large.

These considerations apply also to the Dark quarks sector, and also apply layer by layer to the four normal and Dark layers of quarks.

56.4 Summary of the Other New Gluon – Boson Interactions

The lagrangian term in eq. 56.3 above contains additional boson interactions. Since

$$R''^1{}_{\sigma\mu} = R'_{SU(3)}{}^1{}_{\sigma\mu} + R^1{}_{\sigma\mu}$$
$$R''^2{}_{\sigma\mu} = R'_{SU(3)}{}^2{}_{\sigma\mu} + R^2{}_{\sigma\mu}$$

and since

$${}^{a}\mathbf{A}_I{}^{\mu}(x) = ({}^{a}g_1{}^{a}\mathbf{A}_{SU(3)}{}^{\mu}(x_C),\ {}^{a}g_2{}^{a}\mathbf{W}^{\mu}(x),\ {}^{a}g_3{}^{a}\mathbf{A}_E{}^{\mu}(x),\ {}^{a}g_4{}^{a}\mathbf{W}_D{}^{\mu}(x),\ {}^{a}g_5{}^{a}\mathbf{A}_{DE}{}^{\mu}(x),\ {}^{a}g_6{}^{a}\mathbf{A}_{DSU(3)}{}^{\mu}(x_C),$$
$${}^{a}g_7{}^{a}\mathbf{U}^{\mu}(x),\ {}^{a}g_8{}^{a}\mathbf{V}^{\mu}(x))$$

$$\mathbf{C}_I{}^{\mu}(x) = \Sigma_{a,i}\ {}^{a}\mathbf{A}_{Ii}{}^{\mu}(x) + g_8\mathbf{A}_S{}^{\mu}(x) + g_\Theta\mathbf{A}_\Theta{}^{\mu}(x)$$

$$D_\nu V_\mu = (\partial_\nu + iF_\nu)V_\mu - H^\sigma{}_{\nu\mu}V_\sigma$$
$$= [g^\sigma{}_\mu\partial_\nu + ig^\sigma{}_\mu F_\nu - H^\sigma{}_{\nu\mu}]V_\sigma$$
$$= [g^\sigma{}_\mu\partial_\nu + iD^\sigma{}_{\mu\nu}]V_\sigma$$

$$F^\mu = C_I{}^{1\mu}(x) + C_I{}^{2\mu}(x) + B^{1\mu} + B^{2\mu}$$
$$H^\sigma{}_{\nu\mu} = \Gamma_{GR}{}^\sigma{}_{\nu\mu} + \Gamma_{GR}{}^{2\sigma}{}_{\nu\mu}$$
$$D^\sigma{}_{\mu\nu} = g^\sigma{}_\mu F_\nu + iH^\sigma{}_{\nu\mu}$$

we see the remarkable 'new' 'normal' and Dark gluon – vector boson interactions:

Gluon – Gluon Interaction (chapter 53)
Gluon – Weak Vector Boson Interaction
Gluon – Electromagnetic Interaction
Gluon – Dark Weak Vector Boson Interaction
Gluon – Dark Electromagnetic Interaction
Gluon – Dark Strong Interaction
Gluon – Generation Group Vector Boson Interaction
Gluon – Layer Group Vector Boson Interaction

plus:

Gluon – Graviton Interaction   (Possibly relevant for black holes, and quark and neutron stars)

56.5 New Graviton – Vector Boson Interactions

Eq. 56.2 also contains 'new' graviton – boson interactions:

        Graviton – Weak Vector Boson Interaction
        Graviton – Electromagnetic Interaction
        Graviton – Dark Weak Vector Boson Interaction
        Graviton – Dark Electromagnetic Interaction
        Graviton – Dark Strong Interaction
        Graviton – Generation Group Vector Boson Interaction
        Graviton – Layer Group Vector Boson Interaction

plus an additional

        Graviton – Graviton Interaction

The above new boson – graviton interactions appear to be of cosmological interest in some situations with high gravity such as black holes, and quark and neutron stars.

57. Quantum Gravity and the Wheeler-DeWitt Equation Extended to Complex Coordinates

57.1 Introduction

In this chapter we will consider Quantum Gravity as embodied in the Wheeler-DeWitt[368,369] equation. This equation has many noteworthy points that we will consider below – particularly its extension to complex space-time. It also raises important basic quantum questions: such as "Who is the Observer?" that we addressed in Blaha (2014a) where most of this chapter first appeared.

57.2 Analytically Continued Wheeler-DeWitt Equation to Complex Metrics under a Faddeev-Popov Method Restriction

In this section we extend the Wheeler-DeWitt equation for Quantum Gravity to complex coordinates and metrics <u>by analytic continuation</u> (piece-wise if necessary) and impose the condition that metrics must be real-valued using the Faddeev-Popov Method. Our procedure will be to take the Wheeler-DeWitt equation for real-valued metrics (and coordinates), **analytically continue it to the case of complex metrics and coordinates,**[370] and then impose the condition that physically acceptable metrics must be real-valued through use of a Reality group transformation implemented via the Faddeev-Popov Method. Our motivation is two-fold: we have shown that the form of The Standard Model of Particles can be derived from complex space-time considerations demonstrating that we exist in a "masked" complex space-time; and we

---

[368] DeWitt, B. S., Phys. Rev. **160**, 1113 (1987).

[369] Hartle and Hawking also derive the Wheeler-DeWitt equation from a path integral formalism for quantum gravity.

[370] The piecewise analytic continuation of general relativity to complex coordinates and metrics is described in some detail in Blaha (2004). The gist of the continuation is that all equations have the same form after analytic continuation due to a basic theorem of complex mathematics that the analytic extension of equations to complex values from real values is unique.

have shown the imposition of a restriction on a gauge theory such as gravitation[371] can be implemented using the Faddeev-Popov Method or equivalent.

We start by noting the canonical decomposition of a <u>real-valued</u> metric $g_{\mu\nu}$ is defined by:

$$g_{\mu\nu}(x) = \eta_{\alpha\beta} \, \partial\omega^\alpha/\partial x^\mu \, \partial\omega^\beta/\partial x^\nu \qquad (57.1)$$

with

$$g_{\mu\nu} = g_{\nu\mu}$$

and with inverse

$$g^{\mu\nu} = \eta^{\alpha\beta} \, \partial x^\mu/\partial\omega^\alpha \, \partial x^\nu/\partial\omega^\beta \qquad (57.2)$$

The decomposition of the real-valued metric is

$$g_{\mu\nu} = \begin{bmatrix} -\alpha^2 \beta_k \beta^k & \beta_j \\ \\ \beta_i & \gamma_{ij} \end{bmatrix} \qquad (57.3)$$

$$g^{\mu\nu} = \begin{bmatrix} -\alpha^{-2} & \alpha^{-2}\beta^j \\ \\ \alpha^{-2}\beta^i & \gamma^{ij} - \alpha^2\beta^i\beta^j \end{bmatrix} \qquad (57.4)$$

where

$$\gamma_{ik}\,\gamma^{kj} = \delta_i^{\ j} \qquad\qquad \beta^i = \gamma^{ij}\,\beta_j \qquad (57.5)$$

The Wheeler-DeWitt equation <u>for real-valued metrics</u> is

$$(G_{ijkl}\,\delta/\delta\gamma_{ij}\,\delta/\delta\gamma_{kl} + \gamma^{\frac{1}{2}\,(3)}R + 2\lambda\,\gamma^{\frac{1}{2}\,(3)})\Psi(^{(3)}g) = 0 \qquad (57.6)$$

where λ is the cosmological constant, and where the Wheeler-DeWitt metric is

$$G_{ijkl} = \tfrac{1}{2}\,\gamma^{-\frac{1}{2}}(\,\gamma_{ik}\gamma_{jl} + \gamma_{il}\gamma_{jk} - \gamma_{ij}\gamma_{kl}) \qquad (57.7)$$

---

[371] Such as real-valuedness.

The functional derivatives $\delta/\delta\gamma_{ij}$ have several interpretations that are presumably equivalent. DeWitt characterizes them as coordinate independent specifications of the 3-metric. The wave function $\Psi(^{(3)}\mathcal{G}) = \Psi(\gamma_{ij})$, where $^{(3)}\mathcal{G}$ is a geometry, is *not* coordinate dependent. It is invariant under coordinate changes. $^{(3)}\mathcal{G}$ is a discrete infinity of independent invariants constructed from products of the Riemann tensor and its covariant derivatives.

Hartle and Hawking[372] derive the Wheeler-DeWitt equation from a path integral formalism for quantum gravity. Their path integral can be represented as

$$Z = N \int \delta g(x) \exp(iS_E[g]) \qquad (57.8)$$

where S_E is the classical action for gravity and the functional integral is an integral over all 4-geometries. Changing to DeWitt's notation based on the spatial metric γ_{ij} and expressing eq. 57.8 in a more explicit form for use in conjunction with the Faddeev-Popov Method we have

$$Z = N \int \sum_{i,j} \prod_x d\gamma_{ij}(x) \exp(iS_E[\gamma]) = N \int D\gamma \exp(iS_E[\gamma]) \qquad (57.9)$$

The integrand, being a functional integral over all space, is independent of the coordinates.

The Wheeler-DeWitt equation applies to real-valued metrics $\gamma^{ij}(x)$. We now extend this equation to apply to complex-valued metrics using the local Reality group. In doing this we realize that there are an infinite number of complex-valued metrics in the orbit corresponding to each real-valued metric.

This redundancy can be resolved by realizing that the physical measurement of an invariant interval, and the coordinates from which it is derived, are always real-valued. Yardsticks and clocks can only measure real-valued numbers. Based on this physical principle we can generalize quantum gravity to complex coordinates and metrics by using the Faddeev-Popov method to constrain the set of paths in the quantum

---

[372] Hartle, J. B. and Hawking, S. W., Phys. Rev. D **28**, 2960 (1983).

gravity path integral eq. 57.9. Using the Faddeev-Popov Method the constraint can be expressed in terms of an infinitesimal transformation of the metric to a complex value. Using an infinitesimal Reality group transformation V:

$$[\exp(ia_j(x'',x')U_j)]^\alpha{}_\mu = S(x'',x')^\alpha{}_\mu = \partial x''^\alpha / \partial x'^\mu \qquad (57.10)$$

$$V = \exp(ia_k(x'',x')U_k) \cong I + ia_k(x'',x')U_k \qquad (57.11)$$

$$V^\dagger = [\exp(ia_k(x'',x')U_k)]^\dagger \cong I - ia_k(x'',x')U_k \qquad (57.12)$$

where $a_j(x'',x')$ is the jth real-valued local infinitesimal parameter, and U_k is one of the 16 hermitean generators of the Reality group. (We treat the index on a_k and U_k as lower case and sum on k.) We find the condition fixing the metric to a physical *real* value is:

$$F^{ij}(\gamma(x)) = \text{Im } \gamma^{ij}(x) \equiv -\tfrac{1}{2} i(\gamma^{ij}(x) + \gamma^{ij}(x)^*) = 0 \qquad (57.13)$$

where

$$F^{aij}(\gamma^a(x)) = \text{Im}\{(\delta^i{}_m + ia_k(x',x)U_k{}^i{}_m)(\delta^j{}_p + ia_k(x',x)U_k{}^j{}_p)\gamma^{mp}(x)\} \qquad (57.14)$$

using the infinitesimal form.

The Reality condition eq. 57.13 is implemented within the path integral formalism with the Faddeev-Popov Method identity

$$1 = \int D\gamma \, \Delta(F(\gamma)) \, \delta(\delta F^{aij}(\gamma^a(x))/\delta a_n(x',x)) = \int D\gamma \, \Delta(F(\gamma)) \, \delta(F(\gamma^{ij})) \qquad (57.15)$$

Eq. 57.14 yields

$$
\begin{aligned}
\delta F^{aij}(\gamma^a(x))/\delta a_n(x',x)|_{a=0} &= \text{Im}\{\delta_{kn}iU_k{}^i{}_m\gamma^{mp}(x)\delta^j{}_p + \delta_{kn}\delta^i{}_m i\gamma^{mp}(x)U_k{}^j{}_p]\} \\
&= \text{Re}\{\gamma^{mp}[[U_n{}^i{}_m\delta^j{}_p + \delta^i{}_pU_n{}^j{}_m]\} \\
&= \gamma^{mp}(x)\text{Re}\{U_n{}^i{}_m\delta^j{}_p + \delta^i{}_pU_n{}^j{}_m\} = \gamma^{mp}(x)\xi^{ij}{}_{nmp} \qquad (57.16)
\end{aligned}
$$

since $\gamma^{mp}(x)$ is made real by the $\delta(\text{Im } \gamma^{mp}(x))$ where

$$\xi^{ij}_{nmp} = Re\{U_n{}^i{}_m\delta^j_p + \delta^i_p U_n{}^j{}_m\} \qquad (57.17)$$

Note ξ^{ij}_{nmp} is symmetric in i and j, and becomes effectively symmetric in m and p when combined with $\gamma^{mp}(x)$ in eq. 57.16. Calculating $\Delta(F(\gamma))$ we obtain

$$\Delta(F(\gamma)) = [\det \delta F^{aij}(\gamma^a(x))/\delta a_n(x',x)|_{a=0}]^{-1} \qquad (57.18)$$

We can rewrite this Faddeev-Popov determinant as a path integral over an anti-commuting c-number scalar field χ with a ghost Lagrangian:

$$\Delta(F(\gamma)) = \int D\chi^* D\chi \exp[i\int d^4x \, \gamma^{\frac{1}{2}} \mathscr{L}_\gamma{}^{ghost}(x)] \qquad (57.19)$$

where

$$\mathscr{L}_\gamma{}^{ghost}(x) = \chi^*(x)[U_{nij}\xi^{ij}_{nmp}\gamma^{mp}(x)]\chi(x)$$
$$= \chi^*(x)(Re\, U_{nij})\xi^{ij}_{nmp}\gamma^{mp}(x)]\chi(x) \qquad (57.20)$$
$$= \chi^*(x)\xi_{mp}\gamma^{mp}(x)\chi(x) \qquad (57.21)$$

since $\xi_{mp}\gamma^{mp}(x)$ is a c-number:

$$\xi_{mp} = (Re\, U_{nij})\xi^{ij}_{nmp} = 2\, Re\, U_{npi}\, Re\, U_n{}^i{}_m \qquad (57.22)$$

using $Re\, U_n{}^i{}_m = Re\, U_n{}^m{}_i$ for the U(4) generators in its fundamental representation $\underline{4}$.

We then find ξ_{mp} is a diagonal matrix and has the value

$$\xi_{mp} = \xi_m\delta_{mp} \qquad (57.22a)$$

where the diagonal elements are

$$\xi_0 = 8$$
$$\xi_1 = 8$$
$$\xi_2 = 8$$
$$\xi_3 = 8 \qquad (57.22b)$$

and the non-diagonal elements are zero making ξ_{mp} considered as a matrix a multiple of the Identity matrix..

The Faddeev-Popov generated terms when added to the Einstein Action appear to have important ramifications – particularly with respect to the Cosmological Constant. We explore that issue next.

57.3 Possible Source of the Cosmological Constant in the Complex Space-time – Faddeev-Popov Constraint Term

The Wheeler-DeWitt equation

$$(G_{ijkl}\, \delta/\delta\gamma_{ij}\, \delta/\delta\gamma_{kl} + \gamma^{\frac{1}{2}\,(3)}\, R + 2\lambda\, \gamma^{\frac{1}{2}\,(3)})\Psi(^{(3)}\mathcal{G}) = 0 \qquad (57.23)$$

has the cosmological constant, Λ, as one of its terms. This equation is derived from the Hamiltonian constraint that ultimately follows from the Einstein action. Inserting the Faddeev-Popov term of the previous section in the Einstein lagrangian yields

$$S_E(\gamma) = -(16\pi G)^{-1} \int d^4x\, \gamma^{\frac{1}{2}}\{\, R(x) - 2\lambda + \mathcal{X}^*(x)\mathcal{X}(x)\xi_{mp}\gamma^{mp}(x)\} \qquad (57.24)$$

The constant matrix ξ_{mp} is a product of parts of the generators of U(4) given by eq. 57.22a:

$$\xi_{mp} = 2\,(\text{Re } U_{npi})\,(\text{Re } U_n{}^i{}_m) \qquad (57.25)$$

We will now show that the term

$$\Pi = \gamma^{\frac{1}{2}}\mathcal{X}^*(x)\mathcal{X}(x)\xi_{mp}\gamma^{mp}(x) \qquad (57.26)$$

upon variation of the metric $\delta\gamma_{\mu\nu}$, gives a term which is approximately a constant cosmological term assuming $\mathcal{X}(x)$ is approximately constant (with its implied divergence eliminated by renormalization of the path integral), and assuming an almost flat space-time $\gamma_{\mu\nu} \cong \eta_{\mu\nu}$. Varying the metric for Π yields

$$\delta\Pi = \delta\gamma_{\mu\nu} \{½ \gamma^{½}\chi^*(x)\chi(x)\xi_{mp}\gamma^{mp}(x)\gamma^{\mu\nu} - \gamma^{½}\chi^*(x)\chi(x)\xi_{mp}\gamma^{m\mu}\gamma^{p\nu}\} \qquad (57.27)$$

The resulting modified Einstein field equation is

$$R^{\mu\nu} - ½ g^{\mu\nu}R + \lambda \gamma^{\mu\nu} + ½[\gamma^{mp}(x)\gamma^{\mu\nu} - \gamma^{m\mu}\gamma^{p\nu}]\xi_{mp}\chi^*(x)\chi(x) = -8\pi T^{\mu\nu} \qquad (57.28)$$

The terms $½[\gamma^{mp}(x)\gamma^{\mu\nu} - \gamma^{m\mu}\gamma^{p\nu}]\xi_{mp}\chi^*(x)\chi(x)$ appearing above is, or is a contribution to, the cosmological constant assuming a nearly flat universe as our universe seems to be. Thus $\gamma_{\mu\nu} \cong \eta_{\mu\nu}$ to good approximation and $\chi^*(x)\chi(x)$ can be taken to be constant since the time derivative of $\chi(x)$ does not appear in the lagrangian. Consequently the total cosmological constant term is

$$\lambda_{tot}{}^{\mu\nu} \cong \lambda g^{\mu\nu} + 4 \chi^*(x)\chi(x)g^{\mu\nu} = (\lambda + \lambda_{F-P}) g^{\mu\nu} \qquad (57.29)$$

by eq. 57.22b.

Given the somewhat problematic state of our understanding of the cosmological constant it is not impossible that the complexity of space-time leading to λ_{F-P} may be the sole origin of the cosmological constant. In evaluating eq. 57.29 we may normalize $\chi^*(x)\chi(x) = 1/4$ by adjusting the overall normalization of the path integral since $\chi(x)$ is time independent. Then if the "bare" cosmological constant is zero we obtain

$$\lambda_{tot} = \lambda_{F-P} = 1 \qquad (57.30)$$

by eq. 57.29.

If this is true then we have achieved a space-time origin for the cosmological constant rather than an ad hoc origin. The modified Einstein field equation is then

$$R^{\mu\nu} - ½ g^{\mu\nu}R + \lambda_{tot}\gamma^{\mu\nu} = -8\pi T^{\mu\nu} \qquad (57.31)$$

by eqs. 57.28 and 57.30.

57.4 Impact of the Faddeev-Popov Complexity Term on the Wheeler-DeWitt Equation

The Faddeev-Popov term that arises because of the restriction of the metrics and coordinates to real values also impacts on the Wheeler-DeWitt equation since it is derived from the lagrangian via the Hamiltonian it generates. The Wheeler-DeWitt equation changes to

$$(G_{ijkl} \, \delta/\delta\gamma_{ij} \, \delta/\delta\gamma_{kl} + \gamma^{1/2\,(3)} \, R + 2\lambda \, \gamma^{1/2\,(3)})\Psi(^{(3)}\mathcal{G}) = 0 \qquad (57.32)$$

The functional Wheeler-DeWitt equation of eq. 57.32 resembles a Klein-Gordon equation.[373] Solutions of this equation can be expressed as path integrals:

$$\Psi(^{(3)}\mathcal{G}, \mathcal{L}_F) = N \int_{\mathcal{C}} \delta g(x) \, \exp(-I(g, \mathcal{L}_F)) \qquad (57.33)$$

where $I(g, \mathcal{L}_F)$ is the effective total Euclidean action for the open universe case. See Hartle and Hawking for a detailed study in the case of a symmetric cosmological constant.

It does not appear that the issues of the Wheeler-DeWitt equation are resolved by the extended Wheeler-DeWitt equation presented here:

- Divergences in integrals in inner products, thus requiring renormalization.
- Negative probabilities in inner products,
- Issues with the requirement of space-like surfaces,
- The frontier divergence singularity.

Later we consider another form of the Wheeler-DeWitt equation expressed in terms of Megaverse geometry. We will reconsider the issues of the above formulation again in the Megaverse.

---

[373] Hartle, J. B. and Hawking, S. W., Phys. Rev. D **28**, 2960 (1983).

58. Unified SuperStandard Theory and its Hidden SuperString Aspect

The Unified SuperStandard Theory that we have developed is a Quantum Field Theory that contains the known Standard Model and Gravitation as well as significant new features. It uses extensions of Quantum Field Theory: a Two-Tier PseudoQuantum formulation that makes the theory finite to all orders in perturbation theory, and a formalism that supports quantization in any coordinate system.

In this chapter we will show that the theory has a String aspect that, together with its SuperSymmetric aspect, enables us to characterize it as a hidden variant of SuperString theory.

58.1 String Theory vs. Two-Tier Quantum Field Theory

We begin by comparing the Gervais-Sakita[374] SuperString lagrangian (GS) with a simple Two-Tier Quantum Field Theory lagrangian (TT):[375]

$$\mathcal{L} = \psi(x)(i\gamma^\mu \partial/\partial x^\mu - m)\psi(x) - \partial_a X_\mu \partial^a X^\mu \qquad \text{(GS)}$$
$$\mathcal{L} = \psi(X_T)(i\gamma^\mu \partial/\partial X_T{}^\mu - m)\psi(X_T) + \tfrac{1}{4} M_c{}^4 F^{\mu\nu}(X_T) F_{\mu\nu}(X_T) \qquad \text{(TT)}$$

where

$$X_T{}^\mu(y) = y^\mu + i\, Y^\mu(y)/M_c{}^2$$

and

$$F^{\mu\nu} = \partial X^\mu/\partial y_\nu - \partial X^\nu/\partial y_\mu$$
$$\equiv i\, (\partial Y^\mu/\partial y_\nu - \partial Y^\nu/\partial y_\mu)/M_c{}^2$$

---

[374] Gervais, J. L. and Sakita, B., Nucl. Phys. **B34**, 632 (1971).
[375] Blaha (2002) and (2005a).

The Gervais-Sakita lagrangian differs from the above Two-Tier lagrangian by the introduction of the 'string' X^μ in the fermion part of the Two-Tier lagrangian.

We note that the Two-Tier quantum field Y_μ has two degrees of freedom like the string field. Thus the Two-Tier fermion field is a function of a string[376] whereas the SuperString lagrangian treats the fermion and the string as independent. As a result the Two-Tier quantum field theory embedded in the Unified SuperStandard Theory is a variant on SuperString theory if one also takes account of the SuperSymmetry attributes of the Unified SuperStandard Theory.

58.2 Strings in Particles? or Strings Made into Particles?

The comparison of the two types of theories is made more pointed by the form of Feynman propagators in Two-Tier theories. Fig. 58.1 shows that a 'dressed' free fermion particle has an associated cloud of 'strings' that are the source of the finiteness of perturbation theory calculations. Thus Two-Tier Unified SuperStandard Theory calculations are finite to all orders in perturbation theory by putting 'strings' in particles unlike SuperString theories which make particles from strings.

58.3 Further Evidence of the String-like Substructure of the Unified SuperStandard Theory

Quantum Dimensions $X_T^\mu(y)$ endow a particle with an extended structure that resembles to some extent the extended structure seen in boson string and Superstring theories. For example, Bailin (1994) use the operator[377]

$$V_\Lambda(k) = \int d^2\sigma \sqrt{-h}\, W_\Lambda(\tau, \sigma)\, e^{-ik\cdot X}$$

where X_T^μ is a quantized fourier expansion of the string fields (see eq. 7.22 of Bailin (1994)).

---

[376] Blaha (2002) and (2005a) show there is a form of the Calculus of Variations that supports a canonical derivation of the equations of motion.

[377] D. Bailin and A. Love, *Supersymmetric Gauge Field Theory and String Theory* (Institute of Physics Publishing, Philadelphia, PA, 1994) page 272.

We note our $X_T{}^\mu$ coordinate-field has two transverse degrees of freedom due to gauge invariance, which also invites comparison to the boson string. A point of difference is that we created a well-defined quantum field theoretic formulation in conventional space-time that has the Standard Model as its "large distance" behavior thus introducing a note of reality that is not (yet?) very apparent in Superstring theories. We see that the interacting quantum field theories based on this approach also have good, finite, short distance behavior just as string theories.

Scalar, and other particles, Feynman propagators can be viewed as describing the propagation of a particle cloaked (accompanied) by a cloud of Y particles.

If one considers high energy hadron-hadron collisions then the beclouded quarks within the hadrons can interact in a 'string-like' manner through the degrees of freedom afforded by the Y particle clouds. Then the behavior of the colliding hadrons could well simulate the types of effects that gave rise to String theory in the 1970's.

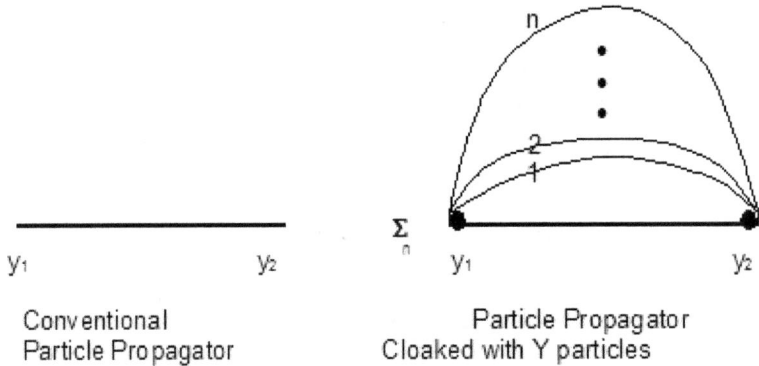

Conventional
Particle Propagator

Particle Propagator
Cloaked with Y particles

Figure 58.1. Feynman diagram for conventional and n^{th} cloaked Two-Tier propagators.

Status of the Derivation at this Point

Since the previous derivation progress report we have extended the derivation to include:

1. The calculation of the Riemann-Christoffel Curvature tensor for bosons.
2. The specification of the vector boson and gravitation terms of the Unified SuperStandard Theory lagrangian.
3. A demonstration of modified gravitation and confining strong interaction consequences of these lagrangian terms as well as other possible effects on the proton spin and radius puzzles.
4. A discussion of the consequences of Quantum Gravity using the Wheeler-DeWitt equation including a Faddeev-Popov Mechanism generation of the Cosmological Constant.
5. A discussion of a SuperString view of the Unified SuperStandard Theory.

Having presented a Unified SuperStandard Theory that successfully accounts for confirmed features of elementary particles and gravitation (possibly excepting new CP-related phenomena) we now turn to evidence for phenomena beyond our universe – The Megaverse.

The Expansion of Our Universe

59. Hubble Constant and Universe Expansion

Our universe is clearly expanding from an initial state called the Big Bang to its current state. The Hubble "Constant" measures the rate of growth. Most of the expansion data on the Hubble Constant is presented below. After reviewing the data we propose a fit to the data in this chapter that explains the apparent growing rate of expansion. Below we derive the form of the fit.

59.1 Hubble Constant Experimental Data

There are a number of astrophysical studies of the universe that suggest that the Hubble Constant is *not* constant. Although there are significant margins of error it appears that the early universe "beginning" epoch around 380,000 years had a Hubble Constant of 67.8 km s^{-1} Mpc^{-1}.[378] More recently, red shift studies of quasars have given a Hubble Constant of 73.2 km s^{-1} Mpc^{-1}.[379] And studies of binary black hole merger gravity waves[380] have given a Hubble Constant of 75.2 km s^{-1} Mpc^{-1} (and earlier of 78 km s^{-1} Mpc^{-1}). Another study of events at 1.8 billion ly yielded a Hubble Constant of 70.0 km s^{-1} Mpc^{-1}.[381] Further studies have given the Hubble Constants: 1) Of variable stars 73.2 km s^{-1} Mpc^{-1}, 2) Of light bent by distant galaxies 72.5 km s^{-1} Mpc^{-1}, 3) Of Magellan Cepheids 74.03 ± 1.42 km s^{-1} Mpc^{-1}, [382] 4) Of distant red giant[383] brightness 69.8 km s^{-1} Mpc^{-1},

---

[378] See, for example, K. Aylor *et al*, arXiv:1811.00537v1 (2018) based on studies of the cosmological sound horizon.

[379] M. Soares-Santos *et al* , arXiv:1901.01540 (2019).

[380] DES and LIGO collaborations *et al*, arXiv:1901.01540 (2019).

[381] B.P. Abbott *et al*, arXiv:1710.05835 (2017).

[382] J. T. Nielsen *et al*, Marginal evidence for cosmic acceleration from Type Ia supernovae, Nature Scientific Reports (2016); arXiv:1506.01354 (2015). A. Riess *et al*, The Astrophysical Journal **875**, 145 (2019) and references therein. A. Riess *et al*, arXiv:1903.07603 (2019).

[383] W. Freedman *et al*, The Astrophysical Journal **880** (July, 2019).

The only apparent conclusion at this time is that there was a Hubble Constant (Constant) H of approximately 67.8 km s^{-1} Mpc^{-1} early in the universe, and ranging up to 75.2 km s^{-1} Mpc^{-1} at the current time. Thus an increasing Hubble Constant.

For the purpose of discussing the apparent increase in H with time, we average the above eight "recent" values of H in the spirit of Bayesian equal probability to obtain a **recent time Hubble average of 73.24** km s^{-1} Mpc^{-1}.[384] Thus there appears to be a 7% - 9% increase in the Hubble Constant over time.

59.2 Fit to the Hubble Constant Data and Scale Factor

It is generally expected that the Hubble Constant will decline with time from the time of the Big Bang. It is generally believed that the Hubble Constant has recently been increasing with time. **The declining value in the past and the current growth of the Hubble Constant imply that it reached a minimum at some time in the past.**

Our fit to the data from Blaha (2019c) and (2019e) was

$$a(t) = (t/t_{now})^{g + ht} \qquad (59.1)$$
$$= \exp[(g + ht)\ln(t/t_{now})]$$

where g and h are constants. (The constant h is *not* the Hubble parameter.) There is an "ht" term in the exponent based on the rise in H(t) suggested by experimental data.

The basis of this choice was:

1. Power law behavior (in part) as in the radiation and matter dominated approximations.

2. The known shape of H(t) at early times, and at present, as described above

3. The simplicity of the fit. Two values of H(t) set the constants g and h.

4. Faster than exponential future growth with no Big Rip.

---

[384] In Blaha (2019c) and (2019e) we used an average estimate of 73.7 km s^{-1} Mpc^{-1}.

The Hubble Constant implied by eq. 59.1 is

$$H(t) = (da/dt)/a = g/t + h(1 + \ln(t/t_{now}) \tag{59.2}$$

We set the value of $H(t)$ by using its value at two values of time determining g and h. Based on experimental data:

$$H(t_c) \equiv H(380{,}000 \text{ yr}) = 67.8 \ \text{km s}^{-1} \ \text{Mpc}^{-1} \tag{59.3}$$
$$H(t_{now}) = 73.24 \ \text{km s}^{-1} \ \text{Mpc}^{-1}$$

and

$$h = (t_c H(t_c) - t_{now}H(t_{now}))[\ t_c - t_{now} + t_c \ln((t_c/t_{now})]^{-1} \tag{59.4}$$
$$g = (H(t_{now}) - h) \ t_{now}$$

where $t_c = 380{,}000$ years after the Big Bang.[385] We obtained

$$h = 2.25983 \times 10^{-18} \ \text{s}^{-1} = 1.49 \times 10^{-33} \ \text{eV} \tag{59.5}$$
$$g = 0.000282377 = 2.82377 \times 10^{-4}$$

59.3 Fluctuating Behavior of a(t) and H(t) – The Big Dip

An examination of the following Figs. 59.1 – 59.4 reveal a Big Dip in $H(t)$ (and also in $a(t)$) which seems to have been unforeseen in astrophysical investigations.

The cause of the Big Dip is the form of the universal scale factor as described in section 59.2. It would be present (although slightly modified) even if the Hubble Constant were truly constant from the 380,000 year point to the present.

Figs. 59.6 through 59.9 plot $H(t)$ (and its logarithm) from the Big Bang to the present and beyond. $H(t)$ has a wide range of values from very large near the Big Bang through the Big Dip to the present and beyond. The recent growth in H is displayed.

59.3.1 Location of the Big Dip in H(t)

The locations in time of the Big Dip events are:

---

[385] Based on the data value of 67.8 km s^{-1} Mpc^{-1} at t = 380,000 years.

Big Dip low point (where H = -445 km s^{-1} Mpc^{-1}) at t = 8.71 × 10^{13} sec.

Big Dip low in a(t) has the value 0.69628 at t = 1.56 ×10^{17} sec.

The decrease in a(t) is to 69.6% of the initial value of 1. The delay from the H(t) Big Dip value to the low point of a(t) may be attributable to the time required to propagate the diminished H value throughout the universe.

59.3.2 Possible Reason for the Big Dip in H(t)

Since the H and Ω_T times match it appears that the Big Dip is a result of massive growth in the mass-energy (and pressure) discuused in a later chapter.

Since the changeover from a radiation-dominated phase to a matter-dominated phase occurs at a slightly earlier time:

Radiation – Matter Domination Transition:[386] t = 1.48 × 10^{12} sec.

it seems reasonable to conclude the transition from radiation-dominated to matter-dominated causes the Big Dip to occur. The matter-dominated phase transition causes shrinkage as shown in a(t) in Fig. 59.4. *The universe contracts by one-third!* [387] We attribute the time delay between the transition and the Big Dip in a(t) to the time required for the transition to occur. (The universe is large at this time after all)

---

[386] In view of our universal scale factor formulation the time of the radiation-matter transition becomes questionable.
[387] Rather like the condensation of water vapor to liquid.

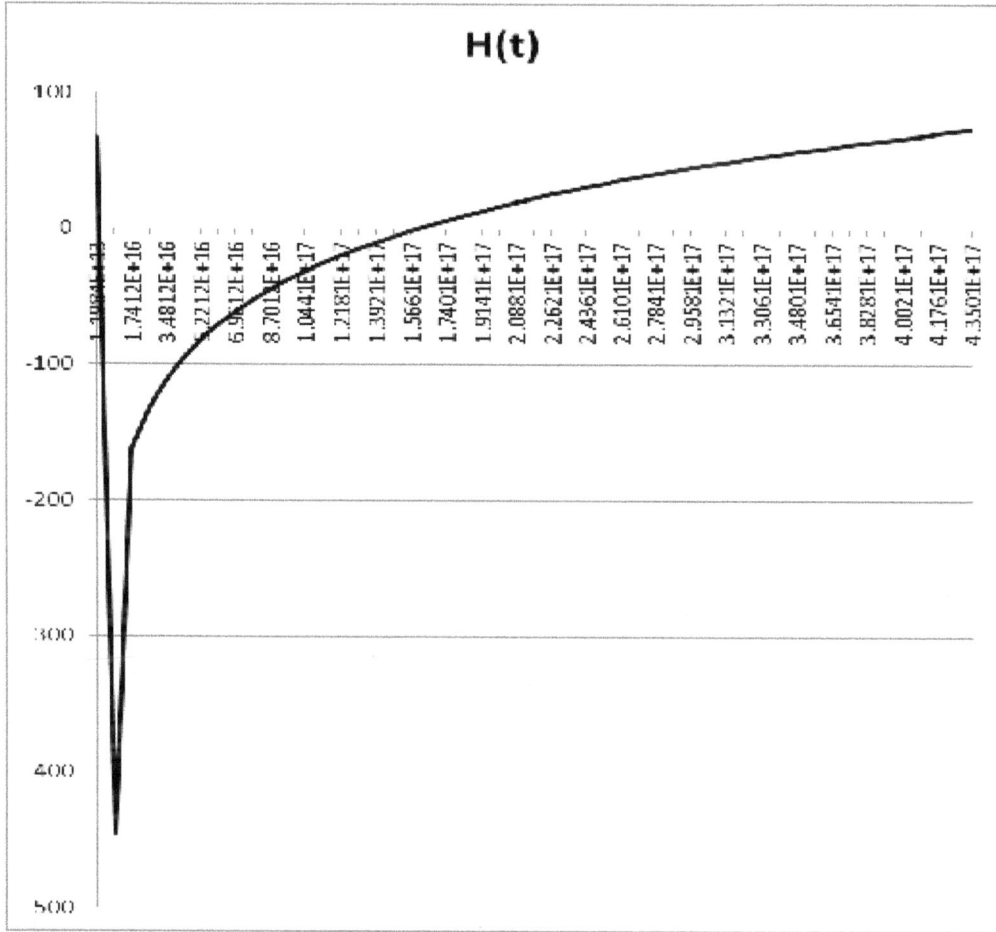

Figure 59.1 Fitted Hubble Constant plotted vs. seconds from t = 1.19 × 10^{13} sec. to 4.35 × 10^{17} sec. The minimum is H = -445 km s^{-1} Mpc^{-1} at t = 8.71 × 10^{13} sec.

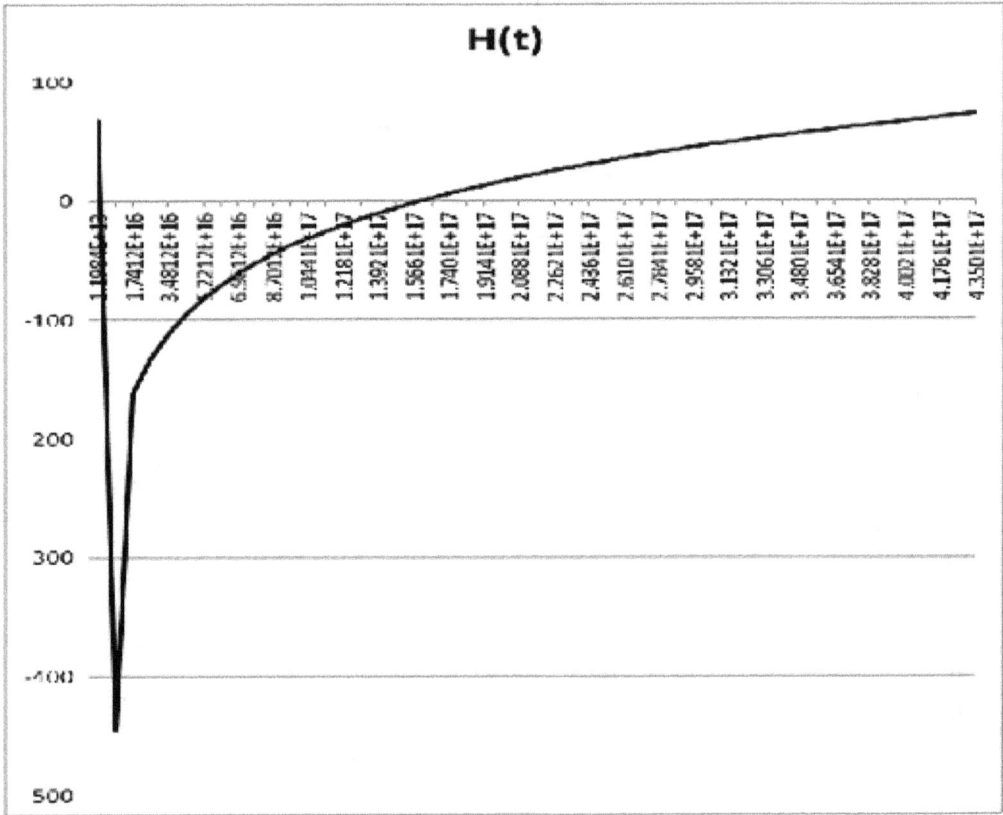

Figure 59.2 A closer view of fitted H(t) plotted vs. seconds from t = 1.19 × 10^{13} sec. to the present 4.35 × 10^{17} sec. The minimum is H = -445 km s^{-1} Mpc^{-1} at t = 8.71 × 10^{13} sec.

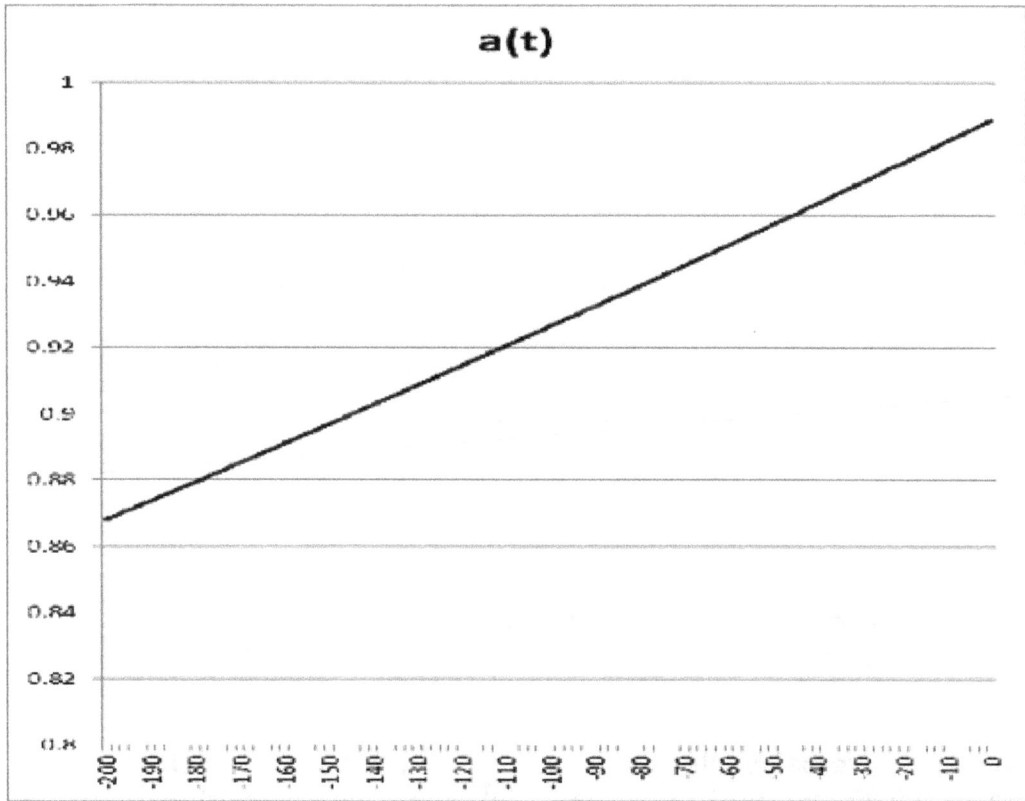

Figure 59.3 Universal Scale Factor a(t) plotted in $\log_{10}$ seconds from t = 10^{-200} sec. to t = 1 sec.

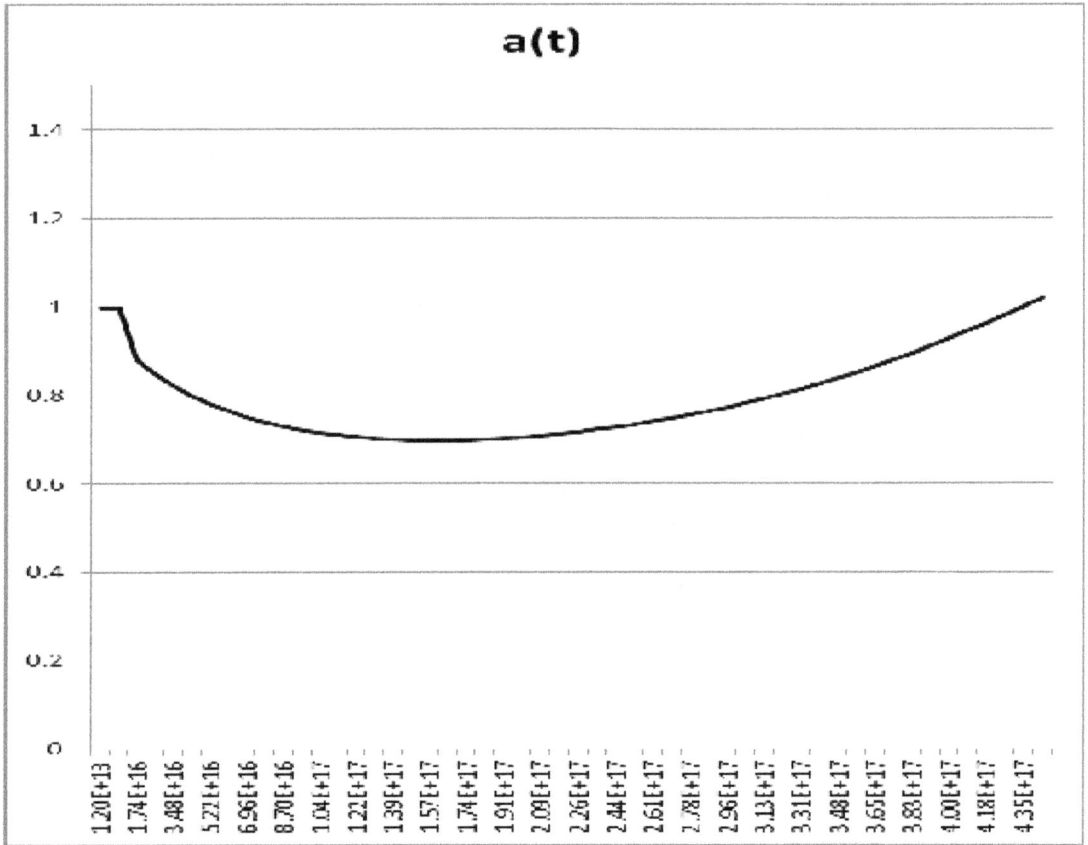

Figure 59.4 Universal Scale Factor a(t) plotted in seconds from t = 1.19 $\times$ 10^{13} sec. (380,000 years) to t = 4.35 $\times$ 10^{17} sec. (the present). Note the Big Dip to a = 0.69628 at t = 1.56 $\times 10^{17}$ sec. The decrease in a(t) is to 69.6% of the initial value of 1. (The almost flat part before 1.20 $\times$ 10^{13} sec is a roundoff effect.)

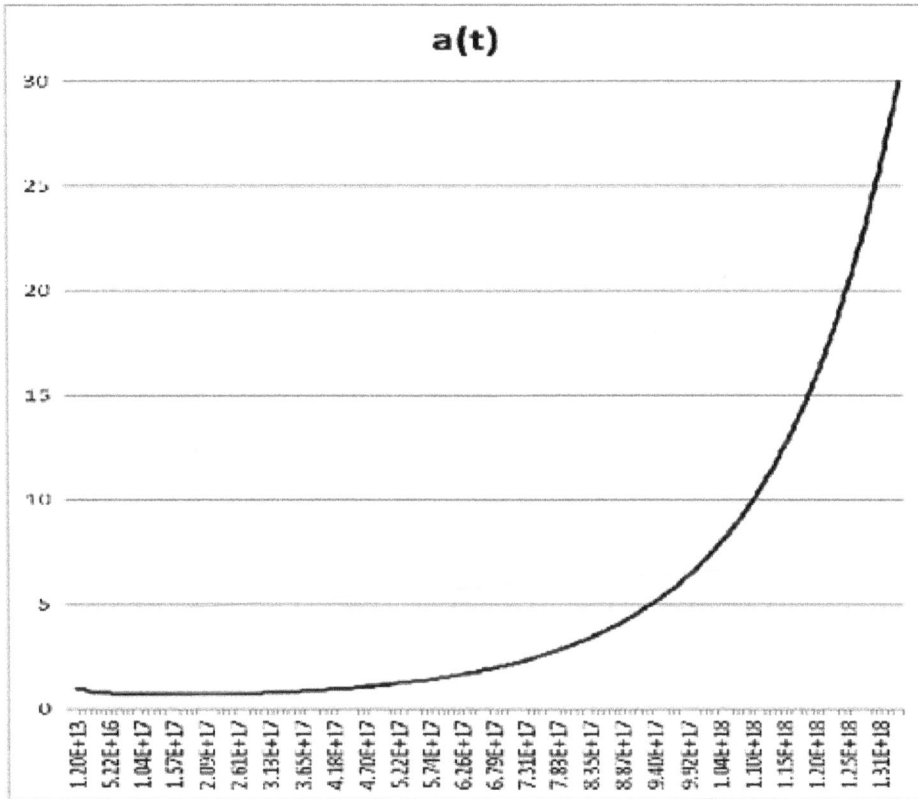

Figure 59.5 Universal Scale Factor a(t) plotted in seconds from t = 1.19 × 10^13 sec. (380,000 years) to t = 1.31 ×10^18 sec. (the distant future). Note the Big Dip appears diminished due to the scale of the plot.

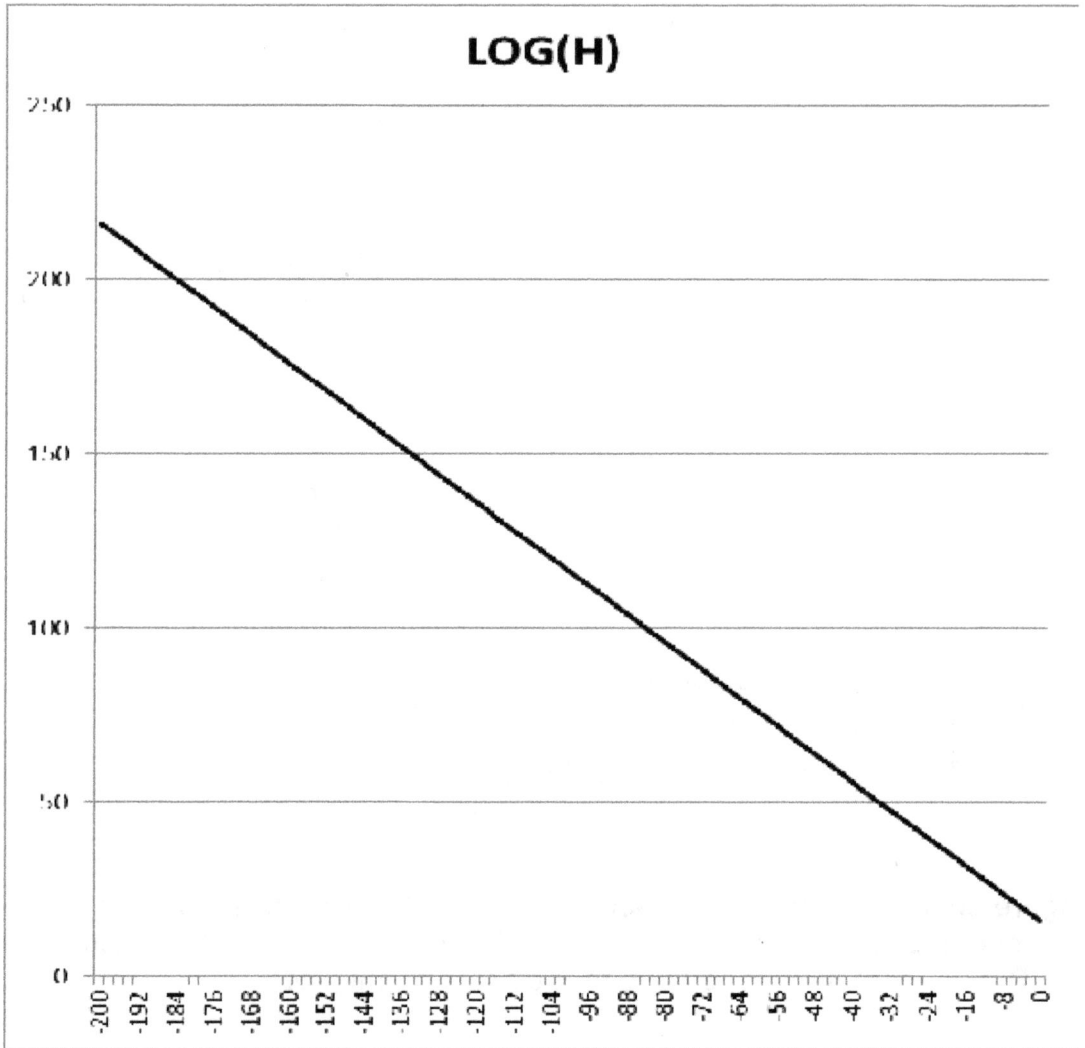

Figure 59.6 Log$_{10}$ H(t) plotted in log$_{10}$ seconds from t = 10^{-200} sec. to 1 sec.

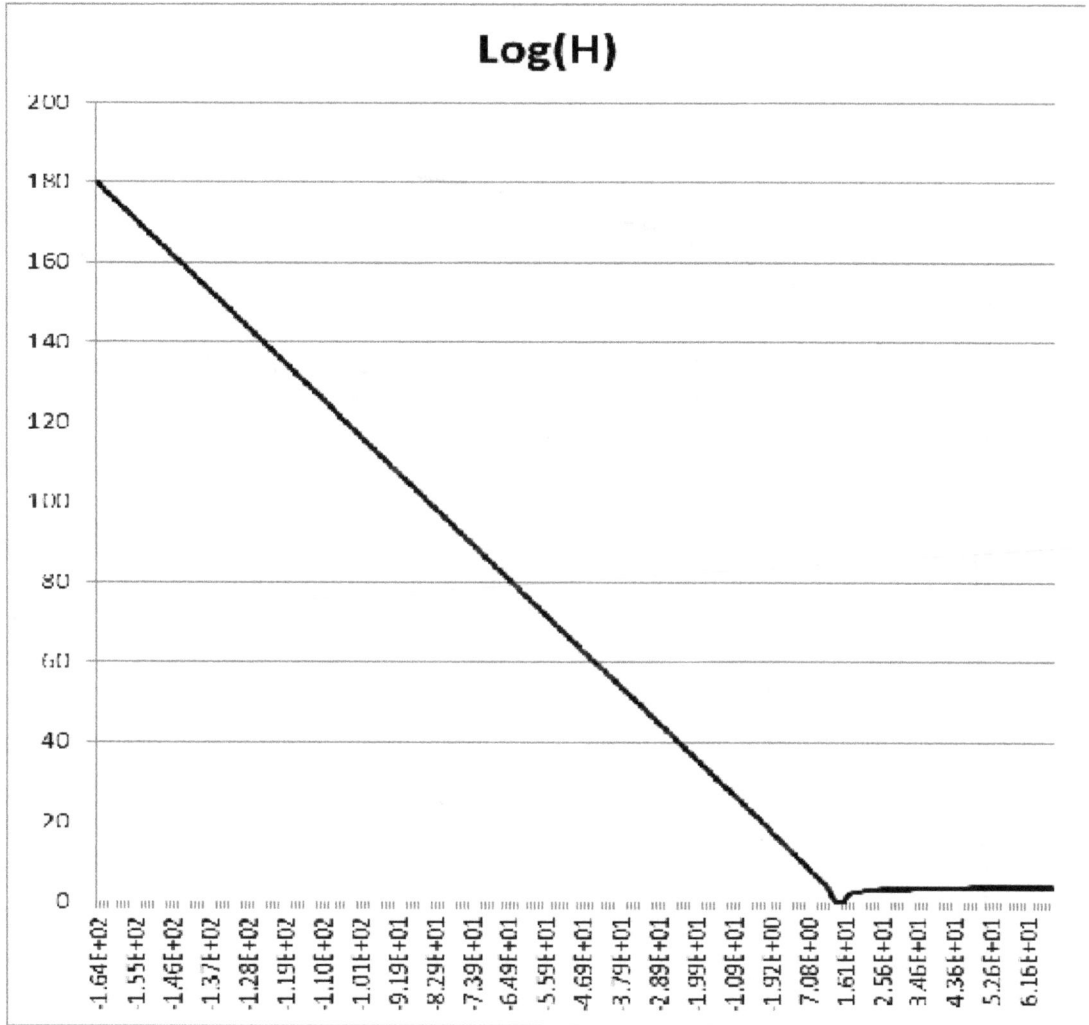

Figure 59.7 Log$_{10}$ H(t) plotted in log$_{10}$ seconds from t = 1.19 $\times$ 10^{-164} sec. to the distant future 4.31 $\times$ 10^{67} sec.

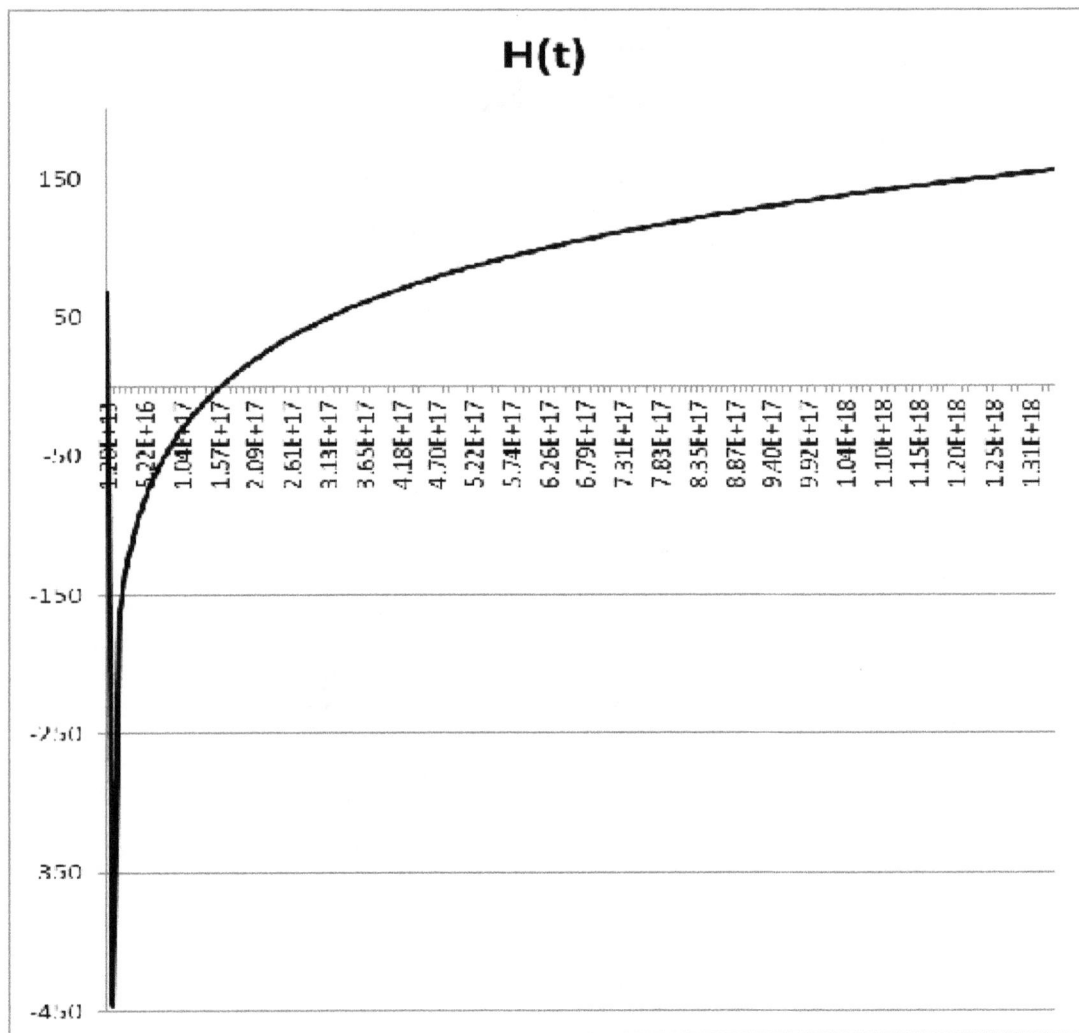

Figure 59.8 H(t) plotted vs. seconds from t = 1.19 × 10¹³ sec. (380,000 years) to 1.31 × 10¹⁸ sec. (three times the present time) showing the Big Dip in H(t).

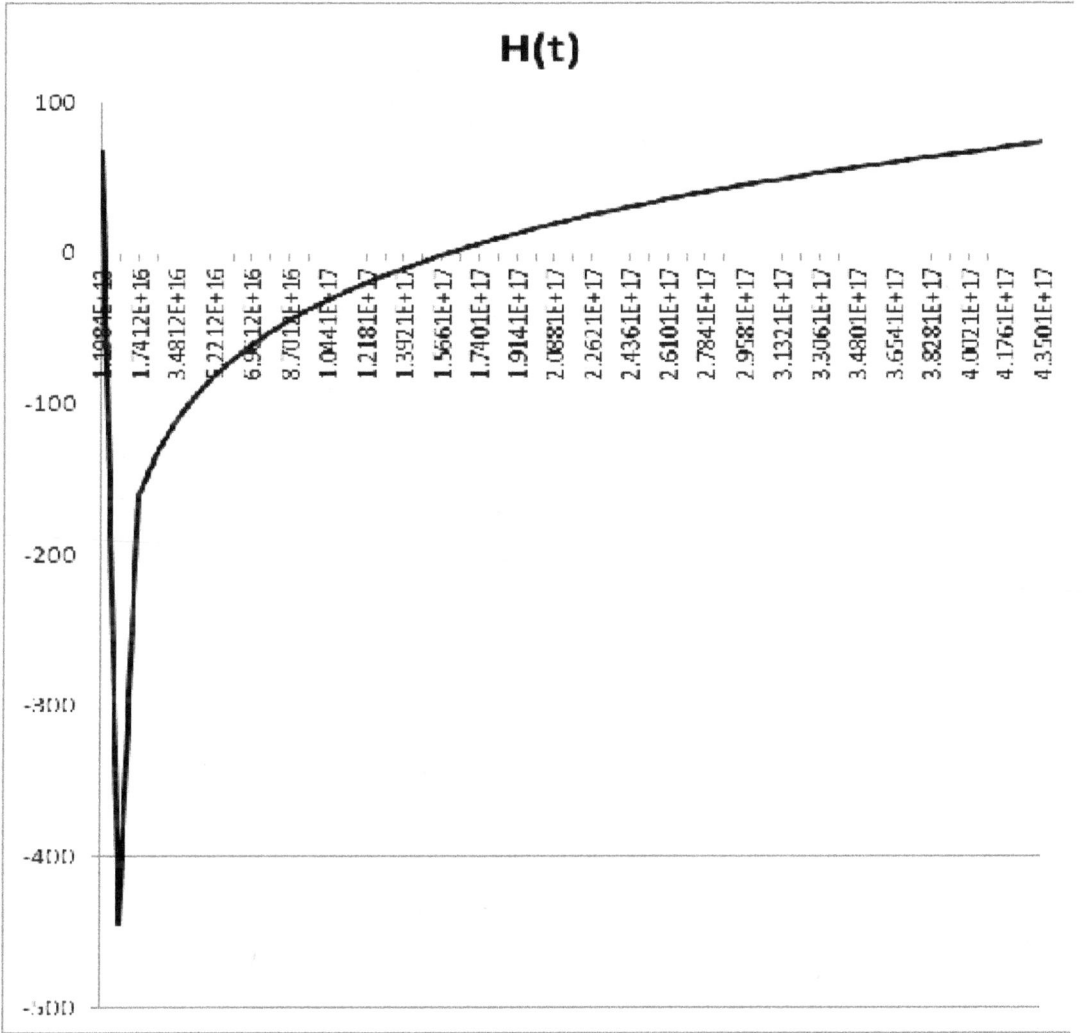

Figure 59.9 A closer view of H(t) plotted vs. seconds from t = 1.19 $\times$ 10^{13} sec. to the present 4.35 $\times$ 10^{17} sec. The minimum is H = -445 km s^{-1} Mpc^{-1} at t = 8.71 $\times$ 10^{13} sec.

59.4 Big Bang Metastate Model Scale Factor and Hubble Constant

In Blaha (2019c), (2019d) and (2019e) and earlier books the author developed a model of a Quantum Big Bang which avoided singularities at $t = 0$ using quantum effects. The results of this model appear below and they can be seen to approximately match at very early times with the scale factor and Hubble Constant values implied by eq. 59.1.

59.4.1 Big Bang Scale Factor a(t) and H(t)

The real part of the complex-valued Big Bang scale factor is

$$\mathrm{Re}\ a_{BBRW}(t_c) \cong \gamma = a(t_c) \tag{59.6}$$

from Blaha (2019c) where

$$\gamma = 1.632 \times 10^{-92}$$
$$t_c = 1.1984 \times 10^{-165}\,\mathrm{s}$$

We can see the removal of the singularity at $t = 0$ by combining the Big Bang metastate scale factor $a_{BBRW}(t)$ with $a(t)$:

For $0 \le t \le t_c$

$$a_{combined}(t) = \gamma/2 + a(t)/2 = \mathrm{Re}\ a_{BBRW}(t) \tag{59.7}$$

For $t_c \le t \le t_{now}$

$$a_{combined}(t) = a(t) \tag{59.8}$$

Note $a(0) = 0$ while $a_{combined}(0) = \gamma/2 = 8.16 \times 10^{-93}$. Thus a Big Bang catastrophe is averted. Fig. 59.10 contains relevant values of the Big Bang Model.

Time	0	t_c
Phase	Big Bang Metastate Beginning	Big Bang Metastate End
Time	0	1.26×10^{-165}
Re a(t)	8.16×10^{-93}	1.632×10^{-92}
Im a(t)	-3.16×10^{-93}	-5.24×10^{-93}
Re radius	4.278×10^{-65} cm	8.5×10^{-65} cm
Im radius	4.278×10^{-65} cm	8.5×10^{-65} cm
Volume	4.37×10^{-192} cm^3	2.6398×10^{-191} cm^3
Central Expansion Energy Density	1.63×10^{218} GeV/cm^3	4.08×10^{217} GeV/cm^3
Edge Expansion Energy Density	8.16×10^{217} GeV/cm^3	2.04×10^{217} GeV/cm^3
Total Expansion Energy[388]	5.34×10^{35} eV	8.07×10^{35} eV
Hubble Constant[389] (km s^{-1} Mpc^{-1})	1.79×10^{218}	1.14×10^{126}

Figure 59.10 Big Bang Model Metastate detailed data.

---

[388] The expansion energy does not include the mass-energy of particles in the Big Bang universe. It only includes Y field black body energy which drives the initial expansion. All particles are massless initially and all interparticle forces are zero.

[389] In the Big Bang metastable state we display the Hubble constant at the "expanding' edge.

59.5 Smooth Fit Connection to Big Bang Metastate

The parameters in eqs. 59.1 and 59.5 were set by the $H(t)$ data at 380,000 years and the present. If we extrapolate back to the end of the Big Bang metastate then we find a good match between the values of $a(t)$ and $H(t)$ of the Big Bang metastate and the extrapolation:

		H(t)
Big Bang Metastate	Big Bang Center	8.95×10^{217} km s^{-1} Mpc^{-1}
	Big Bang Edge	1.14×10^{126} km s^{-1} Mpc^{-1}

$H(t)$ for the universal scale factor 9.149×10^{215} km s^{-1} Mpc^{-1} at t $= 10^{-200}$ sec.
(and much larger as t $\rightarrow$ 0)

where the Big Bang values appear in section 13.6.2 of Blaha (2019c).. Note that the extrapolated value is within the range of values in the Big Bang Metastate and is close (within a factor of 100) to the Big Bang Center value. Thus our $H(t)$ fit (eqs. 59.1 and 59.5) extends smoothly back to the Big Bang. $H(t)$ is in a rather remarkable approximate agreement for the Big Bang Metastate and our fit.

59.6 Universe Contraction – Early Massive Galaxies

The contraction of the universe at the Big Dip has important consequences for galaxy formation and their distribution. It may be relevant to recent studies that suggest synchronous behavior of galaxies and quasars separated by great distances.

The Big Bang contraction would appear to "squeeze" the mass-energy in the universe giving it a "belly" (the Big Belly??? of "squeezed" mass-energy).This mass-energy contraction leads to the early formation of galaxies, and their correlated behavior at large distances. The subsequent expansion creates a "type" of wave that generates massive galaxies (bubbles of mass-energy), and also voids – bubbles of space devoid of galaxies .The galaxies have dispersed in the 13.5 Gyrs that followed.

Evidence[390] has been found for the existence of a huge population of very massive galaxies (39+ have been found so far) that were created within one billion years after the Big Bang. This population of early galaxies is inconsistent with the standard present-day models of galaxy formation. The Big Dip occurs at 2.76 million years – well before one billion years – consistent with the formation of early massive galaxies.

A concentration of mass-energy due to the contraction of the universe appears to present a possible solution. Universe contraction was not considered in the creation of these models of galaxy formation.

Another possible source of universe concentrations (and voids) of energy appears in our Quantum Big Bang Model seen earlier. The cause there is a large difference in expansion rates (Hubble Constant variations) at the center of the Big Bang compared to the outer edge of the Big Bang as shown earlier in the section 2 Big Bang Model.

59.7 An Interlude in the Eons

Based on the above analysis it appears that the universe is currently in an *interlude* following a decline in growth rate after the Big Bang, and a new beginning of major growth.

---

[390] T. Wang *et al,* Nature **572**, 211 (2019).

60. Proof of the Universal Scale Factor for the Expansion of the Universe

There are two approaches to the universal scale factor fit of eq. 59.1. One approach is based on a remarkable coincidence between the power g in the fit and the QED power g seen earlier. It leads to a theory in which the expansion of the universe taken over all time is a vacuum polarization phenomenon. The other approach is based on the Einstein equation for the scale factor. We show that the Universal Scale Factor is consistent with the Einstein equation if additional (dark) energy is properly taken into account.

60.1 Vacuum Polarization Generation of the Early Times Part of the Universal Scale Factor

Perhaps the crowning achievement of our universal scale factor eigenvalue formulation for coupling constants is the successful relation of universe evolution to vacuum polarization due to a vector QED-like interaction between universes.

60.1.1 Recap of Massless QED Vacuum Polarization

In massless QED we found that the vacuum polarization had the form:[391]

$$F_1(\alpha)(p/\Lambda)^{2g_{QED}} \tag{60.1}$$

where $F_1(\alpha)$ is the "eigenvalue function" for the Fine Structure Constant[392] of the Johnson-Baker-Willey model of massless QED, p is the momentum, and Λ is the ultraviolet cutoff. The value of g_{QED} that corresponded to the Fine Structure Constant is

---

[391] Eq. 12 in S. Blaha, Phys Rev **D9**, 2246 (1973).
[392] The author calculated $\alpha = 1/137\ldots$ exactly in Blaha (2019a) and (2019b).

$$g_{QED} = -\ 0.00058053691948 \qquad (60.2)$$

and the Fine Structure Constant was correctly found (well within experimental limits) to be

$$\alpha_{calculated}(g_{QED}) =\ 0.0072973525693 \qquad (60.3)$$

to 13 digit accuracy according to the Particle Data Table of 2019.

Comparing our Universal Scale Factor g value (eq. 27.5) with g_{QED} we find

$$-g\ =\ 0.000282377 \cong -\tfrac{1}{2}g_{QED} = -0.000290268 \qquad (60.4)$$

60.1.2 Comparison of QED Vacuum Polarization Exponent with Universe Vacuum Polarization Exponent

Eq. 60.4 shows the numeric values of the g powers are approximately equal up to a factor of -2. The QED exponent describes high energy vacuum polarization behavior. The universe power g describes the small time universe expansion (near the Big Bang). The relation between the values of g and g_{QED} clearly suggests a close analogy.

Further the low energy (infrared) behavior of the QED vacuum polarization which is mass dependent is analogous to the large time (recent time) behavior of a(t) which is governed by the h term in the exponent of a(t).

The problem now before us is to find the universe vacuum polarization due to a new vector interaction between universes, and show that it is related to the QED vacuum polarization by eq. 60.4.[393]

60.1.3 A New Vector Interaction for Universe Particles

We assume universes can be treated as particles in 4-dimensional space-time.[394] Since experiments appear to have shown that our universe does not rotate (does not

---

[393] The following subsections appeared in Blaha (2019c).
[394] Universes are composite entities but we can treat them as quantum particles in the same manner as physicists treated protons and neutrons etc. as quantum particles before quark theory was accepted. See Blaha (2018e) for a detailed discussion of universe particles.

have spin[395] we will assume the universe is a spin 0 boson. We assume that universes have a vector field interaction similar to QED.

Given this QED-like framework, then universe-antiuniverse pair production and vacuum polarization becomes possible. We assume the QED-like boson lagrangian

$$\mathcal{L} = \tfrac{1}{2}\,(\partial_\mu\varphi^\dagger\partial^\mu\varphi - m^2\varphi^\dagger\varphi) - ie_0\colon \varphi^\dagger(\overrightarrow{\partial_\mu} - \overleftarrow{\partial_\mu})\,\varphi\colon A^\mu + e_0^2\colon A^2\colon \colon\varphi^\dagger\varphi\colon + \delta m^2\colon\varphi^\dagger\varphi\colon$$

$$(60.5)$$

where $\varphi(x)$ is a "charged" quantum universe scalar particle field[396] and A^μ is a QED-like field. We now proceed to calculate the second order vacuum polarization of a universe particle. We will assume the term in $\mathcal{L}$ linear in A^μ is the relevant term since the quadratic term always is negligible compared to the linear term in each order α^n of perturbation theory by a factor of α. The neglected terms will be assumed to not affect the calculated eigenvalue function.

60.1.4 Second Order Vacuum Polarization of a Scalar Universe Particle

The one loop vacuum polarization Feynman diagram is

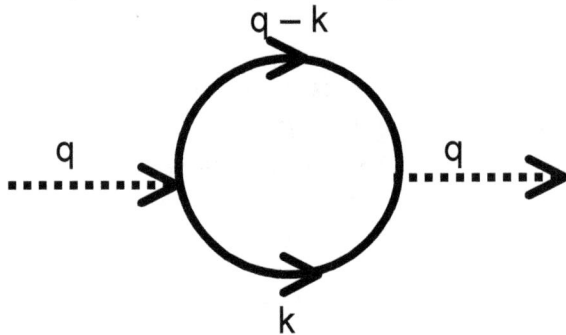

Figure 60.1 One loop vacuum polarization boson Feynman diagram.

---

[395] The lack of universe rotation (spin) is indicated by a study of Cosmic Microwave Background (CMB) by D. Saadeh *et al*, Phys. Rev. Lett. **117**, 313302 (2016).

[396] The charge is not electromagnetic charge.

Its evaluation is

$$I_{\mu\nu} = (-ie_0)^2 \int \frac{d^4k}{(2\pi)^4} \frac{i}{(k^2 - m^2 + i\varepsilon)} \frac{i}{(k^2 - m^2 + i\varepsilon)} (q - 2k)_\mu (q - 2k)_\nu \tag{60.6}$$

$$= \frac{\alpha}{2\pi} \int_0^\infty dz_1 \int_0^\infty dz_2 \frac{g_{\mu\nu} \exp[i(q^2 z_1 z_2/(z_1 + z_2) - (m^2 + i\varepsilon)(z_1 + z_2))]}{(z_1 + z_2)^3} + \text{gauge terms}$$

upon introducing parameters z_1 and z_2 to enable exponentiation and integration over k, where

$$\alpha = e_0^2/4\pi \tag{60.7}$$

Applying $q^2 \partial/\partial q^2$ to $I_{\mu\nu}$ to eliminate the quadratic divergent part, and then using the identity

$$1 = \int_0^\infty d\lambda/\lambda \ \delta(1 - (z_1 + z_2)/\lambda)$$

and letting $z_i = \lambda x_i$ we obtain

$$I_{\mu\nu} = \frac{i\,\alpha}{2\pi} q^2 g_{\mu\nu} \int dx_1 \int dx_2 \int d\lambda/\lambda \ x_1 x_2 \exp[i\lambda(q^2 x_1 x_2 - (m^2 + i\varepsilon))] \ \delta(1 - x_1 - x_2)$$

$$\tag{60.8}$$

up to gauge terms. The λ integration yields a logarithmic divergence which we cut off. Then

$$I_{\mu\nu} = \frac{i\,\alpha}{2\pi} q^2 g_{\mu\nu} \int_0^1 dx \ x(1-x) \ln(q^2 x(1 - x) - m^2) + \dots \tag{60.9}$$

which becomes

$$I_{\mu\nu} = \frac{i\,\alpha}{12\pi} q^2 g_{\mu\nu} \ln(\Lambda^2/m^2) + \dots \tag{60.10}$$

with finite and other gauge terms not shown.

Thus we find the renormalization constant Z_{3U} for a scalar universe particle is

$$Z_{3U} = 1 - \alpha/12\pi \ln(\Lambda^2/m^2) \tag{60.11}$$

If we let

$$\alpha_U = \alpha/4 \tag{60.12}$$

then we obtain the form similar to the one loop value of Z_3 for spin ½ electron QED:

$$Z_{3U} \cong 1 - \alpha_U/3\pi \ln(\Lambda^2/m^2) \tag{60.13}$$

We now *provisionally assume* that α is the QED fine structure constant. We denote it as α_{QED}. We verify this choice later.

Thus the "fine structure constant" α_U for our vector interaction is

$$\alpha_U \equiv \alpha_{QED}/4 = 0.001824338 \tag{60.14}$$

We now turn to the Johnson-Baker-Willey (JBW) model of massless QED since at ultra-high energy our vector interaction theory with lagrangian eq. 60.5 becomes the JBW model for a scalar particle. In the JBW model we calculated α_{QED} and found the corresponding power of the Z_3 divergent factor which we denote g_{QED}.

60.1.5 Finding the Universe g_U

Now we perform the same calculation for universe vacuum polarization and find the g value, which we denote g_U, corresponds to α_U. The value of g_U will be seen to lead to the power g in the universal scale factor almost exactly.

The universe eigenvalue function is[397]

$$F_2(\alpha_U) = F_1(\alpha_U) - [2/3 + \alpha_U/(2\pi) - (1/4)[\alpha_U/(2\pi)]^2] \tag{60.15}$$

For

---

[397] We assume the universe eigenvalue function has the same form as the QED eigenvalue function.

$$\alpha_U \equiv \alpha_{QED}/4 = 0.001824338 \qquad (60.16)$$

we found the eigenfunction value

$$F_2(\alpha_U = 0.001824338) = 5.10824 \times 10^{-12} \cong 0 \qquad (60.17)$$

Examining $F_2(\alpha_U)$[398] as a function of g_U we found the value of g_U corresponding to α_U is

$$g_U = -0.00014525 \qquad (60.18)$$

Thus the universe vacuum polarization is

$$\Gamma_U(p) = (p/\Lambda)^{2g_U} \qquad (60.19)$$

The fourier transform is[399]

$$a(t) = (1/2\pi) \int_0^\infty dp/p \, \exp(-ipt) \, \Gamma_U(p) \qquad (60.20)$$

$$= k \, (t/T)^{-2g_U} \qquad (60.21)$$

where k is a constant and where

$$1/T = \Lambda \qquad (60.22)$$

with Λ being the "momentum space" cutoff mass. Comparing eq. 60.1 and 60.21 we find

$$g = -2g_U$$
$$= 0.0002905 \qquad (60.23)$$

From eq. 60.21 for the power g of a(t) we see the universal scale factor g is

---

[398] $F_2(\alpha_U)$ and $F_2(g_U)$ are alternate notations for the same function.

[399] Those who might object to fourier transforming to time t should remember that inside a Black Hole the "time-like" coordinate is the radius and the time variable t is comparable to a spatial coordinate. The possibility that the universe is a Black Hole is not excluded. This fourier transform appears in Blaha (2019c) in eq. 25.25 with a typographic error—the division by p was omitted.

$$g = 0.000282377 \qquad (60.24)$$

Thus the value of g calculated from the universe vacuum polarization differs from the actual value of g by less than 3%. Given the approximate nature of our JBW calculation of vacuum polarization the agreement is remarkable.[400]

In addition we found the "fine structure constant" for the vector interaction to be given by eq. 60.16 resulting in

$$e_U = (4\pi\alpha_U)^{\frac{1}{2}} = 0.151411 \qquad (60.25)$$

Thus we have shown the universe vacuum polarization $\Gamma_U(p)$ when transformed to time is the universal scale factor a(t) up to a constant. The evolution of our universe is set by universe vacuum polarization. Other 4D universes may be expected to be similar.

The above relation we have found between QED-like vacuum polarization and universe vacuum polarization (Dark Energy) appears to confirm our interpretation of universe Dark Energy as mainly a consequence of universe vacuum polarization due to a universe vector interaction.[401]

60.1.6 Dark Energy is Equivalent to Universe Vacuum Polarization

Dark Energy is elusive both on the experimental and theoretical levels. We know it exists through its effects on our universe. Yet interactions with matter have not been found. Thus it is somewhat of a phantom.

The existence of Dark Energy, which, clearly, strongly affects the evolution of the universe, means that the Einstein equation, usually regarded as central to universe evolution, is incomplete for that purpose. It does not specify the total energy density ρ_{tot}.

---

[400] And may be exact! The value of the Hubble Constant H in recent times varies from about 70 – 75 making the calculation of g also approximate. We chose an average value of 73.24 to obtain the value of g above. If we chose the current value for H to be 75.58 we would have $g = -2g_U$ exactly. Note: studies of binary black hole merger gravity waves have given a Hubble Constant of 75.2 km s^{-1} Mpc^{-1} (and earlier of 78 km s^{-1} Mpc^{-1}), and studies of light bent by distant galaxies give H = 72.5 km s^{-1} Mpc^{-1}. Thus the value H = 75.58 is not unreasonable.

[401] Rather like the discovery of the Ω^- particle in the 1960s confirmed Gell-Mann's SU(3) theory.

$$\dot{a}^2 - 8\pi G\rho_{tot}a^2/3 = -k \tag{60.26}$$

However we can obtain a "handle" on the total energy density by inserting our universal scale factor a(t) in the Einstein equation together with the known radiation density, matter density and Cosmological Constant Λ terms:

$$\rho_{tot}(t) \equiv \rho_{crit}\Omega_{tot}(t) = \rho_{crit}[\Omega_\Gamma(t) + \Omega_M(t) + \Omega_\Lambda + \Omega_T]$$

where the unknown part needed to makes the Einstein equation correct is the elusive Dark Energy $\rho_T(t)$

$$\rho_T(t) = \rho_{crit}\Omega_T(t) \tag{60.27}$$

Then we can calculate energy density $\rho_{Dark}(t)$ as a function of time as well as related quantities as the following plots show. Figs. 29.1 through 29.5 display time plots of $\Omega_T(t)$.

60.1.7 Quasi-Free Universe Particles

Since $F_2 \cong 0$ by eq. 60.17 universe particles are very much like free particles since the vacuum polarization is zero except for a divergence due to the effect of the three subtracted terms displayed in eq. 60.15.

Universe particles are not totally free particles due to gravitation and Standard Model interactions such as electromagnetism. We treated the case of free universe particles in Blaha (2018e).

60.1.8 Doubling Relation Between Coupling Constants

The coupling constants that we have derived show a doubling whose fundamental significance remains to be understood.

INTERACTION	COUPLING CONSTANT[402]
Universe Interaction e_U	0.1514
QED $e_{QED} = (4\pi\alpha_{QED})^{1/2}$	0.303
Weak SU(2) g_W	0.619
Strong SU(3) g_S	1. 22

Figure 60.2 The interaction coupling constants show a regular doubling. A fundamental cause for doubling is not apparent.

60.2 Second Approach to the Universal Scale Factor

In the above proof we addressed only the small time (large momentum) behavior of the Universal Scale Factor exponent g. We now consider the other parameter h (which has no apparent relation to H or the constant h = 0.689.) We will show h has a simple direct interpretation based on the Einstein equation under the assumption that the usually assumed energy density constants Ω_γ, Ω_m, and Ω_Λ may be time dependent:

$$\Omega_\gamma(t), \Omega_m(t), \Omega_\Lambda(t)$$

and the assumption that there is another ultra-energy density $\Omega_T(t)$.

60.2.1 Origin of h

The value of the parameter h is remarkably close to the standard value of the Hubble parameter expressed in eV. It is also remarkably close to t_{now}^{-1}. Thus

$$h = 1.49 \times 10^{-33} \text{ eV} \cong H_0 = 68.9 \text{ km s}^{-1} \text{ Mpc}^{-1} = 1.47 \times 10^{-33} \text{ eV} \cong t_{now}^{-1} \qquad (60.28)$$

We therefore provisionally approximate

$$a(t) \cong (t/t_{now})^{g + H_0 t} \qquad (60.29)$$

---

[402] M. Tanabashi *et al* (Particle Data Group), Phys. Rev. **D98**, 030001 (2018).

We will demonstrate that the $H_0 t$ exponent is due to the existence of an "explosive" phase with energy density $\rho_T(t) = \rho_{cr}\Omega_T(t)$ where ρ_{cr} is the critical energy, which is related to the Hubble parameter by $H_0^2 = 8\pi G\rho_{cr}/3$. Einstein's equation is:

$$\ddot{a}^2 - H_0^2 a^2(t)[\ \Omega_\gamma/a^4(t) + \Omega_m/a^3(t) + \Omega_\Lambda + \Omega_T(t)] = -k \tag{60.30}$$

or

$$H(t) = \dot{a}/a = [H_0^2(\Omega_\gamma/a^4(t) + \Omega_m/a^3(t) + \Omega_\Lambda + \Omega_T) - k/a^2(t)]^{\frac{1}{2}}$$

where we assume the factors are functions of time: $\Omega_\gamma(t)$, $\Omega_m(t)$, $\Omega_\Lambda(t)$ and $\Omega_T(t)$. We keep $\Omega_\Lambda = 0.689$. Conservation laws of the type considered by Weinberg (1972) are not conserved in their simple form but deviate due to large pressures and changes in energy density due to "input/output" from the universe energy. In particular,

$$\Omega_m \neq \text{constant} \tag{60.31}$$
$$\Omega_\Lambda \neq \text{constant}$$
$$\Omega_T \neq \text{constant}$$

60.2.2 Overall Universe Expansion

We now consider the overall universe expansion defining

$$\rho_{tot}(t) \equiv \rho_{crit}\Omega_{tot}(t) = \rho_{crit}[\Omega_\gamma(t) + \Omega_m(t) + \Omega_\Lambda + \Omega_T(t)] \tag{60.32}$$

$$\ddot{a}^2 - H_0^2 a^2(t)\Omega_{tot}(t) = 0 \tag{60.33}$$

We find the solution for $\Omega_{tot}(t)$ where

$$a_{tot}(t) = (t/t_{now})^{g + H_0 t} \tag{60.34}$$

is

$$\Omega_{tot}(t) = [1 + \ln(t/t_{now}) + g/(H_0 t)]^2 \tag{60.35}$$

We note that $\Omega_{tot}(t)$ can be approximately written as

$$\Omega_{tot}(t) = [1 + \ln(H_0 t) + g/(H_0 t)]^2 \qquad (60.36)$$

by eq. 60.28. *Ω_{tot} increases as $(ln\ t)^2$ as t gets large, and increases as $[g/(H_0 t)]^2$ as t →
0.*

The maximum of Ω_{tot} occurs at $g = H_0 t_{max}$ or

$$t_{max} = g/H_0 = 1.32 \times 10^{14} \text{ sec} \qquad (60.37)$$

The time t_{max}, which is approximate due to approximations in eq. 60.28, is relatively
close to the minimum (Big Dip) of H at t = 8.71×10^{13} sec. At the Ω_{tot} maximum

$$\Omega_{tot}(t_{max}) = 37.7 \qquad (60.38)$$

in contrast to $\Omega_\Lambda = 0.689$ showing the general dominance of the Dark Energy in $\Omega_{tot}(t)$
as discussed later.

The minima of Ω_{tot} with values 0.0 occur at

$$t = 1.36 \times 10^{13} \text{ sec} \qquad (60.39)$$
$$t = 1.55 \times 10^{17} \text{ sec} \qquad (60.40)$$

Note both minima of Ω_{tot} are zero. Consequently Ω_T must be negative at those points
since $\Omega_\gamma/a^4(t) + \Omega_m/a^3(t) + \Omega_\Lambda$ is presumably positive. Ω_T is much larger in magnitude
than $\Omega_\gamma/a^4(t) + \Omega_m/a^3(t) + \Omega_\Lambda$ *except* at points in the vicinity of the minima. See Fig.
60.3 for Ω_{tot} details.

60.3 Universe Expansions and Contractions indicated by Ω_{tot}

As the universe expands we expect the total energy density Ω_{tot} to decline, and if
the universe contracts we expect the energy density to increase. Fig. 60.4 shows the
pattern of expansion and contraction of the universe over time. The universe shows a
"Dead Bang Bounce"[403] in reaction to the vast expansion of the universe after the big

---

[403] In Finance it is called a Dead Cat Bounce.

Bang. After the bounce the universe expands again and Ω_{tot} declines. It is possible that another bounce could happen again in the future as Fig. 60.5 indicates. Perhaps the universe is subject to a series of lesser and lesser bounces with corresponding expansions and contractions. The Universal Scale Factor only applies up to the current time and shows only one bounce. It could be modified to handle additional bounces in the future as in Fig. 60.5. This would raise the possibility of an oscillating universe.

60.4 Roles of g and h of the Universal Scale Factor

It is clear from the above considerations that the small time behavior of $a(t)$ is governed by g. Remarkably it corresponds to the small distance behavior of the derivation based on vacuum polarization.

It is also clear that the large time (recent) behavior of $a(t)$ is determined by h, which analogously might be viewed as the infrared (large distance) behavior of the vacuum polarization. Thus time in the universe corresponds to distance in vacuum polarization.

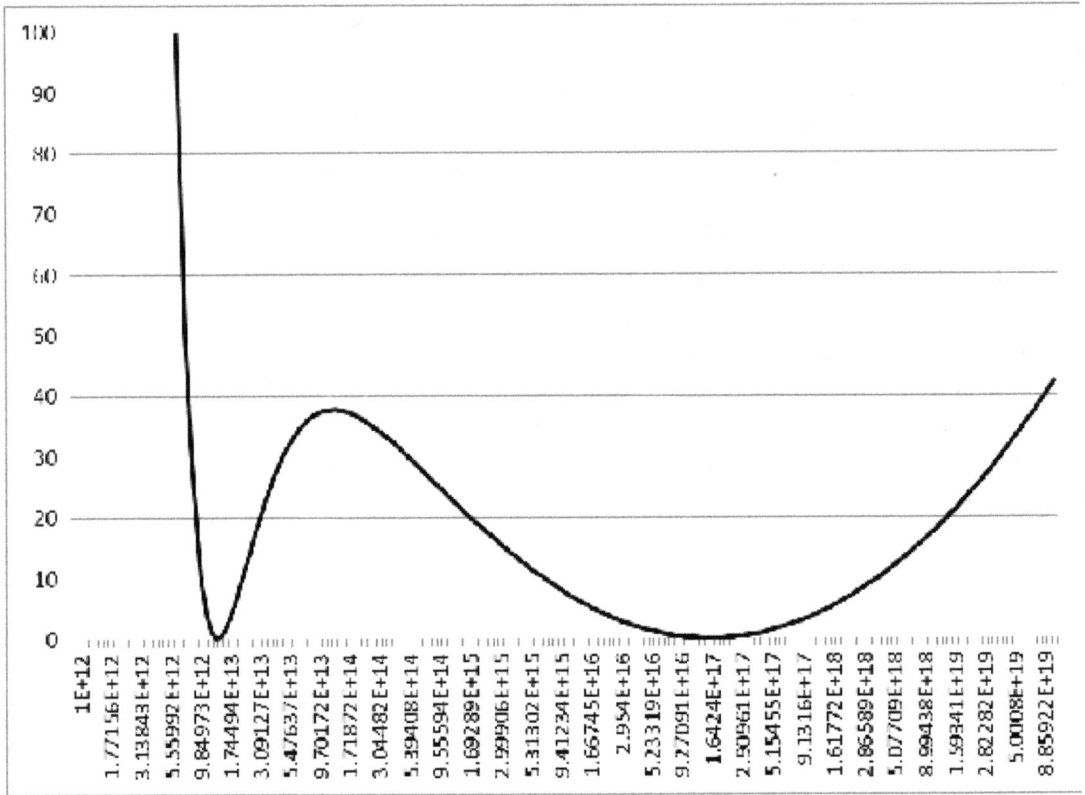

Figure 60.3 The energy density $\Omega_{tot}(t)$ of eq. 60.35 plotted as a function of time in seconds. The minima are at t = 1.36 $\times$ 10^{13} sec with Ω_{tot} = 0.0, and at t = 1.55 $\times$ 10^{17} sec with Ω_{tot} = 0.0. The maximum is at t = 1.32 $\times$ 10^{14} sec with Ω_{tot} = 37.7.

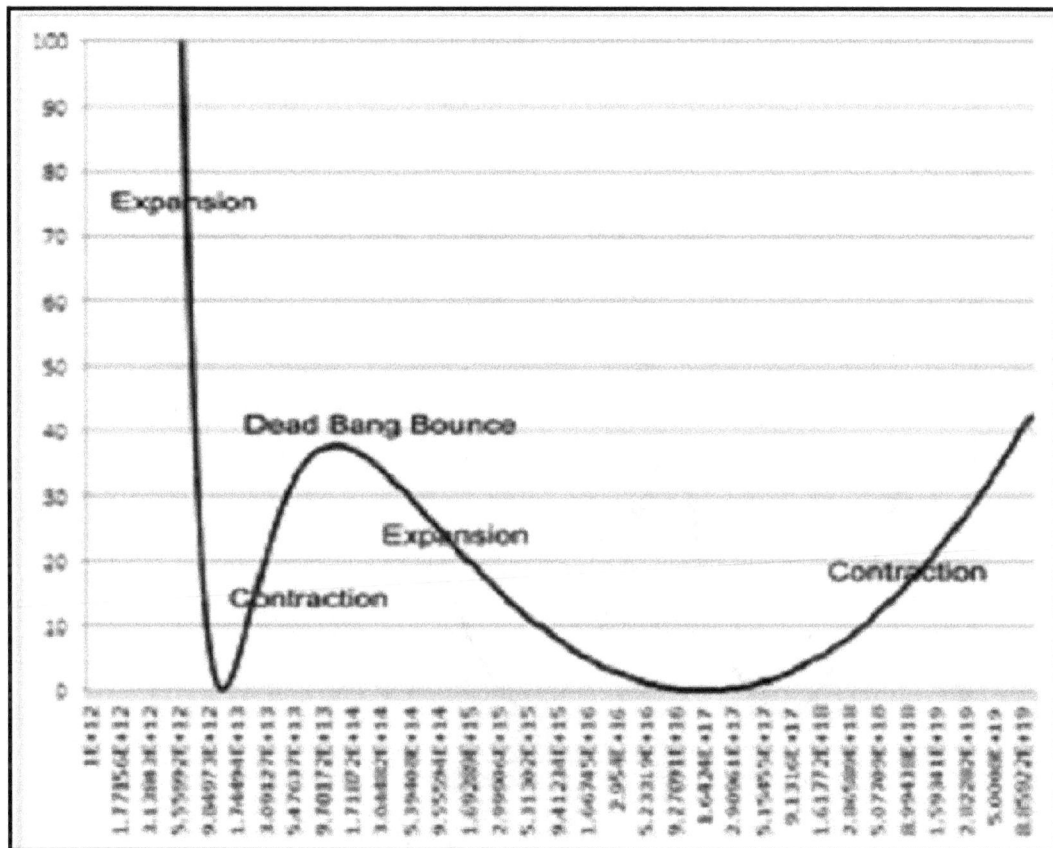

Figure 60.4 The expansions and contractions of the universe as indicated by the changes in total energy density. The diagram shows a "Dead Bang Bounce" in response to the vast expansion of the universe after the Big Bang.

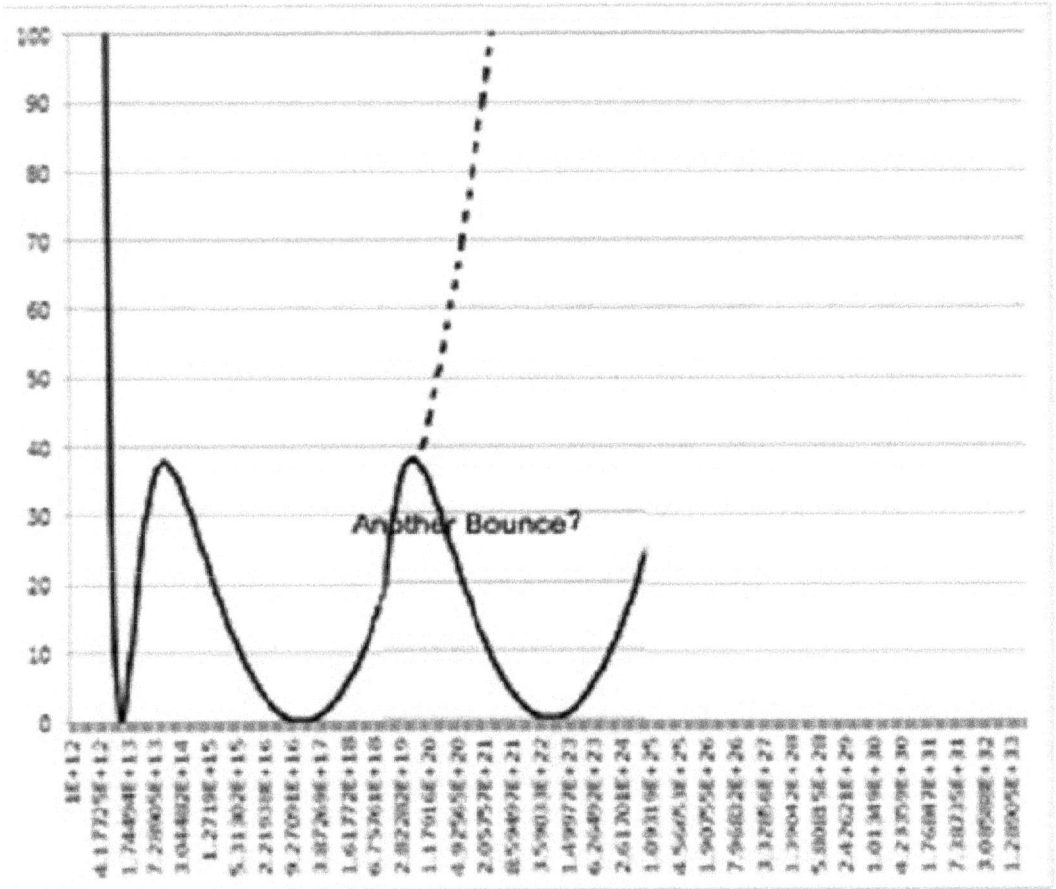

Figure 60.5 The expansions and contractions of the universe as indicated by the changes in total energy density with an illustrative future second bounce added. A second bounce raises the possibility of an oscillating universe. This The dotted line is that of the Universal Scale Factor plot.

60.5 Why $h \cong H_0 \cong 1/t_{now}$?

An important question that emerges from the above discussion is:

Why do we have the approximate equalities $h \cong H_0 \cong 1/t_{now}$

It appears $h \cong H_0$ is understandable to achieve consistency with Einstein equation in above calculation.

One may also understand $h \cong 1/t_{now}$ as simple agreement of the Universal Scale Factor with the available Hubble Constant data.

The resulting form of $\Omega_{tot}(t)$ is very much in agreement with what one might expect:

$$\Omega_{tot}(t) = [1 + \ln(H_0 t) + g/(H_0 t)]^2 \qquad (60.36)$$

Since the form of $\Omega_{tot}(t)$ (which is based on the Einstein equation) depends solely H_0 and the vacuum polarization derived constant g.

61. Energy Density, Pressure, Dark Energy, and Equation of State Implied by the Universal Scale Factor

The universal scale factor defined in chapter 27 implies the time dependence of the universe's Energy Density, Pressure, Dark Energy, and Equation of State. In this chapter we calculate these quantities and then plot their values as a function of time.

We start with the expressions for the universal scale factor and its derivatives:

$$a(t) = (t/t_{now})^{g + ht} \tag{59.1}$$

with

$$da(t)/dt = a[g/t + h + h \ln(t/t_{now})]$$

and

$$d^2a(t)/dt^2 = da/dt \,[g/t + h + h \ln(t/t_{now})] + a(h - g/t)/t$$

61.1 Dark Energy

The Einstein equation

$$\dot{a}^2 - 8\pi G \rho_{tot} a^2/3 = -k + \Lambda \, a^2 c^2/3 \tag{61.3}$$

enables us to determine the dark energy energy) $\Omega_T(t)$ beyond the cosmological constant, using the energy density:

$$\rho_{tot}(t) \equiv \rho_{crit}\Omega_{tot}(t) = \rho_{crit}[\Omega_\gamma(t) + \Omega_m(t) + \Omega_\Lambda + \Omega_T(t)] \tag{61.4}$$

Thus

$$\Omega_T(t) = (H(t)^2 + c^2k/a^2)/H_0^2 - \Omega_\gamma/a^4 - \Omega_m/a^3 - \Omega_\Lambda \tag{61.5}$$

where

$$H_0^2 = 8\pi G \rho_{crit}/3$$

The quantity Ω_T is the excess of energy beyond that specified by Ω_γ, Ω_M, and Ω_Λ in the conventional expressions for the mass-energy density. The calculation of Ω_T which is in part energy density, (and possibly in part a Megaverse energy influx), is based on eq. 61.3 using the Einstein equation in the form

$$H(t) = (da/dt)/a(t) = [H_0^2 \rho_{tot}(t)/\rho_{cri} - c^2 k/a^2(t)]^{\frac{1}{2}}$$

Figs. 61.1, 61.2, 61.4 and 61.5 plot $\Omega_T(t)$ and its derivative. It is clearly much larger than $\Omega_H(t)$ (Fig. 61.2) showing why the standard approaches to calculating a(t) using $\Omega_H(t)$ are faulty. The ratio Ω_H/Ω_T is appreciable only after 380,000 years. See Fig. 61.14.

$$\Omega_H = \Omega_\gamma(t) + \Omega_m(t) + \Omega_\Lambda \qquad (61.6)$$

61.2 Energy Density

The energy density is given by eq. 61.5. It is plotted in Figs. 61.6, 61.7 and 61.8. Fig. 61.6 shows a drop below zero starting at t = 0.12 sec. ($\log_{10}$ t = -0.92) signaling very low values for ρ. There is a minimum at $t = 1.2 \times 10^{16}$ sec. in Fig. 61.7.

Note also the drop below zero in Fig. 61.6 beginning at $t = 5 \times 10^{-16}$ sec. signaling the start of very small values for ρ.

The derivative of total energy density $d\rho_{tot}/dt$ is plotted in Fig. 61.8.

61.3 Pressure

The Friedmann-Lemaître-Robertson-Walker metric equations yield the pressure

$$p = - (c^2\rho_{crit}/(3H_0^2))[2(d^2a/dt^2)/a + (da/dt)^2/a^2 + kc^2/a^2 - \Omega_\Lambda] \qquad (61.7)$$

using eqs. 61.1 and 61.2 above.

An alternate approach to determining the pressure p uses the energy conservation equation

$$d(\rho_{tot}R^3)/dR = -3pR^2 \qquad (61.8)$$

where

$$R = k^{-\frac{1}{2}} a(t) \qquad (61.6)$$

and gives

$$p = - (da/dt)^2\rho_{tot} - a\, da/dt\, d\rho_{tot}/dt \tag{61.7}$$

$$p = -a(t)^2(H(t)^2\rho_{tot} + 1/3\, H(t)\, d\rho_{tot}/dt\,)/[(2.85 \times 10^{37})c^2] \tag{61.8}$$

where $k = 5.56 \times 10^{-57}\, cm^{-2}$, ρ is the total energy-mass density, p is the pressure, and the denominator is required by the dimensions to obtain p in gm/cm^3. See Fig. 61.9, 61.10 and 61.11.

In Fig. 61.9 note the low values of $\log_{10}(-p(t)) \approx -115$ beginning at $\log_{10}(t\ sec.) = 12.1$ ($t = 1.2 \times 10^{12}$ sec.). The pressure is always negative indicating a pressure for expansion. At $t = 1.19 \times 10^{-165}$ sec. we find $\log_{10}(-p(t)) = 593$ or $p(t) = -10^{593}\, gm/cm^3$.

In Fig 61. 11 the maximum pressure in the range of 380,000 years to the present is $-2.62 \times 10^{-124}\, gm/cm^3$ at $t = 1.0 \times 10^{17}$ sec.

61.4 Derivative of Dark Energy

The derivative of the Dark Energy, which we identify as $d\Omega_T/dt$ is

$$d\Omega_T/dt = H(t)[-2g/(H_0^2 t^2) + 2h/(H_0^2 t) - 2c^2 k/a^2 + 4\Omega_\gamma/a^4 + 3\Omega_M/a^3] \tag{61.9}$$

We find that it is well-approximated by

$$d\Omega_T/dt \approx H(t)[-2g/(H_0^2 t^2) + 2h/(H_0^2 t) - 2c^2 k/a^2\,] \tag{61.10}$$

Note the dip below zero in Fig. 61.5 beginning at $t = 1.2 \times 10^{10}$ sec. – somewhat earlier than the Big Dip.

61.5 Dominance of Dark Energy

Similarly we find eq. 61.5 is well approximated by

$$\Omega_T(t) \approx (H(t)^2 + c^2 k/a^2)/H_0^2 \tag{61.3'}$$

due to the relative smallness of Ω_H. Dark Energy dominates.

The total energy density is

$$\rho_{tot}(t) \equiv \rho_{crit}\Omega_{tot}(t) = \rho_{crit}[\Omega_\gamma(t) + \Omega_m(t) + \Omega_\Lambda + \Omega_T(t)] \qquad (61.11)$$

See Fig. 61.6. It is well approximated by

$$\rho_{tot}(t) \approx \rho_{crit}\Omega_T(t) \qquad (61.12)$$

61.6 Equation of State

The equation of state

$$w = p/\rho$$

is plotted in Figs. 61.12 and 61.13 as a function of time. Fig. 61.12 shows an enormous range of values from the Big Bang to the present. Fig. 61.13 shows that w is of the order of 10^{-95} for most of the interval from 380,000 years to the present. At its "peak" Fig. 61.13 shows $w = -1.36 \times 10^{-95}$ at the time $t = 1. \times 10^{17}$ sec.

Since w is approximately zero in these time intervals it can be viewed as describing cold dust or gas.

Since quintessence is viewed as indicated by $w \neq -1$, the theory has quintessence.

61.7 Deceleration Parameter q

The deceleration parameter q is plotted in Figs. 61.15 and 61.16.

$$q = -ad^2a/dt^2/(da/dt)^2 \qquad (61.13)$$

It is proportional to the second derivative of the universal scale factor.

If $q < 0$ then it indicates accelerating expansion of the universe. If $q > 0$ then it indicates a decelerating universe.

Figs. 61.15 and 61.16 for q both have a pronounce maximum and minimum. The maximum occurs at $t = 1.2 \times 10^{13}$ sec. The minimum occurs at 1.6×10^{17} sec. To the left of the maximum (early times) $q > 0$ indicating decelerating expansion. To the right of the maximum at t > 1.7×10^{14} sec. we find q < 0 indicating accelerating expansion The accelerating expansion started 13.78 billion years ago – "just" after the Big bang. At present q = -2.0 – accelerating expansion – as suggested by astrophysical experiments. In the future, at t = 8.2×10^{17} sec. we found q = -1.2. The accelerating expansion will continue.

61.8 Comments on Universal Scale Factor Quantities

The complete universe "life" history presented in the following plots is consistent with a declining pressure, a declining density, and a declining pressure from the Big Bang phase consistent with our physical expectations. The Big Dip and subsequent could be due in part to a sharp influx of energy (Fig. 61.2) possibly from the Megaverse.

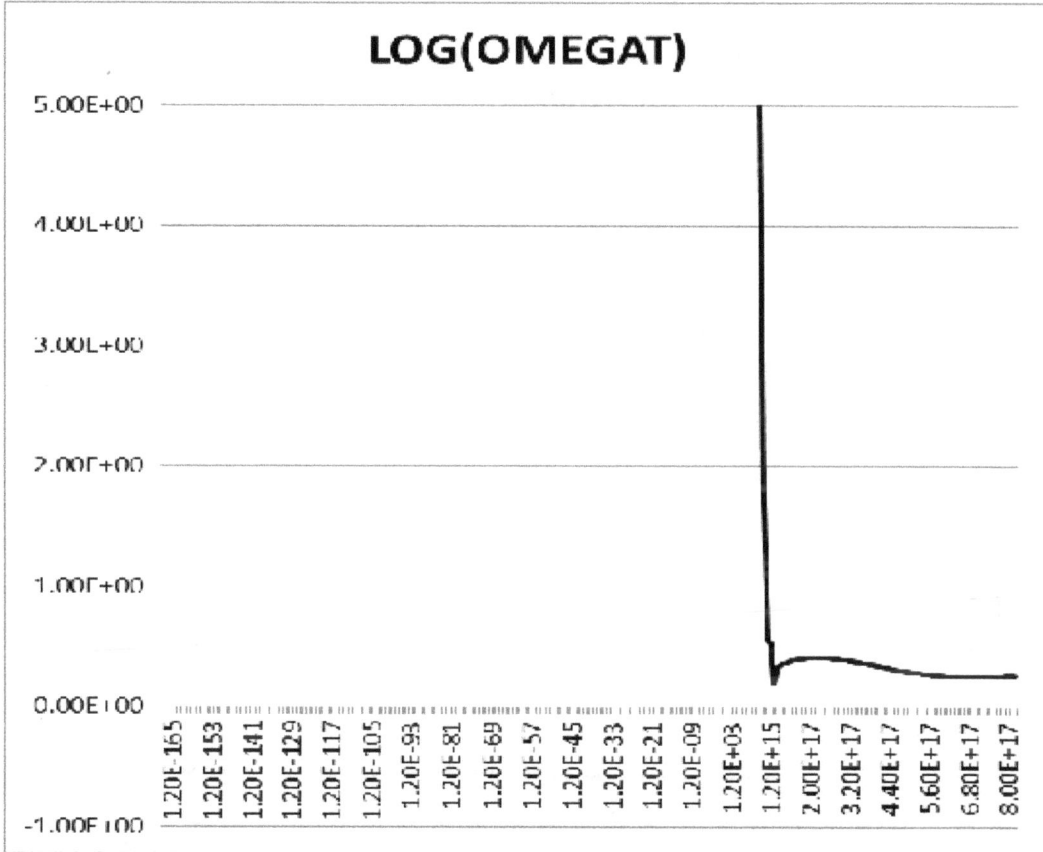

Figure 61.1 Log $\Omega_T(t)$ plotted vs. log time in seconds from 1.19×10^{-165} sec. to the 8.2×10^{17} sec. (almost double the present time). Note the Big Dip in $\Omega_T(t)$ at about t = 8.71×10^{13} sec. followed by a rise then a decline. At t = 1.19×10^{-165} sec. we found (not shown) log $\Omega_T(t) \approx 358$, an enormous value, corresponding to the level of vacuum energy found in quantum field theory. It declines to the plotted data shown above.

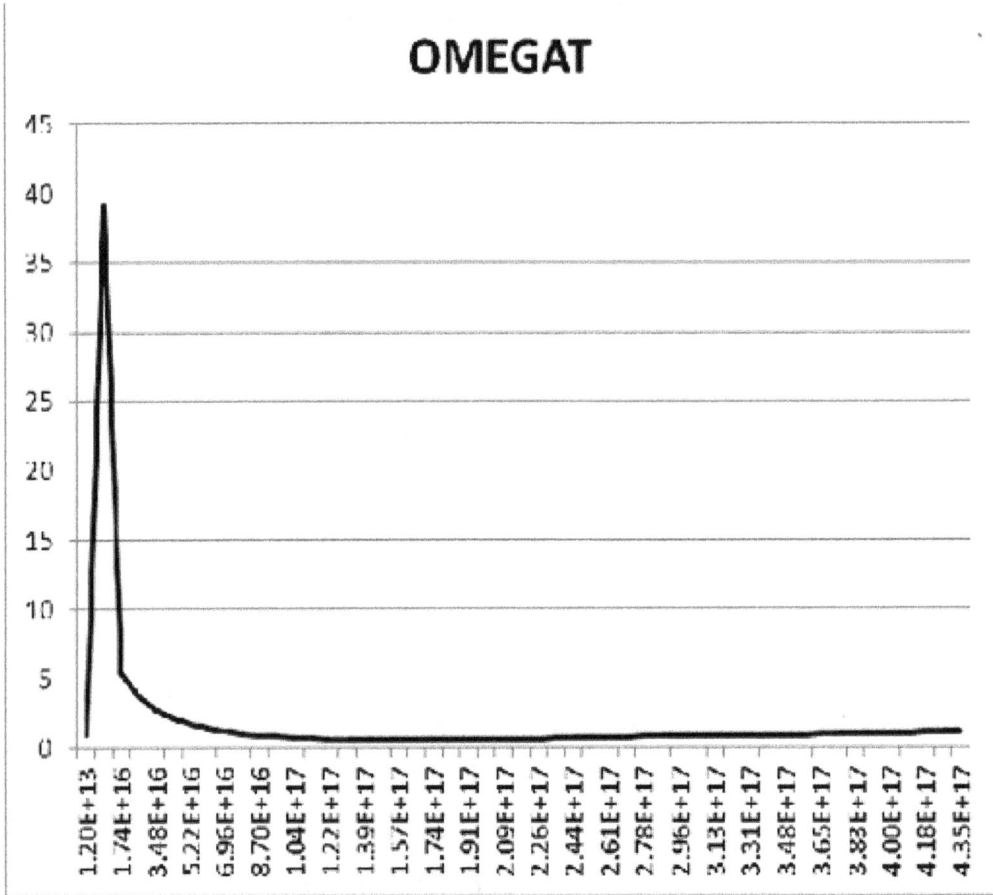

Figure 61.2 $\Omega_T(t)$ plotted vs. time in sec. from the year 380,000 to the present. Note the peak at t = 8.71×10^{13} sec. suggesting an influx into the universe at the transition to the matter-dominated phase, and then a decline followed by a raise. The peak value of $\Omega_T(t)$ is 39.2.

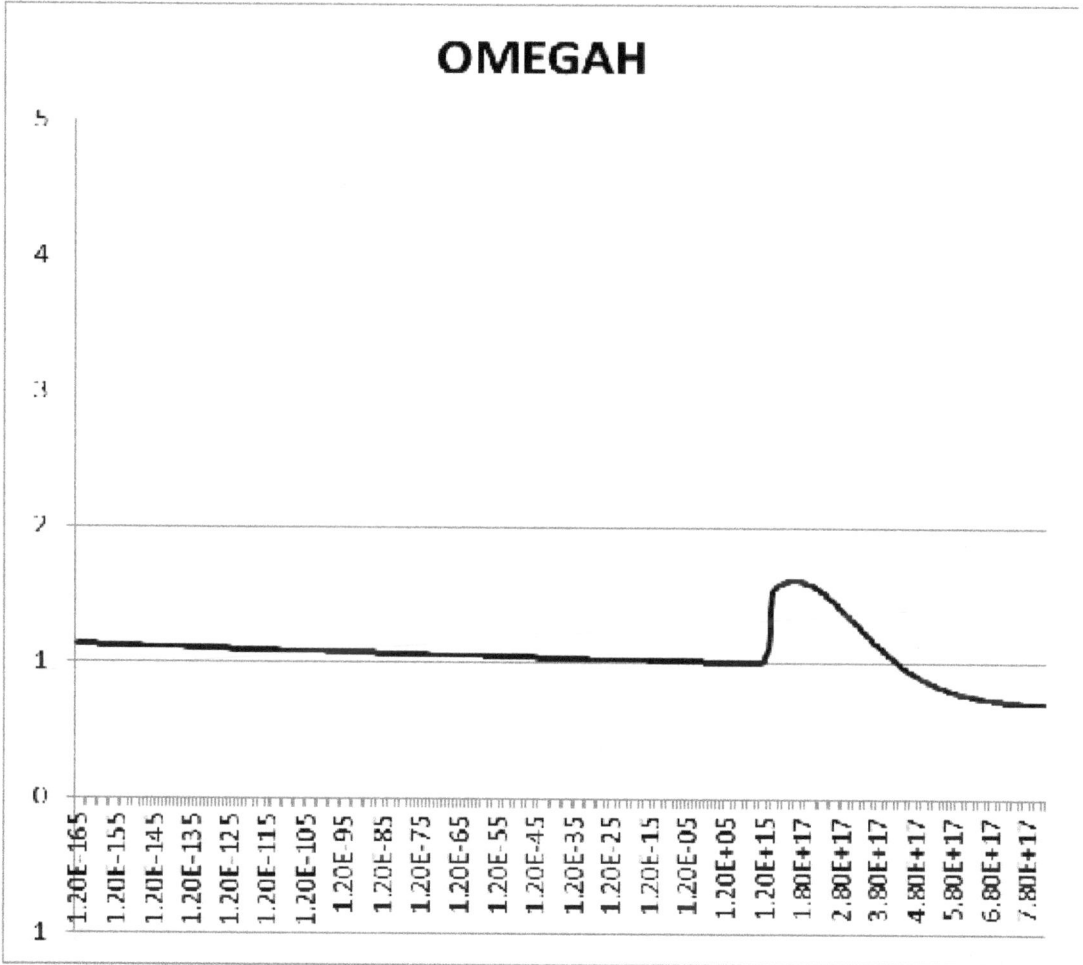

Figure 61.3 $\Omega_H(t)$ plotted vs. log time in seconds from 1.19×10^{-165} sec. to the 8.2×10^{17} sec. (almost double the present time). Note the start of a rise at $t = 1.2 \times 10^{13}$ sec. which coincides with the start of the appearance of atoms at $t_T = 380,000$ years $= 1.19837 \times 10^{13}$ sec.

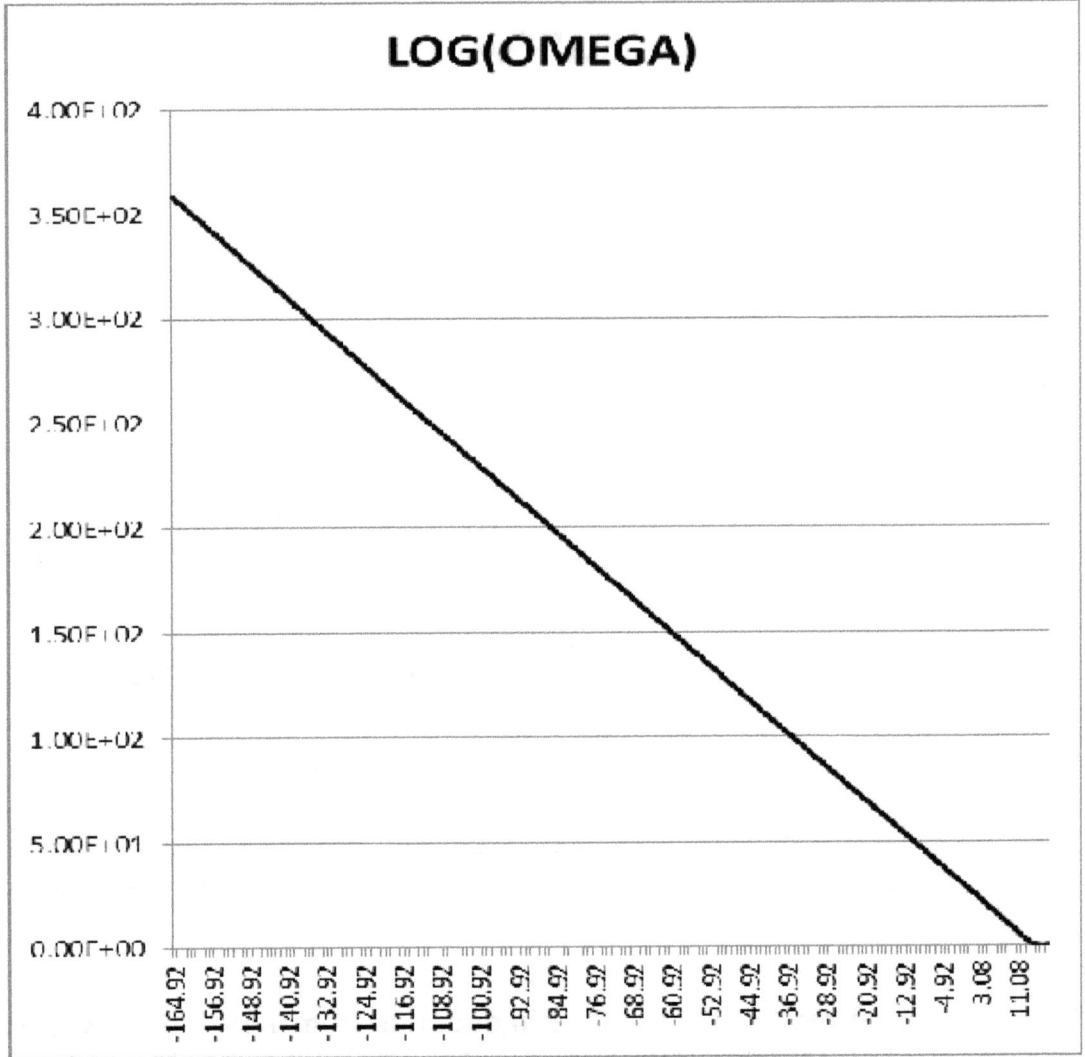

Figure 61.4 Log_{10} $\Omega_T(t)$ plotted vs. log_{10} t sec. from the Big Bang metastate to the present. Note the "dip" below zero of $\Omega_T(t)$ at log_{10} t = 11.

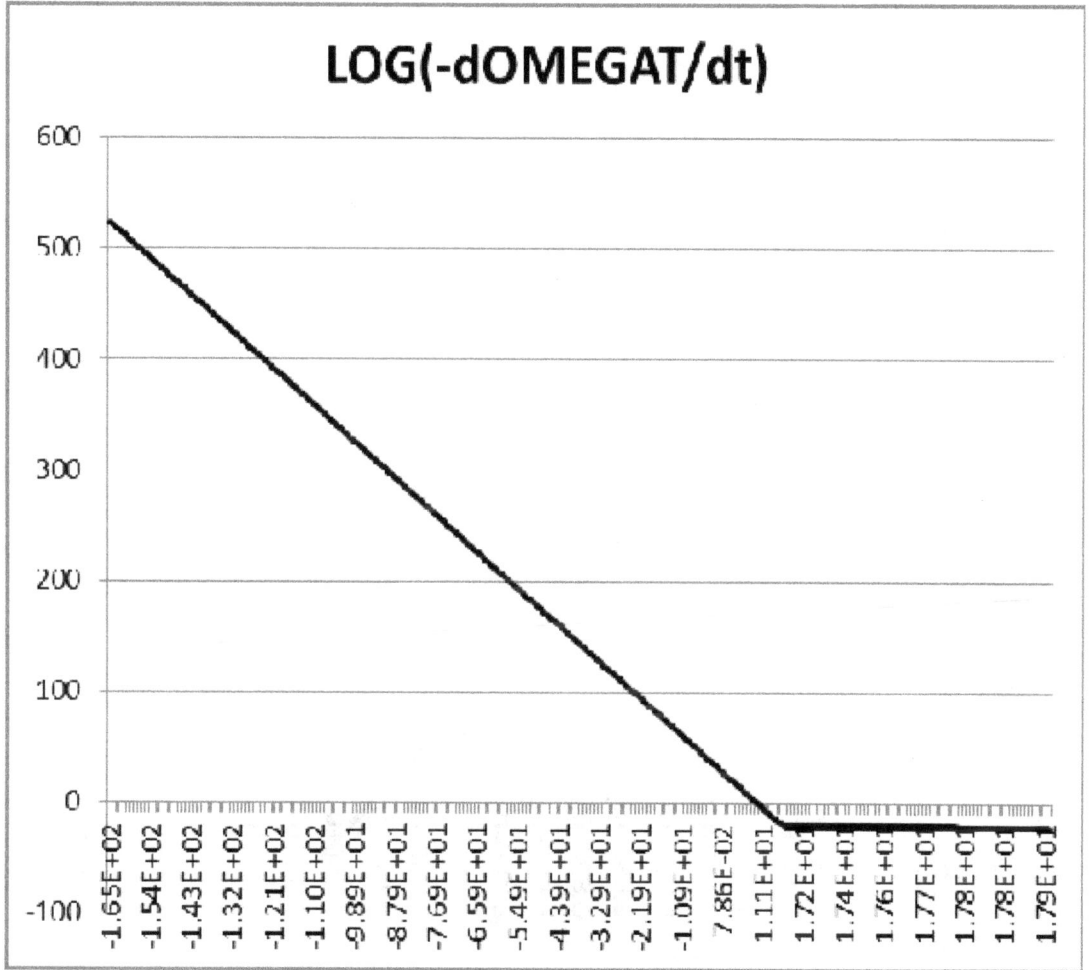

Figure 61.5 $Log_{10}(-d\Omega_T(t)/dt)$ plotted vs. $log_{10}(t)$ sec. from the Big Bang metastate $t = 1.19 \times 10^{-165}$ sec. to the future: $t = 8.2 \times 10^{17}$ sec. Note dip below zero beginning at $t = 1.2 \times 10^{10}$ sec. – somewhat earlier than the Big Dip.

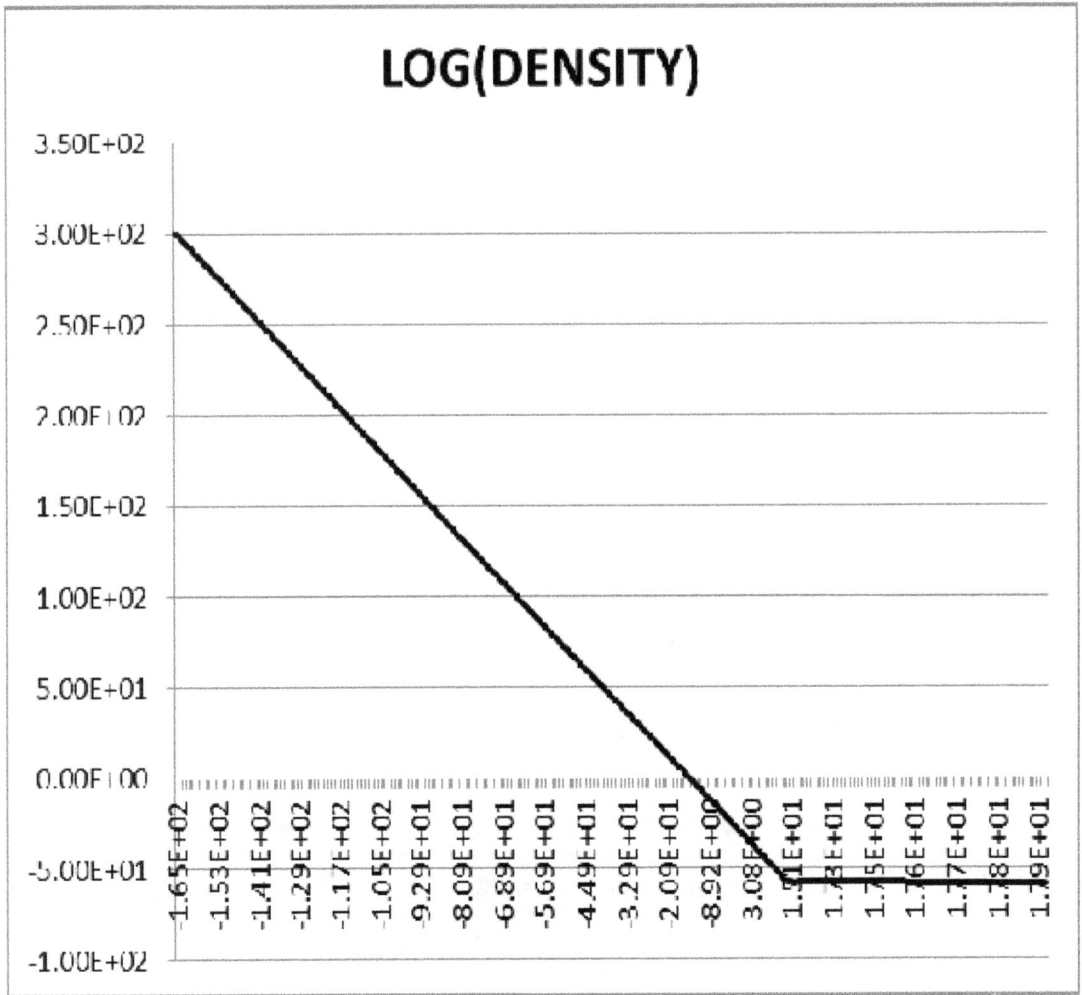

Figure 61.6 Log Density: $\text{Log}_{10}(\rho_{tot}(t) \text{ g/cm}^{-3})$ plotted vs. $\log_{10}(t)$ from the Big Bang metastate at $t = 1.19 \times 10^{-165}$ sec. to the future: $t = 8.2 \times 10^{17}$ sec. Note the drop below zero beginning at $t = 5 \times 10^{-16}$ sec. signaling the start of very small values for ρ.

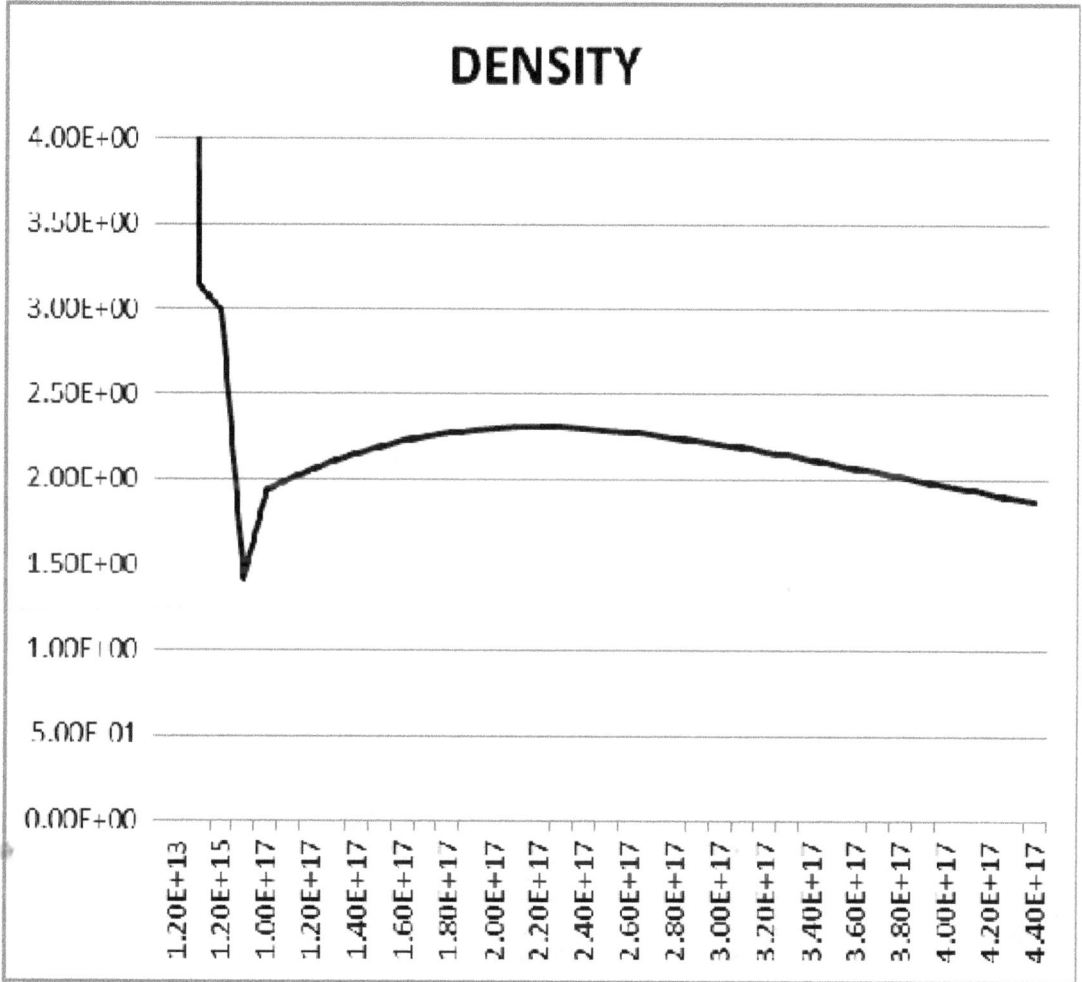

Figure 61.7 Energy density ρ(t) g/cm^{-3} × 10^{29} plotted vs. t from t = 1.19 × 10^{13} sec. (380,000 years) to the present 4.35 × 10^{17} sec. Note the dip at t = 1.2 × 10^{16} sec. with a small value for ρ = 1.42 × 10^{-29} g/cm^{-3}.

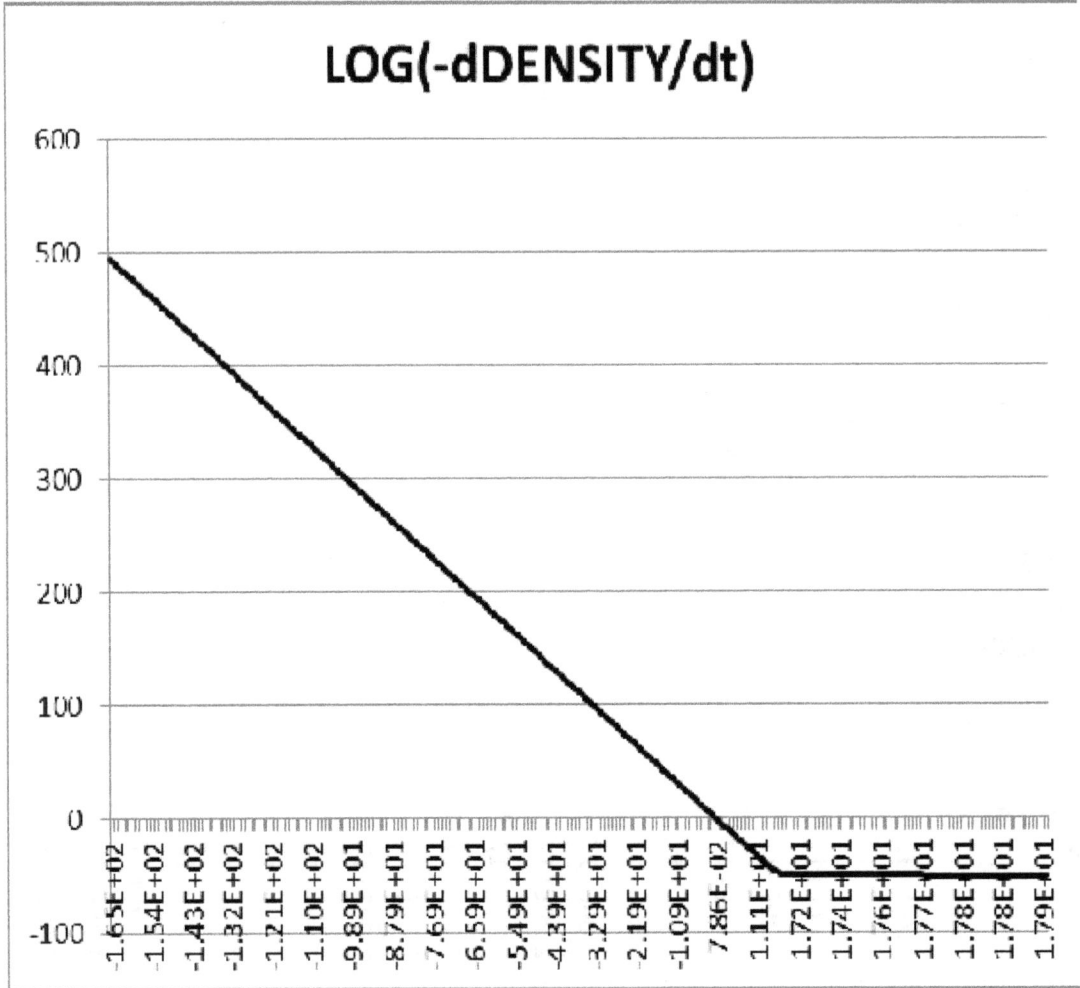

Figure 61.8 $\text{Log}_{10}(-d\rho(t)/dt \text{ g s}^{-1} \text{ cm}^{-3})$ plotted vs. $\log_{10}(t)$ from the Big Bang metastate at $t = 1.19 \times 10^{-165}$ sec. to the future: $t = 8.2 \times 10^{17}$ sec. The plot goes below zero at about $\log_{10}(t \text{ sec.}) = 7.8 \times 10^{-2}$ ($t = 1.198$ sec.) where $\log_{10}(-d\rho/dt \text{ g s}^{-1} \text{ cm}^{-3}) = -0.76$.

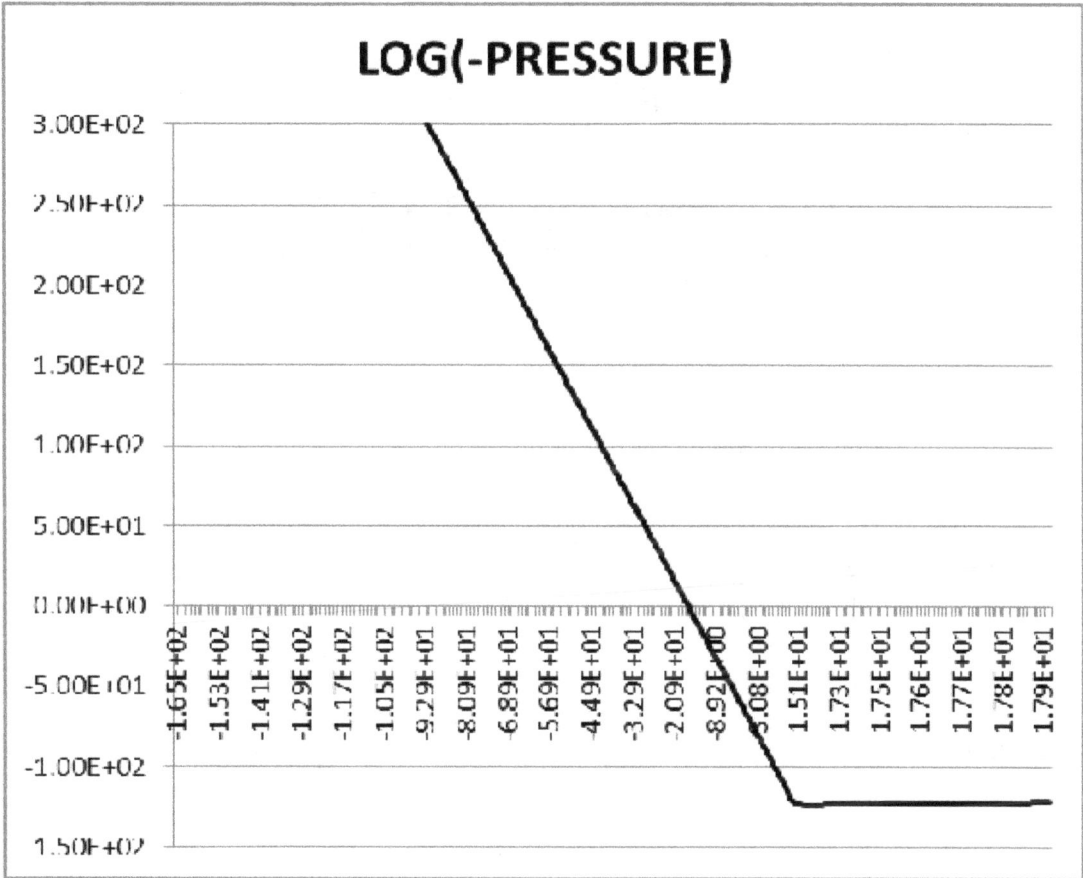

Figure 61.9 Pressure: $Log_{10}(-p(t))$ **plotted vs.** $log_{10}(t)$ **from the Big Bang metastate at** $t = 1.19 \times 10^{-165}$ **sec. to the future time** $t = 8.2 \times 10^{17}$ **sec. Note the low values of** $log_{10}(-p(t)) \approx -115$ **beginning at** $log_{10}(t \text{ sec.}) = 12.1$ **(**$t = 1.2 \times 10^{12}$ **sec.). The pressure is always negative indicating a pressure for expansion. At** $t = 1.19 \times 10^{-165}$ **sec. we find** $log_{10}(-p(t)) = 593$ **or** $p(t) = -10^{593}$.

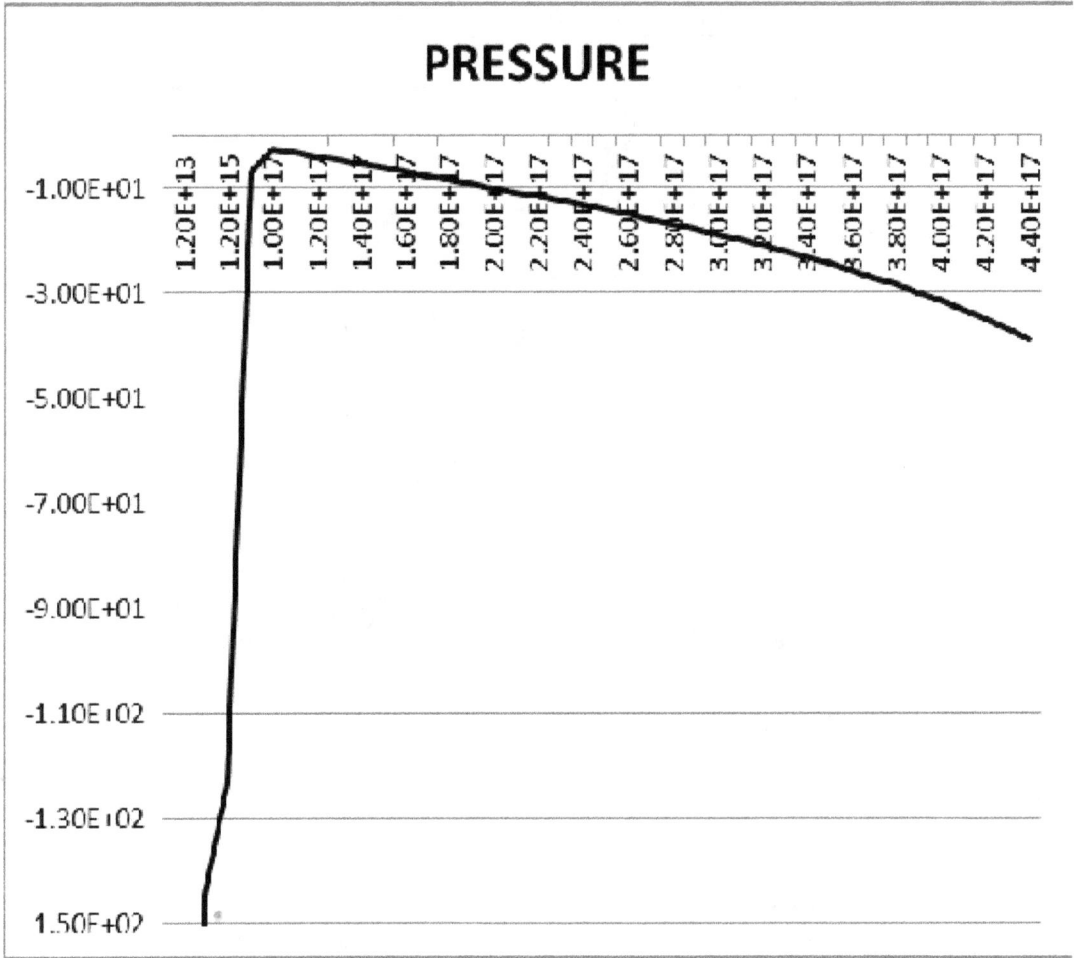

Figure 61.10 Negative Pressure: p(t) × 10^124 plotted vs. t sec. from t = 1.19 × 10^13 sec. (380,000 years) to the present 4.35 × 10^17 sec.

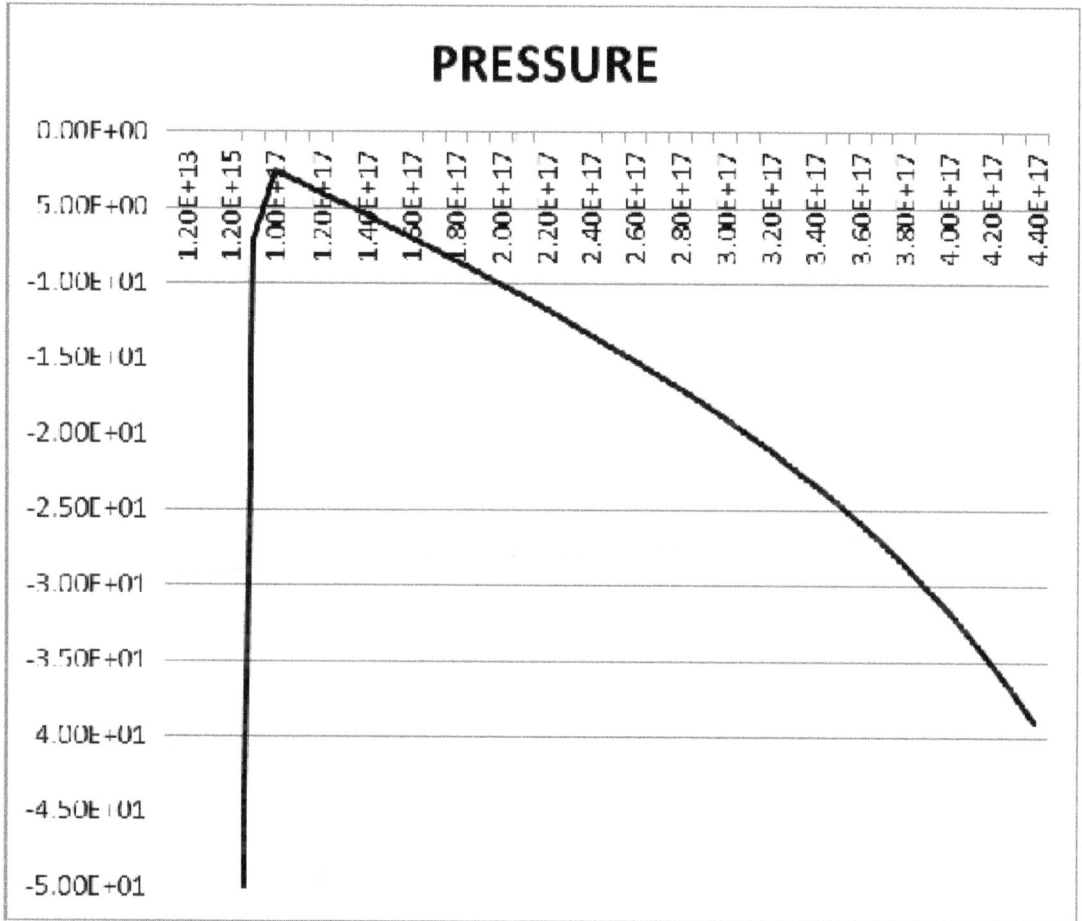

Figure 61.11 Negative Pressure: $p(t) \times 10^{124}$ plotted vs. t sec. from t = 1.19 $\times 10^{13}$ sec. (380,000 years) to the present 4.35×10^{17} sec. The maximum is - 2.62×10^{-124} at t = 1.0×10^{17} sec. A closer view than Fig. 61.10.

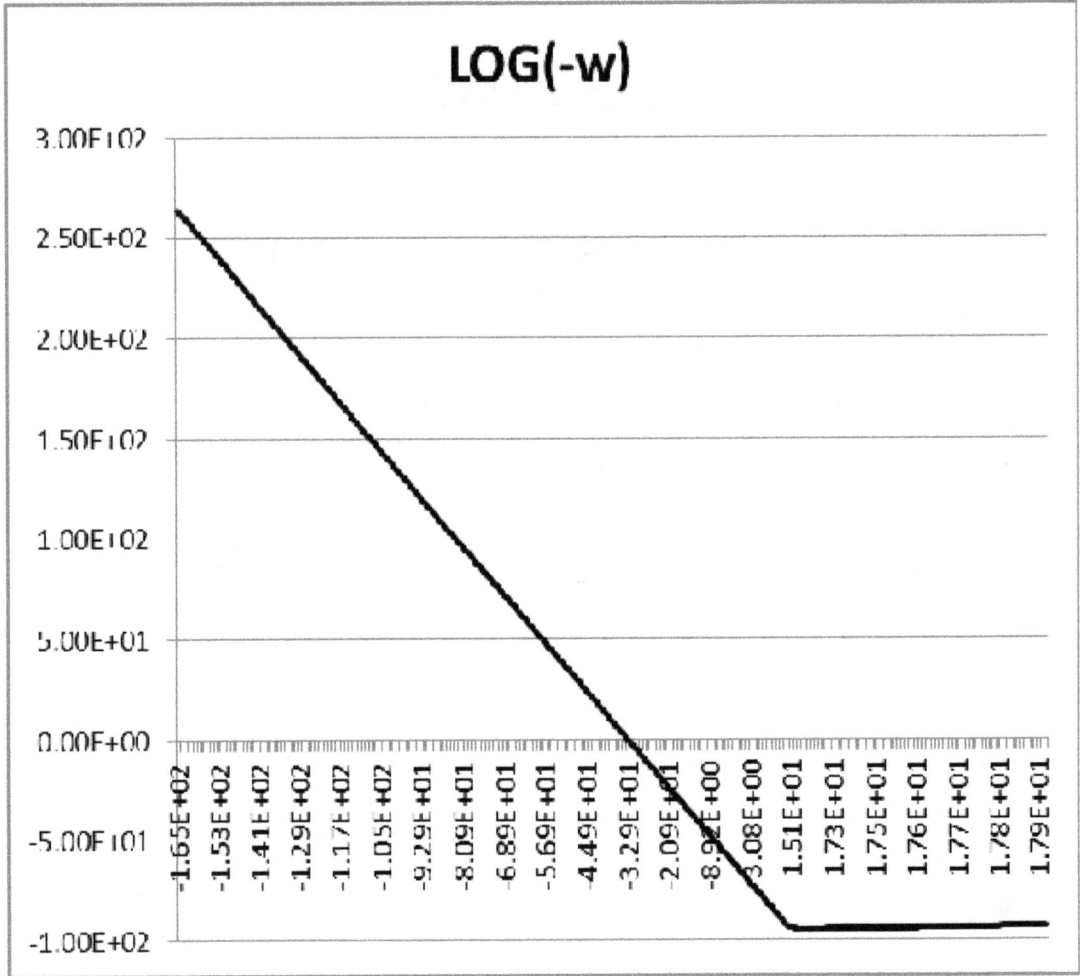

Figure 61.12 The equation of state as a function of time $\log_{10}(-w)$ plotted vs. $\log_{10}(t)$ from the Big Bang metastate at t = 1.19 $\times$ 10^{-165} sec. to the future time t = 8.2 $\times$ 10^{17} sec. Note: w ≠ -1. Quintessence!

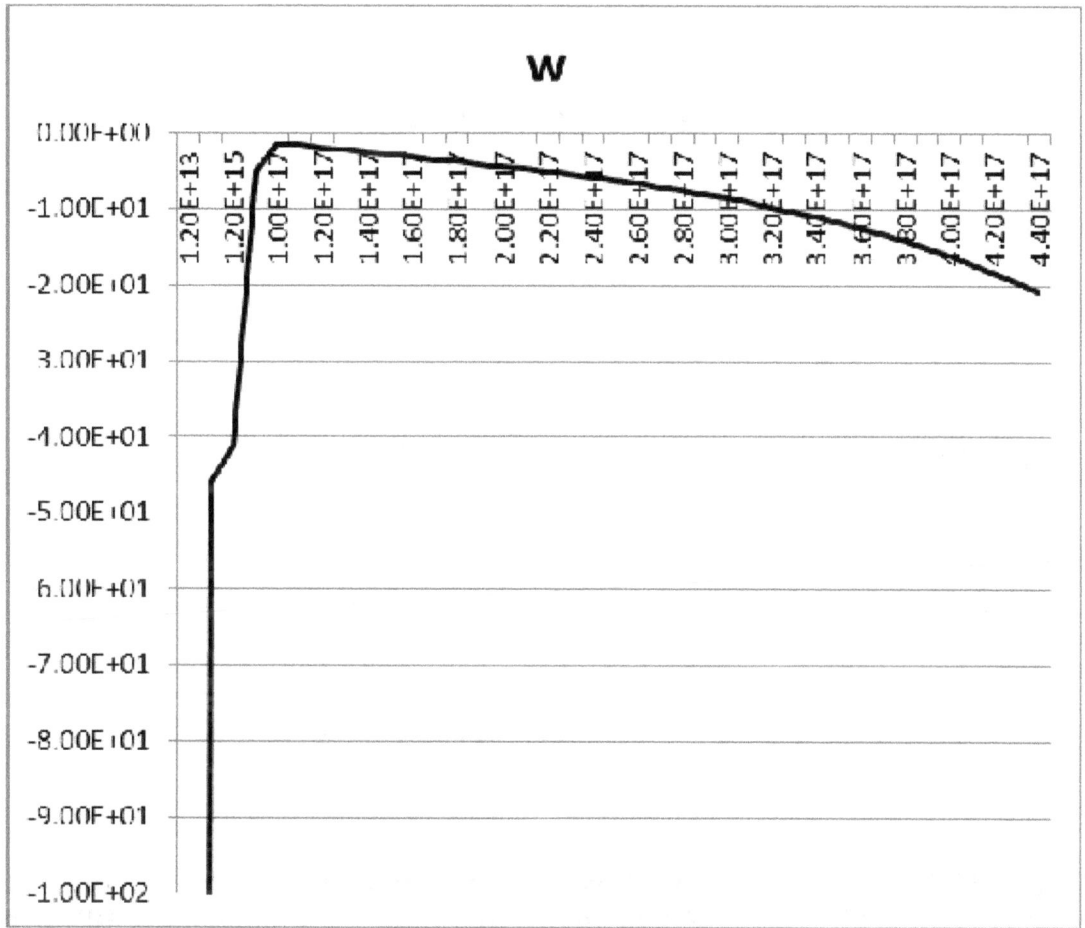

Figure 61.13 The equation of state w $\times 10^{95}$ as a function of time plotted vs. t sec. from t = 1.19×10^{13} sec. (380,000 years) to the present time. Note the peak of w = - 1.36×10^{-95} at t = 1. $\times 10^{17}$ sec. Note: w ≠ -1. Quintessence!

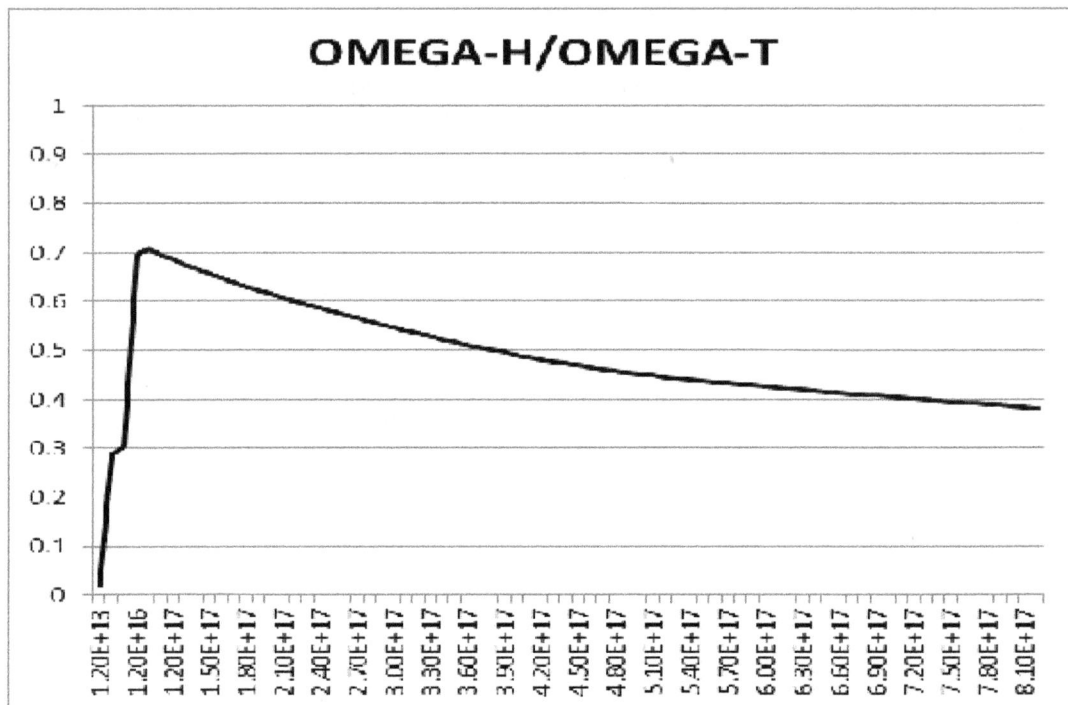

Figure 61.14 The ratio Ω_H/Ω_T showing the dominance of Dark Energy Ω_T from t = 1.19×10^{13} sec. (380,000 years) to the future time t = 8.2×10^{17} sec. Prior to 380,000 the Dark Energy was greater by many orders of magnitude. The peak value of Ω_H/Ω_T is at t = 1.2×10^{16} sec.

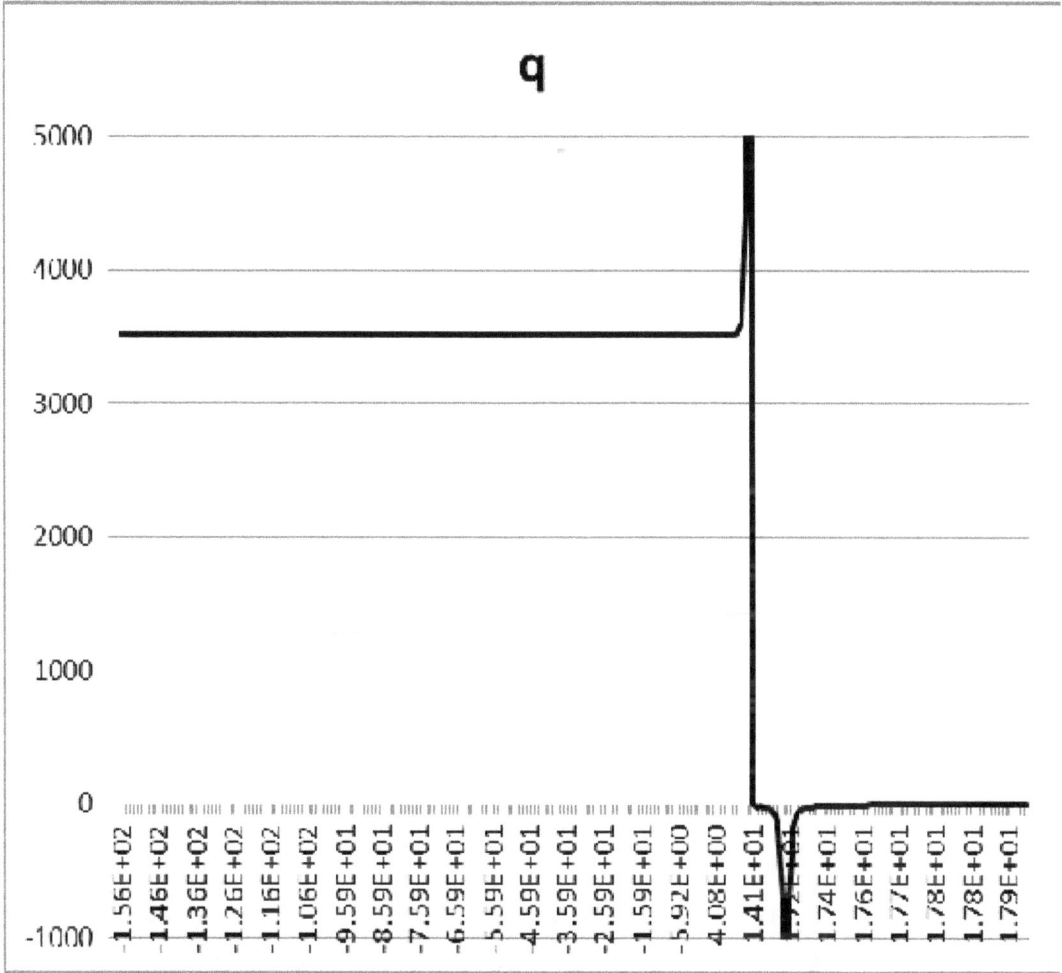

Figure 61.15 The deceleration parameter q vs. log(t sec) from t = 1.2 × 10⁻¹⁵⁶ sec. to the future t = 8.2 × 10¹⁷ sec. The peak occurs at t = 1.2 × 10¹³ sec. The minimum occurs at 1.6 × 10¹⁷ sec. Note: q > 0 for t < 1.7 × 10¹⁴ sec.

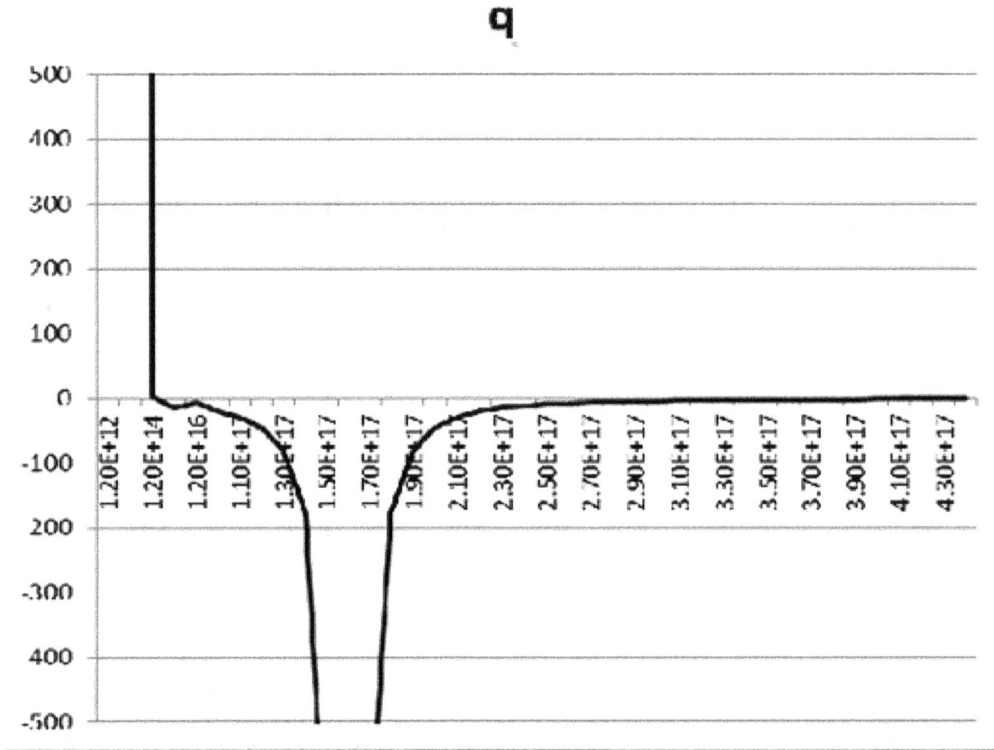

Figure 61.16 The deceleration parameter q vs. log(t sec) from t = 1.2 × 10^{12} sec. to t = 4.4 × 10^{17} sec. The peak occurs at t = 1.2 × 10^{13} sec. The dip occurs at 1.6 × 10^{17} sec. To the left of the peak q > 0 indicating decelerating expansion. To the right of the peak for t > 1.7 × 10^{14} sec. we find q < 0 indicating accelerating expansion starting 13.78 billion years ago – "just" after the Big bang. q = -2.0 presently. In the future, at t = 8.2 × 10^{17} sec. we found q = -1.2.

62. Some Implications of the Universal Scale Factor

In this chapter we consider some implications of our model Quantum Big Bang and Quantum Vacuum Universe based on the time development of relevant parameters presented in chapters 59 through 61.

62.1 Superclusters and Voids

In our study of the Quantum Big Bang[404] we found that the early Big Bang Metastate universe had a wide variation in the Hubble Constant. At its center with radial distance r = 0, we found H = 1.79 $\times$ 10^{218} km s^{-1} Mpc^{-1} while at its periphery H = 1.14 $\times$ 10^{126} km s^{-1} Mpc^{-1}. Thus we see the "explosion" of the energy-mass density of the central region into the outer regions. The result would appear to be a "wave" of mass-energy "emptying" the center and creating a bulge towards the outer region. Thus we anticipate that the primordial universe had variations in mass-energy density favoring the creation of a wave of high mass-energy density and an inner region of low mass-energy. The result is precursors of voids and supercluster bubbles.

In the later Quantum Vacuum Universe period we found a rapid universe contraction to 70% of its previous size. Subsequently the universe expanded – also fairly rapidly. The contraction effectively squeezed the outer portions of the universe again creating a wave of mass-energy. The expanding wave then (as in the case of water waves) developed foam (bubbles) of mass-energy of high density that evolved into the superclusters and voids that we find today.

The number of superclusters is estimated to be ten million. Superclusters seem to contain of the order of 10^5 galaxies or more. Thus the combination of the Quantum

---

[404] See Blaha (2019e).

Big Bang period and the Quantum Vacuum period explain the presence of superclusters and voids found in today's universe.

It appears we have an understanding of the large scale evolution of the universe. Fig. 62.1 presents the large scale pattern of universe evolution.

q rises to a peak value of 408,273 at t = 1.2×10^{13} sec. indicating decelerating expansion.

Dark Energy Ω_T rises to a peak of 39.2 at t = 8.71 $\times 10^{13}$ sec. causing H^2 to simultaneously become very large (Einstein equation), although H (Fig. 22.6) is negative and large with a minimum of - 445 km s^{-1} Mpc^{-1}.

For t > 1.7×10^{14} sec. we find q < 0 indicating accelerating expansion as shown in H

ρ_{tot} declines to a minimum of 1.42×10^{-29} g/cm^{-3} at t = 1.2×10^{16} sec. after which it rises.

At t = 1.0×10^{17} sec. w and –pressure both peak. Note w $\approx$ 0 suggesting a cold dust or gas. The pressure is also quite small. Both peaks are due to the increase in ρ_{tot} after reaching its minimum.

q > 0 for t < 1.7×10^{14} sec. (decelerating). q reaches a minimum at t = 1.6×10^{17} sec.

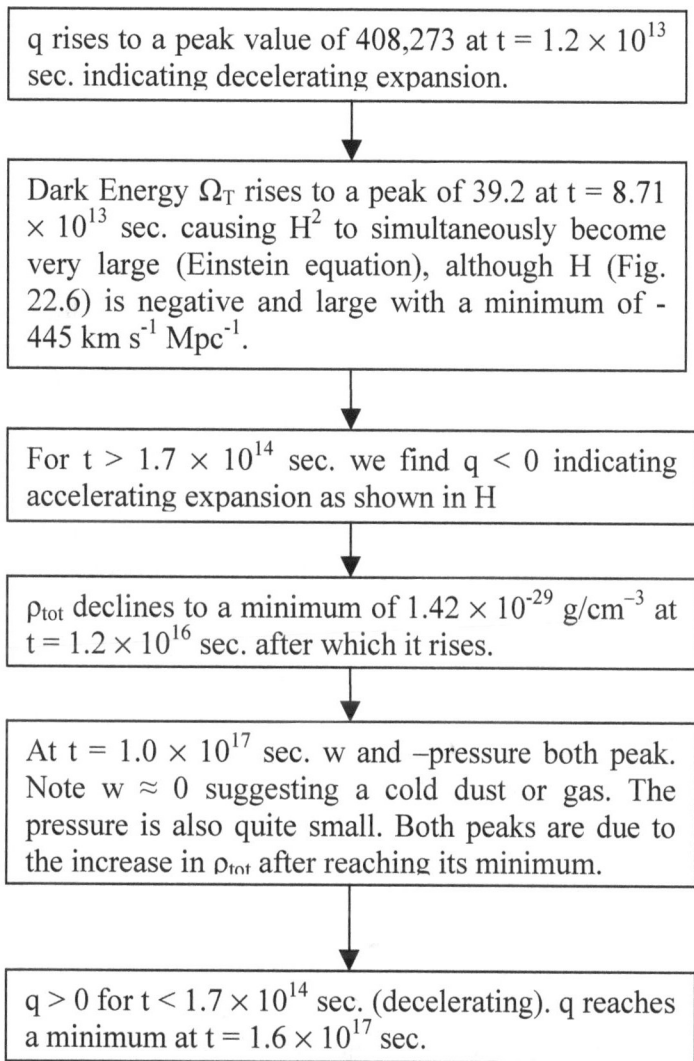

Figure 62.1 Scenario Based on figures of previous chapters. These figures give the evolution in time of the universe.

63. Major Implications For Universes: SuperUniverses

Our ability to calculate the Unified SuperStandard Theory coupling constants and, as we shall see, the universe vector interaction coupling constant (and universe scale factor) raises important issues both within our universe and in possible other universes. (The experimental and theoretical basis of the possibility of other universes is described in earlier books such as Blaha (2018).)

In our universe we see that the Anthropic hypothesis, based on the precise value of the QED Fine Structure Constant, is not compelling. The value of α is set within QED not otherwise. In addition the other known Standard Model coupling constants, which were found to good approximation in our approach, are also *self-determined* in Quantum Field Theory. Since the Standard Model interactions are responsible for the vast majority of the features of Physics, Chemistry and Life we find the universe's detailed structure is set by the Unified SuperStandard Theory.

If, as we have proposed, there are other four-dimensional (4D) universes within a higher dimension space called the Megaverse, then, assuming that the overall physical theory of a four-dimensional (4D) universe[405] satisfies

1. The universe is described by Complex General Relativity.

2. The particles, and particle interactions, in the universe are those of the Unified SuperStandard Theory.

we find 4D universes are based on the same Unified SuperStandard Theory. We call such universes SuperUniverses and their common theoretic underpinning the SuperUniverse Model.

---

[405] And possibly of other universes of different dimension.

The conclusions that follow from this line of reasoning are:

1. Standard Model features are the same in all 4D universes.

2. A SuperUniverse has the same form for the a(t) scale factor, and a similar evolution as other universes and parallels our universe. We therefore expect a SuperUniverse to have Dark Energy dominance, varying Hubble Constants; Big Dips, Superclusters, and voids in other universes. (The total energy of universes may differ.)

3. SuperUniverses have the same Physics, Chemistry and Biology as our universe. Life in other universes is possible but can be expected to take different forms.

4. SuperUniverses can differ in some ways that do not change our overall conclusions: in some SuperUniverses a) anti-particles may dominate; b) left-right symmetry may differ; c) Life may favor either dextrorotary or levorotary molecules;

5. Communication with creatures in other universes may be possible using more advanced quantum entanglement. Quantum entanglement of photons from earth and the sun has been found experimentally.

6. Universes, as a whole, have a composite particle nature and can be described in a second quantized framework. See Blaha (2018e), and comments there by DeWitt in particular on quantum universe features.

The Enclosing Megaverse

64. Evidence for Entities Beyond Our Universe

64.1 Theoretical and Experimental Support

Why are we not content with one universe given its enormous size and variety? It appears that there are important theoretical reasons, and some important experimental observations, that suggest that there is more than our universe 'out there.'

In this chapter[406] we will discuss theoretical reasons and experimental suggestions of a larger space—that we call the *Megaverse*—that contains our universe and, most likely, other universes. The existence of a Megaverse resolves several theoretical issues and may address some important astronomical puzzles that have appeared in recent years.

The theoretical issues, which have been subjects of discussion for many years, are:

1. The need for a 'clock' to measure 'time' knowing that it is to some extent relative and local.
2. The need for a 'quantum observer' to complete the understanding of quantum gravity as described by the Wheeler-DeWitt equation and in other efforts to develop a quantum gravity.
3. The need for other universes to provide theoretical measuring platforms for quantities beyond the charge and mass of the universe. We think here of the other quantum numbers of particles and particle number operators such as Baryon number.
4. The need for an ultimate source of mass and inertia in our universe.

---

[406] Most of this chapter appears in Blaha (2015a) and in earlier books by the author.

In Blaha (2015a) and earlier books we have suggested that there are weighty reasons to believe that other universes exist.[407] The existence of other universes is a solution to these problems.

These problems have a source in Quantum Gravity and the interpretation of the Wheeler-DeWitt equation in particular. We now consider the issues raised above.

64.1.1 Universe Clocks

Asynchronous Logic provides the equivalent of a clock for the synchronization of processes within large electrical systems such as VLSI chips. Similarly there is a need for a universal clock for our universe. As DeWitt[408] points out in his studies of quantum gravity,

'"The variables ... [of the quantized Friedmann model] because of their lack of hermiticity, are not rigorously observable and hence cannot yield a measure of proper time which is valid under all circumstances. ... . It is for this reason that we may say that "time" is only a phenomenological concept ... If the principle of general covariance is truly valid then the quantum mechanics of everyday usage with its dependence on the Schrödinger equations ... is only a phenomenological theory. For the only "time" which a covariant theory can admit is an intrinsic time defined by the contents of the universe itself. Any intrinsically defined time is necessarily non-Hermitean, which is equivalent to saying that there exists no clock, whether geometrical or material, which can yield a measure of time which is operationally valid under *all* circumstances, and hence there exists no operational method for determining the Schrödinger state function with arbitrarily high precision."

The lack of a clock within our universe invalidates quantum mechanics in principle and Quantum Gravity in particular. DeWitt concludes, "Thus [quantum gravity] will say nothing about time unless a clock to measure time is provided."

---

[407] In Blaha (2013a), before the Higgs particle was discovered at CERN we suggested an alternate mechanism was possible if a sister universe existed (making the existence of other universes a reasonable possibility. The Higgs discovery makes the sister universe mechanism unlikely.

[408] DeWitt, B. S., Phys. Rev. **160**, 1113 (1987).

Unruh[409] also has an issue with the source of time:

"One of the key problems is that of time. We see and experience the world in terms of time. We see things grow, develop, and change. However, time does not enter into the Euclidean formulation of quantum gravity directly. In the usual Hamiltonian formulation, the Hamiltonian for quantum gravity is made up of densities which are the generators, not only of spatial coordinate transformations, but also of temporal coordinate transformations. The content of four of Einstein's equations is that some generators are zero. Thus all wave functions are invariant under all spatial and all temporal coordinate transformations. There is nothing in the wave function or the amplitudes which refers to the coordinate t, or the corresponding points of the manifold in any way. How then do we recover the indubitable and ubiquitous experience we have of time? The standard answer is that our experience of time is actually an experience of different correlations between physical quantities in the world. Time is replaced by the readings of clocks. I know that time has changed, not through any direct experience with time, but because the hands of my watch have changed.

Although the implementation of this idea is actually extremely difficult in practice, and although I personally believe that one should formulate one's quantum theory of gravity so as to contain time explicitly, let us nevertheless pursue the consequences of this idea of time as defined internally, as the "reading" of a dynamic variable. For an observer inside the theory, his "time" is not the coordinate t. Rather his time is some one of the given dynamic variables of the theory: y or P. Thus although the coupling to the baby universes via the effective action S is independent of the coordinates t or x, that does not mean that the observer inside the theory will experience the interactions as being independent of time. For him and/or her, time is one of the dynamic variables and so it can depend on the various dynamic variables of the theory, even if it does not depend on the time coordinate t. In general one would expect the observer to see what looks to him like a time-dependent interaction with the baby universes. At one time, some one of the baby universes may couple strongly to the large

---

[409] Unruh, W. G., Phys. Rev. D **40**, 1053 (1989).

universe, while at some other time, another of the baby universes will couple more strongly."

In Blaha (2015a) and earlier books, we suggested the existence of other universes provides a 'clock' in principle for our universe. And being universes, these other universes are excellent clocks. DeWitt points out,

"Because every clock has a "one-sided" energy spectrum, its ultimate accuracy must necessarily be inversely proportional to its rest mass. When the whole universe is cast in the role of a clock, the concept of time can of course be made fantastically accurate (at least in principle) ... "

Setting a mass scale using other universes, also sets[410] a time scale and resolves the issue of a clock for our universe. *In principle the existence of other universes validates the role of time in the Copenhagen interpretation of Quantum Mechanics.*

64.1.2 Quantum Observer

Attempts to create a quantum gravity theory have to confront the need for an *Observer* in any quantum theory within the context of the Copenhagen interpretation. DeWitt points out,

"The Copenhagen view depends on the assumed a priori existence of a classical level to which all questions of observation may ultimately be referred. Here, however, the whole universe is the object of inspection; there is no classical vantage point, and hence the interpretation question must be re-argued from the beginning. While we do not wish to stress this point unduly, since, after all, the Friedmann model ignores the vast complexities of the real universe, it is nevertheless clear that the quantum theory of space-time must ultimately force a deviation from the traditional Copenhagen doctrine." And Unruh states

---

[410] For example the Planck time value is set by the Planck mass.

"One of the key features in the interpretation of such transition amplitudes, or wave functions, is the idea that we, as observers are also a part of the Universe as a whole. We, as physical observers, must be describable from within the theory and not as observers external to the theory as in usual quantum mechanics. In usual quantum mechanics, the interpretation is usually given in terms of observers that are outside of the theory. There one makes a split, with the quantum world at one side of the split, and the observer on the other. von Neumann argued that the predictions of quantum mechanics, at least under certain assumptions, are independent of the exact location of that split, but Bohr argued adamantly for the necessity of such a split (classical observers and quantum world). *There is a great difficulty in setting up such a split for physical observers contained within and influenced by a quantum universe,* [italics added] and for the Universe as a whole, especially including gravity, one cannot argue that the predictions will be independent of where one puts the split. Since all energies interact gravitationally, and our observations are surely energetic phenomenon, the treatment of the energetics of observation as classical would lead to different predictions than if they were treated quantum mechanically. One is therefore forced to devise an interpretation of quantum mechanics in which the observer is part of the quantum system, rather than outside the quantum system.

This means that the interpretation of these transition amplitudes becomes somewhat non-intuitive. One must ask what the system looks like from within, from the viewpoint of an observer who is part of that world, rather than being able to interpret them directly in terms of probabilities for observations made by an external observer."

While the *Observer* question is addressed by a number of authors, the proposed answers are not entirely convincing. *The existence of other universes provides macroscopic Quantum Observers for our universe.* And our universe provides a macroscopic quantum observer for other universes. Thus the quantum observer issue is resolved.

These considerations lead us to view the existence of other universes as a critical solution to the above problems.

64.1.3 The Higgs Mechanism is Explainable by Extra Dimensions

The Higgs Mechanism 'explains' (generates) fermion and boson masses. However the Higgs potential contains a quadratic term with a constant with the dimensions of [mass]. In a sense the Higgs Mechanism trades one mass for another. From where do the Higgs potentials' masses come?

A further explanation is needed is to determine the origin of the "dimensionful" mass terms in the Higgs' particle equations themselves. At present little if any thought has been given to the origin of these terms. We suggested that, excluding a *deus ex machina* source, the only known way to generate these mass terms in the Higgs' equations is through the separation of equations technique of differential equations. This technique requires additional parameters which can only be the coordinates of *extra unknown dimensions*. The best example of the generation of mass terms appears in the Schwarzschild solution of General Relativity where a separation constant, often denoted M, appears that has the dimension of [mass].

Thus extra space-time dimensions would resolve the origin of Higgs potentials' masses. Given extra dimensions it is reasonable to expect that these extra dimensions contain universes. Thus the Megaverse!

64.1.4 Possible Accretion of Megaverse Matter to Fuel Expansion of Our Universe

If matter is distributed outside of universes in the Megaverse, and if this matter can be accreted to universes by gravitational attraction, then the apparent increasing expansion of our universe may be due to this accretion. In chapter 14 of Blaha (2017c) we presented a model in which this possibility is realized. If true, then we would have tangible evidence of the residence of our universe in the Megaverse.

64.1.5 Asynchronous Logic is a Requirement of Universes

By establishing Asynchronous Logic principles[411] as the basis for the existence of universes and for setting the number of dimensions in each universe – four; and basis of fermion particles - qubes – we have found deeper principles of organization for the

---

[411] The basis of this section is described in detail in Blaha (2015a). That book places Physics within a logical framework that is a possible deeper ground for fundamental Physics theory.

foundations of physics. The principles built on this foundation serve to enable the coordination of complex physical processes.

Usually we look at particle processes primarily from a space-time perspective: particles collide and produce new particles. We primarily think of the incoming and outgoing particles in a collision. However, considering the set of fundamental particles – and the particle transforming interactions in themselves – neglecting space-time and momentum considerations – leads us to view particles as constituting an alphabet and their interactions as a type of computer grammar.[412] Then the Asynchronicity Principles enable us to bring in space-time in a way that gives us the maximum complexity with the most minimal assumptions. As Leibniz[413] points out our universe has maximal complexity with minimal assumptions.

64.1.6 The Meaning of Total Quantities of a Universe

The 'external' properties of a universe are normally questioned—for the simple reason that it is assumed that there is no 'outside' of our universe. For example, Misner (1973) asserts:[414]

'There is no such thing as "the energy (or angular momentum, or charge) of a closed universe," according to general relativity, and this for a simple reason. To weigh something one needs a platform on which to stand to do the weighing.'

Misner et al presumes no such platform exists. If there is but one closed universe as most currently believe then one cannot measure any totals of a closed universe (which ours may be to be). Yet if we take a more general view that our universe is only one of many then it becomes possible to measure total mass, charge, angular momentum, baryon number, and many other quantities of interest. Indeed, the existence of other universes (within the encompassing Megaverse) opens the door to an understanding of time, mass, energy, and all the other quantities necessary to develop a dynamical theory of universes.

---

[412] This conceptual approach was first described in Blaha (1998) who went on to characterize our universe as one enormous word evolving in time.
[413] See Rescher (1967).
[414] Pp. 457 - 458.

Later we will also see that one can then treat universes as 'particles', and develop 'universe dynamics', which might explain knotty problems such as the Big Bang and its precursor (if any). We will do this in subsequent chapters after first considering the possible structure of universes in general in the Megaverse.

64.2 Possible Experimental Evidence for the Megaverse

At first glance it would seem impossible to produce evidence for the existence of other universes. However there are subtle means by which we can 'sense' experimentally 'nearby' universes should they exist. The mechanism would appear to be gravitational effects exerted on objects within our universe by unseen objects of enormous mass. Currently there appears to be three experimental suggestions of the existence of 'nearby' universes and one theoretical argument based on an influx of mass-energy from the Megaverse that may cause the expansion of our universe.

64.2.1 Great Attractors

One potential support is the discovery of the Great Attractor (at the center of the Laniakea Galaxy Supercluster), and the more massive Shapley Attractor (centered in the Shapley Supercluster)[415]. These attractors contain massive numbers of galaxies and are drawing galaxies over a distance of millions of light years towards them.

If another universe(s) is 'near' our universe it could act as a 'gravitational magnet' and draw galaxies within our universe towards it to form one or more superclusters which could then act as attractors. Thus attractors might indirectly reveal the presence of other nearby universes—contrary to the expected large scale uniformity of the universe. The only other apparent source of superclusters is chance. Chance seems an unsatisfactory possibility in the present case.

---

[415] Tully, R. Brent; Courtois, Helene; Hoffman, Yehuda; Pomarède, Daniel, "The Laniakea Supercluster of galaxies". Nature (4 September 2014). 513 (7516): 71–73; arXiv:1409.0880.

64.2.2 Bright Bumps in Universe Suggesting Collision with Another Universe

A recent study[416] of the residual brightness of parts of the accessible universe found that bright patches appeared if a model of the CMB (Cosmic Microwave Background) with gases, stars and dust was 'subtracted' from the PLANCK map of the entire sky. After the subtraction one would expect only noise spread throughout the sky. However, bright patches were seen in a certain range of frequencies. These anomalies are thought to be a result of our universe colliding with another object – presumably another universe in the Megaverse.

64.2.3 Cold Spot in Universe Suggesting Collision with Another Universe

Another recent study[417] of a huge cold region of the universe spanning billions of light years revealed that this region is not a relatively empty region but rather is similar to in its distribution of galaxies to the rest of the universe. Previous the Cold Spot (an area where cosmic microwave background radiation – the leftover Big Bang radiation is weak – making it significantly colder (0.00015C colder) than the average temperature of the universe.)

An analysis of 7,000 galaxy redshifts using new high-resolution data has now shown that the Cold Spot is similar to the rest of the universe. The Durham University group suggested that the Cold Spot might have been caused by a collision between our universe and another Universe. They further suggested that there is only a 1 in 50 chance that it could explain by standard cosmology. could produce this feature

Thus we have another important piece of circumstantial evidence in favor of other universes and thus the Megaverse.

64.2.4 Megaverse Energy-Matter Infusion into Our Universe

In chapter 14 of Blaha (2017c) we presented a model for an influx of mass-energy from the Megaverse to support the Bond-Gold-Hoyle-Narlikar Steady State Cosmology, which was originally based on the 'continuous creation of mass-energy' by Hoyle and Narlikar. This model explains why the value of Ω makes the universe close

[416] Ranga-Ram Chary, arXiv.org:/1510.00126 (2015).
[417] T. Shanks et al, Durham University (Australia), Monthly Notices of the Royal Astronomical Society, 2016 .

to flat. If this model is correct then we would have concrete support for a Megaverse with a low mass-energy density leaking mass-energy into our universe. *More generally, it suggests that universes are surfaces of high mass-energy density in a Megaverse of low mass-energy density – with a ratio of mass-energy densities of the other of 10^{30}.*

64.2.5 Conclusion

We conclude that data is beginning to emerge favoring multiple universes and a physical Megaverse in support of the theoretical justifications presented earlier.

64.3 Historical Trend Towards Larger Space-Time Structures

Looking back through the history of Mankind's view of the universe we see a clear progression to a larger and larger view. Before the 16th century the earth was the universe. In the
16th century Giordano Bruno (and possibly others) suggested that the stars were suns with many worlds circling them. So our view of the universe expanded to include stars.

Then over time it was noticed that nebulae existed in space. The astronomer, Edwin Hubble, studied the Andromeda nebula with the 'new' Mt. Wilson telescope and in 1929 announced that it was a galaxy composed of stars. Now the universe was conceptually similar to our current view.

Now we seem to have significant theoretical considerations and some suggestive experimental data that lead us to consider the possibility that our universe is not alone— that our universes is but one of many universes in a space we call the Megaverse. This book (and Blaha (2017c)) pulls together much of our earlier work and adds new insights into the nature of the Megaverse. We shall take the Megaverse as fact, extrapolate the form of our universe into the form of other possible universes, and develop a fairly detailed theory of the Megaverse and its resident universes. We will consider escaping our universe into the Megaverse—mindful that such travel will not happen until the *very* distant future. There are many technical bridges to cross before we can travel to other universes.

Given the vastness of our universe one might ask Why travel? We considered reasons in some detail in Blaha (2017c). For now, it suffices to say, Because they are there. Mankind has always grown and prospered through exploration and exploitation of new territories. Indeed the eminent Historian, Arnold Toynbee, stated that new 'turf' is the source of growth in all civilizations. Eventually, in the very distant future, we may need the new turf in the Megaverse.

64.4 Other Megaverses? Will It Ever End?

Does the trend to larger and larger expanses of space suggest that our Megaverse may be but one of many duplicate Megaverses of the same number of dimensions (but different orientations)? One cannot decide this question in our present, or likely near future, state of knowledge.

However, based on our estimate of the dimension of the Megaverse, and its basis in the geometry and group structure of its interactions, it is reasonable to conjecture duplicate Megaverses would have the same number of dimensions as our Megaverse, and have universes with the same number of dimensions as our universe, and thus have a Physics in each other Megaverses' universes similar to the Physics of our universe.

This scenario would appear to be unlikely in the author's view as it appears uneconomical and gives rise to the question How could a plethora of Megaverses arise?

Another scenario, in which Megaverses appear within Megaverses like the toy Chinese nested boxes, also appears unlikely. For it would require a chain of Megaverses of ever increasing dimension, and raise the question of its origin—a question that would never be answerable. Nor would the nested set of Megaverses be experimentally accessible.

So we are content with one Megaverse.

65. The Embedding of a Universe in a Higher Dimension Megaverse – Surface Tension

65.1 Universes as Mass-Energy Islands

In developing the theory of the Megaverse we view universes as islands of mass-energy that maintain their 'integrity' as surfaces due to gravitational force. Gravity holds universes together rather like molecular forces within a water droplet hold water molecules together. Molecular attractive forces gives droplets cohesion as they (perhaps) descend through the earth's atmosphere. They are the origin of *surface tension* in water.

Similarly gravity holds higher density[418] mass-energy universes together and gives rise to gravity surface tension. In a model presented in Blaha (2017c) for the Big Bang and the expansion of the universe within the Megaverse we found the mass-energy density of the universe was a factor of 10^{30} more than the density in the surrounding Megaverse space.

Thus we have good reason to study the surface tension of universes as a result of gravitational attraction within universes.

65.2 Boundary of a Universe within the Megaverse

In this chapter we describe the embedding of universes within the Megaverse. Much of this chapter appears in several earlier books by the author such as Blaha (2015a).

As stated earlier, we define a universe to be a closed or open surface in the Megaverse with a much higher mass-energy density than the Megaverse.

There are two types of boundaries for a universe embedded in a space of larger dimensions. First there is a boundary of the universe determined by treating the universe

---

[418] Higher density in comparison to the much lower density of the inter-universe space of the Megaverse.

as a surface in the space. Secondly, there is another type of universe boundary defined by the observation that any neighborhood – not strictly within the universe – of every point of the universe has an infinite number of points of the enclosing Megaverse space.[419] Or, every point has neighborhoods with Megaverse points within it. Thus *each point of a universe is on a boundary of the universe due to the larger dimensions of the Megaverse space* within which it resides. Fig. 65.1 schematically illustrates these neighborhoods for any universe point for a universe contained within a higher dimensional Megaverse.

65.3 Confinement of Universes due to 'Surface Tension'

We will assume that other universes have the same physics as our universe with the possible differences that they may have differing interaction coupling constants and particle masses. As we will discuss later, every point of a universe in a higher dimensional Megaverse has Megaverse points in any neighborhood of the point (with the exception of neighborhoods strictly within the universe). Thus we confront the question: what keeps mass-energy at points in a universe or is there leakage from the universe into the Megaverse?

If there is little or no leakage into the Megaverse then, since each point in a universe is part of a Megaverse surface, one can only assume that there is a barrier to movement into the Megaverse. Taking a note from fluid dynamics, and viewing Megaverse space as one 'material' and the universe as a different 'material,'[420] we view the barrier as 'surface tension.'[421] The Megaverse appears to "exert a force" confining the contents of the universe to within itself.[422]

---

[419] Any neighborhood of any point in the universe – with all its points strictly within the universe – has an infinite number of points within the universe. We assume the neighborhood is so small that the curvature of the universe's space can be neglected.

[420] Meaning material with much higher mass-energy density and consequently larger internal gravitational attraction.

[421] See Landau (1987).

[422] Although in actuality it is the universe that holds itself together by gravitation.

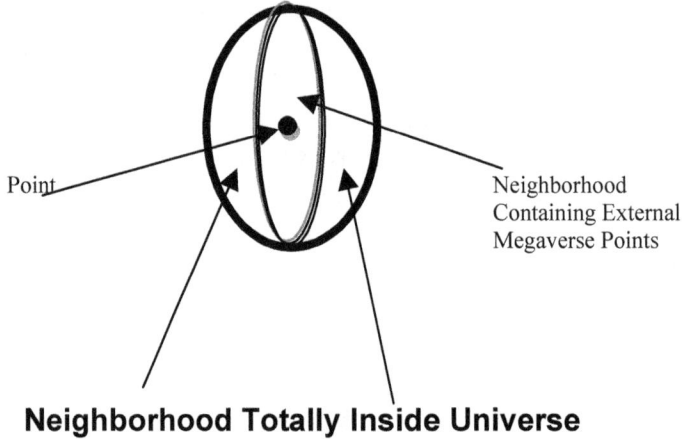

Neighborhood Totally Inside Universe

Figure 65.1. Schematic diagram of a 3-dimensional projection of 'orthogonal' neighborhoods of a point within a universe with one neighborhood strictly within the universe and the other neighborhood containing both universe and external Megaverse points in general. The 'orthogonal' circles around the point differentiate between the two types of neighborhoods.

The surface tension[423] of a universe γ satisfies the relation

$$\gamma = W/\Delta A \tag{65.1}$$

where γ is expressed in erg/cm^2, W is the Work, and ΔA is the Area upon which the work is exerted. The pressure Δp exerted by the surface tension for a 'spherical' surface area is

$$\Delta p = 2\gamma/R \tag{65.2}$$

---

[423] A useful analogy: the Megaverse is a pool of water; a universe is a denser oil bubble within it. Surface tension caused by the cohesiveness of the oil molecules in the bubble makes it spherical (confines it to a spherical shape). Similarly a universe (denser than the Megaverse) is 'confined' within the Megaverse.

where R equals the radius of curvature of the surface. The above equations embody the concept that the surface tension force equals the pressure difference at the surface.

If the universe is flat then the surface pressure approaches ∞ giving confinement of the fields and particles to the universe:

$$R \rightarrow 0 \quad \text{implies } \Delta p \rightarrow \infty \qquad (65.3)$$

Thus we have the theorem:

Theorem: A universe has no leakage of fields or particles into a higher dimensional space if the universe is exactly flat.

This theorem is particularly interesting in the case of our universe. It appears to be flat (or very close to flat). The flatness of our universe may be the reason no leakage of fields or particles from our universe has been detected to high accuracy.

If a universe is found with a non-zero radius of curvature then one can expect that some fields and particles may emerge from it into the Megaverse.

While a zero radius of curvature prevents the exit of fields and particles from a universe, it does not prevent the entry of mass-energy into the universe from the Megaverse. Thus our Continuous Creation Model of chapter 14 in Blaha (2017c) may be relevant. Entry is possible; exit is forbidden in this case.

Eqs. 65.1 – 65.2 are reminiscent of the 'four laws for Black Holes' which are stated later.

65.4 Quantum Fields 'Emanating' from a Universe

If the curvature of the universe is zero (open universe) then no fields emanate from it. If the curvature of the universe is non-zero (closed universe) then fields may 'leak' into the Megaverse. Then continuity conditions between a universe field and its Megaverse counterpart becomes of interest. We discuss this in detail later.

65.5 Universe Confinement by Conservation Laws

Every point in our universe is "infinitely" close to points of the Megaverse.

A universe occupies a region within the Megaverse. However because it is a lower dimension surface within the Megaverse the neighborhood of every point within a universe has an infinite number of Megaverse points that are not within the universe.

One might think that particles within a universe could then 'slip' into the Megaverse outside the universe with ease. However that is not the case. The law of momentum conservation compels particles and interactions within a universe to be confined to the universe. More importantly, the Megaverse surface tension of a flat universe confines particles and fields within a universe.[424]

The only possible ways that a particle could exit from a universe are 1) if the particle collides with a particle with a momentum, some of whose components are in Megaverse dimensions extraneous to the universe's dimensions, or 2) a particle within the universe experiences forces with components in Megaverse dimensions extraneous to the universe. We shall consider the second possibility later when we consider a mechanism for a starship to exit our universe (chapter 34). The first possibility exists if the Megaverse has a very low matter density outside of universes. The fact that this phenomena has not been observed implies the Megaverse matter density is extremely low.

Thus conservation of momentum for particles and interactions, and surface tension, effectively confines particles within a universe even though the neighborhood of every point of a particle's trajectory contains an infinity of Megaverse points exterior to the universe. Similarly every interaction within a universe is confined when expanded in a fourier series (assuming free fields) in universe coordinates.

It is also possible for a Megaverse particle to enter[425] a universe through perhaps a collision that results in the particle being within the universe with momentum also solely within the universe. Thus the 'point boundary' of universes is porous. Particles can enter/exit a universe under appropriate conditions.

---

[424] A useful analogy: the Megaverse is a pool of water; a universe is a denser, oil bubble within it. Surface tension caused by the cohesiveness of the oil molecules in the bubble makes it spherical (confines it to a spherical shape).

[425] In chapter 14 of Blaha (2017c) we consider a model that supplies a mechanism for the Hoyle-Narlikar continuous creation theory for the expansion of our universe. This model 'creates' mass-energy as an inflow from the Megaverse.

We conclude the 'point boundary' of a universe is not a barrier although surface tension force controls the entry/exit of particles and fields. We consider an exit mechanism from a universe later.

65.6 Objects Straddling a Universe-Megaverse Boundary

When a starship, or some other extended object, is entering/exiting a universe at some velocity the question of the state of the object arises It is partially in and partially out of the universe. We know that the object being 4-dimensional will continue to be 4-dimensional, barring effects of forces that might "twist" parts of the object into additional dimensions.

There is also the more subtle quantum effects on the object due to the possibility of different quantizations of the particles in a universe and the Megaverse. Quantizations in different coordinate systems might result in different physical interpretations of matter. We resolve this issue by our PseudoQuantization Method described earlier. In Appendix A we show that one can quantize using a form of PseudoQuantization that preserves (unitarily equivalent) particle interpretations in a universe and the Megaverse.

Thus extended objects can be partly in a universe, and partly in the external Megaverse without issues.

Appendix 65-A. The Shape of Universes and Surface Tension

65-A.1 Shape of a Universe

The shape of a universe[426] either in the case of a Cosmos consisting of one universe or a Megaverse of universes has been a subject of much interest. A universe can be viewed as a 3-surface within a larger space of dimension D > 3.[427]

One interesting possibility, a Poincaré dodecahedral shape, was proposed by J. P. Luminet et al.[428] Experimental evidence from WMAP appears to have ruled out this case.[429]

In this appendix we suggest the surface tension of a universe may make a universe a spherical 3-surface.[430] Earlier we showed that there was an inward directed surface tension force on the surface of a universe due to gravitation between the matter and energy within the universe.

In the absence of other nearby universes, we expect the 3-surface of the universe to be spherical if a uniform distribution of mass-energy in the large is assumed—a common assumption. If there are other nearby universes then the 3-surface of the universe might remain spheroidal but no longer 'perfectly' spherical.

The example of spherical droplets of water in a uniform environment provides a perfect analogy to the effect of surface tension on the shape of a universe. In the case of a water droplet the inward directed surface tension is caused by intermolecular forces between water molecules.

---

[426] See Blaha (2004).

[427] In the case: of D = 4 the map of the 3-surface has been done by R. Lehoucq, J. Weeks, J.-P. Uzan, E. Gausmann, and J. P. Luminet, arXiv:gr-qc/0205009 (2002).

[428] J. P. Luminet, J. R. Weeks, A. Riazuelo, and J.-P. Uzan, Natue **435**, 593 (2003)..

[429] N. J. Cornish, D. N. Spergel, G. D. Starkman, and E. Komatsu, Phys. Rev. Lett. **92**, 201302 (2004).

[430] We note that there appears to be spherical symmetry at the Big Bang point of our universe.

Given a spherical universe one can calculate its surface area and volume. Blaha (2017c), to which the reader is referred, contains estimates of the radius of our expanding universe in various models.

65-A.2 Spheroidal Universes

If a universe has a uniform distribution of galaxies and other mass-energy, as is commonly assumed,, then the universe has a spherical 3-surface in a larger dimensional space such as the Megaverse. We now describe the proof of this assertion.

We note a universe has a well-defined energy in a Megaverse. The energy consists of an internal part $U_{internal}$ and a surface energy part $U_{surface}$.

$$U = U_{internal} + U_{surface}$$

Using the principle of virtual displacements one can calculate surface energy contributions for infinitesimal surface elements δS. We see

$$\delta U_{surface} = -\delta W = \beta \delta S$$

where β is the surface tension, δS is an infinitesimal area element, and δW is the virtual work performed during the virtual displacement. Setting $\delta U_{surface} = 0$ and using Lagrange multipliers one arrives at the equation of a 3-surface sphere.

If the initial assumption of a uniform distribution of matter within the universe does not hold due to the 'nearness' of other universes, then the shape of the universe becomes spheroidal.

66. General Properties of the Megaverse

If one wishes to have a depiction of the Megaverse it seems likely that the universes within it would be scattered in a fashion similar to galaxies within our universe although on a much larger distance scale and in multiple dimensions. In this chapter we overview properties of the Megaverse and its universes. Much of this material previously appeared in Blaha (2017c) and in earlier Physics and starship travel books.

66.1 Megaverse Size, Lifetime, and Universe Separation

The first questions that naturally arise are the age, size, and general of universes within the Megaverse. In the absence of any experimental detail we will assume the relative size of the entities (galaxies) in the universe equals the relative size of entities (universes) in the Megaverse:[431]

(Average Galaxy Size)/(Universe Size) = (Average Universe Size)/(Megaverse Size)

$$(66.1)$$

Taking the average diameter of galaxies to be 400,000 light years, the age of the universe to be 13,800,000,000 light years, and the diameter of the universe to be 91.4 billion light years[432] (the estimated diameter of last scattering surface) we find the diameter of the Megaverse

$$\text{Diameter}_{\text{Megaverse}} = 2 \times 10^{16} \text{ light years} \qquad (66.2)$$
$$= 228{,}500 \times \text{Diameter}_{\text{Universe}}$$

---

[431] We assume the Megaverse is homogeneous at large distance scales. We also assume a proportionality between the average entity size (galaxies in the case of unbiverses; umiverses in the case of the Megaverse) in part based on the relatively long lifetimes of universes and the Megaverse.

[432] There are much larger estimates of the universe's diameter based on the Inflation theory of A. Guth and others.

And, using the 228,500 scale factor, we find the Megaverse age since the Megaverse 'Big Bang' to be

$$\text{Age}_{\text{Megaverse}} = 3 \times 10^{15} \text{ years} \tag{66.3}$$
$$= 3 \text{ million billion years}$$

If the average separation between galaxies in our universe is 3,000,000 light years, then assuming distance scaling by 228,500, the average separation between universes in the Megaverse would be

$$\text{Separation}_{\text{Universes}} = 228{,}500 \times 3{,}000{,}000 \text{ ly} = 7 \times 10^{11} \text{ light years} \tag{66.4}$$
$$= 700 \text{ billion light years}$$

If we now assume the mass of a universe equals the total mass-energy of our universe (including Dark mass and Dark energy) which is estimated to be $m_{\text{universe}} = 3 \times 10^{54}$ kg then the gravitational potential energy between two such universes separated by 700 billion light years is

$$V = G \, m_{\text{universe}}^{2} / \text{Separation}_{\text{Universes}} \tag{66.5}$$
$$= 9 \times 10^{70} \text{ kgm}^2\text{s}^{-2}$$

The gravitational force is

$$F = G \, m_{\text{universe}}^{2} / \text{Separation}_{\text{Universes}}^{2} \tag{66.6}$$
$$= 1.35 \times 10^{43} \text{ kgms}^{-2}$$

and the resulting gravitational acceleration of universes is

$$a = 4.5 \times 10^{-12} \text{ m/s}^2 \tag{66.7}$$

The Baryonic and Leptonic forces associated with the Generation group will slightly modify the force between the universes. We view the small acceleration between universes due to gravity as Physically acceptable. In a billion years the universe velocity would be $v = 1.4 \times 0^5$ m/s with v/c = 0.0005 and the distance traveled equal to 2.2×10^{21}

m = 236,540 light years – negligible compared to the Separation$_{Universes}$. Universes would only make contact after extraordinary long times. Thus we have a coherent view of Megaverse size and distance parameters.

66.2 Likely Features of the Megaverse

There are a number of features of the Megaverse that appear to be true:

66.2.1 Megaverse Curvature

The Megaverse has gravitation. Gravitation appears to be weak in the Megaverse so it is close to a flat space. The sources of Megaverse gravitation are the mass-energy of the universes within it and the density of mass-energy of particles outside of universes. Universes have a larger relative force of gravity due to a higher mass-energy density. Universes therefore have far more curvature.

66.2.2 Megaverse Time Dimension

We will assume that the Megaverse has one complex time dimension denoted y^D for the simple reason that the absence of a time dimension would make the Megaverse static.

66.2.3 Megaverse Forces

In addition to Megaverse gravitation, the Megaverse has the forces in the Unified SuperStandard Theory. These forces satisfy continuity conditions at universe boundaries.

66.2.4 Megaverse Parameters

Megaverse physical constants and particle masses have the same values as in our universe due to continuity.

66.2.5 Megaverse Vacuum Fluctuations

Megaverse Vacuum fluctuations may be a source of the generation of universes and particles. Vacuum fluctuations might account for the Big Bang. The time scale for the persistence of universes generated by a vacuum fluctuation is likely to be an extrapolation of vacuum fluctuation persistence within our universe.

66.2.6 Megaverse Matter and Chemistry

The existence of many more dimensions in the Megaverse suggest that multi-dimensional forms of matter and energy could exist between universes. As a result Megaverse atoms, compounds and Chemistry will be very different and much more varied than in our universe. If such matter exists in the Megaverse then 'mining' such matter for use in our universe—would give us exotic new compounds and Chemistry that would be partially inside, and partially outside, of our universe.

This possibility makes venturing into the Megaverse economically and scientifically desirable since such materials cannot be created within our universe.

66.3 Features of Universes Within the Megaverse

We know of our universe from the 'inside.' However the features of our universe from a Megaverse perspective are not at all certain. In this section we will describe the Megaverse view of a universe's properties.

We shall assume a universe is a closed or open surface within the Megaverse of much higher mass-energy density than the Megaverse's mass-energy density by perhaps as much as a factor of 10^{30}. Chapter 14 of Blaha (2017c) describes a model of Megaverse mass-energy inflow into our universe exemplifying this feature.

66.3.1 Universe Area and Mass

The mass of a universe is an important property since mass is one of the sources of Megaverse gravitation, and interaction between universes.

While a universe is not believed to be a black hole (although Hawking has recently jokingly? suggested that our universe may be a black hole, and even more recently suggested black holes are not quite black holes – grey?), there are general qualitative similarities that lead us to consider the possibility that the four laws of black holes[433] may apply in part (or their entirety) to universes. In particular the 2nd law states

---

[433] Wald, R. M., "The Thermodynamics of Black Holes", *Living Reviews in Relativity* **4** (6): 12119 (2001).

$$dM = \kappa dA/8\pi + \Omega dJ \qquad (66.8)$$

where dM is the change in "mass/energy," A is the area of the Black Hole (universe), Ω is its angular velocity and J is the angular momentum.[434] From eq. 66.8 it appears we can reasonably define a "mass" for a universe in terms of a universe's area:

$$M = \kappa A/8\pi \qquad (66.9)$$

This definition seems to capture the physics of universes that could be used in developing a dynamics of universes as we do later. It allows us to escape the dilemma of having zero total energy for universes that would preclude treating universes as particles in the Megaverse and developing a Megaverse dynamics of universe-particles. Later we will also show how to define a mass for a universe that is time dependent.

66.3.2 Relation Between Universe and Megaverse Vector Fields

Earlier in this book we defined the interactions of The Unified SuperStandard Theory. These interactions and their groups also exist in the Megaverse. The fourier expansions of the fields in the Megaverse are different. They must be expressed in terms of the D coordinates of the Megaverse.

Within a universe, since each point of the universe is surrounded by Megaverse points, a field has both an expression in universe coordinates, x, and an expression in Megaverse coordinates, denoted y. For a vector field in the universe $A_U{}^\mu(x)$ there is an equivalent representation of the field in Megaverse coordinates $A_M{}^i(y)$. These representations are related by a coordinate transformation. If we define a map from universe coordinates to Megaverse coordinates with

$$y^i = f^i(x) \qquad (66.10)$$

then the field representations are related by[435]

---

[434] Although the angular momentum of a universe is not measurable if there is only one universe (as DeWitt argued in a quote earlier), the existence of multiple universes within the Megaverse enables the relative angular momentum of a universe to be determined.

$$A_M^i(y) = \partial y^i/\partial x^\mu \, A_U^\mu(x) \qquad (66.11)$$
$$= \partial y^i/\partial x^\mu \, A_U^\mu(f^{-1}(y))$$

in the domain of the universe. The values of the field at Megaverse points in a neighborhood of a universe point are determined by continuity.

Outside the domain of the universe the Megaverse field value is determined by its sources.

At the boundary of the universe there must be continuity in the expectation values of the Megaverse fields.

66.3.3 Megaverse Gravitation and Free Matter

We assume that a generalization of Einstein's theory of Gravity exists in the Megaverse.

Just as our universe has matter and radiation between galaxies, it seems reasonable to assume that 'free' matter and radiation exists in the Megaverse outside of universes. Such mass-energy would have two roles: to gravitationally affect the dynamics of the Megaverse and the motion of universes within it, and to possibly fuel the expansion of universes. The expansion of our universe may be due to an influx of matter and energy from the external Megaverse. Many years ago Hoyle and Narlikar considered the possibility of 'continuous creation of matter.' We suggest that an influx of Megaverse matter may be the actual source. We consider this possibility in chapter 14 of Blaha (2017c), which contains a paper by this author written approximately seven years ago. (unpublished)

Thus we arrive at a view of the Megaverse of matter and universes that is analogous to our universe of galaxies.

---

[435] Implicit in eq. 33.10 is an inverse relation $x^\mu = f^{-1}(y)$ which is necessarily based on a restriction of the y-coordinates to obtain a 1:1 relation between the y and x coordinates. The restriction is best implemented by requiring the domain of y coordinates be restricted to those y coordinates within the universe surface. The result is a 1:1 relation between the y-domain coordinates and the x universe coordinates.

66.3.4 Expansion of Universes

Our universe expanded from the Big Bang to its current size and is still expanding. It is likely that other universes have undergone similar expansions. According to chapter 32 there is likely an infinite surface tension at the boundary of our universe. How can the universe have expanded, and continue expanding, under such conditions. We see a two phase expansion of the universe.

For a period of time after the Big Bang the universe did not have an infinite surface tension preventing expansion into the Megaverse due to an effect discovered by Eőtvos – a temperature dependence of the surface tension force. Eőtvos pointed out a critical temperature T_c existed that caused the surface tension force to decline as the temperature increased:

$$\gamma V^{2/3} = k(T_c - T) \tag{66.12}$$

where k is the Eőtvos constant, and V is the volume of the universe (the liquid 'drop') Assuming a spherical universe, the volume is

$$V = 4\pi R^3/3. \tag{66.13}$$

Thus

$$\gamma = (4\pi/3)^{-2/3} k(T_c - T)R^{-2} \tag{66.14}$$

For very high temperatures such as existed after the Big Bang $T > T_c$ and thus γ would be negative indicating that there was an outward pressure from the universe into the Megaverse promoting expansion. Thus in the high temperature period after the Big Bang the surface tension force favors expansion of the universe.

After this phase, the surface tension γ is positive. The Megaverse is then superficially 'impeding' expansion. However, the surface tension pressure of the Megaverse causes leakage *into* the universe from the Megaverse causing its mass to increase, and its radius and volume to increase – Expansion! – due to the accretion of Megaverse mass-energy.

The above scenario is supported by the two phase model suggested by section 14.15 consisting of a Big Bang expansion model (chapters 11 – 13), and a mass-energy accretion model (chapter 14) – all of Blaha (2017c). Note as the radius of curvature goes to zero, eq. 66.14 suggests an increasing surface tension pressure 'pushing' particles into the universe.

Thus a complete universe expansion scenario is evident based on surface tension physics.

66.3.5 Universe Generation from Vacuum Fluctuations

Vacuum fluctuations could generate universe-antiuniverse pairs. Antiuniverses would have certain 'negative quantum numbers.' We view universes/antiuniverses, when created (Big Bangs), as 'ultra-small' particles of mass-energy that can proceed to expand to great size. If they are generated by a vacuum fluctuation they will, after possibly a certain time, recombine into the vacuum.

66.3.6 Universes as Black Holes

It is conceivable that a universe could be so dense and confined that it would be effectively a Black Hole. In this case the internal quantum numbers of the Black Hole universe would be inaccessible and only it's mass, velocity and angular momentum would be observables.

66.3.7 Life in Other Universes

It is likely that life exists in some if not all of the universes of the Megaverse. If the physical constants, laws, and masses of the interior of a universe are similar if not identical to ours then the possibility of intelligent species, even human-like species, is very likely. This possibility is an important motivation for humanity to reach for the Megaverse as we discuss later.

66.4 Megaverse Communications

Quantum communication is a rapidly developing field. At this point in time it has only been demonstrated for short ranges. Eventually it should mature as a long distance communication device such as that described in Blaha (2014c). Its great point is its instantaneous nature. Instantaneous communication at very great distances becomes feasible.

Thus one can hope to communicate across the universe. One can also hope for instantaneous communication with ships in the Megaverse. Sending a uniship into the Megaverse should not interfere with quantum communication using a detector at home and another that traversed into the Megaverse on a uniship. The difference in the number of dimensions should not affect the communication between the quantum devices. We expect that quantum states straddling the universe-Megaverse border will not go "out of sync."

Therefore quantum communication offers the possibility of instantaneous communications in our universe, and with uniships/colonies in the Megaverse and other Megaverse universes.

66.4.1 Rapid Interstellar Communication

Recent work on quantum entanglement suggests that instantaneous communication may be possible using this mechanism if an advanced long range form of this laboratory phenomenon can be developed. Quantum entanglement can transcend the borders of universes since it is based on coordinated parts of a quantum state. Its instantaneous nature, which has been verified to great accuracy in recent experiments, makes it the ideal mechanism for communication over trillions of light years. We will describe the application of this concept in more detail in the following sections.

66.4.2 "Instantaneous" Interstellar Communication

Once a uniship capability is achieved it will clearly necessitate a very rapid, if not instantaneous, means of communication. All electromagnetic means of communication are limited by the speed of light and are thus insufficient for multi-million light year communication. If neutrinos are tachyons (faster than light) then they could provide a communications channel except that neutrino detection is very difficult, not reliable, and would require massive detectors that would be an unacceptable addition to the mass of a uniship. More importantly, because neutrinos are extremely light particles their speed is not much more than the speed of light at best.

The only possible method appears to be a quantum entanglement mechanism – currently a subject of intense scientific interest. Based on current thinking about this form of quantum communication it will have the following very desirable features:

1. It is a 1:1 form of communication with no possibility of being intercepted by others.

2. It requires a small amount of power no matter what the distance.

3. It is instantaneous and thus gives direct real time communications over any distance – even trillions of light years.

If history is any guide, the development of inter-universe communications will be similar to the development of telecommunications over the past 150 years, but on a much longer development time scale. Thus we anticipate that it will begin with a primitive Morse code equivalent, and progress eventually to fast digital transmission of images and data. We anticipate bilateral switchboards initially that eventually will lead to communications with uniships beyond our universe in the Megaverse or other universes. Obviously this capability would be needed for exploration – particularly by the initial robot-driven uniships, and for communications between colonies and earth scientific or commercial reasons in other universes.

The basic mechanism will consist of a bilateral quantum entanglement setup that begins as two electrons[436] of opposite spins in a quantum state with total spin zero. Each electron is nudged into a magnetic bottle that does not affect their joint spin state.[437] One bottle is retained on earth; the other bottle is placed in a uniship. As the uniship

---

[436] Protons would be another reasonable alternative.

[437] Several experimental groups have recently been able to detect parts within quantum states without affecting the overall quantum state.

travels the state of the electron spin within its bottle can be periodically sampled but without changing its state.[438] This can also be done on the earth based electron in its bottle. If either electron's spin is flipped the spin of the other electron flips instantaneously no matter what the distance. Thus instantaneous communication of one computer bit takes place.

Eight such bottle pairs allow us by flipping bits to exchange bytes of data. Because of the time contraction associated with much faster than light uniships the byte change must be almost instantaneous for effective communication between a uniship and earth. This fast exchange can be done by ultrafast computers.

Eventually arrays of "bottles" can transmit bytes in bulk in support of large data and image transfer. One can envision electronic switchboards eventually linking arrays of bottles to form a network with a set of uniships and/or colonies. The thought processes and designs are similar to those used in telecommunications.

It is important to note that quantum communications does not require powerful transmitters Thus quantum communications is energy efficient.

66.4.3 Experimental Support for Instantaneous Quantum Entanglement Data Transfer

One might ask if instantaneous quantum data transfer is possible. Both quantum theory and numerous experiments have shown that instantaneous data transfer via entangled pairs works at large distances.[439]

66.4.4 Interstellar Communications and SETI

If our (and others) suggestion that quantum communication is the only reasonable way for communications at large distances, then this might be the reason for SETI's failure to find communications by alien civilizations. Aliens may very well not be communicating by radio or laser waves.

---

[438] 2012 Nobel Prize winner Serge Haroche of France developed ways of detecting the state of particles without disturbing their quantum state.
[439] Matson, John, Quantum Teleportation Achieved Over Record Distances, Nature, 13 August 2012.

It is important to note that quantum communication, as we have proposed it, is inherently private 1:1 communication with no visible manifestations for others to detect of which we are aware.

66.4.5 Quantum Entanglement and the Megaverse

In 1935 Einstein, Podolsky, and Rosen,[440] and shortly afterwards Erwin Schrödinger, began the discussion of what has become known as *quantum entanglement*. It has become a subject of growing, widespread interest since its validity has been demonstrated in numerous experiments.

Quantum entanglement can be described as a phenomenon in which a quantum state of two or more particles evolves to a point where the particles are separated by a distance such that there is no (hitherto) known mechanism by which the particles can communicate. The particles are separated by a space-like distance and thus cannot classically "communicate" at speeds at or below the speed of light. Yet a measurement of a property of one of the particles causes an instantaneous change in the other particles initially entangled with it.

Einstein's unhappiness with quantum entanglement is succinctly expressed with his often quoted description of quantum entanglement, "spooky action at a distance." Yet, since the 1930's, experiments have repeatedly shown that quantum entanglement is correct.

Earlier we showed that quantum entanglement is instantaneous in our universe if we base our SuperStandard Theory on the existence of a particle functional space whose functionals furnish the core of all elementary particles.

Since particle functional space F is not located at any spatial point we can view the universe as a direct product of F with space-time S, and specify the totality of space T as

$$T = S \otimes F \tag{66.15}$$

---

[440] Einstein A, Podolsky B, and Rosen N, "Can Quantum-Mechanical Description of Physical Reality Be Considered Complete?", Phys. Rev. **47**, 777 (1935).

When we consider the Megaverse space M, it is reasonable to generalize space to

$$T_M = M \otimes F \tag{66.16}$$

As we did earlier where T_M is the totality of the Megaverse and particle functional space.[441] (We do not define functionals for universe particles. Each universe particle has a set of functionals for the particles within it.)) Thus the instantaneous nature of Quantum Entangled states also applies to the Megaverse: both to states entirely extraneous to universes as well as states that evolve into parts both within and without a universe.

We conclude that there is no mysterious "spooky action at a distance" in Quantum Entanglement. However the instantaneous nature of Quantum Entanglement extends throughout the Megaverse and its universes.

Returning to our subject: Despite the disappearance of mystery in quantum entanglement, it still remains an exciting topic in view of its potential applications such as communicating instantaneously at interstellar, intergalactic, and Megaverse distances which we have discussed elsewhere.

66.4.6 An Improbable Megaverse Connection to Spirits, the "Spirit World", and UFOs

Ghosts, spirits, and spiritual phenomena have been associated with "the fourth dimension" and other extra dimensions since the 1850s. UFOs are also often described as coming from or through other dimensions. Some scientists have also associated physical phenomena with other dimensions.

We do not believe these characterizations of phenomena as artifacts of other dimensions to be correct in the manner in which they are described.

However, we do think that one can simulate spiritual-like phenomena and UFO-like phenomena using the Megaverse. This section will briefly outline these

---

[441] A person of a philosophic turn of mind might see the ubiquity of functional space throughout the Megaverse as a sign of the unity of all Nature.

possibilities. The hope is to forestall attempts to use our Megaverse theory to bolster support for these phenomena.

66.4.6.1 *"Spiritual" Phenomena and the Megaverse*

Spiritual phenomena have several types: the appearance of visions of people or things of one sort or another; material objects passing through solid objects; unseen voices; and so on. All of these phenomena could be simulated using the Megaverse to make things appear from the Megaverse or to go from within the universe through the Megaverse and reappear in our universe again.

Some examples on Megaverse simulations are:

1. Using a "Sidestep" into the Megaverse to Circumvent Solid Obstacles in our Universe.

2. Going through a Solid One Dimensional Wall. A 1-dimensional example is

------------------|----------------

3. Persons or things going through Solid Three Dimensional Walls.

While these "partly in and partly out of the universe" phenomena can be simulated by a Megaverse agent, the agent's nature and purpose are not determinable from physical considerations of the Megaverse.

66.4.6.2 *UFO Phenomena and the Megaverse*

UFOs have been "seen" in the skies in many regions of the earth. They are often characterized as having high speeds, extremely high accelerations, and the ability to make abrupt changes in direction at high speeds. All of these phenomena are understandable if we are seeing objects in the Megaverse where modest changes in speed, acceleration and direction in Megaverse coordinates can map to the UFO movements that we see in our universe's coordinates.

Beyond stating these phenomena can be understood from a Megaverse perspective we can say no more. They may be real. Their purpose, if they are real, cannot be discerned. Megaverse physics cannot enlighten us on these subjects.

66.5 Traveling and Navigating in the Megaverse

66.5.1 Uniship Baryonic D-Dimensional Observation/Seeing Techniques

If a uniship[442] enters the Megaverse there should be countless universes in view if there is a method for seeing them. However the ability to detect universes is very limited. We cannot detect gravitational waves from universes because their gravity is confined to their universes.

More importantly we cannot detect electromagnetic waves emerging from universes because they are confined to the universe within which they were created.

Consequently, the radiation from univereses that we can expect to encounter in the Megaverse is baryonic or electromagnetic radiation. There are no other potentially known long range types of radiation emitted by universes. This creates a quandary that we can only begin to address at our present state of knowledge. Observing baryonic radiation will be a difficult task – not just because of the weakness of the baryonic field coupling constant but also for several important reasons:

1. We can't see baryonic radiation either visually or through technology at present. No detectors.

2. We cannot focus on the source(s) of baryonic radiation so as to distinguish their direction and distribution.

3. We have no mechanism to magnify the pattern of incoming baryonic radiation. No lenses, telescopes or other viewing mechanisms with "zoom" capabilities. And no technology to amplify baryonic radiation.

---

[442] A starship designed for travel between stars, galaxies and universes.

4. Baryonic radiation is (D – 1)-dimensional. We would have to be able to create 3-dimensional hologram projections that provide an intelligible view. The holographic images would have to be manipulated to enable navigation.

5. Baryonic radiation has (D – 2) polarizations which would provide information on the nature of universes. The detection and analysis of these polarizations is currently beyond our capabilities.

Our inability to detect gravity waves (except in recent studies of the Big Bang – BICEP2) shows the difficulty of detecting and analyzing baryonic radiation.

66.5.2 Relativistic Effects on Radiation

The baryonic view of the Megaverse that a uniship crew "sees", when the uniship is traveling faster than the speed of light, is very different from its view when traveling at low speeds of a few tens of miles per second.

An observer on a uniship traveling at a relativistic speed near, but below, the speed of light will detect universes with baryonic or electromagnetic radiation compressed to within a cone in the frontal direction of the uniship (Fig.34.18). The baryonic radiation cone becomes narrower as the speed of light is approached due to aberration and in the limit as the speed approaches the speed of light becomes a point directly ahead of the uniship.

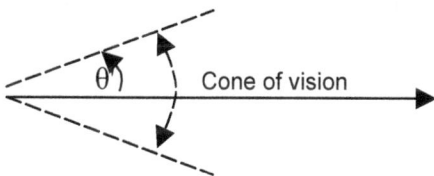

Figure 66.1. Baryonic (and electromagnetic) radiation cone of visibility around direction of uniship motion in the uniship coordinate system with the angle θ' determined by eq. 66.21 for sublight uniship speeds.

The relativistic equation for baryonic radiation aberration is

$$\cos \theta' = (\cos \theta + \beta)/(1 + \beta \cos \theta) \qquad (66.17)$$

where θ is the angle of a universe relative to the uniship's direction of motion as measured in the Megaverse coordinate system and θ' is the angle of a universe relative to the uniship's direction of motion as measured in the uniship's coordinate system.

The inverse relation is

$$\cos \theta = (\cos \theta' - \beta)/(1 - \beta \cos \theta') \qquad (66.18)$$

66.5.2.1.1 SUBLIGHT CASE: B < 1

As $\beta \rightarrow 1$ (the speed of light) eq. 20.1 indicates $\theta' \rightarrow 0°$ showing the entire view of the multi-universe baryonic radiation is compressed to the forward direction. Fig. 34.18 shows the cone of visibility for a uniship traveling near the speed of light at perhaps .6c - .9c. The cone angle θ' satisfies

$$\cos \theta' > \beta \qquad (66.19)$$

The rest of the field of view of the uniship is total baryonic radiation blackness except the point in the directly rearward direction ($\theta' = 180°$) for any object at $\theta = 180°$.

66.5.2.1.2 SUPERLUMINAL CASE: B > 1

For $\beta > 1$ there is a cone of baryonic radiation visibility similar to that depicted in Fig. 66.1. However the cone angle θ' for superluminal speeds, $\beta > 1$, satisfies the relation

$$\cos \theta' > 1/\beta \qquad (66.20)$$

The rest of the field of view of the uniship is total baryonic radiation blackness, as in the sub-light speed case, except the point in the directly rearward direction ($\theta' = 180°$). We note that as β gets very large the cone of visibility becomes larger. At $\beta = \infty$ the cone of visibility becomes the angular region between $\theta' = 0°$ and $\theta' = 90°$ (the forward hemisphere).

66.5.2.1.3 SUPERLUMINAL UNISHIP VISIBILITY

As a result, "visual" baryonic radiation navigation at high superluminal speeds becomes difficult unless we develop an electronic imaging system that "undoes" the effects of aberration and enables "visual" baryonic radiation navigation.

A further problem is the location of a destination. If we send a uniship to a far universe we have to project the location of the universe at the time the uniship arrives based on the universe's current motion. If the motion of the universe is modified by forces exerted by nearby universes as baryonic radiation from the destination universe travels to the uniship, or if the universe's motion is not accurately determined, a uniship could arrive at a point that is still some distance from the destination universe. Thus navigation to a far universe is a significant issue.

66.5.2.1.4 EFFECT OF DOPPLER SHIFT AT SUPERLUMINAL SPEEDS

A uniship traveling at relativistic sublight speeds will see universes having their baryonic radiation spectrum (color) changed significantly due to the Doppler Shift effect. At superluminal speeds the Doppler Shift will also change the "colors" of objects "seen" by the uniship.

This issue is again surmountable if we use electronic imaging techniques to "undo" the Doppler shift and thus display universes with their true baryonic radiation spectrum in the uniship's reference frame.

The relativistic Doppler shift for sublight speeds of a baryonic radiation wave of frequency ν is given by

$$\nu = \nu_0(1 - \beta^2)^{1/2}/(1 - \beta \cos \theta') \tag{66.21}$$

where ν_0 is the frequency of the baryonic radiation emitted by the source universe and θ' is the angle of the source relative to the uniship's velocity.

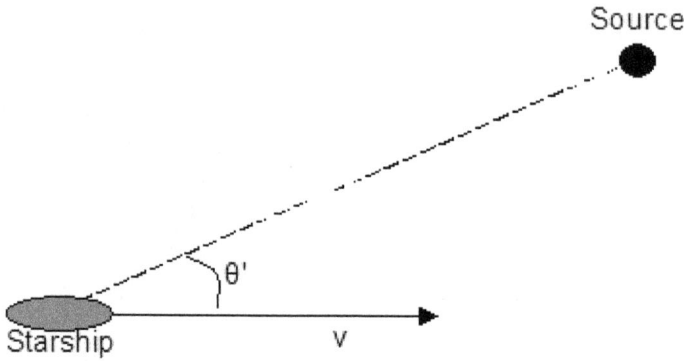

Figure 66.2. The angle of a universe θ' with respect to the uniship's velocity v.

The Doppler shift for superluminal speeds is

$$v = v_0(\beta^2 - 1)^{\frac{1}{2}}/(\beta \cos \theta' - 1) \qquad (66.22)$$

This can be seen by considering a baryonic plane wave, which is a combination of

$$\cos[(k \cdot x - vt)/2\pi] \qquad \text{and} \qquad \sin[(k \cdot x - vt)/2\pi] \qquad (66.23)$$

Upon transforming from the uniship coordinate system, for example, to a coordinate system moving in the "x-direction" at a speed faster than light, both the energy v' (up to a constant) and the time t' obtain a factor of i (that cancel each other) so eq. 66.21 is the correct frequency in the superluminal (faster than light) frame. The sign of the frequency is always positive by convention due to the form of baryonic waves and eq. 66.24 dictates the form of the denominator.

For large $\beta \gg 1$ eq. 66.21 becomes approximately

$$\nu \approx \nu_0/\cos \theta' \qquad (66.24)$$

In the forward direction $\theta' = 0$ the Doppler shift goes to zero. Due to eq. 66.24 the maximum value of the Doppler shift for large β in the field of vision is

$$\nu \approx \beta\nu_0 \qquad (66.25)$$

So the "wide" angle baryonic waves are shifted to large frequency.

The above discussion suggests that frequency shifts will be substantial for extremely fast uniships. The result will be a distorted view of the Megaverse.

However electronic imaging techniques can again be implemented to restore the "correct" view of the baryonic radiation. The combined effects of aberration and the Doppler shift on the view of the Megaverse from the uniship can be electronically corrected to give a "normal" view of the Megaverse. In addition a projection system, probably based on holograms, is required to transform $(D - 1)$-dimensional views of the Megaverse into sets of 3-dimensional depictions of parts of the $(D - 1)$-dimensional view.

66.5.3 Uniship Navigation

Navigating on earth and in space is often a difficult task. First one must know where one is and then one must know where the destination is, and how to get there. In the Megaverse all three items are challenging to discern. For we are in a D-dimensional space where our intuition, based as it is on three dimensional space, fails. We are thus at the mercy of technology to detect these three things with only electronic baryonic eyes to see and guide our motion.

Baryon detectors on board detect other universes through their baryonic radiation just as we detect stars and galaxies by their electromagnetic radiation currently. Within the next 50,000 years we anticipate that baryonic radiation optics will develop and mature to the point where it can provide visual capabilities similar to electromagnetic light that we use at present. Then we can develop universe maps for the Megaverse just as we have star maps currently.

An important issue is the ability to distinguish anti-matter universes from universes dominated by matter such as our universe. We do not want uniships to enter anti-universes unless we can properly shield them from disintegrating under particle-antiparticle interactions.

The navigation system, using 3-dimensional views (holograms) obtained from (D − 1)-dimensional pictures of the Megaverse, can then plot courses to universes of interest for exploration.

The courses selected then direct the (D − 1)-directional thrust system to execute the correct combination of accelerations to travel to the selected universe.

Most of the technology that we have discussed remains to be created. However the rapid progress of technology, if it continues, would seem to be able to provide the needed components in the future.

66.6 Issues for Life on a Megaverse Uniship

Life on a uniship that travels to universes is similar to life on a starship that travels to stars and galaxies except it has significantly more stringent requirements. Travel within the universe is measured in light years ranging to millions of light years. Travel within the Megaverse is most likely in hundreds of billions to trillions of light years in starship time. This results in very large requirements for fuel, materials strength and longevity, and in human suspended animation time, among other requirements.

Starship (uniship) requirements are described in Blaha (2013a), 2014b) and (2014c). In this chapter we will describe some uniship requirements. The following chapters describe other uniship requirements.

66.6.1 Long Distance Starship Requirements for Travel to Far Stars and Galaxies

If we wish to travel to long distances – up to trillions of light years eventually, then critical advances are necessary that could take up to 50,000 years.

We see the uniship effort as a long term exploration and colonization program in an ever widening ring around our universe. In this chapter we discuss many of the advances that would be needed.

A major problem of uniships is the rapid progress of time on a much faster than light uniship. If a uniship has a speed that is much faster than the speed of light, then the progress of time in the uniship is much faster than the progress of time on earth.[443] For example if the uniship is traveling 5,000,000 times the speed of light, then the increase in time on the uniship is 5,000,000 times the increase in time on earth. In an interval of one year of earth time, 5,000,000 years will have passed on the uniship.

The extraordinarily fast passage of time on a very fast uniship requires materials, equipment and engines to continue to work effectively for long periods of uniship time which is just as real on a uniship as earth time is real on earth. One cannot avoid the fact that the distance traveled by a uniship measured in light years is equal to the time of flight to cover that distance measured in years.[444]

Thus we come to the first important long distance uniship requirement – very long lifetime equipment and uniship superstructure. Other requirements follow in this chapter.

66.6.2 Long-Lived Materials

The materials that we use today to build large vehicles such as oil tankers, submarines and aircraft carriers are meant to last up to, at most, a century and often much less. Many of these materials age, deteriorate, rust, migrate within computer chips over time, and actually slowly flow like a liquid in many cases.

Not many materials keep their original characteristics over long periods of time. In the past fifty years there has been much progress in developing new harder, stronger, age-resistant materials and metals. But uniships requirements are extraordinarily larger.

Uniships, in which time moves quickly so that thousands and perhaps millions of years of uniship time elapse, must be composed of materials with a very long stable lifetime. An important part of the R&D for a uniship is the development and use of age tolerant materials. The examination of materials used hundreds of thousands of years ago such as tools and dwellings shows the ravages of age. A uniship should have an initial goal of tens of millions of years of stability without aging. Ultimately one would hope that uniships that don't age in trillions of years could be built to travel to other universes.

---

[443] See p. 15 of Blaha (2011c): *All The Universe.*

[444] Neglecting the time required to accelerate the starship at the beginning and the time required to decelerate back to a "normal" speed of a few miles per second.

These design requirements are far beyond current technology.

66.6.3 Long-Lived Machinery and Electronics

If one has materials that preserve their composition, shape and performance characteristics over millions of years or more, then one can construct machinery and electronic gear such as computers that can last similar periods of time. Long-lived machinery and electronic gear then can be used when a uniship travels the Megaverse.

Long distance uniships need materials and machinery that last "nearly forever" – exactly the opposite of the intent of Earth industries.

66.6.4 Long Shelf Life Nuclear Reactors and Nuclear Shuttles

Several types of nuclear reactors are required for a uniship:

1. A continuous running reactor that can run for up to hundreds of millions of years to provide power to a uniship in flight to a distant location. This reactor may be a low power reactor. It should have a very long lifetime. That this is possible is suggested by the natural nuclear reactor that ran for millions of years about a billion years ago in the Congo.[445]
2. Long shelf life reactors that are not activated until a uniship destination is reached. These would power the uniship inside universes and their solar systems, and nuclear shuttles for travel and landings within a solar system.

66.6.5 Suspended Animation for Long Trips

It is necessary to have suspended animation available for crews on uniship journeys. With suspended animation a crew could go on a journey lasting hundreds of millions of years, or more, of uniship time, and, upon return to earth, have aged physiologically only a short time while out of suspended animation exploring distant universes and their star systems. The round trip travel time will not have aged them.

---

[445] The author suspects that the vast diversity of life in Africa may in part be due to genetic changes caused by this natural reactor. The development of mammalian life may in part also be due to this reactor and the radiation from its waste products and radioactive deposits in the Congo region over the millennia.

When they return to earth they may be some months older, but their families and friends (having aged only by the earth travel time) will still be roughly contemporary with them.

A mechanism for long term suspended animation is thus a major requirement. Any suspended animation mechanism must take account of three important facts: 1) suspended animation must reduce human body temperatures to a low value to "halt" life processes and bodily decay; 2) lowering body temperatures will cause cells to rupture due to the expansion of water upon freezing; 3) the entry into suspended animation and the reentry to a normal bodily state must be rapid and uniform throughout the body.[446]

A mechanism to achieve these goals is not presently known. The current approaches to suspended animation (which all include lowering body temperature) are:

1. Replacing part or all of the blood in an organism with an "antifreeze" solution that will prevent cells and body tissue from bursting when the temperature is lowered. Revival takes place by raising the temperature of the organism while returning blood to the organism's circulatory system. This approach has been successfully applied to dogs that have been put into suspended animation for three hours. Unfortunately some of the dogs had nerve and coordination problems after revival.[447]

2. An organism can have a chemical injected or absorb a chemical while breathing that will counteract the tendency of water to expand when body temperature is lowered and/or lower the metabolic rate of the organism.

3. NASA and other groups have studied the possibility of placing humans into hibernation. Since hibernating organisms do age – perhaps more slowly – this approach is not true suspended animation.

4. A combination of electromagnetic "vibration" of a body having an innocuous chemical dispersed in the body (while awake or in suspended animation) might

---

[446] One cannot "unfreeze" part of a human body and have the rest still frozen.
[447] At the University of Pittsburgh's Safar Center for Resuscitation Research.

allow bodily temperatures to be lowered to a stable "frozen" state without cell rupturing. Turning off the electromagnetic vibration combined with a revival jolt might be an effective way to exit suspended animation procedure.

66.6.6 Robotic Driven Uniships

The initial uniship flights could be manned by robots rather than humans. This approach would be useful to test uniships, and their components, without endangering a crew. The robot guidance systems would, of course, have to be constructed of long-lived components. If it is successful then a robotic trip would also help demonstrate the long term reliability of long-lived computer equipment.

Robotic flights would be especially useful if a method of rapid faster than light, or instantaneous, communication between the uniship and earth existed.

66.6.7 Long-Life Computer Chips

Computers hardened for battle and bad weather conditions currently exist. A long distance uniship would require computers with working lifetimes of between thousands, and hundreds of millions, of years. In time periods of these lengths computer chips would be subject to aging processes such as the intermixing of the metals composing the various chips of the computer and the aging of the wiring of the computer. Since new materials of greater strength and other superior properties are being discovered fairly frequently one can hope that the required types of metals and materials will eventually be found.

66.6.8 Space Dust

The effect of dust and gas molecules in space on uniships are of great importance. These effect should be detectable in "short" distance uniship voyages. If it is important, as it seems to be, then the design of shielding for long distance uniships should incorporate appropriate "armor" to protect the uniship and crew.

66.6.9 Length Dilation Effect

Lengths on a uniship, traveling at high speed much greater than the speed of light, are significantly dilated. A length measured on a uniship will appear to be larger to an observer on earth by a factor of the speed measured in terms of the speed of light than the length on earth. For example if a uniship is moving at 5,000 times the speed of light then a 2 meter long stick on the uniship would appear to be 10,000 meters long to an earth observer.[448]

Does this length dilation phenomenon affect the contents of the speeding uniship? No. It is an illusion that the earth observer "sees." An occupant of the uniship would not notice a difference and would see the stick as still two meters in length. Due to length dilation, a starship traveling at speeds much beyond the speed of light, would appear to be enormously long.

66.6.10 QFT Acceleration Thermal Vacuum Heating – Particle Baths

Recently questions have been raised about spaceships accelerating at a high rate in 'empty' space. Based on an analysis of Unruh and others it has been suggested that the ship would see an incoming wave of particles and a 'heated' vacuum due to its motion. The basis of this claim and the work of Unruh (and others) is standard quantum field theory, which in turn is based on Special Relativity.

Since accelerating reference frame transformations and other exotic reference frame transformations are outside the framework of Special Relativity we suggest that this spaceship phenomena merely reflects a flaw in the choice of second quantization. Earlier we described a more general formulation of quantum field theory in which particle states are unchanged under transformations between accelerating reference systems as well as under transformations to exotic coordinate systems. Thus the particle 'baths' suggested by conventional quantum field theory do not take place for accelerating spaceships nor does the vacuum 'heat up' and spew particles.

---

[448] This discussion assumes that the stick and the starship motion are parallel. For a detailed discussion of this length contraction phenomena see Blaha (2011c) p. 17.

66.7 Uniship Particle Annihilation Drive

This section (from Blaha (2011a)) epitomizes the possibility that the CERN LHC can be used as a *test instrument for the development* of a starship ion drive capable of travel to the stars (after the LHC is retired from its investigation of elementary particle phenomena).

We have shown in Blaha (2017f) that quarks and gluons have complex-valued velocities that enable them to travel faster-than-light (tachyonic), which we showed enables quarks to generate more massive particles after collisions than the initial particles. Recently, new experiments (announced in April, 2017[449]) at the LHC have shown that ultra-high energy proton-proton collisions have produced an abundance of particles similar to that produced in collisions of atomic nuclei. Thus initial quarks in protons effectively 'multiply' to have atomic nuclei characteristics – a feature we attribute to their tachyonic nature.

The tachyonic attributes that seem to explain these new results suggest that a faster-than-light ion starship engine could be created. CERN LHC seems able to do design studies to aid in the development of this new type of starship engine at now accessible energies.

66.8 Voyages into the Megaverse

Traveling in in the Megaverse between universes places extraordinary demands on uniships. Speed, acceleration, and fuel requirements are the primary issues because the distance between universes is vast and human lifetimes are short. We should like to be able to travel fairly quickly between universes. If one wishes colonization, commerce, and timely exploration then the trip between two universes, on average, should be perhaps six months to a year of earth time. (Time on a uniship proceeds much, much more rapidly than earth time. But we can circumvent this potential problem using suspended animation for passengers and very long-live machinery for the uniship and its contents.)

In this section we will consider Uniship distance and speed related requirements for "short" distance travel in the Megaverse.

---

[449] ALICE Collaboration, reported in the journal Nature Physics (April, 2017).

66.8.1 Starship Distance and Speed Requirements in Our Universe

In our book *All the Universe!* (and in earlier books) we developed the theory of faster-than-light starships for travel between stars and galaxies in our universe. For the reader's convenience we reproduce part of *All the Universe!* It appears that speeds of 60,000c give acceptable travel times (up to a year) within our galaxy, and *speeds of a few million c give acceptable travel times to nearby universes where c denotes the speed of light.*

At 3,000,000,000c a universe that is 2,000,000,000 light years away could be reached in eight months (neglecting acceleration and deceleration times) – an acceptable time for trade, exploration and possibly colonization. The fundamental problem is to develop an energy source that can fuel such enormous speeds – a reason for anticipating a 50,000 years of necessary development time.

Destination	Distance (ly)	Approximate Travel Time (years)
To the other end of the Milky Way Galaxy	100,000	3
To the Center of the Milky Way	30,000	1
Large Magellenic Galaxy	150,000	5
Small Magellenic Galaxy	200,000	7
Andromeda Galaxy	2,000,000	70

Figure 34.21. "Coasting" part of travel time to various destinations at a real velocity of 30,000c.

66.8.2 Uniship Distance and Speed Requirements for "Short" Distances in the Megaverse

In this section we will consider some issues associated with uniship "short distance" trips in the Megaverse such as a three trillion light year trip to a nearby universe.[450] The primary issue is the energy (and fuel) required to make a trip at high speed so that the travel time is of the order of months. Another important issue is the acceleration times that are required to attain high speeds.

66.8.2.1 Energy Required to Attain High Velocities

The energy E required to reach a high velocity much greater than the speed of light is not a simple mathematical expression in the speed v because the energy of an object moving much faster than the speed of light approaches zero. (The momentum approaches mc where m is the mass of the object. Thus $E^2 - c^2p^2 = -m^2c^4$ for v > c. The object is tachyonic.)

We will calculate the energy required to reach a speed v > c in the earth's rest frame using the development presented earlier in this book. The energy E expended in the spatial x interval from x_0 to x is defined as

$$E = \int_{x_0}^{x} F dx = \int_{t_0}^{t} F v dt \tag{66.26}$$

where t_0 is the initial time and t is the final time. From eq. 66.26 we obtain

$$E = \int_{t_0}^{t} F v dt = \int_{t_0}^{t} g \gamma v dt \tag{66.27}$$

Eq. 66.27 can be transformed to the form:

$$E = (m/2) \int_{v_0}^{v} dv\, \gamma^{-1}\, d[(\gamma v)^2]/dv \tag{66.28}$$

The exact form of the integral is:

$$E = [m\gamma(v^2 + c^2)/2 + mc^2/(2\gamma)] - [m\gamma_0(v_0^2 + c^2)/2 + mc^2/(2\gamma_0)] \tag{66.29}$$

---

[450] We will use the results found earlier for acceleration and velocity in the x direction. The general case follows directly from this special case.

where $\gamma_0 = (1 - v_0^2/c^2)^{-1/2}$. Since the velocity is in general complex-valued, v and γ are also complex-valued as is E.

Using the Reality group that we introduced in previous books, the physical velocity is the absolute value of v, |v|, and the physical value of E is |E|.

We define

$$E_{part}(v) = m\gamma(v^2 + c^2)/2 + mc^2/(2\gamma) \tag{66.30}$$

The integral in eq. 66.27, although superficially real-valued has complex quantities in the integrand. The result of the integration from a speed below the speed of light to a speed greater than the speed of light can be written as

$$E_{tot} = E_{part}(c - i\varepsilon) - E_{part}(v_0) + E_{part}(v) - E_{part}(c + i\varepsilon) \tag{66.31}$$

as $\varepsilon \to 0$ due to the singularity at v = c. Since we are interested in the energy required to reach ultra-high speeds of the order of tens of thousands to trillions of times the speed of light, we see that

$$E_{tot} \to \lim_{v \to \infty} E_{part}(v) = -imcv/2 \tag{66.32}$$

to leading order in v. Thus

$$|E_{tot}| \to mc|v|/2 \tag{66.33}$$

or, expressed in a more convenient way:

$$|E_{tot}| \to \tfrac{1}{2} E_{rest} |v|/c \tag{66.34}$$

as $v \to \infty$ in the earth's reference frame where $E_{rest} = mc^2$ is the rest mass-energy of the entire ship including the fuel. Knowing the desired cruising speed v of a starship or uniship the ship must have an energy supply equal to $2|E_{tot}| = mc|v|$ when it leaves the earth's reference frame to provide for acceleration *and deceleration* plus the additional energy needed for other ship functions. For a round trip an energy supply of $4|E_{tot}| = 2mc|v|$ would be needed for the ship's engines.

66.8.2.2 Acceleration Time Required to Reach Extremely High Speed

We can determine the acceleration time required to reach velocity v using the previous section:

$$v = c\{1 - 2/(1 + ((c + v_0)/(c - v_0))\exp[2g(t - t_0)/(mc)])\} \qquad (66.35)$$

Inverting eq. 66.35 above yields the acceleration time interval required to achieve a speed v in the earth's reference frame:

$$t(v) - t_0 = (mc/2g)\ln\{[(c - v_0)/(c + v_0)]\,[(c + v)/(c - v)]\} \qquad (66.36)$$

where v_0 is the speed at t_0. For large v in the earth reference frame the time interval approaches

$$|t(v) - t_0| \rightarrow (mc/2g)\ln[(c - v_0)/(c + v_0)] + mc^2/(gv) \qquad (66.37)$$

to leading order in v. As $v \rightarrow \infty$ the interval becomes a constant:

$$|t(v) - t_0| \rightarrow (mc/2g)\ln[(c - v_0)/(c + v_0)] \qquad (66.38)$$

The reason for this limit on the earth reference frame time interval can be understood when one realizes that the corresponding ship time interval approaches infinity. Ship time increases much faster at high speeds greater than $\sqrt{2}c$. We shall see this in the next subsection.

66.8.2.3 Speed, Distance and Acceleration Time in the Starship/Uniship Reference Frame

In this subsection we will calculate dynamical quantities from the point of view of the starship/uniship reference frame.

In case examined in the previous section we found the acceleration in the earth reference frame to be $\gamma g/m$. In the ship reference frame the acceleration is g/m.

The transformation law from the earth reference frame (unprimed coordinates) to the starship/uniship reference frame (primed coordinates) in which the ship is moving at instantaneous speed v in the x direction is

$$t' = \gamma(t - \beta x/c) \tag{66.39}$$
$$x' = \gamma(x - \beta ct)$$
$$y' = y$$
$$z' = z$$

Substituting

$$x = x_0 + (mc^2/g)\ln[(1 - v_0/c)/(1 - v/c)] - c(t - t_0) \tag{66.40}$$

from the previous section and

$$t(v) - t_0 = (mc/2g)\ln\{[(c - v_0)/(c + v_0)] \ [(c + v)/(c - v)]\} \tag{66.41}$$

from this section with $t_0 = 0$ and $x_0 = 0$, we obtain the ship time to be

$$t' = \gamma\{(m(c - v)/2g)\ln\{[(c - v_0)/(c + v_0)] \ [(c + v)/(c - v)]\} - (mv/g)\ln[(c - v_0)/(c - v)]\} \tag{66.42}$$

Eq. 66.42 gives starship time as a function of speed. Thus the faster a ship speeds the more quickly time passes on the ship. As a result ships should have long lifetime equipment and place passengers in suspended animation for long periods of time.

For large v, the absolute value of the ship time approaches

$$|t'| \rightarrow |mc/g)\ln v| \tag{66.43}$$

Thus the ship time approaches infinity as the ship speed approaches infinity. This situation explains the limit on earth time above. It corresponds to ship time approaching infinity. (The ship speed approaches infinity in this limit as well.) Of course the limit is never reached because it would require an infinite amount of thrust and fuel. However it allows very, very large velocities to be reached making starship and uniship travel in reasonable time frames possible.

If the conditions in the above section hold, then it is possible to reach "infinite" speed in a finite time in the earth's reference frame. To reach this limiting speed will require an infinite amount of starship time, and fuel.

67. Unified SuperStandard Theory in the Megaverse

Our view of the Megaverse consists of a vast number of universes in a much vaster space. The space (outside of universes) consists of a very low density of radiation, particles and dust (perhaps accreted into bodies of matter.) Our models[451] suggest that the Megaverse mass-energy density between universes is perhaps 10^{-30} of the average density in our universe.

67.1 Particle Interactions in the Space Between Universes

We assume the interactions between particles in the Megaverse space between universes are described by the Unified SuperStandard Theory, suitably generalized to D dimensions, that was presented earlier in this book. We further assume, based on continuity considerations, that coupling constants, masses and other parameters of the Unified SuperStandard Theory have the same values in the Megaverse as they do in our universe.

Fermion and boson free fields would have fourier expansions in D-dimensional Megaverse coordinates. Using Two-Tier Megaverse coordinates all Megaverse perturbation theory calculations are convergent. There are no infinities in Two-Tier calculations irrespective of the space-time dimension.

67.2 Example of a Megaverse Baryonic Gauge Field – The Planckton Field

The conservation of baryon number has been repeatedly investigated by experimenters and found to be true to extremely high accuracy. For decades theorists have suggested that the conservation law follows from the existence of a gauge field in a manner much

---

[451] See chapters 11 – 14 of Blaha (2017c).

like electric charge conservation follows from the properties of the electromagnetic abelian gauge field.[452]

We will therefore assume a representative baryonic gauge field exists that is similar to the electromagnetic field except for features due to its definition and existence in the D-dimensional Megaverse. This field will couple extremely weakly to individual baryons as well as universe particles with non-zero baryon number. We will call the baryonic gauge field particle a *planckton*. Its electromagnetic analogue is the photon. We described Megaverse baryonic gauge field quantization in Blaha (2014a), (2015a) and earlier books.

Plancktons propagate in the Megaverse, both within universes, and in the Megaverse external to universes. So we will define the planckton field in D-dimensional Megaverse coordinates. They will interact with baryons within a universe with Megaverse coordinates mapped to the curvilinear coordinates in the universe.

Since a planckton field in D-dimensional quantization in conventional coordinates would lead to divergences we will use quantum coordinates:

$$Y^i(y) = y^i + i \, Y_u^{\ i}(y)/M_u^{D/2} \tag{67.1}$$

with quantum coordinate derivatives defined by

$$\partial_i = \partial/\partial Y^i(y) = \partial/\partial(y^i - Y_u^{\ i}(y)/M_u^{D/2}) \tag{67.2}$$

to obtain a completely finite theory of planckton interactions with elementary particles and later with universe particles.

Plancktons, fermion particle fields, gravitation, other gauge fields, Higgs fields, and the $Y_u^{\ i}(y)$ field of quantum coordinates are the only fields in the space between universes in the Megaverse. Universe.

---

[452] See Gell-Mann, M. and Levy, M. *Nuovo Cimento* 16, 705 (1960) for a proof.

67.3 Planckton Second Quantization

We begin by noting that Megaverse quantum coordinates are defined by eqs. 67.1 and 67.2 above. The lagrangian density terms for the free planckton $B_u^i(Y(y))$ fields is

$$\mathscr{L}_{Bu} = -\frac{1}{4}\, F_{Bu}{}^{\mu\nu}(Y(y))F_{Bu_{\mu\nu}}(Y(y)) \tag{67.3}$$

with $Y(y)$ given by eq. 67.1 and the free planckton lagrangian is

$$L_{Bu} = \int d^{D-1}y\, \mathscr{L}_{Bu}(Y(y)) \tag{67.4}$$

with

$$F_{Bu_{\mu\nu}} = \partial B_{u_\mu}(Y(y))/\partial Y^\nu(y) - \partial B_{u_\nu}(Y(y))/\partial Y^\mu(y) \tag{67.5}$$

where the values of μ and ν range from 1 to D.

The equal time commutation relations, derived in the usual way, are:

$$[B_u^\mu(Y(\mathbf{y}, y^0)), B_u^\nu(Y(\mathbf{y}', y^0))] = [\pi_u^\mu(Y(\mathbf{y}, y^0)), \pi_u^\nu(Y(\mathbf{y}', y^0))] = 0 \tag{67.5a}$$
$$[\pi_{uj}(Y(\mathbf{y}, y^0)), B_{uk}(Y(\mathbf{y}', y^0))] = -i\,\delta^{D-1\,tr}{}_{jk}(Y(\mathbf{y},0) - Y(\mathbf{y}',0)) \tag{67.5b}$$

where

$$\pi_u{}^k = \partial\mathscr{L}_{Bu}(B_u(Y(y)))/\partial B_{uk}'(Y(y)) \tag{67.6}$$
$$\pi_u{}^0 = 0 \tag{67.7}$$

and

$$\delta^{D-1\,tr}{}_{jk}(\mathbf{y} - \mathbf{y}') = \int d^{D-1}k\; e^{i\,\mathbf{k}\bullet(Y(y,0)\,-\,Y(y',0))}\,(\delta_{jk} - k_j k_k/\mathbf{k}^2)/(2\pi)^{D-1} \tag{67.8}$$

$$B_{uk}'(Y(y)) = \partial B_{uk}(Y(y))/\partial y^D \tag{67.9}$$

for $j, k = 1, 2, \ldots, D-1$.

If we choose the Coulomb gauge for $B_{uk}(Y(y))$:

$$B_u{}^{16}(Y(y)) = 0$$

$$\partial B_u^{\ j}(Y(y))/\partial Y^j(y) = 0$$

for $j = 1, 2, \ldots, D - 1$ then $D - 2$ degrees of freedom are present in the vector potential.[453] The Fourier expansion of the vector potential $B_u^{\ i}(Y(y))$ is:[454]

$$B_u^{\ i}(Y(y)) = \int d^{D-1}k \ N_{0B}(k) \sum_{\lambda=1}^{D-2} \varepsilon^i(k, \lambda)[a_B(k,\lambda) :e^{-ik\cdot Y(y)}: + a_B^{\dagger}(k,\lambda) :e^{ik\cdot Y(y)}:] \quad (67.10)$$

for $i = 1, \ldots, D - 1$ where

$$N_{0B}(k) = [(2\pi)^{D-1} 2\omega_k]^{-\frac{1}{2}} \quad (67.11)$$

and (since the field is massless)

$$k^D = \omega_k = (\mathbf{k}^2)^{\frac{1}{2}} \quad (67.12)$$

where k^D is the energy, and where the $\varepsilon^i(k, \lambda)$ are the polarization unit vectors for $\lambda = 1, \ldots, D - 2$ and $k^\mu k_\mu = (k^D)^2 - \mathbf{k}^2 = 0$.

The commutation relations of the Fourier coefficient operators are:

$$[a_B(k,\lambda), a_B^{\dagger}(k',\lambda')] = \delta_{\lambda\lambda'}\delta^{D-1}(\mathbf{k} - \mathbf{k}') \quad (67.13)$$
$$[a_B^{\dagger}(k,\lambda), a_B^{\dagger}(k',\lambda')] = [a_B(k,\lambda), a_B(k',\lambda')] = 0 \quad (67.14)$$

and the polarization vectors satisfy

$$\sum_{\lambda=1}^{D-2} \varepsilon_i(k, \lambda)\varepsilon_j(k, \lambda) = (\delta_{ij} - k_i k_j/\mathbf{k}^2) \quad (67.15)$$

The $B_u^{\ \mu}$ Feynman propagator is

$$iD_F^{\text{trTT}}(y_1 - y_2)_{jk} = <0|T(B_{uj}(Y(y_1))B_{uk}(Y(y_2)))|0> \quad (67.16)$$

---

[453] Note we use the Coulomb gauge for $Y(y)$ also.

[454] Normal ordering of the exponentials is required.to avoid spurious infinities. It is analogous to the normal ordering seen in perturbation theory. See Blaha (2005a) for additional details.

$$= -\,ig_{jk} \int \frac{d^D k \; e^{-ik\cdot(y_1 - y_2)} \, R(\mathbf{k}, y_1 - y_2)}{(2\pi)^D \,(k^2 + i\varepsilon)} \qquad (67.17)$$

where g_{jk} is the D-dimensional Lorentz metric and where $R(\mathbf{k}, y_1 - y_2)$ is given by

$$\begin{aligned} R(\mathbf{k}, y_1 - y_2) &= \exp[\,-k^i k^j \Delta_{Tij}(y_1 - y_2)/M_u^{\,D}\,] \qquad (67.18) \\ &= \exp\{\,-k^2[A(v) + B(v)\cos^2\theta] \,/\, [(2\pi)^{D-2} M_u^{\,4} z^2]\,\} \end{aligned}$$

with

$$z^\mu = y_1^{\,\mu} - y_2^{\,\mu}$$
$$z = |\mathbf{z}| = |\mathbf{y_1} - \mathbf{y_2}|$$
$$k = |\mathbf{k}|$$
$$v = |z^0|/z$$
$$A(v) = (1 - v^2)^{-1} + .5v \ln[(v - 1)/(v + 1)]$$
$$B(v) = v^2(1 - v^2)^{-1} - 1.5v \ln[(v - 1)/(v + 1)]$$
$$\mathbf{k}^\cdot\mathbf{z} = kz \cos\theta$$

where $|\mathbf{k}|$ denoting the length of a spatial $(D - 1)$-vector $\mathbf{k}$ while $|z^0|$ is the absolute value of $z^0 \equiv z^D$. Thus the k^2 factor in the exponential damps all spatial components.

As eq. 67.18 indicates, the Gaussian damping factor $R(k, z)$ for all large spatial momentum k^j is the same for both the positive and negative frequency parts of the (two-tier) B_u Feynman propagator. We are assuming the spatial momentum is real-valued in this discussion. It is also important to note that $R(k, z)$ does not depend on $k^0 = k^D$ (in the B_u and Y_u Coulomb gauges) and thus the integration over k^0 proceeds in the usual way to produce time-ordered positive and negative frequency parts.

The Gaussian exponential factor in eqs. 67.17 – 67.18 causes the Feynman propagator to be finite and, together with the Gaussian factor in universe particle propagators, causes all perturbation theory calculations when interactions are introduced to be finite as we have seen earlier in The Unified SuperStandard Theory.

For small momentum much less than M_u then $R(\mathbf{k}, y_1 - y_2) \to 1$ and the Feynman propagator is the "normal" propagator of conventional D-dimensional quantum field theory. For large momentum the corresponding potential approaches $r^{(D-3)}$ due to the k^2/z^2 factor in the exponential in eq. 67.18.[455] Thus the B_u potential is highly non-singular at large energies.

67.4 Planckton Interactions with Universe Particles and Individual Baryons

Section 15.1 of Blaha (2015a) describes the second quantization of plancktons in 16 dimensions. In this section we will develop an interacting theory in D dimensions of universe particles and plancktons from a model lagrangian of universe particles and plancktons using quantum coordinates. The universe particle – planckton lagrangian is:

$$\mathcal{L} = \overline{\psi}(Y(y))[i\gamma^\mu \partial/\partial y^\mu - e_B\gamma^\mu B_{u\mu}(Y(y)) - m(t)]\psi(Y(y)) - \tfrac{1}{4} F_{Bu}^{\ \mu\nu}(Y(y))F_{Bu\mu\nu}(Y(y)) -$$
$$- \tfrac{1}{4} F_u^{\ \mu\nu}(y)F_{u\mu\nu}(y) \tag{67.19}$$

where $\mu, \nu = 1, 2, \ldots, D$ and where

$$\overline{\psi} = \psi^\dagger \gamma^D$$

$$F_{Bu\mu\nu} = \partial B_{u\mu}(Y(y))/\partial Y^\nu(y) - \partial B_{u\nu}(Y(y))/\partial Y^\mu(y) \tag{67.20}$$
$$F_{u\mu\nu} = \partial Y_\mu/\partial y^\nu - \partial Y_\nu/\partial y^\mu$$
$$Y^i(y) = y^i + i\, Y_u^{\ i}(y)/M_u^{D/2}$$

$$e_B = e_{B0}/M_u^{D/2 - 2}$$

with e_{B0} a dimensionless coupling constant, and with μ and ν ranging from 1 through D.

The corresponding lagrangian is

---

[455] See eqs. 4.11 – 4.24 in Blaha (2005a). Blaha (2005a) has a complete discussion of Two-Tier perturbation theory that directly generalizes to the Megaverse.

$$L = \int d^{(D-1)}y \, \mathscr{L} \tag{67.21}$$

Note the dimensions of the fields in the D dimensional space are:

$$
\begin{aligned}
Y^\mu &\sim [\text{mass}]^{D/2-1} \\
B_{u_\mu} &\sim [\text{mass}]^{D/2-1} \\
\psi &\sim [\text{mass}]^{(D-1)/2}
\end{aligned}
\tag{67.22}
$$

as can be seen from the above lagrangian as well as earlier equations. Note also that the mass and thus the size of universe particles is time dependent in general. They can expand or contract with time depending on their internal characteristics (gravitation and effects of elementary particle interactions) which are not embodied in this lagrangian. As a result, this model theory does not conserve energy unless m(t) is constant.

The lagrangian generates baryonic interactions of universe particles using Two–Tier quantum coordinates which prevent infinities in perturbation theory calculations.

The interaction of baryon elementary particles with the baryonic field requires terms in Unified SuperStandard Theory covariant derivatives specifying the baryon field interaction baryons with the form

$$e_B \gamma^\mu B_{u_\mu}(Y(y)) \tag{67.23}$$

67.5 Megaverse Free Scalar Fields

In this section we describe a Megaverse scalar field $\phi(Y)$ (spin 0) which could be viewed as a prototype for a Higgs scalar field. We will use Two-Ticr coordinates defined by eqs. 67.1 and 67.2. The scalar $\phi(Y)$ lagrangian density terms are

$$\mathscr{L}_F = \tfrac{1}{2}[\, (\partial\phi/\partial Y)^2 - m^2\phi^2 \,] \tag{67.24}$$

The free field fourier expansion is

$$\phi(Y) = \int d^{D-1}p\, N_m(p)\, [a(p):e^{-ip \cdot Y}: + a^\dagger(p):e^{ip \cdot Y}:] \qquad (67.25)$$

$$= \int d^{D-1}p\, N_m(p)\, [a(p):e^{-ip \cdot (y + iY/M_u^2)}: + a^\dagger(p):e^{ip \cdot (y + iY/M_u^2)}:]$$

We note the equal time commutation relations of ϕ and the conjugate momentum π_ϕ are the same as the conventional equal time commutation relations of a scalar field despite the fact that Y^μ is itself a quantum field since $[Y^\mu(\mathbf{y}, y^0), Y^\nu(\mathbf{y'}, y^0)] = 0$ for $\mathbf{y} \neq \mathbf{y'}$. In addition, we note the ϕ and π_ϕ fields are not hermitean.

The Fourier expansion of ϕ in eq. 67.25 does require one refinement – the exponential terms in X^μ must be *normal ordered* to avoid infinities in the unequal time commutation relations:

Since the Hamiltonian as well as other quantities are normal ordered in quantum field theory the additional requirement of normal ordering in the field operator is merely an extension of a standard procedure to a more complex situation and is not disturbing. The unequal time commutation relation of the normal ordered ϕ field is:

$$[\phi(Y^\mu(y_1)), \phi(Y^\mu(y_2))] = i\Delta(y_1 - y_2) + O(1/M_u^2) \qquad (67.26)$$

where

$$\Delta(y_1 - y_2) = -i \int d^{D-1}k\, (e^{-ik \cdot (y_1 - y_2)} - e^{ik \cdot (y_1 - y_2)})/[(2\pi)^3 2\omega_k] \qquad (67.27)$$

is a familiar c-number invariant singular function. The additional terms in eq. 67.26 are q-number terms that become significant at very short distances of the order M_u^{-1}. Thus precise measurements of field strengths at larger distances are limited by standard quantum effects as indicated by the commutation relation.

The principle of *microscopic causality* is violated at extremely short distances of the order M_u^{-1} since the commutator is non-zero, in general, for space-like distances of the order of M_u^{-1} due to the q-number terms. This violation is not experimentally measurable now – and for the foreseeable future – and reflects a type of non-locality at extremely short distances.

The short distance behavior of two-tier quantum field theory leads to the elimination of divergences resulting in finite interacting quantum field theories.

67.5.1 Vacuum Fluctuations

While the expectation value of a *conventional* free scalar field $\phi_{conv}(Y)$ is zero in a conventional quantum field theory:

$$<0|\phi_{conv}(Y)|0> = 0 \qquad (67.28)$$

the vacuum fluctuations of *conventional* scalar quantum field theory are quadratically divergent:

$$<0|\phi_{conv}(Y)\phi_{conv}(Y)|0> = \int d^{D-1}p/[(2\pi)^3 2\omega_p] \qquad (67.29)$$

In Two-Tier quantum field theory we find the vacuum expectation value of a free field is zero *and the expectation value of the square of the field is also zero:*

$$<0|\phi(Y)\phi(Y)|0> = \int d^{D-1}p \; e^{-p^i p^j \Delta_{Tij}(0)/Mu^4}/[(2\pi)^3 2\omega_p] = 0 \qquad (67.30)$$

since the exponential factor in the integral is $-\infty$. The exponent contains

$$\Delta_{Tij}(z) = \int d^{D-1}k \; e^{-ik\cdot z} (\delta_{ij} - k_i k_j/\mathbf{k}^2)/[(2\pi)^3 2\omega_k] \qquad (67.31)$$

where "T" is for "Two-Tier". Thus *vacuum fluctuations are zero in Two-Tier quantum field theory.* Correspondingly, we will see that renormalization constants are finite in the Two-Tier Unified SuperStandard Theory .

67.5.2 The Feynman Propagator

The Feynman propagator for a Two-Tier free scalar quantum field is:

$$i\Delta_F{}^{TT}(y_1 - y_2) = <0|T(\phi(Y(y_1)),\phi(Y(y_2)))|0> \qquad (67.32)$$

$$\equiv <0|\phi(Y(y_1))\phi(Y(y_2))|0> \theta(y_1{}^0 - y_2{}^0) + \phi(Y(y_2))\phi(Y(y_1))|0> \theta(y_2{}^0 - y_1{}^0)$$

Since $Y^0 = y^0$ in the Coulomb gauge of the Y^μ field there is no ambiguity in the choice of the relevant time variable. A straightforward calculation shows:

$$i\Delta_F{}^{TT}(y_1 - y_2) = i \int d^4p \, e^{-ip\cdot(y_1 - y_2)} R(\mathbf{p}, y_1 - y_2)/[(2\pi)^4(p^2 - m^2 + i\varepsilon)] \qquad (67.33)$$

where $R(\mathbf{p}, y_1 - y_2)$ is given by eq. 67.18.

67.5.3 Large Distance Behavior of Two-Tier Theories

The large distance behavior of the two-tier Feynman propagator approaches the behavior of the conventional Feynman propagator since

$$R(\mathbf{p}, y_1 - y_2) \to 1 \qquad (67.34)$$

when $(y_1 - y_2)^2$ becomes much larger than $M_u{}^{-2}$. Thus the behavior of a conventional quantum field theory naturally emerges at large distance. We will see that a conventional SuperStandard Theory is the large distance limit of the Two-Tier SuperStandard Theory thus *realizing a form of Correspondence Principle for Quantum Field Theory*. Some features of a conventional SuperStandard Theory that depend specifically on the existence of divergences, such as the axial anomaly, will not be divergent in the Two-Tier SuperStandard Theory since it is a divergence-free theory.

67.5.4 Short Distance Behavior of Two-Tier Theories

At short distances the Gaussian factor dominates and radically changes the behavior of the Feynman propagator eliminating its short distance singular behavior, and thus paving the way to finite quantum field theories. Near the light cone, $M_u{}^{-2} \gg -(y_1 - y_2)^2 \to 0$, we can approximate eq. 67.33 with

$$i\Delta_F{}^{TT}(y_1 - y_2) \approx \int d^{D-1}p \, [N(p)]^2 R(\mathbf{p}, y_1 - y_2) \qquad (67.35)$$

since $e^{-ip \cdot (y_1 - y_2)}$ is approximately unity for small $(y_1 - y_2)$. We assume the mass of the ϕ particle is zero or is negligible at high energies so we set m = 0 to study the high energy behavior of eq. 67.35. Upon performing the integrations in eq. 35.35 for space-like $(y_1 - y_2)^2$ (and analytically continuing to the time-like regions[456,457]) we find that as $(y_1 - y_2)^2 \to 0$ from the space-like or time-like side of the light cone we find eq. 67.35 becomes:

$$i\Delta_F^{TT}(y_1 - y_2) \ \sim \ |(y_1 - y_2)^\mu (y_1 - y_2)_\mu|^{(D-2)/2} \qquad (67.36)$$

Eq. 67.36 has several noteworthy points:

1. The propagator is well behaved on the light cone and approaches zero smoothly from both space-like and time-like directions. This good behavior near the light cone will be seen later for other particle propagators with the net result that the usual infinities found in conventional quantum field theory are absent in two-tier quantum field theories.

2. The quadratic form of the propagator in eq. 67.36 is suggestive of attempts to formulate a relativistic harmonic oscillator model of elementary particles[458] and more recent attempts to achieve quark confinement. The fact that the absolute value of the quadratic term appears in eq. 67.36 neatly avoids the common pitfall seen in fully relativistic harmonic oscillator attempts.

3. The behavior *in coordinate space* of the propagator at short distances (q. 67.36) is equivalent to a high-energy behavior of

$$p^{(2-2D)} \qquad (67.37)$$

---

[456] See S. Blaha, "Relativistic Bound State Models with Quasi-Free Constituent Motion", Phys. Rev. **D12**, 3921 (1975) and references therein.

[457] It should be noted that A and B in eq. 35.18 have the same sign for $0 \le v < 1.1243$ thus making for easy analytic continuation across the light cone (which corresponds to v = 1 in eqs. 35.31 and 35.32).

[458] Yukawa, H., Phys. Rev. **91**, 416 (1953); Y. S. Kim and M. E. Noz, Phys. Rev. **D8**, 3521 (1973) and references therein.

in momentum space. Thus we get the equivalent *of a higher derivative theory* in Two-Tier quantum field theory at high energies while retaining a positive definite energy spectrum. The problems of negative metric states that have plagued conventional higher derivative quantum field theories are avoided.[459]

This concludes the basic description of scalar Unified SuperStandard Theory particles such as Higgs bosons in the Megaverse.

67.6 Megaverse Free Fermion Fields

The fundamental fermions of the Unified SuperStandard Theory that exist in the Megaverse have a basic description very similar to that of fermion universe particles.[460] Fundamental SuperStandard Theory fermions in the space between universes have the same masses, internal symmetries and interactions, which they do in our universe. They have four species just as universe particles. They have corresponding fourier expansions in D dimensions.

Due to the similarity of fundamental SuperStandard Theory fermions to universe particles we will not describe their Megaverse form of SuperStandard Theory fermions here.

---

[459] S. Blaha, Phys.Rev. **D10**, 4268 (1974); S. Blaha, Phys.Rev. **D11**, 2921 (1975); S. Blaha, Nuovo Cim. **A49**, :113 (1979); S. Blaha, "Generalization of Weyl's Unified Theory to Encompass a Non-Abelian Internal Symmetry Group" SLAC-PUB-1799, Aug 1976; S. Blaha, "Quantum Gravity and Quark Confinement" Lett. Nuovo Cim. **18**, 60 (1977); Nakanishi, N., Suppl. Prog. Theo. Phys. **51**, 1 (1972); and references therein.
[460] Universe particle are particle formulations for universes that we describe in chapter 36.

68. Quantum Field Theory of the Megaverse and Universes

In this chapter we describe the Megaverse as a quantum entity within the framework of the Wheeler-DeWitt equation suitably generalized.[461] In addition we describe the relation between a particle Quantum Field Theory in the Megaverse and in a universe within the Megaverse. *We also describe the needed generalizations of Quantum Field Theory to adequately and consistently describe a universe quantum field's relation to its Megaverse counterpart. These generalizations are: the local definition of asymptotic particle states, Bogoliubov transformations between a universe quantum field and its Megaverse counterpart, Two-Tier QFT to eliminate perturbation theory infinities, and PseudoQuantum Field Theory to accommodate Higgs vacuum expectation values and higher derivative interactions.*

68.1 Quantum Gravity in the Megaverse and its Universes

Since our universe is described by Quantum Gravity, and other universes are also, it is reasonable to assume the Megaverse is described by D-dimensional classical and Quantum Gravity. The source of Quantum Gravity in a universe is the mass-energy within the universe including particle masses, the Baryonic and Dark Baryonic gauge fields, the Two-Tier Y^μ field, and interaction gauge fields.

The sources of classical Gravitation and Quantum Gravity in the Megaverse are analogously the "masses" of universe particles, SuperStandard fermions, the Baryonic and Dark Baryonic gauge fields, other gauge fields, Higgs particles, the Θ-group gauge interactions, the mass-energy in the Megaverse outside of universes,[462] and the Megaverse Two-Tier Y^μ field.

---

[461] Most of this chapter appeared in Blaha (2017b), Blaha (2015a) and earlier books by the author.

[462] We believe the Megaverse has a very low density mass-energy between universes.

The surface of a universe in Quantum Gravity is fuzzy and determined by the quantum universe's wave function. This wave function is a solution of the universe Wheeler-Dewitt equation. The Wheeler-DeWitt equation assumes a single 4-dimensional universe with real-valued coordinates and metrics.

Since our theory requires complex coordinates and thus complex-valued metrics, we extended the Wheeler-DeWitt equation to complex metrics. Now we must define a Wheeler-DeWitt for the Megaverse's complex-valued coordinates and metric.

68.2 Megaverse Wheeler-Dewitt Equation

The Megaverse Wheeler-DeWitt equation inside a universe is

$$\left(G_{ijkl} \left\{ \prod_x \sum_m \{ [\delta^m_{\ j} \partial f_n/\partial x^i + \delta^m_{\ i} \partial f_m/\partial x^j](\partial^2 f_n/\partial x^{m2}) \}^{-1} \partial/\partial x^m \right\} \left\{ \prod_x \sum_m \{ [\delta^m_{\ k} \partial f_n/\partial x^l + \right. $$

$$\left. + \delta^m_{\ l} \partial f_n/\partial x^k](\partial^2 f_n/\partial x^{m2}) \}^{-1} \partial/\partial x^m \right\} + \gamma^{\frac{1}{2}\,(3)}R + 2\lambda\gamma^{\frac{1}{2}\,(3)} \right) \Psi(^{(3)}G, L_F) = 0$$

(68.1)

with universe coordinates x. In the Megaverse the Wheeler-DeWitt equation, in terms of Megaverse coordinates y_n is

$$0 = \left(G_{ijkl} \left\{ \prod_x \sum_m \{ [\delta^m_{\ j} \partial y_n/\partial x^i + \delta^m_{\ i} \partial y_m/\partial x^j](\partial^2 y_n/\partial x^{m2}) \}^{-1} \partial/\partial x^m \right\} \left\{ \prod_x \sum_m \{ [\delta^m_{\ k} \partial y_n/\partial x^l + \right. $$

$$\left. + \delta^m_{\ l} \partial y_n/\partial x^k](\partial^2 y_n/\partial x^{m2}) \}^{-1} \partial/\partial x^m \right\} + \gamma^{\frac{1}{2}\,(D-1)}R + 2\lambda\,\gamma^{\frac{1}{2}\,(D-1)} \right) \Psi(^{(D-1)}G, L_F)$$

(68.2)

where the sums over n and m in each pair of {} are done independently. All references to the metric are expressed in terms of Megaverse quantities.

Due to the appearance of products over all coordinates, the Megaverse expression for $\delta/\delta\gamma_{ij}$ is independent of x, and of Megaverse coordinates, in accordance with the space-time independence of the original Wheeler-DeWitt equation.

The Megaverse form of the Wheeler-DeWitt equation also directly relates the metric of a universe to the Megaverse. Every universe has two sets of coordinates:

universe coordinates, usually labeled x, embodying the curvature of the universe induced by gravitation, and Megaverse coordinates usually labeled y.

Outside of universes the Megaverse Wheeler-DeWitt equation becomes

$$(G_{ijkl}\, \delta/\delta\gamma_{ij}\, \delta/\delta\gamma_{kl} + \gamma^{\frac{1}{2}(D-1)}\, R + 2\lambda_M\, \gamma^{\frac{1}{2}(D-1)})\Psi(^{(D-1)}\mathcal{G}) = 0 \qquad (68.3)$$

with a cosmological constant, λ_M which is assumed to be present based on its presence in the universe case above and the uniformity of Nature. We assume a representation of the D-dimensional metric with a "spatial "sub-metric part γ_{ij00}. $\gamma^{\frac{1}{2}(D-1)}$ is the determinant of γ_{ik}.

Due to the products over all coordinates, the Megaverse expression for $\delta/\delta\gamma_{ik}$ is independent of Megaverse coordinates in analogy with the space-time independence of the original Wheeler-DeWitt equation. But the solutions of the Wheeler-DeWitt equations for a universe must be related to the solutions of the Wheeler-DeWitt equations of the Megaverse within the universe. This relationship must constrain the universe solutions at the boundary of a universe.

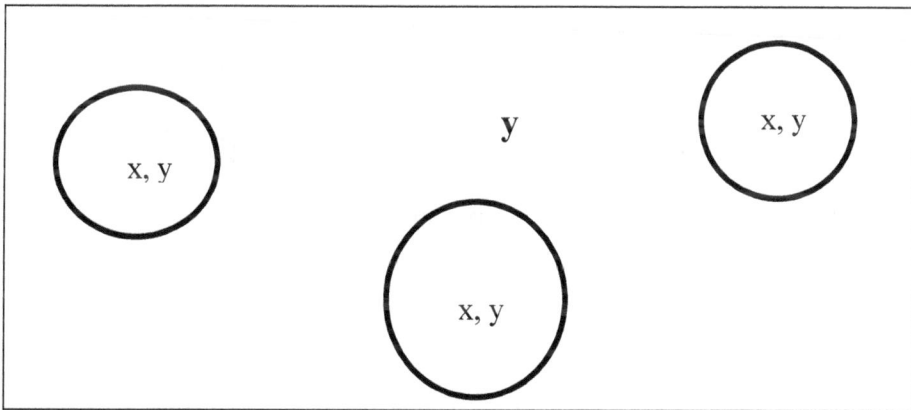

Figure 68.1. A symbolic view of part of the Megaverse with universes depicted as circles. Each universe has its own curvilinear coordinate 4-vector denoted x. The Megaverse coordinates are a D-vector y. A Wheeler-

DeWitt equation applies within each universe and a comprehensive equation describes the entire Megaverse transitioning to each universe's wave function within its domain.

68.3 Megaverse Metric Functional Integral in a Universe

Corresponding to the functional derivative in eq. 68.3 is an explicit Megaverse form of the functional integral for a universe

$$\int D\gamma \equiv \int \prod_x \sum_{i,j} d\gamma_{ij}(x) = \int \prod_x \sum_{i,j,k} d(\partial f_k/\partial x^i \, \partial f_k/\partial x^j) \qquad (68.4)$$

$$= \int \prod_x \sum_{i,j,k} [d(\partial f_k/\partial x^i) \, \partial f_k/\partial x^j + d(\partial f_k/\partial x^j) \, \partial f_k/\partial x^i] \qquad (68.5)$$

$$= 2\int \prod_x \sum_{i,j,k} d(\partial f_k/\partial x^i) \, \partial f_k/\partial x^j$$

$$= 2\int \prod_x \sum_{i,j,k} dx(\partial^2 f_k/\partial x^{i2}) \, \partial f_k/\partial x^j \equiv 2 \int \delta f(x) \qquad (68.6)$$

$$= 2\int \prod_x \sum_{i,j,k} dx(\partial^2 y_k/\partial x^{i2}) \, \partial y_k/\partial x^j \equiv 2 \int \delta f(x) \qquad (68.7)$$

The third line is due to the summation over i and j. The factor of 2 can be absorbed in the normalization factor. Eqs. 68.4 – 68.7, like the Wheeler-DeWitt equation, are independent of the coordinates due to the product over coordinates x. Thus the solution of the Wheeler-DeWitt equation takes the form

$$\Psi(^{(D-1)}g, \, \mathcal{L}_F) = N \int \delta f(x) \exp(-I(g, \, \mathcal{L}_F)) \qquad (68.8)$$

upon absorption of the factor of 2 into the normalization constant N

The Wheeler-DeWitt equation depends on the metric γ_{ij} which is a $(D-1)^2$ components construct with $D \cdot (D-1)/2$ independent components due to its symmetry.

68.4 Megaverse Wheeler-DeWitt Solutions

68.4.1 Tachyonic Solutions of Wheeler-DeWitt Equation

The original Wheeler-DeWitt equation, and its Megaverse equivalent, have a form that is similar in many respects to the Klein-Gordon equation. In particular, as DeWitt[463] noted, it resembles "a Klein-Gordon equation with $-\gamma^{\frac{1}{2}}{}^{(D-1)}R$ (our notation) playing the role of a mass-squared term. An important difference, however, is that ${}^{(D-1)}R$ can be either positive or negative, and hence the wave propagation of the state functional is not confined to time-like directions." DeWitt proceeds later in his paper to exclude consideration of negative "mass" squared terms.

In our view the negative "mass" squared cases represent tachyonic solutions of the Wheeler-DeWitt equation and should be considered as having the same validity as the positive mass squared solutions.

In a universe G_{ijkl} can be regarded as the contravariant metric of a 6-dimensional Riemannian manifold M with hyperbolic signature (-1, 1, 1, 1, 1, 1) with a time-like coordinate. Tachyonic solutions are part of the set of solutions of the Wheeler-DeWitt equation. In the Megaverse formulation the tachyonic solutions are indicators of tachyonic universes in the Megaverse.

If we use the Megaverse form of the Wheeler-DeWitt equation then the changes in time (dilations of γ_{ij} or equivalently dilations of y_n) can be viewed as specifying the overall motion of entire universes. We see now that we can have "normal" or tachyonic motion of universes.

Clearly we are building towards a particle view of universes.[464]

68.4.2 Problems in the Solutions of the Wheeler-DeWitt Equation

There are a number of problem areas associated with the original and our Megaverse Wheeler-DeWitt equations. DeWitt identified most of them in his paper, referenced below. While one could view these apparent problems as negatives, we will

---

[463]   DeWitt, B. S., Phys. Rev. **160**, 1113 (1987).p. 1124.

[464] The Reality groups of 4-dimensional universes, and of the Megaverse, play a role in the physical interpretation of Megaverse phenomena.

take the view that they are indicators of a deeper structure of universes below the Wheeler-DeWitt solutions just as the Dirac equation resolves analogous difficulties with the Klein-Gordon equation.

68.4.2.1 Negative Frequencies and Probabilities – Anti-Universes

DeWitt noted the existence of negative frequencies and negative probabilities associated with the solutions of the Wheeler-DeWitt equation. The rather close analogy to the Klein-Gordon equation in whose solutions similar issues appear is suggestive.

It appears that two general types of universes are embodied in the Wheeler-DeWitt equation. One type, which we call "normal", consists of universes like ours which have an excess of baryons (and consequently electrons to make an overall charge neutral universe). The other type of universe we will call *anti-universes*. These universes have an excess of anti-baryons (and positrons). We suggest such pairs can result from vacuum fluctuations in the Megaverse.

In addition we have "normal" universes and tachyonic universes whose motions in the Megaverse are analogous to the motions of fermions within a universe. In direct analogy with the Klein-Gordon equation the negative frequency and negative probability issues are thereby resolved.

One can speculate that the Wheeler-Dewitt equation, which is effectively second order in the "time" derivative, can be factored into first derivative equations – perhaps through the introduction of more degrees of freedom in a fashion similar to Dirac's introduction of spinors – to achieve first order equations in "time." Then we would have to face the issue of interpreting "Dirac-like" Wheeler-DeWitt equations.[465] We could again fall back on the rationale for introducing spinors for Standard Model fermions as in Blaha (2015a) – Asynchronous Logic – and suggest that universes embody a multi-valued Logic – although the concept of universes as logic values, at first glance, appears strange. We suggest that a universe's logic value is its spin.[466] Tied up with spin is "handedness." Our universe may be 'left-handed' as many phenomena within the universe seem to favor left-handedness. Therefore we can see a meaning for spin in

---

[465] We will not consider factoring the Wheeler-DeWitt equation in this book.
[466] Fermionic universe spin is defined later.

universe internal properties. The "left handedness" of our universe suggests other universes with "right handedness" may exist.

We will be content, for the present, to assume both universes and anti-universes exist. An anti-universe will be presumed to be a universe in the Megaverse where anti-particles predominate. We will postulate that universes, and anti-universes like protons and anti-protons, are always bodies with extension and not point-like. Blaha (2013) and later chapters in this book provide an example of a Big Bang where a universe begins with extension and is not point-like.

Later we describe a proposed dynamics of universes and anti-universes based on Megaverse gauge fields.

68.5 Quantum Aspects of the Megaverse

The Megaverse contains quantized universes. Being quantized, the horizon (surface) of a universe is not precisely defined but undergoes presumably mild quantum smearing. In particular the surface of a quantum universe in the Megaverse must be defined, as particle positions are defined in quantum mechanics, by a wave function whose "square" at each Megaverse point is the probability of a part of the universe being there. Physically we would expect the probability to fall sharply shortly beyond the classical horizon (surface) of the universe.

So in a Megaverse with a low density of universes most of the Megaverse will have zero probability of a universe being present.

There are however four sources of quantum phenomena in the Megaverse: 1) gauge fields that provides interactions between universes as well as quantum fluctuations that create universes; 2) universe particles; 3) Megaverse mass-energy density; and 4) the Y^μ field that appears within universes and in the exterior Megaverse that quantizes the coordinates of each universe. A Y^μ field's existence in the Megaverse makes perturbation theory computations finite to all orders.

68.6 Quantum Field Theory in the Megaverse

The conventional form of Quantum Field Theory does not meet the requirements of a Physical theory in a Megaverse of more than four dimensions. Perturbation theory calculations will yield infinities – not predictions susceptible to physical interpretation. To remedy this flaw in Quantum Field Theory, and for other reasons cited below, we have introduced three enhancements to Quantum Field theory over the past fourteen years. We briefly describe these enhancements within the framework of the Megaverse in this section and refer the reader to previous books that provide detailed descriptions.

68.6.1 Two Tier Quantum Field Theory in the Megaverse

Two-Tier Quantum Field Theory,[467] which was based on a new method in the Calculus of Variations, uses two 'layers' of fields to introduce quantum coordinates. We shall consider this technique, which applies to all fields, for the specific case of a massless vector field $V^i(y)$ analogous to the electromagnetic field.

Since a field, quantized in D-dimensional conventional coordinates (D > 4), would lead to divergences in perturbation theory calculations we use D-dimensional quantum coordinates:

$$Y^i(y) = y^i + i \, Y_u^{\ i}(y)/M_u^{\ D/2} \qquad (68.9)$$

where $Y_u^{\ i}(y)$ for $i = 1, ..., D$ is a D-dimensional free gauge field and M_u is a mass of the order of the Planck mass or greater. The $Y_u^{\ i}(y)$ term adds a quantum field to the D coordinates making them a set of quantum coordinates. Quantum coordinate derivatives are defined by

$$\partial_i = \partial/\partial Y^i(y) = \partial/\partial(y^i - Y_u^{\ i}(y)/M_u^{\ D/2}) \qquad (68.10)$$

The use of these coordinates to quantize particle fields leads to a completely finite perturbation theory. We applied them earlier to create a finite fundamental theory of mater. We will apply them to fields in the Megaverse to achieve a finite theory of Megaverse dynamics for elementary particles and universe particles.

---

[467] See Blaha (2005a), and Blaha (2002), for discussions of this new method to eliminate infinities in quantum field theory calculations.

The second quantization of the vector gauge field, $V^i(y)$ is analogous to the second quantization of the electromagnetic field. The lagrangian density terms for the free $V^i(Y(y))$ fields is

$$\mathscr{L}_{Vu} = -\tfrac{1}{4}\, F_{Vu}{}^{ij}(Y(y))F_{Vuij}(Y(y)) \qquad (68.11)$$

The lagrangian is

$$L_{Vu} = \int d^D y\, \mathscr{L}_{Vu}(Y(y)) \qquad (68.12)$$

with

$$F_{Vuij} = \partial V_i(Y(y))/\partial Y^j(y) - \partial V_j(Y(y))/\partial Y^i(y) \qquad (68.13)$$

where the values of i and j range from 1 to D in this section.

The equal time commutation relations, using the Dth coordinate as the time coordinate, are specified in the usual way:

$$[V^i(Y(\mathbf{y}, y^0)), V^j(Y(\mathbf{y}', y^0))] = [\pi^i(Y(\mathbf{y}, y^0)), \pi^i(Y(\mathbf{y}', y^0))] = 0 \qquad (68.14)$$

$$[\pi_j(Y(\mathbf{y}, y^0)), V_k(Y(\mathbf{y}', y^0))] = -i\, \delta^{(D-1)tr}{}_{jk}(Y(\mathbf{y},0) - Y(\mathbf{y}',0)) \qquad (68.15)$$

where

$$\pi_u{}^k = \partial \mathscr{L}_{Vu}(V(Y(y)))/\partial V_k{}'(Y(y)) \qquad (68.16)$$

$$\pi_u{}^D = 0 \qquad (68.17)$$

for k = 1, ... , (D – 1), and

$$\delta^{(D-1)tr}{}_{jk}(\mathbf{y} - \mathbf{y}') = \int d^{(D-1)}k\, e^{i\,\mathbf{k}\cdot(Y(\mathbf{y},0) - Y(\mathbf{y}',0))} (\delta_{jk} - k_j k_k/\mathbf{k}^2)/(2\pi)^{D-1} \qquad (68.18)$$

$$V_k{}'(Y(y)) = \partial V_k(Y(y))/\partial y^{1D} \qquad (68.19)$$

for j, k = 1, 2, ... , (D – 1).

If we choose the Coulomb gauge for $V_k(Y(y))$:

$$V^D(Y(y)) = 0$$

$$\partial V^j(Y(y))/\partial Y^j(y) = 0$$

for $j = 1, 2, \ldots, (D - 1)$ then $(D - 2)$ degrees of freedom (polarizations) are present in the vector potential.[468] The Fourier expansion of the vector potential $V^i(Y(y))$ is:

$$V^i(Y(y)) = \int d^{(D-1)}k \, N_{0V}(k) \sum_{\lambda=1}^{D-2} \varepsilon^i(k, \lambda)[a_V(k,\lambda) :e^{-ik \cdot Y(y)}: + a_V^\dagger(k,\lambda) :e^{ik \cdot Y(y)}:] \quad (68.20)$$

for $i = 1, \ldots, (D - 2)$ where

$$N_{0V}(k) = [(2\pi)^{(D-1)} 2\omega_k]^{-\frac{1}{2}} \quad (68.21)$$

and (since the field is massless)

$$k^D = \omega_k = (\mathbf{k}^2)^{\frac{1}{2}} \quad (68.22)$$

where k^D is the energy, and where the $\varepsilon^i(k, \lambda)$ are the polarization unit vectors for $\lambda = 1, \ldots, (D - 2)$ and $k^\mu k_\mu = k^{D\,2} - \mathbf{k}^2 = 0$.

The commutation relations of the Fourier coefficient operators are:

$$[a_V(k,\lambda), a_V^\dagger(k',\lambda')] = \delta_{\lambda\lambda'} \delta^{D-1}(\mathbf{k} - \mathbf{k}') \quad (68.23)$$
$$[a_V^\dagger(k,\lambda), a_V^\dagger(k',\lambda')] = [a_V(k,\lambda), a_V(k',\lambda')] = 0 \quad (68.24)$$

and the polarization vectors satisfy

$$\sum_{\lambda=1}^{D-2} \varepsilon_i(k, \lambda)\varepsilon_j(k, \lambda) = (\delta_{ij} - k_i k_j/\mathbf{k}^2) \quad (68.25)$$

The V^μ Feynman propagator is

$$iD_F^{\text{trTT}}(y_1 - y_2)_{jk} = \langle 0|T(V_j(Y(y_1))V_k(Y(y_2)))|0\rangle \quad (68.26)$$

---

[468] Note we use the Coulomb gauge for Y(y) also.

$$= -\, ig_{jk} \int \frac{d^D k\; e^{-ik\cdot(y_1 - y_2)}\; R(\mathbf{k}, y_1 - y_2)}{(2\pi)^{16}\,(k^2 + i\varepsilon)} \qquad (68.27)$$

where g_{jk} is the D-dimensional Lorentz metric and where $R(\mathbf{k}, y_1 - y_2)$ is given by eq. 67.17.

The Gaussian damping factor $R(k, z)$ for *all* large spatial momentum k^j is the same for both the positive and negative frequency parts of the (Two Tier) V Feynman propagator. We are assuming the spatial momentum is real-valued in this discussion. It is also important to note that $R(k, z)$ does not depend on $k^0 = k^D$ (in the V and Y_u Coulomb gauges) and thus the integration over k^0 proceeds in the usual way to produce time-ordered positive and negative frequency parts.

The Gaussian exponential factor in *all* spatial coordinates causes the Feynman propagator to be finite and, together with the Gaussian factor in universe particle propagators, causes all perturbation theory calculations when interactions are introduced to be finite as we have seen in the First Edition.

For small momentum much less than M_u then $R(\mathbf{k}, y_1 - y_2) \to 1$ and the Feynman propagator is the "normal" propagator of conventional D-dimensional field theory. For large momentum, its corresponding potential approaches r^{D-3}. The V potential is highly non-singular at large energies.

Thus using Two-Tier Quantum Field Theory we can perform perturbation theory calculations in the Megaverse that always yield a finite result.[469] This is not true if conventional Quantum Field is used.

---

[469] In particular, the fermion triangle divergence (anomaly) does not occur in our Two-Tier Quantum Field Theory of the fermion sector. Thus there is no requirement for axion-like particles in the Megaverse (or in universes) although the possible existence of this type of particle is not ruled out.

68.6.2 PseudoQuantum Field Theory in the Megaverse

PseudoQuantum Field Theory which we developed in a series of books[470] also can be formulated in the Megaverse. Thus we can use it in the Megaverse to implement the Higgs Mechanism to generate particle masses and symmetry breaking.

In this section we generalize PseudoQuantum field theory to the Megaverse for a scalar field. It can be implemented for other particle fields in an analogous manner.

We will now PseudoQuantize a scalar particle field in the Megaverse that will become a Higgs particle with a non-zero vacuum expectation value.[471] We begin by defining two fields that correspond to the scalar particle:[472] $\varphi_1(x)$ and $\varphi_2(x)$ where x is a D-dimensional vector. These fields will be assumed to have the equal time commutators

$$[\varphi_a(x),\pi_b(y)] = i(1 - \delta_{ab})\delta^{(D-1)}(\mathbf{x} - \mathbf{y}) \qquad (68.28)$$
$$[\varphi_a(x), \varphi_b(y)] = 0$$
$$[\pi_a(x), \pi_b(y)] = 0$$

where δ_{ab} is the Kronecker δ and where $\pi_a(x)$ is the canonically conjugate momentum to $\varphi_a(x)$. The fields $\varphi_1(x)$ and $\pi_1(y)$ will be observable classical fields. The fields $\varphi_2(x)$ and $\pi_2(y)$ will not be observables so that $\varphi_1(x)$ and $\pi_1(y)$ can both be sharp on the set of physical states.

We now specify the lagrangian density for a scalar Megaverse Klein-Gordon particle:

$$\mathcal{L} = \partial\varphi_1/\partial x_\mu \partial\varphi_2/\partial x^\mu \qquad (68.29a)$$

with Hamiltonian density

$$\mathcal{H} = \pi_1 \pi_2 + \partial\varphi_1/\partial x_i \partial\varphi_2/\partial x^i \qquad (68.29b)$$

---

[470] See Blaha (2017f) for the discussion of the PseudoQuantum field theory formalism for Higgs particles in our Extended Standard Model.

[471] Much of this chapter appears in Blaha (2016c), and earlier books, as well as in S. Blaha, Phys. Rev. **D17**, 994 (1978) (Appendix I). The case of fermion Pseudoquantization is also discussed in S. Blaha, Il Nuovo Cimento **49A**, 35 (1979) (See Appendix A).

[472] The subscripts on the fields are not gauge symmetry indices but simply identifiers distinguishing the fields from each other.

where μ labels 16-dimensional coordinates, i labels Megaverse spatial coordinates ((D – 1)-dimensional), and $\pi_1 = \partial\varphi_2/\partial t$ and $\pi_2 = \partial\varphi_1/\partial t$ with $t = x^D$. Eqs. 68.30 are without a potential or mass term.

The lagrangian and Hamiltonian for a massive scalar particle are

$$\mathcal{L} = \partial\varphi_1/\partial x_\mu \partial\varphi_2/\partial x^\mu - m^2 \varphi_1\varphi_2 \qquad (68.30a)$$

with Hamiltonian density

$$\mathcal{H} = \pi_1 \pi_2 + \partial\varphi_1/\partial x_i \partial\varphi_2/\partial x^i + m^2 \varphi_1\varphi_2 \qquad (68.30b)$$

The massless fields can be fourier expanded in terms of creation and annihilation operators:

$$\varphi_i(\mathbf{x}, t) = \int d^{(D-1)}k \, [a_i(k)f_k(x) + a_i^\dagger(k)f_k^*(x)] \qquad (68.31)$$

for i = 1, 2 where

$$f_k(x) = N(k)e^{-ik\cdot x}$$

with N(k) being a normalization factor.

The creation and annihilation operators satisfy the commutation relations:

$$[a_a(k), a_b^\dagger(k')] = (1 - \delta_{ab})\delta^{(D-1)}(\mathbf{k} - \mathbf{k}') \qquad (68.32)$$
$$[a_a(k), a_b(k')] = 0$$
$$[a_a^\dagger(k), a_b^\dagger(k')] = 0$$

for a, b = 1, 2.

In this formulation the defining properties of a physical state are:

$$\varphi_1(x)|\Phi, \Pi> = \Phi(x)|\Phi, \Pi> \qquad (68.33)$$
$$\pi_1(x)|\Phi, \Pi> = \Pi(x)|\Phi, \Pi>$$

where $\Phi(x)$ and $\Pi(x)$ are sharp on the states and thus classical fields with

$$\Phi(\mathbf{x}, t) = \int d^{(D-1)}k \, [\alpha(k)f_k(x) + \alpha^*(k)f_k^*(x)] \qquad (68.34)$$

and correspondingly for $\Pi(x)$.

To implement the mass generation mechanism we set Φ equal to a constant. We can define a set of states satisfying

$$a_1(k)|\alpha\rangle = \alpha(k)|\alpha\rangle$$
$$a_1^\dagger(k)|\alpha\rangle = \alpha^*(k)|\alpha\rangle$$

and correspondingly a set of coherent states

$$|\alpha\rangle = C\exp\left\{\int d^3k \, [\alpha(k)a_2^\dagger(k) + \alpha^*(k)a_2(k)]\right\}|0\rangle \qquad (68.35)$$

where C is a normalization constant and where the vacuum state $|0\rangle$ satisfies

$$a_1(k)|0\rangle = a_1^\dagger(k)|0\rangle = 0 \qquad (68.36a)$$

$$a_2(k)|0\rangle \neq 0 \qquad\qquad a_2^\dagger(k)|0\rangle \neq 0 \qquad (68.36b)$$

The dual vacuum state satisfies

$$\langle 0|a_2(k) = \langle 0|a_2^\dagger(k) = 0 \qquad (68.37a)$$
$$\langle 0|a_1(k) \neq 0 \qquad\qquad \langle 0|a_1^\dagger(k) \neq 0 \qquad (68.37b)$$

With this coherent state formalism, which gives purely classical fields and yet also has quantum fields through the use of φ_2 and its creation and annihilation operators, we now have the machinery to define a mass mechanism without the introduction of a potential whose origin can only be described as dubious.

For we can define a coherent state for some k as

$$|\Phi, \Pi\rangle = C\exp\{[2N(k)]^{-1}\Phi[a_2^\dagger(k) + a_2(k)]\}|0\rangle \qquad (68.38)$$

where C is a normalization constant, that yields a non-zero vacuum expectation value:

$$\varphi_1(x)|\Phi, \Pi> = \Phi| \Phi, \Pi> \qquad (68.39)$$

where Φ is a constant. Evaluating a fermion interaction term we find a mass term emerges[473]

$$\overline{\psi}(\varphi_1 + \varphi_2)\psi \;\; \rightarrow \;\; \overline{\psi}(\Phi + \varphi_2)\psi \qquad (68.40)$$

It generates a mass for an interaction with a gauge field of the form

$$A^{\mu}(\varphi_1 + \varphi_2)^2 A_{\mu} \;\; \rightarrow \;\; A^{\mu}(\Phi + \varphi_2)^2 A_{\mu} \qquad (68.41)$$

It also yields a quantum field theoretic interaction that would result in the production of ElectroWeak particles from these scalar fields. The production of Higgs particles that decay into ElectroWeak gauge particles has recently been found at CERN.

Thus our PseudoQuantum formalism is well-adapted to generate particle masses and symmetry breaking. Section 68.6.3 below shows that we can define Quantum Field Theories that support quantization for arbitrary timelike directions with local definitions of asymptotic states.

68.6.3 Local Definition of Asymptotic Particle States

The local definition of particle states is a significant point of interest for the Megaverse. Given the need for a quantum formulation of particle theory, we must address the issue of particle field quantization in a universe vs. quantization in the Megaverse. A particle state of a field quantized in one coordinate system is, in general, a superposition of particle states if the particle field is quantized in a different coordinate system. This is true within a universe. It is also true if one quantizes a field within a universe's coordinate system, and also quantizes the field in a Megaverse coordinate system. Since a universe can be described in a universe coordinate system, or in Megaverse coordinates, the problem of field quantization in coordinate systems, and the interpretation of particle states, becomes more critical when Megaverse coordinates are brought into consideration.

---

[473] When matrix elements with a "vacuum state" are taken.

Some years ago this author developed a formulation[474] of quantum field theory in which the particle interpretation of particle states was unambiguous: an n particle state in one field quantization's coordinate system was an n particle state for the field quantized in any other coordinate system. Thus the particle interpretation of states was independent of the coordinate system chosen for second quantization.

In this section we overview this method of quantization since it relates to the very real issue of the transition of particles between the Megaverse and a universe. If one envisions a particle (or a starship!) traveling between a universe and the Megaverse, then the fate of the particle (starship) after the transition is directly related to the possible(?) quantum field theoretic change of particle state(s).

The second quantization method described in this section is a generalization of the PseudoQuantization procedure described earlier in this section. The method was developed in the late 1970's by the author to provide a quantization procedure which supports a unique particle interpretation of states in arbitrary non-static space-times where no global timelike coordinate (Killing vector) exists. An N particle state in one quantization is an N particle state in other quantizations. Physical particle states of different quantizations are related by a unitary Bogoliubov transformation that preserves the particle number of the states. (The particle number operator commutes with the operator generating the unitary transformation.) See Appendix A for a detailed discussion.[475] We shall assume the reader has read Appendix A and extend the discussion to Megaverse quantization vs. universe quantization below.

We will consider the case of a scalar particle. Charged scalars, fermions and gauge fields are considered in Appendix A. These other cases are completely analogous.

---

[474] S. Blaha, "The Local Definition of Asymptotic Particle States", IL Nuovo Cimento **49A**, 35 (1979). This paper is reprinted in Appendix A for the reader's convenience. Also the paper S. Blaha, "New Framework for Gauge Field Theories", IL Nuovo Cimento **49A**, 113 (1979) applies this quantization method to non-Abelian gauge theories. It is reprinted in Appendix A.

[475] The discussions in the papers of Appendix A were predicated on an assumption of one universe. The generalization to Megaverse-universe quantizations is straightforward. We note, as Appendix A points out, that differences in the quantizations of two relatively accelerating observers *do cause* different numbers of particles to be evident in corresponding states – both physically, and in the quantized theories particle states. The quantizations described here are for *one* observer using different coordinate systems. The case of *two* relatively accelerating observers is different as noted in Appendix A.

68.6.3.1 Second Quantization and the Definition of Particle States in a Universe

Let us consider the case of a scalar particle that we second quantize in some fashion based on a timelike Killing vector

$$\varphi(x) = \sum_{\alpha} \chi_{\alpha}(x) A_{\alpha} + \chi_{\alpha}^{*}(x) A_{\alpha}^{\dagger} \tag{68.42}$$

where the $\chi_{\alpha}(x)$ are positive frequency with respect to a definition of positive frequency within a universe – following the notation of Appendix A.

68.6.3.2 Second Quantization and the Definition of Particle States in the Megaverse

Consider now the case of the same scalar particle that we second quantize in the Megaverse based on a timelike Megaverse Killing vector

$$\varphi(y) = \sum_{\beta} \psi_{\beta}(y) b_{\beta} + \psi_{\beta}^{*}(y) b_{\beta}^{\dagger} \tag{68.43}$$

where the $\psi_{\beta}(x)$ are positive frequency with respect to a Megaverse definition of positive frequency.

68.6.3.3 Relation Between the Definitions of Quantized Fields and Particle States

Comparing eqs. 68.42 and 68.43 we note the difference in the definition of the coordinates used in the field expansions as well as the implicit difference in the definitions of positive frequency. Therefore, to relate the quantizations to each other *solely within a universe*, we must use the relation between Megaverse coordinates y and universe coordinates x:

$$y_i = f_i(x) \tag{68.44}$$

or, in vector form,

$$y = f(x) \tag{68.45}$$

for i = 1, 2, ... , D. Thus

$$\varphi(f(x)) = \sum_{\beta} \psi_{\beta}(f(x)) b_{\beta} + \psi_{\beta}^{*}(f(x)) b_{\beta}^{\dagger} \tag{68.46}$$

Inverting the above equations to obtain the relation of the fourier coefficient operators we see:

$$A_\alpha = \sum_\beta [C_{\alpha\beta}\, b_\beta + C'_{\alpha\beta}\, b_\beta^\dagger] \qquad (68.47)$$

where $C_{\alpha\beta}$ and $C'_{\alpha\beta}$ are c-number functions of α and β:

$$C_{\alpha\beta} = (\chi_\alpha(x),\, \varphi(f(x))) \qquad (68.48)$$
$$C'_{\alpha\beta} = (\chi_\alpha{}^*(x),\, \varphi(f(x)))$$

with eq. 68.46 substituted for $\varphi(f(x))$ in the inner products, which are integrals over the universe coordinates x.

Eqs. 68.48 imply an N particle state in a universe will appear as a superposition of states of various numbers of particles in Megaverse coordinates IF THE STANDARD QUANTUM FIELD THEORY FORMULATION IS USED. A practical implication of this formalism is that a mouse in a universe is a superposition of protoplasm in the Megaverse – an unpleasant prospect for manned travel out of a universe into the Megaverse.

To REMEDY this situation – which we take to be unphysical[476] – we must reformulate quantum field theory in a manner similar to the PseudoQuantum formulation presented earlier.

The new formulation associates two fields with a particle in a manner very similar to that of PseudoQuantization field theory discussed earlier. The scalar particle case is discussed in the PseudoQuantization paper in Appendix A between eqs. 6 – 31. The reader is directed to read that section.

The conclusions of that section, and the sections following it, in Appendix A are:

---

[476] A mouse is a mouse whether in the universe or Megaverse since the transition between them does not involve a physical change in the mouse. Earlier we suggested the transition is smooth since all neighborhoods of every point of a universe contains an infinite number of exterior Megaverse points.

1. One can define corresponding particle states in a Megaverse quantization or in a universe quantization with the same number of particles.
2. The fourier coefficient operators of the two quantizations are related by Bogoliubov transformations and are unitarily equivalent.
3. The group of the local Bogoliubov transformations is an infinite tensor product of $SU_{1,1}$ groups.
4. The vacuums of the particle are invariant under Bogoliubov transformations that relate the Megaverse and the universe quantizations.
5. Unitarily equivalent perturbation theories of both the Megaverse and the universe quantizations can be defined.

The equations of Appendix A can be taken to apply to a universe quantization vs. a Megaverse quantization with the proviso that a map of Megaverse coordinates to universe coordinates must be used to calculate fourier coefficient operators such as in eqs. 68.46 – 68.48 above.

We thus have shown that our generalized PseudoQuantization formalism can be used to relate Megaverse quantum field theory in Megaverse coordinates to universe quantum field theory in universe coordinates.

68.7 SuperStandard Particles in Universe and Megaverse Coordinates

There are two aspects to the relation of universe fields and Megaverse fields: first the transformation of fields between universe coordinates and Megaverse coordinates; secondly the relation between the fourier expansions of the quantum fields in the respective coordinate systems.

68.7.1 Coordinate Transformation of Fields

The first issue of transforming between universe and Megaverse coordinates must be mindful that universe points are also Megaverse points. The values of a field at Megaverse points in a neighborhood of a universe point are determined by continuity. If we define a map from universe coordinates to Megaverse coordinates with

$$y^i = f^i(x) \tag{68.49}$$

then the vector field representations of a universe vector field $A_U^\mu(x)$ and a Megaverse vector field $A_M^i(y)$ are related by

$$A_M^i(y) = \partial y^i/\partial x^\mu \, A_U^\mu(x) \tag{68.50}$$
$$= \partial y^i/\partial x^\mu \, A_U^\mu(f^{-1}(y))$$

in the domain of the universe. Note the Megaverse vector field has D components.

For scalar fields such as Higgs fields the relationship is

$$\varphi_M^i(y) = A_U^\mu(x) \tag{68.51}$$
$$= \varphi_U^\mu(f^{-1}(y))$$

Fermion field relations are somewhat more intricate. There must be a map from a 4-spinor universe fermion field $\psi_{Ua}(x)$ where a = 1, 2, 3, 4 to the $2^{D/2}$-spinors of a Megaverse $\psi_{Ma}(y)$:

$$\psi_{Ma}(y) = \sum_b D_{ab} \, \psi_{Ub}(x)$$

$$= \sum_b D_{ab} \, \psi_{Ub}(f^{-1}(y)) \tag{68.52}$$

with Megaverse spinor components labeled b. The transformation matrix D_{ab} produces the many spinor component equivalent to a universe fermion field.

68.7.2 Quantum Field Transformation Between Coordinate Systems

The above transformation between universe and Megaverse fields is suitable for classical fields. However, in the case of quantum fields the fourier representation of the fields requires Bogoliubov transformation between the fourier components to guarantee the unitary equivalence of asymptotic particle states in each coordinate system. Chapter 41 describes the PseudoQuantum formalism to implement this change of coordinate systems for the case of scalar particles. Appendix A describes the procedure for the case of vector particles and fermions.

68.8 Unruh Bath for Accelerating Starships?

The preceding discussion in section 68.6.3 eliminates the issue of a starship moving with large acceleration encountering a thermal bath of incoming particles and a 'hot' vacuum.

68.9 Integrity of Starships Passing into the Megaverse

The preceding discussion in section 68.6.3 also eliminates problems with the transition of starships from a universe into the Megaverse. Particles are particles – within and without the Megaverse.

69. Introduction to Universe Particles

69.1 Megaverse Dynamics due to Universe Fields

The particles and fields that we have found in our universe will also exist in the Megaverse but the form of their fourier expansions will be D-dimensional. There will also be universe particles – quantum fields for entire universes.[477]

The universe particles within a Megaverse have dynamical motions in the Megaverse due to forces between them as well as Megaverse gravity.

In Blaha (2017c) as well as earlier books we described features of the Wheeler-DeWitt equation that suggested that universes could be viewed as particles or anti-particles, or tachyons. The solutions of their equations are wave functions on a manifold that are analogous to the solutions of the Klein-Gordon and Dirac equations. The issues of negative probabilities, possible tachyonic solutions, and negative frequency solutions suggest a need for an appropriate particle interpretation of universes that can possibly resolve these problems.

Some physicists have taken the Wheeler-DeWitt equation as the starting point for a theory of a universe as a particle. The Wheeler-DeWitt equation describes the interior of a universe in a quantum framework.

We will take a different approach using the Megaverse as the environment of universe particles that internally have Quantum Gravity, and externally have Megaverse Quantum Gravity.

We view a universe as an extended particle and begin by ignoring the detailed inner structure of universes. This approach is similar to the historical treatment of hadrons such as the proton as particles and developing a theory of them as fundamental particles using form factors, structure functions and so on to approximate their inner structure. Afterwards, as detailed data became available, the detailed investigation of the internal structure of hadrons using quark-parton models followed. We will pursue a

---

[477] Discussed later.

similar theoretical development beginning with a theory of universes as extended particles in the D-dimension Megaverse.[478] The internal structure of the particle universes will eventually be specified by the Wheeler-DeWitt equation expressed in Megaverse coordinates.

The two simplest choices for the nature of universes are "spin 0" *boson universes* and *fermion universes* with odd half integer spin, s_M.[479] We will first consider the possibility of fermion universes, and then consider "spin 0" *boson* universes.

In our study of the Hubble constant earlier we assumed a spin 0 universe whose initial expansion was based on universe particle vacuum polarization.

The first issue of fermion universes (reminiscent of the discussions of spin in the 1920's) is the interpretation of spin states. We suggest that the upper $2^{D/2-1}$ components (with $2^{D/2-2}$ "spin up" and $2^{D/2-2}$ "spin down" states) of a fermion universe wave function represent a left-handed universe with an excess number of baryons. The lower ($2^{D/2-1}$) components lead to right-handed anti-universes where there is an excess of anti-baryons. These associations are analogous to the interpretations of the Dirac electron wave function.[480]

The universe particle "spin up" and "spin down" states are distinguished by their interactions with gauge fields in a manner analogous to quantum electrodynamics.

69.2 "Free Field" Dynamics of Fermion Universe Particles

We now consider fermion universes as extended particles with an odd half integer spin – *fermion universe particles* - in the D-dimensional Megaverse. In the Megaverse there are D 'Dirac' matrices with $2^{D/2}$ rows and $2^{D/2}$ columns that are the equivalent of the four Dirac matrices in four dimensions. We will denote these D matrices as $\gamma_M{}^i$ for i = 1, 2, ... , D. They satisfy the anti-commutation relations:

---

[478] Of course we follow a different course here – looking from within universe particles 'outwards' whereas in our universe we look 'into' particles from 'outside.'

[479] The Megaverse is assumed to be D-dimensional where we have assumed provisionally to be D = 192. The spin of fermionic universe particles was shown to be s_M in Blaha (2017c) in chapter 3.

[480] It is known that phenomena in our universe tend to be left-handed. If this feature of our universe's phenomena is also a property of the universe itself, then, since handedness is an attribute of spin, the treatment of a universe as having spin is not unreasonable.

$$\{\gamma_M{}^i, \gamma_M{}^j\} = 2\,\delta^{ij} \tag{69.1}$$

and thus form a Clifford algebra. We will choose the coordinate y^D to be the time coordinate and thus make it pure imaginary with a Reality group transformation. (The D-dimensional Megaverse space is a complex Euclidean space.) Therefore γ^D will be hermitean $((\gamma^D)^2 = 1)$, and the γ^i matrices for $i = 1, \ldots, (D-1)$ will be anti-hermitean with $(\gamma^i)^2 = -1$. The number of linearly independent matrices in D dimensions is 2^D.

The Megaverse metric is (by using the Reality group) chosen to be

$$g^{ij} = -\delta^{ij}, \qquad g^{D,D} = 1 \tag{69.2}$$

for $i, j = 1, 2, \ldots, (D-1)$; and zero otherwise.

Except for the additional dimensions, fermion dynamics is quite similar to the 4-dimensional case. The free universe particle Dirac equation is

$$(i\gamma^i \partial_i + m)\psi(y) = 0 \tag{69.3}$$

summed over $i = 1, 2, \ldots, D$ where the mass is assumed to be constant, and set below. The derivative operator, is based on the use of quantum coordinates[481]

$$X^i(y) = y^i + i\,Y_u{}^i(y)/M_u{}^{D/2} \tag{69.4}$$

for $i = 1, \ldots, D$ and is defined to be

$$\partial_i = \partial/\partial X^i(y) = \partial/\partial(y_i - Y_{ui}(y)/M_u{}^{D/2}) \tag{69.5}$$

where *we assume $M_u = M_c$ with M_c being a very large mass scale of perhaps the order of the Planck mass.*

$Y_u{}^i$ is a D-dimensional Megaverse gauge field equivalent of the universe $Y^\mu(x)$ used in 4-dimensional Two-Tier renormalization:

---

[481] Giving Two-Tier renormalization without infinities.

$$X^\mu(z) = z^\mu + i\, Y^\mu(z)/M_c^2 \qquad (69.6)$$

where $Y^\mu(z)$ is a free QED-like field. The $Y^i(y)$ quantum coordinates will be used in the Megaverse to eliminate potential divergences, in a manner similar to the case of our universe when universe particle interactions are introduced later. In a higher dimension space the potential divergences in calculations are much greater. *Two-Tier coordinates eliminate these potential divergences.*

69.3 Four Types of Fermion Universe Particles

Assuming universe energies are real-valued,[482] there are four possible types of fermion universe particles in the Megaverse. They are analogous to the four species of fermions. Two of these types are tachyonic. It is important to note that DeWitt points out that the Wheeler-DeWitt equation has tachyonic solutions since the mass-like term dependent on $^{(3)}R$ can be positive or negative.[483] A negative mass is an indication of tachyonic behavior wherein the wave propagation of the state functional is not necessarily in time-like directions and is thus tachyonic.

Eq. 69.3 is a D-dimensional Dirac equation. There are three other general types of universe particle equations. (By assumption fermion universes come in four species like fermions.) The derivation of the four types of universe particles (chapter 38) is similar to the derivation of fermion types discussed previously. We will now consider the D-dimensional equivalent for universe particles in the Megaverse.

The general form of a pure D-dimensional complex Lorentz group[484] boost can be expressed in terms of a complex relative $(D - 1)$-velocity $\mathbf{v_c}$ between inertial reference frames. A D-dimensional coordinate boost has the form

---

[482] The energy of universe particles need not be real-valued since universes can 'decay' – unlike elementary particles which are not subject to decay, by definition, since they are assumed to be *fundamental*. We choose to consider the case of universes with real-valued energies. The case of universes with complex-valued energies is a simple extension of the real-value cases considered here.

[483] DeWitt, B. S., Phys. Rev. **160**, 1113 (1967) p. 1124.

[484] The D-dimensional Complex Lorentz group has similar features to the 4-dimensional Complex Lorentz group. We shall only discuss it to the extent needed for our universe particle type's derivation. See Weinberg (1995) for the 4-dimensional Lorentz group – the D-dimensional Lorentz group generalizes directly from the features of the 4-dimensional Lorentz group.

$$\Lambda_C(\mathbf{v_c}) \equiv \Lambda_C(\omega, \mathbf{v_c}) = \exp[i\omega\hat{\mathbf{w}}{\cdot}\mathbf{K}] \tag{69.7}$$

where

$$\omega = (\omega_r^2 - \omega_i^2 + 2i\omega_r\omega_i\ \hat{\mathbf{u}}_r{\cdot}\hat{\mathbf{u}}_i)^{\frac{1}{2}} \tag{69.8}$$

and

$$\hat{\mathbf{w}} = (\omega_r\hat{\mathbf{u}}_r + i\omega_i\hat{\mathbf{u}}_i)/\omega \tag{69.9}$$

with all vectors being $(D - 1)$-dimensional spatial vectors. We define the real and imaginary unit vectors $\hat{\mathbf{u}}_r{\cdot}\hat{\mathbf{u}}_r = 1 = \hat{\mathbf{u}}_i{\cdot}\hat{\mathbf{u}}_i$ with the result

$$\hat{\mathbf{w}}{\cdot}\hat{\mathbf{w}} = 1 \tag{69.10}$$

The complex relative velocity is

$$\mathbf{v_c} = \hat{\mathbf{w}}\tanh(\omega) \tag{69.11}$$

The free dynamical equations of the four universe particle species will be generated by D-dimensional Lorentz boosts of the free Dirac equation of a universe particle at rest with the *requirement that the time variable* $(t = y^D)$ *and energy are real in the resulting field equations.*[485] The procedure can most easily be performed in D-dimensional momentum space with the Megaverse coordinate space version of the generated equation determined from the momentum space version.

The result is a number of different types of fermion universes that mirror the particle species that we have seen earlier.

- Dirac-like Universe Particle
- Tachyon Universe particle
- Left-Handed Tachyonic Universe Particles
- Right-Handed Tachyonic Universe Particles
- "Up-Quark-like" Universe Particles
- Left-Handed "Down-Quark-like" Tachyonic Universe Particles
- Right-Handed Down-Quark-like Tachyonic Universe Particles

---

[485] The D-dimensional "energy" must be real since it relates to the area of the universe – a real number.

Free lagrangians, which are similar to the corresponding fermion lagrangians seen earlier, follow for these universe particles.

69.4 Embedding Universes Within the Megaverse

Earlier we developed a view of the general nature of universes in the Megaverse. We now examine the general nature of the Megaverse itself. Consistency *suggests that complex-valued coordinates universes, such as our universe, require an embedding in a complex-valued Megaverse* space.

We now assume the Megaverse has an analytic, complex, symmetric metric g_{ik} which satisfies

$$g_{ij} = g_{ji} \tag{69.12}$$

for i, j = 1, 2, ... , D.

The embedding equations for our curved complex-valued universe {x} within the complex, D-dimensional Megaverse {y} are:

$$z_i = f_i(x) \tag{69.13}$$

where x^μ are the complex-valued 4-dimensional coordinates of a universe.
The infinitesimal Megaverse invariant distance is

$$ds^2 = g_{ik}dz^i dz^k \tag{69.14}$$

where the z_i are the complex-valued D-dimensional coordinates of the Megaverse. g_{ik} is the complex-valued Megaverse metric tensor, ds is the D-dimensional invariant Megaverse distance,[486] and the complex-valued, 4-dimensional metric of a universe has the form

$$g_{\mu\nu} = \partial f_j / \partial x^\mu \, \partial f_j / \partial x^\nu \tag{69.15}$$

with an implied sum over the subscript j.

---

[486] We note that ds^2 is complex and this might be found troubling. A Reality group transformation would yield the absolute value of ds^2 and thus $ds_{physical} = ||ds^2||^{1/2}$ would be the physical real-valued invariant distance. This remark also applies to 4-dimensional complex General Relativity.

The picture that we paint of the Megaverse is that of a complex, D-dimensional space containing (perhaps) countless universes, some of which may be (almost) flat like ours, and some of which may be curved, open or closed 4-surfaces. *We assume universes are closed, or curved and open, or flat.*

The Megaverse does have a gravitational field due to universe masses, particle masses, the Two-Tier[487] field Y^μ, the Unified SuperStandard Theory fields. *We will assume all universes are complex, 4-dimensional although universes with a larger number of dimensions are possible and will be discussed below.*

69.5 Possible Dimensions of Universes

In Blaha (2011c) (and in this book) we determined the dimensionality of our universe based on principles of Asynchronous Logic that suggested a 4-valued logic that could be embodied in a 4-dimensional spinor matrix formulation. This 4-dimensional spinor formulation led to a 4-dimensional space-time.

The requirement that the speed of light is the same in all inertial reference frames, and that transformations between reference frames in faster than light relative motion are physically allowed, led to the requirement of complex coordinates and the Complex Lorentz transformation group (as was found necessary in Axiomatic Quantum Field Theory studies.) The reality of all physical time and distance measurements led to the introduction of the Reality group that mapped complex quantities to real physical values. This chain of logic is in accord with Leibniz's minimax principle: nature uses the simplest means to create complex physical phenomena.

While the preceding paragraph applies very nicely to our universe, and presumably to other universes, the question of the existence of other universes within the Megaverse with different dimensions naturally arises. Stars and galaxies have many varieties. Why should all universes be of dimension four?

Having developed the fundamental nature of our universe from Logic (the only sure requirement of any physical theory) it seems reasonable to classify possible universes based on their fundamental logic. In Blaha (2011c) we developed matrix formulations for many-valued logics. Assuming no separate clock mechanism to synchronize parts of complex processes, we developed an n × n matrix formalism for n-valued logic.

Therefore we were able to develop a principal sequence of types of universes based on n-valued logic. We summarize the small n-valued cases in Table 69.1.

Fant (2005) points out that VLSI circuits with spatially separated parts, which require time synchronization of activity without clocks, need a 4-valued logic at minimum. Thus for a complex universe such as ours, the minimum space-time dimensionality is 4. For a smaller number of dimensions the complexity of physical processes is much diminished as the many solvable models of low space-time dimensionality show: easily solved – not very complex phenomena!

Smaller dimensioned universes may well exist – but not with the richness of complexity that leads to our type of universe's phenomena such as life.

Larger dimension universes may well also exist. They would have an excess of phenomena that might preclude life as we know it, or engender new forms of life.

The above classification scheme for universes is based on logic. Another important consideration is size. It appears that universes can have differing sizes and in fact can also grow or diminish in size (expansion or contraction). We have considered the size issue earlier and will again when we consider the expansion/contraction of a universe due to a time-dependent mass. Other possible differentiating factors between universes will also be considered.

70. Fermion and Boson Universe Particles

The Wheeler-DeWitt equation, because of its similarity to the Klein-Gordon equation, has led to numerous proposals to view universes as particles.[488]

In this chapter[489] we will consider a possible particle interpretation of universes that, while consistent with the spirit of the Wheeler-DeWitt equation and the Megaverse, goes far beyond our current experimental knowledge, although some recent astronomical data tends to support it. It can only be justified in this century by its generality and simplicity. It just looks right.

70.1 The Hierarchy of the Cosmos

In our universe we have seen that natural phenomena form a hierarchy ranging from the simplest to the largest/most complex phenomena. One current view of the hierarchy of levels of physical phenomena is:

Elementary particles: leptons, quarks, gluons, gauge bosons, and Higgs particles
Hadrons: protons, neutrons, …
Molecules
Agglomerations of molecules
Macroscopic objects
Planets
Stars
Galaxies
Clusters of galaxies
Supergalaxies
The Universe

---

[488] Some suggestions of this interpretation are: DeWitt, B. S., Phys. Rev. **160**, 1113 (1967); Robles-Perez, S. J., arXiv:1212.4598 (2012); and references therein.
[489] This chapter is obtained from Blaha (2015a).

Each level generally has a set of "simplified" physical laws that describe its phenomena.[490] For example molecules have quantum mechanical laws and regularities that help to understand the phenomena at the molecular level.

Interestingly, while all phenomena at each level should be explainable by the laws at lower levels, and ultimately, all phenomena should be explainable at the level of elementary particles, connecting phenomena at different levels is often quite difficult and, in many cases, impossible.

Consequently, while we believe physical phenomena are ultimately reducible to the lowest level, the problem of relating phenomena at different levels is largely unresolved.

In this book we introduced new levels in the hierarchy of nature: the level of multiple universes, and the level of the all-encompassing Megaverse. In doing this, we seek to maintain what we know of our universe, as embodied in our Extended Standard Model and Quantum Gravity. We will now turn to a discussion of the universes level and a portrayal of universes as extended particles.

70.2 The Particle Interpretation of Extended Wheeler-DeWitt Equation Solutions

In earlier chapters we described features of the Wheeler-DeWitt equation that suggested that universes could be viewed as particles or anti-particles, or tachyons. The solutions of this equation are scalar wave functions on a manifold that are analogous to the solutions of the Klein-Gordon equation. The issues of negative probabilities, possible tachyonic solutions, and negative frequency solutions suggest a need for an appropriate particle interpretation of universes that can possibly resolve these problems.

Some physicists have taken the Wheeler-DeWitt equation as the starting point for a theory of a universe as a particle. The Wheeler-DeWitt equation describes the interior of a universe in a quantum framework.

We will take a different approach using the Megaverse as the environment of universe particles that internally have Quantum Gravity, and externally have Megaverse Quantum Gravity.

---

[490] This point was often made by Nobelist Kenneth Wilson of Cornell and Ohio State Universities.

We view a universe as an extended particle and begin by ignoring the detailed inner structure of universes. This approach is similar to the historical treatment of hadrons such as the proton as particles and developing a theory of them as fundamental particles using form factors, structure functions and so on to approximate their inner structure. Afterwards, as detailed data became available, the detailed investigation of the internal structure of hadrons using quark-parton models followed. We will pursue a similar theoretical development beginning with a theory of universes as extended particles in the D-dimension Megaverse. The internal structure of the particle universes will eventually be specified by the Wheeler-DeWitt equation expressed in Megaverse coordinates.

The two simplest choices for the nature of universes are "spin 0" *boson universes* and *fermion universes* with odd half integer spin, s_M.[491] We will first consider the possibility of fermion universes, and then briefly consider "spin 0" *boson* universes.

The first issue of fermion universes (reminiscent of the discussions of spin in the 1920's) is the interpretation of spin states. We suggest that the upper $2^{D/2 - 1}$ components (with $2^{D/2 - 2}$ "spin up" and $2^{D/2 - 2}$ "spin down" states) of a fermion universe wave function represent a left-handed universe with an excess number of baryons. The lower ($2^{D/2 - 1}$) components lead to right-handed anti-universes where there is an excess of anti-baryons. These associations are analogous to the interpretations of the Dirac electron wave function.[492]

The universe particle "spin up" and "spin down" states are distinguished by their interactions with gauge fields in a manner analogous to quantum electrodynamics.

70.3 "Free Field" Dynamics of Fermion Universe Particles

We now consider universes as extended particles with an odd half integer spin – *fermion universe particles* - in the D-dimensional Megaverse. In the Megaverse there are D 'Dirac' matrices with $2^{D/2}$ rows and $2^{D/2}$ columns that are the equivalent of the

---

[491] Since the Megaverse is D-dimensional, the spin of fermionic universe particles was shown to be s_M in chapter 37.

[492] It is known that phenomena in our universe tend to be left-handed. If this feature of our universe's phenomena is also a property of the universe itself, then, since handedness is an attribute of spin, the treatment of a universe as having spin is not unreasonable.

four Dirac matrices in four dimensions. We will denote these D matrices as $\gamma_M{}^i$ for i = 1, 2, ... , D. They satisfy the anti-commutation relations:

$$\{\gamma_M{}^i, \gamma_M{}^j\} = 2\,\delta^{ij} \qquad (70.1)$$

and thus form a Clifford algebra. We will choose y^D to be the time coordinate and thus make it pure imaginary with a Reality group transformation. (The D-dimensional Megaverse space is a complex Euclidean space.) Therefore γ^D will be hermitean $((\gamma^D)^2 = 1)$, and the γ^i matrices for i = 1, ... , (D – 1) will be anti-hermitean with $(\gamma^i)^2 = -1$. The number of linearly independent matrices in D dimensions is 2^D.

The Megaverse metric is (by use of the Reality group) chosen to be

$$g^{ij} = -\delta^{ij}, \qquad\qquad g^{D,D} = 1 \qquad (70.2)$$

for i, j = 1, 2, ... , (D – 1); and zero otherwise.

Except for the additional dimensions, fermion dynamics is quite similar to the 4-dimensional case. The free universe particle Dirac equation is

$$(i\gamma^i\partial_i + m)\psi(y) = 0 \qquad (70.3)$$

summed over i = 1, 2, ... , D where the mass is assumed to be constant, and set by eq. 70.119 below. The derivative operator, is based on the use of quantum coordinates[493]

$$Y^i(y) = y^i + i\,Y_u{}^i(y)/M_u{}^{D/2} \qquad (70.4)$$

For i = 1, ..., D and is defined to be

$$\partial_i = \partial/\partial Y^i(y) = \partial/\partial(y_i - Y_{ui}(y)/M_u{}^{D/2}) \qquad (70.5)$$

where *we assume $M_u = M_c$ with M_c being a very large mass scale of perhaps the order of the Planck mass.*

---

[493] Giving Two-Tier renormalization.

Y_u^i is a D-dimensional Megaverse gauge field equivalent of the universe $Y^\mu(x)$ used in 4-dimensional Two-Tier renormalization (discussed in I):

$$Y^\mu(y) = y^\mu + i\, Y^\mu(y)/M_c^2$$

where $Y^\mu(y)$ is a free QED-like field. The $Y^i(y)$ quantum coordinates will be used in the Megaverse to eliminate potential divergences, in a manner similar to the case of our universe when universe particle interactions are introduced later.

70.3.1 Four Types of Fermion Universe Particles

Assuming universe energies are real-valued,[494] there are four possible types of fermion universe particles in the Megaverse that are analogous to the four species of fermion described in Blaha (2010b) for The Extended Standard Model. Two of these types are tachyonic. It is important to note that DeWitt points out that the Wheeler-DeWitt equation has tachyonic solutions since the mass-like term dependent on $^{(3)}R$ can be positive or negative.[495] A negative mass is an indication of tachyonic behavior wherein the wave propagation of the state functional is not necessarily in time-like directions and is thus tachyonic.

Eq. 70.3 is a Dirac-type D-dimensional Dirac equation. There are three other general types of universe particle equations. (By assumption fermion universes come in four species like fermions.) The derivation of the four types of universe particles is similar to the derivation of fermion types in the Extended Standard Model in 4-dimensional complex space-time given in Blaha (2010b). We will now consider the D-dimensional equivalent for universe particles in the Megaverse.

---

[494] The energy of universe particles need not be real-valued since universes can 'decay' – unlike elementary particles which are not subject to decay, by definition, since they are assumed to be *fundamental*. We choose to consider the case of universes with real-valued energies. The case of universes with complex-valued energies is a simple extension of the real-value cases considered here.

[495] DeWitt, B. S., Phys. Rev. **160**, 1113 (1967) p. 1124.

The general form of a pure D-dimensional complex Lorentz group[496] boost can be expressed in terms of a complex relative $(D - 1)$-velocity $\mathbf{v_c}$ between inertial reference frames. A D-dimensional coordinate boost has the form

$$\Lambda_C(\mathbf{v_c}) \equiv \Lambda_C(\omega, \mathbf{v_c}) = \exp[i\omega\hat{\mathbf{w}}\cdot\mathbf{K}] \qquad (70.6)$$

where

$$\omega = (\omega_r^2 - \omega_i^2 + 2i\omega_r\omega_i\,\hat{\mathbf{u}}_r\cdot\hat{\mathbf{u}}_i)^{\frac{1}{2}} \qquad (70.7)$$

and

$$\hat{\mathbf{w}} = (\omega_r\hat{\mathbf{u}}_r + i\omega_i\hat{\mathbf{u}}_i)/\omega \qquad (70.8)$$

with all vectors being $(D - 1)$-dimensional spatial vectors. We define the real and imaginary unit vectors $\hat{\mathbf{u}}_r\cdot\hat{\mathbf{u}}_r = 1 = \hat{\mathbf{u}}_i\cdot\hat{\mathbf{u}}_i$ with the result

$$\hat{\mathbf{w}}\cdot\hat{\mathbf{w}} = 1 \qquad (70.9)$$

The complex relative velocity is

$$\mathbf{v_c} = \hat{\mathbf{w}}\tanh(\omega) \qquad (70.10)$$

The free dynamical equations of the four universe particle species will be generated by D-dimensional Lorentz boosts of the free Dirac equation of a universe particle at rest with the *requirement that the time variable* $(t = y^D)$ *and energy are real in the resulting field equations.*[497] The procedure can most easily be performed in D-dimensional momentum space with the Megaverse coordinate space version of the generated equation determined from the momentum space version.

70.3.1.1 Dirac-like Equation – Type I universe Particle

A positive energy plane wave solution of the Dirac equation eq. 70.3 for a universe particle at rest is

---

[496] The D-dimensional complex Lorentz group has similar features to the 4-dimensional complex Lorentz group. We shall only discuss it to the extent needed for our universe particle type's derivation. See Weinberg (1995) for the 4-dimensional Lorentz group – the D-dimensional Lorentz group generalizes directly from the features of the 4-dimensional Lorentz group.

[497] The D-dimensional "energy" must be real since it relates to the area of the universe – a real number.

$$\psi(y) = \exp[-imt]w(0) \tag{70.11}$$

where we set $\partial_t = \partial/\partial y_D$ while temporarily ignoring the $Y_u^i(y)/M_u^{D/2}$ term. $w(0)$ is a $2^{D/2}$ component spinor column vector. The solution $\psi(y)$ satisfies the momentum space Dirac equation for a particle at rest:

$$(m\gamma^D - m)\psi(y) = 0 \tag{70.12}$$

The $2^{D/2} \times 2^{D/2}$ spinor matrix form of a D-dimensional Lorentz boost with a relative real velocity $\mathbf{v}$ of the Dirac matrices is[498]

$$S^{-1}(\Lambda(\mathbf{v}))\gamma^\nu S(\Lambda(\mathbf{v})) = \Lambda^\nu_\mu(\mathbf{v})\gamma^\mu \tag{70.13}$$

where $\Lambda^\nu_\mu(\mathbf{v})$ is a D-dimensional Lorentz boost. $S(\Lambda(\mathbf{v}))$ has the form

$$S(\Lambda(\mathbf{v})) = \exp(-\omega\gamma^D\boldsymbol{\gamma}\cdot\mathbf{v}/(2|\mathbf{v}|))$$

$$= \cosh(\omega/2)I + \sinh(\omega/2)\gamma^D\boldsymbol{\gamma}\cdot\mathbf{p}/|\mathbf{p}| \tag{70.14}$$

with *real* $\omega = \text{arctanh}(|\mathbf{v}|)$ and *real* $\mathbf{v}$. $|\mathbf{p}|$ is the magnitude of the spatial $(D-1)$-vector. Also

$$S^{-1}(\Lambda(\mathbf{v})) = \gamma^D S^\dagger(\Lambda(\mathbf{v}))\gamma^D = \exp(\omega\gamma^D\boldsymbol{\gamma}\cdot\mathbf{v}/(2|\mathbf{v}|))$$

$$= \cosh(\omega/2)I - \sinh(\omega/2)\gamma^D\boldsymbol{\gamma}\cdot\mathbf{p}/|\mathbf{p}| \tag{70.15}$$

If we now apply $S(\Lambda(\mathbf{v}))$ to the momentum space Dirac equation of a particle at rest (eq. 70.12) we find

$$0 = S(\Lambda(\mathbf{v}))(m\gamma^D - m)\,\psi(y)$$
$$= [mS(\Lambda(\mathbf{v}))\gamma^D S^{-1}(\Lambda(\mathbf{v})) - m]S(\Lambda(\mathbf{v}))w(0)$$

---

[498] The indices ν and μ from this point in this chapter have values: 1, 2, ... , D.

A straightforward evaluation shows

$$mS(\Lambda(v))\gamma^D S^{-1}(\Lambda(v)) = g_{D\mu\nu}p^\mu\gamma^\nu = \not{p} \qquad (70.16)$$

where p is a momentum D-vector. In addition we define the D-dimension spinor ($2^{D/2}$ components)

$$S(\Lambda(v))w(0) = w(p) \qquad (70.17)$$

which can be viewed as a "positive energy D Dirac spinor". The Dirac equation in momentum space has the familiar form:

$$(\not{p} - m) \exp[-ip\cdot y]w(p) = 0 \qquad (70.18)$$

Eq. 70.18 implies the free, coordinate space Dirac equation:

$$(i\gamma^\mu \partial/\partial y^\mu - m)\psi(y) = 0 \qquad (70.19)$$

We identify this equation as the dynamical equation of a type 1 universe particle. It corresponds to the free charged lepton elementary particle species Dirac equation in particle physics.

70.3.1.2 Complex Boosts

The form of the D-dimensional spinor boost transformation corresponding to the coordinate transformation eq. 70.6 is:

$$S_C(\omega, \mathbf{v_c}) \equiv S_C = \exp(-\omega\gamma^D\boldsymbol{\gamma}\cdot\hat{\mathbf{w}}/2)$$
$$= \cosh(\omega/2)I + \sinh(\omega/2)\gamma^D\boldsymbol{\gamma}\cdot\hat{\mathbf{w}} \qquad (70.20)$$

with *complex* $\mathbf{v_c}$ and $\hat{\mathbf{w}}$ defined by eqs. 70.10 and 70.8 respectively. The inverse transformation is

$$S_C^{-1}(\omega, \mathbf{v_c}) = \exp(\omega\gamma^D\boldsymbol{\gamma}\cdot\hat{\mathbf{w}}/2)$$

$$= \cosh(\omega/2)I - \sinh(\omega/2)\gamma^D\boldsymbol{\gamma}\cdot\hat{\mathbf{w}} \qquad (70.21)$$

Note that S_C is not unitary just as in the 4-dimensional case.

We now apply a spinor boost to the Dirac equation for a particle at rest in this more general case of complex ω and $\hat{\mathbf{w}}$.

$$0 = S_C(\omega, \mathbf{v_c}))(m\gamma^D - m) \exp[-imt]w(0)$$
$$= [mS_C\gamma^D S_C^{-1} - m] \exp[-imt]S_Cw(0) \qquad (70.22)$$

where $S_C = S_C(\omega, \mathbf{v_c})$. After some algebra we find

$$mS_C\gamma^D S_C^{-1} = m[\cosh(\omega)\gamma^D - \sinh(\omega)\boldsymbol{\gamma}\cdot\hat{\mathbf{w}}] \qquad (70.23)$$

We will use these *complex* boosts to generate the other species' Dirac-like equations.

70.3.1.3 Tachyon Universe particle Dirac Equation

The development of the complex spinor boost transformation (subsection 70.3.1.2 above) leads to two possible forms of the tachyon Dirac-like equation. One form will lead to a lagrangian dynamics for left-handed universe particles. The other form leads to a lagrangian dynamics for right-handed universe particles.

70.3.1.4 Type IIa Case: Left-Handed Tachyonic Universe Particles

If the real and imaginary relative vectors parts of $\hat{\mathbf{w}}$, namely $\hat{\mathbf{u}}_r$ and $\hat{\mathbf{u}}_i$, are parallel, then $\hat{\mathbf{u}}_r\cdot\hat{\mathbf{u}}_i = 1$ and

$$\omega = \omega_r + i\omega_i \qquad (70.24)$$

Eqs. 70.23 and 70.24 then imply

$$mS_C\gamma^D S_C^{-1} = m[\cosh(\omega_r)\cos(\omega_i) + i\sinh(\omega_r)\sin(\omega_i)]\gamma^D -$$
$$- m[\sinh(\omega_r)\cos(\omega_i) + i\cosh(\omega_r)\sin(\omega_i)]\boldsymbol{\gamma}\cdot\hat{\mathbf{u}}_r \qquad (70.25)$$

or

$$mS_C\gamma^D S_C^{-1} = \cos(\omega_i)\gamma \cdot p_r + i\sin(\omega_i)\gamma \cdot p_i \qquad (70.26)$$

where

$$p_r^0 = m\cosh(\omega_r) \qquad p_i^0 = m\sinh(\omega_r) \qquad (70.27)$$

and

$$\mathbf{p_r} = m\hat{\mathbf{u}}_r\sinh(\omega_r) \qquad \mathbf{p_i} = m\hat{\mathbf{u}}_r\cosh(\omega_r) \qquad (70.28)$$

If $\omega_i = 0$, then we recover the momentum space Dirac-like equation. If $\omega_i = \pi/2$, then we obtain the left-handed momentum space tachyon equation:

$$mS_C\gamma^D S_C^{-1} = i\gamma \cdot p_i \qquad (70.29)$$

and the tachyon energy and momentum expressions

$$\mathbf{p} = m\mathbf{v}\gamma_s \qquad E = m\gamma_s \qquad (70.30)$$

where $\sinh(\omega) = \gamma_s = (\beta^2 - 1)^{-\frac{1}{2}}$ with $\beta = v/c > 1$. v is the absolute value of the $(D - 1)$ component spatial velocity. Also

$$S_C w(0) = w_C(p) \qquad (70.31)$$

is a tachyon spinor.

The momentum space tachyonic Dirac-like equation is

$$(i\not{p} - m)\exp[-ip\cdot y]w_T(p) = 0 \qquad (70.32)$$

where $p\cdot y = p^D y^D - \mathbf{p}\cdot\mathbf{y}$ after performing a corresponding boost in the exponential factor. If we apply $i\not{p}$ to eq. 70.32 we find the tachyon mass condition is satisfied

$$-E^2 + \mathbf{p}^2 = m^2 \qquad (70.33)$$

Transforming back to coordinate space we obtain the "left-handed" *tachyonic Dirac-like equation*:

$$(\gamma^\mu \partial/\partial y^\mu - m)\psi_T(y) = 0 \qquad (70.34)$$

70.3.1.5 Type IIb Case: Right-Handed Tachyonic Universe Particles

If the real and imaginary relative vectors parts of $\hat{\mathbf{w}}$, $\hat{\mathbf{u}}_r$ and $\hat{\mathbf{u}}_i$, are anti-parallel $\hat{\mathbf{u}}_r = -\hat{\mathbf{u}}_i$, then $\hat{\mathbf{u}}_r \cdot \hat{\mathbf{u}}_i = -1$ and

$$\omega = \omega_r - i\omega_i \qquad (70.35)$$

then

$$
\begin{aligned}
mS_C\gamma^D S_C^{-1} = {}& m[\cosh(\omega_r)\cos(\omega_i) - i\sinh(\omega_r)\sin(\omega_i)]\gamma^D - \\
& - m[\sinh(\omega_r)\cos(\omega_i) - i\cosh(\omega_r)\sin(\omega_i)]\gamma \cdot \hat{\mathbf{u}}_r \qquad (70.36)
\end{aligned}
$$

or

$$mS_C\gamma^D S_C^{-1} = \cos(\omega_i)\gamma \cdot p_r - i\sin(\omega_i)\gamma \cdot p_i \qquad (70.37)$$

where

$$p_r^D = m\cosh(\omega_r) \qquad p_i^D = m\sinh(\omega_r) \qquad (70.38)$$

and

$$\mathbf{p}_r = m\hat{\mathbf{u}}_r \sinh(\omega_r) \qquad \mathbf{p}_i = m\hat{\mathbf{u}}_r \cosh(\omega_r) \qquad (70.39)$$

If $\omega_i = \pi/2$, then we obtain the right-handed momentum space tachyon equation.[499]

$$(-\gamma^\mu \partial/\partial y^\mu - m)\psi_T(y) = 0 \qquad (70.40)$$

70.3.1.6 Type III Case: "Up-Quark-like" Universe Particles

There are two other cases where we can obtain fermion dynamical equations with a *real* time variable and real energy.[500] In one case we set $\hat{\mathbf{u}}_r \cdot \hat{\mathbf{u}}_i = 0$ and have a real ω.

---

[499] We note that $\gamma_s = (\beta^2 - 1)^{-\frac{1}{2}}$, *if expressed in terms of* ω, *has a branch cut extending from* $<-\infty, +\infty>$ *in the complex* ω *plane. Thus values of* ω *with positive imaginary parts are physically different from values of* ω *with negative imaginary parts.*

[500] The requirement of a real energy for a universe is not strict. For a fundamental free particle the energy must be real or the particle would be subject to decay – contrary to its assumed fundamental nature. Universes can 'decay' to 'smaller' universes. Therefore the requirement for real energy can be violated. Nevertheless the requirement for real

If the real and imaginary relative vectors parts of $\hat{\mathbf{w}}$, namely $\hat{\mathbf{u}}_r$ and $\hat{\mathbf{u}}_i$, are perpendicular, $\hat{\mathbf{u}}_r \cdot \hat{\mathbf{u}}_i = 0$, then

$$\omega = (\omega_r^2 - \omega_i^2)^{\frac{1}{2}} \tag{70.41}$$

Thus ω is either pure real ($\omega_r \geq \omega_i$) or pure imaginary ($\omega_r < \omega_i$).

The momentum space equation generated by the corresponding spinor boost is

$$\{m \cosh(\omega)\gamma^D - m \sinh(\omega)\boldsymbol{\gamma} \cdot (\omega_r\hat{\mathbf{u}}_r + i\omega_i\hat{\mathbf{u}}_i)/\omega - m\} \exp[-imt]w_c(p) = 0 \tag{70.42}$$

Defining the momentum 4-vector

$$p = (p^D, \mathbf{p}) \tag{70.43}$$

where

$$p^D = m \cosh(\omega) \qquad \mathbf{p} = \mathbf{p}_r + i\mathbf{p}_i \tag{70.44}$$

with

$$\mathbf{p}_r = m\omega_r\hat{\mathbf{u}}_r \sinh(\omega)/\omega \quad \mathbf{p}_i = m\omega_i\hat{\mathbf{u}}_i \sinh(\omega)/\omega \tag{70.45}$$

$$\mathbf{p}_r \cdot \mathbf{p}_i = 0 \tag{70.46}$$

then we obtain a positive energy Dirac-like equation

$$[p \cdot \gamma - m]\exp[-imt]w_c(p) = 0$$

or

$$[p^D\gamma^D - (\mathbf{p}_r + i\mathbf{p}_i) \cdot \boldsymbol{\gamma} - m]\exp[-ip \cdot y]w_c(p) = 0 \tag{70.47}$$

with a complex 3-momentum $\mathbf{p}$ and the 4-momentum mass shell condition:

$$p^2 = (p^D)^2 - \mathbf{p}_r \cdot \mathbf{p}_r + \mathbf{p}_i \cdot \mathbf{p}_i = m^2 \tag{70.48}$$

energy is appealing since it leads to four species of universes strengthening the analogy of universes to elementary particles.

Note

$$|\mathbf{v}_c| = |\mathbf{p}|/p^D = [(\mathbf{p_r} + i\mathbf{p_i}) \cdot (\mathbf{p_r} + i\mathbf{p_i})]^{\frac{1}{2}}/p^D = \tanh(\omega) \qquad (70.49)$$

and so the Lorentz factor is

$$\gamma = \cosh(\omega) \qquad (70.50)$$

Eq. 70.47 is the momentum space equivalent of the wave equation[501]

$$[i\gamma^D \partial/\partial t + i\boldsymbol{\gamma} \cdot (\boldsymbol{\nabla_r} + i\boldsymbol{\nabla_i}) - m]\psi_u(t, \mathbf{y_r}, \mathbf{y_i}) = 0 \qquad (70.51)$$

where $\mathbf{y} = \mathbf{y_r} - i\mathbf{y_i}$, and where the grad operators $\boldsymbol{\nabla_r}$ and $\boldsymbol{\nabla_i}$ are with respect to $\mathbf{y_r}$ and $\mathbf{y_i}$ respectively. Since $\hat{\mathbf{u}}_r \cdot \hat{\mathbf{u}}_i = 0$ we see that there is a subsidiary condition on the wave function

$$\boldsymbol{\nabla_r} \cdot \boldsymbol{\nabla_i} \, \psi_u(t, \mathbf{y_r}, \mathbf{y_i}) = 0 \qquad (70.52)$$

We note eq. 70.52 can be put into covariant form as the difference of two vectors squared (which is a real D-dimensional Lorentz group invariant):

$$[\gamma^D \partial/\partial t + i\boldsymbol{\gamma} \cdot (\boldsymbol{\nabla_r} + i\boldsymbol{\nabla_i})]^2 - [\gamma^D \partial/\partial t + i\boldsymbol{\gamma} \cdot (\boldsymbol{\nabla_r} - i\boldsymbol{\nabla_i})]^2 = 4\boldsymbol{\nabla_r} \cdot \boldsymbol{\nabla_i}.$$

We identify eq. 70.51 as the dynamical equation of an "up-quark-like" universe particle.

70.3.1.7 Type IVa Case: Left-Handed "Down-Quark-like" Tachyonic Universe Particles

In this case we set $\hat{\mathbf{u}}_r \cdot \hat{\mathbf{u}}_i = 0$. Then by eq. 70.7

$$\omega = (\omega_r^2 - \omega_i^2)^{\frac{1}{2}}$$

---

[501] The gradient operators $\boldsymbol{\nabla_r}$ and $\boldsymbol{\nabla_i}$ are 191-dimensional spatial gradient operators.

Thus ω again starts out either pure real (if $\omega_r \geq \omega_i$) or pure imaginary (if $\omega_r < \omega_i$). In this case we also choose ω real, and then change ω to

$$\omega = (\omega_r^2 - \omega_i^2)^{\frac{1}{2}} \to \omega' = (\omega_r^2 - \omega_i^2)^{\frac{1}{2}} + i\pi/2 = \omega + i\pi/2$$

by adding $i\pi/2$ to ω since ω is a free parameter. We then proceed as we did in the prior tachyon case.[502]. The resulting Lorentz boost

$$\Lambda_C = \exp[i((\omega_r^2 - \omega_i^2)^{\frac{1}{2}} + i\pi/2)(\omega_r \hat{u}_r + i\omega_i \hat{u}_i)\cdot K/\omega] \qquad (70.53)$$

becomes a left-handed "quark-like" boost. The tachyon dynamical equation is[503]

$$[\gamma^D \partial/\partial t + \gamma \cdot (\nabla_r + i\nabla_i) - m]\psi_d(y) = 0 \qquad (70.54)$$

with the constraint equation

$$\nabla_r \cdot \nabla_i \, \psi_d(t, y_r, y_i) = 0 \qquad (70.55)$$

We will call the universe particles satisfying eqs. 70.54 and 70.55 left-handed *tachyonic quark-like universe particles*.

70.3.1.8 Type IVb Case: Right-Handed Down-Quark-like Tachyonic Universe Particles
In this case we set $\hat{u}_r \cdot \hat{u}_i = 0$. Then by eq. 70.7

$$\omega = (\omega_r^2 - \omega_i^2)^{\frac{1}{2}}$$

Thus ω again starts out either pure real (if $\omega_r \geq \omega_i$) or pure imaginary (if $\omega_r < \omega_i$). In this case we also choose ω real, and then change ω to

---

[502] Here again the choice of ω in eq. 38.53 leads to a "left-handed" universe particle while the choice $\omega' = \omega - i\pi/2$ leads to a right-handed one.
[503] The gradient operators ∇_r and ∇_i are $(D-1)$-dimensional spatial gradient operators.

$$\omega = (\omega_r^{\,2} - \omega_i^{\,2})^{\frac{1}{2}} \rightarrow \omega' = (\omega_r^{\,2} - \omega_i^{\,2})^{\frac{1}{2}} - i\pi/2 = \omega - i\pi/2$$

since ω is a free parameter and proceed as we did in the prior case. The resulting Lorentz boost

$$\Lambda_C = \exp[i((\omega_r^{\,2} - \omega_i^{\,2})^{\frac{1}{2}} - i\pi/2)(\omega_r \hat{\mathbf{u}}_r + i\omega_i \hat{\mathbf{u}}_i) \cdot \mathbf{K}/\omega] \qquad (70.56)$$

becomes a right-handed quark-like boost. The resulting tachyon dynamical equation is

$$[-\gamma^{\mathbf{D}} \partial/\partial t - \gamma \cdot (\nabla_r + i\nabla_i) - m]\psi_d(y) = 0 \qquad (70.57)$$

with the constraint equation

$$\nabla_r \cdot \nabla_i \, \psi_d(t, \mathbf{y}_r, \mathbf{y}_i) = 0 \qquad (70.58)$$

We will call the universe particles satisfying eqs. 70.57 and 70.58 right-handed *tachyonic quark-like universe particles.*

70.3.2 Lagrangians

In this section we will develop a lagrangian formalism for each of the four types of fermion universe particles noting that a tachyonic universe particles have two forms: left-handed and right-handed (discussed later in section 70.3.5).

The various types of universe particles described in section 70.3.1 correspond to universes with differing internal characteristics and motion in the Megaverse. The equations are all free field equations. Internal potentials and interactions must be introduced in these equations to complete the universe dynamical equations. A connection to the Wheeler-DeWitt description of their internal quantum structure also remains to be established (section 70.3.6).

In defining the lagrangians for the four fermion universe types that yield their dynamical equations in a canonical manner, we require the conventional quantum field theory feature that the Hamiltonian derived from the lagrangian is hermitean. We will develop a separate lagrangian for each type.

70.3.2.1 Type I Universe Particle Lagrangian

The Universe particle Dirac equation lagrangian is

$$\mathcal{L}_u = \bar{\psi}(i\gamma^\mu \partial/\partial y^\mu - m)\psi(y) \tag{70.59}$$

where

$$\bar{\psi} = \psi^\dagger \gamma^D$$

and $\psi^\dagger$ is the hermitean conjugate of ψ.

70.3.2.2 Type II Tachyon Universe Particle Lagrangian

This lagrangian includes both left-handed and right-handed cases. It can be separated into lagrangian terms for each case using parity projection operators.

$$\mathcal{L}_{uT} = \psi_T^{\;S}(\gamma^\mu \partial/\partial y^\mu - m)\psi_T(y) \tag{70.60}$$

where

$$\psi_T^{\;S} = \psi_T^{\;\dagger} i\gamma^D \gamma^5 \tag{70.61}$$

with γ^5 being the D-dimensional equivalent for γ^5 in 4 dimensions. The peculiar form of the tachyon universe lagrangian is necessitated by the hermiticity of the Hamiltonian calculated from it.

70.3.2.3 Type III "Up-Quark-like" Universe Particle Lagrangian

The lagrangian density of a free "up-quark-like" universe particle is

$$\mathcal{L}_u = \bar{\psi}_u(i\gamma^\mu D_\mu - m)\psi_u(y) \tag{70.62}$$

where $\bar{\psi}_u = \psi_u^{\;\dagger}\gamma^D$ and

$$\psi_u^{\;\dagger} = [\psi_u(\mathbf{y_r},\, \mathbf{y_i})]^\dagger\,|_{\mathbf{y_i} = -\mathbf{y_i}} \tag{70.63}$$

$$D_D = \partial/\partial y^D$$
$$D_k = \partial/\partial y_r^{\ k} + i\,\partial/\partial y_i^{\ k} \qquad (70.64)$$

for $k = 1, 2, \ldots, (D-1)$. The action

$$I = \int d^{(D-1)}y\,\mathcal{L}_u \qquad (70.65)$$

It is easy to show that this action is also real.

70.3.2.4 Type IV "Down-Quark-like" Tachyon Universe Particle Lagrangian
      The lagrangian density of a free "down-quark-like" universe particle is

$$\mathcal{L}_d = \psi_d^{\ C}(y)(\gamma^D\partial/\partial t + \boldsymbol{\gamma}\cdot(\boldsymbol{\nabla_r} + i\boldsymbol{\nabla_i}) - m)\psi_d(y) \qquad (70.66)$$

where

$$\psi_d^{\ C}(y) = [\psi_d(y)]^\dagger|_{\mathbf{y_i} = -\mathbf{y_i}}\ i\gamma^D\gamma^5 \qquad (70.67)$$

In words, eq. 70.67 states: take the hermitean conjugate of $\psi_d(y)$; change $\mathbf{y_i}$ to $-\mathbf{y_i}$; and then post-multiply by the indicated factors.
      The action is

$$I = \int d^{(D-1)}y\,\mathcal{L}_d \qquad (70.68)$$

The action is real. The lagrangian can also be separated into left-handed and right-handed parts using projection operators.

70.3.3 Form of The Megaverse Quantum Coordinates Gauge Field

The discussions of sections 70.3.1 and 70.3.2 assumed the coordinates were Megaverse coordinates and their derivatives. Prior to those discussions we indicated we would use quantum coordinates in the Megaverse of the form[504]

$$Y^i(y) = y^i + i\, Y_u^i(y)/M_u^{D/2} \tag{70.4}$$

and their derivatives

$$\partial_i = \partial/\partial Y^i(y) = \partial/\partial(y^i - Y_u^i(y)/M_u^{D/2}) \tag{70.5}$$

for $i = 1, 2, \ldots, D$ to eliminate divergences in quantum field theory. The subscript "u" signifies universes. The mass constant for the Megaverse, M_u, may be the same as the mass constant M_c appearing in the Two-Tier mechanism for our universe.

In this section we define the gauge fields $Y_u^i(y)$ of the Megaverse.[505] They are similar to the $Y^\mu(y)$ fields of our New Standard Model.[506] The $Y_u(y)$ D-dimensional vector gauge field, in the absence of external sources, will be defined in a D-dimensional Coulomb gauge:

$$Y_u^D(y) = 0 \tag{70.69}$$
$$\partial Y_u^j(y)/\partial y^j = 0$$

where the sum over j is over the $D - 1$ spatial y coordinates. We follow a procedure similar to Blaha (2003) but for D-dimensional space. The lagrangian density for the free $Y_u^j(y)$ fields is

$$\mathscr{L}_u = -\tfrac{1}{4}\, F_u^{\mu\nu} F_{u\mu\nu} \tag{70.70}$$

and the lagrangian is

$$L_u = \int d^{(D-1)} y\, \mathscr{L}_u \tag{70.71a}$$

with

---

[504] The denominator $M_u^{D/2}$ is necessitated by the dimension of $Y_u^i(y)$ which is $[m]^{D/2-1}$. Eqs. 38.78 and 38.81 below imply this conclusion.

[505] This choice implies that the Megaverse Y mass $M_u = M_C$, its universe mass.

[506] See Blaha (2005a) for details.

$$F_{u\mu\nu} = \partial Y_{u\mu}/\partial y^{\nu} - \partial Y_{u\nu}/\partial y^{\mu} \qquad (70.71b)$$

The equal time commutation relations, derived in the usual way, are:

$$[Y_u^{\mu}(\mathbf{y}, y^0), Y_u^{\nu}(\mathbf{y}', y^0)] = [\pi_u^{\mu}(\mathbf{y}, y^0), \pi_u^{\nu}(\mathbf{y}', y^0)] = 0 \qquad (70.72)$$

$$[\pi_u^{j}(\mathbf{y}, y^0), Y_{uk}(\mathbf{y}', y^0)] = -i\, \delta^{(D-1)\mathrm{tr}}{}_{jk}(\mathbf{y} - \mathbf{y}') \qquad (70.73)$$

for $\mu, \nu, j, k = 1, 2, \ldots, (D-1)$ where

$$\pi_u^{k} = \partial \mathscr{L}_u/\partial Y_{uk}' \qquad (70.74)$$

$$\pi_u^{0} = 0 \qquad (70.75)$$

and

$$\delta^{\mathrm{tr}}{}_{jk}(\mathbf{y} - \mathbf{y}') = \int d^{(D-1)}k\; e^{i\, \mathbf{k}\bullet(\mathbf{y}-\mathbf{y}')}\, (\delta_{jk} - k_j k_k/\mathbf{k}^2)/(2\pi)^{D-1} \qquad (70.76)$$

$$Y_{uk}' = \partial Y_{uk}/\partial y^D \qquad (70.77)$$

The Coulomb gauge indicates $D-2$ degrees of freedom are present in the vector potential. The Fourier expansion of the vector potential is:

$$Y_u^{i}(y) = \int d^{(D-1)}k\; N_0(k) \sum_{\lambda=1}^{D-2} \varepsilon^{i}(k, \lambda)[a(k,\lambda)\, e^{-ik\cdot y} + a^{\dagger}(k,\lambda)\, e^{ik\cdot y}] \qquad (70.78)$$

where

$$N_0(k) = [(2\pi)^{(D-1)}2\omega_k]^{-\frac{1}{2}} \qquad (70.79)$$

and (since the field is massless)

$$k^D = \omega_k = (\mathbf{k}^2)^{\frac{1}{2}} \qquad (70.80)$$

where k^D is the energy, and where the $\varepsilon^{i}(k, \lambda)$ are the polarization unit vectors for $\lambda = 1, \ldots, (D-2)$ and $k^{\mu}k_{\mu} = k^{D\,2} - \mathbf{k}^2 = 0$.

The commutation relations of the Fourier coefficient operators are:

$$[a(k,\lambda), a^\dagger(k',\lambda')] = \delta_{\lambda\lambda'}\delta^{(D-1)}(\mathbf{k} - \mathbf{k}') \tag{70.81}$$

$$[a^\dagger(k,\lambda), a^\dagger(k',\lambda')] = [a(k,\lambda), a(k',\lambda')] = 0 \tag{70.82}$$

and the polarization vectors satisfy

$$\sum_{\lambda=1}^{D-2} \varepsilon_i(k, \lambda)\varepsilon_j(k, \lambda) = (\delta_{ij} - k_i k_j/\mathbf{k}^2) \tag{70.83}$$

It will be convenient to divide the Y field into positive and negative frequency parts:

$$Y_{u}{}^{+}{}_{i}(y) = \int d^{(D-1)}k \, N_0(k) \sum_{\lambda=1}^{D-2} \varepsilon_i(k, \lambda) \, a(k,\lambda) \, e^{-ik\cdot y} \tag{70.84}$$

and

$$Y_{u}{}^{-}{}_{i}(y) = \int d^{(D-1)}k \, N_0(k) \sum_{\lambda=1}^{D-2} \varepsilon_i(k, \lambda) \, a^\dagger(k,\lambda) \, e^{ik\cdot y} \tag{70.85}$$

For later use we note the commutator between the positive and negative frequency parts is:

$$[Y_{u}{}^{-}{}_{j}(y_1), Y_{u}{}^{+}{}_{k}(y_2)] = - \int d^{(D-1)}k \, e^{ik\cdot(y_1 - y_2)} \, (\delta_{jk} - k_j k_k/\mathbf{k}^2)/[(2\pi)^{D-1} 2\omega_k] \tag{70.86}$$

70.3.3.1 Y^μ Fock Space Imaginary Coordinate States

States can also be defines for the quantized Y^μ field. These states will be similar in form to electromagnetic photon states but play a different role in our approach since they are in fact coordinate excitation states for the imaginary part of $Y^i(y)$ (eq. 70.4). Thus universe particles (and other fields) will exist in a real D-dimensional space with quantum excitations into imaginary Quantum Dimensions. These excitations become significant at high energies. At low energies space appears as c-number complex; at very high energies space becomes slightly q-number complex.

There are two types of imaginary coordinate excitations: 1.) Quantum excitations into Fock states consisting of a superposition of states with a definite finite number of Y_u "particles" and 2.) Imaginary coordinate excitations into coherent Y_u

states with an "infinite" number of particles. Coherent states can be viewed as representing "classical" fields.

In this section we will consider Y_u field states with a definite number of excitations ("particles"). The raising and lowering operators of the Y_u field can be used to define free particle states. For example a one particle state can be defined by

$$|k, \lambda> = a^\dagger(k, \lambda)|0>$$
(70.87)

with corresponding bra state

$$<k, \lambda| = <0|a(k, \lambda)$$
(70.88)

where the "coordinate vacuum" is defined as usual:

$$a(k, \lambda)|0> = 0$$
(70.89)
$$<0|a^\dagger(k, \lambda) = 0$$
(70.90)

Multi-particle states can also be defined in the conventional way with products of the raising and lowering operators applied to the vacuum. The set of all states containing a finite number of "particles" constitutes a Fock space.

A state with a finite number of Y_u "particles" represents a quantum fluctuation into imaginary Quantum Dimensions.

70.3.3.2 Y_u Coherent Imaginary Coordinate States

Coherent Y_u states bring us closer what we might consider to be "classical" imaginary dimensions – dimensions that we can, in principle, experience as we do normal dimensions. Let us define the coherent state[507]

$$| y, p> = e^{-\mathbf{p} \cdot \mathbf{Y}u^-(y)/M_u^{D/2}}|0>$$
(70.91)

---

[507] Coherent states are well known in the physics literature. See for example T. W. B. Kibble, J. Math. Phys. **9**, 315 (1968) and references therein; V. Chung, Phys. Rev. **140**, B1110 (1965); J. R. Klauder, J. McKenna, and E. J. Woods, J. Math. Phys. **7**, 822 (1966) and references therein.

This state is an eigenstate of the coordinate operator $Y_u^+(y')$:

$$Y_u^+{}_j(y_1) \, |y_2, p> = -[Y_u^+{}_j(y_1), \, \mathbf{p}\cdot\mathbf{Y}^-(y_2)]/M_u^{D/2}|y, p> \qquad (70.92)$$

$$= -\int d^{D-1}k \, [N_0(k)]^2 \, e^{ik\cdot(y_2 - y_1)} \, (p_j - k_j \mathbf{p}\cdot\mathbf{k}/\mathbf{k}^2)/M_u^{D/2}|y, p>$$

$$= p^i \Delta_{Tij}(y_1 - y_2)/M_u^{D/2}|y, p> \qquad (70.93)$$

where $p^i \Delta_{Tij}(y_1 - y_2)/M_u^{D/2}$ is the eigenvalue of $Y_u^+{}_j(y_1)$. As we will see later, the eigenvalue of Y_u^+ becomes large as $(y_1 - y_2)^2 \rightarrow 0$. Thus the imaginary Quantum Dimensions become significant at very short distances, and then significantly modifies the high-energy behavior of quantum field theories. In particular, Quantum Dimensions have a significant effect when

$$(y_1 - y_2)^2 \lessgtr (2^{D-2}\pi^{D-2}M_u^2)^{-1} \qquad (70.94)$$

We assume the mass scale $M_u = M_C$ is very large – perhaps of the order of the Planck mass (1.221×10^{19} GeV/c^2).

70.3.3.3 Quantization of the Type I Free Universe Particle Dirac Field

The quantization procedure is formally identical to that of a conventional Dirac particle. The standard equal time anti-commutation relations for a D-dimensional fermion field are:

$$\{\psi_\alpha(Y), \, \psi_\beta(Y')\} = \{\pi_{\psi\alpha}(Y), \, \pi_{\psi\beta}(Y')\} = 0 \qquad (70.95)$$

$$\{\pi_{\psi\alpha}(Y), \, \psi_\beta(Y')\} = i\,\delta_{\alpha\beta}\,\delta^{D-1}(\mathbf{Y} - \mathbf{Y}') \qquad (70.96)$$

where α and β are the spinor indices ranging from 1 to $N_{MRC} = 2^{D/2}$ and where

$$\pi_{\psi\alpha}(Y) = i\,\psi_\alpha^\dagger(Y) \qquad (70.97)$$

The field can be expanded in a fourier series:

$$\psi(Y(y)) = \sum_s \int d^{D-1}p \, N^d_m(p) \, [b(p,s)u(p,s) :e^{-ip\cdot(y + iYu/M_u^{D/2})}: \; + $$

$$+ \, d^\dagger(p,s)v(p,s) :e^{ip\cdot(y + iYu/M_u^{D/2})}:] \tag{70.98}$$

$$\psi^\dagger(Y(y)) = \sum_s \int d^{D-1}p \, N^d_m(p) \, [b^\dagger(p,s)\bar{u}(p,s)\gamma^0 :e^{+ip\cdot(y + iYu/M_u^{D/2})}: \; + $$

$$+ \, d(p,s)\bar{v}(p,s)\gamma^0 :e^{-ip\cdot(y + iYu/M_u^{D/2})}:] \tag{70.99}$$

where

$$N^d_m(p) = [m/((2\pi)^{D-1}E_p)]^{\frac{1}{2}} \tag{70.100}$$

and

$$E_p = p^D = (\mathbf{p}^2 + m^2)^{\frac{1}{2}} \tag{70.101}$$

with : ... : signifying normal ordering. The commutation relations of the Fourier coefficient operators are:

$$\{b(p,s), b^\dagger(p',s')\} = \delta_{ss'}\delta^{D-1}(\mathbf{p} - \mathbf{p}') \tag{70.102}$$
$$\{d(p,s), d^\dagger(p',s')\} = \delta_{ss'}\delta^{D-1}(\mathbf{p} - \mathbf{p}') \tag{70.103}$$
$$\{b(p,s), b(p',s')\} = \{d(p,s), d(p',s')\} = 0 \tag{70.104}$$
$$\{b^\dagger(p,s), b^\dagger(p',s')\} = \{d^\dagger(p,s), d^\dagger(p',s')\} = 0 \tag{70.105}$$
$$\{b(p,s), d^\dagger(p',s')\} = \{d(p,s), b^\dagger(p',s')\} = 0 \tag{70.106}$$
$$\{b^\dagger(p,s), d^\dagger(p',s')\} = \{d(p,s), b(p',s')\} = 0 \tag{70.107}$$

The spinors u(p,s) and v(p,s) are defined in a conventional way (as in Bjorken and Drell). However their form is different from the 4-dimensional case. If one takes the $N_{MRC} \times N_{MRC} \equiv 2^{D/2} \times 2^{D/2}$ $\gamma \cdot p$ matrix, then the first $2^{D/2-1}$ columns give u(p,s) up to a normalization for the free particle case, the remaining $2^{D/2-1}$ columns give v(p,s) up to a normalization.

Since there are $2^{D/2-1}$ possible spin values, using the equation 2s + 1 = total number of spin values, we see that the spin of a fermion universe particle is

$$s_M = 2^{D/2 - 2} - \tfrac{1}{2}$$

The possible universe particle spin values are:

Up spin values: $+1/2^{D/2 - 1}$, $+2/2^{D/2 - 1}$, ... , $+2^{D/2 - 2}/2^{D/2 - 1}$
Down spin values: $-2^{D/2 - 2}/2^{D/2 - 1} = -\tfrac{1}{2}$, $-2^{D/2 - 2}/2^{D/2 - 1} + 1$, ... , $-1/2^{D/2 - 1}$

This enormous number of possible spins is reasonable considering the number of dimensions D and the enormous variety in spins one should expect in universes – given their large size and complexity.

70.3.3.4 Feynman Propagators for the Type I Free Universe Particle Dirac Field

The form of the fermion universe particle Feynman propagator differs from a conventional fermion propagator by having a Gaussian factor $R(\mathbf{p}, z)$ in its fourier expansions. This follows from using quantum Megaverse coordinates (eq. 70.4).

$$iS_F^{TT}(y_1 - y_2) = <0|T(\overline{\psi}(Y(y_1))\psi(Y(y_2)))|0> \qquad (70.108)$$

where the time ordering is with respect to $y_1{}^D$ and $y_2{}^D$. Expanding the free fields leads to the fourier representation:

$$iS_F^{TT}(y_1 - y_2) = i \int \frac{d^D p \; e^{-ip\cdot(y_1 - y_2)} (\not{p} + m) \, R(\mathbf{p}, y_1 - y_2)}{(2\pi)^D (p^2 - m^2 + i\varepsilon)} \qquad (70.109)$$

where

$$R(\mathbf{p}, y_1 - y_2) = \exp[-p^i p^j \Delta_{Tij}(y_1 - y_2)/M_u{}^D] \qquad (70.110)$$
$$= \exp\{-p^2[A(v) + B(v)\cos^2\theta] / [(2\pi)^{D-2} M_c{}^4 z^2]\} \qquad (70.111)$$

(Note p^2 is the square of the spatial $(D - 1)$-vector.) with

$$z^\mu = y_1{}^\mu - y_2{}^\mu \qquad (70.112)$$

$$z = |\mathbf{z}| = |\mathbf{y_1} - \mathbf{y_2}| \qquad (70.113)$$

$$p = |\mathbf{p}| \qquad (70.114)$$

$$v = |z^0|/z \qquad (70.115)$$

$$A(v) = (1 - v^2)^{-1} + .5v \, \ln[(v - 1)/(v + 1)] \qquad (70.116)$$

$$B(v) = v^2(1 - v^2)^{-1} - 1.5v \, \ln[(v - 1)/(v + 1)] \qquad (70.117)$$

$$\mathbf{p \cdot z} = pz \cos\theta \qquad (70.118)$$

and $|\mathbf{p}|$ denoting the length of a spatial $(D - 1)$-vector $\mathbf{p}$ while $|z^0|$ is the absolute value of $z^0 \equiv z^D$.

As eq. 70.109 indicates, the Gaussian damping factor[508] $R(p, z)$ for large spatial momentum p is the same for both the positive and negative frequency parts of the Two-Tier Feynman propagator. We are assuming the spatial momentum is real-valued in this discussion. It is also important to note that $R(p, z)$ does not depend on $p^0 = p^D$ (in the Y Coulomb gauge) and thus the integration over p^0 proceeds in the usual way to produce time-ordered positive and negative frequency parts.

70.3.3.5 Feynman Propagators for the Types II, III, and IV Free Universe Particle Dirac Fields

These propagators differ in details from the Type I propagator. The differences modulo the change in dimension appear in Blaha (2011c). See also Blaha (2005a) for a detailed discussion of 4-dimensional spin ½ particle propagators.

70.3.4 Expanding and Contracting Universes: Impact of Time Dependent Universe Particle Masses

Our discussions of the dynamics of universe particles assumed their masses were constant. However the definition of mass in terms of the area of a universe based on the physics of black holes is

$$M = \kappa A/8\pi \qquad (70.119)$$

---

[508] Note the Gaussian damping is for all $D - 1$ spatial momentum integrations.

where A is the area of the black hole shows that *the mass of a universe particle is time dependent* because the area of a universe is generally time dependent. For example, our universe is expanding and its surface area is thus growing with time.

Eqs. 70.11 (and subsequent fermion dynamic equations) must then be modified from

$$\psi(y) = \exp[-imt]w(0) \tag{70.11}$$

to a covariant form:

$$\psi(y) = \exp[-i\int_0^{w\cdot y} m(t')dt']w(0) \tag{70.120}$$

where w is a unit D-vector in the time (y^D) direction ($w^2 = 1$). The lower bound on the integral, 0, is the time of the beginning of the universe particle – its Big Bang. Thus the cumulative change in the mass of the universe particle may be significant. It is interesting to note that the Wheeler-Dewitt equation also has a variable value mass term R that also depends on the evolution of the universe.

Eq. 70.120 satisfies the free covariant Dirac-like universe particle field dynamic equation

$$[i\gamma^i\partial/\partial y^i - m(w\cdot y)]\psi(y) = 0 \tag{70.121}$$

In contrast to the constant mass equation eq. 70.19. Substituting eq. 70.120 in eq. 70.121 we find

$$(\gamma^i w_i\, m(w\cdot y) - m(w\cdot y))\psi(y) = 0 \tag{70.122}$$

or

$$(\gamma^i w_i - 1)\psi(y) = 0 \tag{70.123}$$

Upon performing a D-dimensional Lorentz boost (of the type of eqs. 70.13 – 70.16) on eq. 70.123 we obtain

$$(\gamma_i p^i/m_0 - 1)\psi(y) = 0$$

or

$$(\gamma_i p^i - m_0)\psi(y) = 0 \tag{70.124}$$

where p^i is a momentum D-vector with $p^2 = m_0^2$. Eq. 70.123 is the constant mass momentum space dynamic equation. It determines the spinor in $\psi(y)$. After taking account of the quantum coordinates the quantum Dirac-like universe particle wave function has the form

$$\psi(Y(y)) = \sum_s \int d^{(D-1)}p \; N^d_m(p) \; [b(p,s)u(p,s) : \exp[-iG(p, Y(y))]: + d^\dagger(p,s)v(p,s) \cdot$$
$$\cdot :\exp[+iG(p, Y(y))]:\} \tag{70.125}$$

$$\psi^\dagger(Y(y)) = \sum_s \int d^{(D-1)}p \; N^d_m(p) \; \{b^\dagger(p,s)\bar{u}(p,s)\gamma^0 :\exp[+iG(p, Y(y))]: + d(p,s)\bar{v}(p,s)\gamma^0 \cdot$$
$$\cdot \exp[-iG(p, Y(y))]:\} \tag{70.126}$$

where : ... : denotes normal ordering and

$$G(p, Y(y)) = \int_0^{p \cdot Y(y)/\lambda} m(t')dt' \tag{70.127}$$

with $\lambda = m_0$, and $N^d_m(p)$ a normalization constant. Contrast eqs. 70.125-70.126 to the constant mass case eqs. 70.98-70.101. The *constant mass case* simply sets $m(t') = m_0$.

If we examine the integral eq. 70.127 for a short time interval δt in the particle's rest frame then $G(p, Y(y)) \approx m(0)\delta t$ and so we define $m(0) = m_0$. Based on the formula for universe particle mass (eq. 70.119) we anticipate that m_0 might be as large as the Planck mass or larger – thus an extremely short radius. Blaha (2013) describes a quantum Big Bang model in which the initial radius of the universe is $O(EM_{Planck}^{-2})$ where E is of the order of 1 and has the dimensions of [mass].

Thus we have a closed form definition of a quantum universe particle wave function for universe particles of type I. A similar procedure can be followed for universe particles of types II, III, and IV.

The Feynman propagator for type I quantum fields is *not* eq. 70.109 but now has a form reflecting the Y(y) dependence of the quantum fields in eqs. 70.125 and 70.126:

$$iS_F^{TT}(y_1, y_2) = i \int \frac{d^D p \ \{ <0|\theta(y_{1D} - y_{2D})G(y_1, y_2) + \theta(y_{2D} - y_{1D})G(y_2, y_1)\}0>}{(2\pi)^D \ (p - m_0)} \qquad (70.128)$$

where p^D is the energy and

$$G(y_1, y_2) = \ : \exp[-iG(p, Y(y_1))] : \ :\exp[+iG(p, Y(y_2))] : \qquad (70.129)$$

Let

$$G_{tot}(y_1, y_2) = <0|\theta(y_{1D} - y_{2D})G(y_1, y_2) + \theta(y_{2D} - y_{1D})G(y_2, y_1)\}0> \qquad (70.130)$$
$$= <0|\theta(y_{1D} - y_{2D}):\exp[-iG(p, Y(y_1))]::\exp[+iG(p, Y(y_2))]: +$$
$$+ \ \theta(y_{2D} - y_{1D}) :\exp[-iG(p, Y(y_2))]::\exp[+iG(p, Y(y_1)):]|0>$$

$$= <0|\theta(y^D_1 - y^D_2): \exp[-i\int_0^{p \cdot Y(y1)/\lambda} m(t')dt']::\exp[+i\int_0^{p \cdot Y(y2)/\lambda} m(t')dt']: +$$

$$+ \ \theta(y^D_2 - y^D_1):\exp[-i \exp[+i\int_0^{p \cdot Y(y2)/\lambda} m(t')dt']::\exp[+i\int_0^{p \cdot Y(y1)/\lambda} m(t')dt']:|0>$$

with $\lambda = m_0$ then

$$iS_F^{TT}(y_1, y_2) = i \int \frac{d^D p \ G_{tot}(y_1, y_2)}{(2\pi)^D \ (p - m_0)} \qquad (70.131)$$

Except for the case of a constant mass, where $m(t) = m_0$, the Feynman propagator is not a function of $y_1 - y_2$. The evaluation of eq. 70.130 in the general case of a variable mass is straightforward but cumbersome. For the special case of a linear time dependence of the mass, m(t) = at, we find eq. 70.130 gives

$$G_{tot}(y_1, y_2) = <0|\theta(y^D_1 - y^D_2):exp[-ia(p \cdot Y(y_1)/m_0)^2/2]::exp[+ia(p \cdot Y(y_2)/m_0)^2/2]: +$$
$$+ \theta(y^D_1 - y^D_2):exp[-ia(p \cdot Y(y_2)/m_0)^2/2]::exp[+ia(p \cdot Y(y_1)/m_0)^2/2]:|0>$$

$$(70.132)$$

yielding a complex function of p, y_1, and y_2. *Note that the lower bound of the integrals in the Feynman propagator cancel and thus the need for an understanding of the beginning of a universe is removed in this case.*

We have shown that universe particle theory can handle the case of a variable universe mass m(t). Expanding or contracting (or oscillating) universe particles correspond to expanding and contracting (or oscillating) universes.

70.3.5 Left-Handed and Right-handed Universe Particles

In sections 70.3.1 and 70.3.2 we found that left-handed and right-handed tachyonic universe particles existed. The tachyonic nature of the universe particles indicates that their speed in the universe exceeds the "speed of light" of the Megaverse. The physical meaning of the handedness of these types of universes is an interesting issue. When we consider our universe we see left-handedness in the weak interactions of elementary particles. In addition it appears that organic molecules overwhelmingly favor left-handedness on earth although right-handed molecules exist in outer space and can be created in the laboratory. Right-handed molecules transform into left-handed molecules in watery media through electromagnetic effects.

Why nature favors left-handedness is an open question. It has given rise to speculations that gravitation, especially quantum gravitons, may be left-handed. The European Space Agency's Planck telescope will study polarization effects in the cosmos and may well be able to show that the gravitons starting from the beginning of the universe, and magnified by inflation in the universe's expansion, may be left-handed.

If handedness of gravitation is verified experimentally, then our theory of left-handed/right-handed universe particles would be supported. *Our universe would then be tachyonic and probably left-handed. We, in the universe, would, of course, not know of the velocity of the universe in the Megaverse.*

70.3.6 Internal Structure of Universe Particles

We have treated universes as particles in the preceding discussion taking an extremely large view of Megaverse particles just as elementary particle theory viewed nucleons at low energies (large distances). Now we develop a more detailed view of universe particles in a manner analogous to the high energy view of the internal dynamics of nucleons that led to the quark-parton model of nucleons. In the present case we shall see that high energy Baryonic and other field probes of universe particles can yield a model of the internal structure of universes.

We know that universes are composed of matter and radiation. We believe that there is at least one possible accessible interaction between universes dependent on baryon number – a baryonic, D-dimensional gauge field. There are also other particle number gauge fields – but these are less likely to be significant since Dark matter has yet to be found except through its gravitational effects. In this section we will discuss the use of a baryonic gauge field to probe the baryon structure of universe particles.

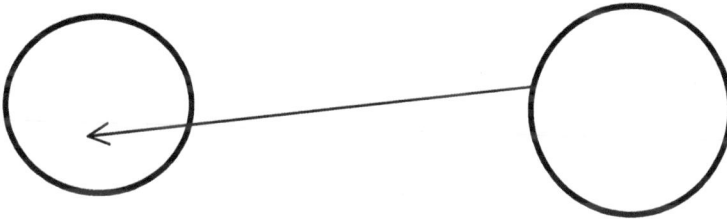

Figure 70.1. A symbolic view of a high resolution (high energy) probe from a universe to a specific baryonic part of another universe.

Figure 70.2. A symbolic view of a low resolution (lower energy) probe from a universe to an entire universe.

There appears to be two types of probes[509] of a universe: 1) a series of high energy probes to specific small regions inside another universe for the purpose of mapping its internal structure; 2) a low energy probe of another universe to get a global view of its structure.[510] The first type of probe corresponds to deep inelastic (high energy) electron-nucleon scattering which led to the quark-parton model of nucleons. The second type of probe corresponds to low energy electron-nucleon scattering to get a "global" view of a nucleon. In both case an electromagnetic (gauge) field particle (photon) was the probe particle.

Besides the inherent scientific interest in such experiments it is possible that they may be of use in the very distant future if Mankind is able to develop Megaverse starships that can travel in the Megaverse to other universes. Then the baryonic gauge field may become the "eyes" of the starship in addition to the electromagnetic field (light) are the eyes of current spaceships. We considered the possibility of universe starships in the book entitled, *All the Megaverse! II* in detail.

70.5 Boson Universe Particles

The previous sections has described fermion universe particles. In this section we will briefly describe aspects of boson (spin 0) universe particles. First it is important to note that the Wheeler-DeWitt equation being second order like the Klein-Gordon equation seems to suggest that universe particles can be boson – like Klein-Gordon equation particles.[511] The Wheeler-DeWitt equation has a mass-like term R that can be positive or negative. If the mass term is negative then the wave-like propagation of the state functional (wave equation solution) can be in space-like directions implying a tachyonic solution. Thus the Wheeler-DeWitt equation supports "normal" state functionals that propagate in time-like directions as well as tachyonic propagation.

---

[509] It would appear that the probes are in Megaverse coordinates, not universe coordinates, in order to 'bridge' the distance between universes. Chapter 35 shows the manner of the mapping between Megaverse coordinates and universe coordinates for quantum fields.

[510] **We note these probes, being limited by the speed of light, are at best theoretical speculations since the travel time between universes is so very large.**

[511] One should remember that the Wheeler-DeWitt equation is not in space but in a 6-dimensional manifold, denoted M, of metrics with one "time" dimension – having hyperbolic signature $- + + + + +$ when the metric is positive definite. See DeWitt's paper.

For this reason we suggest that boson universe particles can be either normal or tachyonic. Tachyonic boson universe particles can fission in a manner similar to tachyonic fermion universe particles. The fission equations of section 70.5.2 also apply to tachyonic boson universe particles.

The quantum field theory of normal and tachyonic boson universe particles is similar to that of ordinary bosons. See Blaha (2005a) for the boson case discussion that is paralleled by our universe particle formalism.

70.6 Physical Meaning of Universe Particle Spin

The physical meaning of spin is a continuing discussion topic. We have suggested[512] that spin states are in essence logic states with changes in spin an analogous to changes in logical values in a discourse or computer program. Since the matrix formalism for spin ½ and higher spin states is formally similar to the formalism for angular momentum, one can combine spin and angular momentum as we do in quantum theory.

In the case of universe particles, one can also associate universe particles with "true" and "false" values. Fermion universe states have $2^{D/2 - 1}$ 'truth' values and correspond to a multi-valued logic. The numerousness of 'truth' values is due to the D-dimensional space within which universe particles reside.

Naturally one would like to know the physical differences between these $2^{D/2 - 1}$ types of universe particles. Does the difference reside in different shapes of the universe particles? Or is the difference somehow a consequence of the global mass-energy distribution of the universe that we have not been able to discern since we only know of one universe?

The physical meaning of spin for elementary particles is also somewhat elusive. It does not reflect the flow of charge within a particle. For if it did reflect physical spinning of a particle, the outer edges of a particle such as an electron would be traveling at a speed faster than light. So spin is not a mechanical property of the internal structure of an elementary particle. We have suggested that it is a truth value (within a

---

[512] Blaha (2011c) and subsequent books.

particle core) in the matrix formulation of a 4-valued logic called Asynchronous Logic. Thus it has no certain tangible physical basis.

In the case of universe particles the situation is unclear at present. It could be taken to be an indirect reflection of the structure of mass-energy within a universe. This view would be contrary to our proposed view of elementary particle spin as truth values. So we can only assert that a logic interpretation is the only sensible one (based on our present knowledge or our lack thereof). The physical role of universe particle spin is only evident in interactions between universe particles via gauge fields. Thus one must simply view it as a construct for the present.

71. Megaverse Interactions

71.1 Unified SuperStandard Theory Interactions in the Megaverse

The particles and interactions of the Unified SuperStandard Theory exist in the Megaverse space between universes. They are, of course, required to be specified in terms of Megaverse coordinates.

71.2 Megaverse Lorentz Group

The Megaverse Lorentz group is SU(D - 1,1) – a direct generalization of the SU(3, 1) Lorentz group for Special Relativity in our universe. The apparently extremely low density of the Megaverse suggests the Megaverse is most likely a flat space-time. In our work we have chosen y^D transformed to a real value as Megaverse time. This assumption is required in order for the Megaverse to be a dynamic entity. We assume the time variable is real since clocks measure real-valued time. A Megaverse Reality group transformation can map complex time coordinates to real values with no loss of generality.

71.3 Megaverse Gravitation

Megaverse Complex General Relativity and Gravitation is also a direct generalization of the 4-dimensional case with one significant exception: the Einstein field equations change to:

$$R_{\mu\nu} - [1/(D - 2)]g_{\mu\nu}R = -8\pi GT_{\mu\nu}$$

where the metric is such that $g_\mu{}^\mu = D-2$, $R_{\mu\nu}$ is the Megaverse Ricci tensor, and R is the Megaverse curvature scalar. (The overall form of these quantities is the same as that of 4-dimensional space-time.)

72. Universe Particle Dynamics

This chapter describes aspects of the interactions of universe particles due to gravitation, baryon number forces, and collisions between universes. It also the describes the genesis of universes due to vacuum fluctuations, the fission of universes, and the internal distortion of universes due to acceleration and the presence of 'nearby' universes..

72.1 The Internal Distortion of Universes

In the absence of external forces universes are considered to be uniform in the large. However the acceleration of a universe in the Megaverse can distort the universe. Also the existence of a nearby universe(s) could cause the uniformity of a universe to be lost due to gravitation and baryonic forces.[513]

72.1.1 Impact of Universe Particle Acceleration – Lopsided Internal Structure of Universe

Universes can accelerate within the Megaverse due to external Megaverse forces. Universe acceleration should be detectable within a universe as a "lopsidedness" – there would be a shift of parts of the universe away from the direction of acceleration resulting in a difference in the features of the universe "in front" compared to those "in back" – an acceleration effect just as one sees when a jet accelerates.

Interestingly new data from the Planck observatory of the European Space Agency confirms and extends earlier data from NASA's WMAP observatory that one side of the universe appears different from the other side. There are temperature differences and mass distribution differences – just as one might expect if the universe were accelerating as a unit.

---

[513] The baryonic forces, the Baryonic force and the Dark Baryonic force, on a universe are large due to their additivity in the universe and nearby universes.

Thus we see the beginning of data suggesting our universe may be accelerating through the Megaverse. Some Planck observatory scientists have suggested their data is a preliminary indication of the Megaverse.

72.1.2 Impact of External Forces on Universe Structure

The presence of a nearby universe could cause a universe to lose its large scale uniformity and the mass-energy of the universe to drift over time to the 'nearby' side of the universe due to gravitation and baryonic forces.

72.2 Universes in Collision

We can assume that the dynamics of universes in collision will be analogous to that of galaxies in collision since gravity is a dominant force in both cases. Colliding galaxies have often been observed. Their dynamics should provide guidance for the case of universes in collision.[514]

It is clear in the case of colliding galaxies, and of colliding large nuclei (gold and lead typically) that there are several types of collisions with differing results. Similarly, the types of universe collisions can be qualitatively classified as:

1. Clean collisions in which universes nudge each other but retain their identity. These are extreme peripheral collisions. If the universes overlap slightly then the typically spherical symmetry of the universes may become distorted and they may become lopsided.[515]

2. Peripheral collisions in which the universes retain their identity but are connected by a trailing string of mass-energy. Eventually the string breaks and

---

[514] The high energy collision of atomic nuclei at Brookhaven, CERN and other laboratories also is analogous in overall detail with universes in collision.

[515] The Wilkinson Microwave Anisotropy Probe (WMAP) and the Planck European Space Agency satellite have been accumulating data since 2001 that suggests the universe may be lopsided with hot and cold spots on opposite sides of the universe differing from those on the other side being hotter and colder respectively. *Perhaps the result of a collision when the universe was young.*

the universes separate. Subsequently the pieces of trailing string in each universe contract due to their universe's gravitational effects.

3. Two universes can collide and produce multiple universes.

4. Two universes can collide in a "central" collision and amalgamate into one universe. They can intermix with both the baryonic, gauge, and gravitational forces causing a redistribution of their masses. They may separate afterwards or may coalesce into a single universe. One result of this may be lopsided universes. Our universe appears to be lopsided. Some cosmologists believe this is due to a near collision of our universe with another shortly after the Big Bang.

72.3 Creation of Universes through Gauge Field Fluctuations

One of the most exciting questions in Cosmology is the origin of our universe. The conventional view is that it originated in a Big Bang from an infinitesimal point in space. The source of the Big Bang and the prior state of the Cosmos, if there was one, is the subject of much speculation. Based on the particle interpretation of the Wheeler-DeWitt equation, the possibility of a baryonic force strongly supported by conservation of baryon number, and the Megaverse concept, it is reasonable to consider the possibility that the universe originated in a vacuum fluctuation.

In this case there would be two Big Bangs one for our universe and one for an anti-universe. One would expect that they would have opposite corresponding features: one with baryon dominance – one with anti-baryon dominance, and one left-handed – one right-handed.[516]

Our formulation of universe particle theory provides for the generation of a universe particle and anti-particle as a vacuum fluctuation. We view a universe particle as having a substantial excess of baryons, N, as we see in our universe. Its anti-universe at the time of creation (the Big Bang point) is its "mirror image" having the "same" number of anti-baryons (baryon number –N) so that baryon number is conserved by the

---

[516] Given our formulation of complex quaternion universes earlier an interesting possibility would be the generation of a pair of complex quaternion universes in a complex octonion Megaverse vacuum.

fluctuation event. Thus the excesses of one universe are compensated by the excesses of the other.

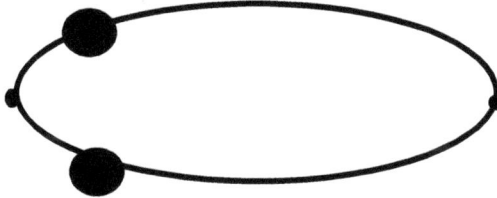

Figure 72.1. Generation of a universe – anti-universe pair as a vacuum fluctuation.

The small value of the coupling constant should lead to an extremely long lifetime for the universes generated by the fluctuation. Thus the 45 billion year life of our universe is not unreasonable. The probability of the creation of universes by vacuum fluctuations should be correspondingly small.

72.4 Fission of Universes

Under certain circumstances the distribution of matter in the universe may lead to the fission of the universe into two separate universes. Our theory supports this possibility for universe particles. The detailed mechanism of the fission process is not specified by the model.

72.4.1 Fission of Normal Universes

The fission of universe particles in our universe particle model is depicted in the Feynman diagram in Fig. 72.2.

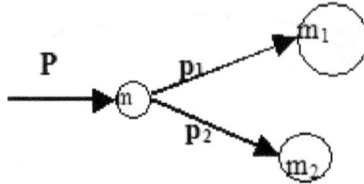

Figure 72.2. Fission of a universe particle into two universe particles.

The sum of the masses of the output universe particles is usually less than the original universe particle mass. However if the fission takes a long time and the masses are time dependent then the produced universe particles combined masses may exceed the original universe's mass.

72.4.2 Tachyon Universe Particle Fission to More Massive Universe Particles

In Blaha (2007a) we showed that a tachyonic (faster than light) particle could fission into particles of larger mass through the conversion of momentum into mass. In this section we show that a tachyonic universe particle may fission into two more massive universe particles.[517] This phenomenon is of particular interest because it enables tachyonic universes to spawn in a new novel way not previously considered in discussions of the origin of universes.

A simple model lagrangian for a tachyonic universe particle is

$$\mathcal{L}_{\parallel} = \psi_T^{\ S}(Y(y))[\gamma^\mu \partial/\partial y^\mu - e_B \gamma^\mu B_{u\mu}(Y(y)) - m(t)]\psi_T(Y(y)) - \tfrac{1}{4}\, F_{Bu}^{\ \mu\nu}(Y(y))F_{Bu\mu\nu}(Y(y)) - \tfrac{1}{4}\, F_u^{\ \mu\nu}(y)F_{u\mu\nu}(y)$$

We assume m(t) is constant.

When a particle or a universe particle fissions (decays) one normally expects that the masses of the particles or universe particles produced by the decay to be smaller

---

[517] We will use the term mass here to denote mass-energy. Since we identified mass as a multiple of area earlier the comments here would appear to apply to universe area as well.

than the mass of the original particle or nucleus. In the case of tachyonic (faster-than-light) elementary particles, or universe particles, a much different possibility is present: a tachyon universe can decay into heavier tachyons (perhaps through a distortion of the universe internally into two 'lumps'.) We will consider the specific case of a tachyon universe particle decaying into two universe particles whose total mass is greater than the original. (See Fig. 72.3.)

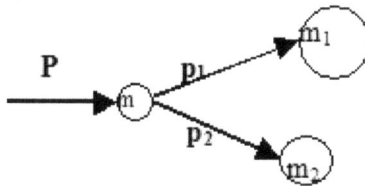

Figure 72.3. Two universe particle decay of a tachyon universe particle.

We will assume the initial tachyon universe particle has zero energy ($p^D = 0$) and thus the tachyons universe particles emerging from the decay also have total universe particle energy zero. The analysis is based on conservation of total universe energy and momentum in Megaverse space outside of universes. The below discussion applies to D-dimensional space with (D – 1)-dimensional spatial coordinates.

Momentum conservation implies

$$\mathbf{P} = \mathbf{p_1} + \mathbf{p_2}$$

Since all energies are zero

$$(c\mathbf{P})^2 = (c\mathbf{P})^2 = m^2$$
$$(c\mathbf{p_1})^2 = (c\mathbf{p_1})^2 = m_1{}^2$$
$$(c\mathbf{p_2})^2 = (c\mathbf{p_2})^2 = m_2{}^2$$

where $P = |\mathbf{P}|$, $p_1 = |\mathbf{p_1}|$, and $p_2 = |\mathbf{p_2}|$. If we now square the above equation for **P** and then use the above three equations we obtain

$$m^2 = m_1^2 + m_2^2 + 2m_1m_2 \cos\theta$$

where θ is the opening angle between the emerging universe particles momenta $\mathbf{p_1}$ and $\mathbf{p_2}$. There are a number of interesting cases:
Case $\theta = 0$:

$$m = m_1 + m_2$$

The masses of the outgoing universe particles sum to the mass of the original tachyon universe particle.

Case $\theta = \pi/2$:

$$m^2 = m_1^2 + m_2^2$$

The masses of each outgoing universe particle tachyon are less than the mass of the original tachyon universe particle.

Case $\theta = \pi$:

$$m^2 = (m_1 - m_2)^2$$

In this case either $m_1 > m$ or $m_2 > m$. Thus one of the outgoing tachyon universe particles has a greater mass than the original tachyon universe particle. Mass is effectively created from the spatial momentum of the initial universe particle. This process is the inverse of normal particle and universe particle fission where the sum of the outgoing masses is always less than the original particle's mass and the difference is mass converted into energy in the form of additional photons.

This last case, where one of the outgoing universe particles is more massive than the original universe particle, is not just for $\theta = \pi$. Since

$$\cos\theta = (m^2 - m_1^2 - m_2^2)/(2m_1m_2)$$

we see that the sum of the outgoing universe particle masses is always greater than the original tachyon universe particle *mass (except when* $\theta = 0$*)* since

$$\cos\theta = 1 + [m^2 - (m_1 + m_2)^2]/(2m_1m_2) \le 1$$

and thus

$$[m^2 - (m_1 + m_2)^2]/(2m_1m_2) \le 0$$

Note $m = m_1 + m_2$ only if $\theta = 0$.

Since we can transform the above discussion to the case of universe particle tachyons having non-zero Megaverse energy using an ordinary D-dimensional Lorentz transformation, the discussion in this subsection is general.

We therefore conclude that when a tachyon universe particle decays into two tachyon universe particles the sum of the masses of the produced tachyon universe particles is greater than the mass of the original tachyon universe particle except if the angle between the momenta of the produced tachyon universe particles is zero. In that case the sum of the masses of the produced tachyon equals the mass of the original tachyon universe particle and the produced universe particles overlap.

73. Universe Particle – Planckton Interactions

While the gravitational force between universe particles is a simple generalization of 4-dimensional gravitation, the baryonic forces, normal and Dark, between universe particles have some points of difference.[518]

They are quite similar to Two-Tier electromagnetic interactions except that universe particles have time-dependent masses, and that the space is D-dimensional.

The time dependence of the universe particle masses is illustrated by Fig. 41.1: the mass of a universe particle after a baryonic interaction vertex is the same as it was before the interaction assuming the point-like interaction specified in the lagrangian.

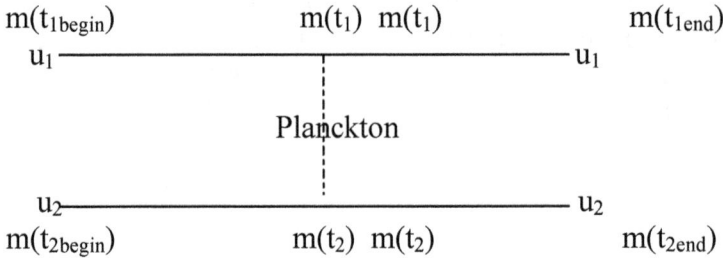

Figure 73.1. A Feynman diagram illustrating the continuity of a universe particle mass through a Planckton interaction.

The reader may verify this by writing the perturbation theory equivalent. A universe particle vertex corresponds to

$$iS_F^{TT}(y_1, y_2)\gamma^\mu iS_F^{TT}(y_2, y_3)$$

---

[518] Since Plancktons travel at the speed of light this section is only of theoretic interest at best.

The universe particle mass is the same on either side of the interaction vertex in lowest order. The weakness of interactions makes higher order corrections negligible.

73.1 Internal Structure of Universe Particles

Planckton field theory gives interactions between baryons. This theory is applicable to universe-universe interactions.[519] It also yields baryon particle – baryon particle interactions as well as baryon particle – universe particle interactions.

It is possible for a planckton to be emitted in one universe and interact with a baryon elementary particle in another universe. This type of "probe" must be a high energy probe just as a photon probe of the internal structure of a nucleon[520] must be a high energy photon to bring out the nucleon's internal structure (parton model).

In this section we will discuss planckton probes of other universes, and the internal structure of a universe as a mass distribution governed by gravitation as it relates to universe particles.

73.1.1 Planckton Probes

Plancktons can be generated in one universe and be used to probe the baryon distribution of another universe. Since the planckton propagator is expressed in Megaverse coordinates the baryon distribution in the target universe will be a distribution in Megaverse coordinates. Megaverse coordinates can be expressed in terms of the curved space-time coordinates of a universe x^μ. However the inversion of the map between universe coordinates and Megaverse coordinates

$$x^\mu = f^{-1\mu}(y)$$

is not 1:1 since x^μ is 4-dimensional and y is a D-dimensional vector. The universe coordinates x^μ are each individually determined up to a subspace. One might be concerned about this situation but the determination of the distribution in Megaverse

---

[519] Again of theoretic interest only due to the vast differences between universes and the limitation of their speed to c.
[520] Deep inelastic electron-nucleon scattering.

coordinates gives a more direct picture not convoluted by the curvature of the target universe.

The detailed probing of a target universe requires high energy plancktons. The similarity of this procedure to deep inelastic electron-nucleon scattering is obvious to the high energy physicist. But in doing this planckton probe experiment one obtains a picture of a different universe – something that is not possible to do with electromagnetic or graviton probes.

The great problem of this approach is the limitation of planckton velocity to the speed of light. A planckton probe between universes has an extremely long transit time that reduces it to only theoretical interest.

73.1.2 Internal Structure of a Universe Particle

The development of the theory of universe particles which resulted in the lagrangian appearing earlier does not fully describe universe particles since it neglects the internal structure of a universe particle. The internal structure of a universe particle is primarily determined by gravitation, electromagnetic effects and nuclear physics.

Consequently the full lagrangian of a universe particle has the form

$$\mathcal{L}_{tot} = \mathcal{L}_{internal} + \mathcal{L}$$

where $\mathcal{L}$ is determined above. As a result the complete quantum wave function of a universe particle has the form

$$\psi_{tot} = \psi_{internal}(Y)\psi_{ext}(Y)$$

where $\psi_{internal}(Y)$ is the internal wave function and $\psi_{ext}(Y)$ is external wave function. It seems reasonable to have a separable equation except when universes collide. In that situation a perturbative mixing of the universes and their wave functions applies and it may be possible to calculate the collision output universes by introducing a further interaction between the internal and external aspects of the universe particles.

73.2 Ultra-high LHC Proton-Proton Collisions Resemble Heavy Nuclei Collisions

Our earlier discussion of tachyonic universes interacting, or decaying, to produce heavier universes is buttressed by new particle interaction data from CERN. New 7 TeV proton-proton collision experiments by the CERN ALICE Collaboration[521] have revealed that the end products of these collisions resemble the end products of heavy nuclei collisions.

Tachyonic particles can transform momentum into mass as pointed out earlier. Since d and s[522] quarks are tachyonic and are present in proton-proton collisions, it is likely that the tachyonic mass production mechanism, at least partly, causes pp collisions to simulate heavy nuclei collisions. The quasi-free nature of the quark-gluon plasma generated in ultra-high energy pp collisions lends weight to this interpretation.

The ALICE results may indicate tachyonic mass creation from the momentum of d and s quarks. Normal u and c quarks do not have a tachyonic nature and thus cannot generate mass from momentum. Their collisions can only generate mass from energy.

---

[521] J. Adam et al, Nature Physics (2017). DOI: 10.1038/nphys4111. Data from LHC run 1.
[522] Enhanced production of s quarks is another feature of ultra-high energy pp collisions according to the ALICE collaboration.

74. Megaverse Hadrons, Atoms, Chemistry, and Materials

The higher dimensionality of the Megaverse causes profound changes in the structure of matter. We assume that the particles of the Unified SuperStandard Theory are the fundamental particles in the very low density Megaverse space between universes. These particles exist in D dimensions and have corresponding fourier expansions. They have the Unified SuperStandard Theory interactions plus additional Megaverse interactions described in chapter 71 Based on these assumptions we can reasonably project the forms of matter that will result over time due to interactions and collisions.

74.1 Hadronic Species

Megaverse quarks must combine in ways satisfying the dictates of SU(3) color confinement. As a result the usual structure of hadrons should be produced as indicated in chapter 18: protons, neutrons, ..., charmonium, ..., pentaquarks, ...

Due to the large Megaverse dimension we can expect that additional varieties of hadrons will exist. The additional spatial dimensions will permit new 'semi-stable' configurations of quarks and antiquarks as these particles disperse in the extra dimensions.

74.2 New Type of Atomic Nuclei

The stable hadrons such as protons, neutrons, and so on will combine using the Megaverse form of the nuclear force to produce atomic nuclei (and using electrons to form atoms.) Atomic nuclei are fairly well modeled by liquid drop models and shell models. The Megaverse equivalents of these models will yield tighter binding and perhaps enhanced stability of isotopes (and a wider variety of isotopes as well.) For example we expect the magic numbers of atomic nuclei shells to increase. The result will be new forms of matter. These new types of matter could yield materials with extraordinary properties of density, hardness, pliability, resilience, conductivity, and so on.

74.3 Megaverse Chemistry

They will be the basis of a vast new Megaverse chemistry with complex molecular configurations of atoms due to the large number of new dimensions.

74.4 The Key Factor in the Megaverse

The key factor that profoundly influences the hadrons, atomic nuclei, chemistry, and life forms in the Megaverse is the enhanced propinquity of the components, which leads to stronger forces amongst them. This feature is simply illustrated by considering a one-dimensional array of four constituents composing an entity in Fig. 74.1 compared to a two-dimensional array in Fig. 74.2.

In the two-dimensional case the average spacing between constituents, and thus the force between them, is stronger resulting in tighter binding and denser entities.

In the case of the D-dimensional Megaverse the strengthening effect can only be much more pronounced.

●   ●   ●   ●

Figure 74.1. One-dimensional array.

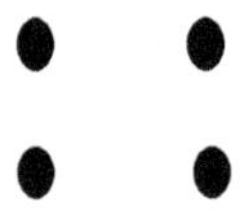

Figure 74.2. Two-dimensional array.

74.5 New Forms of Megaverse Life Beyond Organic

They may also open the possibility of new forms of living creatures beyond organic and hypothesized silicon-based life. See chapter 43.

74.6 Can these New Entities Exist Half in and Half Out of the Universe?

Based on our knowledge of the earth, cosmic rays, and the universe these possibilities do not appear to have been seen. It is conceivable that in regions in the universe, where gravitation and thus spatial curvature is very large, the surface tension of the universe may be near zero. Under these conditions hadrons, atomic nuclei, and atoms might exist with parts in our universe and parts in the Megaverse. Then the question arises: Will the parts that are in our universe be recognizable as in this state? It may be reasonable to search for 'fragments' of these types in spectra and in cosmic rays.

Some spectacular cosmic ray particle events such as the Niu particle event (seen some years ago) and some recent ultra-energetic cosmic ray events may be the result of the entry of a higher dimensional Megaverse particle into our universe.

75. Life Forms in the Megaverse

We have seen that there is evidence for the existence of the Megaverse . We have also seen that it is quite likely that matter and energy exist in Megaverse space between universes.

It is possible that Megaverse matter might have accreted due to gravity and other forces to form stars and planets. If so, then possibly life exists in the Megaverse outside of universes. This possibility raises many provocative questions. We will call Megaverse life *Megaversian*.

We know that life on earth exists in many niches that we would consider hostile: deserts, high altitudes, deep oceans, intense cold and darkness, and even kilometers below the surface of the earth. Life is prevalent on earth.

We have found other solar system bodies have environments that could support life. And we have found planets around distant stars that appear to be capable of supporting life.

Given these facts it seems reasonable to inquire as to the form and nature of life within the D-dimensional Megaverse space.

75.1 Why Life?

If life is prevalent within, and without, our universe then there must be a fundamental reason for Nature's tendency to produce life in just about any 'hospitable' environment. We believe the reason is that "*Life is maximally entropic.*" Entropy almost always increases with time. Sometimes quickly; sometimes more slowly. In any given situation where life exists we believe that entropy increases more quickly than it would if life were not present. This hypothesis, of course, needs comprehensive study. But we will assume it as a working hypothesis in our investigation.

75.2 A Necessary Condition for Life

As part of our fundamental hypothesis for the existence of life, we require that energetic processes must exist in an environment where life appears.[523] Living things need energy to develop and grow as well as for motion should they be mobile.

If one considers the simple fact that giant living objects do not pop into existence, but rather start from the small, we see another requirement for life is that its origin is local – life originates in the small. On earth it went through an evolutionary process starting from viruses to cells to multi-cell creatures – eventually becoming the small and large creatures of our experience.

But one must remember the beginning of all life is in the small and based on local sources of energy that could fuel life in the small and enable it to grow and evolve over time. *We conclude the beginnings of life require a local source of energy. In the earliest stages of the development of life forms, the seed ('virus'?) must utilize an energy source in its immediate vicinity. Only at 'later' stages can life forage for energy over a distance.*

75.3 Types of Known Life

A great deal is known about life on earth. However there is still much to learn. We will pursue the question of Megaverse life based on our knowledge of aspects of earth life.

75.4 Fundamental Aspects of Megaverse Life

It would be nice if the physical laws, coupling constants, and particle masses of the Megaverse were the same as those of our 4-dimensional universe. And it may well be so, since our Continuous Creation Model based on an inflow from the Megaverse (Chapter 14 of Blaha (2017c)) would suggest it.

We shall assume that this is the case but, where possible, not rely on precise details based on phenomena in our universe. Our focus will be on the impact of D dimensions on features of Megaverse life.

---

[523] See Feinberg (1980) for a study of energy processes that could support life.

We will assume, therefore, that the stable nuclei of atoms are the same in our universe and the Megaverse. However the arrangement of electron shells, and their energies, around the nuclei of atoms will be different due to $(D-1)$-dimensional space. As a result the chemistry of molecules and, most particularly, of the equivalent of DNA (if there is one) will differ substantially.

Thus we can expect life in Megaverse space to be quite different from life within our universe. Nevertheless, we can make some 'conclusive' statements based on general considerations.

75.5 Megaversian DNA

DNA is the common denominator of all known forms of life on earth. We assume an equivalent to DNA exists as the basis of life in $(D-1)$-dimensional Megaverse space (Megaverse 'time' is the additional dimension to make the D-dimensional Megaverse.) Based on earth life, we assume Megaversian DNA, *Mega-DNA*, has the following properties:

1. It consists of long strands of linked nucleotides (the equivalents of cytosine, guanine, adenine, and thymine) as well as equivalents of deoxyribose and a phosphate group.

2. There are two strands, one of which may be viewed as a 'back up' copy, to provide stability and for use in DNA replication.

3. The parts of the strands should be tightly bound compared to binding within DNA parts in our universe.

The purpose of the *two* strands suggests that the number of strands is independent of the dimensionality of space.

The large number of dimensions of Megaverse space has a number of important implications:

1. The long Mega-DNA strands can be compacted to better fit into cells as DNA strands are compacted in balls in cells on earth. Earth DNA strands range up to 2 meters in length. Mega-DNA strands, perhaps of length of the order of

$$2(\text{Megaverse-cell radius})^{(D-4)} \text{ meters}$$

can be fit within Megaverse-life cells.

2. The additional Megaverse dimensions suggest more 'flexibility' in the Mega-DNA to provide more epigenetic adaptability and more capacity for mutations.

3. Mega-DNA should support a greater variety of cell and multi-cell life than found on earth.

Thus we can envision 'fantastic' creatures able to rapidly adapt to changing environmental conditions and to possibly have the feature called 'shape shifting' in popular SciFi movies.

75.6 Megaversian Brain Size

Perhaps the most important effect of a large number of dimensions is intelligence level. In our universe, assuming electromagnetic connections in the brain, the size of brains is constrained by the time it takes for signals to go between parts of the brain. (Currently many investigators believe that consciousness, and possibly abstract thought, are a result of total brain 'collaboration.') The maximum brain size is usually estimated to be slightly larger than that of the human brain.

In the Megaverse we do not have three space dimensions but rather $D - 1$ spatial dimensions. Thus the volume of a Megaverse brain is not proportional to r^3 where r is the radius of the brain. Rather it is proportional to $r^{(D-1)}$. Thus a Megaverse brain of the same radius as an earthly brain has incomparably more volume and contents than an earthly brain of the same radius. Thus if a Megaverse intelligent species should exist one can expect it would have massively more memory and analytical power.

Another aspect of Megaversian brain structure is the possibility of distributed brain power. We are familiar with networks of computers uniting to do massive computations. Some earthly species such as the octopus have a distributed brain structure: an octopus has a central brain and a 'sub-brain' in each of its eight limbs. Distributed brain power in Megaverse creatures could result in formidable computational and analytical abilities.

75.7 Megaversian Locomotion

In our universe, particularly on earth, larger animals tend to have two or four legs. Smaller creatures such as insects may have many more legs. Two-legged (and four-legged) animals appear to have two (or four) legs to be able to turn effectively in three dimensions. They do not have three or five legs because the added mobility is outweighed by the added stability requirements placed on the brain and nervous system.

In the Megaverse we would expect that creatures would have $D - 1$ legs (appendages) to maneuver in $D - 1$ spatial dimensions. It is possible that appendages would have 'sub-brains' like the octopus to off load processing and control of limbs. Whether they have hands and arms is open to question since legs could also play the role of arms.

75.8 Megaversian Vision

In $D - 1$ dimensions it would appear reasonable to have compound eyes like insect species on earth. Processing the data coming from the eyes would be a significant burden on the Megaverse brain.

75.9 Composition of Megaversian Life

If the elements present in Megaverse matter are similar to those in our universe the composition of creatures may be similar to that of life in our universe. However the many Megaverse dimensions, and the expected differences in the electronic structure of atoms and compounds might lead to a different chemical composition of life. Perhaps arsenic-based life might exist. Organic chemicals would be different in composition and structure.

These differences might lead to changes in the environments suitable for life: temperature extremes, atmospheres, and food requirements among other things.

75.10 Megaversian Societies and Civilizations

The average density of matter in the Megaverse may be expected to be low (section 14.14) and, consequently, the number of planets and stars per unit volume will be correspondingly small. On some of these objects life may develop and, on a much smaller number, societies and possibly civilizations may appear. After all, on earth we find complex societies of ants, bees, and so on. Their societies are similar to human societies in many respects[524] despite the vast difference between Mankind and insects. *The key factor in the growth of civilizations is the availability of large amounts of surplus energy.*

On this basis we suggest that Megaverse societies and civilizations are a likely possibility given the existence of Megaverse life forms.

75.11 Extension of Human Sight to the Megaverse

If a human entered the Megaverse, his/her sight would be limited to three dimensions since bodily motions would be so limited.[525] To see in all 191 spatial directions a human would need a device (similar to a periscope in concept) that could be oriented to other of the 191 directions. This device would be Megaversian in the sense that part of it would be in the three dimensional subspace defined by the human's orientation, and the rest would be in the remaining 188 dimensions into which the human would want to look. Of course rotating the device to other directions would require force components in those directions. Thus the viewing device would be no simple piece of equipment.

---

[524] See Blaha (2010c) for a comprehensive study of human societies and almost all the known civilizations of Mankind. It shows that the 'ups and downs' of civilizations and societies is based on energetics (Thermodynamics).

[525] One can visualize this situation by considering a 2-dimensional flatworld where the inhabitants could only see things within the flatworld but could not see the things in a third (vertical) dimension. To see things above the flatworld they would need a device that would partly be in the third dimension. These thoughts generalize to the Megaverse case.

Perhaps the most analogous viewing 'device' in our universe is a multi-faceted insect eye. Each facet provides a view of the surroundings. The corresponding Megaverse 'eye' would have one facet for D – 1 spatial directions.

75.12 Megaversian Life

If Megaverse bodies are at all habitable, then one can expect life to exist, and yet to be very different due to the D Megaverse dimensions. Yet despite the differences one can expect certain similarities to life in our universe.

76. Unified SuperStandard Theory Map to Reality

We have derived a finite Unified SuperStandard Theory from basic assumptions and examined some of its consequences. We have gone beyond our universe and considered the possibility of a Megaverse of universes.

Our derivation gave us an unchanging theory of Reality conceptually similar to the concept of Parmenides. The question now becomes: How is that Reality imposed on the material universe that we scc – everywhere with 'infinite precision?' There appear to be three possible answers to that question:

1. Our primitives, axioms and derivation are the only possible choices (modulo minor variations). They are naturally adapted to describe physical reality by construction. This view is similar to that of Professor Hawking who said he did not see a necessity for God in the workings of the universe.

2. There are other possible explanations of the features of the material universe such as a SuperString theory. These explanation(s) may lead to our theory.

3. There is an 'Unmoved Mover' that causes Reality (our theory) to be imposed on the material universe (and Megaverse.) Some see this entity as God.

In any case it is clear that our theory maps directly to the material universe that we see although there are questions that remain to be addressed.

77. Proposals for the Future of Elementary Particle Theory and Experiment

77.1 Theory

The primary theoretical issue facing the Unified SuperStandard Theory is the extension of the theory to yield the symmetry breaking and masses of its elementary fermions and vector bosons. currently there is no theoretical framework available to fully determine their values.

77.2 Experiment Proposals

There are a number of major questions confronting experiment in view of the success of the Unified SuperStandard Theory's determination of the structure of the theory of elementary particles.

77.2.1 Speed of Light

A decisive neutrino experiment should be performed to accurately measure the speed of neutrinos. Existing experiments suggest that neutrinos move faster than the speed of light. However their results are not conclusive. Resolving this question is of great importance not only for the Unified SuperStandard Theory but also for the future expansion of Mankind to the stars.

Einstein's statement on this matter—properly stated—is: particles cannot accelerate up to a speed greater than the speed of light with real-valued accelerations and speeds.

This statement does not exclude particles with complex-valued accelerations and speeds accelerating from below to speeds in excess of the speed of light. It also does not exclude particles such as neutrinos and down-type quarks from always having speeds in excess of the speed of light.

77.2.2 Extremely Massive Particles

The Unified SuperStandard Theory predicts a fourth generation of fermions. It also predicts three additional layers of four generations each. None of these very massive particles have been found although there is new evidence[526] for a possible 4th generation neutrino. In addition the theory predicts four layers of vector bosons. Only one layer is partially known at present. Again there is possible evidence[527] for a layer 2 particle that may be a heavy first generation "electron." There is an important need for much more powerful accelerators to search for these fermions and bosons.

77.2.3 Dark Particles

Dark particles have not as yet been found. However indirect astrophysical data[528] suggests they exist in greater quantity than 'normal' particles. Again more powerful accelerators may find evidence of them. They are predicted in some detail in the Unified SuperStandard Theory.

77.2..4 Left-Right Imaginary Momenta of Quarks

Chapter 26 specifies complex-valued momenta for up-type and down-type quarks. In ultra-high energy experiments at CERN and other laboratories quark-gluon plasmas are generated. Within the quark-gluon plasmas, quarks exist with a distribution of momenta. If one could 'filter' the quark-gluon plasma to create a stream of quarks with the imaginary part of their momenta pointing in the same direction, then one could imagine making a starship engine with complex acceleration and velocity that would

---

[526] The week of September 17, 2018 saw public announcements of a possible fourth generation neutrino seen at the Super-Kamiokande giant particle detector. If so, it would be a potential confirmation of the four fermion generations predicted by our Unified SuperStandard Theory.
[527] New particles have also been found that passed through the earth without generating an interaction shower. They may be second layer, charged leptons (heavier siblings of the electron.) See the paper D. B. Fox *et al*, arXiv:astro-ph.HE 1809.09615 (2018). The fact that these particles traverse earth without an interaction shower indicates they do not have known Standard Model interactions. Thus they are most likely members of a higher fermion layer such as layer 2.
[528] Chapter 46 shows our theory predicts the relative abundances of Dark energy and Dark matter in relatively close agreement with astrophysical data.

enable a starship to travel faster than the speed of light vastly shortening the time of travel to the stars.[529]

Experiments on imaginary momentum filtering of quark-gluon plasmas could have important consequences for interstellar travel.

---

[529] See Blaha (2014b) and (2014c).

Appendix A. The Local Definition of Asymptotic Particle States and New Framework for Gauge Field Theories

S. Blaha, "The Local Definition of Asymptotic Particle States", IL Nuovo Cimento **49A**, 35 (1979). It describes the PseudoQuantization of boson and fermion field theories for use in the quantization of fields in universes and the Megaverse.

See also S. Blaha, "New Framework for Gauge Field Theories", IL Nuovo Cimento **49A**, 113 (1979). It describes the PseudoQuantization of gauge field theories for the purposes of defining higher derivative field theories and for use in the quantization of fields in universes and the Megaverse.

Appendix B. Reality Group of Complex General Relativity

We have seen that Complex Special Relativity is the basis of flat space-time phenomena. Flat space-time coordinates are complex-valued in general. The real-valued coordinates that we experience in everyday life are the result of our measuring instruments: clocks and rulers. Real-valued coordinates are generated from complex-valued coordinates by Reality group transformations.

If flat space-time is governed by Complex Special Relativity then it is clear that curved 'space-time' is governed by Complex General Relativity. Here again there is a Reality group the General Relativistic Reality group – a U(4) group – that maps complex-valued General Relativity coordinates to real-valued curved coordinates.[530] There is a corresponding U(4) Internal Symmetry Reality group that we call the *Species group*. This group rotates fermions as described earlier.

We can isolate the General Relativistic Reality group by factoring complex General Relativistic coordinate transformations into parts that consist of a real-valued General Coordinate transformation and complex-valued coordinate transformations. It will be apparent that the General Relativistic Reality group emerges in this discussion. We begin by defining the tetrad notation.

B.1 Tetrad (Vierbein) Formalism

The *vierbein* formalism begins with the Equivalence Principle that allows us to define an inertial coordinate system in the neighborhood of any point Z in space-time. We will use the notation $\varsigma^\alpha(Z)$ to denote the inertial coordinates at Z. We define a tetrad or vierbein as

$$v^\alpha{}_\mu(x) = (\partial \varsigma^\alpha(x)/\partial x^\mu)_{x=Z} \tag{B.1}$$

---

[530] Much of this chapter appears in Blaha (2016h) and (2017a).

and, in a neighborhood of Z, we can invert the relation between ς and x to define an inverse

$$w^\mu{}_\alpha(x) = (\partial x^\mu(\varsigma)/\partial\varsigma^\alpha)_{x=X} \tag{B.2}$$

such that

$$w^\mu{}_\alpha(x)v^\alpha{}_\nu(x) = \delta^\mu{}_\nu$$
$$w^\mu{}_\beta(x)v^\alpha{}_\mu(x) = \delta^\alpha{}_\beta \tag{B.3}$$

In real-valued General Relativity all *tetrads* are real-valued. In Complex General Relativity a *tetrad* $v^\alpha{}_\mu(x)$ is complex-valued.

The metric at a curved space-time point X is defined in terms of *tetrads* as

$$g_{\rho\sigma}(x) = \eta_{\alpha\beta}\, v^\alpha{}_\rho(x)v^\beta{}_\sigma(x) \tag{B.4}$$
$$g^{\rho\sigma}(x) = \eta^{\alpha\beta}\, w^\rho{}_\alpha(x)w^\sigma{}_\beta(x)$$

The inverse of a *tetrad* transformation can also be expressed as

$$w_\beta{}^\nu(x) = v_\beta{}^\nu(x) = \eta_{\beta\alpha}g^{\nu\mu}(x)v^\alpha{}_\mu(x)$$

Then a *tetrad* and its inverse satisfy

$$v^\alpha{}_\mu(x)v_\beta{}^\mu(x) = \delta^\alpha{}_\beta \tag{B.5}$$

and

$$v^\alpha{}_\mu(x)v_\alpha{}^\nu(x) = \delta^\nu{}_\mu$$

There are two general types of space-time transformations that can be performed on a tetrad.

1. A complex-valued (possibly real-valued) General Relativistic coordinate transformation:

$$v'^\alpha{}_\mu(x) = \partial x^\nu/\partial x'^\mu\, v^\alpha{}_\nu(x)$$

2. A complex-valued, local *Lorentzian transformation*

$$v'^{\beta}{}_{\mu}(x) = \Lambda(x)^{\beta}{}_{\alpha}\, v^{\alpha}{}_{\mu}(x)$$

where $\Lambda(x)^{\beta}{}_{\alpha}$ is an element of a subset of the local Complex Lorentz Group.

The local Lorentzian transformations $\Lambda(x)^{\beta}{}_{\alpha}$ consist of local Lorentz transformations that are real-valued, and complex-valued Lorentz transformations. Both types of transformations satisfy the orthogonality condition:

$$\eta_{\alpha\beta}\Lambda^{\alpha}{}_{\rho}(x)\Lambda^{\beta}{}_{\sigma}(x) = \eta_{\rho\sigma} \tag{B.6}$$

Thus the *tetrad* partakes of both local (position dependent) General Relativistic transformations and local Lorentzian transformations.

B.2 Complex General Relativistic Transformations

The General Relativistic Reality group interaction emerges from complex General Relativistic transformations. We can separate elements of the set of all complex General Coordinate transformations into a product of two factors: a real-valued General Coordinate transformation and a complex-valued General Coordinate transformation. The set of complex factors can be further factored into those that satisfy

$$\Lambda(\omega, \mathbf{u})^{T}G\Lambda(\omega, \mathbf{u}) = G \tag{B.7}$$

and those that do not. We then see that the set of those that do not satisfy the above equation form a curved space representation of the U(4) group under 'multiplication' of transformations.

The elements of the set of real and complex General Coordinate transformations whose flat complex space-time limit satisfy the above equation form the elements of the Complex Lorentz group.[531]

We thus find the set of all 4-dimensional complex, curved space General coordinate transformations can be visualized as in Fig. B.1. The next section describes the interplay of the three parts displayed in Fig. B.1.

B.3 Structure of Complex General Coordinate Transformations

Complex General Coordinate transformations can be uniquely factored into products of two terms, which will later be further factored into three factors. They have the form

$$\partial x''^{\nu}(x)/\partial x^{\mu} = U(x'')^{\nu}{}_{\beta}\, \partial x'^{\beta}(x)/\partial x^{\mu} \tag{B.8}$$

where

$$x''^{\nu}(x) = U(x'')^{\nu}{}_{\beta} x'^{\beta}$$
$$x'^{\mu}(x) = U^{-1\mu}{}_{b}(x'')\, x''^{b}$$

where $U(x')^{\nu}{}_{\beta}$ is complex and where $\partial x'^{\beta}(x)/\partial x^{\mu}$ is a purely real General Coordinate transformation.

We define

$$U(x'')^{\mu}{}_{\nu} = w^{\mu}{}_{a}(x'')\left[\exp\!\left(i \sum_{k} g_{k}\Phi_{k}(x'')\tau_{k}\right)\right]^{a}{}_{b} v^{b}{}_{\nu}(x'') \tag{B.9}$$
$$U^{-1}(x'')^{\mu}{}_{\nu} = w^{\mu}{}_{a}(x'')\left[\exp\!\left(-i\sum_{k} g_{k}\Phi_{k}(x'')\tau_{k}\right)\right]^{a}{}_{b} v^{b}{}_{\nu}(x'')$$

where τ_{k} is a hermitean U(4) generator matrix, where the constants g_{k} are real, and where the Φ_{k} are real. The uniqueness of the factorization follows from the Reality group (and U(4)) property that any complex 4-vector can be uniquely mapped to any specified real 4-vector.

---

[531] It is this part of curved space-time General Relativity that becomes the flat space-time Complex Lorentz group, which leads to the SU(3)⊗SU(2)⊗U(1)⊗SU(2)⊗U(1) Standard Model Reality group.

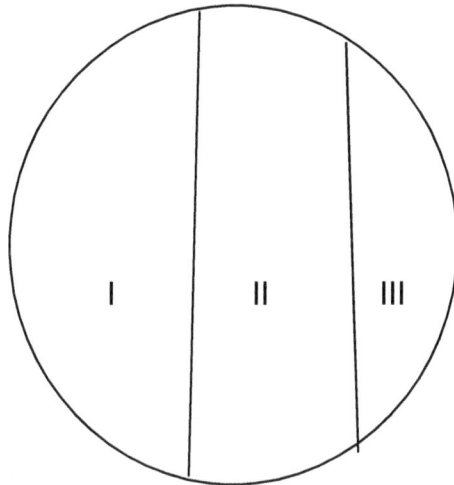

Figure B.1. A visualization of the set of General Coordinate transformations separated into real-valued General coordinate transformations (part I), complex transformations that satisfy $\Lambda(\omega, u)^T G \Lambda(\omega, u) = G$ (part II), and complex transformations that do not satisfy $\Lambda(\omega, u)^T G \Lambda(\omega, u) = G$ (part III). Part I and part II combine in the limit of flat space-time to form the Complex Lorentz group. Parts II and III elements form a U(4) group that we call the General Relativistic Reality group.

Given the factorization above it becomes possible to separate the affine connection correspondingly.

B.4 Complex Affine Connection – General Relativistic Reality Group

The structure of a complex general coordinate transformation enables us to calculate its affine connection for later use in determining the covariant derivative, and the dynamic equations. First the affine connection for the transformation to real-valued x' coordinates from inertial coordinates ς^ρ is

$$\Gamma^{\sigma}{}_{\lambda\mu}(x') = \partial x'^{\sigma}/\partial\varsigma^{\rho}\ \partial^2\varsigma^{\rho}/\partial x'^{\lambda}\partial x'^{\mu} \tag{B.10}$$

Next the Reality group transformation has the affine connection

$$\Gamma^{\sigma}{}_{\lambda\mu}(x'') = \partial x''^{\sigma}/\partial\varsigma^{\rho}\ \partial^2\varsigma^{\rho}/\partial x''^{\lambda}\partial x''^{\mu}$$

which can be re-expressed as

$$\Gamma^{\sigma}{}_{\lambda\mu}(x'') = \partial x''^{\sigma}/\partial x'^{\beta}\ \partial x'^{\beta}(\varsigma)/\partial\varsigma^{\rho}\ \partial/\partial x''^{\mu}[\partial\varsigma^{\rho}/\partial x'^{\alpha}\ \partial x'^{\alpha}/\partial x''^{\lambda}]$$
$$= \partial x''^{\sigma}/\partial x'^{\beta}\ \partial x'^{\alpha}/\partial x''^{\lambda}\ \partial x'^{\gamma}/\partial x''^{\mu}\ \Gamma^{\beta}{}_{\alpha\gamma}(x') + \partial x''^{\sigma}/\partial x'^{\beta}\ \partial^2 x'^{\beta}/\partial x''^{\lambda}\partial x''^{\mu}$$

$$\tag{B.11}$$

Next substituting the General Relativistic Reality group transformation

$$x''^{\nu}(x) = U(x'')^{\nu}{}_{\beta}x'^{\beta}$$
$$x'^{\mu}(x) = U^{-1}(x'')^{\mu}{}_{\beta}\ x''^{\beta}$$

together with

$$\partial x''^{\sigma}/\partial x'^{\beta} = \partial[U(x'')^{\sigma}{}_{\alpha}x'^{\alpha}]/\partial x'^{\beta} = U(x'')^{\sigma}{}_{\beta} + x'^{\alpha}\ \partial U(x'')^{\sigma}{}_{\alpha}/\partial x'^{\beta}$$

$$\partial x'^{\sigma}/\partial x''^{\beta} = \partial[U^{-1}(x'')^{\sigma}{}_{\alpha}x''^{\alpha}]/\partial x''^{\beta} = U^{-1}(x'')^{\sigma}{}_{\beta} + x''^{\alpha}\ \partial U^{-1}(x'')^{\sigma}{}_{\alpha}/\partial x''^{\beta}$$

we find the second term above (eq. B.11) is the Reality fields affine connection

$$\Gamma_R{}^{\sigma}{}_{\lambda\mu}(x'') = \partial[U(x'')^{\sigma}{}_{\alpha}x'^{\alpha}]/\partial x'^{\beta}\ \partial\{\partial[U^{-1}(x'')^{\beta}{}_{\alpha}x''^{\alpha}]/\partial x''^{\lambda}\}/\partial x''^{\mu}$$

and so we find the affine connections are approximately additive. Thus approximately

$$\Gamma^{\sigma}{}_{\lambda\mu}(x'') = \Gamma_{GR}{}^{\sigma}{}_{\lambda\mu}(x') + \Gamma_R{}^{\sigma}{}_{\lambda\mu}(x'')$$

if $x''^{\sigma} \simeq x'^{\sigma}$ where $\Gamma_{GR}{}^{\sigma}{}_{\lambda\mu}(x')$ is the real General Relativistic affine connection.

A complex transformation of types II and III in Fig. B.1 has the form:

$$U(x'')^{\mu}{}_{\nu} = w^{\mu}{}_a(x'')[\exp(i\sum_k \Phi_k(x'')\tau_k)]^a{}_b \, l^b{}_{\nu}(x'')$$

$$U^{-1}(x'')^{\mu}{}_{\nu} = w^{\mu}{}_a(x'')[\exp(-i\sum_k \Phi_k(x'')\tau_k)]^a{}_b \, l^b{}_{\nu}(x'')$$

where τ_k is a U(4) generator matrix. Its infinitesimal transformation is approximately

$$U(x'')^{\nu}{}_{\beta} \approx \delta^{\nu}{}_{\beta} + i\sum_k \Phi_k(x'')[\tau_k]^{\nu}{}_{\beta} \qquad (B.12)$$

$$U^{-1}(x'')^{\nu}{}_{\beta} \approx \delta^{\nu}{}_{\beta} - i\sum_k \Phi_k(x'')[\tau_k]^{\nu}{}_{\beta}$$

using the *vierbein* flat space-time limits

$$w^{\mu}{}_a(x'') \approx \delta^{\mu}{}_a$$
$$l^b{}_{\nu}(x'') \approx \delta^b{}_{\nu}$$

where

$$\Phi_k(x) = \int^x dy_{\lambda} \, A_{Rk}{}^{\lambda}(y) \qquad (B.13)$$

Then

$$\Gamma_R{}^{\sigma}{}_{\lambda\mu} = -\tfrac{1}{2}i\{\sum_k A_{Rk}(x'')_{\mu}[\tau_k]^{\sigma}{}_{\lambda} + \sum_k A_{Rk}(x'')_{\lambda}[\tau_k]^{\sigma}{}_{\mu}\} \qquad (B.14)$$
$$= A_R{}^{\sigma}{}_{\mu\lambda} + A_R{}^{\sigma}{}_{\lambda\mu}$$

(summed over k) with the matrix $A_R{}^{\sigma}{}_{\mu\lambda}$ given by

$$A_R{}^{\sigma}{}_{\mu\lambda} = -\tfrac{1}{2}i\sum_k A_{Rk\mu}[\tau_k]^{\sigma}{}_{\lambda} \qquad (B.15)$$

with $A_R{}^{\sigma}{}_{\mu\lambda}$ transformable to matrix row and column numbers

$$A_{Rflat}{}^{\mu a}{}_b = A_{Rflatk}{}^{\mu}[\tau_k]^{\sigma}{}_{\lambda}\delta_{\sigma}{}^a\delta^{\lambda}{}_b$$

using the flat space-time vierbein values. So $A_{Rflat}{}^a{}_{\mu b}$ may be written in matrix form as

$$A_{R_{flat}\mu} = -\tfrac{1}{2}i\textstyle\sum_k A_{R_{flatk}\mu}\tau_k \qquad (B.16)$$

In the flat space-time limit the $A_{Rk}{}^{\lambda}(y)$ becomes the Coordinate Species group U(4) gauge fields $A_{R_{flatk}}{}^{\lambda}(y)$.

The relevant *quadratic* $A_R{}^{\sigma}{}_{\mu\lambda}$ terms from eq. 11.22 that are needed to find the dynamic equation for the gauge fields $A_{R_{flat}}{}^{i}{}_{\mu}$ are contained in

$$\mathcal{L}_A = \mathrm{Tr}\ \sqrt{g}[M\partial_\nu R^1{}_{\sigma\mu}\partial^\nu R^{2\sigma\mu} + aR^1{}_{\sigma\mu}R^{2\sigma\mu} + bg^{\sigma\mu}(R^1{}_{\sigma\mu} + R^2{}_{\sigma\mu}) + 1/4(g_{\mu\nu} + g^2{}_{\mu\nu})T^{\mu\nu}]$$
$$(B.17)$$

We can let

$$R^i{}_{\sigma\mu} = R^{i\beta}{}_{\sigma\beta\mu} \equiv \partial_\mu(A_R{}^{i\beta}{}_{\sigma\beta} + A_R{}^{i\beta}{}_{\beta\sigma}) - \partial_\beta(A_R{}^{i\beta}{}_{\sigma\mu} + A_R{}^{i\beta}{}_{\mu\sigma}) \qquad (B.18)$$

for i = 1, 2. In the flat space-time limit we chose the Landau gauge

$$\partial_\mu A_{R_{flat}}{}^{i\mu a}{}_b = 0 \qquad (B.19)$$

As a result

$$R^i{}_{\sigma\mu} \equiv \partial_\mu(A_R{}^{i\beta}{}_{\sigma\beta} + A_R{}^{i\beta}{}_{\beta\sigma}) \qquad (B.20)$$

Using

$$A_{R_{flat}}{}^{i\mu\sigma}{}_\lambda = A_{R_{flat}}{}^{i\mu}[\tau_k]^a{}_b\delta^\sigma{}_a\delta_\lambda{}^b \qquad (B.21)$$
$$A_{R_{flat}}{}^{i}{}_\mu = -\tfrac{1}{2}i\textstyle\sum_k A_{R_{flat}}{}^{i}{}_{k\mu}$$

and taking the trace in eq. B.17 we obtain

$$\mathcal{L}_A = \mathrm{Tr}\ \sqrt{g}[8M\partial_\nu\partial_\mu A_{R_{flat}}{}^1{}_\sigma\partial^\nu\partial^\mu A_{R_{flat}}{}^{2\sigma} + 8a\partial_\mu A_{R_{flat}}{}^1{}_\sigma\partial^\mu A_{R_{flat}}{}^{2\sigma} + 1/4(g_{\mu\nu} + g^2{}_{\mu\nu})T^{\mu\nu}]$$
$$(B.22)$$

in the flat space-time limit. Eq. B.17 needs to take account of the complex nature of $(g_{\mu\nu} + g^2{}_{\mu\nu})$ until transformed by the infinitesimal form of the complex Reality transformation:

$$(g_{\beta\alpha} + g^2{}_{\beta\alpha})' \rightarrow U\,(x'')_\beta{}^\mu (g_{\mu\nu} + g^2{}_{\mu\nu}) U^{-1}(x'')\,{}^\nu{}_\alpha$$

$$= (\delta_\beta{}^\mu + i\,\textstyle\sum_k \Phi_k(x'')[\tau_k]_\beta{}^\mu)(g_{\mu\nu} + g^2{}_{\mu\nu})(\delta^\nu{}_\alpha - i\,\textstyle\sum_k \Phi_k(x'')[\tau_k]\,{}^\nu{}_\alpha)$$

$$\cong (g_{\beta\alpha} + g^2{}_{\beta\alpha}) + i\{\textstyle\sum_k \Phi_k(x'')[\tau_k]_\beta{}^\mu - i\,\textstyle\sum_k \Phi_k(x'')[\tau_k]\,{}^\nu{}_\alpha)(g_{\mu\nu} + g^2{}_{\mu\nu})$$

$$\cong (g_{\beta\alpha} + g^2{}_{\beta\alpha}) + i\textstyle\sum_k \Phi_k(x'')\{[\tau_k]_{\beta\alpha} - [\tau_k]_{\beta\alpha}\} \qquad\qquad (B.23)$$

Approximating $\Phi_k(x'')$ with an infinitesimal line we find

$$\textstyle\sum_k \Phi_k(x) \cong \delta x_\lambda A_{R\text{flat}}{}^{1\lambda}(x) \qquad\qquad (B.24)$$

by eq. B.13. Thus

$$1/4(g_{\mu\nu} + g^2{}_{\mu\nu})T^{\mu\nu} \cong \tfrac14\,[(g_{\beta\alpha} + g^2{}_{\beta\alpha}) + i\,\delta x_\lambda A_{R\text{flat}}{}^{1\lambda}(x)\{[\tau_k]_{\beta\alpha} - [\tau_k]_{\beta\alpha}\}]T^{\alpha\beta}$$

Applying the canonical Euler-Lagrange method we obtain the dynamical equations (using integration by parts to handle higher order derivative terms):[532]

$$\square^2 A_{R\text{flat}}{}^1{}_\sigma + (a/M)\square A_{R\text{flat}}{}^1{}_\sigma + i\delta x_\sigma\{[\tau_k]_{\beta\alpha} - [\tau_k]_{\beta\alpha}\}T^{\alpha\beta}/(32M) = 0 \qquad (B.25)$$

$$\square^2 A_{R\text{flat}}{}^2{}_\sigma + (a/M)\square A_{R\text{flat}}{}^2{}_\sigma = 0 \qquad\qquad (B.26)$$

Since the $A_{R\text{flat}}$ gauge field is gravitational in nature it exists, as eq. B.25 shows, as a type of gravitational interaction whose source is the energy-momentum tensor. Following the standard derivation of the gravitational potential we find the Coulomb interaction of $A_{R\text{flat}}{}^{10}$.

B.5 PseudoQuantization of Affine Connections

Having obtained the form of the general affine connection we now PseudoQuantize them for later use in our unification program. We define

---

[532] It is possible that the Reality transformation also depends on $A_{R\text{flat}}{}^2{}_\sigma$. Then eq. A.26 would have an energy-momentum tensor term as well. Consequently there would be an additional interaction of the same form as in eq. A.27.

$$R^{1\beta}{}_{\sigma\nu\mu} = \partial_\mu H^{1\beta}{}_{\sigma\nu} - \partial_\nu H^{1\beta}{}_{\sigma\mu} + H^{1\gamma}{}_{\nu\sigma}H^{1\beta}{}_{\gamma\mu} - H^{1\gamma}{}_{\mu\sigma}H^{1\beta}{}_{\gamma\nu}$$

$$R^{2\beta}{}_{\sigma\nu\mu\rho} = \partial_\mu H^{2\beta}{}_{\sigma\nu} - \partial_\nu H^{2\beta}{}_{\sigma\mu} + H^{2\gamma}{}_{\nu\sigma}H^{2\beta}{}_{\gamma\mu} - H^{2\gamma}{}_{\mu\sigma}H^{2\beta}{}_{\gamma\nu} + \qquad\qquad \text{(B.27)}$$

$$+ H^{1\gamma}{}_{\nu\sigma}H^{2\beta}{}_{\gamma\mu} - H^{1\gamma}{}_{\mu\sigma}H^{2\beta}{}_{\gamma\nu} + H^{2\gamma}{}_{\nu\sigma}H^{1\beta}{}_{\gamma\mu} - H^{2\gamma}{}_{\mu\sigma}H^{1\beta}{}_{\gamma\nu}$$

where

$$H^\sigma{}_{\nu\mu} = \Gamma_{GR}{}^\sigma{}_{\nu\mu} + \Gamma_{GR}{}^{2\sigma}{}_{\nu\mu} + \Gamma_R{}^{1\sigma}{}_{\nu\mu} + \Gamma_R{}^{2\sigma}{}_{\nu\mu} \qquad\qquad \text{(B.28)}$$

and where $\Gamma_{GR}{}^\sigma{}_{\nu\mu}$ and $\Gamma_{GR}{}^{2\sigma}{}_{\nu\mu}$ are affine connections for real-valued General Relativity, and $\Gamma_R{}^{1\sigma}{}_{\nu\mu}$ and $\Gamma_R{}^{2\sigma}{}_{\nu\mu}$ are affine connections for a complex-valued set of transformations embodying a U(4) gauge group that combine with real-valued General Relativistic transformations to yield Complex General Relativistic transformations.

The affine connection is most often viewed as a derived quantity—part of the derivation of the curvature tensor in General Relativity. It is typically derived from manipulations of the metric $g_{\mu\nu}$. However, the affine connection can also be viewed as a set of independent fields that become related to the metric via dynamic equations.

Some years ago A. Einstein and H. Weyl[533] pointed out that the metric and the affine connection should be treated as independent quantities and subject to independent arbitrary infinitesimal variations:

> "In contrast to Einstein's original "metric" conception in terms of the $g_{\nu\mu}$ there was later developed, by Eddington, by Einstein himself, and recently by Schrödinger, an affine field theory operating with the components $\Gamma^\sigma{}_{\nu\mu}$ of an affine connection. But in 1925 Einstein also advocated a "mixed" formulation by means of a lagrangian in which both the $g_{\nu\mu}$ and the $\Gamma^\sigma{}_{\nu\mu}$ are taken as basic field quantities and submitted to independent arbitrary infinitesimal variations.[534] In certain respects this seems to be the most natural procedure."

Following this approach we have introduced the above affine connections for use in the construction of our unification of particle interactions.

[533] H. Weyl, Phys. Rev. **77**, 699 (1950).
[534] A. Einstein, Sitzungsber., Preuss. Akad. Der Wissensch. (1925), p. 414.

Appendix C. The Three Players in Creating a Fundamental Theory of Physical Reality and the Psychology of Reality

When one considers constructing a theory, or model, of fundamental 'real world' physical reality, the usual course of action is to make some assumptions such as a symmetry group, or a construct like strings, and then proceed to develop their implications.[535] Next, the authors of the theory, and experimenters, study the experimental implications of their theory. A tacit assumption, which is sometimes stated, is that their mathematical-mental construct is somehow automatically implemented by Reality. The possibility that their construction is merely a mental exercise, and not Reality, is seldom considered except by the philosophically inclined. And for good reason. No one has put forward a concept that joins thought to Reality in a convincing way except as 'word play.'

In this appendix we will seek to carefully characterize the process of constructing a fundamental physical theory.[536] In doing so, we identify 'three players' that play decisive roles in bringing a fundamental theory from conception to Reality. First is the construction process which is a mathematical-mental effort that leads to a complete theory fundamental theory. Secondly, there is the conceptual mapping of the theory to physical Reality. Thirdly, there is the 'agency' by which Reality is imprinted with the theory. The third 'player', the 'imprinter,' is often alluded to in the statement: "it is surprising how well Reality is described by mathematics." This statement is not good enough! We will examine it in some detail in this appendix together with the other two players in the 'game' of constructing the fundamental theory of everything.

---

[535] In the author's view this approach is satisfactory for non-fundamental theories and models where the underlying physics is presumably well understood. But at the fundamental physics level a more careful analysis is required. This chapter delves into this issue in some detail.
[536] We discuss the metaphysical and logical basis of our approach in more detail in Blaha (2011c).

C.1 Construction of a Fundamental Physics Theory of Everything

The construction process of a fundamental theory is a mental exercise consisting of defining primitive constructs, specifying assumptions (axioms), and then proceeding to develop a theory from the assumptions.

C.1.1 Euclidean Geometry Model

Euclid's theory of geometry is perhaps the best model of the process, and the components of the process. First Euclid identifies primitive unspecified constructs: straight lines, curves, and angles (which are viewed in no need of definition.) He then states five axioms and proceeds to state and prove theorems based on the unspecified, but understood, properties of the constructs. As the set of theorems grows, subsidiary, but unspecified, properties of geometric figures are used in proofs. These properties include the definition of angles, the construction of simple geometric figures such as rectangles, triangles, trapezoids, and so on. The more advanced developments in Euclidean geometry such as the geometry of Ptolemaic astronomy[537] with cycles and epicycles also have implicit principles for construction.

We see from Euclid that the theory construction process requires:

1. The definition of primitive terms.
2. The specification of a set of axioms.
3. The derivation of theorems from the axioms and further ad hoc, but necessary, assumptions.
4. The identification of the 'complete' theory with Reality.

Euclid was fortunate in the creation of Euclidean geometry because he knew the method with which to construct geometry, and its primitive terms with which to define his axioms, and then his theorems. The development of a fundamental physical theory is not so fortunate.

---

[537] Ptolomaic Astronomy is a very-well developed, and mathematically sound, branch of Euclidean geometry. Its flaw is its misidentification with the physical reality of the Solar System. In this regard it very well illustrates our point that the matching of theory with Reality is an important part of the creation of a fundamental theory.

C.1.2 Logic Applied to Physics

Perhaps the most important aspect of constructing a physical theory is the use of Logic to derive results from the fundamental axioms. Questions have been raised about the validity of Logic. Attention is often drawn to logical 'paradoxes' and particularly to Gődel's Theorem. In Blaha (2015a) we showed the problems of these paradoxes and Gődel's Theorem were due to the choice of subject(s) in a statement that were not in the subject domain of the statement. Statements are a verbal form of functions. Like functions they have a domain (set of valid subjects) and a range.

Most scientists, Logicians and Mathematicians are familiar with mathematical functions that have numerical arguments and calculate a numerical value or set of numerical values. However, those familiar with Computer Science know that a function, in general, has two types of output: its numeric value(s), and a status value indicating whether the computation that the function performed was successfully done or failed. Typically the status value is a zero or one, but it is understood as true or false also. (True – computation successful; false – computation failed for some reason) Thus we categorize functions as being in one of three broad categories:

Types of Functions

1	2	3
A Mathematical Function Producing A Value(s) Only;	A Function Producing a Value(s) and a Status Value;	A Function Producing a Status Value Only (T or F);

Examples:

sin function	Programming Function	Logic Statement

Thus Logic is unsullied by paradoxes and, besides being the only way to 'do' Physics theory, is correct and consistent.

C.1.1 Primitive Terms of Fundamental Physics Theories

The primitive terms of the fundamental theory of Physics is an open question. There are various theories that have been proposed: Twistor theory, SuperString theories, SuperStandard Theory, and so on. They differ significantly in their definition of primitive terms and their axioms (or their equivalent to axioms).

Under the circumstances probably the most significant aspects of terms are:

1. Consistency with what we know of Reality – they are somehow evident in Reality.
2. Simplicity.
3. Utility in defining meaningful axioms. This feature implies a forward-looking approach that looks to the development of the theory to guide the choice of primitives.

Euclid's specification of primitive terms agrees with these considerations.

C.1.2 Physical Axioms of a Fundamental Theory

The choice of axioms is very dependent on the eventual theory. Generally it seems that most theories choose a lagrangian and/or symmetry group, and proceed to 'back-engineer' a set of axioms if they consider defining axioms at all.

If one wishes to create a fundamental theory using a top-down approach then the set of axioms, like primitive terms, should satisfy the same three conditions plus Decision Axioms (described below):

1. Consistency with what we know of Reality – they are somehow evident in Reality.
2. Simplicity.
3. Utility in generating the fundamental theory. This feature also implies a forward-looking approach that looks to the development of the theory to guide the choice of axioms.
4. The choice of axioms should be in agreement with the below Decision axioms.

C.1.3 Decision Axioms for a Fundamental Theory

The concept of Decision Axioms does not seem to have been formally considered previously. Decision axioms guide a choice between alternatives in the construction of fundamental theory. Given the complex nature of fundamental theories choices between alternatives frequently occur. SuperString theory, for example, has millions of choices in the determination of the 'correct' SuperString theory.

What rules (Decision Axioms) can guide a choice? In this subsection we identify eight rules that provide guidance for making the proper choice.

C.1.3.1 Ockham's Razor

William of Ockham proposed a Law of Parsimony that is called *Ockham's Razor* which states that the simplest choice is to be preferred in a multiple choice situation. This principle is often stated as 'the simplest solution is usually the correct solution.'[538]

The best rationale for this principle is that it generally reduces the complexity that follows such a choice. Since many physics calculations are extremely difficult, picking the simplest choice would generally tend to make subsequent theorems/and calculations less difficult. This point of view might be thought to be ad hoc or anthropomorphic. But it reflects the reality of scientific calculation and of theory construction.

Thus we will assume Ockham's Law of Parsimony in the construction of our theory with the proviso that Leibniz's Minimax Principle (below) takes precedence if there is a conflict in the implications of the choices.

C.1.3.2 Leibniz's Minimax Principle for Physics Theories

Leibniz[539] developed a Minimax Principle that can be phrased for our purposes as, "The universe is based on the smallest set of properties or features that lead to the greatest variety of phenomena." This principle reflects the spirit of the minimum/maximum criteria of the Calculus

---

[538] William of Ockham – Law of Parsimony – "Pluralitas non est ponenda sine necessitate" or "Plurality should not be posited without necessity." Ockham's Law was first stated by Durand De Saint-Pourçain (1270-1334 A.D.). In simple terms the principle states the simplest solution to a problem is most likely to be the correct solution.
[539] See Rescher (1967).

of Variations[540] that plays a central role in many physics theories. This principle somewhat overlaps Ockham's Law of Parsimony. Given a choice of possible theoretical lines of construction there is a possibility that the Law of Parsimony and the Minimax Principle would suggest different choices. Fortunately, we will see that the construction of the SuperStandard Theory does not seem to present this potential dilemma.

An important, unremarked, aspect of Leibniz's Principle is the decision between a set of choices depends on the future part of the construction or theory. Thus future constructs influence past constructs in minimax decisions.[541]

C.1.3.3 Computationally Minimal

Given a choice between two alternatives that appear equally likely from general considerations and the view of the preceding two Decision Axioms, the preferred choice is the one leading to the most minimal (simplest) computations in the full theory.

C.1.3.4 Experimentally Verifiable Directly or Indirectly

Given a choice between two alternatives, one of which is subject to experimental verification, directly or indirectly although perhaps in the future, and the other choice is not, then the first alternative is preferred.

C.1.3.5 Level of Rigor

Physics is mathematical in nature. Mathematics strives for rigor and will not be satisfied without rigorous results except for conjectures and speculations. Physics prefers rigorously derived results as well. However, there is 'physical rigor,' which is a level of rigor that is rigorous to the extent that it is mathematically rigorous. This circuitous statement reflects the historical fact that Physics has often gone beyond the mathematics of the time. The most significant example is Newtonian physics, which until the mid-19th century used *derivatives* asserting that they were, or would become,

---

[540] Leibniz was one of the founders of the Calculus of Variations.

[541] The knowledgable reader will remember Feynman's speculation that the physical universe may be evolving from the future into the past. Quantum field theories support such an interpretation of their mathematics. The similarity of this fundamental minimax principle's feature with the corresponding feature of physics theories encourages support for the minimax principle as a 'design' law of fundamental physical theory.

rigorous mathematically. After a careful analysis of derivatives, and advances in the understanding of continuity and the various types of infinities, mathematicians were able to put derivatives on a rigorous footing justifying the use of derivatives by physicists for almost four centuries.

Today, the most interesting part of physics in search of mathematical rigor is the path integral formulation in quantum theory and the Faddeev-Popov Method, in particular. The problem here is again an understanding of infinities in functional derivatives and path integrals. Again, mathematical rigor is lacking. Yet physicists use these techniques with the strong belief that their results will be eventually justified rigorously.

In view of the nature of rigor in Physics it seems reasonable to state the Decision Axiom: *In the case of a situation where several possible approaches are possible, one should choose the approach that is most rigorous – should one exist. If all approaches are at an 'equal' level of rigor then the approach that appears 'most' physical should be chosen, taking account of the other Decision Axioms.*

C.1.3.6 Replication Principle - Nature Tends to Repeat Successful Strategies

In many branches of Science one sees that successful strategies and techniques are repeated in several areas – perhaps with some variations and changes. Consequently, when a physical phenomenon is similar to another physical phenomena, and all other axioms that were stated above are not relevant for the phenomena, then one should model the new phenomena in a manner similar to the successfully modeled phenomena.

C.1.3.7 Principle of Maximal Serendipity

A correct physical theory will evolve from fundamental axioms with maximal simplicity to attain better-than-expected ends.

C.1.3.8 Principle of Maximal Descent to a Complete Theory

The development from axioms to a 'complete' theory will take place maximally in the sense that the steps of the derivation lead rapidly to the theory. This principle is analogous to the commonly used mathematical principle of maximal descent.

C.1.4 The Body of the Derived Theory –Jeopardy Points

After the primitives and the axioms have been specified, the body of the theory can be developed mathematically (logically). At points in the development process decisions may have to be made between alternatives. We call these points – *jeopardy points*. The above Decision Axioms can provide guidance if a strictly physical decision is not evident.

Clearly the fewer the jeopardy points in a theoretical development, the better. The existence of many such points raises the question of whether the axiomatic basis needs to be strengthened.

The general lack of jeopardy points suggests the axiomatic basis of the theory is well-founded.

C.1.5 The Fundamental Theory as a Mental Construct

If one creates a fundamental theory it is a mental construct, and although physicists are quick to identify it with Reality, it is not Reality until a map to Reality is specified. Parmenides pointed out this distinction in a way in his only surviving poem *On Nature*. He said the fundamental nature of the universe was an unchanging unity of nature (The Way of Truth); but our view of the universe was that of change and seeming (The Way of Appearance). Translated into the framework of our discussion we see the fundamental theory as an unchanging mental construct, but the Reality to which it maps as ever changing.

In the next section we discuss the mapping realizing that most physical theories make the leap from concept to Reality without effort.

C.2 Conceptual Mapping of a Fundamental Theory to Physical Reality

Generally the fundamental theories of Physics that have been proposed are so close to Reality that the mapping process is straight-forward – pick a symmetry group, specify a lagrangian, find the physical particle spectrum (which was put there by the choice of symmetry group anyway), and calculate experimentally verifiable numbers.

However in some cases the mapping is not so clear. For example, one might find an extremely large particle spectrum with experimentally measurable masses, and the particles are not there! Then one has the dilemma of redesigning the theory.

C.2.1 Mapping of the Fundamental Physical Axioms

One could possibly redesign the set of axioms of the theory if they were at fault and if the axioms had been specified. The new theory that emerges might then be mappable to experimental reality.

C.2.2 Mapping of the Body of the Derived Fundamental Theory

If the body of the fundamental theory does not agree with a mapping to Reality, the creator(s) of the theory often change the parameters or otherwise tinker with its development. One sees this in some current theories.

C.3 The 'Agency' by which Reality is Imprinted with a Fundamental Theory

Mental processes such as the thought embodying a physical theory are clearly different from physical reality. A question that has been a subject of discussion for over 2500 years is How does thought become embodied in Reality? A further question that has been of great interest for the past five hundred years is How does the mathematics of physical theory so well explain/correspond to Reality?

Addressing the second question we note that there are two possibilities: Reality is chaotic or Reality follows physical laws (which are necessarily mathematical). If Reality were chaotic but through some principle of self-organization transformed itself into an ordered system, then we would see order but there would be pockets of chaos in the universe and on earth that we have not seen. Further the transition to order would have to repeat with every change of state of a physical system – an effect which is also not seen. Thus 'self-organization' in any form does not appear to be the source of Physical Law.

Thus we must conclude that nature is ordered and conforms to physical laws. Physical laws are intertwined through mathematics. The fact that we can define theories that embody all physical laws – rather than have a host of unrelated physical laws is another marvel.

So we are left with the issue of who imprinted the concepts of a fundamental physical theory on Reality. The philosophers (Pre-Socratic and more recent) addressed this problem and came to the conclusion that there was an 'Unmoved Mover' – an entity that implemented the imprinting of physical law on Reality.

Most philosophers (and theologians) identified this entity as God. And they called this the proof of the existence of God.

C.4 The Psychology of Physical Reality

C.4.1 Progress in Understanding of Physical Reality

In the past 2,500 years we have progressed in our understanding of the nature of matter and energy from the four elements theories of 500 B.C. to the atomic theory of matter (accepted[542] in only the late 19th Century) and the quantum photon theory of light.

In the 20th Century we saw the joint emergence of Relativity, Quantum Theory, and elementary particles and their interactions. Those developments led to the derivation of the Unified SuperStandard Theory by the author at the beginning of the 21st Century.

The trend illustrated by these milestones ranges from the 'concrete' substances of 500 B.C. to the diaphanous wave-particles that we understand to constitute matter and energy at present. The theory that we have developed in these volumes for wave-particles lends itself to an interpretation as a language description. We have described the language description earlier and in a series of earlier books. It has an alphabet of particles, and production rules embodied in the lagrangian of the Unified SuperStandard Theory, that is particularly evident in the Theory's perturbation theory expansion.

Thus one may say that, in 2,500 years, we have transitioned from 'concrete' matter to an 'evanescent' language of physical reality where matter and energy have the insubstantial form of a language.

---

[542] One of the causes of Boltzmann's suicide was despair over the non-acceptance of atomic theory.

C.4.2 Language and Psychology

Given this linguistic view of physical reality it is reasonable to inquire as to the 'psychology' that the Unified SuperStandard Theory reflects. For a language reflects the culture and concerns of its speakers. The classic example of this 'hidden' aspect of languages is Ancient Greek, which had approximately three hundred words for the various forms of love suggesting a preoccupation for human emotions that we see in the dialogues of Plato and other Pre-Socratic Greek writings. Contrast that with the one word 'Love' that English uses for most forms of love.

In the case of the Unified SuperStandard Theory we may expect to acquire physically relevant information about the physical psychology of the Entity but human-like aspects such as the emotions of love, hate, and so on cannot be determined from the consideration of the physical language of the Unified SuperStandard Theory.

So what does the language of physical reality tell us of the underlying psychology of the putative Creator Entity?

1. It is abstract like mathematical reasoning..
2. It is computationally-oriented. The Unified SuperStandard Theory can be mapped to a type 0 language and grammar production rules.
3. It is rigorously logical.
4. It is oriented toward simplicity in the manner of the Decision Axioms in the book.

C.5 Principle of Vitamorphism

The various coupling constants and mass values embodied in the Unified SuperStandard Theory suggest a profound Vitamorphism on the part of the Entity. For small changes in these values could make life, in all the forms that we can envision, impossible.

Thus we can say the Entity 'designed' the Cosmos for life.[543]

---

[543] If a deeper theory can fully justify the values of these constants so that the Entity had no other choice, then the motive of Vitamorphism is removed.

C.6 Wave-Particles vs. Thought

The wave-particles that are at the basis of the Unified SuperStandard Theory are very much like thoughts within a brain. If wave-particles experienced no interactions, then they would be just as insubstantial as brain waves—the only significant difference being the confinement of thoughts to the brain. So we can make the simple diagram:

$$\text{Reality} = \text{Thoughts} + \text{Interactions}$$

if we regard an interactionless form of the Unified SuperStandard Theory as thought.

Appendix D. New Paradigm: Composition of Extrema in the Calculus of Variations

This Appendix appeared originally in Blaha (2005a) and earlier books.

D.1 A New Paradigm in the Calculus of Variations

The Calculus of Variations has a long and venerable history in Physics and Mathematics. Many problems in Physics and Mathematics have been treated with approaches based on techniques in the Calculus of Variations.[544] In this book we have developed a unified quantum field theory of the known forces of nature based on a new type of problem, or paradigm, in the Calculus of Variations. One way of viewing the spectrum of problems in the calculus of variations is the following progression.

D.1.1 A Classification of Variational Problems

1. Variational problems in a Euclidean, or Minkowski, flat space such as the minimal distance between two points or the extrema of a field theory Lagrangian.

2. Variational problems seeking extrema on a curved surface such as the shortest distance between points on the surface of a sphere.

The development in this book suggests a third and fourth, possibility, which to the author's knowledge, has not been addressed in the literature:

3. Variational problems where the extrema are determined on a surface that is itself defined as an extremum. The discussions in this book exemplify this paradigm.

---

[544] See Akhiezer (1962), Blaha (2003), Gelfand (2000), Giaquinta (1996) and (1998), Jost (1998), and Sagan (1993),

4. Variational problems where the extrema are determined on a surface that is itself defined as an extremum that depends on the extrema on the surface. More simply put the extrema, and the surface upon which they are defined, are jointly determined and are interrelated. Fortunately, our unified theory does not use this paradigm. A future theory might.

In the unified theory that we will develop all particle fields including the graviton field are defined as a mapping of a Minkowski space-time y to a "particle" space-time X with the mapping determined as an extremum of a variation of a fundamental field (a type 3 variational problem in the above classification). Our theory could be generalized to include a back-reaction of the particle fields on the fundamental field (a type 4 variational problem in the above classification). We will not discuss this possibility in this book.

D.2 Simple Physical Example – Strings On Springs

In this section we will describe a simple physical example that illustrates a variational problem of type 3 in the Calculus of Variations. We view it as a composition of extrema. (This problem can be addressed using other calculus of Variations techniques.) The approach used in the solution of this problem is similar to the approach used in Two-Tier quantum field theory.

D.2.1 A Strings on Springs Mechanics Problem

Consider a long string or bar that can oscillate (undulate) in a direction perpendicular to its length. Further assume that one end of this bar or string is attached to a spring that cause the entire bar or string to oscillate back and forth in a direction parallel to its long side. This configuration is illustrated in Fig. D.1.

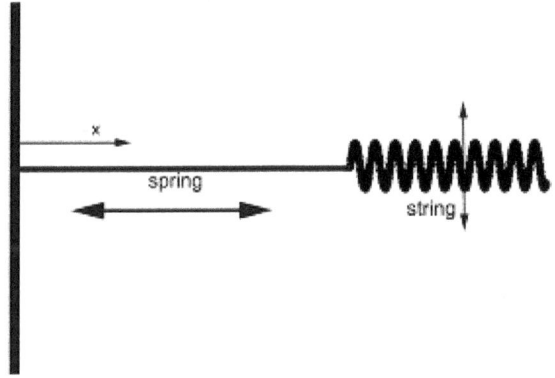

Figure D.1. An oscillating string attached to a spring.

Let x denote the distance to a point on the string when the spring is at equilibrium. If 2π times the frequency of the spring is ω_1, then the location of this point when the spring is oscillating is

$$X(t) = x + A \sin(\omega_1 t + \phi_1) \tag{D.1}$$

where ϕ_1 is a phase, and A is the amplitude of the spring oscillation. Then the vertical displacement of a traveling wave on the *string* can take the form

$$\psi(t) = B \sin(\omega_2 t - k_2(x + A \sin(\omega_1 t + \phi_1)) + \phi_2) \tag{D.2}$$

where B is the amplitude of the string wave, and k_2, ω_2 and ϕ_2 are the parameters of the string wave. These simple mechanical formulae are well known. But they lead to an interesting new application of the ideas of the Calculus of Variations.

Suppose we treat X as an independent variable with X given by eq. (D.1), and with eq. (D.2) written as:

$$\psi(t) = B \sin(\omega_2 t - k_2 X + \phi_2) \tag{D.3}$$

Defining

$$\psi = \psi(X(t), t) \tag{D.4}$$

we can specify the dynamics of the above motion by finding the extrema of

$$I = \int \mathscr{L}_\psi \, dX(t) + \int \mathscr{L}_X \, dt \tag{D.5}$$

where the Lagrangian terms are

$$\mathscr{L}_\psi = \tfrac{1}{2} \{ \mu \, (\partial\psi/\partial t)^2 - Y \, (\partial\psi/\partial X)^2 \} \tag{D.6}$$

with μ and Y being constants, and

$$\mathscr{L}_X = \tfrac{1}{2} \{ m(\partial X/\partial t)^2 - k(X - x)^2 \} \tag{D.7}$$

where m and k are constants, and where x is a parameter. Applying Hamilton's Principle, and performing independent variations of X and ψ yields the Lagrangian equations:

$$\frac{\partial \mathscr{L}_\psi}{\partial \psi} - \frac{\partial}{\partial X} \frac{\partial \mathscr{L}_\psi}{\partial(\partial\psi/\partial X)} - \frac{\partial}{\partial t} \frac{\partial \mathscr{L}_\psi}{\partial(\partial\psi/\partial t)} = 0 \tag{D.8}$$

and

$$\frac{\partial \mathscr{L}_X}{\partial X} - \frac{\partial}{\partial t} \frac{\partial \mathscr{L}_X}{\partial(\partial X/\partial t)} = 0 \tag{D.9}$$

The resulting equations of motion are:

$$\mu \, \partial^2 \psi / \partial t^2 - Y \, \partial^2 \psi / \partial X^2 = 0 \qquad (D.10)$$

and

$$m \, \partial^2 X / \partial t^2 + k(X - x) = 0 \qquad (D.11)$$

with the solutions given in eqs. D.1 and D.2.

The procedure that we use to obtain these results may look a bit strange but they illustrate a type 3 problem in the Calculus of Variations involving the composition of extrema—the composition of an extremum that specifies a manifold in a space (possibly including all of space in a $R^n \rightarrow R^n$ mapping) with an extremum determining a function on that manifold. The procedure is described in detail in the next section.

D.3 The Composition of Extrema – A Lagrangian Formulation

In this section we will explore the general case of the composition of extrema for fields. We will discuss the case of a scalar field ϕ that is a function of a vector field X^μ in a D-dimensional space with coordinate variables that we will denote as y^μ. (The discussion for other types of fields is a straightforward extension of this discussion.) Thus

$$\phi = \phi(X) \qquad (D.12)$$

and

$$X^\mu = X^\mu(y) \qquad (D.13)$$

We assume that the dynamics can be described by a Lagrangian formulation using an extension of Hamilton's principle:

$$I = \int \mathcal{L} d^4 y \qquad (D.14)$$

with

$$\mathcal{L} = \mathcal{L}(\phi(X), \partial\phi/\partial X^\nu, X^\mu(y), \partial X^\mu(y)/\partial y^\nu, y) \qquad (D.15)$$

If we perform a standard variation[545] in ϕ for fixed y (and thus fixed X) we find

$$\delta I = \int [\delta\phi \, \partial\mathscr{L}/\partial\phi + \delta(\partial\phi/\partial X^\nu) \, \partial\mathscr{L}/\partial(\partial\phi/\partial X^\nu)] \, d^4y \qquad (D.16)$$

We can rewrite the variation in the derivative of ϕ as

$$\delta(\partial\phi/\partial X^\nu) = \partial(\delta\phi)/\partial X^\nu \qquad (D.17)$$

$$= \partial y^\mu/\partial X^\nu \, \partial(\delta\phi)/\partial y^\mu \qquad (D.18)$$

with an implied summation over repeated indices. After substituting eq. D.18 in eq. D.16, and performing an integration by parts (and discarding the surface term which is assumed to yield zero in the standard fashion) we obtain:

$$\delta I = \int \delta\phi \, \{\partial\mathscr{L}/\partial\phi - \partial/\partial y^\mu [\partial\mathscr{L}/\partial(\partial\phi/\partial X^\nu) \, \partial y^\mu/\partial X^\nu)]\} \, d^4y$$

Since the variation of $\delta\phi$ is arbitrary we conclude

$$\partial\mathscr{L}/\partial\phi - \partial/\partial y^\mu [\partial\mathscr{L}/\partial(\partial\phi/\partial X^\nu) \, \partial y^\mu/\partial X^\nu)] = 0 \qquad (D.19a)$$

The second term in eq. D.19a shows the effect of the dependence of ϕ on the field X, $\phi = \phi(X)$, rather than directly on the coordinate system y.

Similarly we can perform a variation in X^μ and obtain

$$\partial\mathscr{L}/\partial X^\mu - \partial/\partial y^\nu [\partial\mathscr{L}/\partial(\partial X^\mu/\partial y^\nu)] = 0 \qquad (D.19b)$$

[545] Bogoliubov, N. N., & Shirkov, D. V., Volkoff, G. M. (tr), *Introduction to the Theory of Quantized Fields* (Wiley-Interscience, New York, 1959); Goldstein H., *Classical Mechanics* (Addison-Wesley, Reading, MA 1965).

The X field defines a "manifold" or, more properly, specifies a transformation from R^n $\rightarrow R^n$. If we make standard assumptions about the mapping: that it is continuous and piece-wise invertible, then we can establish the following lemmas:

Lemma 1: *If the transformation $X^\mu = X^\mu(y)$ is a transformation from $R^n \rightarrow R^n$ that is of class C' and piece-wise invertible, then*

$$\frac{\partial}{\partial y^\nu} \frac{\partial y^\nu}{\partial X^\mu} = -\frac{\partial \ln J}{\partial X^\mu} \qquad (D.20)$$

where

$$J = |\partial(X)/\partial(y)| \qquad (D.21)$$

is the absolute value of the Jacobian of the transformation.

Proof:
Consider two equivalent forms of an integral:

$$I = \int \mathscr{L} J \, d^4y = \int \mathscr{L} \, d^4X$$

where $\mathscr{L}$ is specified as in eq. D.15. Then the first expression for I leads to eq. D.19a which can be written in the form

$$\partial \mathscr{L}/\partial\phi - \partial/\partial X^\mu [\partial \mathscr{L}/\partial(\partial\phi/\partial X^\mu)] - \partial \mathscr{L}/\partial(\partial\phi/\partial X^\mu)\{\partial[J\partial y^\nu/\partial X^\mu]/\partial y^\nu\} = 0$$

Using the second expression for I above we obtain the following equation by variation in ϕ:

$$\partial \mathscr{L}/\partial\phi - \partial/\partial X^\mu [\partial \mathscr{L}/\partial(\partial\phi/\partial X^\mu)] = 0$$

Comparing these two expressions and realizing that $\partial[J\partial y^\nu/\partial X^\mu]/\partial y^\nu$ is totally independent of ϕ and its derivatives leads us to conclude

$$\partial\,[\,J\partial y^\nu/\partial X^\mu]/\partial y^\nu\ =0 \tag{D.22}$$

It is a general relationship for a transformation between X and y based on continuity and piece-wise invertibility. After a few elementary manipulations eq. D.22 can be rewritten in the form of eq. D.20. ∎

Lemma 2: *If the transformation $X^\mu = X^\mu(y)$ is a transformation from $R^n \to R^n$ that is of class C' and piece-wise invertible and $\mathscr{L} = \mathscr{L}(\phi(X), \partial\phi/\partial X^\nu, X^\mu(y), \partial X^\mu(y)/\partial y^\nu, y)$, then*

$$\partial\mathscr{L}/\partial(\partial\phi/\partial X^\nu)\,\partial y^\mu/\partial X^\nu = \partial\mathscr{L}/\partial(\partial\phi/\partial y^\mu) \tag{D.23}$$

Proof:

Let us express $\mathscr{L}$ as a power series in derivatives of ϕ:

$$\mathscr{L} = \sum_{n=0} a_{n\mu_1\mu_2\cdots\mu_n}(\phi(X), X^\mu(y), \partial X^\mu(y)/\partial y^\nu, y) \prod_{j=1}^{n} \partial\phi/\partial X^{\mu_j}$$

which can rewritten using piece-wise invertibility as

$$\mathscr{L} = \sum_{n=0} a_{n\mu_1\mu_2\cdots\mu_n}(\phi(X), X^\mu(y), \partial X^\mu(y)/\partial y^\nu, y) \prod_{j=1}^{n} \partial\phi/\partial y^{\nu_j}\,\partial y^{\nu_j}/\partial X^{\mu_j}$$

Taking the derivative of this equation with respect to $\partial\phi/\partial y^\mu$ immediately yields the result. ∎

Eq. D.23 enables us to rewrite eq. D.19a as:

$$\partial \mathscr{L}/\partial\phi - \partial/\partial y^\mu [\partial \mathscr{L}/\partial(\partial\phi/\partial y^\mu)] = 0 \qquad (D.24)$$

which is as one would expect.

In order to get a feeling for the effect of eq. D.19a we will look at a simple example where we specify the relation of the X and y variables directly. Then we will look at the composition of extrema where the transformation between X and y is itself determined as an extremum solution.

D.3.1 Example: a hyperplane

We assume eq. D.19b yields the transformation:

$$X^i = ay^i \qquad \text{for } i = 1,2,3$$

$$X^0 = 0$$

Then eq. D.19a becomes

$$\partial \mathscr{L}/\partial\phi - \partial/\partial y^i [\partial \mathscr{L}/\partial(\partial\phi/\partial y^i)] = 0 \qquad (D.25)$$

with the time derivative disappearing. Effectively the variation of ϕ on the hyperplane $X^0 = 0$ is determined by the differential equation generated by D.25. On this hyperplane the transformation between the X and y variables is invertible.

D.3.2 Coordinate Transformation Determined as an Extremum Solution

We now develop a formalism that determines a mapping from space onto itself as the solution of an extremum problem and also determines the dynamics of one or more fields as a function of this mapping. To this author's knowledge this area in the Calculus of Variations – the determination of an extremum on a manifold where the manifold itself is determined by an extremum – has not been previously explored. We will also develop a Hamiltonian formulation. Then we will proceed to quantize the theory.

D.3.3 Separable Lagrangian Case

Although there are many forms that the composition of extrema could take, one fairly general form that is directly useful in quantum field theory applications is based on a Lagrangian that can be split into two parts which we will call a *separable Lagrangian*:

$$\mathscr{L} = \mathscr{L}_F J + \mathscr{L}_C(X^\mu(y), \partial X^\mu(y)/\partial y^\nu, y) \tag{D.26}$$

where J is defined in eq. D.21, where $\mathscr{L}_F$ contains all the dynamics of the fields and their interactions, and where $\mathscr{L}_C$ defines the coordinate mapping as an extremum solution. The procedure to determine the differential equations that specify the mapping, and the field equations that specify field interactions and evolution, is to vary in the coordinates X^μ and in the fields independently, using Hamilton's Principle. The extrema are to be determined for

$$I = \int \mathscr{L}\, d^4y \tag{D.27}$$

We will begin by considering the case of one scalar field:

$$\mathscr{L}_F = \mathscr{L}_F(\phi(X), \partial\phi/\partial X^\nu) \tag{D.28}$$

and

$$\mathscr{L}_C = \mathscr{L}_C(X^\mu(y), \partial X^\mu(y)/\partial y^\nu, y) \tag{D.29}$$

Eq. D.27 can be written in the form:

$$I = \int \mathscr{L}_F(\phi(X), \partial\phi/\partial X^\nu)\, dX + \int \mathscr{L}_C(X^\mu(y), \partial X^\mu(y)/\partial y^\nu, y)\, d^4y \tag{D.30}$$

using the Jacobian to transform to an integral over dX in the first term. A standard variation of ϕ and the application of Hamilton's Principle yields

$$\partial \mathscr{L}_F / \partial \phi - \partial / \partial X^\mu [\partial \mathscr{L}_F / \partial (\partial \phi / \partial X^\mu)] = 0 \qquad (D.31)$$

reflecting the fact that ϕ is a function of X^μ only, with X^μ a function of the y coordinates.

Next we perform a variation of X^μ determining the mapping from $y \rightarrow X$ as an extremum of the integral in eq. D.27. We note the piece-wise invertibility of the coordinate mapping $X^\mu(y)$ allows us to write the Jacobian J as a function of y^μ only. A standard variation of X^μ and the application of Hamilton's Principle yields

$$\partial \mathscr{L}_C / \partial X^\mu - \partial / \partial y^\nu [\partial \mathscr{L}_C / \partial (\partial X^\mu / \partial y^\nu)] = 0 \qquad (D.32)$$

D.3.4 Klein-Gordon Example

The Klein-Gordon scalar field theory furnishes us with a simple example of the application of the preceding development. The Lagrangian is

$$\mathscr{L}_F = \tfrac{1}{2} [(\partial \phi / \partial X^\nu)^2 - m^2 \phi^2] \qquad (D.33)$$

From eq. D.31 we obtain the field equation:

$$(\Box + m^2) \, \phi(X) = 0 \qquad (D.34)$$

where

$$\Box = \partial / \partial X^\nu \, \partial / \partial X_\nu \qquad (D.34a)$$

A fourier representation of the solution of eq. D.34 is:

$$\phi(X) = \int dp \, \delta(p^2 - m^2) \theta(p^0) \, [A(p) \, e^{-ip \cdot X} + A(p)^* \, e^{ip \cdot X}] \qquad (D.35)$$

where A(k) is a function of k and * indicates complex conjugation.

The determination of $X^\mu(y)$ depends on the Lagrangian $\mathcal{L}_C$ and the solutions of eq. D.3A. If we chose

$$\mathcal{L}_C = -\tfrac{1}{2} (\partial X^\mu / \partial y^\nu)^2 \tag{D.36}$$

Then we obtain the equation

$$\square\, X^\mu = 0 \tag{D.37}$$

with the solution

$$X^\mu = \int dk\, \delta(k^2)\theta(k^0)\, [a^\mu(k)\, e^{-ik\cdot y} + a^\mu(k)^*\, e^{ik\cdot y}] \tag{D.38}$$

where $a^\mu(k)$ are complex vector functions of k in general. (We ignore positivity issues for the moment.) Substitution of eq. D.38 in eq. D.35 yields an expression with a form reminiscent of boson string expressions.[546] We will take up this point later in subsequent sections.

D.4 The Composition of Extrema – Hamiltonian Formulation

The previous section established a Lagrangian formulation of dynamics based on the composition of extrema. In this section we will develop an equivalent Hamiltonian formulation. We will assume a Minkowski space-time with X^0 and y^0 playing the role of the time coordinates in the respective coordinate systems.

Initially, we will assume a scalar field ϕ with a Lagrangian of the form in eq. D.15 and define canonical momenta with

$$\Pi_\phi = \partial\mathcal{L}/\partial\dot\phi \equiv \partial\mathcal{L}/\partial(\partial\phi/\partial X^\mu)\, \partial y^0/\partial X^\mu \tag{D.39}$$

$$\Pi_X^\mu = \partial\mathcal{L}/\partial\dot X_\mu \tag{D.40}$$

---

[546] See for example Polchinski (1998) and Bailin (1994).

where

$$\dot{\phi} = \partial\phi/\partial y^0 \equiv \partial\phi/\partial X^\mu \, \partial X^\mu/\partial y^0 \qquad \text{(D.41)}$$

$$\dot{X}^\mu = \partial X^\mu/\partial y^0 \qquad \text{(D.42)}$$

Then we define the Hamiltonian density as

$$\mathcal{H} = \Pi_\phi \, \dot{\phi} + \Pi_X{}^\mu \, \dot{X}_\mu - \mathcal{L}(\phi(X), \partial\phi/\partial X^\nu, X^\mu(y), \partial X^\mu(y)/\partial y^\nu, y)$$

$$\text{(D.43)}$$

and the Hamiltonian

$$H = \int \mathcal{H} \, d^3y \qquad \text{(D.44)}$$

The Hamiltonian density has the general form

$$\mathcal{H} = \mathcal{H}(\phi(X), \partial\phi/\partial X^i, \Pi_\phi, X^\mu(y), \partial X^\mu(y)/\partial y^j, \Pi_X{}^\mu, y^\nu) \qquad \text{(D.45)}$$

for the case of one scalar field where the indices i and j represent space coordinates; time coordinates are assigned index value 0.

If we calculate the differential change in H using eq. D.45 we obtain

$$dH = \int \left\{ \partial\mathcal{H}/\partial\phi \, d\phi + \partial\mathcal{H}/\partial\Pi_\phi \, d\Pi_\phi - \partial/\partial y^\nu [\partial\mathcal{H}/\partial(\partial\phi/\partial X^i)\partial y^\nu/\partial X^i] d\phi + \right.$$
$$\left. + \partial\mathcal{H}/\partial X^\mu \, dX^\mu + \partial\mathcal{H}/\partial\Pi_X{}^\mu \, d\Pi_X{}^\mu - \partial/\partial y^j \, [\partial\mathcal{H}/\partial(\partial X^\mu/\partial y^j)] \, dX^\mu \right\} d^3y$$
$$\text{(D.46)}$$

after some partial integrations. (Repeated indices indicate summations. Indices labeled i and j indicate space coordinates. Greek indices include all space-time components of a variable.)

Expressing the differential in H using eq. D.43 we obtain

$$dH = \int dy \left\{ \Pi_\phi \, d\dot\phi + \dot\phi \, d\Pi_\phi - \partial\mathcal{L}/\partial\phi \, d\phi - \partial\mathcal{L}/\partial(\partial\phi/\partial X^\mu) d(\partial\phi/\partial X^\mu) + \right.$$

$$\left. + \Pi_X^\mu \, d\dot{X}^\mu + \dot{X}^\mu \, d\Pi_X^\mu - \partial\mathcal{L}/\partial X^\mu \, dX^\mu - \partial\mathcal{L}/\partial(\partial X^\mu/\partial y^j) d(\partial X^\mu/\partial y^j) \right\}$$

(D.47a)

After some manipulations we find

$$dH = \int \left\{ \dot\phi \, d\Pi_\phi + \dot{X}_\mu d\Pi_X^\mu - \partial/\partial y^0 \, \Pi_\phi \, d\phi - \partial/\partial y^0 \, \Pi_X^\mu \, dX_\mu \right\} dy \qquad \textbf{(D47b)}$$

using the equations of motion eqs. D.19a and D.19b.

$$\dot\phi = \partial\mathcal{H}/\partial\Pi_\phi$$

$$\dot\Pi_\phi = -\partial\mathcal{H}/\partial\phi + \partial/\partial y^\nu \, [\partial\mathcal{H}/\partial(\partial\phi/\partial X^j) \, \partial y^\nu/\partial X^j] \qquad \textbf{(D48a)}$$

$$\dot{X}_\mu = \partial\mathcal{H}/\partial\Pi_X^\mu \qquad \textbf{(D48b)}$$

$$\dot\Pi_X^\mu = -\partial\mathcal{H}/\partial X^\mu + \partial/\partial y^j \, [\partial\mathcal{H}/\partial(\partial X^\mu/\partial y^j)] \qquad \textbf{(D.48c)}$$

Comparing eqs D.46 and D.47 we obtain Hamilton's equations in the case of the composition of extrema:

$$\dot\Pi_\phi = \partial \Pi_\phi/\partial y^0$$

$$\dot\Pi_X^\mu = \partial\Pi_X^\mu/\partial y^0 \qquad \textbf{(D.49a)}$$

$$dH = \int dy \left\{ \Pi_\phi \, d\dot{\phi} + \dot{\phi} \, d\Pi_\phi - \partial\mathscr{L}/\partial\phi \, d\phi - \partial\mathscr{L}/\partial(\partial\phi/\partial X^\mu) d(\partial\phi/\partial X^\mu) + \right. $$

$$\left. + \Pi_X{}^\mu \, d\dot{X}^\mu + \dot{X}^\mu d\Pi_X{}^\mu - \partial\mathscr{L}/\partial X^\mu \, dX^\mu - \partial\mathscr{L}/\partial(\partial X^\mu/\partial y^j) d(\partial X^\mu/\partial y^j) \right\} \tag{D.49b}$$

D.5 Translational Invariance

If the Lagrangian of a field theory has no explicit dependence on the coordinates then one expects translational invariance accompanied by a conservation law for an energy-momentum stress tensor. We will show this is the case for Lagrangians implementing the composition of extrema. We assume a Lagrangian without an explicit dependence on the coordinates y^ν:

$$\mathscr{L} = \mathscr{L}(\phi(X), \partial\phi/\partial X^\nu, X^\mu(y), \partial X^\mu(y)/\partial y^\nu) \tag{D.50}$$

Under an infinitesimal displacement,

$$y'^\nu = y^\nu + \epsilon^\nu \tag{D.51a}$$

$$\delta\phi = \phi(X(y + \epsilon)) - \phi(X(y))$$

$$= \epsilon^a \, \partial\phi/\partial y^a \tag{D.51b}$$

$$\delta X^\mu = \epsilon^a \, \partial X^\mu/\partial y^a \tag{D.51c}$$

$$\delta(\partial\phi/\partial X^\mu) = \epsilon^a \, \partial(\partial\phi/\partial y^a)/\partial X^\mu \tag{D.51d}$$

$$\delta(\partial X^\mu/\partial y^\nu) = \epsilon^a \, \partial(\partial X^\mu/\partial y^a)/\partial y^\nu \tag{D.51e}$$

and the Lagrangian changes by

$$\delta \mathcal{L} = \epsilon^a \, \partial \mathcal{L} / \partial y^a \tag{D.52}$$

The change can also be expressed in terms of the changes in the fields, their derivatives and the mapping X^μ:

$$\delta \mathcal{L} = \partial \mathcal{L} / \partial \phi \, \delta \phi + \partial \mathcal{L} / \partial (\partial \phi / \partial X^\mu) \, \delta (\partial \phi / \partial X^\mu) + \partial \mathcal{L} / \partial X^\mu \, \delta X^\mu +$$
$$+ \, \partial \mathcal{L} / \partial (\partial X^\mu / \partial y^\nu) \, \delta (\partial X^\mu / \partial y^\nu) \tag{D.53}$$

Combining eqs. D.51, D.52 and D.53 we obtain (after some manipulations):

$$\epsilon^\nu \, \partial / \partial y_\mu \, \mathcal{T}_{\mu\nu} = 0 \tag{D.54}$$

where

$$\mathcal{T}_{\mu\nu} = -g_{\mu\nu} \mathcal{L} + \partial \mathcal{L} / \partial (\partial \phi / \partial X^\delta) \, \partial y_\mu / \partial X^\delta \, \partial \phi / \partial y^\nu + \partial \mathcal{L} / \partial (\partial X^\delta / \partial y_\mu) \partial X^\delta / \partial y^\nu \tag{D.55a}$$

or, alternately using Lemma 2,

$$\mathcal{T}_{\mu\nu} = -g_{\mu\nu} \mathcal{L} + \partial \mathcal{L} / \partial (\partial \phi / \partial y_\mu) \, \partial \phi / \partial y^\nu + \partial \mathcal{L} / \partial (\partial X^\delta / \partial y_\mu) \, \partial X^\delta / \partial y^\nu \tag{D.55b}$$

Since ϵ^a is an arbitrary displacement we obtain the conservation law:

$$\partial / \partial y_\mu \, \mathcal{T}_{\mu\nu} = 0 \tag{D.56}$$

Eq. D.56 implies the energy-momentum vector

$$P_\beta = \int d^3y \; \mathscr{T}_{0\beta} \tag{D.57}$$

is conserved. We note

$$\partial/\partial y^0 \, P_\beta = 0 \tag{D.58}$$

since eq. D.56 and D.57 can be used to obtain the integral of a divergence, which results in zero.

The Hamiltonian (eqs. D.43-44) is

$$H = P_0 \tag{D.59}$$

We note for later use that the total energy, H, which is conserved, contains a term that represents the energy in the X^μ mapping. Thus energy can be exchanged in principle between the ϕ field sector and the X^μ sector.

D.6 Lorentz Invariance and Angular Momentum Conservation

We can also verify Lorentz invariance and obtain the form of the conserved angular momentum by considering the effect of an infinitesimal Lorentz transformation. We will consider the case of a scalar field ϕ.

Under an infinitesimal Lorentz transformation ($\epsilon_{\mu\nu} = -\epsilon_{\nu\mu}$):

$$y'_\mu = y_\mu + \delta y_\mu = y_\mu + \epsilon_{\mu\nu} y^\nu \tag{D.60a}$$

$$\delta\phi = \phi(X(y')) - \phi(X(y))$$

$$= \epsilon^{\mu\nu} \, y_\nu \, \partial\phi/\partial X^\alpha \, \partial X^\alpha/\partial y^\mu \tag{D.60b}$$

$$\delta X^\mu = S^\mu{}_\alpha X^\alpha(y') - X^\mu(y) \tag{D.60c}$$

$$= e^{\mu}{}_{a} X^{a}(y) + \partial X^{\mu}/\partial y^{\beta} \, \delta y^{\beta} \tag{D.60d}$$

where $S^{\mu}{}_{a}$ is the matrix for the Lorentz transformation of a vector. (If X^{μ} were a gauge field then an additional operator gauge term would have to be added to eq. D.60d.)

The Lagrangian changes by

$$\delta \mathscr{L} = e^{\mu\nu} \, y_{\nu} \, \partial \mathscr{L}/\partial y^{\mu} \tag{D.61}$$

under the infinitesimal Lorentz transformation. The change in the Lagrangian can also be expressed as:

$$\delta \mathscr{L} = \partial \mathscr{L}/\partial \phi \, \delta \phi + \partial \mathscr{L}/\partial(\partial \phi/\partial X^{\mu}) \, \delta(\partial \phi/\partial X^{\mu}) + \partial \mathscr{L}/\partial X^{\mu} \, \delta X^{\mu} + $$
$$+ \, \partial \mathscr{L}/\partial(\partial X^{\mu}/\partial y^{\nu}) \, \delta(\partial X^{\mu}/\partial y^{\nu}) \tag{D.62}$$

Combining eqs. D.61 and D.62, and substituting and simplifying terms leads to:

$$\epsilon_{\mu\nu} \, \partial/\partial y^{\sigma} \, \mathscr{M}^{\sigma\mu\nu} = 0 \tag{D.63}$$

where

$$\mathscr{M}^{\sigma\mu\nu} = (g^{\mu\sigma} y^{\nu} - g^{\nu\sigma} y^{\mu}) \mathscr{L} + \partial \mathscr{L}/\partial(\partial \phi/\partial X^{\alpha}) \, \partial y^{\sigma}/\partial X^{\alpha} \, (y^{\mu} \partial \phi/\partial y_{\nu} - y^{\nu} \partial \phi/\partial y_{\mu}) + $$
$$+ \, \partial \mathscr{L}/\partial(\partial X^{\delta}/\partial y^{\sigma}) \, (g^{\delta\nu} X^{\mu} - g^{\delta\mu} X^{\nu} + y^{\mu} \, \partial X^{\delta}/\partial y_{\nu} - y^{\nu} \, \partial X^{\delta}/\partial y_{\mu}) \tag{D.64}$$

The conserved angular momentum is:

$$M^{\mu\nu} = \int d^{3}y \, \mathscr{M}^{0\mu\nu} \tag{D.65}$$

with

$$\partial M^{\mu\nu}/\partial y^{0} = 0 \tag{D.66}$$

The angular momentum density can be written in the familiar form:

$$\mathcal{M}^{\sigma\mu\nu} = y^{\mu}\,\mathcal{T}^{\sigma\nu} - y^{\nu}\,\mathcal{T}^{\sigma\mu} + \partial\mathcal{L}\big/\partial(\partial X^{\delta}/\partial y^{\sigma})\,(g^{\delta\nu}X^{\mu} - g^{\delta\mu}X^{\nu}) \qquad \text{(D.67)}$$

taking account of the vector nature of X^{μ}. The spatial part of $M^{\mu\nu}$ is the angular momentum.

D.7 Internal Symmetries

We will now consider the case of a set of scalar fields ϕ_r in a Lagrangian with an internal symmetry. Under a local transformation

$$\phi_r(X) \rightarrow \phi_r(X) - i\epsilon\lambda_{rs}\,\phi_s(X) \qquad \text{(D.68)}$$

If the Lagrangian is invariant under this transformation, then

$$\delta\mathcal{L} = 0 = \partial\mathcal{L}\big/\partial\phi_r\,\delta\phi_r + \partial\mathcal{L}\big/\partial(\partial\phi_r/\partial X^{\alpha})\,\delta(\partial\phi_r/\partial X^{\alpha}) \qquad \text{(D.69)}$$

Using the equation of motion eq. D.19a satisfied by all the components ϕ_r we obtain a conserved current:

$$\mathcal{J}^{\nu} = -i\,\partial\mathcal{L}\big/\partial(\partial\phi_r/\partial X^{\delta})\,\partial y^{\nu}/\partial X^{\delta}\,\lambda_{rs}\,\phi_s \qquad \text{(D.70)}$$

which satisfies

$$\partial\mathcal{J}^{\nu}/\partial y^{\nu} = 0 \qquad \text{(D.71)}$$

The conserved charge is

$$Q = \int d^3 y\,\mathcal{J}^0 \qquad \text{(D.72)}$$

$$\partial Q/\partial y^0 = 0 \qquad \text{(D.73)}$$

D.8 Separable Lagrangians

We now consider the case of a separable Lagrangian such as in eq. D.26. Adopting the definitions:

$$\phi' = \partial\phi/\partial X^0 \tag{D.74}$$
$$X_\mu{}' = \partial X_\mu/\partial y^0 \tag{D.75}$$

we define canonical momenta as

$$\pi_\phi = \partial\mathcal{L}/\partial\phi' \equiv \partial\mathcal{L}/\partial(\partial\phi/\partial X^0) \tag{D.76}$$
$$\pi_X{}^\mu = \partial\mathcal{L}/\partial X_\mu{}' \equiv \partial\mathcal{L}/\partial(\partial X_\mu/\partial y^0) \tag{D.77}$$

We now define the separable Hamiltonian density as

$$\mathcal{H}_s = J\pi_\phi\,\phi' + \pi_X{}^\mu\,X_\mu{}' - \mathcal{L}_s \tag{D.78}$$

where J is the Jacobian (eq. D.21) and

$$H_s = \int \mathcal{H}_s\,d^3y \tag{D.79}$$

The separable Lagrangian (from eq. D.26) is:

$$\mathcal{L}_s = \mathcal{L}_F(\phi(X), \partial\phi/\partial X^\mu)\,J + \mathcal{L}_C(X^\mu(y), \partial X^\mu(y)/\partial y^\nu, y) \tag{D.80}$$

In the case of one scalar field the separable Hamiltonian density has the general form

$$\mathcal{H}_s = \mathcal{H}_s(\phi(X), \pi_\phi, \partial\phi/\partial X^i, X^\mu(y), \pi_X{}^\mu, \partial X^\mu(y)/\partial y^j, y^\nu) \tag{D.81}$$

where the indices i and j indicate spatial components. In particular, the terms in the separable Hamiltonian are:

$$\mathcal{H}_s = \mathcal{H}_F J + \mathcal{H}_C \tag{D.82}$$

with

$$\mathcal{H}_F(\phi(X), \pi_\phi, \partial\phi/\partial X^i) = \pi_\phi \phi' - \mathcal{L}_F \tag{D.83}$$

$$\mathcal{H}_C(X^\mu(y), \pi_X{}^\mu, \partial X^\mu(y)/\partial y^j, y^\nu) = \pi_X{}^\mu X_\mu' - \mathcal{L}_C \tag{D.84}$$

where J is the absolute value of the Jacobian defined in D.21.

We now define the time integral of H as we did in eq. D.14 when considering the Lagrangian formulation:

$$G = \int dy^0 H_s \tag{D.85}$$

Thus G is an integral over all space-time coordinates. Using G we can develop a Hamiltonian formulation. First we calculate the differential change in G. Using eqs. D.81-2 and D.85 we obtain

$$dG = \int \{ J \, \partial\mathcal{H}_F / \partial\phi \, d\phi + J \, \partial\mathcal{H}_F / \partial\pi_\phi \, d\pi_\phi + $$
$$+ J \, \partial\mathcal{H}_F / \partial(\partial\phi/\partial X^i) \, d(\partial\phi/\partial X^i) + \partial\mathcal{H}_C / \partial X^\mu \, dX^\mu + $$
$$+ \partial\mathcal{H}_C / \partial\pi_X{}^\mu \, d\pi_X{}^\mu + \partial\mathcal{H}_C / \partial(\partial X^\mu/\partial y^j) \, d(\partial X^\mu/\partial y^j) \} \, d^4y \tag{D.86}$$

with summations implied by repeated indices. (Index labels i and j label spatial coordinates only; Greek indices label space-time coordinates.) Rewriting dG as two integrals and performing partial integrations yields:

$$dG = \int d^4X \{ \partial\mathcal{H}_F / \partial\phi \, d\phi + \partial\mathcal{H}_F / \partial\pi_\phi \, d\pi_\phi - \partial/\partial X^i[\partial\mathcal{H}_F / \partial(\partial\phi/\partial X^i)] \, d\phi \} +$$

$$+ \int d^4 y \left\{ \partial \mathcal{H}_C / \partial X^\mu \, dX^\mu + \partial \mathcal{H}_C / \partial \pi_X{}^\mu \, d\pi_X{}^\mu - \partial / \partial y^j [\partial \mathcal{H}_C / \partial (\partial X^\mu / \partial y^j)] \, dX^\mu \right\}$$

(D.87)

Alternately, expressing the differential in G using eqs. D.82-4 we obtain

$$dG = \int d^4 X \left\{ \pi_\phi \, d\phi' + \phi' d\pi_\phi - \partial \mathcal{L}_F / \partial \phi \, d\phi - \partial \mathcal{L}_F / \partial (\partial \phi / \partial X^\mu) d(\partial \phi / \partial X^\mu) \right\} +$$
$$+ \int d^4 y \left\{ \pi_{X\mu} \, dX^{\mu\prime} + X^{\mu\prime} d\pi_{X\mu} - \partial \mathcal{L}_C / \partial X^\mu \, dX^\mu - \partial \mathcal{L}_C / \partial (\partial X^\mu / \partial y^j) d(\partial X^\mu / \partial y^j) \right\}$$

(D.88)

which becomes

$$dG = \int d^4 X \left\{ -\pi_\phi' \, d\phi + \phi' \, d\pi_\phi \right\} + \int d^4 y \left\{ -\pi_{X\mu}' \, dX^\mu + X^{\mu\prime} d\pi_{X\mu} \right\} \quad (D.89)$$

using the equations of motion eqs. D.31-2.

Comparing eqs D.87 and D.89 we obtain Hamilton's equations for the case of the composition of extrema for a separable Lagrangian:

$$\phi' = \partial \mathcal{H}_F / \partial \pi_\phi \tag{D.90}$$

$$\pi_\phi' = -\partial \mathcal{H}_F / \partial \phi + \partial / \partial X^j [\partial \mathcal{H}_F / \partial (\partial \phi / \partial X^j)] \tag{D.91}$$

$$X_\mu' = \partial \mathcal{H}_C / \partial \pi_X{}^\mu \tag{D.92}$$

$$\pi_{X\mu}' = -\partial \mathcal{H}_C / \partial X^\mu + \partial / \partial y^j [\partial \mathcal{H}_C / \partial (\partial X^\mu / \partial y^j)] \tag{D.93}$$

where

$$\pi_\phi' = \partial \pi_\phi / \partial X^0 \tag{D.94}$$

$$\pi_{X\mu}' = \partial \pi_{X\mu} / \partial X^0 \tag{D.95}$$

Notice that $\mathscr{L}_F$, $\mathscr{H}_F$ and π_ϕ have precisely the same form, as a function of X^μ, as one sees in a conventional field theory formalism. Yet X^μ is a mapping/function of the coordinates y. In reality, it can be viewed as a field as we shall see.

D.9 Separable Lagrangians and Translational Invariance

The general rule for conventional Lagrangians is: if a Lagrangian has no explicit dependence on the coordinates then translational invariance follows accompanied by a conservation law for an energy-momentum tensor. We will show that this rule needs modification for separable Lagrangians that implement the composition of extrema.

Consider the Lagrangian:

$$\mathscr{L}_s = J \, \mathscr{L}_F(\phi(X), \partial\phi/\partial X^\mu) + \mathscr{L}_C(X^\mu(y), \partial X^\mu(y)/\partial y^\nu) \qquad (D.96)$$

in which the X^μ play a dual role as both fields and coordinates. Let us consider a variation in X^μ:

$$X^\mu(y) \rightarrow X^\mu(y) + \delta X^\mu(y) \qquad (D.97)$$

where $\delta X^\mu(y)$ is an arbitrary function of y that vanishes at the endpoints of the integration region of the integral. The action is:

$$I = \int \mathscr{L}_s d^4y \qquad (D.98)$$

We will show that a variation in $X^\mu(y)$ leads to a conserved energy-momentum tensor. But we will use integrals of the Lagrangian density since it provides a simpler derivation of the result. Under the variation of eq. D.97 we find

$$\delta\phi = \phi(X(y) + \delta X^\mu(y)) - \phi(X(y))$$

$$= \delta X^\mu \, \partial\phi/\partial X^\mu \tag{D.99a}$$

$$\delta(\partial\phi/\partial X^\nu) = \delta X^\mu \, \partial(\partial\phi/\partial X^\mu)/\partial X^\nu \tag{D.99b}$$

$$\delta(\partial X^\mu/\partial y^\nu) = \partial(\delta X^\mu)/\partial y^\nu \tag{D.99c}$$

The integral in eq. D.98 changes by

$$\delta I = \int d^4 y \, \delta\mathscr{L}_s = \int d^4 y \, [\, \delta(J\mathscr{L}_F) + \delta\mathscr{L}_C \,] \tag{D.100a}$$

which becomes:

$$\delta I = \int d^4 y \, [\delta X^\mu \, \partial(J\mathscr{L}_F)/\partial X^\mu + \partial(\delta X^\mu \partial\mathscr{L}_C/\partial(\partial X^\mu/\partial y^\nu))/\partial y^\nu] \tag{D.100b}$$

due to the equations of motion of X^μ (eq. D.19b) in X^μs role. Since the second term is a total divergence its contribution to δI is zero. Thus we can express eq. D.100b as:

$$\delta I = \int d^4 y \, [\, J \, \delta\mathscr{L}_F + \mathscr{L}_F \, \delta J \,] \tag{D.101}$$

realizing that the Jacobian J depends on y and thus X:

$$\delta J = \delta X^\mu \, \partial J/\partial X^\mu \tag{D.102}$$

A partial integration gives

$$\mathscr{L}_F \, \delta J = \delta X^\mu \, \partial(J\mathscr{L}_F)/\partial X^\mu - \delta X^\mu \, J \, \partial\mathscr{L}_F/\partial X^\mu \tag{D.103}$$

Evaluating $\delta\mathscr{L}_F$ we find:

$$\delta \mathscr{L}_F = \partial \mathscr{L}_F / \partial \phi \, \delta \phi + \partial \mathscr{L}_F / \partial (\partial \phi / \partial X^\mu) \, \delta (\partial \phi / \partial X^\mu) \qquad (D.104)$$

which gives

$$\delta \mathscr{L}_F = \delta X^\nu \, \partial / \partial X^\mu \, [\partial \mathscr{L}_F / \partial (\partial \phi / \partial X^\mu) \, \partial \phi / \partial X^\nu] \qquad (D.105)$$

using the equations of motion eq. D.31, and using eq. D.99b. Combining eqs. D.100, D.101, D.103 and D.105 we obtain:

$$\int d^4 y \, J \, \delta X^\nu \, \partial / \partial X_\mu \, \mathscr{T}_{F\mu\nu} = \int d^4 X \, \delta X^\nu \, \partial / \partial X_\mu \, \mathscr{T}_{F\mu\nu} = 0 \qquad (D.106)$$

where

$$\mathscr{T}_{F\mu\nu} = - \, g_{\mu\nu} \, \mathscr{L}_F + \partial \mathscr{L}_F / \partial (\partial \phi / \partial X_\mu) \, \partial \phi / \partial X^\nu \qquad (D.107)$$

after some manipulations. Since δX^ν is an arbitrary function of y the differential conservation law follows:

$$\partial / \partial X_\mu \, \mathscr{T}_{F\mu\nu} = 0 \qquad (D.108)$$

Eq. D.108 implies the energy-momentum vector

$$P_{F\beta} = \int d^3 X \, \mathscr{T}_{F0\beta} \qquad (D.109)$$

is conserved:

$$\partial / \partial X^0 \, P_{F\beta} = 0 \qquad (D.110)$$

The Hamiltonian density (eq. D.83) is

$$\mathscr{H}_F = \mathscr{T}_{F0\beta} \tag{D.111}$$

Thus the field energy

$$H_F = P_{F0} = \int d^3X \; \mathscr{T}_{F00} \tag{D.112}$$

is conserved with respect to the "time" X^0. Later we will see that H_F is trivially conserved in the Coulomb gauge of X_μ. (We will also establish an electromagnetic-like quantum field theory for X_μ with gauge invariance.) In other gauges the conservation of H_F is not trivial.

D.10 Separable Lagrangians and Angular Momentum Conservation

We can also verify Lorentz invariance and obtain the form of the conserved angular momentum for a separable Lagrangian by considering the effect of an infinitesimal Lorentz transformation. We will consider the case of a scalar field ϕ.

Under an infinitesimal Lorentz transformation as specified by eqs. D.60a – D.60d the separable Lagrangian changes by

$$\delta\mathscr{L}_s = \epsilon^{\mu\nu} \, y_\nu \partial\mathscr{L}_s / \partial y^\mu \tag{D.113}$$

which can also be expressed as

$$\delta\mathscr{L}_s = \partial\mathscr{L}_s / \partial\phi \; \delta\phi + \partial\mathscr{L}_s / \partial(\partial\phi/\partial X^\mu) \; \delta(\partial\phi/\partial X^\mu) + \partial\mathscr{L}_s / \partial X^\mu \; \delta X^\mu +$$
$$+ \; [\partial\mathscr{L}_s / \partial(\partial X^\mu/\partial y^\nu)] \; \delta(\partial X^\mu/\partial y^\nu) \tag{D.114}$$

Combining eqs. D.113 and D.114 leads to:

$$\epsilon_{\mu\nu} \, \partial/\partial y^\sigma \, \mathscr{M}_s^{\;\sigma\mu\nu} = 0 \tag{D.115}$$

where

$$\mathscr{M}_s^{\;\sigma\mu\nu} = J \, \mathscr{M}_F^{\;\sigma\mu\nu} + \mathscr{M}_C^{\;\sigma\mu\nu} + \mathscr{M}_M^{\;\sigma\mu\nu} \tag{D.116}$$

$$\mathcal{M}_F{}^{\sigma\mu\nu} = (g^{\mu\sigma}y^\nu - g^{\nu\sigma}y^\mu)\mathcal{L}_F + \partial\mathcal{L}_F/\partial(\partial\phi/\partial y_\sigma)\,(y^\mu\partial\phi/\partial y_\nu - y^\nu\partial\phi/\partial y_\mu)$$

$$(D.117)$$

$$\mathcal{M}_C{}^{\sigma\mu\nu} = (g^{\mu\sigma}y^\nu - g^{\nu\sigma}y^\mu)\mathcal{L}_C +$$
$$+ \partial\mathcal{L}_C/\partial(\partial X^\delta/\partial y^\sigma)(g^{\delta\nu}X^\mu - g^{\delta\mu}X^\nu + y^\mu\,\partial X^\delta/\partial y_\nu - y^\nu\,\partial X^\delta/\partial y_\mu)$$

$$(D.118)$$

$$\mathcal{M}_M{}^{\sigma\mu\nu} = \mathcal{L}_F\partial J/\partial(\partial X^\delta/\partial y^\sigma)(g^{\delta\nu}X^\mu - g^{\delta\mu}X^\nu + y^\mu\,\partial X^\delta/\partial y_\nu - y^\nu\,\partial X^\delta/\partial y_\mu)$$

$$(D.119)$$

where the third term originates in the dependence of J on derivatives of X^μ. Eq. D.117 was obtained in part by using the identity:

$$\partial\mathcal{L}/\partial(\partial\phi/\partial y^\sigma) = \partial\mathcal{L}/\partial(\partial\phi/\partial X^\alpha)\,\partial y^\sigma/\partial X^\alpha \qquad (D.120)$$

where $\mathcal{L}$ and ϕ have the form specified in eq. D.15.

The conserved angular momentum is:

$$M_s{}^{\mu\nu} = \int dy\;\mathcal{M}_s{}^{0\mu\nu} \qquad (D.121)$$

with

$$\partial M_s{}^{\mu\nu}/\partial y^0 = 0 \qquad (D.122)$$

D.10.1 Angular Momentum and $\mathcal{L}_F$

An alternate conserved angular momentum can be obtained by considering the "field" part of the Lagrangian $\mathcal{L}_F$ under an infinitesimal Lorentz transformation ($\epsilon_{\mu\nu} = -\epsilon_{\nu\mu}$):

$$X'_\mu = X_\mu + \delta X_\mu \tag{D.123a}$$

$$\delta\phi = \phi(X'(y)) - \phi(X(y))$$
$$= \delta X^\mu \, \partial\phi/\partial X^\mu \tag{D.123b}$$

$$\delta X^\mu = S^\mu{}_a X^a(y) - X^\mu(y) \tag{D.123c}$$

$$= \epsilon^\mu{}_a X^a(y) \tag{D.123d}$$

where $S^\mu{}_a$ is the Lorentz transformation matrix for a vector. (If X^μ is a gauge field then an additional operator gauge term would have to be added to eq. D.123d.)

The Lagrangian changes by

$$\delta\mathscr{L}_F = \epsilon^{\mu\nu} X_\nu \, \partial\mathscr{L}_F/\partial X^\mu \tag{D.124}$$

under an infinitesimal Lorentz transformation. The change can also be expressed as:

$$\delta\mathscr{L}_F = \partial\mathscr{L}_F/\partial\phi \, \delta\phi + \partial\mathscr{L}_F/\partial(\partial\phi/\partial X^\mu) \, \delta(\partial\phi/\partial X^\mu) \tag{D.125}$$

Combining eqs. D.124 and D.125 leads to:

$$\epsilon_{\mu\nu}\partial/\partial X^\sigma \, \mathscr{M}_{FX}{}^{\sigma\mu\nu} = 0 \tag{D.126}$$

where

$$\mathscr{M}_{FX}{}^{\sigma\mu\nu} = (g^{\mu\sigma}X^\nu - g^{\nu\sigma}X^\mu)\mathscr{L}_F + \partial\mathscr{L}_F/\partial(\partial\phi/\partial X^\sigma) \, (X^\mu\partial\phi/\partial X_\nu - X^\nu\partial\phi/\partial X_\mu) \tag{D.127}$$

The conserved angular momentum associated with the X coordinates is:

$$M_{FX}{}^{\mu\nu} = \int d^3X \, \mathscr{M}_{FX}{}^{0\mu\nu} \tag{D.128}$$

with

$$\partial M_{FX}{}^{\mu\nu}/\partial X^0 = 0 \qquad\qquad (D.129)$$

The angular momentum density can be written in the familiar form:

$$\mathcal{M}_{FX}{}^{\sigma\mu\nu} = X^\mu \, \mathcal{T}_F^{\sigma\nu} - X^\nu \, \mathcal{T}_F^{\sigma\mu} \qquad\qquad (D.130)$$

using eq. D.107.

D.11 Separable Lagrangians and Internal Symmetries

We will now consider the case of a set of scalar fields ϕ_r in a separable Lagrangian with an internal symmetry under a local transformation

$$\phi_r(X) \rightarrow \phi_r(X) - i\epsilon\lambda_{rs} \, \phi_s(X) \qquad\qquad (D.131)$$

If the Lagrangian is invariant under this transformation, then

$$\delta\mathcal{L}_S \equiv \delta\mathcal{L}_F = 0 = \partial\mathcal{L}_F/\partial\phi_r \, \delta\phi_r + \partial\mathcal{L}_F/\partial(\partial\phi_r/\partial X^\alpha) \, \delta(\partial\phi_r/\partial X^\alpha)$$
$$(D.132)$$

Using the equation of motion eq. D.31, which is satisfied by all components ϕ_r, we obtain a conserved current:

$$\mathcal{J}^\nu = -i \, \partial\mathcal{L}_F/\partial(\partial\phi_r/\partial X^\nu) \, \lambda_{rs} \, \phi_s \qquad\qquad (D.133)$$

satisfying

$$\partial\mathcal{J}^\nu/\partial X^\nu = 0 \qquad\qquad (D.134)$$

The conserved charge is

$$Q = \int d^3X \, \mathcal{J}^0 \qquad\qquad (D.135)$$
$$\partial Q/\partial X^0 = 0 \qquad\qquad (D.136)$$

We note eq. D.71 provides a corresponding conservation law for the y coordinate system.

Appendix E. CQ Mechanics – PseudoQuantum Mechanics

E.1 Why Quantum Theory?

A question that is not often considered in these days is the reason that Nature 'chose' to be quantum rather than based on classical, deterministic mechanics. In our Theory of Everything presented in Blaha (2015c) and subsequent books in 2015 and 2016 we assumed that al Natural phenomena were ultimately based on quantum field theory.

There is a more fundamental assumption that we could posit that leads to quantum field theory and then to quantum mechanics (which is based on quantum field theory.[547]) If we assume the following postulate:

All entities in the universe are composed of discrete particles, that are integer countable, and all interactions can ultimately be reduced to the interactions of these particles.

Then, when we define field theories, they must be quantum field theories – describe particles (quanta) – and thus the field theories must be second quantized. *Particles are integer countable*[548] whether free or in perturbation theory interaction terms. And the interactions of these theories must be based on the exchange of particles although the particles have a 'cloud' of virtual particles surrounding them. The quantum mechanics

---

[547] Heitler (1954) shows how the Heisenberg Uncertainty Principle is a consequence of quantum field theory using an example from Quantum Electrodynamics. The Uncertainty Principle and the Correspondence Principle lead directly to quantum mechanics. Quantum mechanics is thus a consequence of quantum field theory – not an independent fundamental theory.

[548] Theories with continuous matter have not been shown to exist experimentally.

of the particles constituting atoms (matter) then follows as a consequence of quantum field theory.

Classical mechanics then becomes an approximation to quantum mechanics under certain conditions that turn out to be the common conditions of everyday experience.

In basing the origin of quantum theory on the particulate nature of the entities in the universe we assume that particles exist, and can be defined, in our mostly flat space-time (which itself is generated by amalgamations of graviton particles). We also assume that the particle concept can be extended to unusual space-time coordinates such as non-static space-times. However it became apparent many years ago when accelerating coordinate systems and other non-static[549] coordinate systems were considered, that the definition of particles in quantum field theory is problematic.[550]

Sections 2 and 3 describe the correct definition of particles in quantum field theory. The correct definition of bosons furnishes a basis for a better definition of the Higgs Mechanism. The necessity of higher derivative theories of Gravity and the Strong Interactions[551] to obtain explicit color confinement and to reconcile gravity theory with data on the rotation of stars around galactic centers leads to an extension of the definition of particles to have principal value propagators and thus gives *particulate* action-at-a-distance.[552]

A further issue, that emerges in perturbation theory calculations in quantum field theories, leads to the introduction of a vector field as part of each propagator that eliminates the point-like nature of particle interactions in the high energy (short distance) limit in favor of 'fuzzy' interacting particles. This extension of quantum field theory is called Two-Tier quantum field theory.[553] It is required for the Theory of Everything since a conventional renormalization procedure is not known to exist – and is not likely to exist.

---

[549] A non-static coordinate system mixes space and time coordinates.

[550] S. Blaha, Il Nuovo Cimento **49 A**, 35 (1979). ___, **49 A**, 113 (1979), which appear in appendices A and B and is discussed in the following chapters, describe how to define particles in any coordinate system. The particle definition issued is discussed in papers which they reference.

[551] Unified theory: See Blaha (2016e).

[552] See appendices A and B.

[553] See Blaha (2005a).

The combination of features that we have developed enables us to create a divergence-free[554] Theory of Everything where particles can be defined in any static or non-static coordinate system and where bosons, neglecting interactions, are stable against decay to negative energy states.[555]

The formalism that we present can be applied to quantum and classical dynamics. We define a quantum-classical formalism, that we call *QCMechanics*, that has a fully quantum sector, a classical sector, and an intermediate sector bridging the quantum and classical sectors.

We then proceed to develop the harmonic oscillator theory within this framework. Subsequently we discuss a generalized Feynman path integral formalism, a generalized Schrödinger equation, a generalized Boltzmann equation, the Fokker-Planck equation, a generalized approach to quantum and classical chaos, and to quantum entanglement as well as semi-quantum entanglement. Our formalism applies to both Quantum Field Theory and Quantum Mechanics as well as the path integrals, the Fokker-Planck equation and the Boltzmann equation.

E.2 Boson Particle Formulation

There are two issues confronting the usual approach to the quantization of boson fields that require resolution through a 'new' quantization procedure for boson particles. One problem is the need to quantize boson fields in unconventional coordinate systems such as accelerating coordinate systems and coordinate systems defined for highly curved space-time. The other major problem is the need to quantize boson fields in such a way that bosons of negative energy have a physical interpretation.[556]

In this section we will define a new quantization procedure for bosons that will eliminate both of these problems. In section 3 we will describe the analogous quantization procedure for fermions that will enable us to create well-defined Dirac

---

[554] There are no divergences in perturbation theory calculations and no need for renormalization programs to remove divergences. Physical quantities do get renormalized by finite amounts.
[555] Negative energy boson states are equivalent to classical fields.
[556] There is no Pauli Exclusion Principle for bosons. Negative energy fermions 'fill' their Dirac negative energy sea due to the limitation of fermions to one fermion per state imposed by the Pauli Exclusion Principle.

field particle states in any coordinate system. (A filled Dirac sea of negative energy fermions will exist in this formulation as it does in the conventional formulation.)

In the case of both bosons and fermions we will see that the flat space-time, static coordinate systems that are normally used will remain valid special case approximations to the new formulations of quantum field theory.

Having resolved these problems for quantum field theories of bosons and fermions we will point out in section 5 that 'ordinary' quantum mechanics also has a problem with quantization in unconventional coordinate systems. There are also difficulties in the transition between classical and quantum 'analogues.' For example the transition from the classical Boltzmann equation to a quantum version is uncertain.

Using a framework analogous to our 'new' approach to the quantum field theory of bosons and fermions we will establish a generalization of quantum mechanics that contains both quantum mechanics and classical mechanics, and an intermediate mixed form. This generalization supports a smooth transition between classical mechanics and quantum mechanics. With this generalization we will be able to examine the transition from quantum to classical mechanics in detail without recourse to methods such as expansions in Planck's constant $\hbar$.

E.2.1 Quantization of Boson Fields in Unconventional (Static and Non-Static) Coordinate Systems

The problems associated with the definition of asymptotic particle states in arbitrary coordinate systems have been pointed out by numerous authors.[557] Our 1978 paper (Appendix A) resolves this problem with a consistent procedure for the local definition of asymptotic boson particle states in any coordinate system, which may or may not have a time-like Killing vector.

The general procedure is described in section 2 in Appendix A starting with eq. 6. The boson particle interpretation is described in section 4. In this section we wish to bring out salient details of the procedure which we will be relevant for our consideration of the generalization of quantum mechanics that we will consider later.

---

[557] See appendix A, which contains our 1978 paper Il Nuovo Cimento **49 A**, 35, for references.

The first distinctive feature of this form of boson second quantization is the use of two fields to define a second quantized boson theory. The use of *two* fields enables us to define states which correspond to quantum field particles in any coordinate system. Further they give us the scope to define, not only quantum field particle states, but also classical boson field states. States, which are composites of both classical fields and quantum particles, can also be defined.

In our formulation[558] the simplest lagrangian density for a generic massless, scalar Klein-Gordon particle is:

$$\mathcal{L} = \partial\varphi_1/\partial x_\mu \, \partial\varphi_2/\partial x^\mu \qquad (E.2.1)$$

with Hamiltonian density

$$\mathcal{H} = \pi_1 \pi_2 + \partial\varphi_1/\partial x_i \, \partial\varphi_2/\partial x^i \qquad (E.2.2)$$

where i labels spatial coordinates, and $\pi_1 = \partial\varphi_2/\partial t$ and $\pi_2 = \partial\varphi_1/\partial t$. Eqs. E.2.1 and E.2.2 are without a potential or mass term. Eq. 6, and the discussion following it, in appendix A describe the massive boson case.

The fields can be fourier expanded in terms of creation and annihilation operators:

$$\varphi_i(\mathbf{x}, t) = \int d^3k \, [a_i(k)f_k(x) + a_i^\dagger(k)f_k^*(x)] \qquad (E.2.3)$$

for i = 1, 2 where

$$f_k(x) = e^{-ik\cdot x} /(2\omega_k(2\pi)^3)^{1/2}$$

with $\omega_k = |\mathbf{k}|$ in the massless case and $\omega_k = (|\mathbf{k}|^2 + m^2)^{1/2}$ for a massive boson.

The creation and annihilation operators satisfy the commutation relations:

$$[a_i(k), a_j^\dagger(k')] = (1 - \delta_{ij})\delta^3(\mathbf{k} - \mathbf{k}') \qquad (E.2.4)$$
$$[a_i(k), a_j(k')] = 0$$

---

[558] In earlier books we have called this approach to second quantization *pseudoquantum field theory*.

$$[a_i^\dagger(k), a_j^\dagger(k')] = 0$$

for $i, j = 1, 2$. The vacuum state $|0>$ satisfies

$$a_1(k)|0> = a_1^\dagger(k)|0> = 0 \tag{E.2.5}$$
$$a_2(k)|0> \neq 0 \qquad\qquad a_2^\dagger(k)|0> \neq 0 \tag{E.2.6}$$

The dual vacuum state satisfies

$$<0|a_2(k) = <0|a_2^\dagger(k) = 0 \tag{E.2.7}$$
$$<0|a_1(k) \neq 0 \qquad\qquad <0|a_1^\dagger(k) \neq 0 \tag{E.2.8}$$

Positive energy single particle *ket* states are defined using $a_2^\dagger(k)$ while negative energy ket states are defined using $a_2(k)$. Positive energy single particle *bra* states are defined using $a_1(k)$ while negative energy bra states are defined using $a_1^\dagger(k)$.

E.2.1.1 Transformations to Other (Possibly Non-Static) Coordinate Systems

The preceding discussion applies directly in the rectangular coordinates with which we are familiar. In eqs. 2 – 4, and their discussion, in appendix A we show that the definition of boson field orthonormal sets according to a different definition of positive frequency is related to the definition above in eq. E.2.3 by a local Bogoliubov transformation. The definition of particle states and vacuums are different in general. However, as eqs. 15 – 31 (Appendix A) show, we can define the transformation to preserve the invariance of the particle number operator and thus make the theory under a different definition of positive frequency fully unitarily equivalent to the original theory. Thus the particle interpretation of states is preserved.

The general form of the Bogoliubov transformation (eq. 23) is

$$a_i(k, \lambda_1(k), \lambda_2(k)) = B(\lambda_1(k), \lambda_2(k))a_i(k)B^{-1}(\lambda_1(k), \lambda_2(k)) \tag{23}$$
$$= \exp(i\lambda_1(k))\cosh(\lambda_2(k))a_i(k) + \exp(-i\lambda_1(k))\sinh(\lambda_2(k))a_i^\dagger(k)$$

with $B(\lambda_1(k), \lambda_2(k))$ given by eq. 24 and

$$B^{-1}(\lambda_1(k), \lambda_2(k)) = B^\dagger(\lambda_1(k), \lambda_2(k)) \qquad (E.2.9)$$

where † indicates hermitean conjugate. The text following eq. 23 provide the definition of bra and ket states, inner products, the energy-momentum tensor, equal-time commutation relations, the Green's functions, and the perturbation theory of the 'new' formalism.

Thus PseudoQuantized Field Theory resolves the particle interpretation ambiguities of second quantization in non-static coordinate systems through Bogoliubov rotations.

E.2.1.2 Negative Energy Bosons

Traditional boson second quantization has the problem of the absence of a barrier to the decay of positive energy states to negative energy states since the Pauli Exclusion Principle does not apply to bosons. This problem has been masked ('overcome') by a clever choice of boundary conditions that are embodied in the creation/annihilation momentum space operator conditions:

$$\begin{aligned} a|0> &= 0 \qquad \text{Conventional Approach} \qquad (E.2.10) \\ a^\dagger|0> &\neq 0 \end{aligned}$$

In this conventional approach the creation of negative energy boson states is eliminated *ab initio* by these conditions. Yet boson quantum fields still have a conceptual physical cloud hanging over them that spin ½ fields do not. A spin ½ particle cannot transition to negative energy because there is a filled sea of negative energy particles. No additional particles can fall into the sea due to the Pauli Exclusion Principle that forbids two fermions with the same 4-momentum and quantum numbers.

In the case of scalar particles the Pauli Exclusion Principle does not apply and so, *physically*, a *filled* negative energy sea of bosons is not possible and positive energy bosons should be able to transition to negative energy states. This problem was "resolved" by the above definition of boson vacuums to exclude transitions to negative energy. But the rationale for the definition is lacking. Dirac was once asked about this issue many years ago. He said he had a solution to the problem. However he did not present it – presumably in keeping with his well-known taciturn nature. So the issue remained an open question.

In this book and earlier work[559] we showed that a more physically satisfactory method exists for avoiding the negative energy state problem. This method relies on the use of a larger Fock space in which *negative energy states (or partially negative energy states) are interpreted as states containing classical fields or a mix of classical fields and individual boson particles.* This approach resolves the negative energy boson issue and provides a common framework for boson particles and classical boson fields.

The issue of the spontaneous decay of a positive energy boson into a negative energy state still seems to exist. However all known fundamental scalar bosons are Higgs bosons that have a vacuum expectation value and a 'heavy' quantum field part of positive energy that immediately decays into other particles such as a pair of photons. The decay of a positive energy boson to a negative energy state is precluded by a separation of the formalism into separate positive energy and negative energy sectors as shown in section E.4.5 below.

E.2.2 Classical Field States for Bosons

Classical c-number boson fields exist in our PseudoQuantum Field Theory. A classical c-number field has the form

$$\Phi(\mathbf{x}, t) = \int d^3k \, [\alpha(k)f_k(x) + \alpha^*(k)f_k^*(x)] \qquad (E.2.11)$$

A corresponding classical state is a coherent state with the form

$$| \Phi, \Pi\rangle = C \exp\left\{\int d^3k \, [\alpha(k)a_2^\dagger(k) + \alpha^*(k)a_2(k)]\right\}|0\rangle \qquad (E.2.12)$$

and correspondingly for $\Pi(x)$ where C is a normalization constant.

The defining properties of a classical field state are:

$$\varphi_1(x)|\Phi, \Pi\rangle = \Phi(x)|\Phi, \Pi\rangle \qquad (E.2.13)$$

---

[559] See Appendix 2-A and references therein.

$$\pi_1(x)|\Phi, \Pi\rangle = \Pi(x)|\Phi, \Pi\rangle$$

where $\Phi(x)$ and $\Pi(x)$ are sharp on the states and where $\varphi_i(x)$ is given by eq. E.2.3.

Additional details on coherent states, which differ somewhat from conventional coherent states such as those of Kibble[560] and others, can be found in Appendix C.

E.2.3 The Enigma of Higgs Particles and the Higgs Mechanism

Our PseudoQuantum Field Theory is ideally suited for describing Higgs Mechanism phenomena. In our previous work on the Standard Model, and its generalization to The Unified SuperStandard Theory described in a series of books entitled *Physics is Logic*, we showed that the fermion spectrum results from Complex Special Relativity, the gauge interactions result from the Reality group, the fermion generations result from the Generation group, the layers of fermions result from the U(4) Layer group, and from the combination with Complex General Relativity in our Theory of Everything. Higgs particles and the Higgs Mechanism were inserted *ad hoc* to generate particle masses and symmetry breaking effects.

The apparent recent discovery of Higgs particles at CERN seems to solidify the existence of the Higgs sector of the Standard Model and of our Unified SuperStandard Theory as described in earlier volumes of *Physics is Logic*.[561]

But whence arises Higgs particles? There does not appear to be a more fundamental cause than the need for particle masses obtained through symmetry breaking. And so the Higgs sector was an expedient mechanism. With our method of avoiding divergences in perturbation theory using Two-Tier quantum field theory the need for the Higgs Mechanism appears to have disappeared with the former need for a symmetry breaking mechanism to generate particle masses. The ElectroWeak sector has no divergences in our approach and thus does not need the renormalization program previously developed that was based on symmetry breaking using the Higgs Mechanism.

---

[560] T. W. B. Kibble, Jour. Math. Phys. **2**, 212 (1961).
[561] Blaha (2015a) and (2015b).

In considering the Higgs Mechanism a number of peculiarities appear that diminishes its attractiveness:

1. As remarked above, it is selective in the sense that some gauge fields have associated Higgs particles and utilize the Higgs Mechanism, and some gauge fields do not have associated Higgs particles. In particular, the ElectroWeak gauge fields, the Generation group gauge fields, the Layer group gauge fields, and the complex gravitation fields have associated Higgs particles. The strong interaction (gluon) gauge fields do not.

2. The conventional Higgs potentials have a quadratic mass term of the "wrong" sign plus a quartic interaction term, which together, generate non-zero vacuum expectation values. They obviously accomplish their goal. But the source of these potentials, and why they have their form, is unknown. One suspects a fundamental principle should be operative here.

3. One can imagine creating a Higgs microscope at some super-accelerator. Using this microscope in the presence of a (classical) condensate could enable the Uncertainty Principle to be violated. This possibility, in the case of a microscope using electromagnetic fields, was the source of a heuristic argument for the need to quantize the electromagnetic field.[562]

4. The standard formulation of the Higgs Mechanism uses classical fields under the assumption that a path integral formulation justifies their use. While this may be true, the path integral formulation relies on implicit, unstated boundary conditions that obscure the physics of the quantum field theoretic nature of the mechanism. A direct quantum field theoretic study of the Higgs Mechanism is needed and would further elucidate its character. It is possible, and it has been shown in our earlier books, that the apparently "true"

---

[562] Heitler (1954) p. 86 provides a good discussion of the need to quantize the electromagnetic field.

mechanism described below reveals a number of important new results in a properly formulated version of the Higgs Mechanism.

E.2.4 "True" Origin of an Acceptable Mass Creation Mechanism

In this section we are using *PseudoQuantization*[563] and *PseudoQuantum field theory*. It combines both quantum and classical fields within the same framework. In this extended theory vacuum expectation values appear as coherent ground states that are strictly classical in nature.

This section is based on our 1978 paper that appeared in the peer-reviewed journal *Physical Review D*. The paper is reproduced in appendix C for the reader's convenience.

We suggest the reader skim or read the paper before proceeding. The paper also presents a new formulation of Quantum Theory that incorporates both quantum and classical mechanics within one framework that is of interest in its own right. See section 4 for details. Recently, experimenters have been investigating the possibility of macroscopic and other strange quantum phenomena. The new formulation is ideally suited for tracing the transition from a quantum to a classical regime. For example, it is applicable to "large n atoms" where the outermost electrons approach classical behavior with an almost continuous energy spectrum.

E.2.5 Higgs-Like Vacuum Expectation Value Generation of Masses

The Higgs Mechanism is based on the appearance of non-zero, c-number vacuum expectation values for Higgs fields due to potential terms directly appearing in lagrangians.

E.2.5.1 PseudoQuantization of Higgs Particles

We will now consider the PseudoQuantization of a scalar particle using two fields in a manner shown earlier. It will become a "Higgs" particle with a non-zero vacuum expectation value.

---

[563] This new formalism was first described in S. Blaha, Phys. Rev. D**17**, 994 (1978).

Using the formalism described earlier we define $\varphi_1(x)$ and $\varphi_2(x)$[564] for a generic boson suppressing any internal symmetry indices for simplicity. We define a "vacuum state" containing a coherent superposition that satisfies

$$\varphi_1(x)|\Phi, \Pi> = \Phi|\Phi, \Pi> \qquad (E.2.14)$$

where Φ is a constant. Evaluating a fermion interaction term we find a mass term emerges[565]

$$\bar{\psi}(\varphi_1 + \varphi_2)\psi \;\rightarrow\; \bar{\psi}(\Phi + \varphi_2)\psi \qquad (E.2.15)$$

It can also generate a mass for an interaction with a gauge field of the form
$$A^{\mu}(\varphi_1 + \varphi_2)^2 A_{\mu} \;\rightarrow\; A^{\mu}(\Phi + \varphi_2)^2 A_{\mu} \qquad (E.2.16)$$

for ElectroWeak and other gauge fields. The φ_2 term leads to the production of Higgs particles in interactions. (The production of Higgs particles that decay into ElectroWeak gauge particles has recently been found at CERN.)

The present formalism provides a clean way to separate the vacuum expectation value of a scalar particle from its quantum field part in contrast to the conventional Higgs Mechanism where one has to separate a Higgs field into parts manually.

To obtain both the vacuum expectation value and the interaction with the quantum part of the PseudoQuantum fields we choose to always specify interactions with fermions and gauge fields using $\varphi = \varphi_1 + \varphi_2$ as seen above.

It appears that our formulation of the mass generation mechanism sheds significant light on the reason for the special prominence of inertial frames. Consider massive scalars.[566] Eq. 6 in appendix A describes a massive scalar particle. If the scalar is massive, then the rest frame particle "vacuum" coherent state below yields a non-zero expectation value Φ:

---

[564] The subscripts on the fields are not gauge symmetry indices but simply identifiers distinguishing the fields from one another.
[565] When matrix elements with a "vacuum state" are calculated.
[566] Experiments at CERN have apparently discovered a Higgs particle with a 125 GeV/c mass.

$$|\Phi, \Pi> = C\exp\{[(2\pi)^3 m/2]^{\frac{1}{2}}\Phi[a_2^\dagger(\mathbf{0},m) + a_2(\mathbf{0},m)]\}|0> \qquad (E.2.17)$$

where m is a generic mass. (We note that the conventional Higgs Mechanism also has mass terms.) *Thus our PseudoQuantum formalism allows us to define coherent "vacuum" states that lead to particle masses and Higgs particles.*

E.2.6 PseudoQuantized Non-Abelian Fields

The previous sections have considered scalar boson field theory. PseudoQuantum Field Theory also applies to non-abelian fields. See appendix B, Blaha (2016e) and earlier papers by the author.[567]

E.3 Fermion Quantization

Fermion field quantization is problematic in unconventional coordinate systems such as accelerating coordinate systems and coordinate systems defined for highly curved space-time.

In this section we will define a PseudoQuantization procedure for fermions that supports second quantization in non-static and unusual non-rectangular coordinate systems.

Having resolved these problems for both bosons and fermions using PseudoQuantum field theory we will see in section 4 that 'ordinary' quantum mechanics also has a problem with quantization in unconventional coordinate systems. It also has difficulties in the transition between classical and quantum mechanics. For example the transition from the classical Boltzmann equation to a quantum version is uncertain.

Using a framework analogous to our PseudoQuantum field theory formalisms for bosons and fermions we will establish a generalization of quantum mechanics, PseudoQuantization Mechanics, which contains both quantum mechanics and classical mechanics, and intermediate mixed mechanic states. This generalization supports a smooth transition between classical mechanics and quantum mechanics. With this

---

[567] S. Blaha, Phys. Rev. **D10**, 4268 (1974); Phys. Rev. **D11**, 2921 (1975).

generalization we will be able to examine the transition from quantum to classical mechanics in detail without recourse to methods such as expansions in Planck's constant $\hbar$.

The fermion PseudoQuantization procedure is described in section 3 in Appendix A to which the reader is referred. It is similar to boson PseudoQuantization in that it requires two fermion fields for each fermion particle. Eq. 61 and the following discussion in appendix A show a simple illustrative canonical lagrangian formulation of fermion PseudoQuantization including fourier expansions, equal time commutation relations, and creation and annihilation operators. Eq. 69 shows the general form of fourier expansions while eqs. 77 and 78 show the form restricted by anti-commutation relations and adjointness of the Hamiltonian to have rotations of creation and annihilation operators: b_{1k} and b_{2k}, and d_{1k} and d_{2k} in a manner similar to the analogous boson case in eqs. 40 and 41.

Thus boson and fermion PseudoQuantization field theory both have pairs of fields associated with each particle and utilize rotations to implement unitary equivalence between static and non-static coordinate systems.

E.4 PseudoQuantization Mechanics – Joint Quantum-Classical Mechanics Formalism

Having established the need for paired fields for bosons and fermions using PseudoQuantization field theory to achieve unitary equivalence of quantization in both static and non-static coordinate systems we now turn to the case of quantum mechanics. Here again we find there is a problem associated with the transformations between coordinate systems in certain cases. There is also the problem in determining the transition between quantum mechanical entities and their classical equivalents.

These problems are analogous to those described in previous sections for second quantization. Since quantum mechanics is derived from quantum field theory it is reasonable to suspect that the resolution of quantum mechanics problems will ultimately be found in an analogue of PseudoQuantization which we earlier saw resolved quantum field theory problems.

This section will present PseudoQuantization Mechanics,[568] which contains both fully quantum, and fully classical, sectors as well as an intermediate sector that provides a transition between the quantum and classical regimes. With this formalism we can overcome coordinate system issues as well as the challenge of the correspondence between classical and quantum physics.

E.4.1 Coordinate Systems Problems of Quantum Mechanics

Problems exist in coordinate transformations in certain quantum mechanical situations which are 'fixed' through the use of 'recipes' that patch over the difficulties. One example is the change of coordinates in path integrals.[569] Gutzwiller (1990) points out that there is no simple rule for general canonical transformations of coordinates in path integrals. Given the central role of path integrals in quantum theory the difficulties of canonical transformations in path integrals is of concern. Other problems with canonical transformations in quantum mechanics are also discussed in Gutzwiller (1990).

We shall develop the PseudoQuantization formalism with a view towards facilitating canonical transformations in quantum mechanical studies as well as elucidating the transition from the quantum to the classical regimes.

E.4.2 PseudoQuantization Harmonic Oscillator

The harmonic oscillator plays a central role in classical and quantum mechanics due to its appearance in a variety of physical problems. In this section we will describe the PseudoQuantum formulation of the one-dimensional simple harmonic oscillator as a prelude to the general PseudoQuantum description.

Appendix C contains a paper on the PseudoQuantum harmonic oscillator.[570] In this section we will describe it, in part, with some changes, as a formalism that

---

[568] Much of this chapter was presented in S. Blaha, Phys. Rev. **D17**, 994 (1978) which is reprinted in appendix C. See also S. Blaha, Phys. Rev. **D10**, 4268 (July, 1974) and Phys. Rev. **D11**, 2921 (1974). See appendix D.
[569] See pp. 202-3 in Gutzwiller (1990).
[570] See its description in appendix C – section II of S. Blaha, Phys Rev **D17**, 994 (1978). Excerpts used with the kind permission of Physical Review D.

embodies both classical and quantum sectors, and provides a graceful transition between classical and quantum harmonic oscillator dynamics. Thus we will have an example of a new approach to understanding the classical-quantum transition. In later sections we will apply this approach to physical phenomena where the transition from a classical description to a quantum equivalent is problematic.

We begin with (appendix C) two commuting variables x_1 and p_1, which we augment with two new variables x_2 and p_2, defined by

$$x_i = (m\omega/\hbar)^{-\frac{1}{2}} Q_i \tag{E.4.1}$$
$$p_i = (m\omega\hbar)^{\frac{1}{2}} P_i$$

for i, j = 1, 2 where

$$P_2 = -i\, d/dQ_1 \tag{E.4.2}$$
$$Q_2 = i\, d/dP_1$$

with the commutation relations:

$$[Q_i, P_j] = i(1- \delta_{ij}) \tag{E.4.3}$$

for i, j = 1, 2.

Next we define raising and lowering operators

$$a_i = 2^{-\frac{1}{2}}(Q_i +iP_i) \tag{E.4.4}$$
$$a_i^\dagger = 2^{-\frac{1}{2}}(Q_i - iP_i)$$
$$Q_i = (a_i + a_i^\dagger)/\sqrt{2}$$
$$P_i = (a_i - a_i^\dagger)/(\sqrt{2}i)$$

with

$$[a_i, a_j^\dagger] = (1- \delta_{ij}) \tag{E.4.5}$$
$$[a_i, a_j] = 0$$
$$[a_i^\dagger, a_j^\dagger] = 0$$

for i, j = 1, 2.

We now define an alternate set of raising and lowering operators that will use an angle θ to provide a continuous transition from classical to quantum (and vice versa)[571]

---

[571] This definition differs from that appearing in appendix C.

$$b_1 = Q_1\cos\theta + iP_2\sin\theta \tag{E.4.6}$$
$$b_2 = -Q_2\sin\theta + iP_1\cos\theta$$

$$b_1^\dagger = Q_1\cos\theta - iP_2\sin\theta \tag{E.4.7}$$
$$b_2^\dagger = -Q_2\sin\theta - iP_1\cos\theta$$

Their commutation relations are

$$[b_1, b_1^\dagger] = \sin(2\theta) \tag{E.4.8a}$$
$$[b_2, b_2^\dagger] = -\sin(2\theta)$$
$$[b_1, b_2^\dagger] = [b_2, b_1^\dagger] = 0$$
$$[b_1, b_2] = [b_1^\dagger, b_2^\dagger] = 0$$

The PseudoQuantum Hamiltonian[572] is

$$\hat{H} = p_1 p_2/m + m\omega^2 x_1 x_2 \tag{E.4.8b}$$
$$= \tfrac{1}{2}\omega(\{a_1, a_2^\dagger\} + \{a_2, a_1^\dagger\})$$
$$= \omega(P_1 P_2 + Q_1 Q_2)$$

In terms of the original P and Q variables we find

$$Q_1 = (b_1 + b_1^\dagger)/(2\cos\theta) \tag{E.4.9}$$
$$Q_2 = -(b_2 + b_2^\dagger)/(2\sin\theta)$$
$$P_1 = -i(b_2 - b_2^\dagger)/(2\cos\theta)$$
$$P_2 = -i\sin(\theta)(b_1 - b_1^\dagger)/(2\sin\theta)$$

E.4.2.1 Dirac Metric Operator ζ Transforming From Classical to Quantum Oscillator

At this point we define 'number' states with a_2 and $a_2^\dagger$:

$$|n_+, n_-\rangle = (a_2^\dagger)^{n_+}(a_2)^{n_-}|0,0\rangle \tag{E.4.10}$$

where

---

[572] Eqs. 3, 12, 21 in appendix C with ω made explicit.

$$\hat{H}|n_+, n_-> = \omega(n_+ - n_-)|n_+, n_->$$ (E.4.11)

In view of the commutation relations we wish to transform eq. E.4.10 to

$$|n_+, n_-> = (b_1^\dagger)^{n_+}(b_2^\dagger)^{n_-}|0,0>$$ (E.4.12)

where the vacuum in eqs. E.4.10 and E.4.12 will be seen to be the same:

$$a_1^\dagger|0,0> = a_1|0,0> = 0$$ (E.4.13)
$$b_1|0,0> = b_2|0,0> = 0$$

We define a 'Dirac-like' metric operator ζ. It satisfies

$$\zeta^{-1}a_2^\dagger\zeta = b_1^\dagger$$ (E.4.14)
$$\zeta^{-1}a_2\zeta = b_2^\dagger$$

We provisionally define
$$\zeta = \exp(aP_1Q_1 + bP_1^2 + cQ_2P_1 + dP_2P_1 + eQ_1^2 + fQ_2Q_1 + gP_2Q_1)$$ (E.4.15)

Eqs. E.4.4 and E.4.14 imply the values of the constants in eq. 4.15 so that

$$\zeta = \exp[(-i\cos\theta\, P_1Q_1 - (\cos\theta)/2\, P_1^2 + i\sin\theta\, Q_2P_1 - \sin\theta\, P_2P_1 - (\cos\theta)/2\, Q_1^2 - \sin\theta\, Q_2Q_1 + i\sin\theta\, P_2Q_1)/\sqrt{2}]$$ (E.4.16)

$$= \exp[(\cos\theta\, (a_1^{\dagger 2} - a_1^2)/2 - (\cos\theta)/4\, (a_1 - a_1^\dagger)^2 + (\sin\theta)/2\, (a_2 + a_2^\dagger)(a_1 - a_1^\dagger) +$$
$$+ (\sin\theta)/2\, (a_2 - a_2^\dagger)(a_1 - a_1^\dagger) - (\cos\theta)/4\, (a_1 + a_1^\dagger)^2 - (\sin\theta)/2\, (a_2 + a_2^\dagger)(a_1 + a_1^\dagger) +$$
$$+ (\sin\theta)/2\, (a_2 - a_2^\dagger)(a_1 + a_1^\dagger))/\sqrt{2}]$$

$$\zeta^{-1} = \exp[(\cos\theta\, (a_1^{\dagger 2} - a_1^2)/2 - (\cos\theta)/4\, (a_1 - a_1^\dagger)^2 + (\sin\theta)/2\, (a_2 + a_2^\dagger)(a_1 - a_1^\dagger) +$$ (E.4.17)
$$+ (\sin\theta)/2\, (a_2 - a_2^\dagger)(a_1 - a_1^\dagger) - (\cos\theta)/4\, (a_1 + a_1^\dagger)^2 - (\sin\theta)/2\, (a_2 + a_2^\dagger)(a_1 + a_1^\dagger) +$$
$$+ (\sin\theta)/2\, (a_2 - a_2^\dagger)(a_1 + a_1^\dagger))/\sqrt{2}]$$

We note the ground state (vacuum) explicitly satisfies:

$$|0,0> = \zeta^{-1}|0,0>$$ (E.4.18)

by eq. E.4.13.

We also note

$$\zeta^{-1}[a_2, a_1^\dagger]\zeta = 1$$

and

$$\zeta^{-1}[a_1, a_2^\dagger]\zeta = 1$$

imply

$$\zeta^{-1}a_1\zeta = b_1/\sin(2\theta) \tag{E.4.19}$$
$$\zeta^{-1}a_1^\dagger\zeta = -b_2/\sin(2\theta)$$

using eq. E.4.8.

The transformed Hamiltonian H can be expressed as

$$\hat{H}_\zeta = \zeta^{-1}\hat{H}\zeta = \tfrac{1}{2}\omega(\{b_1, b_1^\dagger\} - \{b_2, b_2^\dagger\})/\sin(2\theta) \tag{E.4.20}$$

E.4.2.2 Classical, Intermediate, and Quantum Wave Functions

The classical $(n_+, n_-)^{th}$ coordinate space wave function (eq. 28 appendix A) has the form:

$$\Psi_{n_-,n_-}(x_1, p_1, x_2, p_2, \theta) = (n_+!n_-!)^{-\frac{1}{2}}<x_1, p_1, x_2, p_2|(a_2^\dagger)^{n_+}(a_2)^{n_-}|0, 0> \tag{E.4.21}$$
$$= (n_+!n_-!)^{-\frac{1}{2}}<x_1, p_1, x_2, p_2|\zeta\zeta^{-1}(a_2^\dagger)^{n_+}\zeta\zeta^{-1}(a_2)^{n_-}\zeta\zeta^{-1}|0, 0>$$
$$= (n_+!n_-!)^{-\frac{1}{2}}<x_1, p_1, x_2, p_2|\zeta^\dagger b_1^{\dagger n_+}b_2^{\dagger n_-}|0,0>$$

using the conventional normalization of states with the form $|n> = (n_-!)^{-\frac{1}{2}}a^{\dagger n}|0>$.

Next we note

$$<x_1, p_1, x_2, p_2|\zeta^\dagger = <x_1, p_1, x_2, p_2| \tag{E.4.22}$$

similarly to eq. E.4.18.

$$b_1^\dagger = Q_1\cos\theta - iP_2\sin\theta \equiv \cos\theta\, Q_1 - \sin\theta\, d/dQ_1 = \cos\theta\, \eta_1 - \sin\theta\, \partial/\partial\eta_1 \tag{E.4.23}$$
$$b_2^\dagger = -Q_2\sin\theta - iP_1\cos\theta \equiv -i(\sin\theta\, d/dP_1 + \cos\theta\, P_1) = -i(\sin\theta\, \partial/\partial\eta_2 + \cos\theta\, \eta_2)$$
$$= \sin\theta\, \partial/\partial\eta_3 - \cos\theta\, \eta_3$$

where $\eta_3 = i\eta_2 = iQ_2$ and $\eta_1 = Q_1$.

Eq. E.4.21 can be expressed as:

$$\Psi_{n_-,n_-}(x_1, p_1, x_2, p_2, \theta) = (n_+!n_-!)^{-\frac{1}{2}}(-1)^{n_-}[\cos\theta \; \eta_1 - \sin\theta \; \partial/\partial\eta_1]^{n_+} \cdot$$
$$\cdot \; [\cos\theta \; \eta_3 - \sin\theta \; \partial/\partial\eta_3]^{n_-}<x_1, p_1, x_2, p_2|0,0> \qquad (E.4.24)$$

The determination of

$$\Psi_{0,0} \equiv <x_1, p_1, x_2, p_2|0,0>$$

begins with noting

$$<x_1, p_1, x_2, p_2|b_1|0,0> = 0$$

or

$$(\cos\theta \; \eta_1 + \sin\theta \; \partial/\partial\eta_1)\Psi_{0,0} = 0$$

and

$$<x_1, p_1, x_2, p_2|b_2|0,0> = 0$$

or

$$(\cos\theta \; \eta_3 + \sin\theta \; \partial/\partial\eta_3)\Psi_{0,0} = 0$$

These conditions require

$$\Psi_{0,0} = C \; \exp[-\tfrac{1}{2}\cot\theta(\eta_1^2 + \eta_3^2)] \qquad (E.4.25)$$

where the normalization $C = [m\omega\cot\theta/(i\pi\hbar)]^{\frac{1}{2}}$ is determined by

$$1 = C^2 \int dx_1 dx_2 \; \exp[-\tfrac{1}{2}\cot\theta(\eta_1^2 + \eta_3^2)] \qquad (E.4.26)$$

Then eq. E.4.24 becomes

$$\Psi_{n_-,n_-}(x_1, p_1, x_2, p_2, \theta) = (n_+!n_-!)^{-\frac{1}{2}}[m\omega\cot\theta/(i\pi\hbar)]^{\frac{1}{2}}(-1)^{n_-}[\cos\theta \; \eta_1 - \sin\theta \; \partial/\partial\eta_1]^{n_+} \cdot$$
$$\cdot \; [\cos\theta \; \eta_3 - \sin\theta \; \partial/\partial\eta_3]^{n_-}\exp[-\tfrac{1}{2}\cot\theta(\eta_1^2 + \eta_3^2)]$$
$$\qquad (E.4.27)$$

$$= (n_+!n_-!)^{-\frac{1}{2}}[m\omega\cot\theta/(i\pi\hbar)]^{\frac{1}{2}}(-1)^{n_-}[\cos\theta \; \eta_1 - \sin\theta \; \partial/\partial\eta_1]^{n_+} \cdot$$
$$\cdot \; [i\cos\theta \; \eta_2 + i\sin\theta \; \partial/\partial\eta_2]^{n_-}\exp[-\tfrac{1}{2}\cot\theta(\eta_1^2 - \eta_2^2)]$$

Note that eq. E.4.27 contains a product of Hermite polynomials if $\theta = \pi/4$. It is not surprising that we obtain quantum harmonic oscillator factors in the wave function since, as eq. E.4.8a shows the b operators have conventional quantum oscillator commutation relations for $\theta = \pi/4$ – thus this value of θ corresponds to the quantum case. We note that Hermite polynomials $H_n(\eta)$ are generated by

$$(\eta - \partial/\partial\eta)^n \exp(-\tfrac{1}{2}\eta^2) = \exp(-\tfrac{1}{2}\eta^2)H_n(\eta) \qquad (E.4.28)$$

We can generalize Hermite polynomials for other values of θ with

$$H_n(\eta, \theta) = \exp[+\tfrac{1}{2}\cot\theta\ \eta^2]\ [\cos\theta\ \eta - \sin\theta\ \partial/\partial\eta]^n\exp[-\tfrac{1}{2}\cot\theta\ \eta^2] \qquad (E.4.29)$$

Then eq. E.4.27 can be expressed by

$$\Psi_{n_+,n_-}(x_1, p_1, x_2, p_2, \theta) = (n_+!n_-!)^{-\frac{1}{2}}[m\omega\cot\theta/(i\pi\hbar)]^{\frac{1}{2}}(-1)^{n_-}H_{n_+}(\eta_1,\theta)H_{n_-}(\eta_3,\theta)\cdot$$
$$\cdot\exp[-\tfrac{1}{2}\cot\theta(\eta_1^2 + \eta_3^2)] \qquad (E.4.30)$$

$$= (n_+!n_-!)^{-\frac{1}{2}}[m\omega\cot\theta/(i\pi\hbar)]^{\frac{1}{2}}(-1)^{n_-}H_{n_+}(\eta_1,\theta)H_{n_-}(i\eta_2,\theta)\exp[-\tfrac{1}{2}\cot\theta(\eta_1^2 - \eta_2^2)]$$
$$= (-1)^{n_-}\Psi_{n_+}(\eta_1, \theta)\Psi_{n_-}(\eta_3, \theta)$$
$$= (-1)^{n_-}\Psi_{n_+}((m\omega)^{\frac{1}{2}} x_1, \theta)\Psi_{n_-}(i(m\omega)^{\frac{1}{2}} x_2, \theta)$$

where

$$\Psi_n(\eta, \theta) = (n!)^{-\frac{1}{2}}[m\omega\cot\theta/(i\pi\hbar)]^{1/4}H_n(\eta,\theta)\ \exp[-\tfrac{1}{2}\cot\theta\eta^2] \qquad (E.4.30a)$$

At $\theta = \pi/4$ the wave function factorizes into a harmonic oscillator wave function times an inverted harmonic oscillator wave function:

$$\Psi_{n_+,n_-}(x_1, p_1, x_2, p_2, \theta=\pi/4) = (n_+!n_-!)^{-\frac{1}{2}}[m\omega/(i\pi\hbar)]^{\frac{1}{2}}(-1)^{n_-}2^{-(n_+ + n_-)/2}H_{n_+}(\eta_1)H_{n_-}(\eta_3)\cdot$$
$$\cdot\exp[-\tfrac{1}{2}(\eta_1^2+\eta_3^2)] \qquad (E.4.30b)$$

$$= (-1)^{n_-}2^{-(n_+ + n_-)/2}\Psi_{n_+}((m\omega)^{\frac{1}{2}} x_1)\Psi_{n_-}(i(m\omega)^{\frac{1}{2}} x_2)$$

where $H_n(\eta)$ is a Hermite polynomial of degree n.

For $\theta = 0$ we find the b commutation relations (eq. 8a) are zero indicating that the wave function is classical in nature. In this case, simply substituting $\theta = 0$ would cause eq. E.4.30 to 'blow up.' However for certain values of n+ and n– the limit $\theta \to 0$ yields a physically interesting result – a wave function that is a delta function similar to that appearing in eq. 43 in appendix C.

Consider first the case n+ = 1 and n– = 0. Then

$$\Psi_{1,0}(x_1, p_1, x_2, p_2, \theta) = [m\omega\cot\theta/(i\pi\hbar)]^{\frac{1}{2}}[\cos\theta \; \eta_1 - \sin\theta \; \partial/\partial\eta_1]\exp[-\frac{1}{2}\cot\theta(\eta_1^2 + \eta_3^2)]$$

As $\theta \to 0$, and for $\eta_3 = 0$, we find

$$\begin{aligned}\Psi_{1,0}(x_1, p_1, x_2, p_2, \theta\to0) &\to [m\omega/(i\hbar)]^{\frac{1}{2}}\eta_1(\pi\sin\theta)^{-\frac{1}{2}}\exp[-\frac{1}{2}\eta_1^2/\sin\theta] \\ &\to [m\omega/(i\hbar)]^{\frac{1}{2}}\eta_1(2\pi\sin\theta)^{-\frac{1}{2}}\exp[-\eta_1^2/(2\sin\theta)] \\ &\to [m\omega/(i\hbar)]^{\frac{1}{2}}\eta_1\delta(\eta_1) = 0 \end{aligned} \qquad (E.4.31)$$

Now consider the case n+ = 0 and n– = 0:

$$\Psi_{0,0}(x_1, p_1, x_2, p_2, \theta) = [m\omega\cot\theta/(i\pi\hbar)]^{\frac{1}{2}}\exp[-\frac{1}{2}\cot\theta(\eta_1^2 + \eta_3^2)]$$

As $\theta \to 0$, and for $\eta_3 = 0$, we find

$$\begin{aligned}\Psi_{0,0}(x_1, p_1, x_2, p_2, \theta\to0) &\to [m\omega/(i\hbar)]^{\frac{1}{2}}(\pi\sin\theta)^{-\frac{1}{2}}\exp[-\frac{1}{2}\cot\theta\eta_1^2/\sin\theta] \\ &\to [m\omega/(i\hbar)]^{\frac{1}{2}}\eta_1(2\pi\sin\theta)^{-\frac{1}{2}}\exp[-\eta_1^2/(2\sin\theta)] \\ &\to [m\omega/(i\hbar)]^{\frac{1}{2}}\delta(\eta_1) = i^{-\frac{1}{2}}\delta(x_1) \neq 0 \end{aligned} \qquad (E.4.32)$$

using

$$\delta(\eta) = \lim_{\varepsilon\to0} (\pi\varepsilon)^{-\frac{1}{2}}\exp[-\eta^2/\varepsilon] \qquad (E.4.33)$$

Thus the Gaussian factor combined with the preceding $(2\pi\sin\theta)^{-\frac{1}{2}}$ grows to a delta-function wave function. *Wave functions corresponding to higher values of n+ and n– go to zero in the limit $\theta \to 0$. Only $\Psi_{0,0}(x_1, p_1, x_2, p_2, \theta\to0)$ is non-zero.*

The introduction of the time dependence and a shift of the location of the minimum of the harmonic oscillator potential to x_0 would lead to a wave function such as:

$$\Psi_{0,0}(x_1, p_1, x_2, p_2, \theta \to 0) = i^{-\frac{1}{2}}\delta(x_1 - x_0 \sin(\omega t)) \tag{E.4.34}$$

A similar behavior may be seen in the case $\eta_1 = 0$. Then we find a wave function with a factor of $\delta(x_2)$.

Lastly, the case of $\theta \to \pi/2$ is of interest. Eq. 4.30 yields

$$\Psi_{n_-,n_-}(x_1, x_2, \theta \to \pi/2) = (n_+! n_-!)^{-\frac{1}{2}}[m\omega\cos\theta/(i\pi\hbar)]^{\frac{1}{2}}[-\partial/\partial\eta_1]^{n_+}[\partial/\partial\eta_3]^{n_-} \cdot$$
$$\cdot \exp[-\tfrac{1}{2}\cos\theta(\eta_1{}^2+\eta_3{}^2)]|_{\theta \to \pi/2}$$
$$= 0 \tag{E.4.35}$$

Figuratively speaking, the wave function progresses from one non-zero 'classical' wave function at $\theta = 0$, to a quantum mechanical wave function at $\theta = \pi/4$, to a zero value wave function at $\theta = \pi/2$. Thus one might say "The good Lord by giving us a quantum universe put us in a position halfway between nothingness and classical mechanics." By implementing Quantum theory we get Second Quantization of particle fields, and thereby, integer countability of particle numbers – a distinct simplification in Nature.

E.4.2.3 Energy Eigenvalues

From eq. E.4.8b, E.4.10, and E.4.11 we see that eq. E.4.11 for the state

$$|n_+, n_-\rangle = (a_2{}^\dagger)^{n_+}(a_2)^{n_-}|0, 0\rangle$$

shows the energy of the wave function (eqs. E.4.27 and E.4.30) to be

$$E_{n_+,n_-} = (n_+ - n_-)\hbar\omega = [n_+ + \tfrac{1}{2} - (n_- + \tfrac{1}{2})]\hbar\omega \tag{E.4.36}$$

Eq. E.4.27 satisfies the PseudoQuantized Schrödinger equation:

$$\hat{H}\Psi_n(x_1, p_1, x_2, p_2, \theta, t) = i\partial\Psi_n(x_1, p_1, x_2, p_2, \theta, t)\partial t \tag{E.4.37}$$

E.4.3 Wave Function as a Function of Position and Momentum

We can define the wave function in terms of x_1 and p_1 with a fourier transform:

$$\Psi_{n+,n-}(x_1, p_1, \theta) = \int dx_2 \, e^{-ip_1x_2} \, \Psi_{n+,n-}(x_1, x_2, \theta) \qquad (E.4.38)$$

$\Psi_{n+,n-}(x_1, p_1, \theta=\pi/4)$, is a wave function for the combined 'normal', and inverted, harmonic oscillators. Thus the full PseudoQuantum theory enables us to define a wave function that is a function of both position and momentum without inconsistency. We discuss this topic in more detail later when we compare it to the Wigner distribution function.

E.4.4 Intermediate Classical-Quantum Wave Functions

For other values of θ in eq. E.4.27 and E.4.30 we obtain wave functions that are intermediate between quantum and classical operator wave functions. Later we will find it of interest to trace the evolution of a classical wave function to a quantum wave function and vice versa in the general case of non-harmonic oscillator dynamics.

It is interesting to note the dependence of the energy level spacing on the angle θ. The transformed energy (eq. E.4.14 implements the transformation to the b operators) has the form:

$$\hat{H}_\zeta = \zeta^{-1}\hat{H}\zeta = \tfrac{1}{2}\omega(\{b_1, b_1^\dagger\} - \{b_2, b_2^\dagger\})/\sin(2\theta) \qquad (E.4.20)$$

If either n_+ and n_- change by one unit, then the energy changes by

$$\Delta E = \tfrac{1}{2}\omega/\sin(2\theta)$$

For $\theta = \pi/4$ (the quantum case)

$$\Delta E = \tfrac{1}{2}\omega \qquad (E.4.39)$$

For $\theta = \pi/8$ (the quantum approaching classical case)

$$\Delta E = \tfrac{1}{2}\omega/0.383 = 1.307\omega \qquad (E.4.40)$$

As $\theta \rightarrow 0$ (the classical case)

$$\Delta E \rightarrow \infty \qquad \text{(E.4.41)}$$

Thus 'higher' (lower) energy states beyond the $n_+ = 0$ and $n_- = 0$ state are inaccessible energy-wise. This corresponds to our above finding that only the wave function $\Psi_{0,0}$ is non-zero in the classical limit.

E.5 General Formalism for a PseudoQuantized System

The basic procedure of our PseudoQuantization Formalism are described in our paper Phys. Rev **D17**, 994 (1978) reprinted in Appendix C.[573] The relevant excerpt is

We shall now briefly outline the procedure for embedding a classical-mechanical system in a quantum system.[6] Consider a classical Hamiltonian system with one degree of freedom, and commuting canonical variables, x_1 and p_1, which have the equations of motion

$$\dot{x}_1 = -i[x_1, \hat{H}], \qquad (1)$$

$$\dot{p}_1 = -i[p_1, \hat{H}], \qquad (2)$$

where defining

$$\hat{H} = -i\left(\frac{\partial H(x_1, p_1)}{\partial p_1}\frac{\partial}{\partial x_1} - \frac{\partial H(x_1, p_1)}{\partial x_1}\frac{\partial}{\partial p_1}\right) \qquad (3)$$

allows us to write Hamilton's equations in com-

---

[573] Excerpt used with the kind permission of Physical Review D.

It is now apparent that we can take the above quantities and equations of motion to describe a quantum mechanical system with two degrees of freedom in the "coordinate" representation where the "coordinates" are (x_1, p_1) and the canonical momenta are $\Pi = (p_2, -x_2)$. As we will see below the linearity of $\hat{H}$ in the momenta is crucial for the maintenance of the classical character of x_1 and p_1, and for the observability of the phase-space trajectory. Since we choose to identify the physical observables with the commutative algebra of the coordinate operators, x_1 and p_1, we are led to impose the superselection condition that the momenta, Π, are unobservable. As a result the Hamiltonian and other generators of canonical transformations, which are all linear in the momenta, are also unobservable. However, in each case there is an associated dynamical quantity which is observable.

The required unobservability of the momenta restricts the form of the interaction between a classical-made-quantum system and an inherently quantum system to

$$H_{int} = \Phi_1 x_2 + \Phi_2 p_2 + X , \qquad (9)$$

where Φ_1, Φ_2, and X are functions of x_1, p_1, and the quantum system variables. The commutation relations of these functions are also constrained[6] by the superselection rule and the commutativity of the classical variables, x_1 and p_1, and their time derivatives. In the next section we will study the simple harmonic oscillator in order to exemplify the quantum-mechanical case described above and also for direct use in the field-theoretic generalizations of subsequent sections.

Based on the above discussion we assume that we start with a conventional Hamiltonian that we express as

$$H = H(x_1, p_1) = \tfrac{1}{2}p_1^2 + V(x_1) \tag{E.5.1}$$

Introducing x_2 and p_2, as in eqs. 4 and 5 above, we can generalize H to a PseudoQuantum Hamiltonian $\hat{H}$:

$$\hat{H} = p_1 p_2 + x_2 \partial V / \partial x_1 \tag{E.5.2}$$

where V is a function of x_1.

We can introduce raising and lowering operators a_i and $a_i^\dagger$ using the procedure of section E.4. Then we can proceed as in the harmonic oscillator case to calculate wave functions. In the next section we apply this procedure to the Boltzmann equation, which has some similarity to the Schrödinger equation.

E.6 PseudoQuantization of the Boltzmann Equation

The Boltzmann equation is a classical dynamics equation that describes the dynamics of a multi-particle system with interactions. The equivalent quantum formulation is not known. However Wigner and others have proposed possible quantum equivalents that of some of the expected features of the quantum Boltzmann function. In this section we will follow a procedure similar to that of section 4 for the Vlasov approximation. For special cases of the collision term of the Boltzmann equation we will obtain quantum equivalents.

E.6.1 Non-Relativistic Boltzmann Equation

The non-relativistic Boltzmann equation for identical particles of one chemical species is

$$[\mathbf{p} \cdot \nabla / m + F \cdot \partial / \partial \mathbf{p}]f = -\partial f / \partial t + (\partial f / \partial t)_{coll}$$

where $f(\mathbf{r}, \mathbf{p}, t)$ is Boltzmann's probability density function. It has often been remarked that this Boltzmann equation strongly resembles the Schrödinger equation.

E.6.2 PseudoQuantum Form of the Boltzmann Equation

We can make the case that it even more strongly resembles the PseudoQuantized Schrödinger equation (see eq. E.5.2) by defining

$$-i[\mathbf{p}_1 \cdot \mathbf{p}_2/m - \mathbf{x}_2 \cdot \mathbf{F}]f = -\partial f/\partial t + (\partial f/\partial t)_{coll} \qquad (E.6.1)$$

where $\mathbf{F} = \mathbf{F}(x_1, t)$ and

$$\mathbf{p}_2 = i\nabla \qquad (E.6.2)$$
$$\mathbf{x}_2 = -i\partial/\partial\mathbf{p}_1$$

Comparing eq. E.6.1 with section E.4, we find we can define

$$\hat{H} = \mathbf{p}_1 \cdot \mathbf{p}_2/m + \mathbf{x}_2 \cdot \partial V/\partial \mathbf{x}_1 \qquad (E.6.3)$$

where

$$\partial V/\partial \mathbf{x}_1 = \mathbf{F}(x_1, t)$$

Given the close similarity of the Boltzmann equation and the Schrödinger equation it is sensible to treat the solution of the equation as a 'wave function' that initially represents a classical state such as the classical harmonic oscillator wave function that we saw in section E.4. Then we will define the Boltzmann distribution in terms of the wave function solution.

The PseudoQuantized Boltzmann wave equation is

$$\hat{H}\psi = -i\partial\psi/\partial t + i(\partial\psi/\partial t)_{coll} \qquad (E.6.4)$$

where

$$\psi = \psi(\mathbf{x}_1, \mathbf{x}_2, \theta)$$

with θ defined later in specific cases. The value of θ determines whether ψ is classical, quantum, or in an intermediate state.

In a manner somewhat analogous to that of Wigner[574] we define a Boltzmann distribution with

---

[574] E. P. Wigner, Phys. Rev. **40**, 749 (1932).

$$f_q(\mathbf{r}_1, \mathbf{p}_1, t, \theta) = \int d^3r_2\, \psi(\mathbf{r}_1, \mathbf{r}_2, t, \theta)\psi^\dagger(\mathbf{r}_1, \mathbf{r}_2, t, \theta)\, exp(-2i\mathbf{r}_2 \cdot \mathbf{p}_1/\hbar) \qquad (E.6.5)$$

in three spatial dimensions where we use the suffix 'q' of f_q to signify the PseudoQuantum Boltzmann probability density function $f_q(\mathbf{r}, \mathbf{p}, t)$. The function f_q can be classical, quantum, or intermediate between classical and quantum depending on the value of θ. We will consider examples that illustrate the dependence of f_q on θ. Later we will also see that our definition of f_q eliminates the problems of the Wigner density function.

E.6.3 PseudoQuantum Form of the Vlasov Equation

The collision-less Boltzmann equation is called the Vlasov equation. It is of interest because of the difficulties associated with solving the full Boltzmann equation. Its PseudoQuantized equivalent is

$$\hat{H}\psi = -i\partial\psi/\partial t \qquad (E.6.6)$$

This equation has 'only' the difficulty of its solution for the various forces $\mathbf{F}(x_1, t)$.

We note that the one-dimensional version of eq. E.6.6 where $V = \frac{1}{2} x_1^2$ is solved in section E.6.4.

E.6.4 PseudoQuantum Vlasov Equation Solution for a Three-dimensional Harmonic Oscillator Force with $\theta = \pi/4$

The choice of $\theta = \pi/4$ gives 'quantum' harmonic oscillator solutions consisting of a harmonic oscillator factor and an inverted harmonic oscillator factor.

The three-dimensional harmonic oscillator PseudoQuantum 'Hamiltonian' equation is

$$(\mathbf{p}_1 \cdot \mathbf{p}_2/m + x_2 \cdot \mathbf{x}_1)\psi = -i\partial\psi/\partial t \qquad (E.6.7)$$

or

$$(\mathbf{p}_1 \cdot \mathbf{p}_2/(2m') + x_2 \cdot \mathbf{x}_1)\,\psi = -i\partial\psi/\partial t \qquad (E.6.8)$$

This equation is fully separable[575] for $\theta = \pi/4$ in rectangular coordinates which we label x, y, and z. The solution is a product of one-dimensional PseudoQuantum harmonic oscillator wave function factors of the form of E.4.30b:

$$\psi_{n+,n-}(\mathbf{r}_1,\mathbf{r}_2,t,\pi/4) = \Psi_{n_{x+},n_{x-}}(x_{1x},p_{1x},x_{2x},p_{2x},t,\theta)\ \Psi_{n_{y+},n_{y-}}(x_{1y},p_{1y},x_{2y},p_{2y},t,\theta)\ \Psi_{n_{z+},n_{z-}}(x_{1z},p_{1z},x_{2z},p_{2z},t,\theta)$$

$$(E.6.9)$$

$$= (-1)^{n_{x-}+n_{y-}+n_{z-}}2^{-(n_{x+}+n_{x-}+n_{y+}+n_{y-}+n_{z+}+n_{z-})/2}\Psi_{n_{x+}}((m\omega)^{\frac{1}{2}}x_1)\Psi_{n_{x-}}(i(m\omega)^{\frac{1}{2}}x_2)\Psi_{n_{y+}}((m\omega)^{\frac{1}{2}}y_1)\cdot$$
$$\cdot\Psi_{n_{y-}}(i(m\omega)^{\frac{1}{2}}y_2)\Psi_{n_{z+}}((m\omega)^{\frac{1}{2}}z_1)\Psi_{n_{z-}}(i(m\omega)^{\frac{1}{2}}z_2)$$

$$= A\Psi_{n_{x+}}((m\omega)^{\frac{1}{2}}x_1)\Psi_{n_{y+}}((m\omega)^{\frac{1}{2}}y_1)\Psi_{n_{z+}}((m\omega)^{\frac{1}{2}}z_1)\Psi_{n_{x-}}(i(m\omega)^{\frac{1}{2}}x_2)\Psi_{n_{y-}}(i(m\omega)^{\frac{1}{2}}y_2)\cdot$$
$$\cdot\Psi_{n_{z-}}(i(m\omega)^{\frac{1}{2}}z_2)$$

$$= A\Psi_1(\mathbf{r}_1)\Psi_2(\mathbf{r}_2) \qquad\qquad (E.6.9a)$$

with the time dependence not displayed and where

$$A = (-1)^{n_{x-}+n_{y-}+n_{z-}}2^{-(n_{x+}+n_{x-}+n_{y+}+n_{y-}+n_{z+}+n_{z-})/2} \qquad (E.6.9c)$$

using eq. E.4.30b.

The energy, which is constant since the Hamiltonian is not explicitly time dependent, is

$$E_{n+,n-} = (n_+ - n_-)\hbar\omega \qquad\qquad (E.6.10)$$

with

$$n_+ = n_{x+} + n_{y+} + n_{z+} \qquad\qquad (E.6.11)$$
$$n_- = n_{x-} + n_{y-} + n_{z-}$$

Following Wigner, a fourier transform for $\theta = \pi/4$ of the eq. E.6.9a factors gives a *quantum* Boltzmann density function:

---

[575] For other values of θ the solutions of the equation do not separate type '1' coordinates from type '2' coordinates. See eq. E.4.30 for the general case.

$$f_q(\mathbf{q}, \mathbf{p}_1, t, \pi/4) = \int d^3Q \, \Psi(\mathbf{q} - \mathbf{Q})\Psi^\dagger(\mathbf{q} + \mathbf{Q}) \exp(-2i\mathbf{Q}\cdot\mathbf{p}_1/\hbar)$$

$$= \int d^3Q \, \Psi_1(\mathbf{q} - \mathbf{Q})\Psi_1^\dagger(\mathbf{q} + \mathbf{Q})\Psi_2(\mathbf{q} - \mathbf{Q})\Psi_2^\dagger(\mathbf{q} + \mathbf{Q}) \exp(-2i\mathbf{Q}\cdot\mathbf{p}_1/\hbar)$$

(E.6.12)

where we let $\mathbf{r}_1 = \mathbf{q} - \mathbf{Q}$ and $\mathbf{r}_2 = \mathbf{q} + \mathbf{Q}$.

If we define the fourier transform of a wave function $\Psi(\mathbf{r})$ by

$$\Phi(\mathbf{p}) = (2\pi\hbar)^{-3/2} \int d^3r \, \Psi(\mathbf{r}) \exp(-i\mathbf{r}\cdot\mathbf{p}/\hbar)$$

then

$$f_{qp}(\mathbf{p}_1, t, \pi/4) = \int d^3q \, f_q(\mathbf{q}, \mathbf{p}_1, t, \pi/4) = \int d^3p \, \Phi(\mathbf{p})\Phi^\dagger(\mathbf{p}) \quad \text{(E.6.13)}$$

yields a projection of the phase space distribution into momentum space. In addition

$$f_{qp}(\mathbf{p}_1, t, \pi/4) = \Psi(\mathbf{q})\Psi^\dagger(\mathbf{q}) \tag{E.6.14}$$

yields a projection of the phase space distribution into coordinate space.

Thus $f_q(\mathbf{q}, \mathbf{p}_1, t, \pi/4)$ can be interpreted as the quantum equivalent of the (classical) Boltzmann distribution. *PseudoQuantization gives us 2n variables just as there are 2n variables in phase space.*

E.6.5 PseudoQuantum Vlasov Equation Solution for a Three-dimensional Harmonic Oscillator Force for Arbitrary θ

The general representation of our PseudoQuantized Vlasov equation solution for any value of θ is given by eq. E.4.30. The 3-dimensional Vlasov representation is

$$\psi_{n_+,n_-}(\mathbf{r}_1, \mathbf{r}_2, t, \theta) = \psi_{n_{x+},n_{x-}}(x_{1x},p_{1x},x_{2x},p_{2x},t,\theta)\psi_{n_{y+},n_{y-}}(x_{1y},p_{1y},x_{2y},p_{2y},t,\theta)\psi_{n_{z+},n_{z-}}(x_{1z},p_{1z},x_{2z},p_{2z},t,\theta)$$

(E.6.15)

$$= (-1)^{n_{x-}+n_{y-}+n_{z-}}2^{-(n_{x+}+n_{x-}+n_{y+}+n_{y-}+n_{z+}+n_{z-})/2}\Psi_{n_{x+}}((m\omega)^{1/2}x_1,t,\theta)\Psi_{n_{x-}}(i(m\omega)^{1/2}x_2,t,\theta)\cdot$$
$$\cdot\Psi_{n_{y+}}((m\omega)^{1/2}y_1,t,\theta)\Psi_{n_{y-}}(i(m\omega)^{1/2}y_2,t,\theta)\Psi_{n_{z+}}((m\omega)^{1/2}z_1,t,\theta)\Psi_{n_{z-}}(i(m\omega)^{1/2}z_2,t,\theta)$$

$$= A\Psi_{n_{x^+}}((m\omega)^{\frac{1}{2}} x_1, t, \theta)\Psi_{n_{y^+}}((m\omega)^{\frac{1}{2}} y_1, t, \theta)\Psi_{n_{z^+}}((m\omega)^{\frac{1}{2}} z_1, t, \theta)\Psi_{n_{x^-}}(i(m\omega)^{\frac{1}{2}} x_2, t, \theta) \cdot$$
$$\cdot \Psi_{n_{y^-}}(i(m\omega)^{\frac{1}{2}} y_2, t, \theta) \ \Psi_{n_{z^-}}(i(m\omega)^{\frac{1}{2}} z_2, t, \theta)$$
$$= A\Psi_1(\mathbf{r}_1, t, \theta)\Psi_2(\mathbf{r}_2, t, \theta)$$

Following similar steps as in the previous section we find

$$f_q(\mathbf{q}, \mathbf{p}_1, t, \theta) = \int d^3Q \ \psi_{n_+,n_-}(\mathbf{q} - \mathbf{Q}, t, \theta) \ \psi_{n_+,n_-}{}^\dagger(\mathbf{q} + \mathbf{Q}, t, \theta) \exp(-2i\mathbf{Q}\cdot\mathbf{p}_1/\hbar)$$

$$= A^2 \int d^3Q \ \Psi_1(\mathbf{q} - \mathbf{Q}, t, \theta)\Psi_1{}^\dagger(\mathbf{q} + \mathbf{Q}, t, \theta)\Psi_2(\mathbf{q} - \mathbf{Q}, t, \theta)\Psi_2{}^\dagger(\mathbf{q} + \mathbf{Q}, t, \theta)\exp(-2i\mathbf{Q}\cdot\mathbf{p}_1/\hbar)$$
$$(E.6.16)$$

where we let $\mathbf{r}_1 = \mathbf{q} - \mathbf{Q}$ and $\mathbf{r}_2 = \mathbf{q} + \mathbf{Q}$.

Following similar steps we can again obtain eqs. E.6.13 and E.6.14 and establish a connection between phase space, and momentum and coordinate space projections.

If we define the fourier transform of a wave function $\Psi(\mathbf{r})$ by

$$\Phi(\mathbf{p}, \theta) = (2\pi\hbar)^{-3/2} \int d^3r \ \Psi(\mathbf{r}, \theta) \exp(-i\mathbf{r}\cdot\mathbf{p}/\hbar)$$

then

$$f_{qp}(\mathbf{p}_1, t, \theta) = \int d^3q \ f_q(\mathbf{q}, \mathbf{p}_1, t, \theta) = \int d^3p \ \Phi(\mathbf{p}, \theta)\Phi^\dagger(\mathbf{p}, \theta) \qquad (E.6.17)$$

yields a projection of the phase space distribution into momentum space. In addition

$$f_{qq}(\mathbf{p}_1, t, \theta) = \Psi(\mathbf{q}, \theta)\Psi^\dagger(\mathbf{q}, \theta) \qquad (E.6.18)$$

yields a projection of the phase space distribution into coordinate space.

Thus $f_q(\mathbf{q}, \mathbf{p}_1, t, \theta)$ can be interpreted as the quantum equivalent of the (classical) Boltzmann distribution. PseudoQuantization gives us 2n variables just as there are 2n variables in phase space.

E.6.6 PseudoQuantum Vlasov Equation Solution for a Three-dimensional Harmonic Oscillator Force for θ = 0 – The Classical Case

In the θ = 0 case, which is the classical mechanics limit, we find that eq. E.4.34 gives a precise expression for the Vlasov Boltzmann equation solution of eq. E.6.16. We note that only the 0-0 wave functions are non-zero. In one dimension we have:

$$\psi = \Psi_{0,0}(x_1, p_1, x_2, p_2, \theta \rightarrow 0) = i^{-\frac{1}{2}}\delta(x_1 - x_0\sin(\omega t)) \qquad (E.4.34)$$

The 3-dimensional case (eq. E.6.16) gives

$$\begin{aligned} f_q(\mathbf{q}, \mathbf{p}_1, t, \theta{=}0) &= \int d^3Q \, \delta^3(\mathbf{q} - \mathbf{Q} - \mathbf{x}_0\sin(\omega t))\delta^3(\mathbf{q} + \mathbf{Q} - \mathbf{x}_0\sin(\omega t))\exp(-2i\mathbf{Q}\cdot\mathbf{p}_1/\hbar) \\ &= \delta^3(\mathbf{q} - \mathbf{x}_0\sin(\omega t)) \qquad (E.6.19) \end{aligned}$$

a classical solution specifying the classical harmonic oscillator trajectory. We note that all other solutions (for other values of n_+ and n_-) are 'pushed' to $E = \infty$ according to eq. E.4.41.

We note $f_q(\mathbf{q}, \mathbf{p}_1, t, \theta{=}0)$ is positive definite as a probability should be. The integral

$$\int d^3q \, f_q(\mathbf{q}, \mathbf{p}_1, t, \theta{=}0) = 1$$

shows the sum of the probabilities of the normalized Boltzmann distribution in coordinate space is unity.

We thus have achieve a quantum-classical Boltzmann distribution in phase space in both coordinates and momenta using PseudoQuantization where the number of phase space parameters is $2n = 6$ in this case—unlike the case of the Wigner density alternative.

E.6.7 Comparison to the Wigner Density Function

The Wigner density function in n dimensions is defined as:

$$\Psi(p, q) = \int d^nQ \, \psi(q - Q) \, \psi^\dagger(q + Q) \, \exp(-2ipQ/\hbar) \qquad (E.6.20)$$

where $\psi(q)$ is the wave function of the system. The interpretation of $\Psi(p, q)$ as the quantum probability in phase space corresponds to $f(p, q)$ – the classical Boltzmann distribution. It is often interpreted in that manner.

However, several concerns are usually expressed about this interpretation:

1. Although real-valued $\Psi(p, q)$ can have a negative value making a probability interpretation problematic.

2. $\Psi(p, q)$ appears to depend on 2n values. However the wave function $\psi(q)$, upon which it is defined, only depends on n variables. Thus the domains of each function are different and $\Psi(p, q)$ can only be viewed as dependent on n variables.

Wigner attempted to overcome these objections by using the quantum mechanics density matrix $\rho(q, q')$ in an attempt to reflect the usual situation that a quantum system is in a mixed state consisting of a superposition of orthogonal states $\psi_k(q)$ with a probability of $\rho_k \geq 0$ with the $\Sigma \rho_k = 1$. The density matrix for this case is defined to be

$$\rho(q, q') = \Sigma \rho_k \psi_k(q)\psi_k^\dagger(q') \qquad (E.6.21)$$

Using the density matrix the Wigner distribution now is

$$\Psi(p, q) = \int d^n Q \, \rho(q - Q, q + Q) \exp(-2ipQ/\hbar) \qquad (E.6.22)$$

The new form of the Wigner distribution is a function of 2n variables and $\Psi(p, q)$ is positive definite. However, the density matrix (eq. E.6.14) has all eigenvalues between 0 and 1 and a trace equal to one. These properties are not shared by every possible Boltzmann probability $f(p, q, t)$. Thus the representation is limited to 'special cases.'

Our form of the *quantum* Boltzmann probability distribution is $f_q(\mathbf{r}_1, \mathbf{p}_1, t, \theta)$ which we have shown overcomes the redundancy of variables in the Wigner quantum generalization of the Boltzmann distribution and gives a sensible result in the classical limit.[576]

E.6.8 PseudoQuantum Form of the BGK Approximation to the Boltzmann Equation

The BKG approximation to the Boltzmann equation is

$$[\mathbf{p} \cdot \nabla/m + F \cdot \partial/\partial \mathbf{p}]f = -\partial f/\partial t + \upsilon(f_0 - f) \qquad (E.6.23)$$

where f_0 is the local Maxwell distribution $f_0 = f_0\,(\mathbf{r}, \mathbf{p})$ and υ is the molecular collision frequency. This model of the collision term due to Bhatnagar, Gross, and Crook[577] has been a much studied approximation.

Before introducing the PseudoQuantum form of the BKG approximation we use the local Maxwell-Boltzmann distribution to re-express the BKG approximation in the form

$$f_0 = n[m/(2\pi kT)]^{3/2} \exp[-m(p - p_0)^2/(2kT)] \qquad (E.6.24)$$

where n is the particle density (assumed constant at equilibrium), k is Boltzmann's constant, T is the temperature, p_0 is the average momentum, and m is the mass of a particle. Inserting f_0 in eq. E.6.23 and letting

$$f = f_0 g \qquad (E.6.25)$$

we obtain

$$[\mathbf{p} \cdot \nabla/m + \mathbf{F} \cdot \partial/\partial \mathbf{p} - (m/kT)\mathbf{F} \cdot (\mathbf{p} - \mathbf{p}_0) + \upsilon]g = -\partial g/\partial t + \upsilon \qquad (E.6.26)$$

The PseudoQuantized equivalent of the expanded BKG approximation (eq. E.6.26) is

---

[576] Our PseudoQuantum equivalent density has the same form as the Wigner density (eq. E.6.21).
[577] P. L. Bhatnagar, E. P. Gross, and M. Crook, Phys. Rev. **94**, 511 (1954).

$$[\mathbf{p_1 \cdot p_2}/m - \mathbf{x_2 \cdot F(x_1)} + (m/kT)\mathbf{F(x_1) \cdot (p_2} + i\mathbf{p_0}) + i\upsilon]g = -i\partial g/\partial t + i\upsilon \qquad \text{(E.6.27)}$$

with the Maxwell-Boltzmann distribution term acting as a 'driving force.'

Given a force F we can proceed to PseudoQuantize using operators that are similar to those of eqs. E.4.1 – E.4.7 but adapted to the force and the Maxwell-Boltzmann distribution 'driving force.' We will not consider BKG examples in this book although a harmonic driving force $\mathbf{F(x_1)} = -m\omega^2\mathbf{x_1}$ is an interesting case to consider.

E.6.9 Relativistic Boltzmann Equation

The Boltzmann equation is non-relativistic and is appropriate in systems that are at rest or moving at non-relativistic velocities. If a system is traveling at relativistic velocities then the relativistic Boltzmann equation must be used. In this section we first generalize the Boltzmann equation to its special relativistic form by making all terms covariant. Then, when we 'go to' a rest frame, the relativistic equation becomes the non-relativistic Boltzmann equation.

E.6.9.1 Relativistic Generalization of the Boltzmann Equation

The relativistic form of the non-relativistic Boltzmann equation of section E.6.1 is

$$[\mathbf{p}^\mu\nabla_\mu/m + F^\mu\partial/\partial\mathbf{p}^\mu]f = (\partial f/\partial t)_{\text{collRelativistic}} \qquad \text{(E.6.28)}$$

where we use indices to transform vectors into 4-vectors: the momentum, derivative operators and the force become Lorentz 4-vectors. The collision term must now be in a relativistic form.

E.6.9.2 PseudoQuantized Relativistic Boltzmann Equation

The PseudoQuantum form of the Boltzmann equation is discussed earlier:

$$-i[\mathbf{p_1 \cdot p_2}/m - \mathbf{x_2 \cdot F}]f = -\partial f/\partial t + (\partial f/\partial t)_{\text{coll}} \qquad \text{(E.6.1)}$$
$$\mathbf{p_2} = i\nabla \qquad \text{(E.6.2)}$$
$$x_2 = -i\partial/\partial\mathbf{p_1}$$

We make it relativistic in a manner similar to the approach in subsection E.6.9.1. The result is the relativistic PseudoQuantum Boltzmann equation:

$$[p_1{}^\mu p_{2\mu} - mx_2{}^\mu F_\mu(x_1{}^\alpha)]f = im(\partial f/\partial t)_{\text{collRelativistic}} \tag{E.6.29}$$

where the collision term is relativistic. This formalism, superficially, has two times. However when the wave functions are calculated only one time $x_1{}^0$ is relevant as the calculations in section 4 in the Vlasov approximation suggest.

E.6.10 Quantum and Classical Entropy

The von Neumann entropy for a system described by a density matrix ρ is defined as

$$S = - \text{tr}[\rho \ln \rho] \tag{E.6.30}$$

Using eigenvectors $|n>$ the density matrix can be expressed as

$$\rho = \sum_i \eta_i |i><i| \tag{E.6.31}$$

Then ρ can be expressed in the information theory Shannon formulation of entropy:

$$S = - \sum_i \eta_i \ln \eta_i \tag{E.6.32}$$

If we use the harmonic oscillator development of section 4 we can express the von Neumann entropy in a form which ranges from classical to quantum as a function of the angle θ. If we define the harmonic oscillator states

$$| n+, n-> = b_1{}^{\dagger n_+} b_2{}^{\dagger n_-} |0,0>$$

as in section 4 where

$$b_1{}^\dagger = Q_1 \cos \theta - iP_2 \sin \theta \tag{E.4.23}$$
$$b_2{}^\dagger = -Q_2 \sin \theta - iP_1 \cos \theta$$

then the density matrix is

$$\rho(\theta) = \sum_{n+,n-} |n+, n-><n+, n-| = \sum_{n+,n-} \eta_{n+,n-} \, b_1^\dagger(\theta)^{n+} b_2^\dagger(\theta)^{n-} |0,0><0,0| b_1(\theta)^{n+} b_2(\theta)^{n-} \quad (E.6.33)$$

In the quantum limit where $\theta = \pi/4$ we see

$$\rho(\pi/4) = \sum_{n+,n-} (\eta_{n+,n-}/2^{n+ + n-})(Q_1 - iP_2)^{n+}(Q_2 + iP_1)^{n-}|0,0><0,0|(Q_1 + iP_2)^{n+}(Q_2 - iP_1)^{n-}$$

$$(E.6.34)$$

yielding a quantum density matrix and thus a von Neumann quantum entropy.
    In the classical limit where $\theta = 0$ we see

$$\rho(0) = \sum_{n+,n-} (\eta_{n+,n-}/2^{n+ + n-})Q_1^{n+}P_1^{n-}|0,0><0,0|Q_1^{n+}P_1^{n-} = \rho(Q_1, P_1) \quad (E.6.35)$$

yielding a purely classical function of Q_1 and P_1 as the density matrix. The von Neumann entropy's classical limit in this case is the classical phase space quantity

$$S = -\{\Sigma(\eta_{n+,n-}/2^{n+ + n-})Q_1^{2n+}P_1^{2n-}\}\ln\{\Sigma(\eta_{n+,n-}/2^{n+ + n-})Q_1^{2n+}P_1^{2n-}\} \quad (E.6.36)$$
$$= S(Q_1, P_1)$$

since Q_1 and P_1 commute.

E.7 PseudoQuantum Path Integral Formulation

The path integral formulation of quantum mechanics (and also of quantum field theory) plays an important role in the understanding of quantum physics. One of its major issues is the transition from a quantum mechanical framework to a classical mechanical framework. We will examine this issue from the point of view of a PseudoQuantum path integral formulation. Earlier we have seen that we can embody both quantum and classical mechanics phenomena within the PseudoQuantum framework and 'rotate' between quantum and classical mechanics solutions.

In order to establish a PseudoQuantum path integral formalism we must first generate a lagrangian from a PseudoQuantum Hamiltonian. In section E.5 we described the general formalism for deriving a PseudoQuantum Hamiltonian from a classical Hamiltonian. We now construct the PseudoQuantum lagrangian. Starting with the equations in section E.5:

$$x_2 = id/dp_1$$
$$p_2 = -id/dx_1$$

$$\hat{H}(x_1, p_1, x_2, p_2) = \partial H(x_1, p_1)/\partial p_1\, p_2 + \partial H(x_1, p_1)/\partial x_1\, x_2$$

we define the velocities[578]

$$x'_1 = \partial\hat{H}(x_1, p_1, x_2, p_2)/\partial p_1 = \partial^2 H(x_1, p_1)/\partial p_1{}^2 p_2 + \partial^2 H(x_1, p_1)/\partial x_1 \partial p_1\, x_2$$

$$\text{(E.7.1)}$$

$$x'_2 = \partial\hat{H}(x_1, p_1, x_2, p_2)/\partial p_2 = \partial H(x_1, p_1)/\partial p_1\big|_{p_2 = p_1} \tag{E.7.2}$$

The lagrangian L is constructed in the canonical way using Legendre transformations

$$L = p_1 x'_1 + p_2\, x'_2 - \hat{H}(x_1, p_1, x_2, p_2) \tag{E.7.3}$$

$$= p_1[\partial^2 H(x_1, p_1)/\partial p_1{}^2 p_2 + \partial^2 H(x_1, p_1)/\partial x_1 \partial p_1\, x_2] - \partial H(x_1, p_1)/\partial x_1\, x_2$$

$$= \partial^2 H(x_1, p_1)/\partial p_1{}^2\, p_1 p_2 + \partial^2 H(x_1, p_1)/\partial x_1 \partial p_1\, p_1 x_2 - \partial H(x_1, p_1)/\partial x_1\, x_2$$
$$\text{(E.7.3a)}$$

where p_1 and p_2 are extracted from eqs. E.7.1 and E.7.2.

We now consider the example of a harmonic oscillator where

$$H = p^2/(2m) + \tfrac{1}{2}\, m\omega^2 x^2$$
$$\hat{H} = p_1 p_2/m + m\omega^2 x_1 x_2 \tag{E.4.8b}$$

Substituting in eq. E.7.3 we find

---

[578] The velocity x'_2 is a defined quantity, which is defined in a manner consistent with the definition of x'_1.

$$L = \partial^2 H(x_1, p_1)/\partial p_1^2 \, p_1 p_2 + \partial^2 H(x_1, p_1)/\partial x_1 \partial p_1 \, p_1 x_2 - \partial H(x_1, p_1)/\partial x_1 \, x_2$$
$$= p_1 p_2/m - m\omega^2 x_1 x_2$$
$$= mx'_1 x'_2 - m\omega^2 x_1 x_2 \tag{E.7.4}$$

with p_1 and p_2 determined, and replaced, as functions of x'_1 and x'_2 by eqs. E.7.1 and E.7.2.

The Lagrange equations of motion determined for $i = 1, 2$ are:

$$d/dt \, (\partial L/\partial x'_i) - \partial L/\partial x_i = 0 \tag{E.7.5}$$

$$mx''_2 + m\omega^2 x_2 = 0 \tag{E.7.6}$$
$$mx''_1 + m\omega^2 x_1 = 0 \tag{E.7.7}$$

E.7.1 Feynman Path Integral formulation

The propagator $K(x - y, t)$ for the Feynman path integral formulation has the form:

$$K(x - y, T) = A \lim_{n \to \infty} \int_{-\infty}^{+\infty}\!\!\!\!\!\int\!\!\int\!\!\int \dots \int dx_0 dx_1 \dots dx_n \exp[i/\hbar \int_t^{t+T} L(x, v, t_a) \, dt_a] \tag{E.7.8}$$

where A is a constant, and the integral over the dx's ranges from $-\infty$ to ∞.

E.7.1.1 Conventional Formulation – Free Particle Case

In the one dimensional free particle case the path integral is a product of n infinitesimal paths of time interval ε:

$$K(x - y, T) = \prod_n G_\varepsilon \tag{E.7.9}$$

where $T = n\delta$. Using $\sim$ to denote proportionality up to a constant we find the fourier transform of an interval of path

$$G_\delta = \int dx \, e^{-ipx} \exp[-ix^2/(2\varepsilon)] \tag{E.7.10}$$
$$\sim \exp[-p^2/(2\varepsilon)]$$

Then the product of the incremental factors that total to time T give

$$K(p, T) \sim \exp[-iTp^2/2] \tag{E.7.11}$$

A fourier transformation yields the free particle propagator

$$K(x - y, T) \sim \int dp \ e^{-ip(x-y)} \exp[-iTp^2/2]$$
$$\sim \exp[-i(x-y)^2/T] \tag{E.7.12}$$

where we normalize the propagator to unity

$$\int dy \ K(x - y, T) = 1 \tag{E.7.13}$$

E.7.1.2 PeseudoQuantum Formulation – Free Particle Case

We will now develop the PseudoQuantum path integral formalism for the case of a free particle. The form of the path integral now is

$$K(x - y, T) =$$

$$= A \lim_{n \to \infty} \int\int\int\int_{-\infty}^{+\infty} \dots \int dx_{10}dx_{11} \dots dx_{1n} \ dx_{20}dx_{21} \dots dx_{2n} \exp[i/\hbar \int_t^{t+T} L(x_1, x_2, x'_1, x'_2, v, t_a) \ dt_a] \tag{E.7.14}$$

where A is a constant, and all integrals over the dx's ranges from $-\infty$ to ∞. Note that we use two sets of coordinates and momenta.

Following a similar path to subsection E.7.1.1 we first we determine the fourier transform on the path integral for an infinitesimal time interval ε:

$$G_\varepsilon = \int\int dx_1 dx_2 \exp[-ip_1x_1 - ip_2x_2] \exp[-imx_1x_2/\varepsilon]$$
$$\sim \exp[-i\varepsilon p_1p_2/m] \tag{E.7.15}$$

Upon combining the intervals to a total time T we obtain the fourier transform of the total path integral

$$K(p, T) \sim \exp[-iTp_1p_2/m] \qquad\qquad (E.7.16)$$

which yields the spatial path integral

$$K(x_1 - y_1, x_2 - y_2, T) \sim \int d\, p_1 dp_2 \, \exp[ip_1(x_1 - y_1) + ip_2(x_2 - y_2)] \, \exp[-iTp_1p_2/m]$$

$$\sim \exp[-im(x_1 - y_1)(x_2 - y_2)/T] \qquad\qquad (E.7.17)$$

which we normalize to unity

$$\int dy_1 dy_2 \, K(x_1 - y_1, x_2 - y_2, T) = 1 \qquad\qquad (E.7.18)$$

with the result

$$K(x_1 - y_1, x_2 - y_2, T) = (m/T)\exp[-im(x_1 - y_1)(x_2 - y_2)/T] \qquad (E.7.19)$$

If we now use the path integral on a free particle wave function we see that it displaces the wave function by the time T:

$$\Psi_0(x_1, x_2, t) = \exp[-ip_1x_1 - ip_2x_2 - iEt] \qquad \text{where } E = p_1p_2/m \qquad (E.7.20)$$

$$\Psi(y_1, y_2, t + T) = \iint dx_1 dx_2 \, K(x_1 - y_1, x_2 - y_2, T)\Psi_0(x_1, x_2, t) \qquad (E.7.21)$$
$$= \exp[-ip_1y_1 - ip_2y_2 - iE(t + T)]$$
$$= \Psi_0(y_1, y_2, t + T) \qquad\qquad (E.7.22)$$

E.7.1.3 Introducing the Rotation Between the Quantum and Classical Cases of the Path Integral

We begin by expressing the coordinates in terms of new 'rotated' coordinates:

$$x_1 = u_1 \cos \theta + u_2 \sin \theta \qquad\qquad (E.7.23)$$
$$x_2 = -u_1 \sin \theta + u_2 \cos \theta$$
$$p_1 = -p_{u1} \sin \theta + p_{u2} \cos \theta$$
$$p_2 = p_{u1} \cos \theta + p_{u2} \sin \theta$$

Then the quantities of interest are now expressed as

$$\Psi_0(x_1, x_2, t) = \exp[-ip_1x_1 - ip_2x_2 - iEt] \qquad \text{where} \quad E = p_1p_2/m \qquad (E.7.24)$$

$$= \Psi_0(u_1, u_2, t) = \exp\{-i[(p_{u2}u_2 - p_{u1}u_1)\sin(2\theta) + (p_{u2}u_1 + p_{u1}u_2)\cos(2\theta)] - iEt\} \qquad (E.7.25)$$

where the energy E now having the form

$$E = (-p_{u1}\sin\theta + p_{u2}\cos\theta)(p_{u1}\cos\theta + p_{u2}\sin\theta)/m$$
$$= [(p_{u2}^2 - p_{u1}^2)\sin(2\theta)]/(2m) + [p_{u2}p_{u1}\cos(2\theta)]/m \qquad (E.7.26)$$

We will now examine the two special cases: $\theta = 0$ corresponding to classical mechanics and $\theta = \pi/4$ corresponding to quantum mechanics.

$\underline{\theta = 0}$

The wave equation in this case is

$$\Psi_0(u_1, u_2, t) = \exp\{-i(p_{u2}u_1 + p_{u1}u_2) - iEt\} \qquad (E.7.27)$$

with

$$E = p_{u2}p_{u1}/m$$

The wave function has a 'classical' form as we showed in eq. 47 in appendix C.

$\underline{\theta = \pi/4}$

The wave equation in this case is

$$\Psi_0(u_1, u_2, t) = \exp\{-i(p_{u2}u_2 - p_{u1}u_1) - iEt\} \qquad (E.7.28)$$

with

$$E = (p_{u2}^2 - p_{u1}^2)/(2m)$$

This wave function has a quantum form with a positive energy part and a negative energy part:

E.7.1.4 Free Path Integral with Rotation Between Classical and Quantum Mechanics

The incremental path integral factor, expressed in terms of u_1 and u_2, and then fourier transformed is:

$$G_\varepsilon (p_{u1}, p_{u2}) = \iint du_1 du_2 \exp\{-i[(p_{u2}u_2 - p_{u1}u_1)\sin(2\theta) + (p_{u2}u_1 + p_{u1}u_2)\cos(2\theta)]\} \cdot$$
$$\cdot \exp\{-im[(u_2{}^2 - u_1{}^2)\sin(2\theta)/2 + u_2u_1\cos(2\theta)]\}/\varepsilon\}$$
$$(E.7.29)$$
$$\sim \exp\{-i\varepsilon[(p_{u2}{}^2 - p_{u1}{}^2)\sin(2\theta)/2 + p_{u1}p_{u2}\cos(2\theta)]/m\}$$

yielding the cumulative product for the time interval T

$$K(p_{u1}, p_{u2}, T) \sim \exp[-iT[(p_{u2}{}^2 - p_{u1}{}^2)\sin(2\theta)/2 + p_{u1}p_{u2}\cos(2\theta)]/m] \qquad (E.7.30)$$

Upon fourier transforming to coordinate space we find

$$K(u_1 - v_1, u_2 - v_2, T) \sim \int dp_{u1}dp_{u2} \exp\{i[(p_{u2}w_2 - p_{u1}w_1)\sin(2\theta) + (p_{u2}w_1 + p_{u1}w_2)\cos(2\theta)]\}K(p_{u1}, p_{u2}, T)$$

$$\sim \exp[im[(w_2{}^2 - w_1{}^2)\sin(2\theta)/2 + w_2w_1\cos(2\theta)]/T] \qquad (E.7.31)$$

where
$$w_i = u_i - v_i$$

The special cases of interest are:

$\theta = 0$
$$K(u_1 - v_1, u_2 - v_2, T) \sim \exp[imw_2w_1]/T] \qquad (E.7.32)$$

This gives the *Classical* path integral without use of any limiting or approximation procedure such as one often finds in the literature.

$\theta = \pi/4$

This case yields the familiar free particle path integral – but with a part for a positive energy particle and a part for a negative energy particle. The negative energy part can be removed easily yielding the conventional free particle quantum path integral.

$$K(u_1 - v_1, u_2 - v_2, T) \sim \exp[im(w_2{}^2 - w_1{}^2)/(2T)] \tag{E.7.33}$$

E.7.2 General PseudoQuantum Formulation

The propagator $K(x - y, t)$ for the conventional Feynman path integral formulation has the form:

$$K(x - y, T) = A \lim_{n \to \infty} \int\!\!\int\!\!\int\limits_{-\infty}^{+\infty} \dots \int dx_0 dx_1 \dots dx_n \exp[i/\hbar \int\limits_{t}^{t+T} L(x, v, t_a)\, dt_a] \tag{E.7.34}$$

where A is a constant, and the integral over the dx's ranges from $-\infty$ to ∞.

The PseudoQuantum form of the path integral formalism is based on the PseudoQuantum lagrangian

$$L(x_1, v_1, x_2, v_2) = \partial^2 H(x_1, p_1)/\partial p_1{}^2\, p_1 p_2 + \partial^2 H(x_1, p_1)/\partial x_1 \partial p_1\, p_1 x_2 - \partial H(x_1, p_1)/\partial x_1\, x_2 \tag{E.7.35}$$

$$K(x_1 - y_1, x_2 - y_2, T) = A \lim_{n \to \infty} \int\!\!\int\!\!\int\limits_{-\infty}^{+\infty} \dots \int dx_{10} dx_{11} \dots dx_{1n}\, dx_{20} dx_{21} \dots dx_{2n} \exp[i/\hbar \int\limits_{t}^{t+T} L(x_1, v_1, x_2, v_2)\, dt_a]$$
$$\tag{E.7.36}$$

From subsection E.7.1.2, where

$$G_\varepsilon = \int\!\!\int dx_1 dx_2 \exp[-ip_1 x_1 - ip_2 x_2]\, \exp[-imx_1 x_2/\varepsilon]$$
$$\sim \exp[-i\varepsilon p_1 p_2/m] \tag{E.7.37}$$

we can perform the x_2 integration in G_ε using

$$x'_2 \cong x_2(t + \varepsilon) - x_2(t)\, /\varepsilon \tag{E.7.38}$$

if

$$p_2 = mx'_2$$

Then

$$G_\varepsilon = \iint dx_1 dx_2 \exp[-ip_1x_1 - ip_2x_2] \exp\{-i\varepsilon[\partial^2 H(x_1, p_1)/\partial p_1{}^2 \, p_1 mx_2/\varepsilon + \partial^2 H(x_1, p_1)/\partial x_1 \partial p_1 \, p_1 x_2 - \partial H(x_1, p_1)/\partial x_1 \, x_2]]$$

$$= \int dx_1 \exp[-ip_1x_1] \, \delta\big(p_2 - (m\partial^2 H(x_1, p_1)/\partial p_1{}^2 \, p_1 + \varepsilon\partial^2 H(x_1, p_1)/\partial x_1 \partial p_1 \, p_1 - \varepsilon\partial H(x_1, p_1)/\partial x_1)\big)$$

$$= \int dx_1 \exp[-ip_1x_1] \, \delta\big(p_2 - (m\partial^2 H(x_1, p_1)/\partial p_1{}^2 \, p_1 + \varepsilon\partial^2 H(x_1, p_1)/\partial x_1 \partial p_1 \, p_1 - \varepsilon\partial H(x_1, p_1)/\partial x_1)\big) \qquad \text{(E.7.39)}$$

E.7.2.1 PseudoQuantum Wave Functions and Schrödinger equation

In the presence of a potential, the path integral formulation leads us to transform it into a Schrödinger equation. The incremental time displacement form of the wave equation is

$$\Psi(y_{1k+1}, y_{2k+1}, t + \varepsilon) = \iint dx_{1k} dx_{2k} \exp\{(i\varepsilon/\hbar)[m(x_{1k+1} - x_{1k})(x_{2k+1} - y_{2k})/\varepsilon^2 -$$
$$- x_{2k+1}(\partial H(x_1, p_1)/\partial x_1)|_{x_1=x_{1k+1}}]\}\Psi_0(x_{1k}, x_{2k}, t) \qquad \text{(E.7.40)}$$

$$= \iint dx_{1k} dx_{2k} \exp\{(i\varepsilon/\hbar)[m(x_{1k+1} - x_{1k})(x_{2k+1} - y_{2k})/\varepsilon^2 -$$
$$- x_{2k+1}(\partial V(x)/\partial x)|_{x=x_{1k+1}}]\}\Psi_0(x_{1k}, x_{2k}, t)$$

for a potential $V(x)$. In the case of the harmonic oscillator the PseudoQuantum potential term is

$$x_{2k+1}\partial V(x)/\partial x|_{x=x_{1k+1}} = m\omega^2 x_{1k+1}x_{2k+1} \qquad \text{(E.7.41)}$$

The PseudoQuantum Schrödinger equation that results is

$$i\hbar\partial\Psi(x_1, x_2, t)/\partial t = (-\hbar^2/m)\partial^2\Psi(x_1, x_2, t)/\partial x_1\partial x_2 + V(x_1, x_2)\Psi(x_1, x_2, t) \qquad \text{(E.7.42)}$$

E.7.2.2 Decomposition of PseudoQuantum Schrödinger Equation into Quantum and Classical Parts

The Schrödinger equation can be decomposed into a quantum and a classical part. Starting from eq. E.7.42:

$$i\hbar\partial\Psi(x_1, x_2, t)/\partial t = (-\hbar^2/m)\partial^2\Psi(x_1, x_2, t)/\partial x_1\partial x_2 + V(x_1, x_2)\Psi(x_1, x_2, t) \qquad \text{(E.7.43)}$$

we find

$$i\hbar\partial\Psi(u_1, u_2, t, \theta)/\partial t = (-\hbar^2/m)[\sin(2\theta)\,(\partial^2/\partial u_2{}^2 - \partial^2/\partial u_1{}^2)\,/2 + \cos(2\theta)\partial^2/\partial u_1\partial u_2]\cdot$$
$$\cdot\Psi(u_1, u_2, t, \theta) + V(u_1\cos\theta + u_2\sin\theta, -u_1\sin\theta + u_2\cos\theta)\Psi(u_1, u_2, t, \theta)$$
$$(\text{E.7.44})$$

using the relation to u_1 and u_2:

$$x_1 = u_1\cos\theta + u_2\sin\theta$$
$$x_2 = -u_1\sin\theta + u_2\cos\theta$$

Then

$$\Psi(u_1, u_2, t, \theta) = \exp\{-i[(p_{u2}u_2 - p_{u1}u_1)\sin(2\theta) + (p_{u2}u_1 + p_{u1}u_2)\cos(2\theta)] - iEt\} \quad (\text{E.7.45})$$

and the energy is

$$E = [(p_{u2}{}^2 - p_{u1}{}^2)\sin(2\theta)]/(2m) + [p_{u2}p_{u1}\cos(2\theta)]/m \quad (\text{E.7.46})$$

Again there are two cases of interest:

The Quantum part $\theta = \pi/4$

$$i\hbar\partial\Psi(u_1, u_2, t, \theta = \pi/4)/\partial t = (-\hbar^2/m)[(\partial^2/\partial u_2{}^2 - \partial^2/\partial u_1{}^2)\,/2]\Psi(u_1, u_2, t, \theta = \pi/4) +$$
$$+ V((u_1 + u_2)/\sqrt{2}, (-u_1 + u_2)/\sqrt{2})\Psi(u_1, u_2, t, \theta = \pi/4)$$
$$(\text{E.7.47})$$

In the quantum free particle case

$$\Psi(u_1, u_2, t, \theta = \pi/4) = \exp\{-i(p_{u2}u_2 - p_{u1}u_1) - iEt\} \quad (\text{E.7.48})$$

where

$$E = (p_{u2}{}^2 - p_{u1}{}^2)/(2m)$$

The Classical part $\theta = 0$

$$i\hbar\partial\Psi(u_1, u_2, t, \theta = 0)/\partial t = (-\hbar^2/m)\partial^2/\partial u_1\partial u_2\,\Psi(u_1, u_2, t, \theta = 0) + V(u_1, u_2)\Psi(u_1, u_2, t, \theta = 0)$$

In the classical free particle case

$$\Psi(u_1, u_2, t, \theta = 0) = \exp\{-i(p_{u2}u_1 + p_{u1}u_2) - iEt\} \tag{E.7.49}$$

where

$$E = p_{u2}p_{u1}/m$$

E.7.3 Classical Part of Free Particle PseudoQuantum Feynman Path Propagator

The classical part of the free particle PseudoQuantum path integral is described as follows. First the time incremental factor is

$$G_\varepsilon(p_{u1}, p_{u2}) = \iint du_1 du_2 \exp\{-i[(p_{u2}u_2 - p_{u1}u_1)\sin(2\theta) + (p_{u2}u_1 + p_{u1}u_2)\cos(2\theta)]\} \cdot$$
$$\cdot \exp\{-im[(u_2^2 - u_1^2)\sin(2\theta)/2 + u_2u_1\cos(2\theta)]/\varepsilon\}$$

$$\sim \exp\{-i\varepsilon[(p_{u2}^2 - p_{u1}^2)\sin(2\theta)/2 + p_{u1}p_{u2}\cos(2\theta)]/m\} \tag{E.7.50}$$

The product of the incremental terms is

$$K(p_{u1}, p_{u2}, T) \sim \exp[-iT[(p_{u2}^2 - p_{u1}^2)\sin(2\theta)/2 + p_{u1}p_{u2}\cos(2\theta)]/m] \tag{E.7.51}$$

which yields the *classical* path integral

$$K(u_1 - v_1, u_2 - v_2, T) \sim \int dp_{u1} dp_{u2} \exp\{i[(p_{u2}w_2 - p_{u1}w_1)\sin(2\theta) + (p_{u2}w_1 + p_{u1}w_2)\cos(2\theta)]\}K(p_{u1}, p_{u2}, T)$$
$$\sim \exp[im[(w_2^2 - w_1^2)\sin(2\theta)/2 + w_2w_1\cos(2\theta)]/T] \tag{E.7.52}$$

where

$$w_i = u_i - v_i$$

For $\theta = 0$ the classical path integral steps are:

$$G_\varepsilon(p_{u1}, p_{u2}) \sim \exp\{-i\varepsilon p_{u1}p_{u2}/m\} \tag{E.7.53}$$

$$K(p_{u1}, p_{u2}, T) \sim \exp[-iTp_{u1}p_{u2}/m] \tag{E.7.54}$$

$$K(u_1 - v_1, u_2 - v_2, T) \sim \exp[im(u_1 - v_1)(u_2 - v_2)/T] \qquad (E.7.55)$$

This gives us a classical path integral formulation that avoids approximation techniques which have hitherto been used. The above development can be rewritten in terms of the x and p variables.

E.7.4 Fokker-Planck Equation

The Feynman path integral formulation and equations can be transformed into similar forms by letting $i\hbar$ be changed to a positive constant. The Fokker-Planck equation gives the probability density of a particle velocity's time evolution under the impact of forces. This equation is also known as the Smoluchowski equation—named after its originator.

In its formulation a variable x_2 is introduced as the 'response' variable to the 'primary' variable x_1. The Fokker-Planck equation for the probability density is

$$p(x_1', t + \varepsilon) = (1/2\pi i) \int_{-\infty}^{\infty} dx_1 \int_{-i\infty}^{i\infty} dx_2 \exp\{\varepsilon[-x_2(x_1' - x_1)/\varepsilon + x_2 D_1(x_1, t) + x_2{}^2 D_2(x_1, t)]\} p(x_1, t)$$

$$(E.7.56)$$

The origin of x_2 in our formalism, and in the Fokker-Planck equation, are very different. The Fokker-Planck equation has the lagrangian:

$$L = \int dt \, [x_2 D_1(x_1, t) + x_2{}^2 D_2(x_1, t) - x_2 \, \partial x_1/\partial t]$$

A comparison of this formulation with the PseudoQuantum path integral formulation shows an apparently remarkable similarity. Thus one might view the Fokker-Planck equation as a precursor of the PseudoQuantum formulation. The papers appearing in the appendices of this book show that the origin of our formalism and the Fokker-Planck formalism are very different.

E.8 The Transition Between Classical and Quantum Chaos

Chaos has become an increasingly important field of activity. While classical chaos has been the more studied aspect of chaos there has been an increasingly larger interest in quantum chaos. In this section we wish to show that the PseudoQuantum formalism appears to be a useful means of relating classical chaos to quantum chaos for many systems. It makes it possible to trace the transition from a classical chaos situation to quantum chaos. It also offers the possibility to determine the quantum analogue of a classical chaos phenomenon.

Given a quantum system it is often difficult to determine whether it has a chaotic regime. Frequently extensive numerical analysis is needed for this determination.

In this section we show that the PseudoQuantum formalism enables us to determine a classical Hamiltonian from a quantum formalism where chaos is known to occur based on extensive numerical investigations.

A well-studied[579,580] Hamiltonian for a quantum theory, known to have chaotic regions, is

$$H = (p_x^2 + p_y^2/2 + x^2 y^2 + \beta(x^4 + y^4)/4 \qquad (E.8.1)$$

Creating the equivalent PseudoQuantum Hamiltonian we obtain

$$\hat{H} = p_{x1}p_{x2} + p_{y1}p_{y2} + y_1^2 x_1 x_2 + x_1^2 y_1 y_2 + \beta(x_1^3 x_2 + y_1^3 y_2) \qquad (E.8.2)$$

Introducing new variables we can develop a form of eq. E.8.2 that allows us to trace the transition from a quantum theory to a classical theory, which also should have chaotic regimes.

$$x_1 = u_{x1} \cos \theta + u_{x2} \sin \theta \qquad (E.8.3)$$
$$x_2 = -u_{x1} \sin \theta + u_{x2} \cos \theta$$

$$p_{x1} = p_{ux1} \cos \theta + p_{ux2} \sin \theta$$
$$p_{x2} = -p_{ux1} \sin \theta + p_{ux2} \cos \theta$$

---

[579] The Hamiltonian model above has been studied by: Y. Y. Bai, G. Hose, K. Stefański, and H. S. Taylor, Phys. Rev. **A31**, 2821 (1985), R. L. Waterland, J.-M. Yuan, C. C. Martens, R. E. Gillilan, and W. P. Reinhardt, Phys. Rev. Lett. **61**, 2733 (1988, and other papers..

[580] Another much studied model—the 2-Dimensional stadium, Quantum billiard ball model has classical chaotic dynamics, and a quantum approximation that is chaotic: S. W. McDonald and A. N. Kaufman, Phys. Rev. **A37**, 3067 (1988); _____, Phys. Rev. Lett., **42**, 1189 (1979), and other papers.

$$y_1 = u_{y1} \cos\theta + u_{y2} \sin\theta$$
$$y_2 = -u_{y1} \sin\theta + u_{y2} \cos\theta$$

$$p_{y1} = p_{uy1} \cos\theta + p_{uy2} \sin\theta$$
$$p_{y2} = -p_{uy1} \sin\theta + p_{uy2} \cos\theta$$

Then we obtain the PseudoQuantum Hamiltonian

$$\hat{H}(\theta) = p_{x1}p_{x2} + p_{y1}p_{y2} + y_1^2 x_1 x_2 + x_1^2 y_1 y_2 + \beta(x_1^3 x_2 + y_1^3 y_2)$$

$$
\begin{aligned}
= & (p_{ux2}^2 - p_{ux1}^2 + p_{uy2}^2 - p_{uy1}^2)\sin(2\theta)/2 + (p_{uy1}p_{uy2} + p_{ux1}p_{ux2})\cos(2\theta) + \\
& + (u_{y1}\cos\theta + u_{y2}\sin\theta)^2[(u_{x2}^2 - u_{x1}^2)\sin(2\theta)/2 + u_{x1}u_{x2}\cos(2\theta)] + \\
& + (u_{x1}\cos\theta + u_{x2}\sin\theta)^2[(u_{y2}^2 - u_{y1}^2)\sin(2\theta)/2 + u_{y1}u_{y2}\cos(2\theta)] + \\
& + \beta\{(u_{x1}\cos\theta + u_{x2}\sin\theta)^2[(u_{x2}^2 - u_{x1}^2)\sin(2\theta)/2 + u_{x1}u_{x2}\cos(2\theta)] + \\
& + (u_{y1}\cos\theta + u_{y2}\sin\theta)^2[(u_{y2}^2 - u_{y1}^2)\sin(2\theta)/2 + u_{y1}u_{y2}\cos(2\theta)]\}
\end{aligned}
$$

(E.8.4)

The classical Hamiltonian that results from this analysis is

$$\hat{H}(0) = (p_{uy1}p_{uy2} + p_{ux1}p_{ux2}) + u_{y1}^2 u_{x1}u_{x2} + u_{x1}^2 u_{y1}u_{y2} + \beta\{u_{x1}^2 u_{x1}u_{x2} + u_{y1}^2 u_{y1}u_{y2}\}$$

(E.8.5)

The Quantum Hamiltonian that emerges is

$$
\begin{aligned}
\hat{H}(\pi/4) = & (p_{ux2}^2 - p_{ux1}^2 + p_{uy2}^2 - p_{uy1}^2)/2 + u_{y2}^2(u_{x2}^2 - u_{x1}^2)/2 + u_{x2}^2(u_{y2}^2 - u_{y1}^2)/2 + \\
& + \beta\{u_{x2}^2(u_{x2}^2 - u_{x1}^2)/2 + u_{y2}^2(u_{y2}^2 - u_{y1}^2)/2\}
\end{aligned}
$$

(E.8.6)

While these Hamiltonians require numerical analysis to understand their chaotic features, they offer the possibilities of comparative studies of quantum and classical chaos.

The study of other models of a similar character would appear to be of importance in elucidating quantum chaos.

E.9 The Transition Between Classical & Quantum Entanglement Dynamics

Quantum Entanglement has become of great importance judging from its increasing number of papers. It offers the possibility of new forms of communication that might be of great value in interstellar communication should Mankind reach the stars.

In this section we will study a prototype example of quantum entanglement with a view towards investigating the transition from quantum entanglement through 'semi-classical' quantum entanglement to a 'classical' limit.

The example that we consider will use a combination of a positive energy single particle state entangled with a negative energy particle state to simulate the more commonly studied case of entangled spins. The particles, in a superposed state, are assumed to separate, and one particle will be measured thus determining the state of the other particle due to entanglement. We define the 'entangled' NOON-type state:

$$\Psi = (|n+ = 1, n- = 0> + | n+ = 0, n- = 1>)/\sqrt{2} \qquad (E.9.1)$$

where one pure state $|n+ = 1, n- = 0>$ is a one particle state of positive energy and the other state $| n+ = 0, n- = 1>$ has a one negative energy particle.

We define projection operators of the form:[581]

$$\rho(\theta) = ||n+(\theta),n-(\theta)><n+(\theta),n-(\theta)| \qquad (E.9.2)$$

using an angle θ as we have done previously to specify the quantum-classical content of the projection.

More generally we will define harmonic oscillator states:

$$| n+, n-> = b_1^{\dagger n+} b_2^{\dagger n-} |0,0> \qquad (E.9.3)$$

with a density operator

---

[581] We will be using the harmonic oscillator formalism of E.4 although the conclusions will be more far reaching.

$$\rho(\theta) = \sum_{n+,n-} |n+, n-><n+, n-| = \sum_{n+,n-} b_1^{\dagger}(\theta)^{n+}b_2^{\dagger}(\theta)^{n-}|0,0><0,0|b_1(\theta)^{n+}b_2(\theta)^{n-} \qquad (E.9.4)$$

We will consider the particular projection:

$$P(\theta) = |1(\theta),0><1(\theta),0| \qquad (E.9.5)$$

which we will apply to Ψ:

$$P(\theta)\Psi = |1(\theta), 0>/\sqrt{2} = \sin(2\theta)b_1^{\dagger}(\theta)|0, 0>/\sqrt{2} = \sin(2\theta)(Q_1\cos\theta - iP_2\sin\theta)|0, 0>/\sqrt{2}$$
$$(E.9.6)$$

Using the familiar relations:

$$b_1 = Q_1\cos\theta + iP_2\sin\theta \qquad (E.4.6)$$
$$b_2 = -Q_2\sin\theta + iP_1\cos\theta$$
$$b_1^{\dagger} = Q_1\cos\theta - iP_2\sin\theta \qquad (E.4.7)$$
$$b_2^{\dagger} = -Q_2\sin\theta - iP_1\cos\theta$$

with commutation relations

$$[b_1, b_1^{\dagger}] = \sin(2\theta) \qquad (E.4.8a)$$
$$[b_2, b_2^{\dagger}] = -\sin(2\theta)$$
$$[b_1, b_2^{\dagger}] = [b_2, b_1^{\dagger}] = 0$$
$$[b_1, b_2] = [b_1^{\dagger}, b_2^{\dagger}] = 0$$

We find

$$\Psi' = P(\theta)\Psi \qquad (E.9.7)$$

has the following forms for $\theta = \pi/4$ and $\theta=0$:

<u>$\theta = \pi/4$</u>

$$\Psi' = (Q_1 - iP_2)|0, 0>/2 \qquad (E.9.8)$$

gives a quantum state.

θ = 0

$$\Psi' = 0 \qquad\qquad (E.9.9)$$

We note that we showed in section E.4 that 'classical' states containing particles have infinite energy—This above result, $\Psi' = 0$, reflects the infinite energy of states containing one or more particles. The value of $\Psi' = 0$ leaves the positivity or negativity of the other particle undetermined since entanglement is a purely quantum phenomena.

Other values of θ

$$\Psi' = \sin(2\theta)(Q_1\cos\theta - iP_2\sin\theta)|0, 0\rangle/\sqrt{2} \qquad (E.9.10)$$
$$= \sin(2\theta)b_1^\dagger(\theta)|0, 0\rangle/\sqrt{2} = \sin(2\theta)|1(\theta),0\rangle/\sqrt{2}$$

An intermediate result occurs but the other entangled separated particle state's energy is determined to be negative.

E.10 PseudoQuantum Transition Between Quantum and [582]Classical Dynamics

The preceding sections have shown that the use of the PseudoQuantum framework, which contains a purely quantum sector (albeit with both positive and negative energy parts that are separable), a purely classical sector, and an intermediate sector that is partly quantum and partly classical, enables us to

1. Relate the corresponding quantum and classical dynamics of a physical phenomenon.

2. Study the transition from quantum to classical behavior without recourse to approximations or limits such as $\hbar \rightarrow 0$.

3. Determine the classical equivalent of a quantum dynamical system.

---

[582] Gutzwiller (1990) points out the use of harmonic oscillator wave functions in several studies of the quantum-classical connection.

4. Determine the quantum equivalent of a classical dynamical system.

These advantages appear to be fairly general in nature—as evidenced by the harmonic oscillator case studied in section E.4—since the harmonic oscillator plays such a prominent role in many physical situations.

We conclude that the PseudoQuantum formalism, which is not only relevant for quantum-classical mechanics dynamics, but is also of importance in Quantum Field Theories as shown in our earlier sections and in our papers in the appendices. It has also recently been used in a new GraviStrong unified theory that relates quark confinement to deviations from Newtonian gravitation at galactic distance scales using a canonical PseudoQuantum formulation of a higher derivative Quantum Field Theory. (See Blaha (2016e).)

Appendix F. Complex Lorentz Group Details

F.1 Transformations Between Coordinate Systems

The measurement of time and space is simple in practice but raises weighty questions when their underlying basis is examined. We shall begin by measuring spatial distances with a ruler, and by measuring time with a clock. Earlier we determined that four dimensions: one space dimension and three space dimensions were required. We now define rectangular coordinate systems with x, y, and z axes as pictured in Fig. F.1 below. We then postulate:

Postulate F.1. Any observer can define a set of time and space coordinates called a coordinate system in which the observer is at rest. One can define a transformation that relates the coordinate systems of two observers traveling at a constant velocity with respect to each other.

One can always relate the coordinates of two coordinate systems by having an observer in each coordinate system specify the coordinates of objects located at each spatial point, and then creating a map between the coordinates of corresponding spatial locations.

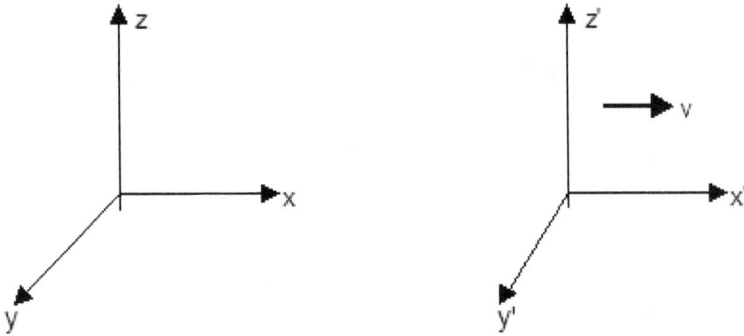

Figure F.1. Depiction of two coordinate systems. The "primed" coordinate system is moving with velocity **v** in the positive x direction with respect to the "unprimed" coordinate system. We choose parallel axes for convenience.

If space is flat the relation between the respective coordinates is linear. (One could reverse the logic of that statement by defining a flat space to be one in which the coordinates of a point in any coordinate system are linearly related to the coordinates of any other coordinate system moving at a constant velocity with respect to it.) Thus we can express the relation between the coordinates in the "unprimed" system to the coordinates in the "primed" system as a transformation between coordinate systems:

$$\mathbf{a'} = A\mathbf{a} + \mathbf{B}t + \mathbf{C}$$
$$t' = Dt + \mathbf{E} \cdot \mathbf{a}$$

(F.1)

where A is a 3×3 matrix, **B**, **C** and **E** are 3-vectors, and D is a number (scalar value).

Having restricted the set of transformations between coordinate systems to the form of eq. F.1 we now assert postulates that restrict the form of the transformation to Lorentz transformations and transformations similar to Lorentz transformations.

Postulate F.2. The speed of light, c, is the same in all coordinate systems.

Postulate F.3. The invariant interval or distance $d\tau$ is defined by

$$d\tau^2 = g_{\mu\nu}dx^\mu dx^\nu \qquad (F.2)$$

It is invariant under a change of coordinate systems. The 16 quantities $g_{\mu\nu}$ are known as the metric tensor.[583] *The four quantities dx^μ are infinitesimal displacements in space and time.*

If we expand eq. F.2 in rectangular coordinates it is equivalent to

$$d\tau^2 = g_{00}dx^0dx^0 + g_{11}dx^1dx^1 + g_{22}dx^2dx^2 + g_{33}dx^3dx^3 \qquad (F.3)$$

which equals

$$d\tau^2 = c^2dt^2 - dx^2 - dy^2 - dz^2 \qquad (F.4)$$

using the familiar form of the time and rectangular space coordinates.

F.2 The Lorentz Group

The metric tensor $g_{\mu\nu}$ for rectangular coordinates has the matrix form $G = \text{diag}(1, -1, -1, -1)$:

$$G = \begin{bmatrix} 1 & 0 & 0 & 0 \\ 0 & -1 & 0 & 0 \\ 0 & 0 & -1 & 0 \\ 0 & 0 & 0 & -1 \end{bmatrix} \qquad (F.5)$$

---

[583] The repeated indices indicate a summation. In this case from 0 to 3 as shown in eq. 4-A.3

The invariant interval under a transformation between rectangular coordinate systems (with "primed" and "unprimed" coordinates) has the form of eq. F.4 for the unprimed coordinates and the same form for the primed coordinates:

$$d\tau^2 = c^2 dt'^2 - dx'^2 - dy'^2 - dz'^2 \tag{F.6}$$

In matrix form we can define an "unprimed" coordinate column vector with

$$a = \begin{bmatrix} t \\ x \\ y \\ z \end{bmatrix} \tag{F.7a}$$

and its corresponding "primed" coordinate with

$$a' = \begin{bmatrix} t' \\ x' \\ y' \\ z' \end{bmatrix} \tag{F.7b}$$

If a and a' are the coordinates of the same point in the respective coordinate systems then, by postulates F.2 and F.3, they are related by a boost Lorentz transformation $\Lambda(\mathbf{v})$ with the form

$$a' = \Lambda(\mathbf{v})a \tag{F.8}$$

(and possibly a spatial rotation matrix factor), where $\mathbf{v}$ is the relative velocity of the coordinate systems. The form of the transformation eq. F.8, which is called a Lorentz *boost*, is constrained by postulates F.2 and F.3 to be[584]

---

[584] We shall consider only the proper, orthochronous Lorentz group at this point. We assume that the primed and unprimed coordinate systems have parallel axes. So there is no rotation of axes embodied in eq. 15.9.

$$\Lambda(\mathbf{v}) = \begin{bmatrix} \gamma & -\gamma v_x & -\gamma v_y & -\gamma v_z \\ -\gamma v_x & 1 + (\gamma - 1)v_x^2/v^2 & (\gamma - 1)v_x v_y/v^2 & (\gamma - 1)v_x v_z/ \\ -\gamma v_y & (\gamma - 1)v_x v_y/v^2 & 1 + (\gamma - 1)v_y^2/v^2 & (\gamma - 1)v_y v_z/v^2 \\ -\gamma v_z & (\gamma - 1)v_x v_z/v^2 & (\gamma - 1)v_y v_z/v^2 & 1 + (\gamma - 1)v_z^2/v^2 \end{bmatrix} \qquad \text{(F.9)}$$

where $\gamma = (1 - v^2)^{-\frac{1}{2}}$, $\mathbf{v} = (v_x, v_y, v_z)$, $v = |\mathbf{v}|$ and we set $c = 1$ for convenience.[585] The set of all matrices of the form of $\Lambda(\mathbf{v})$, or $\Lambda(\mathbf{v})\mathcal{R}(\boldsymbol{\theta})$ or $\mathcal{R}(\boldsymbol{\theta})\Lambda(\mathbf{v})$ where $\mathcal{R}(\boldsymbol{\theta})$ is a spatial rotation with angle vector $\boldsymbol{\theta}$, for $v < c$ form a matrix representation of the Lorentz group. Elements, $\Lambda(\mathbf{v}, \boldsymbol{\theta})$, of the Lorentz group satisfy the defining relation of the Lorentz group:

$$\Lambda(\mathbf{v}, \boldsymbol{\theta})^T G \Lambda(\mathbf{v}, \boldsymbol{\theta}) = G \qquad \text{(F.10)}$$

where the superscript ^{T} specifies the transpose of the matrix.

The Lorentz group, with which we are familiar, relates the coordinates of an event in two coordinate systems that differ by a spatial rotation, and a relative velocity whose magnitude is less than the speed of light. The inhomogeneous Lorentz group includes coordinate displacements.[586]

The group elements of the homogeneous Lorentz group can be expressed in terms of the generators $\mathbf{K}$ of boosts to coordinate systems moving at a constant velocity $\mathbf{v}$ and the generators $\mathbf{J}$ of purely spatial rotations by

$$\Lambda(\mathbf{v}, \boldsymbol{\theta}) = \exp[i\omega\hat{\mathbf{u}}\cdot\mathbf{K} + i\boldsymbol{\theta}\cdot\mathbf{J}] \qquad \text{(F.11)}$$

---

[585] One can set $c = 1$ by an appropriate choice of time and spatial distance scales. The demonstration that $\Lambda(\mathbf{v})$ has the form given by eq. 15.9 can be found in many textbooks.
[586] See Weinberg (1995) for a discussion of the inhomogeneous Lorentz group.

where the vector $\boldsymbol{\theta}$ is a 3-vector specifying the rotation angles, and where $\mathbf{v} = \hat{\mathbf{u}}\tanh\omega$, $\hat{\mathbf{u}}\cdot\hat{\mathbf{u}} = 1$.

The boost transformation $\Lambda(\mathbf{v}) = \Lambda(\mathbf{v}, \mathbf{0})$ has the form

$$\Lambda(\mathbf{v}) = \exp[i\omega\hat{\mathbf{u}}\cdot\mathbf{K}] \tag{F.12}$$

Its matrix form is eq. F.9. The matrix form can be expressed in terms of the unit normalized velocity vector $\mathbf{u} = (u_x, u_y, u_z)$ and ω as

$$\Lambda(\omega, \mathbf{u}) = \Lambda(\mathbf{v}) \tag{F.13}$$

$$= \begin{bmatrix} \cosh(\omega) & -\sinh(\omega)u_x & -\sinh(\omega)u_y & -\sinh(\omega)u_z \\ -\sinh(\omega)u_x & 1 + (\cosh(\omega)-1)u_x^2 & (\cosh(\omega)-1)u_xu_y & (\cosh(\omega)-1)u_xu_z \\ -\sinh(\omega)u_y & (\cosh(\omega)-1)u_xu_y & 1 + (\cosh(\omega)-1)u_y^2 & (\cosh(\omega)-1)u_yu_z \\ -\sinh(\omega)u_z & (\cosh(\omega)-1)u_xu_z & (\cosh(\omega)-1)u_yu_z & 1 + (\cosh(\omega)-1)u_z^2 \end{bmatrix}$$

where $\Lambda(\omega, \mathbf{u}) = \Lambda(\omega, \mathbf{u}, \boldsymbol{\theta} = \mathbf{0})$ in the previous notation. This definition of the general form of proper, orthochronous, Lorentz boost matrices $\Lambda(\omega, \mathbf{u})$ will be used in subsequent sections to define faster-than-light boost transformations.

The vector form of a Lorentz boost transformation is

$$\begin{aligned} \mathbf{x}' &= \mathbf{x} + (\gamma-1)\mathbf{x}\cdot\mathbf{v}\,\mathbf{v}/v^2 - \gamma\mathbf{v}t \\ t' &= \gamma(t - \mathbf{v}\cdot\mathbf{x}/c^2) \end{aligned} \tag{F.14}$$

where $\gamma = (1-\beta^2)^{-\frac{1}{2}}$ with $\beta = v/c = v$ (since we set $c = 1$).

F.3 The Nature of $\Lambda(\omega, \mathbf{u})$ for Complex ω

We now turn to the case of complex ω which includes superluminal (faster-than-light) Lorentz transformations as well as conventional Lorentz transformations. Since, for any complex value z

$$\cosh^2(z) - \sinh^2(z) = 1 \tag{F.15}$$

it follows that for any complex value of ω, $\Lambda(\omega, \mathbf{u})$ is a member of the Lorentz group, and/or of the complex Lorentz group[587] for complex ω:

$$\Lambda(\omega, \mathbf{u})^T G \Lambda(\omega, \mathbf{u}) = G \tag{F.16}$$

For certain values of the imaginary part of ω the matrix $\Lambda(\omega, \mathbf{u})$ has a particularly simple form, similar to that of $\Lambda(\omega, \mathbf{u})$ for real ω, but which generates boosts to relative velocities greater than the speed of light. Among these values are:

$$\omega = \omega_\pm = \omega \pm i\pi/2 \tag{F.17}$$

Later we will see that these alternate choices $\omega_\pm$ correspond to specific choices of parity.

F.4 Complex Lorentz Group

In the preceding section we saw that the parameter ω can be complex and the boost transformation will still satisfy the Lorentz condition eq. F.10. More generally we can consider complex homogeneous Lorentz transformations $\Lambda(\mathbf{v}, \boldsymbol{\theta})$ which can be represented by eq. F.11 with complex parameters ω, $\hat{\mathbf{u}}$, and $\boldsymbol{\theta}$ where $\hat{\mathbf{u}}$ and $\boldsymbol{\theta}$ are complex 3-vectors. $\boldsymbol{\theta}$ specifies a rotation angle.

In general $\Lambda(\mathbf{v}, \boldsymbol{\theta})$ is then a transformation between coordinate systems that have complex coordinates. One coordinate system is moving at a constant complex velocity

---

[587] The complex Lorentz group is defined as the group of all complex transformations that satisfy eq. 15.16.

with respect to the other. Coordinate systems do not necessarily have parallel spatial axes in general.

Within the complex Lorentz group, denoted L(C),[588] there are subsets of boosts that play important physical roles in the derivation of the form of The Standard Model. In particular we will see that certain classes of boosts generate faster-than-light transformations. These transformations can be further divided into subclasses of "left-handed" and "right-handed" transformations based on the quantum field theories to which they lead. Further within each subclass there are subclasses of transformations that naturally lead to Dirac-like free field equations that can be described as lepton-like and quark-like.

Thus these boosts are a key ingredient to understanding the form of The Standard Model.

F.5 Faster-than-Light Transformations

In this section we will substitute $\omega_\pm$ for ω in $\Lambda(\omega, \mathbf{u})$ and then show that we obtain two sets of possible transformations from sublight reference frames to faster-than-light reference frames. One set of transformations, where $\omega_L = \omega + i\pi/2$, will be called *left-handed superluminal boosts*. They eventually lead to the "left-handed" part of The Standard Model. We denote members of this set, $\Lambda_L(\omega, \mathbf{u})$, with the subscript "L" for left-handed.

The other set of boosts where $\omega_R = \omega - i\pi/2$ will be called *right-handed superluminal boosts*. They eventually lead to a right-handed, unphysical,[589] version of The Standard Model. We denote members of this set of boosts, $\Lambda_R(\omega, \mathbf{u})$, with the subscript "R" for right-handed.

Before considering faster-than-light boosts we note the relation between a real-valued ω in a *conventional* Lorentz boost $\Lambda(\omega, \mathbf{u})$, and the magnitude of the relative velocity v for v < 1, is

$$\mathbf{v} = \hat{\mathbf{u}} \tanh\omega \qquad \text{with} \qquad \hat{\mathbf{u}} \cdot \hat{\mathbf{u}} = 1$$

---

[588] Streater (2000) points out that the complex Lorentz group is essential to the proof of the CPT theorem.

[589] Currently the case. If a right-handed counterpart to the current Standard Model surfaces at higher energies then the features emerging from right-handed superluminal boosts then become physically important.

$$\cosh(\omega) = \gamma = (1 - v^2)^{-\frac{1}{2}} \qquad \text{(F.18)}$$
$$\sinh(\omega) = v\gamma = \beta\gamma$$

where $\beta = v = |\mathbf{v}|$.

F.6 Left-Handed Superluminal Transformations

Left-handed (proper orthochronous) superluminal boost transformations $\Lambda_L(\mathbf{v})$ have the same form as eq. F.9 for ordinary (proper orthochronous) Lorentz boost transformations. However the magnitude of the relative velocity $\mathbf{v}$ is greater than the speed of light. Thus $\gamma = (1 - v^2)^{-\frac{1}{2}}$ is pure imaginary and $\Lambda_L(\mathbf{v})$ is complex.

$$\Lambda_L(\mathbf{v}) = \begin{bmatrix} \gamma & -\gamma v_x & -\gamma v_y & -\gamma v_z \\ -\gamma v_x & 1 + (\gamma - 1)v_x^2/v^2 & (\gamma - 1)v_x v_y/v^2 & (\gamma - 1)v_x v_z/v^2 \\ -\gamma v_y & (\gamma - 1)v_x v_y/v^2 & 1 + (\gamma - 1)v_y^2/v^2 & (\gamma - 1)v_y v_z/v^2 \\ -\gamma v_z & (\gamma - 1)v_x v_z/v^2 & (\gamma - 1)v_y v_z/v^2 & 1 + (\gamma - 1)v_z^2/v^2 \end{bmatrix} \qquad \text{(F.19)}$$

This transformation raises several issues – the most prominent of which is the interpretation of the imaginary coordinates generated by the transformation. Imaginary coordinates would appear at first glance to be unphysical. However we view the measurement of these quantities operationally: an observer measures distances with "rulers", and time with clocks, which both give real numeric values. Thus an observer *in any coordinate system* will always measure real numbers for time and space distances. However an observer *in another coordinate system* that is related to the first coordinate system by a superluminal transformation will view the coordinates in the first system as complex as eq. F.19 indicates.

The reconciliation of these points of view requires the introduction of a new transformation, called a Reality group transformation, in addition to a superluminal Lorentz transformation for the case of faster than light transformations. Reality group transformations maps the complex coordinates generated by a Lorentz transformation to the real coordinates seen by the observer[590] in the "faster than light" reference frame. We describe the Reality group transformations in detail earlier. We show how they imply the Reality group is $SU(3) \otimes SU(2) \otimes U(1) \otimes SU(2) \otimes U(1)$.

F.6.1 Cosh-Sinh Representation of Left-Handed Superluminal Boosts

We will now develop the representation of left-handed superluminal boost transformations in terms of $\cosh(\omega)$ and $\sinh(\omega)$ for later use in our discussion of tachyons. We find that we must use a complex $\omega_L \equiv \omega + i\pi/2$ to properly describe left-handed superluminal boosts. The relation between ω_L and v is different from eq. F.18 for the case of left-handed superluminal boosts:

$$\cosh(\omega_L) = i \sinh(\omega) = -\gamma = i\gamma_s \tag{F.20}$$
$$\sinh(\omega_L) = i \cosh(\omega) = -\beta\gamma = i\beta\gamma_s$$

where $\beta = v > 1$, $\boldsymbol{\omega \geq 0}$, and
$$\gamma_s = (\beta^2 - 1)^{-\frac{1}{2}} \tag{F.21}$$

Eq. F.20 implies
$$\sinh(\omega) = \gamma_s \tag{F.22}$$
$$\cosh(\omega) = \beta\gamma_s$$

Upon substituting $\boldsymbol{\omega_L}$ for ω in eq. F.13 we obtain another form for a left-handed superluminal transformation (equivalent to that of eq. F.19):

---

[590] The linearity of the superluminal transformation makes this secondary transformation physically possible.

$\Lambda_L(\omega, \mathbf{u}) = \Lambda(\omega + i\pi/2, \mathbf{u})$

$$= \begin{bmatrix} \cosh(\omega_L) & -\sinh(\omega_L)u_x & -\sinh(\omega_L)u_y & -\sinh(\omega_L)u_z \\ -\sinh(\omega_L)u_x & 1+(\cosh(\omega_L)-1)u_x{}^2 & (\cosh(\omega_L)-1)u_xu_y & (\cosh(\omega_L)-1)u_xu \\ -\sinh(\omega_L)u_y & (\cosh(\omega_L)-1)u_xu_y & 1+(\cosh(\omega_L)-1)u_y{}^2 & (\cosh(\omega_L)-1)u_yu_z \\ -\sinh(\omega_L)u_z & (\cosh(\omega_L)-1)u_xu_z & (\cosh(\omega_L)-1)u_yu_z & 1+(\cosh(\omega_L)-1)u_z{}^2 \end{bmatrix}$$

$$= \begin{bmatrix} i\gamma_s & -i\beta\gamma_su_x & -i\beta\gamma_su_y & -i\beta\gamma_su_z \\ -i\beta\gamma_su_x & 1+(i\gamma_s-1)u_x{}^2 & (i\gamma_s-1)u_xu_y & (i\gamma_s-1)u_xu_z \\ -i\beta\gamma_su_y & (i\gamma_s-1)u_xu_y & 1+(i\gamma_s-1)u_y{}^2 & (i\gamma_s-1)u_yu_z \\ -i\beta\gamma_su_z & (i\gamma_s-1)u_xu_z & (i\gamma_s-1)u_yu_z & 1+(i\gamma_s-1)u_z{}^2 \end{bmatrix} = \Lambda_L(\mathbf{v})$$

$$(F.23)$$

A simple case that illustrates a left-handed superluminal boost is to assume the relative velocity is in the x direction. Then eq. F.23 becomes

$$\Lambda_L(\omega, \mathbf{u} = (1,0,0)) = \begin{bmatrix} i\gamma_s & -i\beta\gamma_s & 0 & 0 \\ -i\beta\gamma_s & i\gamma_s & 0 & 0 \\ 0 & 0 & 1 & 0 \\ 0 & 0 & 0 & 1 \end{bmatrix} \qquad (F.24)$$

implementing the coordinate transformation:

$$X' = \Lambda_L(\omega, \mathbf{u} = (1,0,0))X$$

or

$$\begin{aligned} t' &= i\gamma_s(t - \beta x) \\ x' &= i\gamma_s(x - \beta t) \\ y' &= y \\ z' &= z \end{aligned} \qquad (F.25)$$

The addition rule for the x-component of velocity can be computed for infinitesimal displacements in space and time:

$$v_x' = \Delta x' / \Delta t' = (\Delta x \, \gamma_s - \Delta t \, \beta \gamma_s)/(\Delta t \, \gamma_s - \Delta x \, \beta \gamma_s)$$
$$= (v_x - \beta)/(1 - \beta v_x) \tag{F.26}$$

in the limit $\Delta t \rightarrow 0$ where the x component of a particle's velocity in the unprimed frame is $v_x = \Delta x / \Delta t$. $\Delta t'$ is determined by

$$\Delta t' = i \Delta t \, \gamma_s (1 - \beta v_x) \tag{F.27}$$

Note the velocity of light is the same in the primed and unprimed reference frames. (If $v_x = 1$ then $v_x' = 1$.) *Thus left-handed superluminal transformations preserve the constancy of the speed of light in all reference frames.* (Postulate F.2)

Further note that increasing the value of ω in $\Lambda_L(\omega, \mathbf{u})$ corresponds to decreasing the magnitude of the relative velocity v since

$$v = \coth(\omega) \tag{F.28}$$

by eq. F.22. Thus when $\omega = 0$ then $v = \infty$, and when $\omega = \infty$ then $v = 1$. This is the reverse of the sublight case: by eq. F.18 $v = \tanh(\omega)$. Thus when $\omega = 0$ then $v = 0$, and when $\omega = \infty$ then $v = \infty$.

F.6.2 General Velocity Transformation Law – Left-Handed Superluminal Boosts

The general velocity transformation law for a particle moving with velocity $\mathbf{v}$ in the unprimed reference frame and velocity $\mathbf{v}'$ in the primed reference frame is

$$\mathbf{v}' = [\mathbf{v} + (\gamma - 1)\mathbf{w}{\cdot}\mathbf{v} \, \mathbf{w}/w^2 - \gamma\mathbf{w}]/[\, \gamma(1 - \mathbf{w}{\cdot}\mathbf{v})] \tag{F.29}$$

where $\mathbf{w}$ is the relative velocity of the primed reference frame with respect to the unprimed reference frame, and $\gamma = (1 - w^2)^{-\frac{1}{2}}$. Eq. F.29 is obtained by calculating the derivative $d\mathbf{x}'/dt'$ using eqs. F.14. The relative velocity $\mathbf{w}$ can be greater or less than the speed of light. Eq. F.29 implies

$$v'^2 = 1 + (v^2 - 1)(1 - w^2)/(1 - \mathbf{w\cdot v})^2 \tag{F.30}$$

The relation of the velocities (eq. F.30) will be used to determine the multiplication rules for subluminal and superluminal Lorentz transformations (next subsection).

F.6.3 Left-Handed Transformations Multiplication Rules

In this subsection we will determine the multiplication rules of left-handed subluminal and superluminal Lorentz boosts. To do this we will consider three reference frames: an "unprimed" frame, a "primed" frame moving with velocity **w** with respect to the unprimed frame, and a "double-primed" frame moving with velocity **v** with respect to the unprimed frame and velocity **v'** with respect to the primed frame. (See Fig.F.2.)

The velocity **v'** is related to **v** by eqs. F.29 and F.30. Think of the double-primed coordinate system as attached to a particle. In addition note that the transformation law from the unprimed to the double-primed reference frame can be viewed as the product of consecutive transformations (boosts) from the unprimed to the primed reference frames and then from the primed to the double-primed reference frames.

Thus the transformations have the general form:

$$\Lambda_?(\mathbf{v}) = \Lambda_?(\mathbf{v'})\Lambda_?(\mathbf{w}) \tag{F.31}$$

where the "?" subscripts indicate subluminal or superluminal transformations (boosts) depending on the magnitude of the relative velocity in the transformation's parentheses.

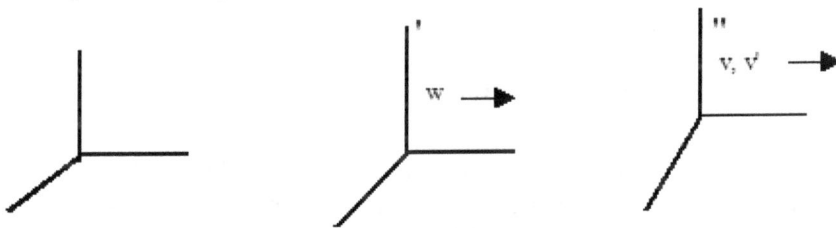

Figure F.2. Three reference frames used to establish transformation multiplication rules.

We now consider the various cases using eq. F.30:

1) If $w > 1$ and $v' > 1$

   then eq. F.30 implies $v < 1$ and thus the left $\Lambda_?(\mathbf{v})$ is a subluminal transformation

$$\Lambda(\mathbf{v}) = \Lambda_L(\mathbf{v'})\Lambda_L(\mathbf{w}) \tag{F.32}$$

2) If $w > 1$, $v' < 1$

   then eq. F.30 implies $v > 1$ and thus the left $\Lambda_?(\mathbf{v})$ is a superluminal transformation

$$\Lambda_L(\mathbf{v}) = \Lambda(\mathbf{v'})\Lambda_L(\mathbf{w}) \tag{F.33}$$

3) If $w < 1$, $v' > 1$

   then eq. F.30 implies $v > 1$ and thus the left $\Lambda_?(\mathbf{v})$ is a superluminal transformation

$$\Lambda_L(\mathbf{v}) = \Lambda_L(\mathbf{v'})\Lambda(\mathbf{w}) \tag{F.34}$$

4) If $w < 1$, $v' < 1$

   then eq. F.30 implies $v < 1$ and thus the left $\Lambda_?(\mathbf{v})$ is a Lorentz transformation

$$\Lambda(\mathbf{v}) = \Lambda(\mathbf{v'})\Lambda(\mathbf{w}) \tag{F.35}$$

where, in each above case, the transformation on the left side of the equation may be a boost or a combination of a boost and a spatial rotation. Thus we have obtained the multiplication rules for left-handed subluminal and superluminal Lorentz transformations.

F.6.4 Inverse of Left-Handed Transformations

The inverse of a Lorentz boost is

$$\Lambda^{-1}(\omega, \hat{\mathbf{u}}) = \exp[-i\omega\hat{\mathbf{u}}\cdot\mathbf{K}] \tag{F.36}$$

where $\omega \geq 0$. Thus the inverse is generated by letting $\omega \to -\omega$. Note that since $v = \tanh\omega$, the effect of $\omega \to -\omega$ is to let $v \to -v$. In the case of superluminal left-handed boosts, since

$$\Lambda_L(\omega, \mathbf{u}) = \Lambda(\omega + i\pi/2, \mathbf{u}) = \exp[i(\omega + i\pi/2)\hat{\mathbf{u}}\cdot\mathbf{K}] \tag{F.37}$$

we find the inverse is

$$\Lambda_L^{-1}(\omega, \mathbf{u}) = \Lambda(-(\omega + i\pi/2), \mathbf{u}) = \exp[-i(\omega + i\pi/2)\hat{\mathbf{u}}\cdot\mathbf{K}] \tag{F.38}$$

where $\omega \geq 0$. Since $\Lambda_L^{-1}(\omega, \mathbf{u})$ is not the hermitean conjugate of $\Lambda_L(\omega, \mathbf{u})$, superluminal boosts are not unitary. However unitarity is not required since complex Lorentz group elements satisfy the defining relation of the Lorentz group (eq. F.10).

F.7 Right-Handed Superluminal Transformations

When we transform between reference frames using a *right-handed*[591] superluminal boost the relation between ω and v is different. The variable ω becomes $\omega_R = \omega - i\pi/2$ and

$$\cosh(\omega_R) = -i\sinh(\omega) = \gamma = -i\gamma_s \tag{F.39}$$
$$\sinh(\omega_R) = -i\cosh(\omega) = \beta\gamma = -i\beta\gamma_s \tag{F.40}$$

where $\beta = v > 1$ and $\omega \geq 0$. Note that $\omega = \text{Re } \omega_R$

---

[591] We call these transformations right-handed because they lead eventually to an alternate right-handed Standard Model This alternate right-handed Standard Model does not appear to correspond to current experimental reality.

$$\sinh(\omega) = \gamma_s \tag{F.41}$$
$$\cosh(\omega) = \beta\gamma_s \tag{F.42}$$

with

$$\gamma_s = (\beta^2 - 1)^{-\frac{1}{2}} \tag{F.43}$$

Upon substituting ω_R for ω in eq. F.13 we obtain the form of the right-handed superluminal boost:[592]

$$\Lambda_R(\omega, \mathbf{u}) = \Lambda(\omega - i\pi/2, \mathbf{u}) \tag{F.44}$$

$$= \begin{bmatrix} -i\gamma_s & i\beta\gamma_s u_x & i\beta\gamma_s u_y & i\beta\gamma_s u_z \\ i\beta\gamma_s u_x & 1 + (-i\gamma_s - 1)u_x^2 & (-i\gamma_s - 1)u_x u_y & (-i\gamma_s - 1)u_x u_z \\ i\beta\gamma_s u_y & (-i\gamma_s - 1)u_x u_y & 1 + (-i\gamma_s - 1)u_y^2 & (-i\gamma_s - 1)u_y u_z \\ i\beta\gamma_s u_z & (-i\gamma_s - 1)u_x u_z & (-i\gamma_s - 1)u_y u_z & 1 + (-i\gamma_s - 1)u_z^2 \end{bmatrix}$$

A simple case that illustrates right-handed superluminal transformations is to assume a relative velocity in the x direction. Then eq. F.44 becomes

$$\Lambda_R(\omega, \mathbf{u} = (1,0,0)) = \begin{bmatrix} -i\gamma_s & i\beta\gamma_s & 0 & 0 \\ i\beta\gamma_s & -i\gamma_s & 0 & 0 \\ 0 & 0 & 1 & 0 \\ 0 & 0 & 0 & 1 \end{bmatrix} \tag{F.45}$$

implementing the coordinate transformation:

$$X' = \Lambda_R(\omega, \mathbf{u})X$$

or

$$t' = -i\gamma_s(t - \beta x)$$

---

[592] We note the singularities at $\beta = \pm 1$ or $\omega = \pm\infty$. **As a result we have a branch cut in the complex ω-plane consisting of the entire real ω axis. Therefore three left-handed boosts are not equivalent to a right-handed boost but rather appear on a different Riemann sheet.**

$$x' = -i\gamma_s(x - \beta t) \qquad (F.46)$$
$$y' = y$$
$$z' = z$$

Comparing eq. F.45 with eq. F.24 for a left-handed superluminal boost we see that

$$PT\Lambda_L(\omega, \mathbf{u} = (1,0,0)) = \begin{bmatrix} -i\gamma_s & i\beta\gamma_s & 0 & 0 \\ i\beta\gamma_s & -i\gamma_s & 0 & 0 \\ 0 & 0 & -1 & 0 \\ 0 & 0 & 0 & -1 \end{bmatrix}$$

where P is the parity operator and T is the time reversal operator. If we now apply a spatial rotation $\mathcal{R}$ of π radians around the x axis then we obtain

$$\mathcal{R}PT\Lambda_L(\omega, \mathbf{u} = (1,0,0))\mathcal{R}^{-1} = \begin{bmatrix} -i\gamma_s & i\beta\gamma_s & 0 & 0 \\ i\beta\gamma_s & -i\gamma_s & 0 & 0 \\ 0 & 0 & 1 & 0 \\ 0 & 0 & 0 & 1 \end{bmatrix} \qquad (F.47)$$
$$= \Lambda_R(\omega, \mathbf{u} = (1,0,0))$$

Since P and T commute with spatial rotations we find

$$\Lambda_R(\omega, \mathbf{u} = (1,0,0)) = PT\mathcal{R}\Lambda_L(\omega, \mathbf{u} = (1,0,0))\mathcal{R}^{-1} \qquad (F.48)$$

or, more generally, performing additional spatial rotations:

$$\Lambda_R(\omega, \mathbf{u}) = PT\mathcal{R}_u\mathcal{R}\mathcal{R}_w\Lambda_L(\omega, \mathbf{w})\mathcal{R}_w^{-1}\mathcal{R}^{-1}\mathcal{R}_u^{-1} \qquad (F.49)$$

or,

$$\Lambda_R(\omega, \mathbf{u}) = PT\mathcal{R}_{tot}\Lambda_L(\omega, \mathbf{w})\mathcal{R}_{tot}^{-1} \qquad (F.50)$$

where $\mathbf{u}$ and $\mathbf{w}$ are unit vectors, and $\mathcal{R}_{tot} = \mathcal{R}_u\mathcal{R}\mathcal{R}_w$. Alternately,

$$\Lambda_L(\omega, \mathbf{w}) = PT\mathscr{R}_{tot}^{-1}\Lambda_R(\omega, \mathbf{u})\mathscr{R}_{tot} \tag{F.51}$$

or

$$\Lambda_L(\omega, \mathbf{w}) = PT\Lambda_R(\omega, \mathbf{u'}) \tag{F.52}$$

for some unit vector $\mathbf{u'}$.

Thus we have shown that PT can be used to relate left-handed and right-handed boosts in a one-to-one fashion. *The appearance of the* parity *operator P takes on great significance when we derive features of the Standard Model. The appearance of left-handed form of The Standard Model stems directly from the implicit parity dependence of the left-handed sector of the superluminal part of the complex Lorentz group.*

For a right-handed boost the addition rule for the x-component of velocity can be computed for infinitesimal displacements in space and time:

$$\begin{aligned} v_x' = \Delta x' / \Delta t' &= (\Delta x\, \gamma_s - \Delta t\, \beta\gamma_s)/(\Delta t\, \gamma_s - \Delta x\, \beta\gamma_s) \\ &= (v_x - \beta)/(1 - \beta v_x) \end{aligned} \tag{F.53}$$

in the limit $\Delta t \to 0$ where the x component of a particle's velocity in the unprimed frame is $v_x = \Delta x/\Delta t$. Note if $v_x = 1$ then $v_x' = 1$. *Thus right-handed superluminal transformations also preserve the constancy of the speed of light in all reference frames.*

F.8 Inhomogeneous Left-Handed Lorentz Group Transformations

The *Left-Handed transformations of the complex Lorentz group* consist of the elements of the real Lorentz group plus left-handed superluminal boost transformations, and combinations of boosts and spatial rotations. Thus the homogeneous left-handed superluminal transformations have the general form:

$$\Lambda_L(\mathbf{v}, \boldsymbol{\theta}) = \exp[i\omega_L \hat{\mathbf{u}}\cdot\mathbf{K} + i\boldsymbol{\theta}\cdot\mathbf{J}] \tag{F.54}$$

where $\omega_L' = \omega + i\pi/2$, $\boldsymbol{\theta}$ is the angular vector, and $\mathbf{J}$ is the angular momentum operator vector. Inhomogeneous left-handed superluminal transformations, which include displacements, can be expressed as

$$\Lambda_L(\mathbf{v}, \boldsymbol{\theta}, \mathbf{d}) = \exp[i\omega_L\hat{\mathbf{u}}\cdot\mathbf{K} + i\boldsymbol{\theta}\cdot\mathbf{J} - i\mathbf{d}\cdot\mathbf{P}] \tag{F.55}$$

where $\mathbf{P}$ is the momentum operator vector and $\mathbf{d}$ is a displacement vector.

We note

$$\det \Lambda_L(\omega, \mathbf{u}) = \pm 1 \tag{F.56}$$

The ordinary Lorentz group is divided into four disjoint subgroups that are often denoted:

$$
\begin{array}{ll}
L_+^\uparrow: & \det \Lambda(\omega, \mathbf{u}) = +1; \ \ \mathrm{sgn}\ \Lambda(\omega, \mathbf{u})^0{}_0 = +1 \\
L_-^\uparrow: & \det \Lambda(\omega, \mathbf{u}) = -1; \ \ \mathrm{sgn}\ \Lambda(\omega, \mathbf{u})^0{}_0 = +1 \\
L_+^\downarrow: & \det \Lambda(\omega, \mathbf{u}) = +1; \ \ \mathrm{sgn}\ \Lambda(\omega, \mathbf{u})^0{}_0 = -1 \\
L_-^\downarrow: & \det \Lambda(\omega, \mathbf{u}) = -1; \ \ \mathrm{sgn}\ \Lambda(\omega, \mathbf{u})^0{}_0 = -1
\end{array}
\tag{F.57}
$$

where $\mathrm{sgn}\ \Lambda(\omega, \mathbf{u})^0{}_0$ is the sign of the 00 component of the $\Lambda(\omega, \mathbf{u})$ matrix. The various subgroups are related by the discrete transformations of parity P and time reversal T:

$$
\begin{array}{ccc}
& P & \\
L_+^\uparrow & \xrightarrow{} & L_-^\uparrow \\
& PT & \\
L_+^\uparrow & \xrightarrow{} & L_+^\downarrow \\
& T & \\
L_+^\uparrow & \xrightarrow{} & L_-^\downarrow
\end{array}
$$

The left-handed superluminal transformations are disjoint in a somewhat different way. By eq. F.56 the determinants are ± 1. However the 0-0 matrix element of eq. F.16 gives

$$\Lambda_L{}^0{}_0{}^2 - \Sigma_i\, (\Lambda_L{}^i{}_0)^2 = 1 \tag{F.58}$$

The representation of superluminal boosts shows that each factor in eq. F.58 is imaginary. Thus eq. F.58 implies

$$\Sigma_i\, |\Lambda_L{}^i{}_0|^2 \geq 1 \tag{F.59}$$

$$|\Lambda_L{}^0{}_0| \geq 0 \qquad (\text{not} \geq 1) \qquad (\text{F.60})$$

where $\|$ indicates absolute value since the quantities in eq. F.58 are squares – not in absolute value. Thus the magnitude of $\Lambda_L{}^0{}_0$ does not have a gap. Therefore left-handed superluminal transformations can be divided into two categories:

$$\begin{aligned} {}_L L_+: &\quad \det \Lambda_L(\omega, \mathbf{u}) = +1 \\ {}_L L_-: &\quad \det \Lambda_L(\omega, \mathbf{u}) = -1 \end{aligned} \qquad (\text{F.61})$$

as one expects for complex Lorentz group transformations.[593]

Earlier we saw that under a PT transformation a left-handed superluminal transformation becomes a right handed superluminal transformation. Again, as in the left-handed case, the various disjoint pieces are related by the discrete transformations of parity P and time reversal T:

$$\begin{aligned} {}_L L_+ &\xrightarrow[T]{P} {}_L L_- \\ {}_L L_+ &\rightarrow {}_L L_- \end{aligned} \qquad (\text{F.62})$$

F.9 Inhomogeneous Right-Handed Extended Lorentz Group

The inhomogeneous right-handed part of the complex Lorentz group[594] consists of the real Lorentz group plus right-handed superluminal transformations plus rotations and displacements that have the form:

$$\Lambda_R(\mathbf{v}, \boldsymbol{\theta}, \mathbf{d}) = \exp[i\omega_R \hat{\mathbf{u}}\cdot\mathbf{K} + i\boldsymbol{\theta}\cdot\mathbf{J} - i\mathbf{d}\cdot\mathbf{P}] \qquad (\text{F.63})$$

in general where $\omega_R = \omega - i\pi/2$.

---

[593] Streater (2000) p. 13.
[594] Since γ_s has branch points at $v = \pm 1$ (which corresponds to $\omega = \pm\infty$ for both the left-handed and right-handed groups) there is a cut along the real ω axis between $-\infty$ and $+\infty$ in the ω complex plane. Therefore, we note, the product of three left-handed Lorentz transformations does not yield a right-handed transformation (as might be supposed from eqs. 2.43 and 2.51) but rather a left-handed transformation on the second sheet. A transformation with $\omega + 3i\pi/2$ is not equivalent to a transformation with $\omega - i\pi/2$.

F.10 General Forms of Superluminal Boosts

The group elements of the homogeneous complex Lorentz group L(C) can be expressed in terms of the group generators as

$$\Lambda_C = \exp[i(\omega_r \hat{\mathbf{u}}_r + i\omega_i \hat{\mathbf{u}}_i)\cdot\mathbf{K} + i\boldsymbol{\theta}_c\cdot\mathbf{J}] \tag{F.64}$$

where the vector $\boldsymbol{\theta}_c$ is a complex 3-vector, $\omega_r \geq 0$ and $\omega_i \geq 0$ are real numbers, and $\hat{\mathbf{u}}_r$ and $\hat{\mathbf{u}}_i$ are real normalized 3-vectors such that $\hat{\mathbf{u}}_r\cdot\hat{\mathbf{u}}_r = 1 = \hat{\mathbf{u}}_i\cdot\hat{\mathbf{u}}_i$. The generators of the homogeneous complex Lorentz group are $\mathbf{K}$, and $\mathbf{J}$ just as for the homogeneous real Lorentz group.

We now focus on boosts because they will be crucial in the determination of the equations of motion of various types of spin ½ particles. A boost has the form

$$\Lambda_C(\mathbf{v_c}) = \exp[i\omega\hat{\mathbf{w}}\cdot\mathbf{K}] \tag{F.65}$$

where

$$\omega = (\omega_r^2 - \omega_i^2 + 2i\omega_r\omega_i\,\hat{\mathbf{u}}_r\cdot\hat{\mathbf{u}}_i)^{\frac{1}{2}} \tag{F.66}$$

and

$$\hat{\mathbf{w}} = (\omega_r\hat{\mathbf{u}}_r + i\omega_i\hat{\mathbf{u}}_i)/\omega \tag{F.67}$$

Since $\hat{\mathbf{u}}_r\cdot\hat{\mathbf{u}}_r = 1 = \hat{\mathbf{u}}_i\cdot\hat{\mathbf{u}}_i$ we see

$$\hat{\mathbf{w}}\cdot\hat{\mathbf{w}} = 1 \tag{F.68}$$

The complex relative velocity is

$$\mathbf{v_c} = \hat{\mathbf{w}}\,\tanh(\omega) \tag{F.69}$$

Having placed boost transformations in the form of eq. F.12 we can take advantage of the form of real proper orthochronous Lorentz boost transformations, eq. F.13, and analytically continue to complex ω and complex unit vectors $\hat{\mathbf{w}}$ provided eq. F.69 is satisfied. The resulting complex generalization will be the matrix form of proper boosts:

$$\Lambda_C(\mathbf{v_c}) = \exp[i\omega\hat{\mathbf{w}}\cdot\mathbf{K}] \equiv \Lambda_C(\omega, \hat{\mathbf{w}})$$

$$= \begin{bmatrix} \cosh(\omega) & -\sinh(\omega)\hat{w}_x & -\sinh(\omega)\hat{w}_y & -\sinh(\omega)\hat{w}_z \\ -\sinh(\omega)\hat{w}_x & 1+(\cosh(\omega)-1)\hat{w}_x^{\ 2} & (\cosh(\omega)-1)\hat{w}_x\hat{w}_y & (\cosh(\omega)-1)\hat{w}_x\hat{w}_z \\ -\sinh(\omega)\hat{w}_y & (\cosh(\omega)-1)\hat{w}_x\hat{w}_y & 1+(\cosh(\omega)-1)\hat{w}_y^{\ 2} & (\cosh(\omega)-1)\hat{w}_y\hat{w}_z \\ -\sinh(\omega)\hat{w}_z & (\cosh(\omega)-1)\hat{w}_x\hat{w}_z & (\cosh(\omega)-1)\hat{w}_y\hat{w}_z & 1+(\cosh(\omega)-1)\hat{w}_z^{\ 2} \end{bmatrix}$$

$$(F.70)$$

Since analytic continuations are unique, the above form for $\Lambda_C(\mathbf{v_c})$ is well-defined and unique. It spans the complete set of proper complex Lorentz boosts.

We now will study six classes of boosts that have the property that they boost from a coordinate system with real time and space coordinates to a coordinate system with either a purely real or purely imaginary time, and real, imaginary or complex spatial coordinates. These boosts produce left-handed lepton-like and "quark-like" free Dirac-like equations. They also produce right-handed lepton-like and "quark-like" free Dirac-like equations. We will discuss these Dirac-like equations in detail later. First we describe the four categories of boosts that have the property that they transform the reference frame of a particle at rest to a reference frame where the energy is either purely real or purely imaginary – the distinguishing feature of these four sets of transformations.

F.10.1 "Lepton-like" Left-Handed Boosts

If we let

$$\hat{\mathbf{u}}_i = \hat{\mathbf{u}}_r \equiv \hat{\mathbf{u}} \tag{F.71}$$

so that the vector $\hat{\mathbf{u}}_i$ is parallel to $\hat{\mathbf{u}}_r$, and let

$$\omega_i = \pi/2 \tag{F.72}$$

then $\Lambda_C(\mathbf{v_c})$ becomes a lepton-like left-handed boost:[595]

$$\Lambda_C = \exp[i(\omega_r + i\,\pi/2)\hat{\mathbf{u}}_r\cdot\mathbf{K}] \tag{F.73}$$

F.10.2 "Lepton-like" Right-Handed Boosts

If we let

$$\hat{\mathbf{u}}_i = -\hat{\mathbf{u}}_r \equiv -\hat{\mathbf{u}} \tag{F.74}$$

so that the vector $\hat{\mathbf{u}}_i$ is anti-parallel to $\hat{\mathbf{u}}_r$, and

$$\omega_i = -\pi/2 \tag{F.75}$$

then $\Lambda_C(\mathbf{v_c})$ becomes a right-handed boost:

$$\Lambda_C = \exp[i(\omega_r - i\,\pi/2)\hat{\mathbf{u}}_r\cdot\mathbf{K}] \tag{F.76}$$

F.10.3 "Quark-like" Left-Handed Boosts

If the real and imaginary relative vectors parts of $\hat{\mathbf{w}}$, namely $\hat{\mathbf{u}}_r$ and $\hat{\mathbf{u}}_i$, are perpendicular, $\hat{\mathbf{u}}_r\cdot\hat{\mathbf{u}}_i = 0$, then by eq. F.66

$$\omega = (\omega_r^2 - \omega_i^2)^{\frac{1}{2}} \tag{F.77}$$

Thus ω is either pure real ($\omega_r \geq \omega_i$) or pure imaginary ($\omega_r < \omega_i$). We choose ω real, and then reset

$$\omega = (\omega_r^2 - \omega_i^2)^{\frac{1}{2}} \rightarrow \omega' = (\omega_r^2 - \omega_i^2)^{\frac{1}{2}} + i\pi/2 = \omega + i\pi/2 \tag{F.78}$$

by adding $i\pi/2$ to the ω factor in eq. F.65 since ω is a free parameter. Then the resulting Lorentz transformation then becomes a "quark-like" left-handed boost:[596]

---

[595] We say "lepton-like" because we obtain a lepton-like Dirac-like equation using these boosts later. Similarly for "quark-like.'

$$\Lambda_C = \exp[i((\omega_r^2 - \omega_i^2)^{\frac{1}{2}} + i\pi/2)(\omega_r \hat{\mathbf{u}}_r + i\omega_i \hat{\mathbf{u}}_i) \cdot \mathbf{K}/\omega] \qquad (F.79)$$

F.10.4 "Quark-like" Right-Handed Boosts

If the real and imaginary relative vectors parts of $\hat{\mathbf{w}}$, namely $\hat{\mathbf{u}}_r$ and $\hat{\mathbf{u}}_i$, are perpendicular, $\hat{\mathbf{u}}_r \cdot \hat{\mathbf{u}}_i = 0$, then by eq. F.66

$$\omega = (\omega_r^2 - \omega_i^2)^{\frac{1}{2}} \qquad (F.80)$$

Thus ω again starts out either pure real ($\omega_r \geq \omega_i$) or pure imaginary ($\omega_r < \omega_i$). In this case we also choose ω real, and then reset

$$\omega = (\omega_r^2 - \omega_i^2)^{\frac{1}{2}} \rightarrow \omega' = (\omega_r^2 - \omega_i^2)^{\frac{1}{2}} - i\pi/2 \qquad (F.81)$$

by subtracting $i\pi/2$ from ω in eq. F.65 since ω is a free parameter. The resulting Lorentz boost

$$\Lambda_C = \exp[i((\omega_r^2 - \omega_i^2)^{\frac{1}{2}} - i\pi/2)(\omega_r \hat{\mathbf{u}}_r + i\omega_i \hat{\mathbf{u}}_i) \cdot \mathbf{K}/\omega] \qquad (F.82)$$

becomes a quark-like right-handed boost.[597]

F.10.5 "Quark-like" Boosts

If the real and imaginary relative vectors parts of $\hat{\mathbf{w}}$, namely $\hat{\mathbf{u}}_r$ and $\hat{\mathbf{u}}_i$, are perpendicular, $\hat{\mathbf{u}}_r \cdot \hat{\mathbf{u}}_i = 0$, then by eq. F.66

$$\omega = (\omega_r^2 - \omega_i^2)^{\frac{1}{2}} \qquad (F.83)$$

Thus ω again starts out either pure real ($\omega_r \geq \omega_i$) or pure imaginary ($\omega_r < \omega_i$). In this case choose ω_r real and use ω as defined by eq. F.83.

---

[596] We say "quark-like" because we will later obtain a quark-like left-handed Dirac-like equation with complex spatial momentum terms using these boosts.
[597] We say "quark-like" because we obtain a quark-like right-handed Dirac-like equation with complex spatial momentum terms using these boosts later.

Then the resulting Lorentz boost

$$\Lambda_C = \exp[i(\omega_r^2 - \omega_i^2)^{\frac{1}{2}}(\omega_r\hat{\mathbf{u}}_r + i\omega_i\hat{\mathbf{u}}_i)\cdot\mathbf{K}/\omega] \tag{F.84}$$

becomes a quark-like boost without handedness.[598]

F.10.6 Conventional "Dirac" Boosts

If we let

$$\hat{\mathbf{u}}_i = \hat{\mathbf{u}}_r \equiv \hat{\mathbf{u}} \tag{F.85}$$

so that the vector $\hat{\mathbf{u}}_i$ is parallel to $\hat{\mathbf{u}}_r$, and let

$$\omega_i = 0 \tag{F.86}$$

then $\Lambda_C(\mathbf{v}_c)$ becomes a Dirac boost:[599]

$$\Lambda = \exp[i\omega_r\hat{\mathbf{u}}_r\cdot\mathbf{K}] \tag{F.87}$$

This boost can be used to generate the free Dirac equation.

---

[598] We again say "quark-like" because we obtain a quark-like Dirac equation with complex spatial momentum terms using these boosts later.

[599] We say "Dirac" because we obtain a Dirac equation using this boost later.

Appendix G. Phenomena Beyond the Light Barrier

G.1 Superluminal (Faster-than-Light) Transformations

In this Appendix we will briefly survey some of the very different features of faster-than light physical phenomena. We will frame our discussion in terms of the two simple reference frames depicted in Fig. G.1. The prime frame is moving at a speed v > c (the speed of light) in the positive x direction with respect to the unprimed reference frame.

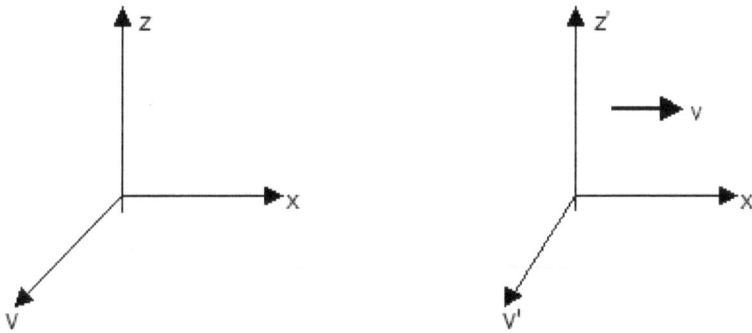

Figure G.1. Two coordinate systems having a relative speed v in the x direction.

As shown later in the text we define a superluminal (faster-than-light) transformation between coordinates in these reference frames with (Eqs. 2.16 and 2.13 are in Blaha (2007b))

$$t' = \gamma_s(t - \beta x/c)$$
$$x' = \gamma_s(x - \beta ct) \qquad (2.16)$$

$$y' = iy$$
$$z' = iz$$

where

$$\gamma_s = (\beta^2 - 1)^{-\frac{1}{2}} \tag{2.13}$$

and $\beta = v/c > 1$. The appearance of imaginary values for y' and z' is not a cause for alarm. An observer resident in the prime coordinate system will measure real y and z distances with a ruler. The only purpose of the factors of i is to relate the y and z coordinates to y' and z'. An observer in either coordinate system will view his/her coordinates as real.

The energy and momentum of a tachyon (faster-than-light) particle of mass m traveling at a speed $v > c$ is

$$E = \gamma_s mc^2 \tag{G.1}$$

and

$$\mathbf{p} = m\gamma_s \mathbf{v} \tag{G.2}$$

Note that the tachyon defining condition is satisfied:

$$E^2 - c^2 \mathbf{p}^2 = -m^2 c^4 \tag{G.3}$$

Also note that in the limit $\beta \to \infty$ that

$$E = 0 \tag{G.4}$$

and

$$p = mc \tag{G.5}$$

where $p = |\mathbf{p}|$. Tachyons are always in motion. The minimal momentum of a tachyon is given by eq. G.5. It corresponds to zero energy. It is the tachyon equivalent of Einstein's famous $E = mc^2$.

G.2 Length Dilations and Time Contractions

In ordinary Lorentz transformations a moving ruler will appear to be shorter in the direction of its motion when measured in another reference frame. This phenomenon is called *Lorentz contraction.*

Superluminal Length Dilation/Contraction

In the case of a superluminal transformation we find precisely the opposite effect, *superluminal length dilation,* is a possibility. Consider the case of the transformation of eq. 2.16 above (corresponding to Fig. G.1), which relates the prime reference frame traveling at speed v in the positive x direction to the unprimed reference frame. A ruler perpendicular to the x-axis will have the same length in both reference frames if its endpoints are simultaneously measured – perhaps by photographing it. The y and z equations in eqs. 2.16 specify this fact up to an extraneous factor of i.

If the ruler is at rest in the prime reference frame and parallel to the x' axis, then a simultaneous measurement of its endpoints at the same time t_0 by an observer in the unprimed reference frame (perhaps by photographing it) will reveal both *length contraction and dilation* depending on the value of β. If the length is $L' = x'_2 - x'_1$ in the prime frame and $L = x_2 - x_1$ in the unprimed frame, then the equations:

$$x'_1 = \gamma_s(x_1 - \beta ct_0) \tag{G.7}$$
$$x'_2 = \gamma_s(x_2 - \beta ct_0) \tag{G.8}$$

imply

$$L' = \gamma_s L = (\beta^2 - 1)^{-\frac{1}{2}} L \tag{G.9}$$

Thus we have three cases:

Case 1: $\beta \in <1, \sqrt{2}>$:	$L < L'$	Contraction	(G.10)
Case 2: $\beta = \sqrt{2}$:	$L = L'$	Equality	(G.11)
Case 3: $\beta \in <\sqrt{2}, \infty>$:	$L > L'$	Dilation	(G.12)

Superluminal Time Contraction/Dilation

In the case of a superluminal transformation we find *superluminal time contraction* is a possibility. Consider again the case of the transformation of eq. 2.16 above corresponding to Fig. G.1 relating the prime reference frame traveling at speed v in the positive x direction to the unprimed reference frame. Consider the time interval between two events occurring at the same point x'_0 in the prime reference frame. From the viewpoint of an observer in the unprimed frame the events take place at different points x_1 and x_2. If the time interval is $T' = t'_2 - t'_1$ in the prime frame and $T = t_2 - t_1$ in the unprimed frame, then the inverse of eqs. 2.16 give:

$$t_1 = \gamma_s(t'_1 + \beta x'_0/c) \tag{G.13}$$
$$t_2 = \gamma_s(t'_2 + \beta x'_0/c) \tag{G.14}$$

and imply

$$T = \gamma_s T' = (\beta^2 - 1)^{-\frac{1}{2}} T' \tag{G.15}$$

Again we have three cases:

Case 1: $\beta \in <1, \sqrt{2}>$: $\quad\quad T > T' \quad$ Dilation $\tag{G.16}$

Case 2: $\beta = \sqrt{2}$: $\quad\quad T = T' \quad$ Equality $\tag{G.17}$

Case 3: $\beta \in <\sqrt{2}, \infty>$: $\quad\quad T < T' \quad$ Contraction $\tag{G.18}$

The time interval in the unprimed frame can be less than, equal to, or greater than the time interval in the frame where the events take place at the same spatial point.

Thus superluminal transformations are more complex than Lorentz transformations with respect to space and time, dilation and contraction.

G.3 Tachyon Fission to More Massive Particles – Reverse Fission

Another way in which faster-than-light phenomena differ from sublight phenomena is particle fission. Normally when a particle or nucleus decays or fissions the masses of the particles produced by the decay are smaller than the mass of the original particle or nucleus. And energy is released. We are familiar with fission as the source of nuclear energy.

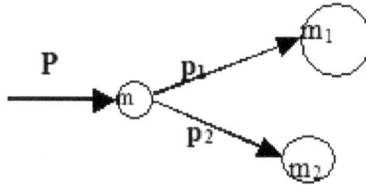

Figure G.2. Two particle decay of a tachyon.

In the case of faster-than-light particles, tachyons, a much different possibility is present: a tachyon can decay into heavier tachyons: *a particle's spatial 3-momentum can be transformed into mass*. We will consider the specific case of a tachyon decaying into two particles to illustrate this possibility. (See Fig. G.2.)

We will assume the initial tachyon has zero energy[600] and thus the tachyons emerging from the decay also have zero energy. The analysis is based on conservation of total energy and momentum.

Momentum conservation implies

$$\mathbf{P} = \mathbf{p_1} + \mathbf{p_2} \tag{G.19}$$

Since all energies are zero

$$(cP)^2 = (c\mathbf{P})^2 = m^2$$
$$(cp_1)^2 = (c\mathbf{p_1})^2 = m_1^2 \tag{G.20}$$
$$(cp_2)^2 = (c\mathbf{p_2})^2 = m_2^2$$

---

[600] If a particle has zero energy its velocity is infinite so the case considered is somewhat artificial. However the results would still be approximately true for very large velocities. The simplicity of the kinematics led us to consider this case.

where $P = |\mathbf{P}|$, $p_1 = |\mathbf{p_1}|$, and $p_2 = |\mathbf{p_2}|$. If we now square eq. G.19 and use eqs. G.20 we obtain

$$m^2 = m_1^2 + m_2^2 + 2m_1m_2 \cos \theta \qquad \text{(G.21)}$$

where θ is the angle between the emerging particles momenta $\mathbf{p_1}$ and $\mathbf{p_2}$.

Eq. G.21 has a number of interesting cases:

Case $\theta = 0$:

$$m = m_1 + m_2 \qquad \text{(G.22)}$$

The masses of the outgoing tachyons sum to the mass of the original tachyon.

Case $\theta = \pi/2$:

$$m^2 = m_1^2 + m_2^2 \qquad \text{(G.23)}$$

The masses of each outgoing tachyon is less than the mass of the original tachyon.

Case $\theta = \pi$:

$$m^2 = (m_1 - m_2)^2 \qquad \text{(G.24)}$$

In this case either $m_1 > m$ or $m_2 > m$. Thus one of the outgoing tachyons has a greater mass than the original tachyon. Mass is effectively created from the spatial momentum of the particle. This process is the inverse of normal particle decay or fission where the sum of the outgoing masses is always less than the original particle's mass and the difference is mass converted into energy in the form of additional photons via $E = mc^2$.

This last case, where one of the outgoing particles is more massive than the original particle, is not just for $\theta = \pi$. Since

$$\cos\theta = (m^2 - m_1^2 - m_2^2)/(2m_1 m_2) \qquad (G.25)$$

we see that *the sum of the outgoing tachyon masses is always greater than the original tachyon mass (except when θ = 0)* since

$$\cos\theta = 1 + [m^2 - (m_1 + m_2)^2]/(2m_1 m_2) \leq 1 \qquad (G.26)$$

and thus

$$[m^2 - (m_1 + m_2)^2]/(2m_1 m_2) \leq 0 \qquad (G.27)$$

Note $m = m_1 + m_2$ only if $\theta = 0$.

Since we can transform the above discussion to the case of tachyons with a non-zero energy using an ordinary Lorentz transformation, the above discussion in this subsection is general.

We therefore conclude that when a tachyon decays into two tachyons the sum of the masses of the produced tachyons is greater than the mass of the original tachyon except if the angle between the momenta of the produced tachyons is zero. In that case the sum of the masses of the produced tachyon equals the mass of the original tachyon.

*Thus tachyons can engage in **reverse fission** in which **momentum is converted into mass so the outgoing particles have a total mass greater than the incoming particle**.* In the case of "normal" fission part of the mass of a particle can be converted to energy and the sum of the masses of the decay product particles is less than the mass of the original particle.

G.4 Light Chasing Faster-than-Light Particles?

Einstein told a story that he imagined positioning himself in a (Galilean) reference frame moving at the speed of light and seeing electromagnetic waves "frozen" in time so that they were no longer vibrating. This vision inspired him to reconsider the transformation laws between coordinate systems and to derive the theory of Special Relativity. In Special Relativity the speed of light is the same in all reference frames.

In this subsection we will consider a light pulse from the points of view of two reference frames whose relative speed v is greater than the speed of light. We will use the example considered earlier and add a pulse of light traveling in the positive x direction. (See Fig. G.3.)

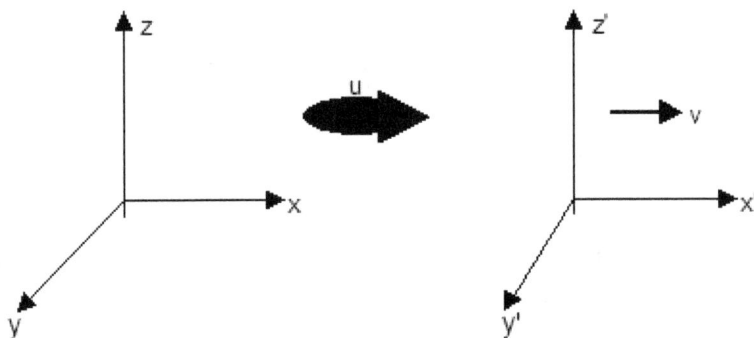

Figure G.3. Two coordinate systems having a relative speed v in the x direction. A pulse of light is displayed as a thick arrow.

The general law for the addition of velocities in a situation such as depicted in Fig. G.3 is well known. If we adapt it to the present example and let u be the speed of the pulse in the unprimed frame (temporarily forgetting it is a light pulse) we find it implies

$$u' = (u - \beta c)/(1 - \beta u/c) \qquad (2.17a)$$

where $\beta = v/c > 1$, and u' is the speed of the pulse in the prime frame. Then if we set u = c we see that u' = c as well. *Thus our superluminal transformations preserve the constancy of the speed of light just like Lorentz transformations.*

As a result the pulse of light will intersect the z' axis eventually. However if superluminal transformations did not preserve the speed of light in all frames the pulse might never reach the z' axis. For example under a Galilean transformation the speed of the pulse would be u' = u – v = c – v and the pulse would actually be falling further and further behind the z' axis.

G.5 Electromagnetic Field of a Charged Tachyon – A Pancake Effect?

The electric field of a charge q at rest in a reference frame is:

$$\mathbf{E} = (q/(4\pi\varepsilon_0)\check{\mathbf{r}}/r^2 \qquad (G.28)$$

in spherical coordinates where $\check{\mathbf{r}}$ is a unit vector in the radial direction.

Sublight Charged Particle

The electric and magnetic fields of a charge q moving in the positive x direction with speed v < c are

$$\mathbf{E} = (q/(4\pi\varepsilon_0)\check{\mathbf{r}}(1 - \beta^2)/[r^2(1 - \beta^2\sin^2\theta)^{\frac{3}{2}}] \qquad (G.29)$$
$$\mathbf{B} = (q/(4\pi\varepsilon_0)\check{\mathbf{r}}\beta(1 - \beta^2)\sin\theta/[r^2(1 - \beta^2\sin^2\theta)^{\frac{3}{2}}] \qquad (G.30)$$

where $\check{\mathbf{r}}$ is the radial unit vector, $\beta = v/c$, and θ is measured with respect to the polar axis which is taken to be the x axis. As $\beta \rightarrow 1$ the electric and magnetic fields develop a "pancake" form with large field strengths in the directions perpendicular to the direction of motion similar to the transverse fields of electromagnetic quanta. This feature is the basis of the Weizsäcker-Williams method of virtual quanta.

Charged Tachyon

The electric and magnetic fields of a tachyon of charge q moving in the positive x direction with speed v > c are

$$\mathbf{E} = (q/(4\pi\varepsilon_0)\check{\mathbf{r}}(\beta^2 - 1)/[r^2(\beta^2\sin^2\theta - 1)^{\frac{3}{2}}] \qquad (G.31)$$
$$\mathbf{B} = (q/(4\pi\varepsilon_0)\check{\mathbf{r}}\beta(\beta^2 - 1)\sin\theta/[r^2(\beta^2\sin^2\theta - 1)^{\frac{3}{2}}] \qquad (G.32)$$

where $\beta = v/c > 1$, and θ is again measured with respect to the polar axis which is taken to be the x axis. In the case of tachyons there are three cases of interest.

Case $\beta^2\sin^2\theta - 1 < 0$:

The electric and magnetic fields are pure imaginary and are excluded from the forward and backward cones surrounding the x axis defined by $|\sin\theta| < \beta^{-1}$.

Case $\beta^2\sin^2\theta - 1 = 0$:

The electric and magnetic fields are infinite. Thus the field strengths are infinite on a cone at the angle θ with respect to the x-axis. By comparison, a magnetic monopole only has a one-dimensional, singularity line extending from the monopole to infinity.

Case $\beta^2\sin^2\theta - 1 > 0$:

The electric and magnetic fields decrease in strength as $\sin^2\theta$ increases. Thus the region of maximum field strength are the forward and backward cones where $|\sin\theta|$ is greater than but near β^{-1} in value. The pancake picture of the sublight charged particle does not hold for charged tachyons.

Are There Tachyon Cones in the Au-Au Scattering Quark-Gluon Plasma?

Cones have been observed in high energy Au-Au scattering in which a quark-gluon plasma is created. These cones have been attributed to a variety of causes such as hydrodynamically generated Mach cones, and Cherenkov radiation. The possibility exists that tachyon excitations may transiently exist in the quark-gluon plasma and may, in part, explain the observed cones and dips. The above described cones in the case of a moving charged tachyon are remarkably similar in character. See the CERES collaboration paper arXiv:nucl-ex/0701023, and references therein, for experimental findings.

G.6 Superluminal (Tachyon) Physics is Different

The simple classical examples presented in this appendix demonstrate that superluminal physics has many interesting new features that are worthy of interest. Since tachyons exist in Black Holes, and, perhaps, in other contexts, their study is a worthwhile endeavor.

Appendix H. Superluminal (Faster Than Light) Kinetic Theory and Thermodynamics

This appendix[601] changes the flow of topics of these volumes from gravitation and particle theory to superluminal many particle dynamics. We will progress from superluminal Kinetic theory to Thermodynamics. We will see that there are strong similarities with non-relativistic Thermodynamics.

H.1 Superluminal Kinetic Theory

Assemblages of large numbers of particles embody the Maxwell-Boltzmann distribution. The Boltzmann H theorem is the beginning point for derivations of the non-relativistic Maxwell-Boltzmann distribution. The non-relativistic Maxwell-Boltzmann distribution has the form

$$f(\mathbf{v}, \mathbf{r}) = n(m/(2\pi kT))^{3/2}\exp\{-[m(\mathbf{v} - \mathbf{v}_0)^2/2 + V(r)]/(kT)\} \qquad (H.1)$$

where n is the particle density, T is the temperature, $\mathbf{v}_0$ is the average velocity, m is the particle mass, V(r) is an external conservative force, and k is Boltzmann's constant. In terms of a Hamiltonian

$$H(\mathbf{v}, \mathbf{r}) = m\mathbf{v}^2/2 + V(r) \qquad (H.2)$$

we can express the Maxwell-Boltzmann distribution as

$$f(\mathbf{v}, \mathbf{r}) = n(m/(2\pi kT))^{3/2}\exp\{-H(\mathbf{v} - \mathbf{v}_0, \mathbf{r})/(kT)\} \qquad (H.3)$$

---

[601]From Blaha (2012a).

H.1.1 Relativistic Form of the Maxwell-Boltzmann Distribution

If we assume that we have a container containing a distribution of relativistic (sublight) particles with an average velocity $\mathbf{v}_0 = 0$, and no external force, then the form of eq. H.3 generalizes to the relativistic Maxwell-Boltzmann distribution

$$f_R(\mathbf{v}) = C_R \exp\{-H/(kT)\} \qquad (H.4)$$

where C_R is a normalization constant and H is the relativistic Hamiltonian for a free particle:

$$H = c(m^2c^2 + \mathbf{p}^2)^{\frac{1}{2}} \qquad (H.5)$$

with $\mathbf{p} = \gamma m\mathbf{v}$ and $\gamma = (1 - v^2/c^2)^{-\frac{1}{2}}$. C_R is determined by the condition

$$\int d^3v f_R(\mathbf{v}) = 1 \qquad (H.6)$$

H.1.2 Superluminal Form of the Maxwell-Boltzmann Distribution

The superluminal form of Maxwell-Boltzmann distribution is based on the form of the mass shell condition for superluminal particles:

$$E^2 - c^2\mathbf{p}^2 = m^2c^4 \qquad (H.7)$$

which implies a free Hamiltonian

$$H_S = c(\mathbf{p}^2 - m^2c^2)^{\frac{1}{2}} \qquad (H.8)$$

where

$$\mathbf{p} = \gamma_s m\mathbf{v} \qquad (H.9)$$

and

$$\gamma_s = (v^2/c^2 - 1)^{-\frac{1}{2}}$$

The seemingly slight difference between eqs. H.8 and H.9, and eq. H.5 causes major differences between superluminal and relativistic kinetic theory and thermodynamics. On the other hand relativistic kinetic theory and thermodynamics are qualitatively similar in many ways with their non-relativistic counterparts.

One major difference is the behavior of kinematic variables near the speed of light:

As $v \rightarrow c$ Below the Speed of Light

$$p \rightarrow \infty$$
$$H \rightarrow \infty$$

As $v \rightarrow c$ From Above the Speed of Light

$$p \rightarrow \infty$$
$$H_S \rightarrow \infty$$

As $v \rightarrow \infty$

$$p \rightarrow mc$$
$$H_S \rightarrow 0$$

Thus as v ranges from c to ∞, H_S decreases monotonically from ∞ to zero and p decreases from ∞ to mc. This behavior contrasts with H in eq. H.5, which increases monotonically with p as v increases from 0 to c. Thus the sublight Maxwell-Boltzmann distribution decreases with v as v increases from 0 to c.

The superluminal Maxwell-Boltzmann distribution *increases* with v as v increases from c to ∞ as we see below. The superluminal Maxwell-Boltzmann distribution decreases with p as p increases from mc to ∞. *As a result the natural physical parameterization of the Maxwell-Boltzmann distribution should be in terms of the momentum rather than the velocity.* Thus Boltzmann's H function which normally is

$$H_B(t) = \int d^3v \, f(\mathbf{v}, t) \log f(\mathbf{v}, t)$$

must be replaced with[602]

$$H_{BS}(t) = \int d^3p \; f_S(\mathbf{p}, t) \log f_S(\mathbf{p}, t)$$

The equilibrium superluminal Maxwell-Boltzmann distribution can be derived from $H_S(t)$. It has the same general form as the relativistic distribution

$$f_S(\mathbf{p}) = C_S \exp\{-H_S/(kT)\} \tag{H.10}$$

where C_S is a normalization constant and H_S is the superluminal Hamiltonian for a free particle.

We now apply the normalization condition[603]

$$n = N/V = \int d^3p f_S(\mathbf{p}) = C_S \int d^3p \exp\{-H_S/(kT)\} \tag{H.11}$$

where n is the particle density, N is the number of particles in the system, and V is the volume of the system. We calculate C_S by evaluating the integral:

$$n = 4\pi C_S \int_m^\infty dp \; p^2 \exp\{-H_S/(kT)\} \tag{H.12}$$

Letting $x = p/(mc)$ and $\alpha = mc^2/(kT)$ we see eq. H.12 becomes

$$n = 4\pi m^3 c^3 C_S \int_1^\infty dx \; x^2 \exp\{-\alpha(x^2 - 1)^{\frac{1}{2}}\} \tag{H.13}$$

Then letting $y^2 = x^2 - 1$ we find

$$n = 4\pi m^3 c^3 C_S \int_0^\infty dy \; y(y^2 + 1)^{\frac{1}{2}} \exp(-\alpha y)$$

---

[602] Note the additional factor of m^3 in $\int d^3p$ will be absorbed in the normalization (eq. H.11).
[603] We note that using $\int d^3v$ rather than $\int d^3p$ in eq. H.11 would result in a divergence – another reason for our choice of integration parameter.

$$= -m^3c^3C_SG^{31}_{13}(\alpha^2/4 \mid {}^{0}_{-3/2,0,\,\frac{1}{2}}) \qquad (H.14)$$

where $G^{31}_{13}(\dots)$ is Meijer's G-Function.[604] Therefore

$$C_S = -[m^3c^3G^{31}_{13}((mc^2/(2kT))^2 \mid {}^{0}_{-3/2,0,\,\frac{1}{2}})/n]^{-1} \qquad (H.15)$$

The most probable momentum of a particle p_p is the maximum of

$$p_p = \text{Max}\{p^2\exp[-H_S/(kT)]\}$$

$$= \{(2(kT)^2/c^2)[1 + (1 - m^2c^4/(kT)^2)^{\frac{1}{2}}]\}^{\frac{1}{2}} \qquad (H.16)$$

For large T or small T the maximum is

$$p_p \approx 2kT/c \; > mc$$

The velocity v_p corresponding to the maximum in the momentum is

$$v_p = cp_p/(p_p^{\,2} - m^2c^2)^{\frac{1}{2}}$$

For large T or small T, the velocity v_p corresponding to the maximum in the momentum is approximately

$$v_p \approx c + \tfrac{1}{2}\, m^2c^5/(2kT)^2$$

H.2 Superluminal Thermodynamics

Turning now to the thermodynamics of a dilute superluminal gas implied by the superluminal Maxwell-Boltzmann distribution we begin by calculating the average energy per particle

---

[604] See Gradshteyn (1965) integral 3.389.2 and p. 1068 for the properties of Meijer's G-Function.

$$\varepsilon = C_S \int d^3p \; H_S \; exp[-H_S/(2kT)]/\int d^3p \; C_S \; exp[-H_S/(kT)] \tag{H.17}$$

$$= (C_S/n) \int d^3p \; H_S \; exp[-H_S/(kT)]$$

$$= (C_S/n)2kT\alpha \; 4\pi m^2 c^3 \int_0^\infty dy \; y^2(y^2+1)^{\frac{1}{2}} \; exp(-\alpha y)$$

$$= -(C_S/n)m^3 c^5 G^{31}{}_{13}(\alpha^2/4 \; |^{-\frac{1}{2}}{}_{-2,0,\;\frac{1}{2}})$$

$$= mc^2 \; G^{31}{}_{13}((mc^2/(2kT))^2|^{-\frac{1}{2}}{}_{-2,0,\;\frac{1}{2}})/G^{31}{}_{13}((mc^2/(2kT))^2|^0{}_{-3/2,0,\;\frac{1}{2}}) \tag{H.18}$$

The Maxwell-Boltzmann normalization factor is related to the energy per particle by

$$C_S = -n\varepsilon/(m^3 c^5 G^{31}{}_{13}(\alpha^2/4 \; |^{-\frac{1}{2}}{}_{-2,0,\;\frac{1}{2}})) \tag{H.19}$$

Note that C_S is proportional to the energy in contrast to the non-relativistic case where the Maxwell-Boltzmann normalization factor $C = (3m/(4\pi\varepsilon))^{3/2}$.

We now calculate the superluminal pressure for the case of a distribution of superluminal particles bouncing on a wall perpendicular to the z-axis. The wall is assumed to be a perfectly reflecting plane. The pressure is the average force per unit area due to the gas of superluminal particles. The number of particles bombarding the wall per second is with $v_z > 0$ is $v_z f_S(\mathbf{p})d^3p$. Thus the pressure is

$$P = \int d^3p \, 2p_z v_z f_S(\mathbf{p}) \tag{H.20}$$

where the particle momentum changes by $2p_z$ due to reflection. Due to spherical symmetry one expects the average values for the various components of $\mathbf{v}$ to be equal. Consequently we can re-express eq. H.20 as

$$P = 1/3 \int d^3p \, 2m\gamma_s v^2 f_S(\mathbf{p}) \tag{H.21}$$
$$= 1/3 \int d^3p \, 2pv f_S(\mathbf{p})$$

Since

$$v = cp/(p^2 - m^2c^2)^{1/2} \qquad\qquad (H.22)$$

we see

$$P = 8\pi c/3 \int_m^\infty dp\, p^4 f_S(\mathbf{p})/(p^2 - m^2 c^2)^{1/2}$$

Following steps similar to eqs. H.12 – H.15 leads to

$$P = m^4 c^4\, C_S G^{31}_{13}((mc^2/(2kT))^2|{}^{1/2}_{-2,0,\,1/2}) \qquad\qquad (H.23)$$

The *equation of state* relating the pressure and energy is

$$P = -(m/c)\{G^{31}_{13}((mc^2/(4kT))^2|{}^{1/2}_{-2,0,\,1/2})/G^{31}_{13}(\alpha^2/4\,|{}^{-1/2}_{-2,0,\,1/2}))\}n\varepsilon \qquad (H.24)$$

Substituting for ε we find

$$P = -(nm^2 c)\{G^{31}_{13}(\rho\,|{}^{1/2}_{-2,0,\,1/2})/\,G^{31}_{13}(\rho\,|{}^{0}_{-3/2,0,\,1/2})\} \qquad\qquad (H.25)$$

where

$$\rho = (mc^2/(2kT))^2 \qquad\qquad (H.26)$$

Turning now to the consideration of a dilute gas the internal energy of the gas can be defined to be[605]

$$U(t) = N\varepsilon \qquad\qquad (H.27)$$

We note that the work done by the superluminal gas if its volume increases by dV is PdV. Then the superluminal (and usual) form of the first law of thermodynamics is

$$dQ = dU + PdV \qquad\qquad (H.28)$$

---

[605] The internal energy of a gas of non-interacting non-relativistic particles is $U(t) = 3NkT/2$. In the superluminal case it appears that it is eq. H.18.

where Q is the heat absorbed. The heat capacity of the system for constant volume is

$$C_V = (\partial U/\partial T)_V \qquad (H.29)$$

The second law of thermodynamics, Boltzmann's H theorem, is based on

$$H = -S/kV \qquad (H.30)$$

where H is the negative of the entropy divided k times the volume V. In systems where there are no superluminal particles, the H theorem states that the entropy never decreases for an isolated gas of fixed volume.

We can calculate H for a superluminal system under equilibrium conditions, H_e, from[606]

$$H_e = \int d^3p f_S(\mathbf{p})\ln(f_S(\mathbf{p})) \qquad (H.31)$$

$$= \int d^3p f_S(\mathbf{p})[\ln C_S - H_S/(kT)] \qquad (H.32)$$
$$= n \ln C_S - \int d^3p f_S(\mathbf{p})H_S/(kT)$$

$$= n \ln C_S - n\varepsilon/(kT) \qquad (H.33)$$

by eqs. H.11 and H.17. Therefore

$$S = -kVH_{Se} = -kN \ln C_S + N\varepsilon/T \qquad (H.34)$$

Consequently we obtain the superluminal *and* standard non-relativistic result

$$1/T = (\partial S/\partial U)_x \qquad (H.35)$$

---

[606] We consistently assume that integrals over the momentum $\int d^3p$ are the proper integration (rather than integrations over velocity $\int d^3v$) because, for example, the calculation of the normalization constant eq. H.11 would diverge if the integration were over $\int d^3v$.

where x represents all other extensive variables.

H.3 Approximate Calculation of Kinetic and Thermodynamic Quantities

We can obtain more tractable expressions for kinetic and thermodynamic quantities by assuming $\mathbf{p}^2 \gg m^2c^2$ and approximating the Hamiltonian (eq. H.8) with

$$H_{Sa} = cp \tag{H.36}$$

The approximate normalization condition is

$$n = N/V = \int d^3 p f_{Sa}(\mathbf{p}) = C_{Sa} \int d^3 p \, \exp\{-H_{Sa}/(kT)\} \tag{H.37}$$

where n is the particle density, N is the number of particles in the system, and V is the volume of the system. C_S is determined by

$$n = 4\pi C_{Sa} \int_{m_c}^{\infty} dp \, p^2 \exp\{-cp/(kT)\} \tag{H.38}$$

Letting $\alpha = c/(kT)$ we see eq. H.38 becomes

$$n = 4\pi C_{Sa} \, d^2/d\alpha^2 \int_{m_c}^{\infty} dp \, \exp(-\alpha p)$$

$$= 4\pi C_{Sa} \, d^2/d\alpha^2 \, [(1/\alpha) \exp(-\alpha mc)] \tag{H.39}$$

Therefore the normalization factor is

$$C_{Sa} = n/\{4\pi \, d^2/d\alpha^2 \, [(1/\alpha)\exp(-\alpha mc)]\} \tag{H.40}$$

The most probable momentum of a particle p_p is the maximum of

$$p_{pa} = \text{Max}\{p^2\exp[-H_{Sa}/(kT)]\}$$
$$= 2kT/c \tag{H.41}$$

The velocity v_{pa} corresponding to the maximum in the momentum is

$$v_{pa} = cp_{pa}/(p_{pa}^2 - m^2c^2)^{1/2}$$

For large T or small T, the velocity v_{pa} corresponding to the maximum in the momentum is approximately

$$v_{pa} \approx c + \tfrac{1}{2}\, m^2c^5/(2kT)^2$$

Turning now to the thermodynamics implied by the superluminal Maxwell-Boltzmann distribution we begin by calculating the average energy per particle

$$\varepsilon_a = \int d^3p\; H_{Sa}\exp[-H_{Sa}/(kT)]/\int d^3p\; \exp[-H_{Sa}/(kT)] \tag{H.42}$$
$$= (C_{Sa}/n)\int d^3p\; H_{Sa}\exp[-H_{Sa}/(kT)]$$
$$= -(4\pi c C_{Sa}/n)\, d^3/d\alpha^3[(1/\alpha)\exp(-\alpha mc)] \rightarrow 3kT \text{ for } T \gg mc$$

where $\alpha = c/(kT)$.[607]

The Maxwell-Boltzmann normalization factor is related to the energy per particle by

$$C_{Sa} = -n\varepsilon_a/\{4\pi c\; d^3/d\alpha^3[(1/\alpha)\exp(-\alpha mc)]\} \tag{H.43}$$

Note that C_{Sa} is proportional to the energy ε_a in contrast to the non-relativistic case where the Maxwell-Boltzmann normalization factor $C = (3m/(4\pi\varepsilon))^{3/2}$.

We now calculate the superluminal pressure for the case of a distribution of superluminal particles bouncing on a wall perpendicular to the z-axis. The wall is

---

[607] The Superluminal case differs from the non-relativistic case: $\varepsilon_a = 3kT/2$. An example of $\varepsilon_a = 3kT$ is a crystal with a potential energy of compression. See p. 192 Morse (1964).

assumed to be a perfectly reflecting plane. The pressure is the average force per unit area due to the gas of superluminal particles. The number of particles bombarding the wall per second is with $v_z > 0$ is $v_z f_{Sa}(\mathbf{p}) d^3 p$. Thus the pressure is

$$P_a = \int d^3p \, 2p_z v_z f_{Sa}(\mathbf{p}) \tag{H.44}$$

where the particle momentum changes by $2p_z$ due to reflection. Due to spherical symmetry one expects the average values for the various components of $\mathbf{v}$ to be equal. Consequently we can re-express eq. H.44 as

$$P_a = 1/3 \int d^3p \, 2m\gamma_s v^2 f_{Sa}(\mathbf{p}) \tag{H.45}$$
$$= 1/3 \int d^3p \, 2pv f_{Sa}(\mathbf{p})$$

Since

$$v = cp/(p^2 - m^2c^2)^{\frac{1}{2}} \tag{H.46}$$

we see

$$P_a = 8\pi c/3 \int_{m_c}^{\infty} dp \, p^4 f_{Sa}(\mathbf{p})/(p^2 - m^2c^2)^{\frac{1}{2}} \tag{H.47}$$

$$\cong 8\pi c/3 \int_{m_c}^{\infty} dp \, p^3 f_{Sa}(\mathbf{p})$$

Evaluating eq. H.47 yields

$$P_a = -(8\pi c/3) \, C_{Sa} \, d^3/d\alpha^3 [(1/\alpha)\exp(-\alpha m_c)] \tag{H.48}$$

The *equation of state* relating the pressure and energy is[608]

$$P_a = 2/3 \, n\varepsilon_a \tag{H.49}$$

For $T \gg mc$ we found[609]

---

[608] The same equation of state as non-relativistic kinetic theory. See p. 72 Huang (1965).
[609] Later we will define temperature in terms of the entropy S as $1/T = (\partial S/\partial U)_x$ where x is all other extensive variables.

$$\varepsilon_a \rightarrow 3kT \tag{H.50}$$

then, contrary to non-relativistic kinetic theory, we find ($T \gg mc$)

$$P_a = 2nkT \tag{H.51}$$

Turning now to the consideration of a dilute gas the internal energy of the gas for $T \gg mc$ is

$$U(t) = N\varepsilon \rightarrow 3NkT \tag{H.52}$$

We note again that the work done by the superluminal gas if its volume increases by dV is PdV. Then the superluminal (and usual) form of the first law of thermodynamics is

$$dQ = dU + PdV \tag{H.53}$$

where Q is the heat absorbed. The heat capacity of the system for constant volume is (T $\gg$ mc)

$$C_V \rightarrow 3Nk \tag{H.54}$$

The second law of thermodynamics, Boltzmann's H theorem, is based on

$$H = -S/kV \tag{H.55}$$

where H is the negative of the entropy divided k times the volume V. In systems where there are no superluminal particles, the H theorem states that the entropy never decreases for an isolated gas of fixed volume.

We can calculate H_{BS} for a superluminal system under equilibrium conditions, H_{BSea}, from[610]

$$H_{BSea} = \int d^3 p f_{Sa}(\mathbf{p}) \ln(f_{Sa}(\mathbf{p})) \tag{H.56}$$
$$= \int d^3 p f_{Sa}(\mathbf{p})[\ln C_{Sa} - H_{Sa}/(kT)] \tag{H.57}$$
$$= n \ln C_{Sa} - \int d^3 p f_{Sa}(\mathbf{p}) H_{Sa}/(kT)$$
$$= n \ln C_{Sa} - n\varepsilon_a/(kT) \tag{H.58}$$

by eqs. H.11 and H.17. Therefore

$$S_a = -kV H_{BSea} = -kN \ln C_{Sa} + N\varepsilon_a/T \tag{H.59}$$

The superluminal *and* standard non-relativistic result still holds

$$1/T = (\partial S/\partial U)_x \tag{H.60}$$

where x represents all other extensive variables.

H.4 Superluminal Kinetics and Thermodynamics Are Similar to the Non-Relativistic Case

In the previous sections we have shown that kinetic theory and the laws of thermodynamics are usually similar in the superluminal and non-relativistic cases modulo detail differences in the values of the various quantities due to differences between superluminal kinematics and non-relativistic kinematics.

---

[610] We consistently assume that integrals over the momentum $\int d^3 p$ are the proper integration (rather than integrations over velocity $\int d^3 v$) because, for example, the calculation of the normalization constant eq. H.11 would diverge if the integration were over $\int d^3 v$.

Appendix I. PseudoQuantization – Embedding Classical Fields in Quantum Field Theories

S. Blaha, Phys. Rev. **D17**, 994 (1978).

Appendix J. PseudoQuantum – Second Quantized Non-Abelian Field Theory for Hadrons with Quark Confinement …

S. Blaha, Phys. Rev. D**11**, 2921 (1975).

Appendix K. Towards a Field Theory of Hadron Binding

S. Blaha, Phys. Rev. **D10**, 4268 (1974).

REFERENCES

Akhiezer, N. I., Frink, A. H. (tr), 1962, *The Calculus of Variations* (Blaisdell Publishing, New York, 1962).

Bjorken, J. D., Drell, S. D., 1964, *Relativistic Quantum Mechanics* (McGraw-Hill, New York, 1965).

Bjorken, J. D., Drell, S. D., 1965, *Relativistic Quantum Fields* (McGraw-Hill, New York, 1965).

Blaha, S., 1998, *Cosmos and Consciousness* (Pingree-Hill Publishing, Auburn, NH, 1998).

_____, 2002, *A Finite Unified Quantum Field Theory of the Elementary Particle Standard Model and Quantum Gravity Based on New Quantum Dimensions™ & a New Paradigm in the Calculus of Variations* (Pingree-Hill Publishing, Auburn, NH, 2002).

_____, 2003, *A Finite Unified Quantum Field Theory of the Elementary Particle Standard Model and Quantum Gravity Based on New Quantum Dimensions™ and a New Paradigm in the Calculus of Variations* (Pingree-Hill Publishing, Auburn, NH, 2003).

_____, 2004, *Quantum Big Bang Cosmology: Complex Space-time General Relativity, Quantum Coordinates™Dodecahedral Universe, Inflation, and New Spin 0, ½, 1 & 2 Tachyons & Imagyons* (Pingree-Hill Publishing, Auburn, NH, 2004).

_____, 2005a, *Quantum Theory of the Third Kind: A New Type of Divergence-free Quantum Field Theory Supporting a Unified Standard Model of Elementary Particles and Quantum Gravity based on a New Method in the Calculus of Variations* (Pingree-Hill Publishing, Auburn, NH, 2005).

_____, 2005b, *The Metatheory of Physics Theories, and the Theory of Everything as a Quantum Computer Language* (Pingree-Hill Publishing, Auburn, NH, 2005).

_____, 2005c, *The Equivalence of Elementary Particle Theories and Computer Languages: Quantum Computers, Turing Machines, Standard Model, Superstring Theory, and a Proof that Gödel's Theorem Implies Nature Must Be Quantum* (Pingree-Hill Publishing, Auburn, NH, 2005).

_____, 2006a, *The Foundation of the Forces of Nature* (Pingree-Hill Publishing, Auburn, NH, 2006).

_____, 2006b, *A Derivation of ElectroWeak Theory based on an Extension of Special Relativity; Black Hole Tachyons; & Tachyons of Any Spin.* (Pingree-Hill Publishing, Auburn, NH, 2006).

_____, 2007a, *Physics Beyond the Light Barrier: The Source of Parity Violation, Tachyons, and A Derivation of Standard Model Features* (Pingree-Hill Publishing, Auburn, NH, 2007).

_____, 2007b, *The Origin of the Standard Model: The Genesis of Four Quark and Lepton Species, Parity Violation, the ElectroWeak Sector, Color SU(3), Three Visible Generations of Fermions, and One Generation of Dark Matter with Dark Energy* (Pingree-Hill Publishing, Auburn, NH, 2007).

_____, 2008a, *A Direct Derivation of the Form of the Standard Model From GL(16) (Pingree-Hill Publishing, Auburn, NH, 2008).*

_____, 2008b, *A Complete Derivation of the Form of the Standard Model With a New Method to Generate Particle Masses Second Edition* (Pingree-Hill Publishing, Auburn, NH, 2008)

_____, 2009, *The Algebra of Thought & Reality: The Mathematical Basis for Plato's Theory of Ideas, and Reality Extended to Include A Priori Observers and Space-Time Second Edition* (Pingree-Hill Publishing, Auburn, NH, 2009).

_____, 2010a, *Operator Metaphysics: A New Metaphysics Based on a New Operator Logic and a New Quantum Operator Logic that Lead to a Mathematical Basis for Plato's Theory of Ideas and Reality* (Pingree-Hill Publishing, Auburn, NH, 2010).

_____, 2010b, *The Standard Model's Form Derived from Operator Logic, Superluminal Transformations and GL(16)* (Pingree-Hill Publishing, Auburn, NH, 2010).

_____, 2010c, *SuperCivilizations: Civilizations as Superorganisms* (McMann-Fisher Publishing, Auburn, NH, 2010).

_____, 2011a, *21st Century Natural Philosophy Of Ultimate Physical Reality* (McMann-Fisher Publishing, Auburn, NH, 2011).

_____, 2011b, *All the Universe! Faster Than Light Tachyon Quark Starships & Particle Accelerators with the LHC as a Prototype Starship Drive Scientific Edition* (Pingree-Hill Publishing, Auburn, NH, 2011).

_____, 2011c, *From Asynchronous Logic to The Standard Model to Superflight to the Stars* (Blaha Research, Auburn, NH, 2011).

_____, 2012a, *From Asynchronous Logic to The Standard Model to Superflight to the Stars volume 2: Superluminal CP and CPT, U(4) Complex General Relativity and The Standard Model, Complex Vierbein General Relativity, Kinetic Theory, Thermodynamics* (Blaha Research, Auburn, NH, 2012).

_____, 2012b, *Standard Model Symmetries, And Four And Sixteen Dimension Complex Relativity; The Origin Of Higgs Mass Terms* (Blaha Reasearch, Auburn, NH, 2012).

_____, 2013a, *Multi-Stage Space Guns, Micro-Pulse Nuclear Rockets, and Faster-Than-Light Quark-Gluon Ion Drive Starships* (Blaha Research, Auburn, NH, 2013).

_____, 2013b, *The Bridge to Dark Matter; A New Sister Universe; Dark Energy; Inflatons; Quantum Big Bang; Superluminal Physics; An Extended Standard Model Based on Geometry* (Blaha Reasearch, Auburn, NH, 2013).

_____, 2014a, *Universes and Megaverses: From a New Standard Model to a Physical Megaverse; The Big Bang; Our Sister Universe's Wormhole; Origin of the Cosmological Constant, Spatial Asymmetry of the Universe, and its Web of Galaxies; A Baryonic Field* between Universes and Particles; Megaverse Extended Wheeler-DeWitt Equation (Blaha Reasearch, Auburn, NH, 2014).

_____, 2014b, *All the Megaverse! Starships Exploring the Endless Universes of the Cosmos Using the Baryonic Force* (Blaha Research, Auburn, NH, 2014).

_____, 2014c, *All the Megaverse! II Between Megaverse Universes: Quantum Entanglement Explained by the Megaverse Coherent Baryonic Radiation Devices – PHASERs Neutron Star Megaverse Slingshot Dynamics Spiritual and UFO Events, and the Megaverse Microscopic Entry into the Megaverse* (Blaha Research, Auburn, NH, 2014).

_____, 2015a, *PHYSICS IS LOGIC PAINTED ON THE VOID: Origin of Bare Masses and The Standard Model in Logic, U(4) Origin of the Generations, Normal and Dark Baryonic Forces, Dark Matter, Dark Energy, The Big Bang, Complex General Relativity, A Megaverse of Universe Particles* (Blaha Research, Auburn, NH, 2015).

_____, 2015b, *PHYSICS IS LOGIC Part II: The Theory of Everything, The Megaverse Theory of Everything, U(4)⊗U(4) Grand Unified Theory (GUT), Inertial Mass = Gravitational Mass, Unified Extended Standard Model and a New Complex General Relativity with Higgs Particles, Generation Group Higgs Particles* (Blaha Research, Auburn, NH, 2015).

_____, 2015c, *The Origin of Higgs ("God") Particles and the Higgs Mechanism: Physics is Logic III, Beyond Higgs – A Revamped Theory With a Local Arrow of Time, The Theory of Everything Enhanced, Why Inertial Frames are Special, Universes of the Mind* (Blaha Research, Auburn, NH, 2015).

_____, 2015d, *The Origin of the Eight Coupling Constants of The Theory of Everything: U(8) Grand Unified Theory of Everything (GUTE), S^8 Coupling Constant Symmetry, Space-Time Dependent Coupling Constants, Big Bang Vacuum Coupling Constants, Physics is Logic IV* (Blaha Research, Auburn, NH, 2015).

_____, 2016a, *New Types of Dark Matter, Big Bang Equipartition, and A New U(4) Symmetry in the Theory of Everything: Equipartition Principle for Fermions, Matter is 83.33% Dark,*

Penetrating the Veil of the Big Bang, Explicit QFT Quark Confinement and Charmonium, Physics is Logic V (Blaha Research, Auburn, NH, 2016).

_____, 2016b, *The Periodic Table of the 192 Quarks and Leptons in The Theory of Everything: The U(4) Layer Group, Physics is Logic VI* (Blaha Research, Auburn, NH, 2016).

_____, 2016c, *New Boson Quantum Field Theory, Dark Matter Dynamics, Dark Matter Fermion Layer Mixing, Genesis of Higgs Particles, New Layer Higgs Masses, Higgs Coupling Constants, Non-Abelian Higgs Gauge Fields, Physics is Logic VII* (Blaha Research, Auburn, NH, 2016).

_____, 2016d, *Unification of the Strong Interactions and Gravitation: Quark Confinement Linked to Modified Short-Distance Gravity; Physics is Logic VIII* (Blaha Research, Auburn, NH, 2016).

_____, 2016e, *MoND: Unification of the Strong Interactions and Gravitation II, Quark Confinement Linked to Large-Scale Gravity, Physics is Logic IX* (Blaha Research, Auburn, NH, 2016).

_____, 2016f, *CQ Mechanics: A Unification of Quantum & Classical Mechanics, Quantum/Semi-Classical Entanglement, Quantum/Classical Path Integrals, Quantum/Classical Chaos* (Blaha Research, Auburn, NH, 2016).

_____, 2016g, *GEMS: Unified Gravity, ElectroMagnetic and Strong Interactions: Manifest Quark Confinement, A Solution for the Proton Spin Puzzle, Modified Gravity on the Galactic Scale* (Pingree Hill Publishing, Auburn, NH, 2016).

_____, 2016h, *Unification of the Seven Boson Interactions based on the Riemann-Christoffel Curvature Tensor* (Pingree Hill Publishing, Auburn, NH, 2016).

_____, 2017a, *Unification of the Eleven Boson Interactions based on 'Rotations of Interactions'* (Pingree Hill Publishing, Auburn, NH, 2017).

_____, 2017b, *The Origin of Fermions and Bosons, and Their Unification* (Pingree Hill Publishing, Auburn, NH, 2017).

_____, 2017c, *Megaverse: The Universe of Universes* (Pingree Hill Publishing, Auburn, NH, 2017).

_____, 2017d, *SuperSymmetry and the Unified SuperStandard Model* (Pingree Hill Publishing, Auburn, NH, 2017).

_____, 2017e, *From Qubits to the Unified SuperStandard Model with Embedded SuperStrings: A Derivation* (Pingree Hill Publishing, Auburn, NH, 2017).

_____, 2017f, *The Unified SuperStandard Model in Our Universe and the Megaverse: Quarks, ... ,* (Pingree Hill Publishing, Auburn, NH, 2017).

_____, 2018a, *The Unified SuperStandard Model and the Megaverse SECOND EDITION A Deeper Theory based on a New Particle Functional Space that Explicates Quantum Entanglement Spookiness (Volume 1)* (Pingree Hill Publishing, Auburn, NH, 2018).

_____, 2018b, *Cosmos Creation: The Unified SuperStandard Model, Volume 2, SECOND EDITION* (Pingree Hill Publishing, Auburn, NH, 2018).

_____, 2018c, *God Theory (*Pingree Hill Publishing, Auburn, NH, 2018).

_____, 2018d, *Immortal Eye: God Theory: Second Edition* (Pingree Hill Publishing, Auburn, NH, 2018).

_____, 2018e, *Unification of God Theory and Unified SuperStandard Model THIRD EDITION* (Pingree Hill Publishing, Auburn, NH, 2018).

_____, 2019a, *Calculation of: QED α = 1/137, and Other Coupling Constants of the Unified SuperStandard Theory* (Pingree Hill Publishing, Auburn, NH, 2019).

_____, 2019b, *Coupling Constants of the Unified SuperStandard Theory SECOND EDITION* (Pingree Hill Publishing, Auburn, NH, 2019).

_____, 2019c, *New Hybrid Quantum Big_Bang–Megaverse_Driven Universe with a Finite Big Bang and an Increasing Hubble Constant* (Pingree Hill Publishing, Auburn, NH, 2019).

_____, 2019d, *The Universe, The Electron and The Vacuum* (Pingree Hill Publishing, Auburn, NH, 2019).

_____, 2019e, *Quantum Big Bang – Quantum Vacuum Universes (Particles)* (Pingree Hill Publishing, Auburn, NH, 2019).

_____, 2019f, *The Exact QED Calculation of the Fine Structure Constant Implies ALL 4D Universes have the Same Physics/Life Prospects* (Pingree Hill Publishing, Auburn, NH, 2019).

_____, 2019g, *Unified SuperStandard Theory and the SuperUniverse Model: The Foundation of Science* (Pingree Hill Publishing, Auburn, NH, 2019).

_____, 2020a, *Quaternion Unified SuperStandard Theory (The QUeST) and Megaverse Octonion SuperStandard Theory (MOST)* (Pingree Hill Publishing, Auburn, NH, 2020).

_____, 2020b, *United Universes Quaternion Universe - Octonion Megaverse* (Pingree Hill Publishing, Auburn, NH, 2020).

Eddington, A. S., 1952, *The Mathematical Theory of Relativity* (Cambridge University Press, Cambridge, U.K., 1952).

Fant, Karl M., 2005, *Logically Determined Design: Clockless System Design With NULL Convention Logic* (John Wiley and Sons, Hoboken, NJ, 2005).

Feinberg, G. and Shapiro, R., 1980, *Life Beyond Earth: The Intelligent Earthlings Guide to Life in the Universe* (William Morrow and Company, New York, 1980).

Gelfand, I. M., Fomin, S. V., Silverman, R. A. (tr), 2000, *Calculus of Variations* (Dover Publications, Mineola, NY, 2000).

Giaquinta, M., Modica, G., Souchek, J., 1998, *Cartesian Coordinates in the Calculus of Variations* Volumes I and II (Springer-Verlag, New York, 1998).

Giaquinta, M., Hildebrandt, S., 1996, *Calculus of Variations* Volumes I and II (Springer-Verlag, New York, 1996).

Gradshteyn, I. S. and Ryzhik, I. M., 1965, *Table of Integrals, Series, and Products* (Academic Press, New York, 1965).

Heitler, W., 1954, *The Quantum Theory of Radiation* (Claendon Press, Oxford, UK, 1954).

Huang, Kerson, 1992, *Quarks, Leptons & Gauge Fields 2nd Edition* (World Scientific Publishing Company, Singapore, 1992).

Jost, J., Li-Jost, X., 1998, *Calculus of Variations* (Cambridge University Press, New York, 1998).

Kaku, Michio, 1993, *Quantum Field Theory*, (Oxford University Press, New York, 1993).

Kirk, G. S. and Raven, J. E., 1962, *The Presocratic Philosophers* (Cambridge University Press, New York, 1962).

Landau, L. D. and Lifshitz, E. M., 1987, *Fluid Mechanics 2nd Edition*, (Pergamon Press, Elmsford, NY, 1987).

Misner, C. W., Thorne, K. S., and Wheeler, J. A., 1973, *Gravitation* (W. H. Freeman, New York, 1973).

Rescher, N., 1967, *The Philosophy of Leibniz* (Prentice-Hall, Englewood Cliffs, NJ, 1967).

Rieffel, Eleanor and Polak, Wolfgang, 2014, *Quantum Computing* (MIT Press, Cambridge, MA, 2014).

Riesz, Frigyes and Sz.-Nagy, Béla, 1990, *Functional Analysis* (Dover Publications, New York, 1990).
Sagan, H., 1993, *Introduction to the Calculus of Variations* (Dover Publications, Mineola, NY, 1993).

Sakurai, J. J., 1964, *Invariance Principles and Elementary Particles* (Princeton University Press, Princeton, NJ, 1964).

Sorokin, Pitirim, 1941, *Social and Cultural Dynamics* (Porter Sargent Publishers, Boston, MA, 1941).

Streater, R. F. and Wightman, A. S., 2000, *PCT, Spin, Statistics, and All That* (Princeton University Press, Princeton, NJ 2000).

Weinberg, S., 1972, *Gravitation and Cosmology* (John Wiley and Sons, New York, 1972).

Weinberg, S., 1995, *The Quantum Theory of Fields Volume I* (Cambridge University Press, New York, 1995).

Weinberg, S., 2000, *The Quantum Theory of Fields Volume III Supersymmetry* (Cambridge University Press, New York, 2000).

Weyl, H., 1950, *Space, Time, Matter* (Dover, New York, 1950).

Weyl, H., (Tr. S. Pollard et al), 1987, *The Continuum* (Dover Publications, New York, 1987).

INDEX

About the Author

Stephen Blaha is a well-known Physicist and Man of Letters with interests in Science, Society and civilization, the Arts, and Technology. He had an Alfred P. Sloan Foundation scholarship in college. He received his Ph.D. in Physics from Rockefeller University. He has served on the faculties of several major universities. He was also a Member of the Technical Staff at Bell Laboratories, a manager at the Boston Globe Newspaper, a Director at Wang Laboratories, and President of Blaha Software Inc. and of Janus Associates Inc. (NH).

Among other achievements he was a co-discoverer of the "r potential" for heavy quark binding developing the first (and still the only demonstrable) non-Aeolian gauge theory with an "r" potential; first suggested the existence of topological structures in superfluid He-3; first proposed Yang-Mills theories would appear in condensed matter phenomena with non-scalar order parameters; first developed a grammar-based formalism for quantum computers and applied it to elementary particle theories; first developed a new form of quantum field theory without divergences (thus solving a major 60 year old problem that enabled a unified theory of the Standard Model and Quantum Gravity without divergences to be developed); first developed a formulation of complex General Relativity based on analytic continuation from real space-time; first developed a generalized non-homogeneous Robertson-Walker metric that enabled a quantum theory of the Big Bang to be developed without singularities at t = 0; first generalized Cauchy's theorem and Gauss' theorem to complex, curved multi-dimensional spaces; received Honorable Mention in the Gravity Research Foundation Essay Competition in 1978; first developed a physically acceptable theory of faster-than-light particles; first derived a composition of extremums method in the Calculus of Variations; first quantitatively suggested that inflationary periods in the history of the universe were not needed; first proved Gödel's Theorem implies Nature must be quantum; provided a new alternative to the Higgs Mechanism, and Higgs particles, to generate masses; first showed how to resolve logical paradoxes including Gödel's Undecidability Theorem by developing Operator Logic and Quantum Operator Logic; first developed a quantitative harmonic oscillator-like model of the life cycle, and interactions, of civilizations; first showed how equations describing superorganisms also apply to civilizations. A recent book shows his theory applies successfully to the past 14 years of history and to *new* archaeological data on Andean and Mayan civilizations as well as Early Anatolian and Egyptian civilizations.

He first developed an axiomatic derivation of the form of The Standard Model from geometry – space-time properties – The Unified SuperStandard Model. It unifies all the known forces of Nature. It also has a Dark Matter sector that includes a Dark ElectroWeak sector with Dark doublets and Dark gauge interactions. It uses quantum coordinates to remove infinities that crop up in most interacting quantum field theories and additionally to remove the infinities that appear in the Big Bang and generate inflationary growth of the universe. It shows gravity has a MOND-like form without sacrificing Newton's Laws. It relates the interactions of the MOND-like sector of gravity with the r-potential of Quark Confinement. The axioms of the theory lead to the question of their origin. We suggest in the preceding edition of this book it can be attributed to an entity with God-like properties. We explore these properties in "God Theory" and show they predict that the Cosmos exists forever although individual universes (or incarnations of our universe) "come and go." Several other important results emerge from God Theory such a functionally triune God. The Unified SuperStandard Theory has many other important parts described in the Current Edition of *The Unified Superstandard Theory* and expanded in subsequent volumes.

Blaha has had a major impact on a succession of elementary particle theories: his Ph.D. thesis (1970), and papers, showed that quantum field theory calculations to all orders in ladder approximations could not give scaling deep inelastic electron-nucleon scattering. He later showed the eigenvalue equation for the fine structure constant α in Johnson-Baker-Willey QED had a zero at $\alpha = 1$ not 1/137 by solving the Schwinger-Dyson equations to all orders in an approximation that agreed with exact results to 4th order in α thus ending interest in this theory. In 1979 at Prof. Ken Johnson's (MIT) suggestion he calculated the proton-neutron mass difference in the MIT bag model and found the result had the wrong sign reducing interest in the bag model. These results all appear in Physical Review papers. In the 2000's he repeatedly pointed out the shortcomings of SuperString theory and showed that The Standard Model's form could be derived from space-time geometry by an extension of Lorentz transformations to faster than light transformations. This deeper space-time basis greatly increases the possibility that it is part of THE fundamental theory. Recently, Blaha showed that the Weak interactions differed significantly from the Strong, electromagnetic and gravitation interactions in important respects while these interactions had similar features, and suggested that ElectroWeak theory, which is essentially a glued union of the Weak interactions and Electromagnetism, possibly modulo unknown Higgs particle features, be replaced by a unified theory of the other interactions combined with a stand-alone Weak interaction theory. Blaha also showed that, if Charmonium calculations are taken seriously, the Strong interaction coupling constant is only a factor of five larger than the electromagnetic coupling constant, and thus Strong interaction perturbation theory would make sense and yield physically meaningful results.

In graduate school (1965-71) he wrote substantial papers in elementary particles and group theory: The Inelastic E- P Structure Functions in a Gluon Model. Phys. Lett. B40:501-502,1972; Deep-Inelastic E-P Structure Functions In A Ladder Model With Spin 1/2 Nucleons, Phys.Rev. D3:510-523,1971; Continuum Contributions To The Pion Radius, Phys. Rev. 178:2167-2169,1969; Character Analysis of U(N) and SU(N), J. Math. Phys. 10, 2156 (1969); and The Calculation of the Irreducible Characters of the Symmetric Group in Terms of the Compound Characters, (Published as Blaha's Lemma in D. E. Knuth's book: *The Art of Computer Programming Vols. 1 – 4*).

In the early 1980's Blaha was also a pioneer in the development of UNIX for financial, scientific and Internet applications: benchmarked UNIX versions showing that block size was critical for UNIX performance, developing financial modeling software, starting database benchmarking comparison studies, developing Internet-like UNIX networking (1982) and developing a hybrid shell programming technique (1982) that was a precursor to the PERL programming language. He was also the manager of the AT&T ten-year future products development database. His work helped lead to commercial UNIX on computers such as Sun Micros, IBM AIX minis, and Apple computers.

In the 1980's he pioneered the development of PC Desktop Publishing on laser printers. and was nominated for three "Awards for Technical Excellence" in 1987 by PC Magazine for PC software products that he designed and developed.

Recently he has developed a theory of Megaverses – actual universes of which our universe is one – with quantum particle-like properties based on the Wheeler-DeWitt equation of Quantum Gravity. He has developed a theory of a baryonic force, which had been conjectured many years ago, and estimated the strength of the force based on discrepancies in measurements of the gravitational constant G. This force, operative in D-dimensional space, can be used to escape from our universe in "uniships" which are the equivalent of the faster-than-light starships proposed in the author's earlier books. Thus travel to other universes, as well as to other stars is possible.

Blaha also considered the complexified Wheeler-DeWitt equation and showed that its limitation to real-valued coordinates and metrics generated a Cosmological Constant in the Einstein equations.

The author has also recently written a series of books on the serious problems of the United States and their solution as well as a book on the decline of Mankind that will follow from current social and genetic trends in Mankind.

In the past twelve years Dr. Blaha has written over 40 books on a wide range of topics. Some recent major works are: *From Asynchronous Logic to The Standard Model to Superflight to the Stars*, *All the Universe!*, *SuperCivilizations: Civilizations as Superorganisms*, *America's Future: an Islamic Surge, ISIS, al Qaeda, World Epidemics, Ukraine, Russia-China Pact, US Leadership Crisis, The Rises and Falls of Man – Destiny – 3000 AD: New Support for a Superorganism MACRO-THEORY of CIVILIZATIONS From CURRENT WORLD TRENDS and NEW Peruvian, Pre-Mayan, Mayan, Anatolian, and Early Egyptian Data, with a Projection to 3000 AD*, and *Mankind in Decline: Genetic Disasters, Human-Animal Hybrids, Overpopulation, Pollution, Global Warming, Food and Water Shortages, Desertification, Poverty, Rising Violence, Genocide, Epidemics, Wars, Leadership Failure.*

He has taught approximately 4,000 students in undergraduate, graduate, and postgraduate corporate education courses primarily in major universities, and large companies and government agencies.